THE TIDES OF THE PLANET EARTH

Second Edition

Related Pergamon Titles of Interest

Books

GORSHKOV:
World Ocean Atlas Volume 1: Pacific Ocean
Volume 2: Atlantic and Indian Oceans
Volume 3: Antarctic Ocean

JORDAN:
The Expanding Earth

MARCHUK & KAGAN:
Ocean Tides (Mathematical Models and Numerical Experiments)

PICKARD & EMERY:
Descriptive Physical Oceanography, 4th edition

POND & PICKARD:
Introductory Dynamic Oceanography, 2nd edition

STILLER & SAGDEEV:
Planetary Interiors

UDINTSEV:
Geological/Geophysical Atlas of the Indian Ocean

Journals

Continental Shelf Research

Deep-Sea Research

Vistas in Astronomy

THE TIDES OF THE PLANET EARTH

By

PAUL MELCHIOR

Observatoire Royal de Belgique, Bruxelles

Second Edition

PERGAMON PRESS

OXFORD · NEW YORK · TORONTO · SYDNEY · PARIS · FRANKFURT

U.K. Pergamon Press Ltd., Headington Hill Hall, Oxford OX3 0BW, England

U.S.A. Pergamon Press Inc., Maxwell House, Fairview Park, Elmsford, New York 10523, U.S.A.

CANADA Pergamon Press Canada Ltd., Suite 104, 150 Consumers Rd., Willowdale, Ontario M2J 1P9, Canada

AUSTRALIA Pergamon Press (Aust.) Pty. Ltd., P.O. Box 544, Potts Point, N.S.W. 2011, Australia

FRANCE Pergamon Press SARL, 24 rue des Ecoles, 75240 Paris, Cedex 05, France

FEDERAL REPUBLIC OF GERMANY Pergamon Press GmbH, Hammerweg 6, D-6242 Kronberg-Taunus, Federal Republic of Germany

---

First edition 1978
Second edition 1983

Library of Congress Cataloging in Publication Data

Melchior, Paul J.
The tides of the planet earth.
Bibliography: p.
Includes indexes.
1. Earth tides I. Title.
QC809.E2M48 1982 551.1'4 82-16567

British Library Cataloguing in Publication Data

Melchior, Paul
The tides of the planet Earth.—2nd ed.
1. Earth tides
I. Title
525'.6 QC809.E2

ISBN 0-08-026248-1

In order to make this volume available as economically and as rapidly as possible the author's typescript has been reproduced in its original form. This method unfortunately has its typographical limitations but it is hoped that they in no way distract the reader.

Printed in Great Britain by A. Wheaton & Co. Ltd., Exeter

Preface to the Second Edition

Essential progress has been realized in the field of tidal research during the short interval of the past five years. New penetrating theoretical approaches, the extension of high-precision observational data, especially by establishing a great many temporary stations in the tropical areas and the southern hemisphere, through Trans World tidal gravity profiles and the simultaneous development of new noteworthy oceanic cotidal maps, have allowed us a deeper insight into the problems related to the tidal deformations of the earth's body and to the tidal interactions between oceans and solid earth.

A new edition of *The Tides of the Planet Earth* thus offers an opportunity to include these important results in a revised text. The chapter dealing with observational results had to be completely rewritten.

On the other hand, the development of new powerful means of measurements like VLBI, satellite and moon laser ranging devices and superconducting gravimeters makes precise tidal knowledge a fundamental need and not just an esoteric subject. In particular, these developments could offer in the near future new ways of investigating the motions inside the earth's core, for which such a new tool would be most helpful to improve the understanding of the behaviour of this essential piece of our Earth.

Many thanks are due to Mrs. Jean-Barthélemy who (again) prepared the typing for direct reproduction with such great care and skill and to Mr. F. Cumps and Mr. H. Lauwers who prepared the drawings.

I also wish to thank Pergamon Press for its continued support and interest in having a new edition of this book.

PAUL MELCHIOR

Preface to the First Edition

The present book is not intended as a second edition of *The Earth Tides* which was published under my authorship by Pergamon Press in 1966.

Looking back to the earlier book after more than ten years, I would like to state that *The Earth Tides* was simply the result of a detailed analysis of all previous work published on the subject of earth tides and of some preliminary, still very imperfect and uncertain, new results obtained during the International Geophysical Year and the few years following.

Methods at that time were somewhat inaccurate but quickly evolving, while our ideas were quite elementary and rather old-fashioned.

Progress achieved since 1966, theoretical as well as experimental, has been so considerable that a simple revision of the 1966 book appeared difficult if not impossible.

I preferred to retain the historical character of the earlier book intact with the hope that it would thus preserve some value in this respect and I wrote the present book under a new title, although the reader will certainly find traces of the former work to a greater or lesser extent.

I am indebted to many colleagues for their assistance in the preparation of this book. I particularly wish to thank Dr. B. Ducarme and Professor R. Lecolazet who read the manuscript and suggested many improvements. I have used their own developments in several chapters.

Many thanks are due to Mrs. Jean-Barthélemy who prepared the typing for direct reproduction with such great care and skill and to Mr. F. Cumps and Mr. H. Lauwers who prepared all the drawings.

I also wish to thank Pergamon Press for its support and its interest in having a new book on the earth tides published in 1978.

PAUL MELCHIOR

Contents

Introduction

Earth Tides is a subject which holds a key position in relation to geophysics, geodesy and astronomy, since it is a phenomenon which consists of an elastico-viscous deformation of the terrestrial globe, and this is caused by the gravitational action of the Moon and the Sun. Every point on the surface of the Earth is subject to two forces: the force of gravity due to the newtonian attraction of the whole mass of the Earth, and the centrifugal force due to the rotation of the Earth. The resultant of these two forces is a vector, directed towards the interior of the Earth, whose length represents the intensity of the gravity at the point considered, and whose direction defines the direction of the vertical at the point. These elements cannot be considered as constants because both the Sun and the Moon attract the point under consideration; this attraction varies with time and with the paths of these two bodies. This is indeed the cause of oceanic tides, which result from the fact that the free surface of the sea constantly adapts itself to the level surface perpendicular to the pencil of disturbed verticals. The special circumstances which apply to the oceans, fluidity and the size of the basins, causes a resonance with the disturbing force (even though in an extremely varied manner) and have made immediate and simple observation possible but the theory more complex.

If the Earth were perfectly rigid it would be possible to observe, by means of very sensitive instruments, small periodic deviations of the vertical (with an amplitude of approximately 0ʺ05) and variation of the intensity of gravity (with an amplitude of approximately 2.4×10^{-7} or approximately 0.24 mgal). The laws governing these variations and their instantaneous amplitude may be calculated, in the various components and with high accuracy (this mainly depends upon the accuracy of our knowledge of the mass of the Moon), from elements of terrestrial and lunar orbits and the values of the mass of the Moon and the Sun. But because the Earth is anything but an ideal body, it does change its shape. Indeed, the Earth has physical properties that obey very complex, and as yet little known, laws whose study is the subject of rheology and which combine parameters of elasticity, viscosity and plasticity.

These deformations of the entire Earth, which evidently are governed by the same law of variation as the lunar-solar forces that cause them, sensibly modify the amplitude and in a lesser way the phase of the phenomena we are going to measure; they moreover produce periodic internal strains and cubic dilatations. The interest of the measurements lies, therefore, in a comparison of the observed phenomenon with the corresponding phenomenon calculated for a model Earth: the amplitude ratio and phase difference for each of the main waves that the instrument's accuracy enables us to detect constitute the basic elements of geophysical study by allowing a refinement of our models mainly in respect to the liquid core of the Earth. This principle of comparison, introduced by Kelvin, has been the method of investigating Earth tides for some hundred years. It is to be noted that Earth tides are the Earth's only deformation phenomenon for which we are able to calculate a priori the forces at work. On the other hand, it is worth while keeping in mind that the frequencies of the tidal phenomena are extremely low frequencies quite unusual in physics. The classical twelve-hour period indeed corresponds to 0.000 046 hertz only.

Prior to 1957 remarkable efforts had been made to observe these phenomena, thereby revealing their essential characteristics. But this sporadic study, conducted on a very modest scale, could never rely upon a network of permanent stations. Several new factors have determined the considerable development recently made in this study, and it must be said right here that it is because of the organisation of the International Geophysical Year that a whole program of research has been conceived and conducted. But the stimulus given by the IGY could not, alone, have been sufficient to focus on this problem the interest it arouses today. Precisely at the very moment when the IGY was being organised, new techniques were evolved at every stage of the study of Earth tides, eventually leading to a thorough re-examination of the problem and to a complete renovation of all its aspects.

The points that, in this evolution, seem decisive to us are as follows:

(1) Construction of recording gravimeters capable of measuring variations of gravity with a precision of 1 μgal (that is 10^{-9} of g).
(2) Construction of horizontal pendulums capable of following oscillations of the vertical with a precision comparable to 0".0002 in one measurement; this involves new techniques of installation.
(3) Improvements in the methods of calibration of the instruments ensuring a precision of at least 0.5%.
(4) The development with electronic computers of new methods of analysis better adapted to the study of Earth tides.
(5) The importance of Earth-tide effects on satellite orbit computations and in the other space techniques (lunar laser, VLBI).

Moreover, as a result of a continual increase in precision of measurements in the experimental sciences the existence of Earth tides is now observed in many phenomena which at first glance appear dissimilar. In actual fact every instrument attached to the Earth's crust is disturbed by the deformations of the Earth and, when it is sufficiently sensitive, the measurements appear altered by some periodicities and must be duly corrected before being suitable for interpretation. It is therefore important to investigate this phenomenon so as to be able to calculate the necessary corrections.

SHORT HISTORY

The story of the discovery of Earth tides is not long, but it nevertheless dates back to the beginning of the Christian era when Pliny the Elder states in his *Historia Naturalis* that at Cadiz, near to the temple of Hercules,

> "there is a closed source similar to a well which occasionally rises and falls with the ocean, but at other times does the opposite".

He also gives other examples, one on the banks of the Guadalquivir, the other near Seville, where in both cases the wells are lowest when the tide is high and begin to refill when the tide ebbs. These phenomena are obviously due to the cubic dilatations in the crust (Chapter 10).

According to Galitzin, the mathematician Abel was the first to indicate, in 1824, that the direction of the vertical does not remain constant but varies under the influence of the luni-solar attraction. Yet it seems that Ximenes in 1757 had already attempted to calculate this effect. It was, however, C. A. Peters who in 1844 published the first correct calculation.

The concept that the globe is not entirely rigid, but deformable, began to be accepted at the beginning of the nineteenth century. Simultaneously, some astronomers (Brioschi at Naples) began to suspect that there was a periodic variation in the latitudes and to study the oscillations of the direction of the vertical and the local deformations of the Earth's crust. Studies begun in France in 1831, carried out by d'Abbadie, using baths of mercury, demonstrated that there were rather irregular variations which, near the Gulf of Gascony in particular, appeared to be linked up with the ocean tides, although in a quite complex manner.

The first instrument to remain one of the basic instruments in the study of Earth tides and Seismology as well, the horizontal pendulum, was invented in 1832. The original idea, with its ingenious suspension, is due to Hengler of Munich. The instrument was introduced into geophysics by the German Zöllner, who was able to undertake its construction and to introduce it into practice. As is often the case, it was his name and not that of the inventor, Hengler, that was given to the instrument, but it must be recorded that Zöllner played an important part in the development of the instrument. The imperfections in the suspensory wires available at that time led another German, von Rebeur Paschwitz, to conceive a suspension on metallic points, and this was the first instrument to record deviations of the vertical caused by Earth tides. These records of very minute amplitudes now have only a qualitative value, but they were extremely important in the development of the research. Three stations - Potsdam, Strasbourg and Teneriffe - equipped with von Rebeur Paschwitz's instruments, demonstrated the existence of periodic oscillations of the vertical in 1890 at the same time as two Germans, Küstner and Marcuse, were experimentally demonstrating the actual existence of the periodic displacements of the instantaneous axis of rotation of the Earth inside the globe. The two phenomena of the variation of latitude and Earth tides which are governed by the same theory were thus discovered at the same epoch and in the same country.

Other attempts made by G. H. Darwin with a bifilar pendulum remained unsuccessful because of the too low sensitivity of this first instrument and of the interference due to various disturbances, mainly solar heating effects, but also local effects caused by oceanic tides, defects in the installation of the apparatus, etc., of which we are well aware today.

About 1876, Lord Kelvin drew attention to the effect of deformations of the Earth itself, showing that it was no longer acceptable to consider the Earth as being completely rigid, and from that time it has been accepted that the body of the Earth is deformed as a result of the tides, in the same way as for the oceans but to a lesser degree. Kelvin then showed that the amplitudes observed at the surface of the Earth, for each phenomenon deriving from the tidal potential (oceanic tides, deviations of the vertical, variations in the force of gravity), would be affected by the deformation of the surface, on which all our measurements are made.

The easiest observational method which led to the clear demonstration of the existence of the bodily tides was based upon an extremely simple reasoning: ocean tides are observed in relation to marks "fixed" on the crust with the help of tide gauges. These marks would be effectively fixed if the globe was perfectly rigid and the amplitude of the ocean tide observed would then be equal to that calculated. On the contrary, if the solid part is also deformed, the measured amplitude will be equal to the difference between oceanic and Earth tides.

G. Darwin first applied this method to observations of long period oceanic tides (monthly and semi-monthly lunar tides) which can be considered as static tides at least for a first approach to the problem. He found that the amplitude was only two-thirds of the theoretical, that is to say the Earth's crust has tides with an amplitude three times less than those of the oceans that cover it. This is the historical result (1883) that led to the often repeated statement that the rigidity of the Earth (supposed homogeneous) was the same as that of steel. For more than half a century, observations of many different types have confirmed the views of Kelvin, and together with the data from seismology and the polar motion, they made possible the first investigations concerning the elastic properties of the Earth.

It is indeed correct to apply the Newton static theory to the bodily tides: for the case of a solid globe, the rigid links between the molecules (the modulus of rigidity being indeed about the same as that of steel) does not allow currents to be established, the particles are only displaced a few decimetres and, therefore, equilibrium is quickly established.

The periods of the tides (8 h, 12 h, 24 h and more) are long compared to the periods of free oscillation of the Earth (the longest being about an hour) so that the resonance phenomena do not have a chance to occur. We will, however, have to correct this view in the case when tesseral tidal forces are applied to the liquid core of the Earth (Chapter 6).

1. ASTRONOMY

(a) The effect of Earth tides is obvious in all fundamental astronomical observations. The astronomical latitude of an observatory is the angle the vertical makes with the celestial equator defined by the adopted fundamental catalogue of star declinations, it therefore undergoes periodical variations corresponding to the deviations of the vertical. Therefore Peters engaged in an interesting study of deviations of the vertical in the Astronomische Nachrichten of 1845. T. Shida was the first to draw attention to the existence of such variations in the observations of the International Latitude Service. Longitudes are also disturbed and corrections for that have been included in the Bureau International de l'Heure procedures for several years.

(b) The period of oscillation of pendulum clocks is a function of the instantaneous value of the gravity; they are therefore subject to the tidal variations. This was investigated initially by Brown and Brouwer (1930) but is of no more concern to us as such clocks were abandonned in about 1950 in favor of quartz clocks and, later on, atomic clocks.

(c) A certain class of tides corresponding to a zonal distribution of the deformations causes variations of the largest moment of inertia of the Earth and as a result changes its speed of rotation. Predicted in 1928 by H. Jeffreys, these periodic variations of speed have been proved experimentally by many authors since atomic standards became available.

(d) The dissipation of energy due to the tidal deformations of the Earth's body explains to some extent the secular retardation of the rotation of the Earth (Chapter 14).

(e) A most important relation between Earth tides and fundamental astronomy is the fact that the same external potential is simultaneously responsible for the diurnal Earth tides and for the phenomenon of precession and nutation of the axis of rotation and axis of inertia (Chapter 2) in space: each term in the development of precession-nutation corresponds to two components in the diurnal tide. A similar relation must exist between the Moon's librations and the tidal deformations of the Moon. However, Poincaré, Jeffreys and Molodensky have shown that, owing to the fluidity of the Earth's core, it is necessary to introduce inertial terms in the equations of motion for phenomena which are represented by a tesseral function, that is phenomena producing oscillations of the principal axis of inertia This includes the eulerian free motion of the axis of rotation (responsible for the variation of latitude). The theoretical result is that the amplitude of each component near to an existing resonance is modified as a function of the speed of the nutation n (the speeds ω-n and ω+n are the speeds of the corresponding tides). The experimental results obtained from Earth-tide observations here provide an important contribution to the theory of the rotation of the Earth since they contribute to a more accurate knowledge of nutation phenomenon.

(f) Finally, it is to be emphasized that the determination of the irregularities of the rotation of the Earth, including the polar motion, which may now be derived from laser measurements of Earth-Moon distance or by very long base interferometry implies a very precise determination of Earth tide.

2. SPACE DYNAMICS

(a) The perturbing effects of tidal variations of the Earth's potential on artificial satellite orbits have been derived from observations by R. Newton and by Y. Kozai. They amount to as much as 50 metres on a satellite position and therefore must be modelled and included in the computer programs for orbit determination.

(b) As laser distance measurements to the Moon and satellites will have ratings of some centimetres and the radial tidal deformation reach an amplitude of 30 to 40 cm, the latter must be taken into account. But a simple analytical elastic Earth-model being insufficient to calculate exactly this tidal deformation, experimental measurements are needed at the sites concerned.

3. GEODESY

The relation is obvious as Earth tides can produce periodic systematic effects on very high precision levelling-distance and gravity measurements.

(a) The hydrostatic levellings of Nörlund have established that the phenomenon was observable, even though it had not then been isolated in operations on land.

(b) Geodesy is interested not only in the periodic deformations of the level surfaces and of the geoid in particular, but also in the variations of the intensity of gravity, which must be taken into account in the gravimetric surveys.

(c) Absolute determinations of gravity are made now with a precision of nearly 1 microgal (10^{-9}) (Sakuma, Faller, Istituto di Metrologia Colonetti, Torino). The comparison of such determinations made at different epochs also depends upon the precision with which the tidal correction can be given. Special high-precision gravity profiles presently developed (Scandinavia) require a comparable precision. Here again a simple analytical model is not sufficient.

(d) On the other hand, a better knowledge of Earth-tide linear strains will allow improvements in the accuracy of long base interferometers as well as other sensitive instruments which use Earth-based length standards. The International Association of Geodesy has shown a deep interest in the problem of Earth tides since the time it was first raised, and has devoted a General Report to this subject at each of its triennial or quadriennial meetings. In 1957 the Association has founded and has since then supported an International Centre for Earth Tides (ICET) affiliated to the Federation of Astronomical and Geophysical Services (FAGS) of the International Council of Scientific Unions (ICSU).

4. PHYSICS OF THE EARTH'S INTERIOR

Love introduced in 1909 special parameters that we call the "Love numbers". These are most useful for the interpretation of the experimental measurements and their comparison with theoretical models. This is because, like the moments of inertia, they represent integral properties of the planet as they are observed outside of it (or on its surface). The theory of Love is strictly only true for a spherical symmetry of the Earth's properties but is an excellent and sufficient first approximation. The numerical values of the Love numbers have been calculated for a great number of Earth models (Chapter 5). The determination of the Love numbers is evidently a fundamental objective in the Earth's tide study. They give an estimate of the maximum rigidity that the core may have ($10^9 < \mu < 10^{10}$ as obtained for the first time by Takeuchi in 1951), information which it is difficult for seismology to obtain as we do not see any transverse waves in the core.

As indicated by Poincaré and later on by Jeffreys, Jeffreys-Vicente and Molodensky, a resonance due to dynamical effects in the liquid core causes a perturbation in the Love numbers values for waves having their frequency near this resonance. We therefore will have to separate carefully these waves from the others by harmonic analysis.

On the other hand, the determination of the eventual lag of the tides with respect to the acting potential should give important information on the viscosity of the Earth. The present results show clear regional distributions, but an extension of the world net is necessary to understand this effect correctly.

5. CRUSTAL STRUCTURE DERIVED FROM OCEANIC LOADING EFFECTS

Oceanic tides produce important displacements of water masses which besides a direct variable attraction effect on the instruments installed in the continental areas produce loading tilts of the crust which extend their effects far away in the middle of the continents. They are usually called the "indirect effects" (Chapter 11).

The indirect effects are difficult to distinguish from the "body force" tide because their frequencies are exactly the same, being derived from the same astronomical input. However, the spatial behaviour of the two tidal effects is quite different. The bodily tide varies smoothly over the Earth's surface, the load tide is more irregular because of the discontinuity in the forcing function at the coast line and because ocean tides form localized circulations around the amphidromes. The only difference in the associated boundary conditions is that on the free surface the latter exerts a normal stress which is missing in the former.

The various acceptable Earth models differ only slightly in their response to the tidal body force so that simple subtraction of the calculated body tide from the observed tide gives a good estimate of the load tide.

Our knowledge of the elastic properties of the Earth's crust and upper mantle is provided almost exclusively by seismology. Seismic studies determine compressional and shear wave velocities which, together with density, completely describe the elastic parameters of a region in the Earth.

The study of crustal loading differs from seismology in three important ways (as noted by Lambert and Beaumont). First, the observed deformation is sensitive to the Earth's elastic properties averaged over the area including the load and the observation site, whereas the travel times of seismic waves depend on the elastic properties and density averaged over the path between source and receiver. Second,tilt - unlike seismic wave velocities - is insensitive to the density distribution in the Earth providing the load is not too extensive.

Third, the frequencies of loading are intermediate between those of seismic wave transmission, a nearly elastic process, and tectonic deformation, a very anelastic process. Therefore, loading observations are potentially useful in determining the frequency at which crustal and upper mantle rocks start to respond anelastically.

6. OCEANOGRAPHY

Ocean-continent interactions are investigated in marine geodesy. Trans World tidal gravity profiles are now under way between Europe and Polynesia (about 35 ICET stations) as well as in North America, in order to determine with precision the loading effects produced by the oceanic tides on the upper crust. The European profile is more difficult to interpret than the US profile because of the complication of the loading effect due to oceanic tides. Indeed, the distribution of oceanic tides along European coasts is very complicated. Similar difficulties arise in South East Asia and the South Pacific area.

7. HYDROLOGY

The elastic deformations of the Earth involve periodic cubic expansions and compressions. They in turn produce oscillations in the water wells with opposite phase (Chapter 10).

Analyses of the water-level fluctuations caused by Earth tide can be used to compute the specific storage and the porosity of the aquifer, parameters that greatly interest groundwater hydrologists.

The importance of tidal variations of underground fluids and gases is emphasized as recent studies suggest that pore fluids play an important role in tectonic processes and that their behaviour can be diagnostic of them. It is important therefore to know the response spectrum of well systems to solid strain.

8. TECTONICS

In addition to that under the heading of Hydrology, we may expect that the harmonic constants (amplitude and phase) of the main tidal waves determined at many places could give new information on the limits of the plates constituting the Earth's crust.

List of Symbols

1- Astronomical parameters

Symbol	Description	Dimensions
G	Newtonian gravitation constant (6 672 ± 4.1) 10^{-14} m^3 s^{-2} kg^{-1}	M^{-1} L^3 T^{-2}
ϕ	latitude	
θ	colatitude = $\frac{\pi}{2} - \phi$	
λ	longitude	
AH	hour angle (local)	
H	hour angle (zero meridian)	
t	universal time (UT)	T
t'	sidereal time	T
$t_{\odot}$	solar time	T
τ	lunar time	T
s	mean longitude of the Moon	
h	mean longitude of the Sun	
p	mean longitude of the Moon perigee	
N	mean longitude of the node of the Moon's orbit	
N' = -N		
p_s	mean longitude of the Sun's perigee	
ε	obliquity of the ecliptic	
i	inclination of the orbit	
μ	ratio of the mass of the Moon to the mass of the Earth 1/81.3007 ± 0.0003	
z	zenithal distance	
α	right ascension	
δ	declination	

2- Geodetical parameters

Symbol	Description	Dimensions
a	equatorial radius of the Earth (6 378 140 ± 5)m	L
c	polar radius of the Earth 6 356 755 m	L
$e = \frac{a-c}{a}$	Earth's flattening 1/e=(298 257±1.5)X10^{-3}	no
A,B,C	principal moments of inertia	$M L^2$
D,E,F	products of inertia	$M L^2$
D	Doodson's constant 2.627 723 m^2 s^{-2}	L^2 T^{-2}
h,k,ℓ,f	Love numbers	no

h', k', ℓ'	Load deformation numbers	no
g	gravity	$L\ T^{-2}$
GM	geocentric gravitational constant including the atmosphere $(3\ 986\ 005 \pm 3)\ 10^{8}\ m^{3}\ s^{-2}$	$L^{3}\ T^{-2}$
r	radius vector	L
r_o	radius of the core 0.545 13a	L
V	Earth gravitational potential	$L^{2}\ T^{-2}$
W	Tidal potential	$L^{2}\ T^{-2}$
$\vec{\omega}(pqr)$	instantaneous rotation vector of the Earth	
ω	rotational angular velocity of the Earth $7\ 292\ 115 \times 10^{-11}$ rad s^{-1}	T^{-1}
ω_i	tidal frequencies	T^{-1}

3- Rheological paremeters

ρ	density	$M\ L^{-3}$
p	pressure	$M\ L^{-1}\ T^{-2}$
λ	compressibility, Lamé constant	$M\ L^{-1}\ T^{-2}$
μ	modulus of rigidity (shear modulus), Lamé constant	$M\ L^{-1}\ T^{-2}$
$k=\lambda+\frac{2}{3}\mu$	bulk modulus	$M\ L^{-1}\ T^{-2}$
ν	Poisson coefficient $\lambda/2(\lambda+\mu)$	no
E	Young modulus $\mu(3\lambda+\mu)/(\lambda+\mu)$	$M\ L^{-1}\ T^{-2}$
Q	quality factor	no
ε	phase-lag	no
$e_{rr}, e_{\theta\theta}, e_{\lambda\lambda}$	strain components	$M\ L^{-1}\ T^{-2}$
$\sigma_{rr}, \sigma_{\theta\theta}, \sigma_{\lambda\lambda}$	stress components	no
θ	cubical dilatation	no
k_f	modulus of compressibility of the fluid	$M\ L^{-1}\ T^{-2}$
k_r	modulus of compressibility of the rock	$M\ L^{-1}\ T^{-2}$
k_p	permeability of the rock	L^{2}
ϕ	volume porosity	no
S	storage coefficient or hydraulic capacitivity	$M^{-1}\ L\ T^{-2}$
C	hydraulic conductivity	T
ν	kinematic viscosity	$L^{2}\ T^{-1}$

CHAPTER 1

Tidal Potential

1.1. INTRODUCTION

We consider a planet whose figure is an ellipsoid of revolution and whose internal distribution of mechanical (density ρ) and rheological (elasticity, viscosity, plasticity) parameters present the same ellipsoidal symmetry of revolution.

Let $W(r,\phi,\lambda)$ (r, radius vector; ϕ, latitude; λ, longitude; $\theta=\pi/2-\phi$ is the co-latitude) be the differential gravitational potential exerted at point $A(r,\phi,\lambda)$ by an external body at a distance d from the centre of mass of the planet which is taken as the origin of the reference system.

$$W(A) = GM \sum_{n=2}^{\infty} \frac{r^n}{d^{n+1}} P_n \ (\cos z), \qquad (1.1)$$

where M is the mass of the disturbing body, P_n is the Legendre's polynomial of order n, and z is the geocentric zenithal distance of the external body at the considered point A.

If we limit ourselves to the first terms of second and third order, then we have

$$\left.\begin{aligned} W_2(A) &= \frac{GM}{2} \left[\frac{r^2}{d^3} (3 \cos^2 z - 1) \right], \\ W_3(A) &= \frac{GM}{2} \left[\frac{r^3}{d^4} (5 \cos^3 z - 3 \cos z) \right]. \end{aligned}\right\} \qquad (1.2)$$

This local coordinate is inconvenient in analytical development. Using the fundamental formula of the position triangle of spherical astronomy (Fig.1.1),

$$\cos z = \sin \phi \sin \delta + \cos \phi \cos \delta \cos H(A). \qquad (1.3)$$

This permits the appearance of the geocentric coordinates (ϕ, λ positive towards West) of the point A and the equatorial coordinates (right ascension α and declination δ) of the perturbing body. The local hour angle H(A) is given by

$$H(A) = H - \lambda(A) = \omega t' - \alpha - \lambda(A), \qquad (1.4)$$

H and t' being respectively the hour angle of the external body and the sidereal time for a fixed meridian of the planet, arbitrarily chosen as the origin of the longitudes. In the case of the Earth this is the international zero meridian. ω is the sidereal velocity of rotation of the Earth.

Then we get the general expression

$$W(A) = \sum_{\ell=2}^{\infty} \sum_{m=0}^{\ell} W_{\ell m}\, r^{\ell}\, P_{\ell}^{m} (\sin \delta)\, P_{\ell}^{m} (\sin \phi) \cos mH[A], \qquad (1.5)$$

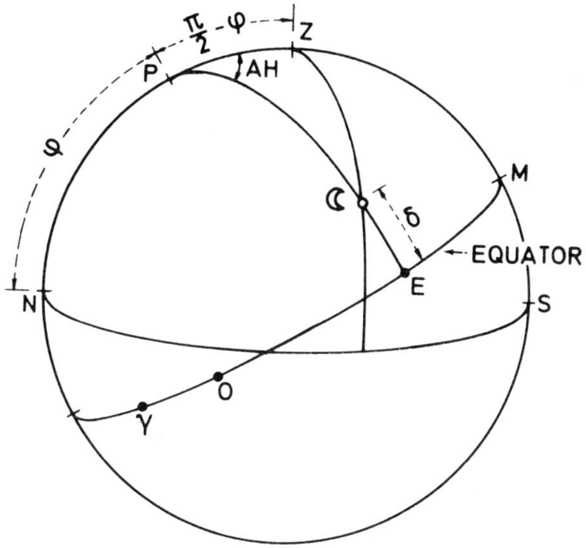

Fig. 1.1. O is the conventional origin of the longitudes defined by BIH in the equator; γ is the vernal equinox; E γE = α, γO is the sidereal time of O[t'(o)]; γM is the local sidereal time t'; EM is the H(A) (hour angle of the Moon); OM is the -λ (λ positive towards the West); EM = γO + OM - γE; H(A) = t' (o) - λ - α and H(A) = ωt' - λ - α.

where

$$W_{\ell m} = \frac{2(\ell-m)!\ GM}{(\ell+m)!\ d^{\ell+1}} \qquad (m = 1 \text{ to } \ell) \tag{1.6}$$

and

$$W_{\ell o} = \frac{GM}{d^{\ell+1}} \tag{1.7}$$

and, further,

$$\cos m\, H(A) = \cos m\, H \cos m\lambda + \sin mH \sin m\lambda. \tag{1.8}$$

Generally limited to second order, this expression furnishes the classical development of Laplace in three families of spherical harmonics (Fig. 1.2):

$$W_2(A) = \frac{3}{4}\, GM\, \frac{r^2}{d^3} \begin{cases} \cos^2\phi \cos^2\delta \cos 2H(A) & \text{sectorial,} \\ + \sin 2\phi \sin 2\delta \cos H(A) & \text{tesseral,} \\ + 3(\sin^2\phi - \frac{1}{3})\ (\sin^2\delta - \frac{1}{3}) & \text{zonal,} \end{cases} \tag{1.9}$$

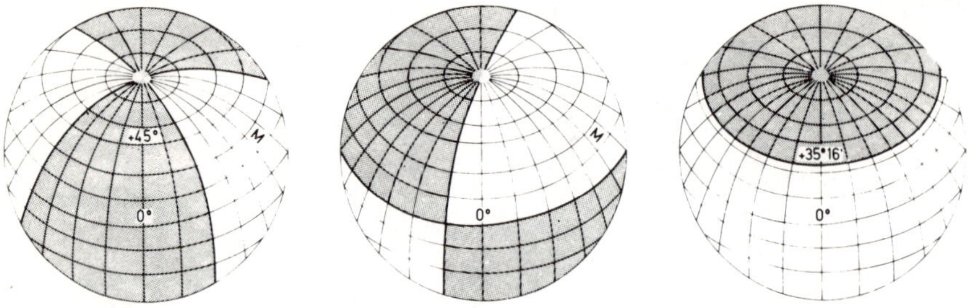

(a) (b) (c)

Fig. 1.2. The three kinds of tides

We owe this analysis of the potential into three terms to Laplace, who was the first to call attention to its remarkable meaning and geometric characteristics. These three terms represent the three types of spherical harmonic functions of the second order (Fig. 1.2).

Figure 1.2a represents the first of these functions which as nodal lines (lines where the function is zero) has only the meridians situated 45° to either side of the meridian of the perturbing body. These lines divide the spherical surface into four sectors where the function is alternately positive and negative. The areas where W is positive are those of high tides, the negative areas are those of low tides.

This function is termed the *sectorial function*, the period of the corresponding tides is *semi-diurnal*, and their amplitude has a maximum at the equator when the declination of the perturbing body is zero. Amplitudes are zero at the poles. Laplace called tides of this type "marées de troisième espèce" (tides of the third kind). It should be noted that variations of mass distribution at the Earth's surface subject to the sectorial distribution do not modify either the position of the pole of inertia or the major moment of inertia C (which determines the speed of rotation of the Earth).

However, these tides are responsible at least partly for the secular retardation of the Earth's speed of rotation owing to internal friction and energy dissipation (see Chapter 14).

As nodal lines, the second function represented on Fig. 1.2b has a meridian (90° from the meridian of the perturbing body) and a parallel, namely the equator. This is a *tesseral function* dividing the sphere into areas which change sign with the declination of the perturbing body. The corresponding tidal period is *diurnal*, and the amplitude is maximum at latitude 45°North and 45° South when the declination of the perturbing body is maximum; the amplitude is always zero at the equator and at the poles. Laplace referred to tides of this type as "marées de deuxième espèce" (tides of the second kind). The variations of mass distribution at the surface of the Earth following the tesseral distribution produce oscillations of the principal axis of inertia but no variations of the principal moment of inertia C. The perturbing potential due to the polar motion has the same form. This distribution corresponds to the precession-nutational torque which, acting on the Earth's equatorial bulge, tends to tilt the equatorial plane towards the ecliptic. Thus a diurnal tesseral wave in the harmonic development of the tidal potential corresponds to a term in the development of the precession-nutation. The effect of this torque may be that the fluid core rotates relative to the mantle.

The third function which is represented on Fig. 1.2c depends only on the latitude, this is a *zonal function*; its nodal lines are the parallels +35°16' and -35°16'. Since it is a squared sine function of the declination of the perturbing body, its fundamental period will be *fourteen days* for the Moon and *six months* for the Sun. These are Laplace's tides of the first kind. The variations of the mass distribution inside the Earth corresponding to the zonal distribution do not produce displacement of the inertial pole, but do affect the principal moment of inertia C. We may therefore expect some fluctuations in the Earth's speed of rotation corresponding to the periods given above which are effectively detected by the time services equipped with high-precision instruments. Moreover, the constant part of this function has as a consequence that the equipotential surface is lowered 28 cm at the pole and raised 14 cm at the equator. The effect of this permanent tide is a slight increase of the Earth's flattening.

1.2. COMPONENTS OF THE TIDAL FORCE

To measure the components of the tidal forces at a given place one normally has to set up three instruments directed along the three axes of a local reference system which are usually chosen as follows (Fig. 1.3):

a) The direction of the vertical towards the zenith as the Oz axis along which acts terrestrial gravity (gravity + centrifugal force). The instrument utilized for this measurement will necessarily be a gravimeter.

b) Two directions in the horizontal plane which will be that of the meridian (Ox) towards the South, and that of the prime vertical (Oy) towards the East. The instrument will be a horizontal pendulum for each direction.

The components of the tidal force along these three directions are obtained by taking the derivative of the potential with respect to the corresponding variables:

OZ: $-\dfrac{\partial W}{\partial r}$ vertical component (variation of the acceleration of gravity),

OX: $-\dfrac{\partial W}{r\partial \phi}$ North-South horizontal component,

OY: $-\dfrac{\partial W}{r\cos\phi\;\partial \lambda}$ East-West horizontal component.

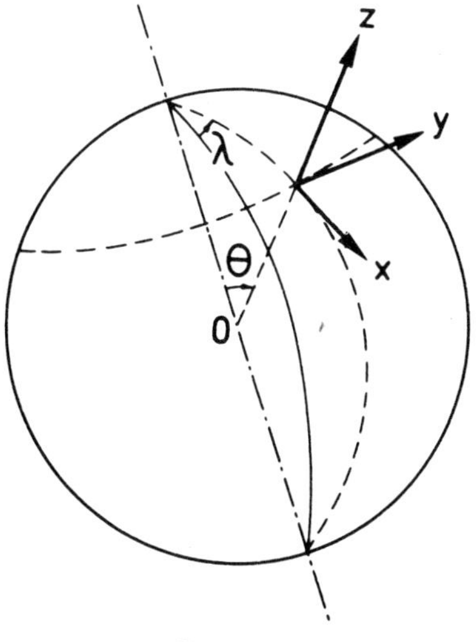

Fig. 1.3

This gives three trigonometric functions for each tidal family as shown in Table 1.1.

A representation of the forces as functions of latitude is given on Figs. 1.4, 1.5 and 1.6.

TABLE 1.1. Trigonometric part in the expression of the different families of Tides in the three local components

Function	Potential and vertical component	North-South component	East-West component
Zonal	$3(\sin^2\phi-\frac{1}{3})\ (\sin^2\delta-\frac{1}{3})$	$-\ 3\ \sin 2\phi\ (\sin^2\delta-\frac{1}{3})$	0
Tesseral	$\sin 2\phi\ \sin 2\delta\ \cos H$	$-\ 2\ \cos 2\phi\ \sin 2\delta\ \cos H$	$+2\sin\phi\ \sin 2\delta\ \sin H$
Sectorial	$\cos^2\phi\ \cos^2\delta\ \cos 2H$	$+\ \sin 2\phi\ \cos^2\delta\ \cos 2H$	$+\ \cos\ \phi\ \cos^2\delta\ \sin 2H$

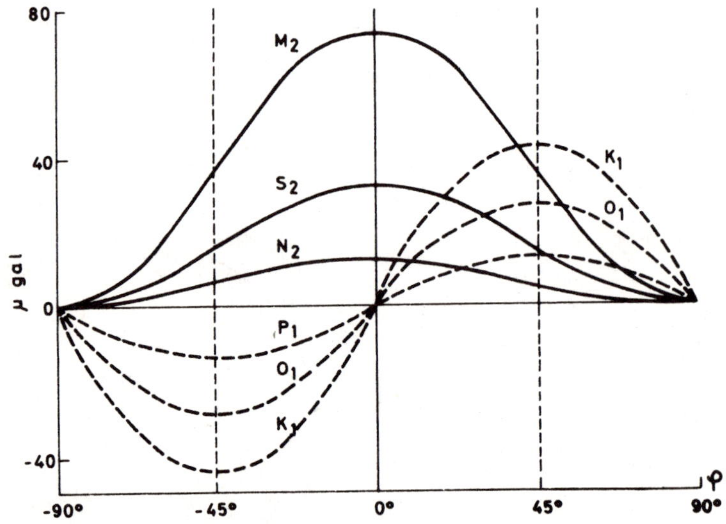

Fig. 1.4. Amplitude variation of the principal waves as a function of latitude for the vertical component of tidal force.
K_1, O_1, P_1: tesseral diurnal waves;
M_2, S_2, N_2: sectorial semi-diurnal waves.

The direction of the vertical is the resultant of the forces applied to the point under consideration, and hence of the composition of this perturbational force with gravity.

Thus the deviation of the vertical (which is to be taken in the opposite direction of the plumb line deviation) is given at the surface of the Earth (r = a) by its components:

$$\left.\begin{aligned} n_1 &\simeq \tan n_1 = \frac{-1}{ag}\frac{\partial W}{\partial \theta} && \text{along the meridian} \\ n_2 &\simeq \tan n_2 = \frac{1}{ag \sin\theta}\frac{\partial W}{\partial \lambda} && \text{along the prime vertical} \end{aligned}\right\} \quad (1.10)$$

Figure 1.6 relating to the tesseral part of the tidal potential, shows that the direction of the vertical describes a circular cone at the pole. This is evident *a priori*: at this singular point the directions North-South and East-West are meaningless, hence the vertical can only be fixed (in the case of the semi-diurnal sectorial waves) or describe a circle (in the case of the tesseral diurnal waves).

It will be noted that the East-West component does not contain any long-period terms, which was evident *a priori* since these terms are derivatives of a *zonal* harmonic function.

1.3. UNITS

We should adopt the SI units which are the radian for the angular measurements and the metre per second squared for the accelerations. Unfortunately these units are far too big and thus inconvenient and out of touch with normal usage.

Since time immemorial astronomers have used a sexagesimal system which has many arithmetical advantages. It is obviously unrealistic to ask them to measure now in nanoradians. Therefore we shall continue to use the sexagesimal second of degree, keeping in mind that

$$0''.001 = 4.848 \times 10^{-9} \text{ radian or about 5 nanoradians.}$$

Similarly, we shall keep the unit of the gravimetrists

$$1 \text{ gal} = 1 \text{ cm s}^{-2},$$

with its submultiples milligal and microgal for the accelerations.

1.4. GEOID TIDES

An equipotential surface is a surface upon which any displacement does not involve any work. We can define it as normal, at each point P, to the resultant of all forces acting on that point, i. e. the gravity, the centrifugal force and the luni-solar forces deriving from the potential W.

Let

$$V = h_1$$

be the equation of one of these surfaces.

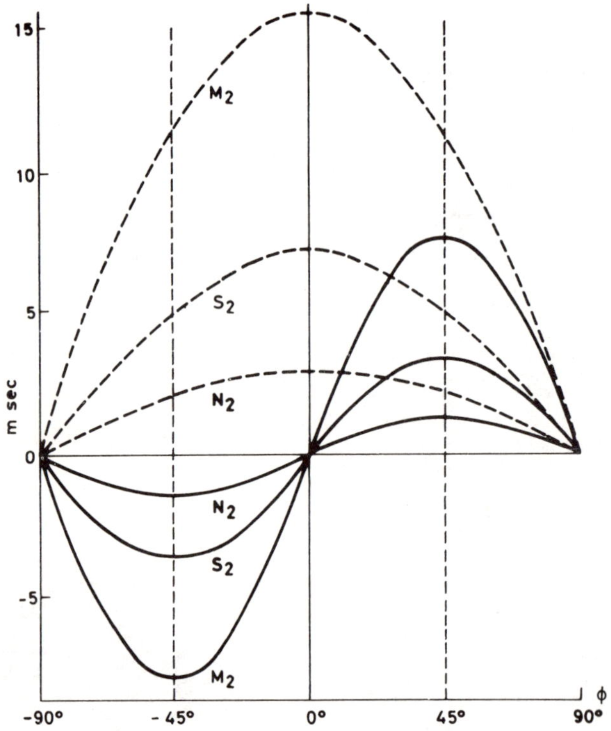

Fig. 1.5. Amplitude variation of the principal semi-diurnal sectorial waves (M_2, S_2, N_2) as a function of the latitude for the horizontal components of tidal force. Dashed line: East-West component.
Solid line : North-South component.

h_1 can define the distance (geopotential height, expressed in metres kilogals) to a reference surface which can be the geoid of which the equation will thus be

$$V_0 = 0.$$

The acceleration of gravity is

$$g = -\frac{\partial V}{\partial r}. \tag{1.11}$$

Now let us apply the tidal forces. The equipotential surface is deformed: the point P(r) is transported to P(r+ξ) where the Earth's potential is given by a Taylor development limited to the first order:

$$V(r + \xi) = V(r) + \xi \frac{\partial V(r)}{\partial r}$$

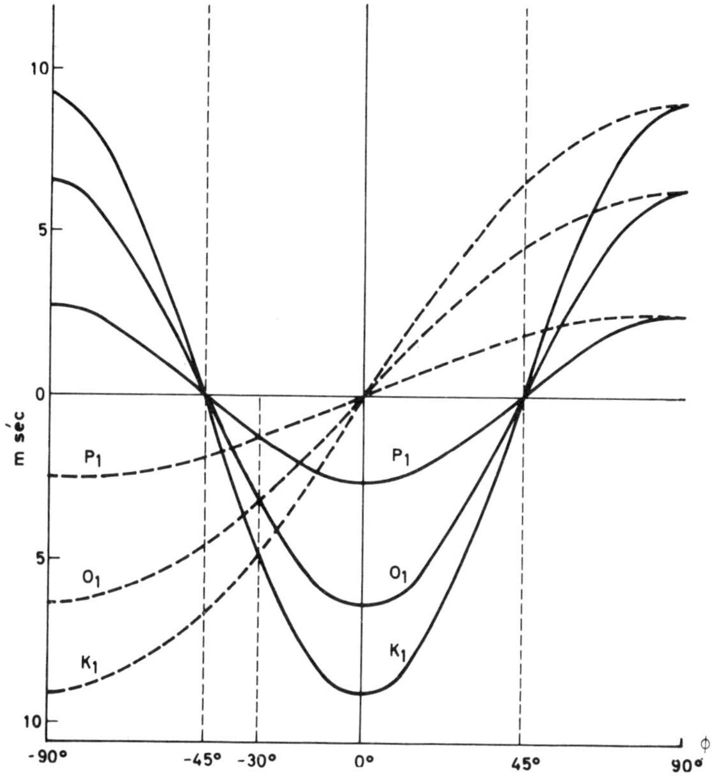

Fig. 1.6. Amplitude variation of the principal diurnal tesseral waves (K_1, O_1, P_1) as a function of latitude for the horizontal components of tidal force. Solid line: North-South component. Dashed line:East-West component.

and the equation of the equipotential surface of level h_1 becomes

$$V + \xi \frac{\partial V}{\partial r} + W = h_1.$$

As it is an equipotential, the variable part of it must be zero, thus

$$\xi \frac{\partial V}{\partial r} = - W$$

or

$$\xi = \frac{W}{g}, \tag{1.12}$$

which will be called here the "geoïd tide".

This conclusion may be reached directly by stating that the work accomplished in transporting the unitary mass from the initial level to the perturbed level is ξg, which must be equal to the potėntial variation W.

1.5. NUMERICAL ESTIMATION OF THE TIDAL EFFECTS

In the expression (1.1) of the potential let us introduce as constants the mean distance of the Moon c and the radius of the sphere of volume equal to that of the Earth:

$$\bar{a} = \sqrt[3]{a^2 b}.$$

We can define a general constant for tidal phenomena in the form

$$D = \frac{3}{4} GM \frac{\bar{a}^2}{c^3}, \qquad (1.13)$$

which is named Doodson's constant. This constant has the dimensions of a work ($L^2\ T^{-2}$). Thus

$$W_2 = 2D \left(\frac{c}{d}\right)^3 \left(\frac{r}{\bar{a}}\right)^2 \left(\cos^2 z - \frac{1}{3}\right),$$

$$W_3 = \frac{2}{3} D \left(\frac{c}{d}\right)^4 \left(\frac{r}{\bar{a}}\right)^3 \left(5 \cos^3 z - 3 \cos z\right),$$

Numerical Value of Doodson's Constant

From its definition we get

$$D = \frac{3}{4} \left(\frac{\bar{g}\ \bar{a}^2}{E}\right) M_☾ \frac{\bar{a}^2}{c^3}, \qquad (1.14)$$

where $M_☾$ is the mass of the Moon, E is the mass of the Earth, $\bar{g} = GE/\bar{a}^2$ is the so called "mean gravity" (mean value of g on a sphere having the same mass and volume as the Earth), sin π is the sine of the lunar parallax (*), and a_E is the equatorial radius of the Earth, giving

$$D = \frac{3}{4} \left(\frac{M}{E}\right) \bar{g}\ \bar{a}^4 \left(\frac{\sin^3 \pi}{a_E^3}\right). \qquad (1.15)$$

Then, comparing the old and new fundamental constants of astronomy we can establish Table 1.2.

TABLE 1.2. The Fundamental Constants used in Tidal Research

Parameter	Numerical value before 1967	Numerical value since 1967	Ratio
M/E	1/81.53	1/81.30	1.002 829
$\bar{a}$	6 371 221.00	6 371 023.60	0.999 875
$\bar{g}$	9.820 4	9.820 24	0.999 984
a_E	6 378 388.00	6 378 160.00	1.000 107
sin π	0.016 593 71	0.016 592 51	0.999 774
D($m^2\ s^{-2}$)	2.620 629	2.627 723	1.002 57

(*) One must take care to distinguish here the sine equatorial parallax defined as a/c = sin π = 3 422"451 = 0.016 592 51 radian from the equatorial parallax π which is π = 3 422"608 = arc sin 3 422"451.

The value of D for the Sun should be

$$D' = \frac{M_\odot}{M_☾} \frac{c_☾^3}{c_\odot^3} D, \tag{1.16}$$

$$D' = 0.459\ 90\ D. \tag{1.17}$$

The order of magnitude of the main tidal phenomena can now easily be estimated.

The purely numerical coefficient in expression (1.12) of the geoid tide is

$$\frac{D}{g} = \frac{2.627\ 723}{9.820\ 240} = 0.267\ 6\ \text{m}, \tag{1.18}$$

while its trigonometrical part can vary from -1 to +1.

The total range of the lunar tides can thus reach 53.52 cm. The solar contribution will reach 24.61 cm and the total could in exceptional occasions reach 78.13 cm.

The time variation of gravity being given at the surface by

$$\Delta g = -\left(\frac{\partial W_2}{\partial r}\right)_{r=a} = -\frac{2}{a} W_2\ (a), \tag{1.19}$$

its numerical coefficient is

$$\frac{2D}{a} = 0.082\ 49\ \text{mgal}, \tag{1.20}$$

Thus the Moon effect has a total range of 0.1650 mgal and the Sun effect has 0.075 9 mgal. The periodic deviation of the vertical

$$n \sim \tan n = -\frac{1}{ag}\frac{\partial W}{\partial z} = \frac{2D}{ag} \sin 2z \tag{1.21}$$

has a coefficient

$$C_{xy} = \frac{2D}{ag} = 8.399 \times 10^{-8}\ \text{radian} = 0''\!.017\ 326 \tag{1.22}$$

for the Moon and 0".007 968 for the Sun. This makes a total (exceptional) range of 0".050 6.

1.6. BASIC ASTRONOMICAL DEVELOPMENTS

The expression of the potential obtained by Laplace (1.9) and its derivatives (Table 1.1) are not suitable for the analysis of tidal phenomena because the term $(c/d)^3$ as well as the trigonometric functions containing δ and H exhibit very complicated time variations due to the complexity of the orbital motions of the Earth around the Sun and of the Moon around the Earth.

The components of these motions must be separated carefully in order to describe the tidal potential and its derivatives as a sum of purely sinusoidal waves, i. e. waves having as argument purely linear functions of the time.

Such a development was given for the first time by Ferrel in 1874, but it contained only a very limited number of principal waves, while Doodson published in 1922 a purely harmonic expansion containing the 386 components having an amplitude coefficient greater than (or equal to) 0.000 1 D.

Recently, Cartwright, Tayler and Edden have published a new extended development obtained from a computer program. This development is in general in excellent agreement with Doodson.

The choice of six variables - which are, over an interval of a century or so, practically linear increasing functions of the time - was made by Doodson on the basis of the results accumulated in fundamental astronomy.
These variables are:

- τ — mean lunar time = H + 180° (it is counted from the lower transit of the Moon).
- s — mean tropic longitude of the Moon.
- h — mean tropic longitude of the Sun.
- p — mean tropic longitude of the lunar perigee.
- $N' = -N$ — mean tropic longitude of the ascending lunar node which is changed in sign because N is the only one of the six variables which is decreasing with time (retrogression of the node).
- p_s — mean tropic longitude of the perihelion.

These angles are measured with respect to the mean instantaneous vernal equinox (the aberration not being taken into account) and are therefore called *mean* longitudes.

One has immediately the relationship

$$\tau + s = t + h = t' \qquad \text{(sidereal time)}, \tag{1.23}$$

t being the mean solar time; thus

$$t = \tau + s - h. \tag{1.24}$$

Taking advantage of the work performed by all the astronomers of the last century, it has been possible to determine with extraordinary precision the coefficients in the quasi-linear expressions of these variables by utilizing the extensive observations.

From 40 000 observations of the Sun, covering an interval of 140 years, Newcomb derived in 1895 the following formulas for the solar elements:

$$h = 279\overset{\circ}{.}696\ 68 + 36\ 000\overset{\circ}{.}768\ 92\ T + 0\overset{\circ}{.}000\ 30\ T^2, \tag{1.25}$$

$$p_s = 281\overset{\circ}{.}220\ 83 + 1\overset{\circ}{.}719\ 02\ T + 0\overset{\circ}{.}000\ 45\ T^2 + 0\overset{\circ}{.}000\ 003\ T^3 \tag{1.26}$$

the obliquity of the ecliptic:

$$\varepsilon = 23°27'8\overset{''}{.}261 - 46\overset{''}{.}845\ T - 0\overset{''}{.}005\ 9\ T^2 + 0\overset{''}{.}001\ 83\ T^3 \tag{1.27}$$

and the eccentricity of the orbit:

$$e = 0.016\ 751\ 04 - 0.000\ 041\ 80\ T - 0.000\ 000\ 126\ T^2, \tag{1.28}$$

T being the time expressed in Julian centuries of 36 525 mean solar days, starting with

T = 0 at 1900 January 0.5 ET (ET: ephemeris time)
or 1900 January 0 at 12h ET,
that is 1899 December 31 at 12h 0m 0s ET,
that is 2 415 020.00 Julian date.

In 1919 Brown gave similar expressions for the lunar elements:

$$s = 270°436\ 59 + 481\ 267°890\ 57\ T + 0°001\ 98\ T^2 + 0°000\ 002\ T^3 \quad (1.29)$$

$$p = 334°329\ 56 + 4\ 069°034\ 03\ T - 0°010\ 32\ T^2 - 0°000\ 01\ T^3 \quad (1.30)$$

$$N' = 259°183\ 28 + 1\ 934°142\ 01\ T + 0°002\ 08\ T^2 + 0°000\ 002\ T^3 \quad (1.31)$$

Remarks

One has, indeed, h=0 on March 21, h=180° on September 23, and h=270° on December 21. It follows that January 0 at 12h corresponds, indeed, to 279°7 as the Sun's longitude increase is about 0°98 per day.

Moreover, formula (1.25) gives

$$x = \frac{36\ 000.768\ 92}{36\ 000} = 1.000\ 021\ 359$$

$$\frac{365.25}{x} = 365.242\ 198\ 79, \text{ duration of the tropic year.}$$

The longitude of the Sun is 279°41'48"048 and its linear coefficient (36 000°768 92 = 129 602 768"13 per Julian century = 3548"330 407 per mean solar day) define the epoch 1900 January 0, 12h ET.

The tropic year J = 365.242 198 79 days contains J X 86400 = 31 556 925. 9747 seconds of ET.

This is the definition of the second of ET (International Astronomical Union, Hamburg, 1964).

Let us observe that the term $+0°001\ 98\ T^2 = +7"128\ T^2$ represents the so-called secular acceleration of the Moon(*) on its orbit which, later on (see Chapter 14), will be an important correction to be taken into account for the evaluation of the secular retardation of the Earth's rotation.

In the improved lunar ephemeris (1954), this term has been now corrected for the observed lunar acceleration $1/2\ \dot{n}\ T^2$; the value of $\dot{n}$ as determined by Spencer Jones from lunar eclipses and occultations (-22"44) was adopted.

This gives

$$-4"09\ T^2 = -0°001\ 136\ T^2$$

Recent investigations based upon a very large amount of data have shown that the Spencer Jones coefficient is too small by a factor of about 2, the concluded value being -40"75 ± 2"25 which should give a term

$$-13"25\ T^2 = -0°003\ 680\ 5\ T^2$$

The precision of these numerical coefficients is very high as they are currently used to define ET which is uniform to 10^{-9} and has been compared carefully with atomic time (TA). This means that tidal frequencies will be determined from astronomy with a corresponding precision: this remark has a bearing on the construction of numerical methods of analysis of tides.

From the coefficients of the first-order terms we can easily derive the hourly speed of the six fundamental variables.

(*) In the formula the acceleration has to appear as $1/2\ \gamma\ T^2$ and therefore the acceleration is twice the coefficient, i. e. + 14"256.

Example for $\dot{s}$: $\frac{481\ 267^{\circ}.890\ 57}{24 \times 365\ 25} = 0^{\circ}.549\ 016\ 5.$

In this way we obtain the hourly speeds and corresponding periods given in Table 1.3.

The synodic month (29.5 days) gives approximately the commensurability of the lunar and solar main periods (see Table 1.2) and will be a fundamental interval for an harmonic analysis.

1.6.1. The Orbit of the Moon

The major perturbations produced by the Sun on the orbit of the Moon are evection and variation, both of which change periodically the ellipticity of the orbit.

It may be easily understood that if the Moon's perigee remained fixed, the projection of the major axis in the ecliptic should contain the Sun on it twice a year, and this should result in an increase of the ellipticity, the frequency of which is evidently $(\dot{s}-\dot{p}) - 2(\dot{h}-\dot{p}) = (\dot{s}-2\dot{h}+\dot{p})$. This is called *evection*.

Another effect of the Sun is also to modify the ellipticity at each conjunction (new moon) or opposition (full moon) and the argument is $2(\dot{s}-\dot{h})$. This is called *variation*.

True Longitude

The longitude variation can easily be deduced from the ellipse formula as follows.

Let c be the semi-major axis of the lunar orbit.

Then

$$d = c\,[1-e\cos(\dot{s}-\dot{p})t] \qquad (1.32)$$

and

$$\frac{c}{d} = 1 + e\cos(\dot{s}-\dot{p})t. \qquad (1.33)$$

The second order of e may be provisionally neglected as

$$e = 0.054\ 900\ 489. \qquad (1.34)$$

Now Kepler's second law can be expressed as

$$\frac{1}{2}\dot{s}\,d^2 = \text{constant} = \frac{1}{2}\dot{s}_o\,c^2,$$

where $\dot{s}_o$ is the mean angular speed corresponding to the mean distance c.

Therefore with

$$d^2 = c^2\,[1-2e\cos(\dot{s}-\dot{p})t + e^2\cos^2(\dot{s}-\dot{p})t]$$

we obtain

$$\dot{s}c^2\,[1-2e\cos(\dot{s}-\dot{p})t] = \dot{s}_o\,c^2$$

and

$$\dot{s} = \dot{s}_o\,[1+2e\cos(\dot{s}-\dot{p})t]. \qquad (1.35)$$

The true longitude of the Moon at the instant t is

$$\ell = \int_0^t \dot{s}\,dt = \dot{s}_o t + \left(\frac{2\,e\,\dot{s}_o}{\dot{s}-\dot{p}}\right)\sin(\dot{s}-\dot{p})t \qquad (1.36)$$

TABLE 1.3. Periods of variation of the six variables with corresponding hourly speeds

Variable	Interval defined	Hourly speed	Period
$\dot{t} = \dot{t}' - \dot{h}$	Mean solar day	15°.000 000 0	1.000 000 m.s.d.
$\dot{t}' = \dot{t} + \dot{h}$	Sidereal day	15°.041 068 6	0.997 270 m.s.d. $(t-4^m)$
$\dot{\tau} = \dot{t}' - \dot{s}$	Mean lunar day	14°.492 052 1	1.035 050 m.s.d. = 24h 50.47m
$\dot{s}$	Tropic month	0°.549 016 5	27.321 582 day. Period of variation of the declination
$\dot{h}$	Tropic year	0°.041 068 6	365.242 199 day (oscillations in longitude)
$\dot{p}$		0°.004 641 8	8.847 year. Period of revolution of the mean perigee of the Moon
$\dot{N}'$		0°.002 206 4	18.613 year. Period of revolution of lunar nodes (Saros)
$\dot{p}_s$		0°.000 002 0	20.940 year. Period of revolution of the solar perihelion
$\dot{s} - \dot{N}$	Mean draconitic month	0°.551 222 9	27.212 22 day. Oscillations of the Moon in latitude
$\dot{s} - \dot{p}$	Mean anomalistic month	0°.544 374 7	27.554 55 day. Interval between two passages of the Moon at the perigee
$\dot{s} - \dot{h}$	Mean synodic month	0°.507 947 9	29.530 59 day. Return of the lunar phases
$\dot{s} - 2\dot{h} + \dot{p}$		0°.471 521 1	31.812 day. Evection period
$\dot{h} - \dot{p}_s$		0°.041 066 7	365.259 64 day. Mean anomalistic year
$\dot{h} - \dot{p}$		0°.036 426 8	411.784 71 day
$2(\dot{s}-\dot{h})$		1°.015 895 8	14.765 30 day. Period of variation

or

$$\ell = \dot{s}_0\, t + 0.110\,809\,9 \sin(\dot{s}-\dot{p})t \quad \text{in radians,}$$

$$\ell = \dot{s}_0\, t + 22\,856''18 \sin(\dot{s}-\dot{p})t. \qquad (1.37)$$

By the same procedure we can introduce evection and variation and from

$$\frac{c}{d} = 1 + \underset{\text{ellipticity}}{0.054\,9 \cos(\dot{s}-\dot{p})t} + \underset{\text{evection}}{0.010 \cos(\dot{s}-2\dot{h}+\dot{p})t} + \underset{\text{variation}}{0.008 \cos(2\dot{s}-2\dot{h})t} \qquad (1.38)$$

we obtain

$$\ell = \dot{s}_0 t + 0.110\,8 \sin(\dot{s}-\dot{p})t + 0.023 \sin(\dot{s}-2\dot{h}+\dot{p})t + 0.011 \sin(2\dot{s}-2\dot{h})t \qquad (1.39)$$

The complete development of ℓ is given in Woolard (1953), table 4, page 52(*).

(*)Astronomers use the following arguments: $\ell=s-p$ $\ell'=h-p_s$ $D=s-h$ $F=s-N$.

One easily establishes the first terms of the development of the trigonometric functions entering in the development of the potential:

$$\sin\delta = \sin\varepsilon \sin \ell = 0.397\ 98 \sin \ell,$$

$$\sin^2\delta = 0.079\ 196\ (1-\cos 2\ell),$$

and consequently

$$\cos^2\delta = 0.920\ 80 + 0.079\ 196 \cos 2\dot{s}t - 0.036 \cos \dot{N}t - 0.036 \cos(2\dot{s}-\dot{N})t + \ldots \qquad (1.40)$$

$$\sin 2\delta = 2 \sin\delta \cos\delta = 0.763\ 79 \sin \dot{s}t + \ldots \qquad (1.41)$$

From (1.38) one has also

$$\left(\frac{c}{d}\right)^3 = 1 + 0.164\ 7 \cos(\dot{s}-\dot{p})t + 0.030 \cos(\dot{s}-2\dot{h}+\dot{p})t + 0.024 \cos(2\dot{s}-2\dot{h})t. \qquad (1.42)$$

1.6.2. The Orbit of the Earth

We will consider here only the principal term

$$\frac{c}{d} = 1 + e \cos(\dot{h}-\dot{p}_s)t \qquad (1.43)$$

with

$$e = 0.016\ 730\ 1 \text{ at } 1950.0 \ [T = \tfrac{1}{2} \text{ in } (1.28)]$$

which for the longitude of the Sun gives

$$L = \dot{h}_0 t + \left(\frac{2e\dot{h}_0}{\dot{h}-\dot{p}_s}\right) \sin(\dot{h}-\dot{p}_s)t. \qquad (1.44)$$

Thus $L = \dot{h}_0 t + 0.033\ 461\ 7 \sin(\dot{h}-\dot{p}_s)t$ radian (1.45)

or $L = \dot{h}_0 t + 6\ 901''.97 \sin(\dot{h}-\dot{p}_s)t = \dot{h}_0 t + 460\text{s} \sin(\dot{h}-\dot{p}_s)t.$

This coefficient 460s of time is called the "equation of the centre" which is one of the three components of the famous "equation of time".

One has here

$$\left(\frac{c}{d}\right)^3 = 1 + 0.050\ 19 \cos(\dot{h}-\dot{p}_s)t + \ldots \qquad (1.46)$$

1.7. TIDAL SPECTRUM

1.7.1. Sectorial Function - Semi-diurnal Waves

The general expression is

$$D \left(\frac{c}{d}\right)^3 \cos^2\phi \cos^2\delta \cos 2H. \qquad (1.47)$$

If we consider first the lunar effect we write it as follows:

$$D \cos^2\phi\,[1 + 0.164\ 7 \cos(\dot{s}-\dot{p})t + 0.030 \cos(\dot{s}-2\dot{h}+\dot{p})t + 0.024 \cos(2\dot{s}-2\dot{h})t] \times$$

$$(0.920\ 8 + 0.079\ 196 \cos 2\dot{s}t + \ldots) \cos 2\dot{\tau}t \qquad (1.48)$$

which, developed, gives a main wave

$$0.920\ 8 \cos 2\dot{\tau}t \qquad \text{called } M_2$$

and then a series of waves grouped by pairs and resulting from the combination of $\cos 2\dot{\tau}t$ with the other arguments.

For example, $0.164\ 7 \cos(\dot{s}-\dot{p})t \cos 2\dot{\tau}t$ gives the pair

$$0.082\ 3 \cos [2\dot{\tau} + (\dot{s}-\dot{p})]t \qquad \text{called } L_2,$$

$$0.082\ 3 \cos [2\dot{\tau} - (\dot{s}-\dot{p})]t \qquad \text{called } N_2,$$

which are obviously the result of the ellipticity of the Moon's orbit.

Similarly, we have a pair

$$0.039\ 6 \cos(2\dot{\tau} + 2\dot{s})t \qquad \text{called } K_2,$$

$$0.039\ 6 \cos(2\dot{\tau} - 2\dot{s})t,$$

which are "declinational" waves.

The procedure, extended to the complete development of $(c/d)^3$ and $\cos^2\delta$ generates an infinite number of components which have their frequencies symmetrically distributed on both sides of the half-lunar day frequency $2\dot{\tau}$. Arguments and periods can be deduced easily from Table 1.2. For example:

M_2: $2\dot{\tau}$ = 28°.984 104 2 per hour; period = 12h 25m 14s.

N_2: $2\dot{\tau} - \dot{s} + \dot{p}$ = 28°.439 729 5 per hour; period = 12h 39m 30s.

K_2: $2\dot{\tau} + 2\dot{s}$ = 30°.082 137 2 per hour; period = 11h 58m 2s.

Let us now consider the solar effect with (1.46):

$$D' \cos^2\phi\ [1 + 0.050\ 19 \cos(\dot{h}-\dot{p}_s)t + \ldots] \times$$

$$(0.920\ 8 + 0.079\ 196 \cos 2\dot{h}t + \ldots) \cos 2\dot{\tau}t \text{ with } \dot{t} = \dot{\tau}+\dot{s}-\dot{h}.$$

The same elementary procedure generates the waves:

S_2 argument $(2\dot{\tau} + 2\dot{s} - 2\dot{h})$ = 30°.000 00 per hour;
period 12h 0m 0s:

two elliptic waves:

R_2 argument $(2\dot{\tau} + 2\dot{s} - 2\dot{h}) - (\dot{h} - \dot{p}_s)$ = 29°.958 933 per hour;
period 12h 0m 59s,

T_2 argument $(2\dot{\tau} + 2\dot{s} - 2\dot{h}) + (\dot{h} - \dot{p}_s)$ = 30°.041 066 7 per hour;
period 11h 59m 0s;

two declinational waves:

K_2 argument $(2\dot{\tau} + 2\dot{s} - 2\dot{h}) + 2\dot{h}$ = 30°.082 137 2 per hour;
period 11h 58m 2s,

and argument $(2\dot{\tau} + 2\dot{s} - 2\dot{h}) - 2\dot{h}$.

It is immediately obvious that the same pulsation which corresponds to a one-half sidereal day period is produced simultaneously by the Moon and Sun. Respective contributions of the two bodies are in the ratio 0.459 7 but cannot be distinguished or separated within the tidal phenomena. That combined wave will be called K_2.

The evection waves are called λ_2 (argument $2\dot{\tau} + \dot{s} - 2\dot{h} + \dot{p}$) and ν_2 (argument $2\dot{t} - \dot{s} + 2\dot{h} - \dot{p}$), while the main variation wave is called μ_2 (argument $2\dot{\tau} - 2\dot{s} + 2\dot{h}$). Figure 1.10 and Table 1.4 give a pictorial and tabulated view of the sectorial tidal spectrum.

1.7.2. Tesseral Function-Diurnal Waves

Their general expression is

$$D \left(\frac{c}{d}\right)^3 \sin 2\phi \sin 2\delta \cos H \qquad (1.49)$$

which, for the Moon, can be written from (1.41), (1.42) as:

$$-D \sin 2\phi \,[1 + 0.164\,7 \cos(\dot{s}-\dot{p})t + \ldots]\ (0 + 0.763\,79 \sin \dot{s}t + \ldots) \cos \dot{\tau}t \qquad (1.50)$$

[since $\cos H = \cos(\tau - 180°)$].

One immediately sees that the principal term (corresponding to M_2) with argument τ does not exist as the mean value of $\sin 2\delta$ is zero. Thus one has, as principal terms, the declinational waves

$K_1^{☾}$ of argument $\dot{\tau} + \dot{s}$ = 15°.041 068 6; period 23h 56m 4s

and

O_1 of argument $\dot{\tau} - \dot{s}$ = 13°.943 035 6; period 25h 49m 10s.

Similarly, the Sun will generate two declinational waves:

$K_1^{\odot}$ of argument $\dot{t} + \dot{h}$ = 15°.041 068 6; period 23h 56m 4s.

P_1 of argument $\dot{t} - \dot{h}$ = 14°.958 931 4; period 24h 3m 57s.

Again, two waves of exactly the same frequency are generated by the two bodies. This frequency is that of the sidereal rotation of the Earth. This very important luni-solar wave is called K_1: practically two-thirds of its amplitude is furnished by the Moon and one-third by the Sun. $K_1^{☾}$ and O_1 have practically the same amplitude, while the same occurs for $K_1^{\odot}$ and P_1 so that they cancel themselves when the corresponding body is in the equator (δ=0).

Figure 1.7 shows the interference of the wave O_1 with the lunar part of the wave K_1. Their amplitudes cancel themselves when the Moon is in the equator (δ=0).

Therefore we have approximately:

K_1 amplitude $\triangleq$ O_1 amplitude + P_1 amplitude

and

P_1 amplitude $\triangleq$ 0.46 O_1 amplitude.

Let us mention the main lunar elliptical waves:

Q_1 of argument $(\dot{\tau}-\dot{s}) - (\dot{s}-\dot{p})$ = 13°.398 66; period 26h 52m 6s,

J_1 of argument $(\dot{\tau}+\dot{s}) + (\dot{s}-\dot{p})$ = 15°.585 44; period 23h 5m 54s,

TABLE 1.4. Principal Tidal Waves

Symbol	Argument number	Argument	Frequency in degrees per hour	Amplitude	Origin (L, lunar; S, solar)
		Long-period components			
M_0	055.555	0	0°000 000	+50458	L constant flattening
S_0	055.555	0	0°000 000	+23411	S constant flattening
S_a	056.554	$h - p_s$	0°041 067	+ 1176	S elliptic wave
Ss_a	057.555	2h	0°082 137	+ 7287	S declinational wave
M_m	065.455	s - p	0°544 375	+ 8254	L elliptic wave
M_f	075.555	2s	1°098 033	+15642	L declinational wave
		Diurnal components			
Q_1	135.655	$(\tau-s) - (s-p)$	13°398 661	+ 7216	L elliptic wave of O_1
O_1	145.555	$\tau - s$	13°943 036	+37689	L principal lunar wave
M_1	155.655	$(\tau+s) - (s-p)$	14°496 694	- 2964	L elliptic wave of $^mK_1$
π_1	162.556	$(t-h) - (h-p_s)$	14°917 865	+ 1029	S elliptic wave of P_1
P_1	163.555	t - h	14°958 931	+17554	S solar principal wave
S_1	164.556	$(t+h) - (h-p_s)$	15°000 002	- 423	S elliptic wave of $^sK_1$
$^mK_1$	165.555	$\tau + s = t'$	15°041 069	-36233	L declinational wave
$^sK_1$	165.555	t + h = t'	15°041 069	-16817	S declinational wave
ψ_1	166.554	$(t+h) + (h-p_s)$	15°082 135	- 423	S elliptic wave of $^sK_1$
ϕ_1	167.555	t + 3h	15°123 206	- 756	S declinational wave
J_1	175.455	$(\tau+s) + (s-p)$	15°585 443	- 2964	L elliptic wave of $^mK_1$
OO_1	185.555	$\tau + 3s$	16°139 102	- 1623	L declinational wave
		Semi-diurnal components			
$2N_2$	235.755	$2\tau - 2(s-p)$	27°895 355	+ 2301	L elliptic wave of M_2
μ_2	237.555	$2\tau - 2(s-h)$	27°968 208	+ 2777	L variation wave
N_2	245.655	$2\tau - (s-p)$	28°439 730	+17387	L major elliptic wave of M_2
ν_2	247.455	$2\tau - (s-2h+p)$	28°512 583	+ 3303	L evection wave
M_2	255.555	2τ	28°984 104	+90812	L principal wave
λ_2	263.655	$2\tau + (s-2h+p)$	29°455 625	- 670	L evection wave
L_2	265.455	$2\tau + (s-p)$	29°528 479	- 2567	L minor elliptic wave of M_2
T_2	272.556	$2t - (h-p_s)$	29°958 933	+ 2479	S major elliptic wave of S_2
S_2	273.555	2t	30°000 000	+42286	S principal wave
R_2	274.554	$2t + (h-p_s)$	30°041 067	- 354	S minor elliptic wave of S_2
$^mK_2$	275.555	$2(\tau+s) = 2t'$	30°082 137	+ 7858	L declinational wave
$^sK_2$	275.555	$2(t+h) = 2t'$	30°082 137	+ 3648	S declinational wave
		Ter-diurnal component			
M_3	355.555	3τ	43°476 156	- 1188	L principal wave

derived respectively from O_1 and K_1, and two solar elliptical waves:

π_1 of argument $(\dot{t}-\dot{h}) - (\dot{h}-\dot{p}_s) = 14°917\ 86$; period 24h 7m 56s

ψ_1 of argument $(\dot{t}+\dot{h}) + (\dot{h}-\dot{p}_s) = 15°082\ 14$; period 23h 52m 9s.

Hydrodynamical considerations in the Earth's core will emphasize their importance later on (Chapter 6).

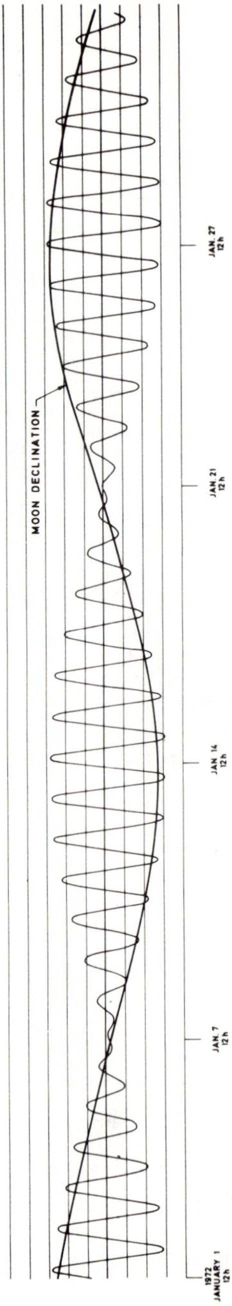

Fig. 1.7. Combination of waves $K_1^{☾}$ and O_1

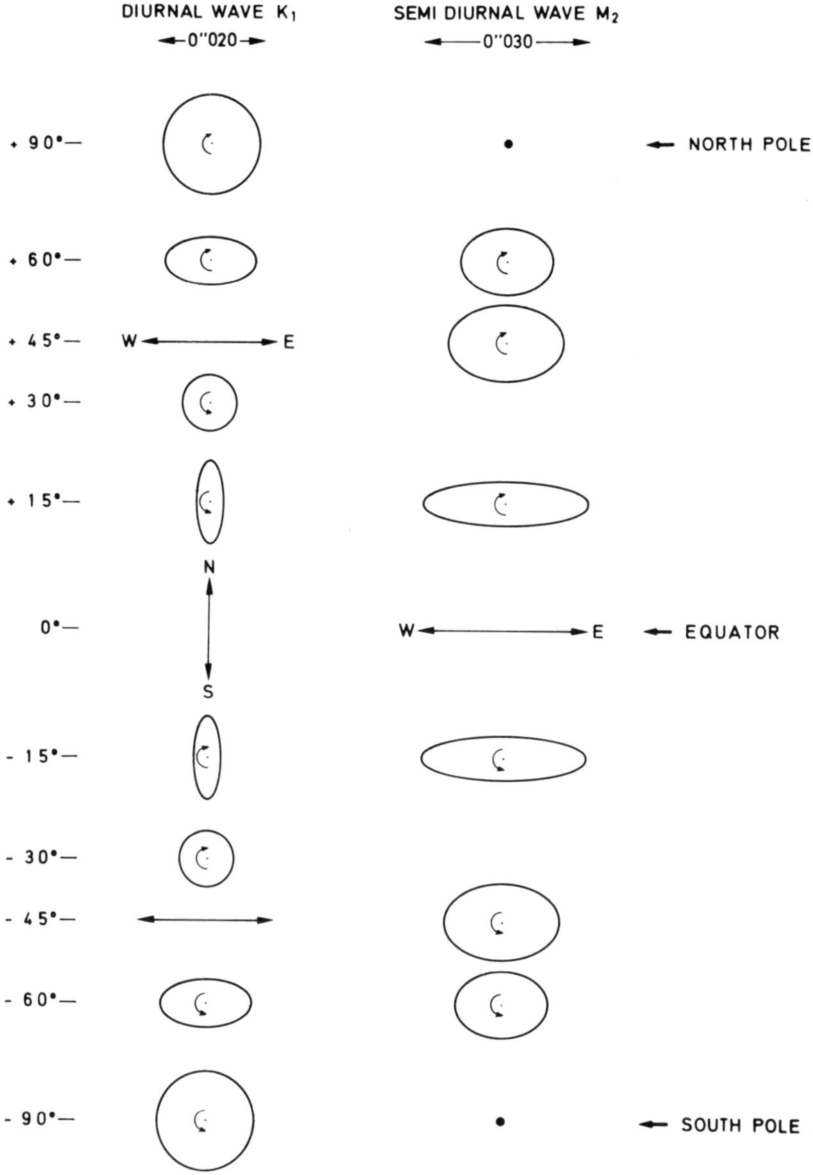

Fig.1.8. Curves described by the direction of the vertical at different latitudes.

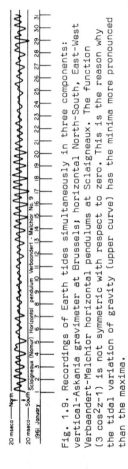

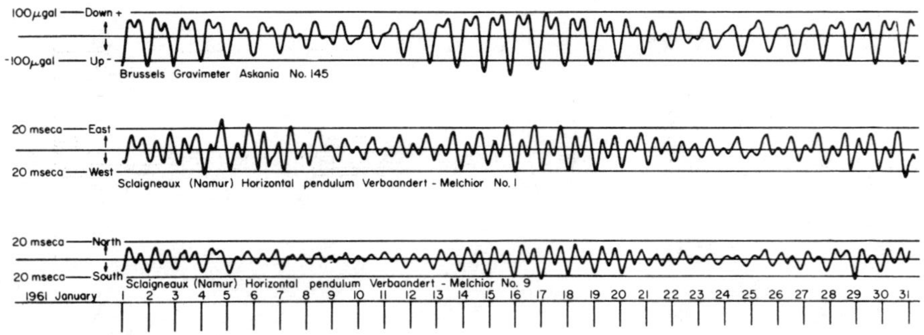

Fig. 1.9. Recordings of Earth tides simultaneously in three components: vertical-Askania gravimeter at Brussels; horizontal North-South, East-West Verbaandert-Melchior horizontal pendulums at Sclaigneaux. The function $(3 \cos^2 z - 1)$ is not symmetric with respect to zero. This is the reason why the tidal variation of gravity (upper curve) has the minima more pronounced than the maxima.

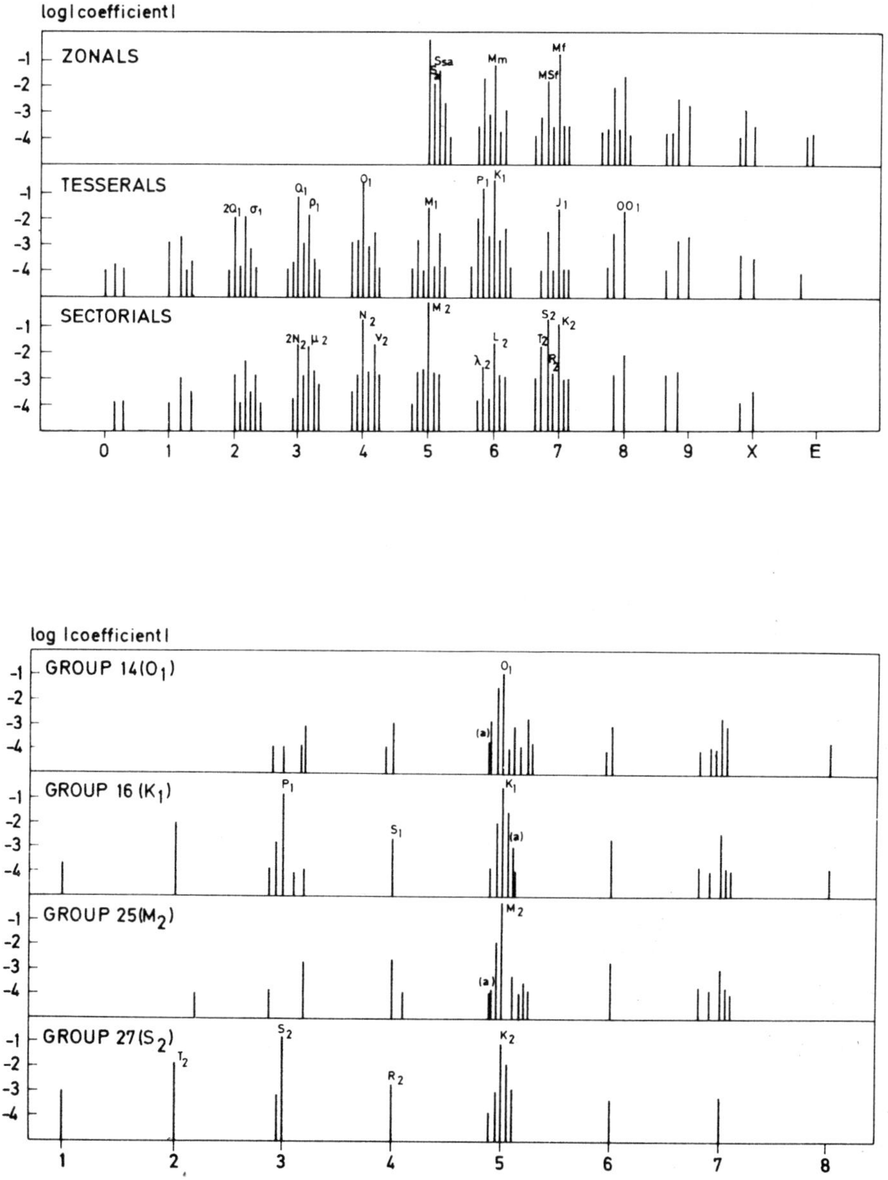

Fig. 1.10. Tidal spectrum

The effects of evection and variation give rise to diurnal waves called ρ_1, θ_1, χ_1, σ_1, SO1, and τ_1 which do not play, at present, any role in the normal analysis of Earth tides.

All the main waves are tabulated in Table 1.4 and represented on Fig 1.10.

1.7.3. Zonal Function - Long-Period Waves

$$30(\sin^2\phi - 1/3)(\sin^2\delta - 1/3) = 30\,(\frac{c}{d})^3(\sin^2\phi - 1/3)(\frac{2}{3} - \cos^2\delta) \qquad (1.51)$$

Introducing the developments (1.40), (1.42) and (1.46) we obtain for the Moon:

$$\left.\begin{aligned}(\frac{c}{d})^3\,(\frac{2}{3} - \cos^2\delta) &= -0.254\,13 - 0.079\,196\cos 2\dot{s}t + 0.036\,0\cos \dot{N}t \\ &+0.036\,0\cos(2\dot{s}-\dot{N})t \\ &-0.041\,85\cos(\dot{s}-\dot{p})t + \ldots\end{aligned}\right\} \qquad (1.52)$$

The main lunar wave has a period of 13.66 days; it is called M_f (Moon, fortnightly), its speed is 1°.098 per mean solar hour and its argument 2s. Elliptic waves join with the constant term and with M_f and the most important is

M_m (Moon, monthly) of argument $\dot{s}-\dot{p}$ = 0°.544 4.

The main solar waves are Ssa (Sun, semi-annual) of argument $2\dot{h}$ and Sa (Sun, annual) of argument $(\dot{h}-\dot{p}_s)$:

$$(\frac{c}{d})^3(\frac{2}{3} - \cos^2\delta) = -0.254\,13 - 0.079\,196\cos 2\dot{h}t - 0.012\,754\cos(\dot{h}-\dot{p}_s)t + \ldots \qquad (1.53)$$

1.7.4. The Constant Terms in the Zonal Function (Honkasalo correction)

Honkasalo pointed out that the tidal corrections applied to gravity measurements imply a constant low tide at the pole and a permanent high tide at the equator of about 40cm due to the constant terms appearing in (1.52) and (1.53). If such a correction, involved in the current tidal formulae, is applied to the measurements, particularly to the world gravity net and to the gravity calibration lines, it means that we are not using the real mean sea level as geoid but a geoid which must be defined so that the Moon and Sun are considered as being at an infinite distance.

The existence of such a permanent flattening caused by external bodies raises a very special problem for Geodesy.

The geoïd, defined as the equipotential surface fitting with the mean sea level, has a global flattening which includes this effect (about 40cm over a total of 21 km). Similarly the experimental determination of $J_2 = (C - A) / M a^2$ from satellite orbits also contains this contribution which is indeed a real geophysical effect and should thus remain as a component in our reference surface. This was the reason of Honkasalo proposal.

Moreover it would be practical not to have to evaluate it for a correction because we are not sure which factor to apply to a zero frequency effect to allow for the earth's rheology. No earth model gives clear informations. On the other hand Heikkinen (1979) has reminded that if two external fictitious bodies are kept outside the earth's surface (S_o and M_o fictitious circular rings which generate the stationary parts of the tidal potential) the Stokes' formula is no more applicable as it presupposes that all the masses involved are located *within* the geoid.

Indeed, by applying the Stokes formula with these two external masses, one arrives to a contradiction on the heights between equator and pole which amounts to approximately 90 cm as shown by Heikkinen.

This of course is not acceptable for modern Geodesy.

A proposal made by Groten should be to eliminate only the external circular ring masses S_0 and M_0 without considering the earth's response to these attractions which avoids any hypothesis upon its rheological response.

However a removal of the secular indirect effect using any short period elastic response should not cause any relevant error and should have the advantage of consistency with all other operations.

It is to be expected that a similar problem will soon be raised for chord measurements by Very Long Base Interferometry methods.

1.8. A SYSTEMATIC CLASSIFICATION FOR TIDAL WAVES

The "names" of tidal waves were given by George Darwin to provide a mnemotechnic way to memorize them. One can see, indeed, that the letters chosen are arranged in the alphabet as pairs with respect to M (Moon) and S (Sun). But this is obviously limited to a few main components.

Doodson (1922) introduced a very clever notation which makes possible the automatic classification of all tidal waves, deduced from theory, in the order of their increasing speed. The parameter used, which he called the argument number, can be deduced from the mathematical expression of the argument by the following rule: we write the argument as a function of the six independent variables:

$$a\dot{\tau} + b\dot{s} + c\dot{h} + d\dot{p} + e\dot{N}' + f\dot{p}_s, \tag{1.54}$$

the variables always being arranged in the same order. We then obtain the argument number by the combination of six successive ciphers:

$$a,\ (b+5),\ (c+5),\ (d+5),\ (e+5),\ (f+5). \tag{1.55}$$

This results from the statement that the values taken by a are 0, 1, 2, 3, ..., always positive, whereas the other coefficients generally vary from -4 to +4. The change of sign of N is very useful here.

The first three variables have the greatest velocity; we shall separate the first three numbers from the last three numbers by a point.

Taking some examples we have:

for M_2 of argument	$2\dot{\tau}$	i = 255.555
S_2	$2\dot{\tau} - 2\dot{h} + 2\dot{s}$	273.555
N_2	$2\dot{\tau} - \dot{s} + \dot{p}$	245.655
R_2	$2\dot{\tau} - \dot{h} + 2\dot{s} - \dot{p}_s$	274.554
Q_1	$\dot{\tau} - 2\dot{s} + \dot{p}$	135.655
π_1	$\dot{\tau} + \dot{s} - 3\dot{h} + \dot{p}_s$	162.556
J_1	$\dot{\tau} + 2\dot{s} - \dot{p}$	175.455
M_f	$2\dot{s}$	075.555

and so on.

Table 1.4 gives the argument number for each wave mentioned here, and as the indices are classified in increasing order, the waves are classified in this table according to their increasing velocity.

Because the last three indices only represent effects of a very slow variation, the waves whose arguments can only be distinguished in this part of the argument will only be separable if we have available extensive continuous observations. Thus it is impossible to separate the two waves forming S_1, which have indices of 164.556 and 164.554.

We shall therefore call the first section of three digits of the argument number the constituent number because it differentiates the tidal waves which we can separate providing we have at least one year's observations. The two first digits of the argument number is the group number which characterizes the tidal waves separable from one month's observations. Then the first digit is precisely the species number in the sense of Laplace; the two types of semi-diurnal and diurnal tide are separable after only a few days'observations. Therefore, from 29 days'observations we will be able to separate the following groups of waves:

Group	wave	Group	wave
13	Q_1, ρ_1	18	OO_1
14	O_1	23	2 N_2, μ_2
15	M_1	24	N_2, ν_2
16	π_1, P_1, S_1, K_1, ψ_1, ϕ_1	25	M_2
17	J_1	26	λ_2, L_2

Group 27 T_2, S_2, R_2, K_2

Group 16 contains in fact very many constituents, but the main six are:

π_1	162.556
P_1	163.555
S_1	164.554
	164.556
K_1	165.555
ψ_1	166.554
ϕ_1	167.555

From 6 months'observations we can separate P_1 from K_1 but not from π_1; from one-year data we can separate all these six main components.

CHAPTER 2

Relation between the Tidal Theory and the Precession - Nutations Theory

2.1. INTRODUCTION

The mechanical disturbances induced by an external body on a planet produce several associated phenomena:

1) A precession and nutations of the axis of inertia and the axis of rotation in space.
2) Almost diurnal nutations of the axis of rotation with respect to the Earth itself.
3) Periodic deviations of the vertical and variations of the intensity of the gravity field in every part of the planet (phenomena producing the tides).

The amplitudes and periods of these associated phenomena are interrelated in a simple way, but this does not appear clearly in the classical treatises of celestial mechanics: indeed, the torques producing precession and nutations are those exerted by the tesseral diurnal tidal forces. However, precession and nutations are movements of the axis of figure of the Earth described in an inertial system of fixed axes, while tides are observed at points fixed with respect to the Earth, rotating with the angular velocity

$$\omega = \frac{2\pi}{t'} = 15^{\circ}.041\ 069 \text{ per hour}$$

From this follows a first theorem:

Theorem 1: *The frequency of a nutation can be directly deduced from the frequency of the corresponding tide by simple subtraction of the "sidereal frequency" (15°041 per hour of universal time).*

Similarly, one can calculate the amplitudes of the nutations from the amplitudes of the corresponding tides.

Then, it will be shown that the diurnal nutations of the axis of rotation with respect to the Earth have the same frequency as the component of the corresponding tide and practically the same amplitude as the deviation of the vertical at the pole itself.

We have observed on Fig. 1.6 that the North-South component of the tesseral diurnal force of the tide is zero at latitudes +45° and -45° and that it has the same amplitude at the equator and pole but opposite sign. This is illustrated by Fig. 2.1 which shows that, because of the flattening of the Earth the equatorial torque is greater than the polar one. This produces a rotation which tends to tilt the terrestrial equator towards the ecliptic: this is precisely the precession-nutation phenomenon which would not exist if the Earth was a sphere or if the Sun and the Moon were moving in the equatorial plane (δ=0 in formula 1.9 cancels the tesseral part of the tidal potential). One passes from this schematic representation of two torques to a volume integral to include the sum of all the acting torques.

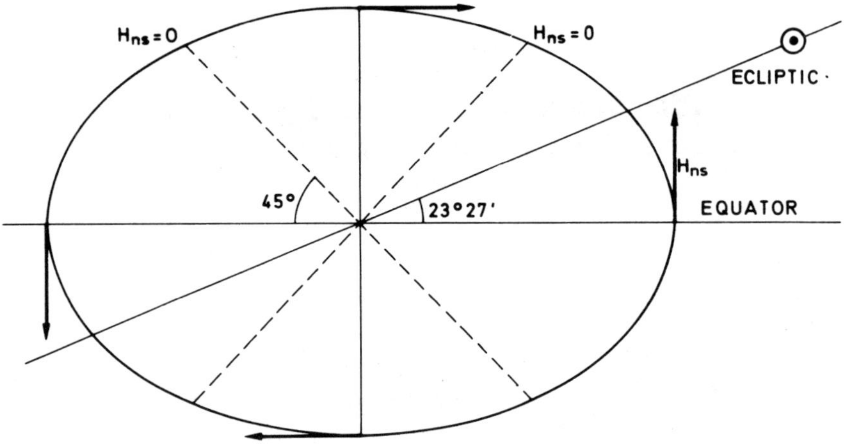

Fig. 2.1. Torques produced on an elliptic Earth by the North-South component of the tidal forces.

We also have to consider the torques exerted by the East-West component of the tesseral forces.

2.2. CALCULATION OF THE PRECESSION-NUTATION TORQUE FROM THE TIDAL POTENTIAL

The total torque exerted on the planet is

$$\mathbf{N} = \iiint_V (\mathbf{r} \wedge \mathbf{grad}\ W)\rho\ dv. \tag{2.1}$$

This integral, extended to the entire volume of the planet, is transformed (*) to

$$\mathbf{N} = -\iiint_V \mathbf{rot}\,(\rho W \mathbf{r})\ dv - \iiint_V (\mathbf{r} \wedge \mathbf{grad}\ \rho)\ W\ dv \tag{2.2}$$

and, using the Ostrogradsky theorem:

$$\mathbf{N} = -\oiint_S (\mathbf{n} \wedge \mathbf{R})\rho W\ dS - \iiint_V (\mathbf{r} \wedge \mathbf{grad}\rho) W\ dv,$$

where $\mathbf{R}$ is the vectorial radius at the external surface and $\mathbf{n}$ is the external normal (Fig. 2.2).
The first term is zero in the case of a spherical Earth ($\mathbf{n} // \mathbf{R}$)(geometrical ellipticity) while the second term is zero for a density distribution with spherical symmetry ($\mathbf{r} // \mathbf{grad}\ \rho$)(dynamical ellipticity). A surface integral term exists for every surface of discontinuity of ρ.

The development of the tidal potential has to be introduced in this expression of the torque and, to integrate with respect to λ, one has to take into account the following relations (Fig. 2.2):

(*) $\mathbf{rot}\,(\rho W \mathbf{r}) = \rho W\, \mathbf{rot}\ \mathbf{r} - (\mathbf{r} \wedge \mathbf{grad}\ \rho W) = -(\mathbf{r} \wedge \mathbf{grad} W)\rho - (\mathbf{r} \wedge \mathbf{grad}\rho) W$, $\mathbf{rot}\ \mathbf{r} = 0$.

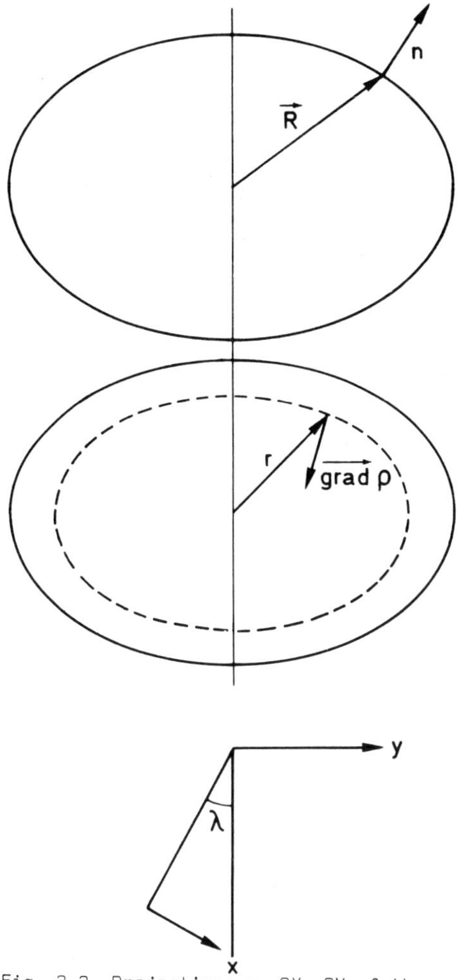

Fig. 2.2. Projections on OX, OY of the vectorial products (**n** Λ **r**) and (**r** Λ **grad** ρ)

$$(\mathbf{n} \wedge \mathbf{r}).\mathbf{e}_x = \sin\lambda \,\|\mathbf{n} \wedge \mathbf{r}\|$$
$$(\mathbf{n} \wedge \mathbf{r}).\mathbf{e}_y = \cos\lambda \,\|\mathbf{n} \wedge \mathbf{r}\| \tag{2.3}$$

$$(\mathbf{r} \wedge \mathbf{grad}\rho).\mathbf{e}_x = \sin\lambda \,\|\mathbf{r} \wedge \mathbf{grad}\rho\|$$
$$(\mathbf{r} \wedge \mathbf{grad}\rho).\mathbf{e}_y = \cos\lambda \,\|\mathbf{r} \wedge \mathbf{grad}\rho\| \tag{2.4}$$

and

$$dS = \frac{R}{(\bar{\mathbf{n}}.\bar{\mathbf{R}})} R^2 \cos\phi \, d\phi \, d\lambda; \; (*) \; dv = dr \, dS \tag{2.5}$$

$(\mathbf{e}_x, \mathbf{e}_y)$ is the equatorial plane; $(\mathbf{e}_x, \mathbf{e}_z)$ is the meridian plane taken as origin of longitudes.
Then from equations (1.5) and (1.6) we have

$$\left.\begin{aligned}
N_x = &- \sum_\ell \sum_m W_{\ell m} P_\ell^m(\sin\delta) \int_{-\frac{1}{2}\pi}^{\frac{1}{2}\pi} d\phi \frac{\|\mathbf{n} \wedge \mathbf{R}\|}{(\mathbf{n}.\mathbf{R})} R^{\ell+3} \rho \cos\phi \, P_\ell^m (\sin\phi) \\
&\times \int_0^\pi d\lambda \sin\lambda \, (\cos mH \cos m\lambda + \sin mH \sin m\lambda) \\
&- \sum_\ell \sum_m W_{\ell m} P_\ell^m (\sin\delta) \int_{-\frac{1}{2}\pi}^{\frac{1}{2}\pi} d\phi \cos\phi \, P_\ell^m(\sin\phi) \int_0^R dr \|\vec{\mathbf{r}} \wedge \vec{\mathbf{grad}}\rho\| \, \mathbf{r}^{\ell+2} \\
&\times \int_0^\pi d\lambda \sin\lambda \, (\cos mH \cos m\lambda + \sin mH \sin m\lambda)
\end{aligned}\right\} \tag{2.6}$$

and a similar equation for N_y, where $R = R(\phi)$ since the Earth is ellipsoidal. Figure 1.2 shows that the sectorial forces have no effect if the Earth has symmetry about its rotational axis. The integration of formula (2.6) with respect to λ leads to the disappearance of the sectorial and zonal functions and brings the index m to the single value m = 1.

After this integration, which is easily done, we put

$$J'_\ell = \pi \int_{-\frac{1}{2}\pi}^{\frac{1}{2}\pi} d\phi \cos\phi \, P_\ell^1 (\sin\phi) \left[\rho \frac{\|\mathbf{n} \wedge \mathbf{R}\|}{(\mathbf{n}.\mathbf{R})} R^{\ell+3} + \int_0^R dr \|\mathbf{r} \wedge \mathbf{grad}\rho\| r^{\ell+2}\right]. \tag{2.7}$$

This allows us to write simply:

$$\left.\begin{aligned}
N_x &= - \sum_\ell J'_\ell W_{\ell 1} P_\ell^1 (\sin\delta) \sin H, \\
N_y &= - \sum_\ell J'_\ell W_{\ell 1} P_\ell^1 (\sin\delta) \cos H.
\end{aligned}\right\} \tag{2.8}$$

The potential exerted by the Earth at the gravity centre of a satellite is

$$V = \frac{G}{d}\left[1 - \sum_{\ell=2}^{\infty} \frac{a^\ell}{d^\ell} J_\ell P_\ell (\sin\delta)\right] \tag{2.9}$$

(*) (La Vallée Poussin, Traité d'Analyse Mathématique, Vol., I, p. 387, N°293).

and the torque with respect to the centre of the Earth is

$$d\left(M\,\frac{\partial V}{d\partial\delta}\right) = M\,\frac{\partial V}{\partial\delta}. \tag{2.10}$$

V does not depend on H(A).

Then the total torque exerted by the satellite on the Earth is

$$\left.\begin{aligned} N_y &= M\,\frac{\partial V}{\partial\delta} = -\,\frac{GM}{d}\sum_{\ell=2}^{\infty}\frac{a^{\ell}}{d^{\ell}}\,J_\ell\,\frac{\partial}{\partial\delta}P_\ell\,(\sin\delta) \\ &= -\,\frac{GM}{d}\sum_{\ell=2}^{\infty}\frac{a^{\ell}}{d^{\ell}}\,J_\ell\,P_\ell^1\,(\sin\delta) \end{aligned}\right\} \tag{2.11}$$

and it follows from (1.6) and (2.8) that

$$J'_\ell = -\frac{1}{2}\,\ell(\ell+1)a^{\ell}\,J_\ell \tag{2.12}$$

as

$$W_{\ell 1} = \frac{2(\ell-1)!}{(\ell+1)!}\,\frac{GM}{d^{\ell+1}}.$$

Thus

$$J'_2 = 3a^2J_2 = 3a^2\,\frac{C-A}{Ma^2}. \tag{2.13}$$

If the axis Ox_0 is taken in the direction of the vernal equinox, and the axis Oz_0 in the direction of the North pole of the Earth, we have

$H = -\alpha$, right ascension of the perturbing body and

$$\left.\begin{aligned} N_{xo} &= +\sum_\ell J'_\ell\,W_{\ell 1}\,P_\ell^1\,(\sin\delta)\,\sin\alpha, \\ N_{yo} &= -\sum_\ell J'_\ell\,W_{\ell 1}\,P_\ell^1\,(\sin\delta)\,\cos\alpha. \end{aligned}\right\} \tag{2.14}$$

Introducing the variations of declination and right ascension of the external body with time, we can develop the perturbing potential in the form of a sum of simple periodic terms (see equation (1.6)):

$$W_{\ell m}\,P_\ell^m\,(\sin\delta)\,\cos mH = K_\ell\sum_i A_{\ell mi}\,\cos[\omega_i t+\tfrac{1}{2}(\ell-m)\pi], \tag{2.15}$$

with

$$K_2 = \frac{1}{2}\,\frac{GM}{c^3} = \frac{2}{3}\,\frac{D}{\bar{a}^2} \tag{2.16}$$

The tesseral tidal frequency spectrum is symmetric with respect to the central sidereal frequency ω: there are n lines on the left and n lines on the right of ω (see Fig. 1.10). Thus let us put

$$\omega_i = \omega + \Delta\omega_i = 15^\circ\!.041\,069 + \Delta\omega_i \tag{2.17}$$

with

$$\Delta\omega_i = -\Delta\omega_{-i}. \tag{2.18}$$

Hence we have

$$H = \omega t - \alpha \tag{2.19}$$

and consequently

$$W_{\ell 1} P_\ell^1 (\sin \delta) \cos \alpha = K_\ell \sum_i A_{\ell 1 i} \cos[\Delta\omega_i t + \tfrac{1}{2}(\ell-1)\pi],$$

$$-W_{\ell 1} P_\ell^1 (\sin \delta) \sin \alpha = K_\ell \sum_i A_{\ell 1 i} \cos[\Delta\omega_i t + \tfrac{1}{2}(\ell-2)\pi] \tag{2.20}$$

In the expression of the precession torque, only the differences $\Delta\omega_i$ of the frequency with respect to the "sidereal" frequency appear (Theorem 1):

$$\left.\begin{aligned} N_{xo} &= - \sum_\ell \sum_i K_\ell J'_\ell A_{\ell i} \cos[\Delta\omega_i t + \tfrac{1}{2}(\ell-2)\pi], \\ N_{yo} &= - \sum_\ell \sum_i K_\ell J'_\ell A_{\ell i} \cos[\Delta\omega_i t + \tfrac{1}{2}(\ell-1)\pi], \end{aligned}\right\} \tag{2.21}$$

in which the index m=1 has been dropped.

Let us consider the development limited only to order 2. If $\ell=2$ and dropping this index again to simplify the notation, we have

$$\left.\begin{aligned} N_{xo} &= - \sum_{i=-n}^{i=+n} K J' A_i \cos(\Delta\omega_i t), \\ N_{yo} &= + \sum_{i=-n}^{i=+n} K J' A_i \sin(\Delta\omega_i t). \end{aligned}\right\} \tag{2.22}$$

This permits the statement of a second theorem:

Theorem 2: *Two waves of symmetric frequency with respect to the sidereal frequency form only one and the same wave of nutation; the sum of their amplitudes (major axis) appears in N_{xo} and their difference (minor axis) in N_{yo}.*

$$\left.\begin{aligned} N_{xo} &= - \sum_{i=-o}^{i=+n} K J' (A_i + A_{-i}) \cos(\Delta\omega_i t), \\ N_{yo} &= + \sum_{i=-o}^{i=+n} K J' (A_i - A_{-i}) \sin(\Delta\omega_i t). \end{aligned}\right\} \tag{2.23}$$

This theorem remains valid if $\ell \neq 2$.

Conversely we may consider an elliptic nutation as equivalent to two circular nutations of equal and opposite velocity corresponding to the two symmetrical tidal waves.

Introducing the harmonic development obtained in Chapter 1, we immediately observe a number of interesting features. Applying Theorem 1, a nutation of zero frequency, i. e. a secular one, corresponds to the luni-solar wave K_1. This is the luni-solar precession, two-thirds of which are due to the Moon and one-third to the Sun.

The waves J_1 and M_1 are symmetrical with respect to the sidereal frequency 15°041. Applying Theorem 2 we find only one nutation, the period of which will be 27.6 days. Likewise the waves ψ_1 and S_1 give rise to the annual nutation of 365 days.

From the complete development the existence of a wave 001 symmetrical to O_1 and that of a wave ϕ_1, symmetrical to P_1 is apparent.

Table 2.1 gives the principal diurnal tidal waves and the nutations associated with them.

2.3. EXPLICIT CALCULATION OF THE PRECESSION-NUTATION TORQUE LIMITED TO THE SECOND ORDER

Let the intersection of the equator and the meridian plane of the external body be the Ox_1 axis and the polar axis of inertia of the Earth the Oz_1 axis (Fig. 2.3).

The torque due to the North-South component of the tesseral forces is

$$\mathbf{L} = (r\mathbf{n} \wedge \mathbf{F}_{NS})\ dm; \tag{2.24}$$

its projections are:

$$\left.\begin{aligned} L_{x_1} &= \|\mathbf{L}\| \sin H(A), \\ L_{y_1} &= \|\mathbf{L}\| \cos H(A). \end{aligned}\right\} \tag{2.25}$$

$$\|\mathbf{L}\| = r\ F_{NS}\ dm, \tag{2.26}$$

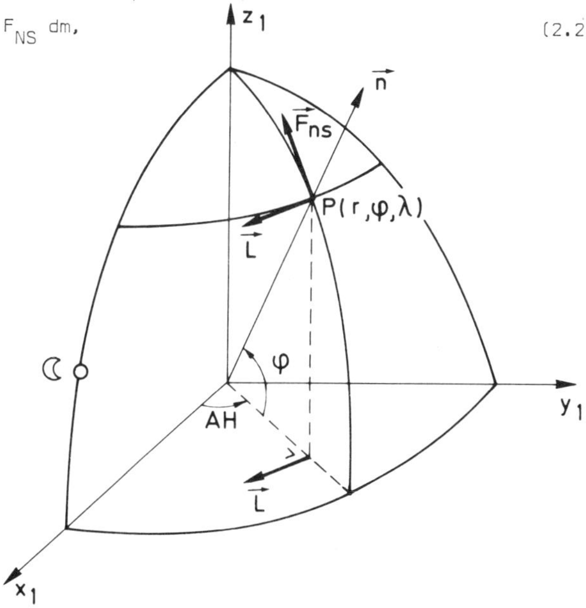

Fig. 2.3. Torque produced by the North-South component

$$\left.\begin{aligned} L_{x_1} &= \int_{-\pi}^{+\pi} d\lambda \int_{-\frac{\pi}{2}}^{+\frac{\pi}{2}} d\phi \int_0^{R(\phi)} dr \left(\frac{3}{2} GM \frac{r}{d^3} \cos 2\phi \sin 2\delta \cos H(A) \ X \right.\\ &\quad \left. (r \sin H(A)\right) \rho\ (r, \phi)\ r^2 \cos \phi, \\ L_{y_1} &= -\int_{-\pi}^{+\pi} d\lambda \int_{-\frac{\pi}{2}}^{+\frac{\pi}{2}} d\phi \int_0^{R(\phi)} dr \left(\frac{3}{2} GM \frac{r}{d^3} \cos 2\phi \sin 2\delta \cos H(A) \ X \right.\\ &\quad \left. (r \cos H(A)\right) \rho\ (r, \phi)\ r^2 \cos \phi, \end{aligned}\right\} \quad (2.27)$$

where $H(A) = (H-\lambda)$, H being the hour angle of the body with respect to a fixed point on the equator which is the origin of the geographical longitudes. Likewise the torque due to the East-West component of the tesseral forces is:

$$\mathbf{M} = (r\,\mathbf{n} \wedge \mathbf{F}_{EW})\ dm. \quad (2.28)$$

Its projections are (Fig. 2.4):

$$\left.\begin{aligned} M_{x_1} &= \|\mathbf{M}\| \sin \phi \cos H(A), \\ M_{y_1} &= \|\mathbf{M}\| \sin \phi \sin H(A), \\ \|\mathbf{M}\| &= r\ F_{EW}\ dm. \end{aligned}\right\} \quad (2.29)$$

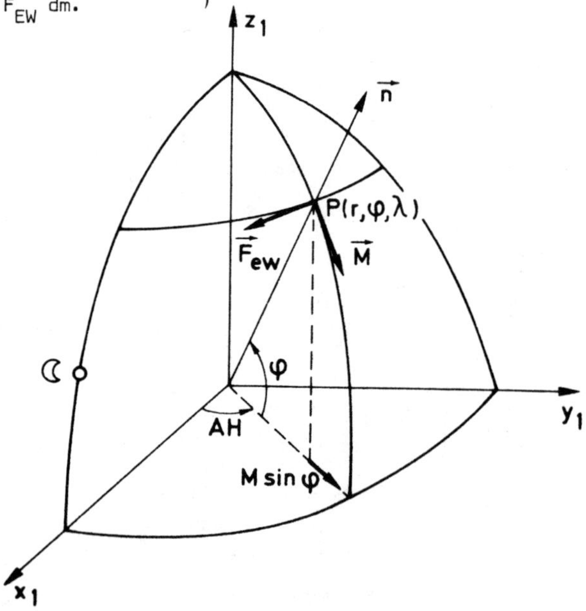

Fig. 2.4. Torque produced by the East West component

$$\left.\begin{aligned} M_{x_1} &= \int_{-\pi}^{+\pi} d\lambda \int_{-\frac{\pi}{2}}^{+\frac{\pi}{2}} d\phi \int_0^{R(\phi)} dr \left(\frac{3}{2}\, GM \frac{r}{d^3} \sin\phi \sin H(A) \sin 2\delta\right) \times \\ &\quad (r \sin\phi \cos H(A))\, \rho\, (r,\, \phi)\; r^2 \cos\phi, \\ M_{y_1} &= \int_{-\pi}^{+\pi} d\lambda \int_{-\frac{\pi}{2}}^{+\frac{\pi}{2}} d\phi \int_0^{R(\phi)} dr \left(\frac{3}{2}\, GM \frac{r}{d^3} \sin\phi \sin H(A) \sin 2\delta\right) \times \\ &\quad (r \sin\phi \sin H(A))\, \rho\, (r,\, \phi)\; r^2 \cos\phi. \end{aligned}\right\} \quad (2.30)$$

Integrating with respect to λ (assuming a symmetry of revolution about the z_1 axis) we find:

$$L_{y_1} = -\frac{3}{2}\pi \frac{GM}{d^3} \sin 2\delta \int_{-\frac{1}{2}\pi}^{\frac{1}{2}\pi} d\phi \int_0^{R(\phi)} dr\; r^4 \cos 2\phi \cos\phi\; \rho\,(r,\phi), \quad (2.31)$$

$$M_{y_1} = +\frac{3}{2}\pi \frac{GM}{d^3} \sin 2\delta \int_{-\frac{1}{2}\pi}^{\frac{1}{2}\pi} d\phi \int_0^{R(\phi)} dr\; r^4 \sin^2\phi \cos(\phi)\; \rho\,(r,\phi), \quad (2.32)$$

$$L_{x_1} = 0, \qquad M_{x_1} = 0. \quad (2.33)$$

The moments of inertia of the Earth being:

$$\begin{aligned} C &= \int_{-\pi}^{\pi} d\lambda \int_{-\frac{1}{2}\pi}^{\frac{1}{2}\pi} d\phi \int_0^{R(\phi)} dr\; (r^2 \cos^2\phi)\; \rho\,(r,\phi) r^2 \cos\phi \\ &= \pi \int_{-\frac{1}{2}\pi}^{\frac{1}{2}\pi} d\phi \int_0^{R(\phi)} dr\; 2r^4 \cos^3(\phi)\; \rho\,(r,\phi), \end{aligned} \quad (2.34)$$

$$\begin{aligned} A &= \int_{-\pi}^{\pi} d\lambda \int_{-\frac{1}{2}\pi}^{\frac{1}{2}\pi} d\phi \int_0^{R(\phi)} dr\; r^2\, (\sin^2\phi + \cos^2\phi \sin^2\lambda) \rho(r,\phi) r^2 \cos\phi \\ &= \pi \int_{-\frac{1}{2}\pi}^{\frac{1}{2}\pi} d\phi \int_0^{R(\phi)} dr\; r^4 (2\sin^2\phi \cos\phi + \cos^3\phi) \rho(r,\phi). \end{aligned} \quad (2.35)$$

Hence

$$(C-A) = \pi \int_{-\frac{1}{2}\pi}^{\frac{1}{2}\pi} d\phi \int_0^{R(\phi)} dr\; r^4\, \rho(r,\phi) \cos\phi (1-3\sin^3\phi). \quad (2.36)$$

The total precession-nutation torque is written:

$$N_{y_1} = M_{y_1} + L_{y_1} = -\frac{3}{2}\pi \frac{GM}{d^3} \sin 2\delta \int_{-\frac{1}{2}\pi}^{\frac{1}{2}\pi} d\phi \int_0^{R(\phi)} dr\; r^4\, \rho(r,\phi) \cos\phi (1-3\sin^2\phi), \quad (2.37)$$

whence immediately

$$N_{y_1} = -\frac{3}{2}\frac{GM}{d^3}\sin 2\delta\,(C-A). \qquad (2.38)$$

Let us return to the coordinate system of Fig. 2.5, where the Ox_0 axis in the direction of the vernal equinox and the z axis is left invariant; we then have

$$\left.\begin{aligned} N_{xo} &= \frac{3}{2}\frac{GM}{d^3}(C-A)\sin 2\delta\sin\alpha, \\ N_{yo} &= -\frac{3}{2}\frac{GM}{d^3}(C-A)\sin 2\delta\cos\alpha, \end{aligned}\right\} \qquad (2.39)$$

obtained by projecting N_{y1} as given by (2.38) on the Ox_0, Oy_0 axis.

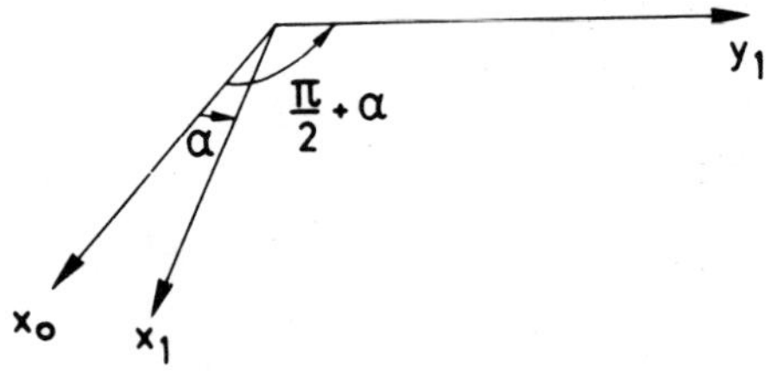

Fig. 2.5. The used axes systems: Ox_0 is directed towards the vernal equinox while Ox_1 is in the meridian plane of the external body.

2.4. EQUATIONS OF THE PRECESSION-NUTATION (FIRST APPROXIMATION)

This first approximation will neglect the mean motion of the body with respect to ω, the angular velocity of rotation of the Earth.

The coordinate system (x_0, y_0, z_0) is variable because of the precession-nutation; therefore let ω_0 be the rotation vector of this trihedral with components (p_0, q_0, r_0). These three quantities are small and we can neglect their products.

Thus the components of the moment of momentum are

$$\mathcal{M}_1 = Ap_0, \qquad \mathcal{M}_2 = Aq_0, \qquad \mathcal{M}_3 = C(r_0+\omega), \qquad (2.40)$$

and the equations of the movement are:

$$\left.\begin{aligned} A\dot{p}_0 + Cq_0(r_0+\omega) - Ar_0q_0 &= N_{xo} = -N_{y1}\sin\alpha, \\ A\dot{q}_0 + Ar_0p_0 - Cp_0(r_0+\omega) &= N_{yo} = N_{y1}\cos\alpha, \\ C\dot{r}_o + Ap_oq_o - Aq_op_o &= 0; \end{aligned}\right\} \qquad (2.41)$$

and neglecting the small terms:

$$A\dot{p}_0 + Cq_0\omega = -N_{y1}\sin\alpha, \qquad A\dot{q}_0 - Cp_0\omega = N_{y1}\cos\alpha, \qquad C\dot{r}_0 = 0. \qquad (2.42)$$

Omitting the mean motion of the body with respect to ω, by supposing that

$$\frac{d}{dt}\left[N_{y1}\begin{pmatrix}\cos\alpha\\ \sin\alpha\end{pmatrix}\right] = 0, \qquad (2.43)$$

we have immediately

$$q_o = -\frac{N_{y1}}{C\omega}\sin\alpha, \qquad p_o = -\frac{N_{y1}}{C\omega}\cos\alpha. \qquad (2.44)$$

Let us introduce the Eulerian angles ϕ, θ, ψ of the moving trihedral with respect to a trihedral fixed in space, the z axis of which is the pole of the ecliptic.

Figure 2.6 shows that

$$p_o = -\dot{\theta}, \qquad q_o = +\dot{\psi}\sin\theta, \qquad r_o = -\dot{\psi}\cos\theta, \qquad (2.45)$$

$\phi = 0$ (by hypothesis if the Earth is non-viscous).
From this it can be deduced that

$$\left.\begin{aligned}\dot{\theta} &= -\frac{3}{2}\frac{GM}{d^3}\frac{C-A}{C\omega}\sin 2\delta\cos\alpha,\\ \dot{\psi} &= +\frac{3}{2}\frac{GM}{d^3}\frac{C-A}{C\omega\sin\theta}\sin 2\delta\sin\alpha.\end{aligned}\right\} \qquad (2.46)$$

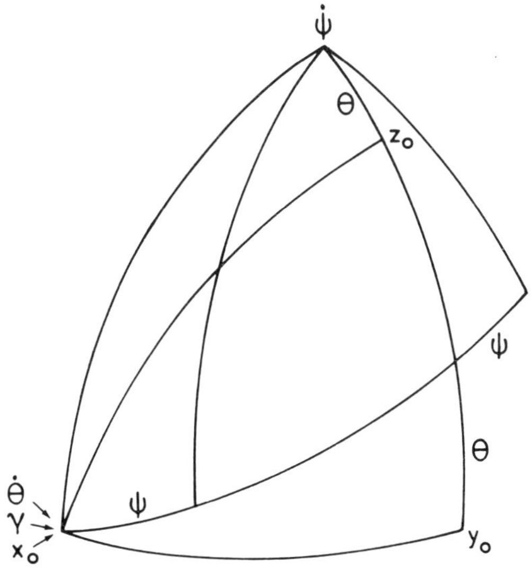

Fig. 2.6. Projections on the celestial sphere.

Transforming to coordinates relative to the ecliptic (β,λ) and introducing the known numerical values,

$$\left.\begin{aligned}&\theta = 23^\circ\!.27'; \ \sin\theta = 0.397\,985; \ \cos\theta = 0.917\,392\\ &\frac{1-2\sin^2\theta}{\sin\theta} = 1.716\,684,\end{aligned}\right\} \qquad (2.47)$$

we obtain

$$\dot{\theta} = -3\frac{GM}{d^3}\frac{C-A}{C\omega}(0.917\,392 \sin\beta\cos\beta\cos\lambda + 0.397\,985\cos^2\beta\sin\lambda\cos\lambda),$$

$$\dot{\psi} = +3\frac{GM}{d^3}\frac{C-A}{C\omega}[1.716\,684\sin\beta\cos\beta\sin\lambda + 0.917\,392(\cos^2\beta\sin^2\lambda - \sin^2\beta)]. \qquad (2.48)$$

These are the terms of Woolard's development (Woolard, 1953, pages 47, 48) which are independent of the movement of the ecliptic itself, due to planetary perturbations. But if we consider formulas (2.39), we note that equations (2.46) could be written as

$$\dot{\theta} = +\frac{N_{yo}}{C\omega}, \qquad \sin\theta.\dot{\psi} = +\frac{N_{xo}}{C\omega}, \qquad (2.49)$$

and using (2.23) we obtain the nutations in terms of the development in tidal waves:

$$\dot{\theta} = \sum_i \frac{K}{C\omega}J'(A_i - A_{-i})\sin(\Delta\omega_i.t),$$

$$\sin\theta.\dot{\psi} = -\sum_i \frac{K}{C\omega}J'(A_i + A_{-i})\cos(\Delta\omega_i.t). \qquad (2.50)$$

Let us introduce a new dimensionless constant:

$$E = \frac{2}{a^2}D\left(\frac{C-A}{C\omega^2}\right) = \frac{3}{2}\frac{GM}{c^3}\frac{C-A}{C\omega^2} = \frac{KJ'}{C\omega^2}. \qquad (2.51)$$

Its value, expressed in seconds of arc, for the Moon is

$$E_{☾} = 0''016\,443 \qquad (2.52)$$

Then

$$\dot{\theta} = +E_{☾}\,\omega\sum_i(A_i - A_{-i})\sin(\Delta\omega_i t),$$

$$\dot{\psi}\sin\theta = -E_{☾}\,\omega\sum_i(A_i + A_{-i})\cos(\Delta\omega_i t). \qquad (2.53)$$

The K_1 tidal field of forces is distributed according to the $\cos(\tau+s)$ function, i. e. the cosine of the sidereal time or hour angle of the vernal equinox. Its maximum therefore permanently points towards the vernal equinox ($\dot{\theta}$ axis), and the torques produced have no resultant component along $\dot{\theta}$ axis but along the direction 90° apart, i. e. $\dot{\psi}\sin\theta$.

The equations for K_1 give

$$\dot{\psi} = -E_{☾}\frac{\omega A(K_1)}{\sin\theta}, \qquad \dot{\theta} = 0, \qquad (2.54)$$

and from

$$E = 0''016\,443, \qquad \omega = 7.292\times10^{-5}\ s^{-1},$$

$$\sin^{-1}\theta = 2.512, \qquad A = 0.530\,5,$$

we obtain

$$\dot{\psi} = -50''38 \text{ per year, the luni-solar precession constant.} \qquad (2.55)$$

The nutations are obtained by integration of equations (2.53):

$$\left.\begin{aligned}\Delta\theta &= - E_{☾} \sum_i \frac{\omega}{\Delta\omega_i} (A_i - A_{-i}) \cos(\Delta\omega_i t),\\ \Delta\psi &= - \frac{E_{☾}}{\sin\theta} \sum_i \frac{\omega}{\Delta\omega_i} (A_i + A_{-i}) \sin(\Delta\omega_i t),\end{aligned}\right\} \quad (2.56)$$

The presence of $\Delta\omega_i$ in the denominator shows that the waves give rise to nutations of an amplitude which becomes lower as their frequency diverges from that of the sidereal day (wave K_1), even when the amplitude of the tide is comparable to that of K_1 (this is the case with O_1 versus P_1).

In the first approximation we observe that tidal waves symmetrical with respect to K_1 and of equal amplitude ($A_i = A_{-i}$) do not cause nutations in obliquity ($\Delta\theta = 0$) but only nutations in longitude. This is the case for waves generated by the ellipticity of the orbits:

NO_1 and J_1 for the Moon,

S_1 and ψ_1 for the Sun.

The periods of nutations associated with the ellipticity of the orbits are evidently a month and a year as is shown by Table 2.1.

To a second approximation a very small nutation in obliquity will be found. Woolard's expressions of the nutation given in table 26, p. 153 of his paper, leave the amplitudes in obliquity of these nutations undetermined. Indeed, this development, in agreement with the recommendations of the International Astronomical Union, includes only the coefficients attaining at least 0".000 2.

In the sense of mechanics it seems unsuitable to classify these components among the "short period nutations", as they do not practically alter the angle θ and show only a variation of ψ, that is to say a precession.

The two components produced by the ellipticity of the orbits should logically have been named "short period precessions".

2.5. DIURNAL NUTATIONS OF THE AXIS OF ROTATION WITHIN THE EARTH TO A FIRST APPROXIMATION

Let Oxyz be a trihedral rigidly bound to the Earth, the axis Oz being the principal polar axis of inertia, and let (p,q,r) be the components of the rotation ω along these axes (Fig. 2.7).

The classical relations of Euler:

$$\left.\begin{aligned}p &= - \dot{\theta} \cos\phi + \dot{\psi} \sin\theta \sin\phi \qquad (\phi = \omega t = t_s),\\ q &= + \dot{\theta} \sin\phi + \dot{\psi} \sin\theta \cos\phi,\\ r &= - \dot{\psi} \cos\theta + \dot{\phi},\end{aligned}\right\} \quad (2.57)$$

enables us to calculate the position of the axis of the rotation with respect to the Earth, starting from the expressions for $\dot{\theta}$, $\dot{\psi}$, $\dot{\phi}$:

$$\left.\begin{aligned} p &= + E_{☾}\, \omega \left(\frac{c}{d}\right)^3 \sin 2\delta\ (\cos\alpha\cos\phi + \sin\alpha\sin\phi), \\ q &= + E_{☾}\, \omega \left(\frac{c}{d}\right)^3 \sin 2\delta\ (-\cos\alpha\sin\phi + \sin\alpha\cos\phi). \end{aligned}\right\} \quad (2.58)$$

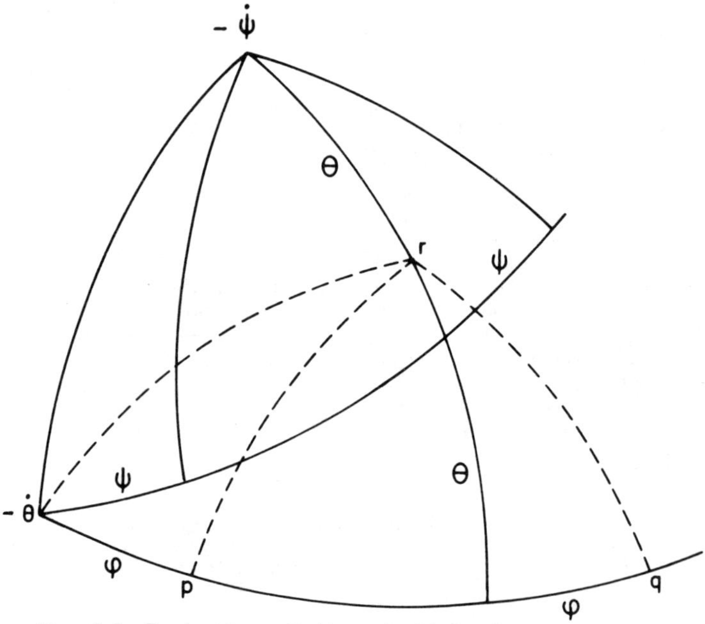

Fig. 2.7. Projections on the celestial sphere.

The coordinates of the pole on the surface of the Earth are:

$$\left.\begin{aligned} x &= \frac{p}{\omega} = E_{☾}\left(\frac{c}{d}\right)^3 \sin 2\delta\ \cos(t_s - \alpha), \\ y &= \frac{q}{\omega} = - E_{☾}\left(\frac{c}{d}\right)^3 \sin 2\delta\ \sin(t_s - \alpha). \end{aligned}\right\} \quad (2.59)$$

We find again the development in tidal waves, each diurnal wave giving rise to a forced circular nutation of the same period:

$$\left.\begin{aligned} x &= - E_{☾} \sum_i A_i \sin \omega_i t, \\ y &= - E_{☾} \sum_i A_i \cos \omega_i t. \end{aligned}\right\} \quad (2.60)$$

It is interesting to compare the amplitudes of these nutations with those of the corresponding deviations of the vertical.

For this purpose we will return to the coordinate system formed by the local trihedral.

OZ: Direction of the vertical towards the nadir.
OX: Line tangent to the meridian, directed toward South.
OY: Line tangent to the prime vertical, directed toward East.

The formula for the deviation of the vertical is

$$\left.\begin{aligned} e_\lambda &= -\frac{1}{g}\frac{\partial W}{r\cos\phi\partial\lambda} = -\frac{3}{2}\frac{GM}{d^3}\frac{r}{g}\sin\phi\sin 2\delta\sin(t_s-\alpha),\\ e_\phi &= -\frac{1}{g}\frac{\partial W}{r\partial\phi} = -\frac{3}{2}\frac{GM}{d^3}\frac{r}{g}\cos 2\phi\sin 2\delta\cos(t_s-\alpha). \end{aligned}\right\} \tag{2.61}$$

In particular, at the pole (ϕ=90°) we obtain an oscillation which can only be circular:

$$\left.\begin{aligned} e_x &= +\frac{3}{2}\frac{GM}{d^3}\frac{r}{g}\sin 2\delta\cos(t_s-\alpha),\\ e_y &= -\frac{3}{2}\frac{GM}{d^3}\frac{r}{g}\sin 2\delta\sin(t_s-\alpha). \end{aligned}\right\} \tag{2.62}$$

Let us put

$$C_{xy} = \frac{3}{2}\frac{GM}{c^3}\frac{r}{g} = \frac{2}{a}\frac{D}{g}. \tag{2.63}$$

C_{xy} is a dimensionless constant (subject to a very small variation as a function of the latitude).

For the Moon and for a mean Earth radius:

$$\left.\begin{aligned} C_{xy} &= 0.017\ 317\ 2'',\\ e_x &= -C_{xy}\cos 2\phi\,(c/d)^3\sin 2\delta\cos(t_s-\alpha),\\ e_y &= -C_{xy}\sin\phi\,(c/d)^3\sin 2\delta\sin(t_s-\alpha). \end{aligned}\right\} \tag{2.64}$$

These equations have the same form as those for the nutations of the axis of rotation in the Earth. Approximately, we have:

$$\frac{E_☾}{C_{xy}} = \frac{C-A}{Cq}, \tag{2.65}$$

where q is the geodynamic parameter, the ratio of the centrifugal force to the gravity at the equator (ω^2a^3/GM).

We have

$$e_x = +C_{xy}\cos 2\phi\sum_i A_i\sin\omega_i t, \qquad e_y = -C_{xy}\sin\phi\sum_i A_i\cos\omega_i t, \tag{2.66}$$

and at the pole

$$e_x = -C_{xy}\sum_i A_i\sin\omega_i t, \qquad e_y = -C_{xy}\sum_i A_i\cos\omega_i t. \tag{2.67}$$

TABLE 2.1. Correspondence between Tides and Nutations

Diurnal tides					
Argument	Origin Symbol	Doodson argument number	Frequencies f_i in degrees per hour	Period	Amplitude of North-South component at the equator in microgals
$\tau-2s+p=(\tau+s)-(3s-p)$	L Q_1	135.655	13°398 660 9	26h52m48s	+ 5.942 8
$(\tau+s)+(3s-p)$	L -	195.455	16°683 476 3	21h34m42s	- 0.256 1
$\tau-s=(\tau+s)-2s$	L O_1	145.555	13°943 035 6	25h49m 8s	+31.039 1
$\tau+3s=(\tau+s)+2s$	L OO_1	185.555	16°139 101 7	22h18m22s	- 1.336 6
$(\tau+s)-(s-p)$	eL NO_1	155.655	14°496 694 0	24h49m59s	- 2.441 0
$(\tau+s)+(s-p)$	eL J_1	175.455	15°585 443 3	23h 5m53s	- 2.441 0
$t-2h+p_s=(t+h)-(3h-p_s)$	S π_1	162.556	14°917 864 7	24h 7m55s	+ 0.847 4
$(t+h)+(3h-p_s)$	S -	168.554	15°164 272 4	23h44m24s	- 0.036 2
$t-h=(t+h)-2h$	S P_1	163.555	14°958 931 4	24h 3m54s	+14.481 5
$t+3h=(t+h)+2h$	S ϕ_1	167.555	15°123 205 9	23h48m14s	- 0.622 6
$(t+h)-(h-p_s)$	eS S_1	164.556	15°000 002 0	23h59m56s	- 0.348 4
$(t+h)+(h-p_s)$	eS ψ_1	166.554	15°082 135 3	23h52m 8s	- 0.348 4
$t+h-2N$	L -	165.575	15°045 481 4	23h56m39s	+ 0.126 8
$(t-h)+N$	L -	165.545	15°038 862 2	23h56m16s	- 0.864 7
$(t+h)-N$	L -	165.565	15°043 275 1	23h55m51s	+ 5.914 8
$(t+h)$	LS K_1	165.555	15°041 068 6	23h56m 3s	-43.689 8

L: Lunar tides   eL: elliptic lunar tides
S: Solar tides   eS: elliptic solar tides

TABLE 2.1. Correspondence between Tides and Nutations

Nutations				
Frequency in degrees per hour $\frac{1}{2}(f_2-f_1)$	Period in days	Amplitude		Name
		in the longitude	in the obliquity	
$3\dot{s}-\dot{p}=\ell+2F+2\Omega=1^{\circ}\!.642\ 408$	9.1	$-\ 0''\!.026\ 1$	$+0''\!.011\ 3$	
$2\dot{s}=2F+2\Omega=1^{\circ}\!.098\ 033\ 1$	13.7	$-\ 0''\!.203\ 7$	$+0''\!.088\ 4$	fortnightly
$\dot{s}-\dot{p}=\ell=0^{\circ}\!.544\ 374\ 594$	27.6	$+\ 0''\!.067\ 5$	0	monthly
$3\dot{h}-\dot{p}_s=\ell'+2F-2D+2\Omega=0^{\circ}\!.123\ 203\ 9$	122	$-\ 0''\!.049\ 7$	$-0''\!.021\ 6$	
$2\dot{h}=2F-2D+2\Omega=0^{\circ}\!.082\ 137\ 3$	183	$-\ 1''\!.272\ 9$	$+0''\!.552\ 2$	semi-annual
$\dot{h}-\dot{p}_s=\ell'=0^{\circ}\!.041\ 066\ 678$	365	$+\ 0''\!.126\ 1$	0	annual
$2\dot{N}=2\dot{\Omega}=0^{\circ}\!.004\ 412\ 8$	3 399	$+\ 0''\!.208\ 8$	$-0''\!.090\ 4$	
$\dot{N}=\dot{\Omega}=0^{\circ}\!.002\ 206\ 4$	6 798	$-17''\!.232\ 7$	$+9''\!.210\ 0$	principal nutation of 18. years 66
0	secular		0	precession

The vertical turns in the same direction as the axis of rotation of the Earth and in phase.

Table 2.2 shows a comparison of the diurnal nutations with the deviations of the vertical at the pole.

Consequently the corresponding effects in the observation of the variations of the latitude are added up for an observatory with latitude $\phi > 45°$ and are subtracted for the observatories of the International Polar Motion Service ($\phi = 39°8'$).

Table 2.3 gives the amplitude of the deviations of the vertical due to the wave K_1 in the two directions North-South and East-West. Figure 1.8 shows the oscillation of the vertical at different latitudes.

Equations of the precession-nutation to a second approximation have been developed by Melchior and Georis (1968).

TABLE 2.2. Diurnal nutations and diurnal tides

Principal diurnal tidal waves		Amplitude A_i	Amplitude of the cones of diurnal nutations in the Earth	Deviations of the vertical at the pole
LS	K_1	-0.530 50	0"008 7 = 27 cm	0"009 19
L	O_1	+0.376 89	0"006 2 = 19 cm	0"006 53
S	P_1	+0.175 54	0"002 9 = 9 cm	0"003 04
L	Q_1	+0.072 16	0"001 2 = 4 cm	0"001 25
eL	J_1	-0.029 64	0"000 5 = 1.5 cm	0"000 53
eL	M_1	-0.029 64	0"000 5 = 1.5 cm	0"000 53
cf. Tisserand, 1891, Tome II:		p 494	p 420	

L: lunar, S: solar, e: elliptic.

TABLE 2.3. Amplitude of the wave K_1

Latitude	North-South component	East-West component
0°	-0"009 19	0
30°	-0"004 60	0"004 60
39°8'	-0"001 87	0"005 71
45°	0	0"006 50
50°	+0"001 60	0"007 05
60°	+0"004 60	0"007 97
90°	+0"009 19	0"009 19
factor:	$\cos 2\phi$	$\sin \phi$

2.6. CONCLUSION

The techniques of positional astronomy and the "geophysical" study of Earth tides do not differ fundamentally: the astronomers determine geometrically the characteristics of the nutations by angular measurements, i. e. the oscillations of a trihedral fixed into the Earth, while the geophysicists, using dynamometers (gravimeters, pendulums), directly measure the forces producing these oscillations.

In certain aspects, the direct measurement of the tidal forces presents some advantages compared with determinations of the astronomical nutations:

1) it enables continuous measurements to be made: this cannot be done by optical astronomy;
2) it is possible to display small components, principally the waves O_1 and P_1, the associated nutations of which are only accessible to astronomers with some difficulty.

This will be very valuable in a thorough study of the influence of the internal structure of the Earth on these phenomena.

TABLE 2.4. Periods of the nutations corresponding to each diurnal tesseral tidal wave as deduced by application of theorem 1 or formulas (2.17) and (2.18)

	Argument			Amplitude	Frequency ω_i	Nutation Period Sid. day	Nutation Period Solar day
1	M 105.955			0.00011	− 11.76553679	4.591947	4.579409
2	M 107.755			0.00046	− 11.83839039	4.696403	4.683580
3	M 109.555			0.00028	− 11.91124400	4.805722	4.792600
5	M 115.845			0.00021	− 12.30770508	5.502769	5.487744
6	M 115.855	N		0.00108	− 12.30991149	5.507214	5.492177
8	M 117.645			0.00053	− 12.38055867	5.653453	5.638016
9	M 117.655	N		0.00278	− 12.38276508	5.658145	5.642696
10	M 118.654			0.00021	− 12.42383176	5.746926	5.731235
11	M 119.445			0.00010	− 12.45341229	5.812622	5.796751
12	M 119.455			0.00054	− 12.45561970	5.817584	5.801700
13	M 124.756			− 0.00013	− 12.81321951	6.751385	6.732951
16	M 125.745	N		0.00180	− 12.85207977	6.871240	6.852478
17	M 125.755	N	2Q1	0.00955	− 12.85428618	6.878173	6.859392
18	M 126.556			− 0.00016	− 12.88607310	6.979628	6.960571
19	M 126.655			− 0.00011	− 12.89071297	6.994688	6.975590
20	M 126.754			0.00015	− 12.89535285	7.009814	6.990674
22	M 127.545	N		0.00218	− 12.92493337	7.107801	7.088393
23	M 127.555	N	SIG1	0.01153	− 12.92713978	7.115219	7.095792
24	M 128.544			0.00014	− 12.96600005	7.248468	7.228676
25	M 128.554	N		0.00079	− 12.96820646	7.256183	7.236371
26	M 129.355	W		0.00035	− 12.99999339	7.369188	7.349067
27	M 133.855	W		− 0.00023	− 13.32580728	8.768966	8.745023
28	M 134.656	N		− 0.00061	− 13.35759420	8.934539	8.910144
29	M 135.435	W		− 0.00028	− 13.38496439	9.082199	9.057400
32	M 135.635	N		− 0.00042*	− 13.39424805	9.133398	9.108460
33	M 135.645	N		0.01360	− 13.39645446	9.145651	9.120680
34	M 135.655	N	Q1	0.07216	− 13.39866087	9.157938	9.132932
36	M 135.855	W		− 0.00019*	− 13.40794455	9.209997	9.184850
37	M 136.456			− 0.00013	− 13.43044780	9.338677	9.313178
38	M 136.555	W		− 0.00039	− 13.43508767	9.365658	9.340085
39	M 136.644			0.00011	− 13.43752113	9.379870	9.354259
40	M 136.654	N		0.00068	− 13.43972754	9.392795	9.367148
41	M 137.445			0.00258	− 13.46930807	9.569567	9.543438
42	M 137.455	N	RO1	0.01371	− 13.47151448	9.583019	9.556854
44	M 137.655	N		− 0.00078*	− 13.48079814	9.640039	9.613717
45	M 137.665	W		0.00024	− 13.48300455	9.653690	9.627331

	Argument			Amplitude	Frequency ω_i	Nutation Period Sid. day	Solar day
46	M 138.444			0.00011	−13.51037475	9.826307	9.799477
47	M 138.454	N		0.00064	−13.51258116	9.840491	9.813623
48	M 139.455	W		−0.00014	−13.55365176	10.112207	10.084597
49	M 143.535	W		−0.00017	−13.85648548	12.697351	12.662682
50	M 143.745	W		−0.00020	−13.86797556	12.821718	12.786709
51	M 143.755	N		−0.00113	−13.87018179	12.845897	12.810804
52	M 144.546	W		−0.00015	−13.89976249	13.178820	13.142836
53	M 144.556	N		−0.00130	−13.90196890	13.204347	13.168293
55	M 145.535	N		−0.00218*	−13.93862275	13.643362	13.606110
56	M 145.545	N		0.07105	−13.94082916	13.670722	13.633395
57	M 145.555	N	O1	0.37689	−13.94303557	13.698192	13.660790
61	M 145.755	N		−0.00243*	−13.95231924	13.814996	13.777275
62	M 145.765	N		−0.00040	−13.95452565	13.843049	13.805252
63	M 146.544	W		0.00012	−13.98189583	14.200769	14.161995
64	M 146.554	N		0.00115	−13.98410224	14.230413	14.191558
65	M 147.355	W		−0.00491*	−14.01588917	14.671644	14.631584
67	M 147.545	N		−0.00021	−14.02296643	14.773633	14.733294
68	M 147.555	N	TO1	0.00014	−14.02517284	14.805720	14.765293
69	M 147.565	N		0.00107	−14.02737925	14.837946	14.797432
70	M 148.554	N		−0.00033*	−14.06623952	15.429441	15.387312
71	M 152.656	W		−0.00014	−14.37349898	22.530781	22.469262
72	M 153.645	N		−0.00063	−14.41235025	23.923379	23.858058
73	M 153.655	N		−0.00278	−14.41455666	24.007631	23.942080
74	M 154.656	N		0.00015*	−14.45562726	25.691844	25.621694
75	M 155.435	N		0.00017*	−14.48299745	26.951882	26.878291
76	M 155.445	N		−0.00197	−14.48520386	27.058862	26.984980
77	M 155.455	N		−0.01065	−14.48741027	27.166696	27.092519
81	M 155.645	N		0.00085	−14.49448752	27.518456	27.443319
82	M 155.655	N	M1	−0.02964*	−14.49669393	27.629992	27.554550
83	M 155.665	N		−0.00594	−14.49890034	27.742435	27.666686
84	M 155.675	N		0.00017	−14.50110675	27.855797	27.779738
85	M 156.555	N		0.00016*	−14.53312073	29.611439	29.530587
86	M 156.654	N		−0.00018*	−14.53776060	29.884420	29.802822
87	M 157.445	N		0.00016	−14.56734113	31.750465	31.663772
88	M 157.455	N	KI1	−0.00566*	−14.56954754	31.899036	31.811938
89	M 157.465	N		−0.00124	−14.57175395	32.049005	31.961497
90	M 158.454	N		−0.00024*	−14.61061422	34.942303	34.846895
91	S 161.557	N		0.00042	−14.87679800	91.562737	91.312731
92	S 162.556	N	PI1	0.01029	−14.91786469	122.082691	121.749353
93	M 163.535	N		0.00014*	−14.95451854	173.784552	173.310044
94	M 163.545	N		−0.00199	−14.95672495	178.330713	177.843793
95	SM163.555	N	P1	0.17584	−14.95893136	183.121117	182.621116
96	S 163.557	N		−0.00011*	−14.95893528	183.129856	182.629832
97	M 163.755	N		−0.00026*	−14.96821503	206.456079	205.892364
98	S 164.554	N		−0.00147	−14.99999804	366.224800	365.224848
99	S 164.556	N	S1	−0.00423*	−15.00000196	366.259758	365.259710
101	M 165.545	N		0.01050	−15.03886222	6816.987155	6798.373824
102	MS165.555	P	K1	−0.53050	−15.04106863	∞	∞
103	M 165.565	N		−0.07182	−15.04327504	−6816.987155	−6798.373824
104	M 165.575	N		0.00154	−15.04548145	−3408.493577	−3399.186912
106	S 166.554	N	PSI1	−0.00423*	−15.08213530	−366.259758	−365.259710
107	M 167.355	N		−0.00026*	−15.11392223	−206.456079	−205.892364
108	S 167.553	N		−0.00011*	−15.12320198	−183.129856	−182.629832
109	S 167.555	N	FI1	−0.00756	−15.12320590	−183.121117	−182.621116
110	M 167.565	N		0.00029	−15.12541231	−178.330713	−177.843793
111	M 167.575	N		0.00014*	−15.12761872	−173.784552	−173.310044
112	S 168.554	N		−0.00044	−15.16427258	−122.082681	−121.749343
113	M 172.656	N		−0.00024*	−15.47152304	−34.942303	−34.846895
114	M 173.445	N		−0.00017	−15.50109965	−32.695770	−32.606496
115	M 173.645	N		0.00018	−15.51038331	−32.049005	−31.961497
116	M 173.655	N	TT1	−0.00566*	−15.51258972	−31.899036	−31.811938
117	M 173.665	N		−0.00111	−15.51479613	−31.750465	−31.663772
119	M 174.456	N		−0.00018*	−15.54437666	−29.884420	−29.802822
120	M 174.555	N		0.00016*	−15.54901653	−29.611439	−29.530587

	Argument			Amplitude	Frequency ω_t	Nutation Period Sid. day	Solar day
121	M 175.445	N		0.00087	−15.58323692	−27.742435	−27.666686
122	M 175.455	N	J1	−0.02964*	−15.58544333	−27.629992	−27.554550
123	M 175.465	N		−0.00587	−15.58764974	−27.518456	−27.443319
124	M 175.475	W		0.00013	−15.58985615	−27.407818	−27.332983
126	M 175.655	N		0.00046	−15.59472699	−27.166696	−27.092519
127	M 175.665	N		0.00029	−15.59693340	−27.058862	−26.984980
128	M 175.675	N		0.00017*	−15.59913981	−26.951882	−26.878291
129	M 176.454	N		0.00015*	−15.62651000	−25.691844	−25.621694
130	M 177.455	N		0.00012	−15.66758060	−24.007631	−23.942080
131	M 182.556	N		−0.00032*	−16.01589774	−15.429441	−15.387312
132	M 183.545	N		−0.00016	−16.05475801	−14.837946	−14.797432
133	M 183.555	N	SO1	−0.00492*	−16.05696442	−14.805720	−14.765293
134	M 183.565	N		−0.00096	−16.05917083	−14.773633	−14.733294
135	M 185.355	N		−0.00240*	−16.12981802	−13.814996	−13.777275
136	M 185.365	N		−0.00048	−16.13202443	−13.787055	−13.749411
139	M 185.555	N	OO1	−0.01623	−16.13910169	−13.698192	−13.660790
140	M 185.565	N		−0.01039	−16.14130810	−13.670722	−13.633395
141	M 185.575	N		−0.00218*	−16.14351451	−13.643362	−13.606110
142	M 185.585	W		−0.00014	−16.14572092	−13.616111	−13.578933
143	M 191.655	W		−0.00015	−16.52848550	−10.112207	−10.084597
144	M 193.455	N		−0.00078*	−16.60133912	−9.640039	−9.613717
145	M 193.465	N		−0.00015	−16.60354553	−9.626426	−9.600141
146	M 193.655	N		−0.00059	−16.61062278	−9.583019	−9.556854
147	M 193.665			−0.00038	−16.61282919	−9.569567	−9.543438
148	M 195.255	W		−0.00019*	−16.67419271	−9.209997	−9.184850
149	M 195.455	N	NU1	−0.00311	−16.68347639	−9.157938	−9.132932
150	M 195.465	N		−0.00199	−16.68568280	−9.145651	−9.120680
151	M 195.475	N		−0.00042*	−16.68788921	−9.133398	−9.108460
152	M 1X3.555	N		−0.00050	−17.15499748	−7.115219	−7.095792
153	M 1X3.565	N		−0.00032	−17.15720389	−7.107801	−7.088393
154	M 1X5.355	N		−0.00041	−17.22785108	−6.878173	−6.859392
155	M 1X5.365	N		−0.00027	−17.23005749	−6.871240	−6.852478
156	M 1E3.455	N		−0.00012	−17.69937218	−5.658145	−5.642696

(X = 10, E = 11, if appearing in place of digits)

P Precession
N IAU nutation terms
W Woolard's nutation terms
* Symmetrical waves of K1/precession/and of equal amplitudes - no nutation in obliquity
M Lunar waves
S Solar waves
3 Waves deriving from 3th order potential

TABLE 2.5. Construction of the nutation tables for a rigid earth by using equations (2.56)

$E_{(} = 0.0164427$

Associated Tides					Period	$\sin\theta \cdot \Delta\psi$	$\Delta\theta$	Nut. arg.
9	156	M	117.655	1E3.455	5.658145	−0.000247	0.000269	X3.455
16	155	M	125.745	1X5.365	6.871240	−0.000172	0.000233	95.365
17	154	M	125.755	1X5.355	6.878173	−0.001033	0.001126	95.355
22	153	M	127.545	1X3.565	7.107801	−0.000217	0.000292	93.565
23	152	M	127.555	1X3.555	7.115219	−0.001290	0.001407	93.555
32	151	M	135.635	195.475	9.133398	0.000126	0.000000	85.475
33	150	M	135.645	195.465	9.145651	−0.001745	0.002344	85.465
34	149	M	135.655	195.455	9.157938	−0.010397	0.011334	85.455
36	148	M	135.855	195.255	9.209997	0.000057	0.000000	85.255
41	147	M	137.445	193.665	9.569567	−0.000346	0.000465	83.665
42	146	M	137.455	193.655	9.583019	−0.002067	0.002253	83.655
44	144	M	137.655	193.455	9.640039	0.000247	0.000000	83.455
48	143	M	139.455	191.655	10.112207	0.000048	0.000001	81.655
55	141	M	145.535	185.575	13.643362	0.000978	0.000000	75.575
56	140	M	145.545	185.565	13.670722	−0.013635	0.018306	75.565
57	139	M	145.555	185.555	13.698192	−0.081233	0.088544	75.555
61	135	M	145.755	185.355	13.814996	0.001097	−0.000006	75.355
67	134	M	147.545	183.565	14.773633	0.000199	0.000267	73.565
68	133	M	147.555	183.555	14.805720	0.002393	0.000002	73.555
69	132	M	147.565	183.545	14.837946	−0.000222	0.000300	73.545
70	131	M	148.554	182.556	15.429441	0.000164	−0.000002	72.556
73	130	M	153.655	177.455	24.007631	0.001050	−0.001144	67.455
74	129	M	154.656	176.454	25.691844	−0.000126	0.000000	66.454
75	128	M	155.435	175.675	26.951882	−0.000150	0.000000	65.675
76	127	M	155.445	175.665	27.058862	0.000747	−0.001005	65.665
77	126	M	155.455	175.655	27.166696	0.004551	−0.004962	65.655
81	123	M	155.645	175.465	27.518456	0.002271	0.003040	65.465
82	122	M	155.655	175.455	27.629992	0.026931	0.000000	65.455
83	121	M	155.665	175.445	27.742435	0.002312	−0.003106	65.445
85	120	M	156.555	174.555	29.611439	−0.000155	0.000000	64.555
86	119	M	156.654	174.456	29.884420	0.000176	0.000000	64.456
87	117	M	157.445	173.665	31.750465	0.000501	0.000668	63.665
88	116	M	157.455	173.655	31.899036	0.005937	0.000000	63.655
89	115	M	157.465	173.645	32.049005	0.000558	−0.000748	63.645
90	113	M	158.454	172.656	34.942303	0.000275	0.000000	62.656
92	112	S	162.556	168.554	122.082681	−0.019772	0.021539	58.554
93	111	M	163.535	167.575	173.784552	−0.000800	0.000000	57.575
94	110	M	163.545	167.565	178.330713	0.004984	−0.006685	57.565
95	109	SM	163.555	167.555	183.121117	−0.506692	0.552218	57.555
96	108	S	163.557	167.553	183.129856	0.000662	0.000000	57.553
97	107	M	163.755	167.355	206.456079	0.001765	0.000000	57.355
99	106	S	164.556	166.554	366.259758	0.050948	0.000000	56.554
101	103	M	165.545	165.565	6816.987155	6.873338	9.227222	55.565

. (X = 10, E = 11, if appearing in place of digits)

The nutation argument (last column) is obtained by subtracting 110.000 from the highest Doodson argument of the pair of associated tidal waves.

CHAPTER 3

Love Numbers and the Description of Tidal Deformations

3.1. INTRODUCTION

In 1909 Love introduced two new parameters in spherical elasticity, h and k. These parameters are dimensionless numbers and bear the name of Love. In 1912 Shida introduced a third number ℓ. These three numbers allow a very practical representation of all deformation phenomena produced by a potential which can be developed in spherical polynomials. As these mathematical functions are orthogonal, each characteristic deformation effect (e. g. pressure, cubic expansion, linear displacement components, additional potential) may be developed in the same way as a series of spherical polynomials, and each term of this development is proportional to the corresponding term of the perturbing potential as was shown by Kelvin (Chapter 4).

The coefficient of proportionality is a "Love number" or a fairly simple arithmetic combination of Love numbers.

We shall assume here a spherical symmetry for physical properties in the Earth's interior:

density: $\rho = \rho(r)$
compressibility: $\lambda = \lambda(r)$
rigidity: $\mu = \mu(r)$

Then Love numbers characterizing the deformation are defined by the following relations:

components of the displacement:

$$u_r = \sum_{n=2}^{\infty} H_n(r)\,\frac{W_n}{g} = \xi, \tag{3.1}$$

$$u_\theta = \frac{1}{g}\sum_{n=2}^{\infty} L_n(r)\,\frac{\partial W_n}{\partial\theta}, \tag{3.2}$$

$$u_\lambda = \frac{1}{g}\sum_{n=2}^{\infty} L_n(r)\,\frac{\partial W_n}{\sin\theta\,\partial\lambda}, \tag{3.3}$$

cubic expansion:

$$\theta = \sum_{n=2}^{\infty} f_n(r)\,\frac{W_n}{g}. \tag{3.4}$$

Likewise the potential caused by the deformation itself and the density variation which accompanies cubic dilatation and superficial displacement of matter will be expressed in the form

$$V = V_0 + \sum_{n=2}^{\infty} K_n(r)\,W_n = V_0 + \sum_{n=0}^{\infty} V_n. \tag{3.5}$$

At the surface (r=a) one puts:

$$H_n(a) = h_n, \qquad L_n(a) = \ell_n, \qquad K_n(a) = k_n. \qquad (3.6)$$

3.2. POTENTIAL INDUCED BY A BULGE. RELATION BETWEEN LOVE NUMBERS h AND k

Let A be a point of observation at the surface of the Earth,

$$OA = a,$$

and consider an internal element M of coordinates $[r<a,\ \theta=\pi/2 - \phi,\ \lambda,\ \mu=\cos\theta]$, the mass element at M is

$$dm = \rho r^2 \cos\phi\, d\phi\, d\lambda\, dr = -\rho r^2 \sin\theta\, d\theta\, d\lambda\, dr = +\rho r^2\, dr\, d\mu\, d\lambda. \qquad (3.7)$$

Its potential at external point A is thus

$$V = G\rho \left(\sum_n \frac{r^{n+2}}{a^{n+1}} P_n\right) dr\, d\mu\, d\lambda. \qquad (3.8)$$

When the Earth is deformed, particles which were on the surface of radius r are displaced on a surface:

$$r(1+\varepsilon) = r(1+\sum_i^n Y_i), \qquad (3.9)$$

where ε is developed in spherical harmonics Y_i.

This deformation produces a potential at A:

$$V^* = G\rho \int_{-1}^{+1}\int_0^{2\pi}\int_r^{r(1+\varepsilon)} \left(\sum_n \frac{r^{n+2}}{a^{n+1}} P_n\right) dr\, d\mu\, d\lambda, \qquad (3.10)$$

and neglecting terms of the order ε^2 or higher,

$$V^* = \sum_n G\rho \frac{r^{n+3}}{a^{n+1}} \int_{-1}^{+1}\int_0^{2\pi} \varepsilon P_n\, d\mu\, d\lambda, \qquad (3.11)$$

ε is developed in spherical harmonics Y_i (3.9). Because of the orthogonality property, we have to consider only those terms for which i=n:

$$V^* = \sum_n G\rho \frac{r^{n+3}}{a^{n+1}} \int_{-1}^{+1}\int_0^{2\pi} Y'_n P_n\, d\mu\, d\lambda = \sum_n G\rho \frac{4\pi}{2n+1} \cdot \frac{r^{n+3}}{a^{n+1}} Y_n, \qquad (3.12)$$

thus considering the effect of the deformation of an infinitesimal shell of thickness dr within which the density may be considered as constant:

$$dV^* = \frac{4\pi G\rho}{a} \frac{\partial}{\partial r}\left(\sum_n \frac{r^{n+3}}{(2n+1)\, a^n} Y_n\right) dr. \qquad (3.13)$$

The global effect produced by the deformation of the Earth is obtained by integration from the centre to the surface:

$$V = \frac{4\pi G}{a}\int_0^a \rho \frac{\partial}{\partial r}\left(\sum_n \frac{r^{n+3}}{(2n+1)\, a^n} Y_n\right) dr. \qquad (3.14)$$

Theory for the Effects of Second and Third Order

In Earth tides the term Y_2 is by far the most important. However, investigations on the term of third-order Y_3 have been made as soon as a sufficiently long series was available (Melchior and Venedikov, 1967).

Let us put the luni-solar potential as

$$W = W_2 + W_3 = r^2\, S_2(\phi, \lambda) + r^3\, S_3(\phi, \lambda), \qquad (3.15)$$

S_2 and S_3 being surface spherical harmonics.

The Earth tide at the surface of the Earth is:

$$h_2 \frac{W_2}{g} = h_2 \frac{a^2 S_2}{g} \quad \text{for the second order,} \qquad (3.16)$$

$$h_3 \frac{W_3}{g} = h_3 \frac{a^3 S_3}{g} \quad \text{for the third order.} \qquad (3.17)$$

Within the Earth, on a surface of radius r, one has deformations like

$$\xi(r) = H_2(r) \frac{r^2 S_2}{g} \quad \text{and} \quad H_3(r) \frac{r^3 S_3}{g}. \qquad (3.18)$$

Let us put

$$\xi(r) = \xi(a)\, \chi(r), \qquad (3.19)$$

i. e.

$$H_2(r) \frac{r^2}{g} = \frac{h_2 a^2}{g} \chi(r). \qquad (3.20)$$

The parametric equation (limited to the second order) of the deformed surface becomes

$$r(1+\varepsilon) = r\left[1 + \frac{h_2 a}{g} \frac{a}{r} \chi(r)\, S_2\right], \qquad (3.21)$$

$$\varepsilon(r) = \frac{\xi(r)}{r} = \frac{h_2 a}{g} \frac{a}{r} \chi(r)\, S_2, \qquad (3.22)$$

and (3.11 and 3.14) give

$$k_2\, a^2 = \frac{4\pi G}{a}\int_0^a \rho \frac{\partial}{\partial r}\left[\frac{r^5}{5a^2} \frac{h_2 a}{g} \left[\frac{a}{r} \chi(r)\right]\right] dr. \qquad (3.23)$$

An interesting simplification is obtained for this expression when we consider the deformations as homothetic with respect to the centre:

$$\frac{\xi(r)}{r} = \frac{\xi(a)}{a} \qquad (3.24)$$

which corresponds to

$$\chi(r) = \frac{r}{a}. \tag{3.25}$$

This means that $\frac{[rH(r)]}{g}$ is independent of r.

As
$$g = \frac{4\pi G}{a^2}\int_0^a \rho\, r^2\, dr, \tag{3.26}$$

one obtains

$$k_2 = h_2 \frac{\int_0^a \rho\, r^4\, dr}{a^2 \int_0^a \rho\, r^2\, dr} = \frac{3}{5} h_2 \frac{\int_0^a 5\, \rho\, r^4\, dr}{a^2 \int_0^a 3\, \rho\, r^2\, dr} \tag{3.27}$$

a relation given by Melchior in 1950.

Using the d'Alembert parameter one can write

$$\lambda k = h. \tag{3.28}$$

λ is related to the fundamental parameters of geodynamics in such a way that

$$\frac{k}{h} = \frac{1}{\lambda} = \frac{e - \frac{q}{2}}{\frac{C-A}{C}} = 1 - \frac{2}{5}\sqrt{1 + \eta} = \frac{3}{2}\frac{C}{Ma^2} \tag{3.29}$$

where η is the Radau parameter, e the geometrical flattening, A and C the principal moments of inertia, and $q = \omega^2 a^3 / GM$. The restrictions for the validity of these relations are evidently important: 1) the hydrostatic equilibrium is realized, and 2) deformations are homothetic with respect to the centre. But a comparison with the most recent experimental results shows a fairly good agreement.

Equation (3.29) gives the following relationships:

η	C/Ma^2	k/h
0.543	0.335 50	0.500 3
0.566	0.332 96	0.498 9
0.585	0.330 96	0.496 4

while the reduction of more than 30 000 days' observations (Melchior, 1975) has given

$$k/h = 0.496. \tag{3.30}$$

Jobert (1951) has shown that if we do not introduce the condition of homothety we cannot, of course, obtain the numerical value of k/h from the law of density, but we can still put an upper limit on its value.

$$k/h < 1/\lambda. \tag{3.31}$$

For the deformations of third order one evidently has

$$k_3 = \frac{3}{7} h_3 \frac{\int_0^1 6 \rho r^5 dr}{\int_0^1 3 \rho r^2 dr} \tag{3.32}$$

and generally

$$k_n = \frac{3}{2n+1} h_n \frac{\int_0^1 (n+3) \rho r^{n+2} dr}{\int_0^1 3 \rho r^2 dr}. \tag{3.33}$$

Case of Homogeneity

In this case, (3.27), (3.32) and (3.33) give immediately

$$k_2 = \frac{3}{5} h_2, \qquad k_3 = \frac{3}{7} h_3, \qquad k_n = \frac{3}{2n+1} h_n. \tag{3.34}$$

Remark. This very clearly demonstrates that Love numbers are different for the deformations of the Earth corresponding to the different orders in the development of the tidal potential.

Due to its form ($W_n = r^n S_n$) this potential introduces the internal rheological properties of the Earth according to the power n of the radius.

Therefore the terms of higher order give more and more importance to the external parts of the planet.

3.3. SOME DIRECT OBSERVATIONS OF THE SOLID EARTH TIDAL EFFECTS

3.3.1. Amplitude of Oceanic and Lake Tides

As the potential of the deformed surface is $W_n^* = (1+k_n) W_n$ the static tide in an ocean or a lake will be

$$\zeta_0 = (1 + k_n) \frac{W_n}{g}$$

while the tide of the Earth's crust is by definition

$$\zeta_c = h_n \frac{W_n}{g}.$$

A mareograph attached to the crust therefore measures the differential displacement

$$\zeta = \zeta_0 - \zeta_c = (1 + k_n - h_n) \frac{W_n}{g} \tag{3.35}$$

and, in principle, should allow a measure of the coefficient

$$\gamma_n = 1 + k_n - h_n. \tag{3.36}$$

3.3.2. Deviations of the Vertical with Respect to the Crust

The equation of the deformed surface being

$$r = a + h_n \frac{W_n}{g} = a + dr,$$

the angle, with respect to the undeformed surface in the meridian plane at the point P(r, θ), is (Fig. 3.1)

$$i = \frac{dr}{a d\theta} = \frac{h_n}{g} \frac{\partial W_n}{a \partial\theta}.$$

On the other hand, the deviation of a pendulum having its suspension support fixed to the crust at point P will be

$$\frac{\partial W^*}{g a \partial\theta} = \frac{1 + k_n}{g} \frac{\partial W_n}{a \partial\theta}.$$

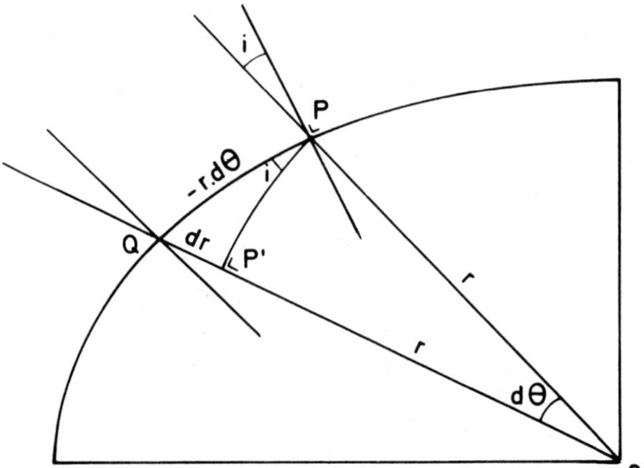

Fig. 3.1.

This instrument measures the difference

$$\left.\begin{aligned} i &= \frac{1 + k_n - h_n}{g} \frac{\partial W_n}{a \partial\theta} \text{ in the meridian,} \\ &\text{and similarly} \\ j &= \frac{1 + k_n - h_n}{g} \frac{\partial W_n}{a \sin\theta \partial\lambda} \text{ in the prime vertical} \end{aligned}\right\} \qquad (3.37)$$

both give the same coefficient

$$\gamma_n = 1 + k_n - h_n. \qquad (3.38)$$

3.3.3. Variations of the Intensity of Gravity

A gravimeter installed at the point P, on the crust, is submitted to an acceleration deriving from the total potential at this point,

$$V = V_0 + \zeta_c \frac{dV}{da} + W_n + V_n , \qquad (3.39)$$

where

$$\zeta_c = h_n \frac{W_n}{g}$$

and, at the surface, from (3.5) and (3.14),

$$V_n = \frac{4\pi G}{a^{n+1}} \int_0^a \rho \frac{\partial}{\partial r} \left(\frac{r^{n+3}}{2n+1} Y_n\right) dr. \qquad (3.40)$$

Let us calculate the force acting on the gravimeter's beam by derivation of the potential (3.39) with respect to r and making thereafter r = a:

$$\frac{dV}{da} = \frac{dV_0}{da} + \zeta_c \frac{d^2V}{da^2} + \frac{dW_n}{da} + \frac{dV_n}{da}, \qquad (3.41)$$

where

$$-g = -g_0 + \frac{2g}{a} \zeta_c + \frac{nW_n}{a} - (n+1) \frac{V_n}{a}$$

or

$$-\Delta g = \frac{2g}{a} h_n \frac{W_n}{g} + \frac{nW_n}{a} - \frac{n+1}{a} k_n W_n \qquad (3.42)$$

or

$$\Delta g = - [n + 2h_n - (n+1) k_n] \frac{W_n}{a}.$$

This instrument measures

$$\Delta g = - \left(1 + \frac{2}{n} h_n - \frac{n+1}{n} k_n\right) \frac{dW_n}{da} \qquad (3.43)$$

and allows a determination of the coefficient

$$\delta_n = 1 + \frac{2}{n} h_n - \frac{n+1}{n} k_n \qquad (3.44)$$

which, for orders n=2, n=3, is

$$\delta_2 = 1 + h_2 - \frac{3}{2} k_2 , \qquad (3.45)$$

$$\delta_3 = 1 + \frac{2}{3} h_3 - \frac{4}{3} k_3. \qquad (3.46)$$

3.3.4. Deviations of the Vertical as Measured with Astronomical Instruments

The deflexions of the vertical determined with horizontal pendulums or levels are related to the crust itself; in a horizontal pendulum the deviations of a beam of light reflected by a mirror attached to a moving arm are compared with a fixed point on the recording drum, i. e. to the deformed crust.

The situation is quite different for astronomical instruments. All deflexions of the vertical at a point cause the astronomic coordinates - and especially the latitude - to vary. The latitude is determined by comparing the direction of the vertical measured by two spirit-levels (Talcott's method), or with a bath of mercury (zenith tube, astrolabe) to the observed directions of a series of fixed fundamental stars.

In this method the movement of the vertical is not related to the crust but to the direction of the axis of rotation of the Earth (called the Earth's axis) in space. If the Earth was perfectly rigid there would not be any difference between the results of the two methods of measurement, but it is clear that if it changes shape the observed phenomena will be different for the two methods.

The astronomical observations are affected by Earth tides as follows: one observes a deviation of the vertical produced by the perturbed potential $W^* = (1+k_n)W_n$ with an instrument which is displaced by (u_θ, u_λ), and this produces a geometrical shift of the vertical with respect to the axis of rotation which amounts to $(u_\theta/a,\ u_\lambda/a)$.

Then one measures:

$$\text{in the meridian} \qquad (1 + k - \ell)\,\frac{1}{ag}\,\frac{\partial W}{\partial \theta}\,n,$$

$$\text{in the prime vertical} \qquad (1 + k - \ell)\,\frac{1}{ag\,\sin\theta}\,\frac{\partial W_n}{\partial \lambda}.$$

The coefficient occurring here is

$$\Lambda = 1 + k - \ell. \tag{3.47}$$

3.3.5. Linear, Areal and Cubic Expansions

Instruments called extensometers (described in Chapter 10) allow the direct measurements of linear, areal and cubic expansion.

From formulas (4.10), established in the Chapter 4, we can immediately write:

$$\left.\begin{aligned}
\varepsilon_{33} &= e_{rr} = \frac{\partial u_r}{\partial r} = (ah' + nh)\,\frac{W_n}{gr} = \eta\,\frac{W_n}{gr},\\
\varepsilon_{11} &= e_{\theta\theta} = \frac{1}{r}\,\frac{\partial u_\theta}{\partial \theta} + \frac{u_r}{r} = \frac{\ell}{gr}\,\frac{\partial^2 W_n}{\partial \theta^2} + \frac{h}{gr}\,W_n\,,\\
\varepsilon_{22} &= e_{\lambda\lambda} = \frac{1}{r\sin\theta}\,\frac{\partial u_\lambda}{\partial \lambda} + \frac{u_\theta}{r}\cot\theta + \frac{u_r}{r}\\
&= \frac{\ell}{gr\sin^2\theta}\,\frac{\partial^2 W_n}{\partial \lambda^2} + \frac{\ell}{gr}\cdot\frac{1}{\sin\theta}\cos\theta\,\frac{\partial W_n}{\partial \theta} + \frac{h}{gr}\,W_n,\\
\varepsilon_{12} &= e_{\theta\lambda} = \frac{1}{r}\left(\frac{\partial u_\lambda}{\partial \theta} - u_\lambda\cot\theta\right) + \frac{1}{r\sin\theta}\,\frac{\partial u_\theta}{\partial \lambda}\\
&= \frac{2\ell}{gr\sin\theta}\,\frac{\partial^2 W_n}{\partial \theta\,\partial \lambda} - \frac{2\ell}{gr\sin\theta}\cot\theta\,\frac{\partial W_n}{\partial \lambda},\\
\varepsilon_{23} &= e_{\lambda r} = \frac{1}{r\sin\theta}\,\frac{\partial u_r}{\partial \lambda} + \frac{\partial u_\lambda}{\partial r} - \frac{u_\lambda}{r}\\
&= \frac{h}{gr\sin\theta}\,\frac{\partial W_n}{\partial \lambda} + \frac{\ell}{gr\sin\theta}\,\frac{\partial W_n}{\partial \lambda} + \frac{\ell'}{g\sin\theta}\,\frac{\partial W_n}{\partial \lambda},\\
\varepsilon_{31} &= e_{r\theta} = \frac{\partial u_\theta}{\partial r} - \frac{u_\theta}{r} + \frac{1}{r}\,\frac{\partial u_r}{\partial \theta}\\
&= \frac{\ell}{gr}\,\frac{\partial W_n}{\partial \theta} + \frac{h}{gr}\,\frac{\partial W_n}{\partial \theta} + \frac{\ell'}{g}\,\frac{\partial W_n}{\partial \theta},
\end{aligned}\right\} \tag{3.48}$$

formulas correctly given by Lecolazet in 1963.

having put

$$h' = \left(\frac{dH(r)}{dr}\right)_{r=a} \qquad \ell' = \left(\frac{\partial L(r)}{\partial r}\right)_{r=a}. \tag{3.49}$$

The deformation along any one direction of the direction cosines $(\alpha_1, \alpha_2, \alpha_3)$ is given by

$$d = \alpha_1^2 e_{rr} + \alpha_2^2 e_{\theta\theta} + \alpha_3^2 e_{\lambda\lambda} + \alpha_1\alpha_2 e_{r\theta} + \alpha_2\alpha_3 e_{\theta\lambda} + \alpha_3\alpha_1 e_{\lambda r}. \tag{3.50}$$

Considerable interest attaches to

$$\Sigma = e_{\theta\theta} + e_{\lambda\lambda} = \frac{2h}{ag} W_n + \frac{\ell}{ag}\left(\frac{1}{\sin^2\theta}\frac{\partial^2 W_n}{\partial\lambda^2} + \frac{1}{\sin\theta}\frac{\partial}{\partial\theta}\sin\theta\frac{\partial W_n}{\partial\theta}\right), \tag{3.51}$$

which approximately to the second order represents the horizontal areal deformation, i. e. the increase of the horizontal surface per unit of surface or surface expansion. This expression is considerably simplified by application of the fundamental property of harmonic functions, namely of having a zero Laplacian. In polar coordinates

$$\frac{1}{\sin\theta}\frac{\partial}{\partial\theta}\left(\sin\theta\frac{\partial S_n}{\partial\theta}\right) + \frac{1}{\sin^2\theta}\frac{\partial^2 S_n}{\partial\lambda^2} = -n(n+1)\, S_n \tag{3.52}$$

which, introduced in (3.51), gives

$$\Sigma_n = \left[2\, h_n - n(n+1)\, \ell_n\right]\frac{W_n}{ag}. \tag{3.53}$$

For order n=2 this gives

$$\Sigma_2 = 2(h_2 - 3\ell_2)\,\frac{W_2}{ag}. \tag{3.54}$$

Moreover, the difference $e_{\theta\theta} - e_{\lambda\lambda}$ allows direct determination of the number ℓ:

$$e_{\theta\theta} - e_{\lambda\lambda} = \frac{\ell}{ag\sin\theta}\left(\sin\theta\frac{\partial^2 W_n}{\partial\theta^2} - \cos\theta\frac{\partial W_n}{\partial\theta} - \sec\theta\frac{\partial^2 W_n}{\partial\lambda^2}\right). \tag{3.55}$$

Equations (3.48), (3.50) and (3.51) will be used to interpret the results yielded by an array of extensometers. Six extensometers will clearly be needed to obtain a complete representation of the phenomenon.

Expression (3.53) is particularly noteworthy because it reveals the simple combination $(h-3\ell)$, which may also be experimentally determined by two suitably arranged extensometers.

From formulas (3.48) and (3.50) it becomes quite clear that the arithmetic combination of Love numbers appearing as the coefficient in any linear displacement depends upon the azimuth in which this displacement is measured.

But because of the different combinations of the derivatives of the potential with respect to λ and to θ, it also depends upon the geometrical properties of the Legendre polynomial as this can be sectorial, tesseral or zonal.

Conditions at the Free Surface (Boundary Conditions)

The depth at which our instruments are buried (the present maximum is 1500 m) is always much less than the wavelengths of the tidal strain. We can thus consider that our measurements are made at the free surface.

At the free surface one has, of course, the conditions

$$\sigma_{rr} = 0, \qquad \sigma_{r\theta} = \sigma_{\lambda r} = 0 \tag{3.56}$$

and in a Hooke body [see (4.22)],

$$e_{\lambda r} = e_{r\theta} = 0, \tag{3.57}$$

but also [see (4.21) and (4.22)]

$$e_{\theta\theta} = \frac{1}{E}\sigma_{\theta\theta} - \frac{\nu}{E}\sigma_{\lambda\lambda}, \quad e_{\lambda\lambda} = \frac{1}{E}\sigma_{\lambda\lambda} - \frac{\nu}{E}\sigma_{\theta\theta}, \quad e_{rr} = -\frac{\nu}{E}(\sigma_{\theta\theta} + \sigma_{\lambda\lambda}), \tag{3.58}$$

which gives

$$(\sigma_{\theta\theta} + \sigma_{\lambda\lambda}) = \frac{E}{1-\nu}(e_{\theta\theta} + e_{\lambda\lambda}) \tag{3.59}$$

and as a consequence

$$e_{rr} = -\frac{\nu}{1-\nu}(e_{\theta\theta} + e_{\lambda\lambda}). \tag{3.60}$$

The definition of Poisson's coefficient ν [see (4.23)]

$$\nu = \frac{\lambda}{2(\lambda+\mu)}$$

gives

$$e_{rr} = -\frac{\lambda}{\lambda+2\mu}(e_{\theta\theta} + e_{\lambda\lambda}). \tag{3.61}$$

and in the case $\lambda = \mu$ $(\nu = 0.25)$,

$$e_{rr} = -\frac{1}{3}(e_{\theta\theta} + e_{\lambda\lambda}) \tag{3.62}$$

a useful relation to check numerical computations.

With these conditions in mind let us consider the different cases.

Tesseral Waves

For the general form of tesseral waves of the second order, let us put the potential as

$$\mathscr{T}_2 = D\left(\frac{c}{d}\right)^3 \left(\frac{r}{a}\right)^2 \sin 2\theta \sin 2\delta \cos H \tag{3.63}$$

with

$$H = H_i - \lambda. \tag{3.64}$$

H_i being the hour angle for the international meridian one obtains at the surface

$$\left.\begin{aligned}
e_{\theta\theta} &= \frac{h - 4\ell}{g}\frac{\mathcal{T}_2}{a},\\
e_{\lambda\lambda} &= \frac{h - 2\ell}{g}\frac{\mathcal{T}_2}{a},\\
e_{\theta\lambda} &= -\frac{2\ell}{g}\left(\frac{\mathrm{tg}\ H}{\cos\theta}\right)\frac{\mathcal{T}_2}{a},\\
e_{\lambda r} &= (h+\ell+a\ell')\frac{\partial\mathcal{T}_2}{ag\sin\theta\partial\lambda} = 0 \quad \text{(condition 3.57)},\\
e_{r\theta} &= (\ell+h+a\ell')\frac{\partial\mathcal{T}_2}{ag\partial\theta} = 0 \quad \text{(condition 3.57)},\\
e_{rr} &= \left\{a\left(\frac{dh(r)}{dr}\right)_{r=a} + 2h\right\}\frac{\mathcal{T}_2}{ag} = \eta\frac{\mathcal{T}_2}{ag},\\
\Sigma &= \frac{(2h-6\ell)}{g}\frac{\mathcal{T}_2}{a},\\
e_{\lambda\lambda} - e_{\theta\theta} &= \frac{2\ell}{g}\frac{\mathcal{T}_2}{a}.
\end{aligned}\right\} \qquad (3.65)$$

Thus for an arbitrary direction of direction cosine $(\alpha_1\ \alpha_2\ \alpha_3)$ one has at the surface:

$$\left.\begin{aligned}
d &= \alpha_1^2\, e_{rr} + \alpha_2^2\, e_{\theta\theta} + \alpha_3^2\, e_{\lambda\lambda} + \alpha_2\,\alpha_3\, e_{\theta\lambda},\\
&\alpha_1^2 + \alpha_2^2 + \alpha_3^2 = 1.
\end{aligned}\right\} \qquad (3.66)$$

In the horizontal plane $(\alpha_1 = 0)$,

$$d_1 = \left[(h-4\ell)\alpha_2^2 + (h-2\ell)(1-\alpha_2^2) - 2\ell\,\frac{\mathrm{tg}\ H}{\cos\theta}\,\alpha_2\sqrt{1-\alpha_2^2}\right]\frac{\mathcal{T}_2}{ag} \qquad (3.67)$$

gives a distribution of deformations according to the azimuth α_2 as shown by Fig. 3.2.

In the meridian plane $(\alpha_3 = 0)$,

$$d_3 = \left\{\alpha_1^2\left[a\left(\frac{dh(r)}{dr}\right)_{r=a} + 2h\right] + (h-4\ell)(1-\alpha_1^2)\right\}\frac{\mathcal{T}_2}{ag} \qquad (3.68)$$

and in the prime vertical $(\alpha_2 = 0)$,

$$d_2 = \left\{\alpha_1^2\left[a\left(\frac{dh(r)}{dr}\right)_{r=a} + 2h\right] + (h-2\ell)(1-\alpha_1^2)\right\}\frac{\mathcal{T}_2}{ag}. \qquad (3.69)$$

The deformations in function of the dip angle α_1 are also represented on Fig. 3.2. To draw Fig. 3.2 we have adopted the following numerical values:

$$h = 0.638; \qquad \ell = 0.088; \qquad \eta = a\left(\frac{dh(r)}{dr}\right)_{r=a} + 2h = -0.250;$$

and the scale value

$$\frac{\mathcal{T}_2}{ag} = 1.$$

The null deformation dip direction depends obviously upon the relative values of $h - 2\ell$ (0.462), $h - 4\ell$ (0.286) and η (-0.250). With these values it is nearly 53° from the horizontal plane in the prime vertical and nearly 47° from the horizontal plane in the meridian.

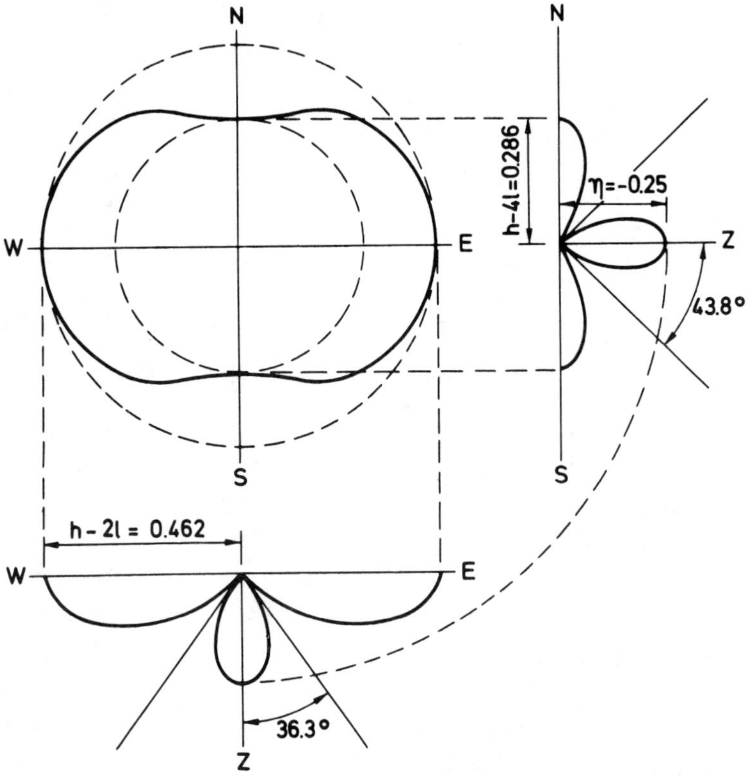

Fig.3.2. Distribution of the deformations as a function of the azimuth and dip at the Earth's surface: wave O_1, latitude 45°

Sectorial Waves

For the general form of sectorial waves of the second order:

$$\mathcal{S}_2 = D\left(\frac{c}{d}\right)^3\left(\frac{r}{\bar{a}}\right)^2 \sin^2\theta \cos^2\delta \cos 2H. \qquad (3.70)$$

With $\quad 2H = 2H_i - 2\lambda \qquad (3.71)$

one obtains at the surface:

$$\left.\begin{aligned}
e_{\theta\theta} &= \left[h + 2\left(\frac{1-2\sin^2\theta}{\sin^2\theta}\right)\ell\right]\frac{\mathcal{S}_2}{ag},\\
e_{\lambda\lambda} &= \left[h - 2\left(\frac{1+\sin^2\theta}{\sin^2\theta}\right)\ell\right]\frac{\mathcal{S}_2}{ag},\\
e_{\theta\lambda} &= 4\,\ell\left(\frac{\cos\theta}{\sin^2\theta}\ \tan 2H\right)\frac{\mathcal{S}_2}{ag},\\
e_{\lambda r} &= 0,\\
e_{r\theta} &= 0,\\
e_{rr} &= \left[a\left(\frac{dh(r)}{dr}\right)_{r=a} + 2h\right]\frac{\mathcal{S}_2}{ag},
\end{aligned}\right\} \qquad (3.72)$$

and once again formula (3.66) is used.

For $\theta = 45°$, $1 - 2\sin^2\theta = 0$, $2\,\frac{1+\sin^2\theta}{\sin^2\theta} = 6$ so that the extreme deformations in the horizon are determined by $h = 0.638$ and $h - 6\ell = 0{,}110$.

One obtains the configurations represented on Figure 3.3. With these numerical values for h and ℓ the East-West component $e_{\lambda\lambda}$ vanishes at the latitude ϕ=52° (which will be seen later on Figure 11.10).

Zonal Waves

For the general form of zonal waves of the second order

$$\mathcal{Z}_2 = D\left(\frac{c}{d}\right)^3\left(\frac{r}{a}\right)^2(\cos^2\theta - \tfrac{1}{3})(\sin^2\delta - \tfrac{1}{3}) \qquad (3.73)$$

one has

$$\left.\begin{aligned}
e_{\theta\theta} &= \left(h - \frac{2\cos 2\theta}{\cos^2\theta-\frac{1}{3}}\,\ell\right)\frac{\mathcal{Z}_2}{ag},\\
e_{\lambda\lambda} &= \left(h - \frac{2\cos^2\theta}{\cos^2\theta-\frac{1}{3}}\,\ell\right)\frac{\mathcal{Z}_2}{ag},\\
e_{\theta\lambda} &= 0,\\
e_{\lambda r} &= 0,\\
e_{r\theta} &= 0,\\
e_{rr} &= \left[a\left(\frac{dh(r)}{dr}\right)_{r=a} + 2h\right]\frac{\mathcal{Z}_2}{ag},
\end{aligned}\right\} \qquad (3.74)$$

at

$$\left.\begin{aligned} \theta &= 45°, \\ e_{\theta\theta} &= h\,\frac{\mathscr{W}_2}{ag}, \\ e_{\lambda\lambda} &= (h - 6\ell)\,\frac{\mathscr{W}_2}{ag} . \end{aligned}\right\} \tag{3.75}$$

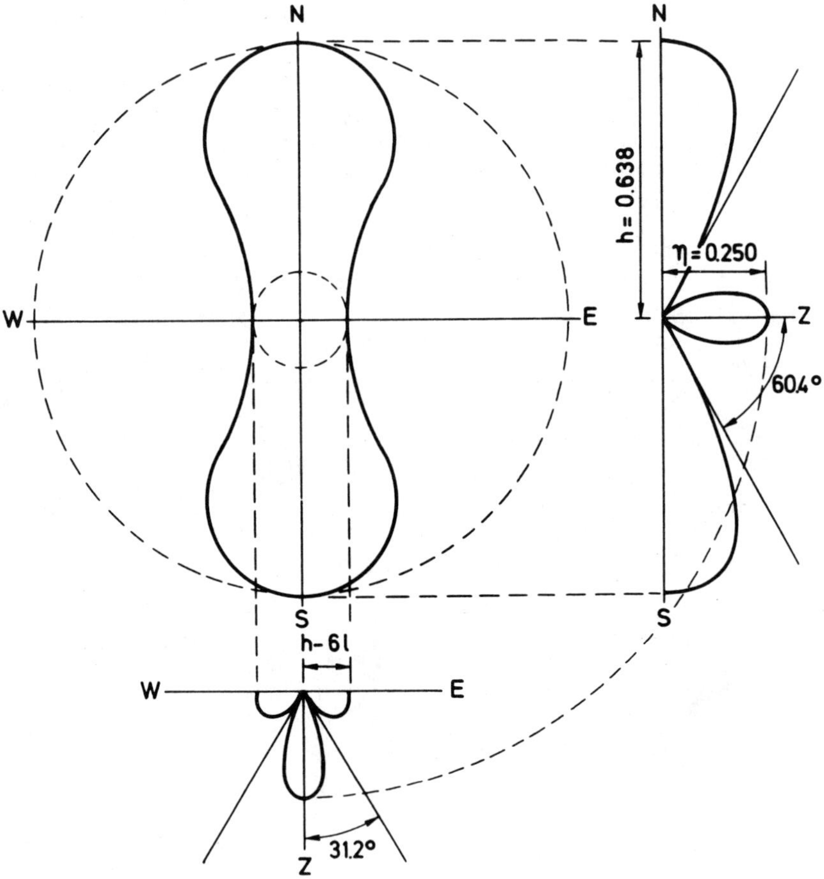

Fig.3.3. Distribution of deformations as a function of the azimuth and dip at the Earth's surface: wave M_2. (latitude 45°)

One observes that for all tidal families

$$e_{\lambda\lambda} + e_{\theta\theta} = (2h - 6\ell)\,\frac{W_2}{ag} \tag{3.76}$$

with

$$W_2 = \mathcal{T}_2 + \mathcal{S}_2 + \mathcal{Z}_2.$$

This is just a check of (3.54) and results from the general property (3.52) of harmonic functions.

Determination of Ratio ℓ/h from only one Extensometer

When an extensometer is not installed in one of the two principal directions North-South or East-West, the term $e_{\lambda\theta}$ plays a role which is maximum when the azimuth α is 45°. This term contains the first derivative of W_2 with respect to λ instead of its second derivative as in $e_{\lambda\lambda}$. Therefore a sine term of the hour angle is introduced which can be easily and correctly separated in the harmonic analysis of the registered curve if the time marks are accurate.

Then if we take the amplitude ratio of the cosine to the sine term, which by the way also eliminates the difficult problem of calibration, one obtains, putting $x = \ell/h$:

for the *diurnal waves*

$$\frac{\text{Cosine term}}{\text{-Sine term}} = \frac{h - 2\ell - 2\ell \cos^2\alpha}{\ell \sin 2\alpha} \cos\theta = \frac{1 - 2x(1+\cos^2\alpha)}{x \sin 2\alpha} \cos\theta; \quad (3.77)$$

for the *semi-diurnal waves*,

$$\frac{\text{Cosine term}}{\text{Sine term}} = \frac{(h - 2\ell)\sin^2\theta + 2\ell\,(\cos 2\alpha - \sin^2\theta\cos^2\alpha)}{2\ell\cos\theta\sin 2\alpha}$$

$$= \frac{(1 - 2x)\sin^2\theta + 2x\,(\cos 2\alpha - \sin^2\theta\cos^2\alpha)}{2x\cos\theta\sin 2\alpha}. \quad (3.78)$$

Because of the different response of the Earth's liquid core and of the Earth's crust to the tesseral and sectorial waves, it is not permissible to combine waves of these two different kinds to derive information about h, ℓ or k.

Cubic Expansion

The cubic dilatation of an infinitesimal volume is given by

$$\theta = e_{rr} + e_{\theta\theta} + e_{\lambda\lambda} = e_{rr} + \Sigma \quad (3.79)$$

which gives, using (3.48) and (3.53),

$$\theta_n = [a\,h'_n + (n+2)\,h_n - n\,(n+1)\,\ell_n]\,\frac{W_n}{ag} \quad (3.80)$$

and, for n = 2,

$$\theta_2 = (ah' + 4h - 6\ell)\,\frac{W_2}{ag}. \quad (3.81)$$

At the surface, using (3.62),

$$\theta = \frac{1-2\nu}{1-\nu}(e_{\theta\theta} + e_{\lambda\lambda}) = \frac{2}{3}(e_{\theta\theta} + e_{\lambda\lambda}) \quad \text{if } \nu=0.25 \quad (3.82)$$

The order of magnitude of the tidal cubic dilatations is 2.10^{-8}.

Directions of Zero Deformations in Vertical Planes

Diurnal Waves

For any direction (α_1, α_2, α_3) at the Earth's surface the deformation is

$$d = (ah' + 2h)\,\alpha_1^2 + (h - 4\ell)\,\alpha_2^2 + (h - 2\ell)\,\alpha_3^2 - 2\ell\left(\frac{\tan H}{\cos\theta}\right)\alpha_2\,\alpha_3 \qquad (3.83)$$

with

$$\alpha_1 = \cos z, \qquad \alpha_2 = \sin z\cos\alpha, \qquad \alpha_3 = \sin z\sin\alpha \qquad (3.84)$$

z being, as usual, the zenith distance and α the azimuth counted from the North, it follows that

$$d = (ah' + 2h) - (ah' + h + 2\ell + 2\ell\cos^2\alpha + \ell\frac{\tan H}{\cos\theta}\sin 2\alpha)\sin^2 z \quad (3.85)$$

and the deformation vanishes when

$$\sin z = \sqrt{\frac{ah' + 2h}{ah' + h + 2\ell + 2\ell\cos^2\alpha + \ell\dfrac{\tan H}{\cos\theta}\sin 2\alpha}}\,. \qquad (3.86)$$

With the numerical values (see Chapter 13),

$$ah' + 2h = -0.23, \qquad h = 0.638, \qquad \ell = 0.088$$

one has, when H=0, for the meridian plane (α=0),

$$z_0 = 43°81$$

and for the prime vertical ($\alpha=\frac{\pi}{2}$),

$$z_0 = 36°31.$$

This is illustrated by Fig. 3.2.

Semi-diurnal Waves

In the case of sectorial waves one has, similarly,

$$d = (ah' + 2h)\,\alpha_1^2 + \left(h - 2\ell\,\frac{1-2\sin^2\theta}{\sin^2\theta}\right)\alpha_2^2 + \left(h - 2\ell\,\frac{1+\sin^2\theta}{\sin^2\theta}\right)\alpha_3^2$$

$$+ 4\,\ell\,\chi\,\alpha_2\,\alpha_3, \qquad (3.87)$$

χ being a function of tan H

or

$$d = (ah' + 2h) - [ah' + h + 2\ell\,(1-3\cos^2\alpha + \frac{1}{\sin^2\theta} - 2\chi\sin 2\alpha)]\sin^2 z. \quad (3.88)$$

For H = 0 this deformation vanishes for the following values of z:

	East-West	North-South
at the equator (θ = 90°)	26°15	36°31
at latitude 45° (θ = 45°)	31°17	60°44

which is represented on Fig.3.3.

Linear, Areal and Cubic Expansions Deriving from the Third-order Potential

One easily deduces from the formulas of spherical elasticity (4.10) the following relations:

$$\left.\begin{aligned}
e_{rr} &= \left[a\left(\frac{dH_3}{dr}\right)_{r=a} + 3h_3\right]\frac{W_3}{ag},\\
e_{\theta\theta} &= \left[\ell_3(6 \sin\theta \cos^2\theta - 3 \sin^3\theta) + h_3 \sin^3\theta\right]\frac{W_3}{ag \sin^3\theta},\\
e_{\lambda\lambda} &= \left(-9\ell_3 \sin\theta + 3\ell_3 \sin\theta \cos^2\theta + h_3 \sin^3\theta\right)\frac{W_3}{ag \sin^3\theta},\\
e_{\theta\lambda} &= \left(3\ell_3 \sin 2\theta\right)\frac{W_3}{ag \sin^3\theta},\\
e_{r\theta} &= e_{r\lambda} = 0,\\
\Sigma &= e_{\theta\theta} + e_{\lambda\lambda} = 2(h_3 - 6\ell_3)\frac{W_3}{ag},\\
e_{\theta\theta} - e_{\lambda\lambda} &= 6\ell_3 (1+\cos\theta)\frac{W_3}{ag \sin^2\theta},\\
\theta &= e_{rr} + e_{\theta\theta} + e_{\lambda\lambda} = \left[a\left(\frac{dH_3}{dr}\right)_{r=a} + 5h_3 - 12\ell_3\right]\frac{W_3}{ag}.
\end{aligned}\right\} \quad (3.89)$$

Curl of the Tidal Displacement

If we define the displacement by

$$u_r = H(r)\, S_n(\theta\lambda), \quad u_\theta = T(r)\,\frac{\partial S_n(\theta\lambda)}{\partial\theta}, \quad u_\lambda = T(r)\,\frac{\partial S_n(\theta\lambda)}{\sin\theta\,\partial\lambda}, \qquad (3.90)$$

separating completely the variable r from the variables (θ, λ) as W_n is a homogeneous function of degree n, equations (4.15) immediately give the curl components:

$$\omega_{r\theta} = -\frac{1}{2r}(H - r\dot{T} - T)\frac{\partial S_n}{\partial\theta}, \quad \omega_{\theta\lambda} = 0, \quad \omega_{\lambda r} = \frac{1}{2r\sin\theta}(H - r\dot{T} - T)\frac{\partial S_n(\theta\lambda)}{\partial\lambda}. \quad (3.91)$$

if $H = r\dot{T} + T$ (see 5.91) the tidal displacement is irrotational.

3.4. THE CHANDLERIAN MOTION OF THE POLE

The Earth's rotation can be described easily if we take the principal axis of inertia as the axis of reference. Let $\vec{\omega}$ be the instantaneous axis of rotation and (p, q, r) its components with respect to these axes and let (L, M, N) be the projections on the same axis of the torque exerted by the luni-solar external forces. If (A, B, C) are the principal moments of inertia, the equations governing the Earth's rotation are the famous Euler equations:

$$\left.\begin{aligned} A\frac{dp}{dt} + (C - B)\,qr &= L, \\ B\frac{dq}{dt} + (A - C)\,rp &= M, \\ C\frac{dr}{dt} + (B - A)\,pq &= N. \end{aligned}\right\} \tag{3.92}$$

The free motion is obtained by putting L=M=N=0, and as we can accept as an excellent approximation that A=B, the integration of the equations gives the equations of the free Euler motion of the pole:

$$\left.\begin{aligned} p &= R\cos\left(\frac{C-A}{A}\,\omega t - \beta\right), \\ q &= R\sin\left(\frac{C-A}{A}\,\omega t - \beta\right), \\ r &= \text{constant} = \omega, \end{aligned}\right\} \tag{3.93}$$

(R, β) being the integration constants.

The coordinates of the moving pole of rotation are usually denoted as

$$x = \frac{p}{r}, \qquad y = \frac{q}{r}. \tag{3.94}$$

Such a displacement of the axis of rotation with respect to the Earth itself produces a perturbation of the centrifugal force

$$V = \frac{1}{2}\,\omega^2\, r^2 \sin^2\theta, \tag{3.95}$$

as the colatitude of any point P(θ, λ) is modified as follows:

$$d\theta = -\,(x\cos\lambda + y\sin\lambda). \tag{3.96}$$

The perturbing potential is then

$$W_2 = dV = \frac{1}{2}\,\omega^2\, r^2 \sin 2\theta\, d\theta = -\frac{1}{2}\,\omega^2\, r^2\,(x\cos\lambda + y\sin\lambda)\sin 2\theta, \tag{3.97}$$

a tesseral harmonic spherical function of the second order which produces tidal deformations just like the tesseral luni-solar potential. The response of the elastic Earth to this centrifugal potential can evidently be described by the same Love numbers as the luni-solar tidal deformations.

Thus if the radial deformation is U and the cubic expansion θ, the density at any point P becomes

$$\rho_o - U\,\frac{d\rho_o}{dr} - \rho_o\,\theta \tag{3.98}$$

as a denser particle comes at P if U > 0.

From (3.8) we can write

$$V_o = G \iiint \rho \sum_n \frac{P_n}{a^{n+1}} \frac{\partial}{\partial r}\left(\frac{r^{n+3}}{n+3}\right) dr\, d\mu\, d\lambda \tag{3.99}$$

with

$$r(1+\varepsilon) = r + r^2\, H(r)\, \frac{S_2}{g}. \tag{3.100}$$

Neglecting all terms of an order higher than the first order of the deformations we have:

$$V = G \iiint (\rho_o - \rho_o\, \theta) \left(\sum_n \frac{r^{n+2}}{a^{n+1}} P_n\right) dr\, d\mu\, d\lambda$$
$$+ G \iiint \rho_o \sum_n \frac{P_n}{a^{n+1}} \frac{\partial}{\partial r}\left(r^{n+4}\, H(r)\, \frac{S_2}{g}\right) dr\, d\mu\, d\lambda$$

or

$$\left.\begin{aligned} V = V_o &- G \iiint \rho_o \frac{1}{g} r^2 f(r)\, S_2 \left(\Sigma \frac{r^{n+2}}{a^{n+1}} P_n\right) dr\, d\mu\, d\lambda \\ &+ G \iiint \rho_o \frac{1}{g} S_2\, \Sigma \frac{\partial}{\partial r}\left(H(r)\, r^{n+4}\right) \frac{P_n}{a^{n+1}}\, dr\, d\mu\, d\lambda \end{aligned}\right\} \tag{3.101}$$

According to (3.5):

$$V - V_o = K_n(r)\, W_n.$$

For the order n=2 this gives

$$K(a) = \frac{3}{5\bar{\rho} a^6} \int_0^a \rho_o \left\{\frac{d}{dr}\left[r^6\, H(r)\right] - r^6\, f(r)\right\} dr. \tag{3.102}$$

Now, the displacement of the axis of inertia due to this tesseral deformation generates non-zero products of inertia as

$$D = \int_M YZ\, dm = \int_M (\rho_o - \rho_o\, \theta) \frac{\partial}{\partial r}\left(\frac{r^5}{5}\right) \sin^2\theta \cos\theta \sin\lambda\, d\theta\, d\lambda\, dr \tag{3.103}$$

or

$$D = \int_0^a \rho_o \frac{1}{g} \left\{\frac{d}{dr}\left[r^6\, H(r)\right] - r^6\, f(r)\right\} dr \int_0^{2\pi}\int_0^{2\pi} \sin^2\theta \cos\theta \sin\lambda\, S_2\, d\theta\, d\lambda \tag{3.104}$$

with, from (3.97)

$$S_2 = -\frac{1}{2}\omega^2 (x\cos\lambda + y\sin\lambda) \sin 2\theta. \tag{3.105}$$

The x term disappears in the integration. Finally, one has

$$\left.\begin{aligned} D &= -\frac{8\pi}{15}\, y\, \frac{\omega^2}{2g} \int_0^a \rho_o \left\{\frac{d}{dr}\left[r^6\, H(r)\right] - r^6\, f(r)\right\} dr, \\ E &= -\frac{8\pi}{15}\, x\, \frac{\omega^2}{2g} \int_0^a \rho_o \left\{\frac{d}{dr}\left[r^6\, H(r)\right] - r^6\, f(r)\right\} dr, \\ F &= 0. \end{aligned}\right\} \tag{3.106}$$

Let us put

$$D = -\,yw\,(C-A), \qquad E = -\,xw\,(C-A), \qquad F = 0. \tag{3.107}$$

Then Euler's equations can be written:

$$\left.\begin{aligned} A\frac{dp}{dt} + \omega q\,(C-A)\,(1-w) &= 0,\\ A\frac{dq}{dt} - \omega p\,(C-A)\,(1-w) &= 0,\end{aligned}\right\} \tag{3.108}$$

and the real period of the polar motion is Chandler's period

$$\tau = \frac{2\pi}{\omega}\,\frac{A}{C-A}\,\frac{1}{1-w}, \tag{3.109}$$

while for an indeformable Earth it is Euler's period

$$\tau_o = \frac{2\pi}{\omega}\,\frac{A}{C-A}. \tag{3.110}$$

Thus

$$w = 1 - \frac{\tau_o}{\tau}. \tag{3.111}$$

Equations (3.106), (3.107) and (3.102) give

$$w = \frac{8\pi}{15}\,\frac{\omega^2}{2g}\,\frac{5\bar{\rho}a^6}{3}\,\frac{K(a)}{C-A}. \tag{3.112}$$

Using the hydrostatic equilibrium equation

$$C - A = \frac{2}{3}\,(e - \frac{q}{2})\,Ma^2 \tag{3.113}$$

with

$$q = \omega^2\,a^3\,/\,GM$$

we obtain the relation between Chandler's period and the number k:

$$1 - \frac{\tau o}{\tau} = k\,\frac{\omega^2\,a/2g}{e-\omega^2\,a/2g} = k\,\frac{q}{2e-q}. \tag{3.114}$$

τ = 432d corresponds to k = 0.278

τ = 442d corresponds to k = 0.289

3.5. PERIODIC VARIATIONS OF THE SPEED OF ROTATION OF THE EARTH

The deformations produced by the zonal part of the tidal potential result in variations of the principal moments of inertia.

This zonal potential has been written

$$\mathcal{W}_2 = 3\,D\,(\frac{c}{d})^3\,(\sin^2\phi - \frac{1}{3})\,(\sin^2\delta - \frac{1}{3}) \tag{3.115}$$

and the bulge produced by it has a potential

$$\Delta\mathcal{W}_2 = k_2\mathcal{W}_2\,(\frac{a}{r})^3 \tag{3.116}$$

at the distance r.

On the other hand, as it is a purely zonal deformation ($dA = dB$), MacCullagh's formula gives by differentiation

$$\Delta \mathcal{H}_2 = G \frac{dC - dA}{2a^3} (3 \cos^2\theta - 1) (\frac{a}{r})^3 \qquad (3.117)$$

and by identification

$$\frac{3}{4} G \mu \left(\frac{a^2}{d^3}\right) \left(\sin^2\delta - \frac{1}{3}\right) k_2 = \frac{GE}{2a} \left(\frac{dC - dA}{Ma^2}\right) \qquad (3.118)$$

(E having been taken as the unit of mass).

The central moment of inertia of the Earth is

$$I = \frac{1}{2} (A + B + C) = \iiint r^2 \, dm. \qquad (3.119)$$

As the zonal forces do not change the revolution shape of the body ($dA = dB$), the variation of I is

$$dI = dA + \frac{dC}{2} = 2 \iiint r \, dr \, dm, \qquad (3.120)$$

where

$$dr = \sum_{n=2}^{\infty} h_n (r) S_n^{\circ} (\theta), \quad dm = \rho \, r^2 \sin\theta \, dr \, d\theta \, d\lambda.$$

Thus

$$dI = 2 \sum_n \int r^3 h_n (r) \rho(r) \, dr \iint S_n^{\circ} (\theta) \sin\theta \, d\theta \, d\lambda \qquad (3.121)$$

the surface integral gives zero (orthogonality theorem) and therefore

$$dA = - \frac{dC}{2}. \qquad (3.122)$$

The same property appears if we introduce zonal variations of the density $\rho(r)$. This demonstration has been given by Parysky and Pertsev (1964). Then (3.118) becomes

$$k_2 \mu (\frac{a}{d})^3 \left(\sin^2\delta - \frac{1}{3}\right) = \frac{dC}{Ma^2}. \qquad (3.123)$$

Euler's third equation is

$$C\omega = \text{constant}.$$

Thus

$$\frac{dC}{C} = - \frac{d\omega}{\omega} \qquad (3.124)$$

and we obtain

$$\frac{d\omega}{\omega} = - \frac{k_2 \mu}{0.334} (\frac{a}{d})^3 (\sin^2\delta - \frac{1}{3}) \qquad (3.125)$$

as

$$C = 0.334 \, M \, a^2.$$

This gives, with $\mu = 1/81.3$ and $(a/c)^3 = 4.57 \; 10^{-6}$,

$$\frac{d\omega}{\omega} = - 1.683\ k_2\ 10^{-7}\ (\frac{c}{d})^3\ (\frac{2}{3} - \cos^2\delta), \tag{3.126}$$

where we introduce (1.52) and

k_2 = 0.30 and ω = 86.400 sidereal seconds per sidereal day:

$$d\omega = 0.3447 \cos 2\dot{s}t - 0.1571 \cos \dot{N}t + \ldots + 0.1605 \cos 2\dot{h}t + \ldots \tag{3.127}$$

The difference between a true sidereal hour and a mean sidereal hour will be given by

$$\frac{dT}{T} = - \frac{d\omega}{\omega}. \tag{3.128}$$

Thus

$$\int dT.dt = - \frac{T}{\omega} \int d\omega.dt.$$

Because s, h and N are linear functions of t, we have $\frac{ds}{dt} = \frac{s}{t}$, $\frac{dh}{dt} = \frac{h}{t}$ and $\frac{dN}{dt} = \frac{N}{t}$, i. e. $dt = \frac{t}{s} ds$, $2\dot{s}t = 2s$, and so on.

Therefore,

$$\int dT.dt = - \frac{T}{\omega} (0.3447 \frac{t}{s} \int \cos 2s\ ds - 0.1571 \frac{t}{N} \int \cos N\ dN + \ldots + 0.1605 \frac{t}{h} \int \cos 2h\ dh). \tag{3.129}$$

Now

$$\frac{t}{\omega s} = \frac{27.3}{2\pi}, \qquad \frac{t}{\omega N} = \frac{6794}{2\pi}, \qquad \frac{t}{\omega h} = \frac{366}{2\pi}$$

and

$$\int dT\ dt = 1.4977 \int \cos 2s\ ds - 169.8721 \int \cos N\ dN + 9.3492 \int \cos 2h\ dh + \ldots \text{ milliseconds.} \tag{3.130}$$

This integration emphasizes the cumulative effect in the very long-period tides as

$$\int_{-\frac{\pi}{4}}^{+\frac{\pi}{4}} \cos 2x\ dx = 1, \qquad \int_{-\frac{\pi}{2}}^{+\frac{\pi}{2}} \cos N\ dN = 2.$$

Therefore we will observe a variation in the universal time which reaches

1.50 ms in 7 days,
9.35 ms in 91 days,
339.74 ms in 9 years.

Parysky and Pertsev pointed out that the variation of the centrifugal force resulting from this variation in the rotation speed itself has little effect on the moments of inertia, and this should be taken into account in such a determination of k_2. They estimate this effect as lower than 0.44% of the numerical value of k_2.

3.6. THE TRUE PHASE-LAG OF THE TIDAL DEFORMATION

The analysis of a tidal curve provides as "observed phase-lag" a global value resulting from the vectorial combination of the external tidal force itself and of the elastico-viscous tidal deformation effect.

Figure 3.4 represents such a combination for the vertical component and for the horizontal component.

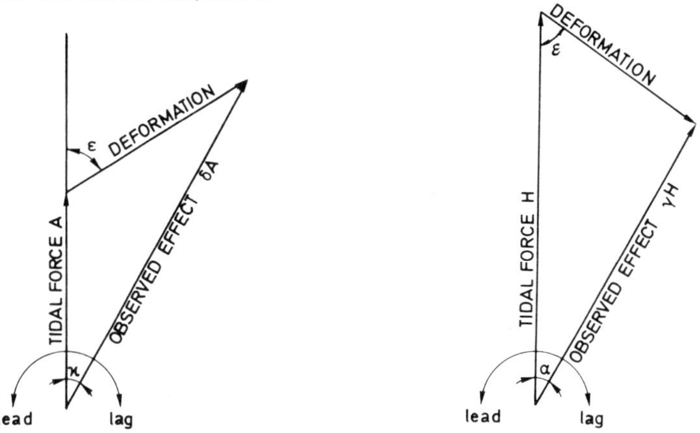

Polar diagram for the Vertical component
the result of analysis is κ
the real phase of the deformation is ε

Polar diagram for the Horizontal component
the result of analysis is α
the real phase of the deformation is ε

Fig. 3.4

Obviously, for the vertical component, one has

$$\frac{\sin(\varepsilon-\kappa)}{A} = \frac{\sin(\pi-\varepsilon)}{\delta A}$$

or

$$\text{tg } \varepsilon = \frac{\delta \sin \kappa}{\delta \cos \kappa - 1} \tag{3.131}$$

while for the horizontal component

$$\frac{\sin \varepsilon}{\gamma H} = \frac{\sin(\pi-\varepsilon-\alpha)}{H}$$

or

$$\text{tg } \varepsilon = \frac{\gamma \sin \alpha}{1 - \cos \alpha} \tag{3.132}$$

For current analysis results such as

$$\delta = 1.16, \qquad \kappa = -0.20°,$$

$$\gamma = 0.68, \qquad \alpha = -0.69°,$$

one should conclude therefore, that

$$\varepsilon = -1.45°. \tag{3.133}$$

which is in good agreement with the theoretical calculations of Molodensky.

(see also page 432)

CHAPTER 4

Kelvin's and Herglotz's Theories of Earth Tides

4.1. INTRODUCTION

The theoretical problem of Earth tides consists in the calculation of the deformations produced within the Earth by forces deriving from the luni-solar attraction potential. It falls into three historically well-separated phases:

1) Before 1890, i. e. before the development of experimental seismology, nobody could have had a clear idea of the internal structure of the Earth. It was therefore impossible to approach the problem with any chance of success. However, fundamental work was performed by Lord Kelvin who gave a theory of the solid tides for a homogeneous and incompressible Earth.

2) After 1890, the discovery of the polar motion and the first conclusions derived from its observation led two theoreticians, Sloudsky (1896) and Poincaré (1911) to elaborate a theory for a planet having a liquid core. Later on, with the progress of seismology and the discovery of the core by Wiechert, Love and Herglotz developed a theory for a heterogeneous Earth. However, the means at that time available for numerical calculus were so limited that they had to satisfy themselves with very simple Earth models, using binomial laws for the internal distribution of density and rigidity, which was far from being true. Schweydar, Prey, Boaga, and Jeffreys therefore reduced the system to one differential equation of the sixth order in the radial displacement, eliminating the other unknowns which, similar to pressure variation, could not have been observed experimentally. However, in 1950 Takeuchi succeeded in performing the first numerical integration of the system for some models more compatible with the real structure of the Earth.

3) After 1960, when modern computers became available, the orientation of this theoretical work was completely modified and, instead of eliminating one by one the unknowns, one prefers to build a system of six linear differential equations of the first order, a system which can be solved numerically by the Runge Kutta method. But the choice of the six variables must be made with care in order that the derivation of experimental functions like $\rho(r)$, $\lambda(r)$ and $\mu(r)$ can be avoided.

4.2. SPHERICAL ELASTICITY

We need now to resume the fundamental equations used in the study of an elastic sphere. The equilibrium equations are:

$$\text{div}\ \bar{\bar{\Sigma}} + \bar{F} - \rho\bar{\Gamma} = 0 \qquad (4.1)$$

where $\bar{\bar{\Sigma}}$ is the tensor of constraints, $\bar{F}$ the body force and $\bar{\Gamma}$ the acceleration.

The equilibrium of momentum shows that the tensor of constraints is a symmetric tensor.

Writing these equations in a suitable form for a spherical body one has to introduce the spherical coordinates r, θ, λ defined by:

$$x = r \sin\theta \cos\lambda, \quad y = r \sin\theta \sin\lambda, \quad z = r \cos\theta. \tag{4.2}$$

In this transformation the fundamental metric tensor has as components:

$$g_{11} = 1, \qquad g_{22} = r^2, \qquad g_{33} = r^2 \sin^2\theta, \tag{4.3}$$

the Jacobian being:

$$r^2 \sin\theta. \tag{4.4}$$

Christoffel symbols of the first kind are (with particular convention)

$$\Gamma_{ijk} = \frac{1}{2} (\partial_i g_{jk} + \partial_j g_{ki} - \partial_k g_{ij}) \tag{4.5}$$

and here give:

$$\left.\begin{aligned} \Gamma_{122} &= -\Gamma_{221} = \Gamma_{212} = r, \\ \Gamma_{133} &= -\Gamma_{331} = \Gamma_{313} = r \sin^2\theta, \\ \Gamma_{233} &= -\Gamma_{332} = \Gamma_{323} = r^2 \sin\theta \cos\theta, \end{aligned}\right\} \tag{4.6}$$

while Christoffel symbols of the second kind are:

$$\left.\begin{aligned} &\Gamma_{12}^{\ 2} = \frac{1}{r}, &&\Gamma_{21}^{\ 2} = \frac{1}{r}, &&\Gamma_{31}^{\ 3} = \frac{1}{r}, \\ &\Gamma_{13}^{\ 3} = \frac{1}{r}, &&\Gamma_{22}^{\ 1} = -r, &&\Gamma_{32}^{\ 3} = \cot\theta, \\ & &&\Gamma_{23}^{\ 3} = \cot\theta, &&\Gamma_{33}^{\ 1} = -r \sin^2\theta, \\ & && &&\Gamma_{33}^{\ 2} = -\sin\theta\cos\theta, \end{aligned}\right\} \tag{4.7}$$

all the others being zero.

The experimental geophysical measurements give the physical elements. These elements are precisely those appearing in the equations of elasticity. The physical components are expressed in a base obtained by normalizing the vectors of the natural base. However, differentiation as well as derivation operations cannot be performed except on the natural components.

One has therefore to proceed as follows:

(a) determine the natural components from the physical components;
(b) differentiation or derivation performed upon the natural components;
(c) determine the physical components obtained by operation (b).

This is formulated as follows:

$$t^*_{ij} = \frac{1}{e_i\, e_j} \left[\partial_i \,(e_j\, u^*_j) - \Gamma_{ij}^{\ k} \,(e_k\, u^*_k) \right]. \tag{4.8}$$

where $e_i = \sqrt{g_{ii}}$.

Let $(u_r, u_\theta, u_\lambda)$ be the physical components of the displacement vector **U** at point M. Following (4.8) the deformations are given by:

$$
\left.
\begin{aligned}
\varepsilon_{11} &= \varepsilon_{rr} = \frac{1}{e_1^2}\left[\partial_1(e_1\, u_r) - \Gamma_{11}^{\ \ k}\,(e_k\, u_k^*)\right],\\
\varepsilon_{22} &= \varepsilon_{\theta\theta} = \frac{1}{e_2^2}\left[\partial_2(e_2\, u_\theta) - \Gamma_{22}^{\ \ k}\,(e_k\, u_k^*)\right],\\
\varepsilon_{33} &= \varepsilon_{\lambda\lambda} = \frac{1}{e_3^2}\left[\partial_3(e_3\, u_\lambda) - \Gamma_{33}^{\ \ k}\,(e_k\, u_k^*)\right],\\
\varepsilon_{12} &= \varepsilon_{r\theta} = \frac{1}{2}\,\frac{1}{e_1 e_2}\left\{\left[\partial_1(e_2\, u_\theta) - \Gamma_{12}^{\ \ k}\,(e_k\, u_k^*)\right] + \right.\\
&\qquad \left.\left[\partial_2(e_1\, u_r^*) - \Gamma_{21}^{\ \ k}(e_k\, u_k^*)\right]\right\},\\
\varepsilon_{13} &= \varepsilon_{r\lambda} = \frac{1}{2}\,\frac{1}{e_1 e_3}\left\{\left[\partial_1(e_3\, u_\lambda) - \Gamma_{13}^{\ \ k}(e_k\, u_k^*)\right] + \right.\\
&\qquad \left.\left[\partial_3\,(e_1\, u_r^*) - \Gamma_{31}^{\ \ k}(e_k\, u_k^*)\right]\right\},\\
\varepsilon_{23} &= \varepsilon_{\theta\lambda} = \frac{1}{2}\,\frac{1}{e_2 e_3}\left\{\left[\partial_2(e_3\, u_\lambda) - \Gamma_{23}^{\ \ k}(e_k\, u_k^*)\right] + \right.\\
&\qquad \left.\left[\partial_3(e_2\, u_\theta^*) - \Gamma_{32}^{\ \ k}(e_k\, u_k^*)\right]\right\},
\end{aligned}
\right\}
\quad (4.9)
$$

which, using (4.7), give the physical components of the strain tensor expressed in function of the physical components of the displacement

$$
\left.
\begin{aligned}
\varepsilon_{rr} &= \frac{\partial u_r}{\partial r},\\
\varepsilon_{\theta\theta} &= \frac{1}{r}\,\frac{\partial u_\theta}{\partial \theta} + \frac{u_r}{r},\\
\varepsilon_{\lambda\lambda} &= \frac{1}{r\sin\theta}\,\frac{\partial u_\lambda}{\partial \lambda} + \frac{u_r}{r} + \cot\theta\,\frac{u_\theta}{r},\\
\varepsilon_{r\theta} &= \frac{1}{2}\left(\frac{1}{r}\,\frac{\partial u_r}{\partial \theta} + \frac{\partial u_\theta}{\partial r} - \frac{u_\theta}{r}\right),\\
\varepsilon_{r\lambda} &= \frac{1}{2}\left(\frac{\partial u_\lambda}{\partial r} + \frac{1}{r\sin\theta}\,\frac{\partial u_r}{\partial \lambda} - \frac{u_\lambda}{r}\right),\\
\varepsilon_{\theta\lambda} &= \frac{1}{2}\left(\frac{1}{r}\,\frac{\partial u_\lambda}{\partial \theta} + \frac{1}{r\sin\theta}\,\frac{\partial u_\theta}{\partial \lambda} - \cot\theta\,\frac{u_\lambda}{r}\right),
\end{aligned}
\right\}
\quad (4.10)
$$

and the cubic expansion

$$
\Theta = \varepsilon_{rr} + \varepsilon_{\theta\theta} + \varepsilon_{\lambda\lambda} = \frac{\partial u_r}{\partial r} + \frac{1}{r}\,\frac{\partial u_\theta}{\partial \theta} + \frac{1}{r\sin\theta}\,\frac{\partial u_\lambda}{\partial \lambda} + \frac{2}{r}\,u_r + \cot\theta\,\frac{u_\theta}{r}. \quad (4.11)
$$

It is also convenient to write the differential operators in spherical coordinates.

The *gradient* has the physical components:

$$\left.\begin{aligned} \text{grad}_r\ V &= \text{grad}_1\ V = \frac{\partial V}{\partial r}, \\ \text{grad}_\theta\ V &= \frac{1}{r}\,\text{grad}_2\ V = \frac{1}{r}\frac{\partial V}{\partial \theta}, \\ \text{grad}_\lambda\ V &= \frac{1}{r\sin\theta}\,\text{grad}_3\ V = \frac{1}{r\sin\theta}\frac{\partial V}{\partial \lambda}. \end{aligned}\right\} \tag{4.12}$$

The *curl* is the antisymmetric tensor

$$\omega_{ij} = \frac{1}{2}\,(\partial_i\, u_j - \partial_j\, u_i) \tag{4.13}$$

which in curvilinear coordinates becomes

$$\frac{1}{2}\,(D_i\, u_j - D_j\, u_i) = \frac{1}{2}\,(\partial_i\, u_j - \partial_j\, u_i - \Gamma_{ij}^{\ \ k}\, u_k + \Gamma_{ji}^{\ \ k}\, u_k) \tag{4.14}$$

or explicitly:

$$\left.\begin{aligned} \omega_{r\theta} &= \frac{1}{2r}\,(u_\theta + r\,\frac{\partial u_\theta}{\partial r} - \frac{\partial u_r}{\partial \theta}), \\ \omega_{\theta\lambda} &= \frac{1}{2\,r\sin\theta}\,(u_\lambda\cos\theta + \sin\theta\,\frac{\partial u_\lambda}{\partial \theta} - \frac{\partial u_\theta}{\partial \lambda}), \\ \omega_{\lambda r} &= \frac{1}{2\,r\sin\theta}\,(\frac{\partial u_r}{\partial \lambda} - u_\lambda \sin\theta - r\sin\theta\,\frac{\partial u_\lambda}{\partial r}). \end{aligned}\right\} \tag{4.15}$$

The *divergence* is

$$\text{div}^i\, \overset{=}{\Sigma} = D_j\, \sigma^{ij} = \partial_j\, \sigma^{ij} + \Gamma_{j\ell}^{\ \ i}\, \sigma^{j\ell} + \Gamma_{j\ell}^{\ \ j}\, \sigma^{i\ell}, \tag{4.16}$$

that is

$$\left.\begin{aligned} \text{div}^1\, \overset{=}{\Sigma} &= \frac{\partial\sigma^{11}}{\partial r} + \frac{\partial\sigma^{12}}{\partial \theta} + \frac{\partial\sigma^{13}}{\partial \lambda} - r\sigma^{22} - r\sin^2\theta\,\sigma^{33} + \frac{2}{r}\,\sigma^{11} + \cot\theta\,\sigma^{12}, \\ \text{div}^2\, \overset{=}{\Sigma} &= \frac{\partial\sigma^{12}}{\partial r} + \frac{\partial\sigma^{22}}{\partial \theta} + \frac{\partial\sigma^{23}}{\partial \lambda} + \frac{2}{r}\,\sigma^{12} - \sin\theta\cos\theta\,\sigma^{33} + \frac{2}{r}\,\sigma^{12} + \cot\theta\,\sigma^{22}, \\ \text{div}^3\, \overset{=}{\Sigma} &= \frac{\partial\sigma^{13}}{\partial r} + \frac{\partial\sigma^{23}}{\partial \theta} + \frac{\partial\sigma^{33}}{\partial \lambda} + \frac{2}{r}\,\sigma^{13} + 2\cot\theta\,\sigma^{23} + \frac{2}{r}\,\sigma^{13} + \cot\theta\,\sigma^{23}, \end{aligned}\right\} \tag{4.17}$$

with

$$\left.\begin{aligned} u_r &= u^1 = u_1, \\ u_\theta &= r\,u^2 = \frac{1}{r}\,u_2, \\ u_\lambda &= r\sin\theta\,u^3 = \frac{1}{r\sin\theta}\,u_3, \end{aligned}\right\} \tag{4.18}$$

and from

$$\left.\begin{aligned} &\sigma_{rr} = \sigma_{11} = \sigma^{11}, && \sigma_{\theta\theta} = r^2\,\sigma^{22} = \frac{1}{r^2}\,\sigma_{22}, \\ &\sigma_{r\theta} = r\sigma^{12} = \frac{1}{r}\,\sigma_{12}, && \sigma_{\lambda\lambda} = r^2\sin^2\theta\,\sigma^{33} = \frac{1}{r^2\sin^2\theta}\,\sigma_{33}, \\ &\sigma_{\lambda\theta} = r^2\sin\theta\,\sigma^{32} = \frac{1}{r^2\sin\theta}\,\sigma_{32}, && \sigma_{r\lambda} = r\sin\theta\,\sigma^{13} = \frac{1}{r\sin\theta}\,\sigma_{13}, \end{aligned}\right\} \tag{4.19}$$

we get the expressions:

$$\left.\begin{aligned}
\operatorname{div}_r \bar{\bar{\Sigma}} &= \frac{\partial \sigma_{rr}}{\partial r} + \frac{1}{r}\frac{\partial \sigma_{r\theta}}{\partial \theta} + \frac{\partial \sigma_{r\lambda}}{r \sin\theta\, \partial\lambda} + \frac{1}{r}(2\sigma_{rr} - \sigma_{\theta\theta} - \sigma_{\lambda\lambda} + \cot\theta\, \sigma_{r\theta}),\\
\operatorname{div}_\theta \bar{\bar{\Sigma}} &= \frac{\partial \sigma_{r\theta}}{\partial r} + \frac{1}{r}\frac{\partial \sigma_{\theta\theta}}{\partial \theta} + \frac{1}{r \sin\theta}\frac{\partial \sigma_{\theta\lambda}}{\partial\lambda} + \frac{1}{r}\left[3\sigma_{r\theta} + \cot\theta\,(\sigma_{\theta\theta} - \sigma_{\lambda\lambda})\right],\\
\operatorname{div}_\lambda \bar{\bar{\Sigma}} &= \frac{\partial \sigma_{r\lambda}}{\partial r} + \frac{1}{r}\frac{\partial \sigma_{\theta\lambda}}{\partial \theta} + \frac{1}{r \sin\theta}\frac{\partial \sigma_{\lambda\lambda}}{\partial\lambda} + \frac{1}{r}(3\sigma_{r\lambda} + 2\cot\theta\, \sigma_{\theta\lambda}).
\end{aligned}\right\} \quad (4.20)$$

4.3. INTRODUCTION OF A RHEOLOGICAL LAW

At this stage we have obtained a system of three equations of equilibrium with six unknowns, the stress tensor being a symmetric tensor of the second order. If we can express this stress tensor as a function of the strain tensor through some rheological law, we will be able to reduce our system to one of three equations in three unknowns as the strain tensor components are expressed as a function of the displacement vector by (4.9) or (4.10).

Let us assume now that the behaviour of the Earth is that of a perfect elastic Hooke body, i. e.

$$\bar{\bar{\Sigma}} = \lambda\, \theta\, \bar{\bar{E}} + 2\,\mu\, \bar{\bar{\epsilon}} \qquad (4.21)$$

or

$$\sigma_{ij} = \lambda\, \theta\, \delta_{ij} + 2\mu\, \varepsilon_{ij} \qquad (4.22)$$

where

$$\lambda = \frac{\nu E}{(1+\nu)\,(1-2\nu)}, \qquad \mu = \frac{E}{2\,(1+\nu)}, \qquad (4.23)$$

λ and μ are the Lamé coefficients, E being Young's modulus and ν Poisson's coefficient.
Introducing these relations in the fundamental equations (4.1) we easily obtain three equations of the second order in the displacements:

$$(\lambda+\mu)\, \mathbf{grad}\, \theta + \mu\, \Delta\, \mathbf{U} + \mathbf{F} - \rho\, \mathbf{\Gamma} = 0 \qquad (4.24)$$

or, in spherical coordinates and considering more generally a non homogeneous body (ρ, λ, μ being functions of r)

$$\left.\begin{aligned}
\frac{\partial}{\partial r}(\lambda\theta + 2\mu\, \varepsilon_{rr}) + \frac{\mu}{r}\frac{\partial \varepsilon_{r\theta}}{\partial\theta} + \frac{\mu}{r\sin\theta}\frac{\partial \varepsilon_{r\lambda}}{\partial\lambda} + \frac{\mu}{r}(4\varepsilon_{rr} - 2\varepsilon_{\theta\theta} - 2\varepsilon_{\lambda\lambda} + \operatorname{cotg}\theta\, \varepsilon_{r\theta})\\
+ \rho\, F_r = \rho\, \frac{\partial^2 u_r}{\partial t^2}
\end{aligned}\right\} \quad (4.25)$$

$$\left.\begin{aligned}
&\frac{\partial}{\partial r}(\mu\,\varepsilon_{r\theta}) + \frac{1}{r}\frac{\partial}{\partial\theta}(\lambda\Theta+2\mu\,\varepsilon_{\theta\theta}) + \frac{\mu}{r\sin\theta}\frac{\partial\varepsilon_{\theta\lambda}}{\partial\lambda} + \\
&\qquad \frac{\mu}{r}\left[2\,\mathrm{cotg}\,\theta\left(\frac{1}{r}\frac{\partial\varepsilon_{\theta\theta}}{\partial\theta} - \frac{\varepsilon_{\theta\theta}}{r}\,\mathrm{cotg}\,\theta - \frac{1}{r\sin\theta}\frac{\partial\varepsilon_{\lambda\lambda}}{\partial\lambda}\right) + 3\varepsilon_{r\theta}\right] + \\
&\qquad\qquad \rho\,F_\theta = \rho\,\frac{\partial^2 u_\theta}{\partial t^2}, \\
&\frac{\partial}{\partial r}(\mu\,\varepsilon_{r\lambda}) + \frac{\mu}{r}\frac{\partial\varepsilon_{\theta\lambda}}{\partial\theta} + \frac{1}{r\sin\theta}\frac{\partial}{\partial\lambda}(\lambda\Theta+2\mu\,\varepsilon_{\lambda\lambda}) + \\
&\qquad 3\,\frac{\mu}{r}\,\varepsilon_{r\lambda} + 2\,\frac{\mu}{r}\,\varepsilon_{\theta\lambda}\,\mathrm{cotg}\,\theta + \rho\,F_\lambda = \rho\,\frac{\partial^2 u_\lambda}{\partial t^2}
\end{aligned}\right\} \quad (4.25)$$

Superficial Traction

The expression of the superficial traction will be needed to express the boundary conditions. It can be written as

$$\mathbf{T} = \overline{\overline{\boldsymbol{\Sigma}}}\,.\,\frac{\mathbf{r}}{r} \qquad (4.26)$$

and, using (4.21), becomes

$$X_r = \lambda\,\Theta\,\frac{x}{r} + \frac{\mu}{r}\left(\frac{\partial\xi}{\partial x} + r\,\frac{\partial u}{\partial r} - u\right) \qquad (4.27)$$

with

$$\xi = \mathbf{r}\,\,.\,\mathbf{U} = r\,u_r = r\zeta. \qquad (4.28)$$

4.4. PARTICULAR SOLUTIONS OF THE HOMOGENEOUS EQUATION

This equation is

$$(\lambda+\mu)\,\mathbf{grad}\,\Theta + \mu\,\Delta\,\mathbf{U} = 0 \qquad (4.29)$$

By means of spherical harmonics designated as ϕ_n, χ_n and ω_n, Love obtained three kinds of solutions:

first: $\mathbf{U} = \mathbf{grad}\,\phi_n$, (4.30)

second: $\mathbf{U} = \mathbf{r}\,\wedge\,\mathbf{grad}\,\chi_n$, (4.31)

which obviously satisfy (4.29), and

third: $\mathbf{U} = r^2\,\mathbf{grad}\,\omega_n + \alpha_n\,\omega_n\,\mathbf{r}$, (4.32)

which satisfies (4.29) if

$$\alpha_n = -2\,\frac{n\lambda + (3n+1)\mu}{(n+3)\,\lambda + (n+5)\,\mu}. \qquad (4.33)$$

4.5. KELVIN'S THEORY FOR AN HOMOGENEOUS INCOMPRESSIBLE EARTH

The difficulty with the application of the theory of elasticity to the Earth is that it supposes that the exerted constraints are very small. However, in the case of the Earth the mutual gravitational attraction exerts extremely high constraints. To avoid that difficulty, Kelvin assumes that these initial constraints do not produce any deformation and that they reduce themselves to an initial hydrostatic pressure with

$$\mathbf{grad}\ p_o = \rho\ \mathbf{grad}\ V, \tag{4.34}$$

which can be written

$$\frac{dp_o}{dr} = -\ \rho g\ (r) = -g\ \rho\ \frac{r}{a}$$

g being the gravity at the surface of the homogeneous Earth, i. e.

$$g = \frac{4\pi G\rho a}{3}. \tag{4.36}$$

The integration of (4.35) gives

$$p_o = -\ \frac{g\rho}{2a}\ r^2 + \mathcal{E}, \tag{4.37}$$

and as $p_o = 0$ at the free surface ($r=a$) this gives

$$p_o = \frac{g\rho}{2a}\ (a^2 - r^2). \tag{4.38}$$

This initial stress balances the self-gravitation. Then, when the sphere is deformed by external forces, we can measure its deformation with respect to the initial state considered as an undeformed state and we can suppose that the deformation at any point is associated with an additional stress superposed on the initial stress p_o.

$$\lim \lambda\ \theta = -p,$$

which is the additional pressure due to the deformation.

As ρ, λ and μ are independent of r, the equilibrium equations are now

$$-\ \mathbf{grad}\ p + \mu\ \Delta\ \mathbf{U} + \rho\ \mathbf{grad}\ V = 0, \qquad \operatorname{div} \mathbf{U} = 0, \tag{4.39}$$

V being the potential of the external forces (X, Y, Z).

Let us write the equation of the deformed surface as

$$r = a + \mu_r\ (a) = a + \varepsilon_n\ S_n\ (\theta, \lambda). \tag{4.40}$$

The traction exerted on the surface $r=a$, after the deformation, is equal to the weight of the column of matter contained between the undeformed and the deformed surface and is directed towards the interior (boundary condition):

$$(\mathbf{T})_{r=a} = -\ g\rho\ \varepsilon_n\ S_n\ \frac{\mathbf{r}}{r} \tag{4.41}$$

On the other hand, the potential of the thin spherical shell which is created by the deformation and which can be represented by the density distribution

$$\rho^* = \rho \sum_{n=0}^{\infty} \varepsilon_n S_n (\theta, \lambda) \tag{4.42}$$

is for an internal point

$$V_i = \frac{4\pi G\rho}{2n+1} a \left(\frac{r}{a}\right)^n \varepsilon_n S_n = \frac{3g}{2n+1} \left(\frac{r}{a}\right)^n \varepsilon_n S_n \tag{4.43}$$

and for an external point

$$V_e = \frac{4\pi G\rho}{2n+1} \frac{a^2}{r} \left(\frac{a}{r}\right)^n \varepsilon_n S_n = \frac{3g}{2n+1} \left(\frac{a}{r}\right)^{n+1} \varepsilon_n S_n . \tag{4.44}$$

Then the total perturbing potential acting upon any internal point can be written

$$V_n = W_n + \frac{3g}{2n+1} \left(\frac{r}{a}\right)^n \varepsilon_n S_n . \tag{4.45}$$

Now, before solving the system (4.39) let us apply the divergence operator to its first equation: we observe that p is harmonic.

Kelvin then searches for a solution of the form

$$p = \sum_n \rho V_n + \sum_n p_n , \tag{4.46}$$

$$\mathbf{U} = \sum_n A_n r^2 \,\mathbf{grad}\, p_n + \left(\sum_n B_n p_n\right) \mathbf{r} , \tag{4.47}$$

which, when introduced in the system, easily gives

$$\left.\begin{aligned} \mu \left[2(2n+1) A_n + 2B_n\right] &= 1, \\ 2n A_n + (n+3) B_n &= 0. \end{aligned}\right\} \tag{4.48}$$

This is a particular solution of the heterogeneous equations. One has to combine it with a solution of the homogeneous equations:

$$\mathbf{U} = \sum_n \mathbf{grad}\, \phi_n , \tag{4.49}$$

which is a Love solution of the first class, and we must satisfy the boundary conditions for the traction

$$\mathbf{T}_r (a) = - \rho g \sum_n \varepsilon_n S_n \frac{\mathbf{r}}{r} \tag{4.50}$$

and for the radial displacement

$$u_r (a) = \sum_n \varepsilon_n S_n . \tag{4.51}$$

The radial component of the displacement being

$$\zeta = u_r = \frac{\mathbf{r}}{r} \cdot \mathbf{U} = \frac{1}{r}\left[\sum (nA_n + B_n) r^2 p_n + \sum n\phi_n\right] \tag{4.52}$$

allows condition (4.51) to be written in the form

$$(n A_n + B_n) a^2 p_n + n\phi_n = a \varepsilon_n S_n , \tag{4.53}$$

while the traction at the surface (4.27), after some simple developments utilising the classical properties of the spherical harmonics (as Euler's theorem for homogeneous functions), becomes

$$\mathbf{T}_r = \mu\,\Sigma\left\{-\frac{\rho}{\mu} V_n \frac{\mathbf{r}}{r} + \left[2nA_n + (n+2)\,B_n - \frac{1}{\mu}\right] p_n \frac{\mathbf{r}}{r} + (2nA_n + B_n)\, r\,\mathbf{grad}\, p_n + \frac{2(n-1)}{r}\,\mathbf{grad}\,\phi_n\right\}. \qquad (4.54)$$

In this expression only the first two terms represent the purely radial traction which must satisfy the boundary condition (4.50). The other terms must therefore be identically zero. Thus, with (4.45) we obtain two additional conditions:

$$\left.\begin{aligned} &\left\{\mu\left[2nA_n + (n+2)B_n\right] - 1\right\} p_n + g\rho\left(1 - \frac{3}{2n+1}\right)\varepsilon_n S_n = \rho\, W_n, \\ &(2nA_n + B_n)\, a^2 p_n + 2(n-1)\,\phi_n = 0. \end{aligned}\right\} \qquad (4.55)$$

The systems (4.48), (4.53) and (4.55) allow the five unknowns to be calculated as follows:

$$\left.\begin{aligned} A_n &= \frac{(n+3)}{2(2n^2 + 5n + 3)\,\mu}, \\ B_n &= -\frac{n}{(2n^2 + 5n + 3)\,\mu}, \end{aligned}\right\} \qquad (4.56)$$

$$\varepsilon_n S_n = \frac{n(2n+1)\,\rho\, a\, W_n}{2\left[\mu\,(2n^2 + 4n + 3) + \rho\, g\, a\, n\right](n-1)}, \qquad (4.57)$$

$$p_n = \frac{-(2n^2 + 5n + 3)\,\mu\,\rho\, W_n}{\mu(2n^2 + 4n + 3) + \rho\, g\, a\, n} \qquad (4.58)$$

$$\phi_n = \frac{a^2\, n(n+2)\,\rho\, W_n}{2\left[\mu\,(2n^2 + 4n + 3) + \rho\, g\, a\, n\right](n-1)} \qquad (4.59)$$

Let us put, following Love, at the free surface (r=a)

$$\varepsilon_n S_n = h_n \frac{W_n}{g}. \qquad (4.60)$$

This gives

$$h_n = \frac{(2n+1)}{2(n-1)} \cdot \left[1 + \frac{(2n^2+4n+3)\,\mu}{n\, g\, \rho\, a}\right]^{-1}. \qquad (4.61)$$

That is for the second order n=2

$$h_2 = \frac{5}{2} \cdot \left[1 + \frac{19\,\mu}{2\, g\, \rho\, a}\right]^{-1} \qquad (4.62)$$

and for the third order n=3

$$h_3 = \frac{7}{4} \cdot \left[1 + \frac{33\,\mu}{3\, g\, \rho\, a}\right]^{-1}. \qquad (4.63)$$

In a similar way, Love puts

$$k_n W_n = \frac{3g}{2n + 1} \varepsilon_n S_n \tag{4.64}$$

for the additional potential which gives

$$k_2 = \frac{3}{5} h_2 = \frac{3}{2} \left[1 + \frac{19 \mu}{2 g \rho a} \right]^{-1} \tag{4.65}$$

and

$$k_3 = \frac{3}{7} h_3 = \frac{3}{4} \left[1 + \frac{33 \mu}{3 g \rho a} \right]^{-1}. \tag{4.66}$$

Tangential Components of the Displacement-Number ℓ

We can write

$$\left.\begin{aligned} u_r &= u \sin \theta \cos \lambda + v \sin \theta \sin \lambda + w \cos \theta, \\ u_\theta &= u \cos \theta \cos \lambda + v \cos \theta \sin \lambda - w \sin \theta, \\ u_\lambda &= - u \sin \lambda + v \cos \lambda, \end{aligned}\right\} \tag{4.67}$$

with

$$\operatorname{tg} \lambda = \frac{y}{x}, \quad \operatorname{tg} \theta = \frac{\sqrt{x^2 + y^2}}{z}, \quad r = \sqrt{x^2 + y^2 + z^2}. \tag{4.68}$$

Thus

$$\left.\begin{aligned} u_\theta &= \frac{1}{r \sqrt{x^2 + y^2}} \left[xzu + yzv - (x^2 + y^2) w \right], \\ u_\lambda &= \frac{1}{\sqrt{x^2 + y^2}} (xv - yu). \end{aligned}\right\} \tag{4.69}$$

Introducing the general solution

$$\mathbf{U} = A_n r^2 \,\mathbf{grad}\, p_n + B_n p_n \mathbf{r} + \mathbf{grad}\, \phi_n \tag{4.70}$$

from which we may write

$$u_\theta = A_n r^2 \frac{\partial p_n}{r \partial \theta} + \frac{\partial \phi_n}{r \partial \theta} \tag{4.71}$$

we can replace A_n, p_n and ϕ_n respectively by using (4.56), (4.58) and (4.59) and obtain immediately

$$\ell_n = \frac{3}{2n(n-1)} \cdot \left[1 + \frac{2n^2 + 4n + 3) \mu}{n g \rho a} \right]^{-1}. \tag{4.72}$$

Thus

$$\ell_2 = \frac{3}{4} \cdot \left[1 + \frac{19 \mu}{2 g \rho a} \right]^{-1}, \tag{4.73}$$

$$\ell_3 = \frac{3}{12} \cdot \left[1 + \frac{33 \mu}{3 g \rho a} \right]^{-1}, \tag{4.74}$$

and, consequently,

$$\ell_2 = \frac{3}{10} h_2 , \qquad (4.75)$$

$$\ell_3 = \frac{1}{7} h_3 . \qquad (4.76)$$

4.6. HETEROGENEOUS AND COMPRESSIBLE EARTH

The hypotheses are as follows:

1) The Earth is a Hooke body.

2) The rheological parameters ρ, λ and μ are constants on an equipotential surface which in first approximation can be considered to be a sphere. Thus

$$\rho = \rho(r), \qquad \lambda = \lambda(r), \qquad \mu = \mu(r). \qquad (4.77)$$

3) The initial stress was an hydrostatic pressure. This hypothesis furnishes an additional equation

$$\mathbf{grad}\ p_0 = \rho_0\ \mathbf{grad}\ V_0 \qquad (4.78)$$

which allows the elimination of p_0 and V_0.

4) The external forces are weak with respect to the Earth's gravity and may be developed in a series of spherical harmonic functions.

Let ρ_p^o be the initial density at a point P.

Due to the deformation, this density becomes

$$\rho_p = (\rho_p^o - \zeta \frac{d\rho_p^o}{dr})\ (1-\theta) = \rho_p^o - \frac{d\rho_p^o}{dr} - \rho_p^o\ \theta. \qquad (4.79)$$

Then, Poisson's equation, which in the initial state is

$$\Delta\ V_p^o = -\ 4\ \pi\ G\ \rho_p^o\ , \qquad (4.80)$$

becomes, in the deformed state,

$$\Delta(V_p^o + V_p) = -4\ \pi\ G\ (\rho_p^o - \zeta \frac{d\rho_p^o}{dr} - \rho_p^o\ \theta), \qquad (4.81)$$

which, by difference, gives

$$\Delta\ V_p = 4\ \pi\ G \left(\rho_p^o\ \theta + \zeta \frac{d\rho_p^o}{dr}\right), \qquad (4.82)$$

which describes the gravity effects due to the new distribution of the internal masses in the deformed body.

On the other hand, the internal pressure becomes

$$p'_0 = p_0 + \zeta \frac{dp_0}{dr} \qquad (4.83)$$

and the stress tensor components are

$$X_x = -\ p'_0 + \lambda\theta + 2\mu \frac{\partial u}{\partial x} \qquad (4.84)$$

with

$$\frac{\partial \mu}{\partial x} = \frac{d\mu}{dr}\frac{dr}{dx} = \frac{x}{r}\frac{d\mu}{dr} . \tag{4.85}$$

Using the expression of superficial traction (4.27), we obtain the general form of the equations of elasticity:

$$\left.\begin{aligned} &(\lambda+\mu)\frac{\partial \theta}{\partial x} + \mu\,\Delta\,u - \frac{\partial}{\partial x}\left(\underline{p_o} + \zeta\frac{dp_o}{dr}\right) + \theta\frac{\partial \lambda}{\partial x} + \frac{1}{r}\frac{d\mu}{dr}\left(\frac{\partial \xi}{\partial x} + r\frac{du}{dr} - u\right) + \\ &\qquad \rho_o\,(\underline{1}-\theta)\frac{\partial V_o}{\partial x} - \zeta\frac{d\rho}{dr}\frac{\partial V_o}{\partial x} + \rho_o\frac{\partial V}{\partial x} = \rho_o\frac{\partial^2 u}{\partial t^2}. \end{aligned}\right\} \tag{4.86}$$

Hypothesis (4.78) allows cancellation of the underlined terms. Moreover, the term

$$-\frac{\partial}{\partial x}\left(\zeta\frac{dp_o}{dr}\right)$$

can be written

$$-\frac{\partial}{\partial x}\left(\zeta\,\rho_o\frac{dV_o}{dr}\right) = +\,\rho_o\frac{\partial}{\partial x}\,(g_o\,e_{rr}),$$

while

$$-\,\rho_o\,\theta\frac{\partial V_o}{\partial x} = +\,\frac{4\,\pi\,G\,\rho_o^2}{3}\times\theta. \tag{4.87}$$

Considering that the independent terms from u in equations (4.86) contain either xW_n or $\partial W_n/\partial x$, Herglotz (1905) looked for a solution of the type

$$u = F_n\,(r)\frac{\partial W_n}{\partial x} + G_n\,(r)\times W_n\,, \tag{4.88}$$

which is sufficiently general to satisfy the boundary conditions.

The Tangential Displacements

$$\left.\begin{aligned} r u_\theta &= F_n(r)\left(\frac{xz}{\sqrt{x^2+y^2}}\frac{\partial W}{\partial x} + \frac{zy}{\sqrt{x^2+y^2}}\frac{\partial W}{\partial y} - \sqrt{x^2+y^2}\frac{\partial W}{\partial z}\right), \\ u_\lambda &= F_n(r)\left(-\frac{y}{\sqrt{x^2+y^2}}\frac{\partial W}{\partial x} + \frac{x}{\sqrt{x^2+y^2}}\frac{\partial W}{\partial z}\right), \end{aligned}\right\} \tag{4.89}$$

where

$$\left.\begin{aligned} \frac{\partial W}{\partial \theta} &= r\cos\theta\cos\lambda\frac{\partial W}{\partial x} + r\cos\theta\sin\lambda\frac{\partial W}{\partial y} - r\sin\theta\frac{\partial W}{\partial z}, \\ \frac{\partial W}{\partial \lambda} &= -\,r\sin\theta\sin\lambda\frac{\partial W}{\partial x} + r\sin\theta\cos\lambda\frac{\partial W}{\partial y}, \end{aligned}\right\} \tag{4.90}$$

can be written

$$u_\theta = F_n(r)\frac{\partial W_n}{r\partial\theta}, \quad u_\lambda = F_n(r)\frac{\partial W_n}{r\sin\theta\,\partial\,\lambda}. \tag{4.91}$$

The function $F_n(r)$ is thus related to the third Love number L_n:

$$\frac{L_n(r)}{g} = \frac{F_n(r)}{r}. \tag{4.92}$$

The radial displacement is

$$u_r = \zeta = \frac{x}{r} u + \frac{y}{r} v + \frac{z}{r} w = [\frac{n}{r} F_n(r) + rG_n(r)] W_n .$$

Thus the first Love number is

$$h_n(r) = \frac{n F_n(r)}{r} + r G_n(r). \tag{4.93}$$

The cubic expansion is

$$\theta = \frac{\partial u}{\partial x} + \frac{\partial v}{\partial y} + \frac{\partial w}{\partial z} = \frac{n}{r} \frac{dF_n}{dr} W_n + r \frac{dG_n}{dr} W_n + (n+3) G_n W_n .$$

Therefore

$$f_n(r) = \frac{n}{r} \frac{dF_n}{dr} + r \frac{dG_n}{dr} + (n+3) G_n . \tag{4.94}$$

Finally, Poisson's equation

$$\Delta V = \Delta (K_n (r) W_n) = 4 \pi G \rho_0 \theta,$$

neglecting ζ $(d\rho^\circ_p/dr)$ with respect to $\rho_0\theta$ because of

$$\Delta V = \left(\frac{d^2K_n}{dr^2} + \frac{2(n+1)}{r} \frac{dK_n}{dr} \right) W_n , \tag{4.95}$$

gives another relation

$$f_n(r) = \frac{1}{4\pi G\rho_0} \left(\frac{d^2K_n}{dr^2} + \frac{2(n+1)}{r} \frac{dK_n}{dr} \right). \tag{4.96}$$

We define Poincaré's operator

$$D \equiv \frac{d^2}{dr^2} + \frac{2(n+1)}{r} \frac{d}{dr}. \tag{4.97}$$

Then Poisson's equation can be written

$$f_n(r) = \frac{DK_n (r)}{4 \pi G \rho_0}. \tag{4.98}$$

Herglotz has considered an incompressible Earth ($\theta = 0$) and neglected the inertial effects ($\rho\, \partial^2 u/\partial t^2$). Then he combined the three equations (4.86) in two ways: a) multiplying them respectively by x, y, z, and adding, b) deriving them respectively with respect to x, y, z and adding.

He obtained in this way one equation in Δp and one equation in $\Delta\xi$. As ξ is interesting only from the experimental point of view, he used the combination

$$\Delta \text{ (equ. } \Delta\zeta) + \left(r \frac{\partial}{\partial r} + 2 \right) \text{(equ. } \Delta p) = 0, \tag{4.99}$$

which eliminates p and gives a fourth-order equation containing K_n as well as h_n. Then an application of Poincaré's operator eliminates K_n because

$$DK_n = \frac{d^2K_n}{dr^2} + \frac{2(n+1)}{r} \frac{dK_n}{dr} = 4 \pi G \frac{d\rho^\circ_p}{dr} h_n \tag{4.100}$$

when $\theta = 0$.

The equation obtained is the differential equation of Herglotz and it is a sixth-order equation:

$$\left.\begin{array}{l} D\left[\frac{r_1}{\rho'} D\,(\mu\, Dh_n)\right] - 2D\left(\frac{\mu'}{\rho'} Dh_n\right) - 2D\left[\frac{d}{dr_1}\left(\frac{\mu'}{r_1}\right)\frac{r_1^{n^2}}{\rho'}\frac{d}{dr_1}\left(r_1^{1-n^2} h_n\right)\right] \\ \qquad = n(n+1)\, a^4\, D\,(\frac{r_1}{\rho'}\gamma\, h_n) - 4\,\pi\, G\, n(n+1)\, a^2 \frac{\rho'}{r_1} h_n\,, \end{array}\right\} \quad (4.101)$$

where

$$r_1 = \frac{r}{a}$$
$$\gamma\, r^2 = -\frac{1}{r^2}\frac{d\rho}{dr}\frac{dV_0}{dr} = \frac{4\pi G}{r^2}\rho' \int_0^r \rho r^2\, dr = \frac{4\pi G}{r^2}\rho'\, M. \qquad (4.102)$$

This equation has been used by Herglotz, Jeffreys, Schweydar, Prey and Boaga, who tried to integrate it introducing binomial or polynomial laws for rigidity.

It is quite interesting to point out that making $\mu=0$, Herglotz's equation reduces to Clairaut's famous equation of the ellipticities of equipotential surfaces of a rotating fluid.

4.7. METHOD OF INTEGRATION OF TAKEUCHI

This method is described in a fundamental paper published in 1951, where, for the first time, a numerical integration of the equations had been performed with success.

Putting

$$\xi = \frac{r}{a} \qquad (4.103)$$

where ξ is a variable between 0 and 1

$$F(\xi) = F_n\,(r)/a^2 \qquad G_n(\xi) = G_n\,(r),$$
$$K_n(\xi) = K_n\,(r)/4\pi G a^2 \qquad (4.104)$$

Takeuchi has written all the equations as a function of that variable and obtains three differential equations of the second order. He integrates these equations numerically for n=2. Therefore he has to calculate the coefficients of F_2'', F_2', F_2, G_2'', G_2', G_2, K_2'', K_2', K_2. The first step consists in the calculation of ρ, λ, μ, $d\rho/d\xi$, $d\lambda/d\xi$, $d\mu/d\xi$ at a sufficient number of points in the mantle. Takeuchi selected points at the following depths: 500, 800, 1100, 1400, 1700, 2000, 2300, 2600, 2900 km.

Having determined ρ, λ and μ at each of these points by using the Wadati and Gutenberg tables of the velocities of the P and S seismic waves, Takeuchi obtains numerical values at 650, 950, ... and 2750 km by a linear interpolation and thereafter the mean gradient on each interval (650-800), (800-950) ... . He adopts for numerical values of the derivatives the mean of two consecutive gradients as shown in Table 4.1.

Then, performing the computations one needs the numerical values of $dV_0/d\xi$ and $dP_0/d\xi$. Considering that g does not vary by more than 1% between the surface and the core mantle boundary, Takeuchi adopts a constant value g = 982 and obtains in c. g. s. units:

$$\frac{dV_o}{d\xi} = -\,ag = -\,0.625\ 53 \times 10^{12}, \qquad (4.105)$$

$$\frac{d^2V_o}{d\xi^2} = 0, \qquad (4.106)$$

$$\frac{dP_o}{d\xi} = \rho_o \frac{dV_o}{d\xi} = 0.625\ 53 \times 10^{12}\ \rho_o, \qquad (4.107)$$

then it is possible to calculate at each point numerical values of the coefficients of F" F' F G' G K in three second-order equations.

TABLE 4.1. Rheologic constants and their gradients, as adopted by Takeuchi for the Earth's mantle (expressed in 10^{12} cgs)

Depth (km)	ρ	μ	λ	$\frac{d\rho_o}{d\xi}$	$\frac{d\mu}{d\xi}$	$\frac{d\lambda}{d\xi}$
	3.30	0.600	0.680			
250	3.51	0.772	0.936	-4.9686	-4.1558	-6.5815
500	3.69	0.926	1.197			
500	4.22	1.217	1.644	-4.6730	-9.9371	-8.4296
800	4.44	1.685	2.041	-4.4590	-8.2172	-6.4443
1100	4.64	1.991	2.251	-3.7158	-4.5970	-8.5570
1400	4.79	2.118	2.847	-3.1850	-3.6203	-8.3977
1700	4.94	2.332	3.042	-3.0788	-4.2466	-4.6713
2000	5.08	2.518	3.287	-3.0788	-3.0788	-8.3340
2300	5.23	2.622	3.827	-3.1850	-2.7072	-10.9350
2600	5.38	2.773	4.317	-3.1850	-3.2805	-8.2916
2900	5.53	2.931	4.608	-3.1850	-3.3549	-6.1789

In the core Takeuchi used a Bullen law of density: $\rho = 12.284\ (1-0.64014\ r^2)$

Takeuchi constructed tables of these values, separating the terms which are a function of n from those which are not. This allows a separate treatment of the different orders. He resolves then, at each point, the three equations with respect to F", G" and K" under such a form as the following examples obtained at $\xi = 0.544\ 74$ (see Table 4.2):

$$\begin{aligned} F'' &= -\,15.641F' + 8.5353F - 1.4012G' - 13.594G - 0.64169K \\ K'' &= 20.302F' - 11.693F + 3.0124G' + 25.915G - 11.014K' \end{aligned} \qquad (4.108)$$

etc. which could be solved by the Runge-Kutta method. Takeuchi has preferred a method due to Frazer, Duncan and Collar which he describes in his paper (1951).

Results Obtained by Takeuchi for Five Different Models

These models have the same mantle but differ by the adopted value for the core rigidity which is considered as homogeneous (Table 4.3).

TABLE 4.2. Numerical integration by Takeuchi

EQUATION in F"

ξ	F'	G'	K'	F	G	K
0.54474	-15.641	-1.4012	0.0	8.5353	-13.594	-0.64169
0.59184	-14.215	-1.5132	0.0	8.0988	-13.366	-0.65986
0.63893	-12.927	-1.5715	0.0	7.1377	-12.841	-0.67842
0.68603	-11.329	-1.5816	0.0	7.2438	-11.822	-0.68615
0.73313	-9.9216	-1.6895	0.0	8.5828	-11.216	-0.72046
0.78022	-9.4267	-1.8290	0.0	8.0080	-11.284	-0.76916
0.82732	-7.6765	-1.7627	0.0	9.1055	-9.5369	-0.79262
0.87441	-4.7554	-1.9336	0.0	14.924	-7.3507	-0.89620
0.92151	-1.2778	-2.1663	0.0	22.430	-4.2311	-1.1793
0.92151	-4.8281	-2.1120	0.0	15.147	-7.0284	-1.3550
0.96075	-3.3890	-2.1255	0.0	17.115	-5.1623	-1.5453
1.0	-1.3403	-2.1333	0.0	20.734	-2.2999	-1.8706

EQUATION in G"

ξ	F'	G'	K'	F	G	K
0.54474	90.952	-4.0110	-0.32975	-45.509	71.315	3.1141
0.59184	71.979	-3.3714	-0.31345	-36.533	61.905	2.7084
0.63893	57.844	-3.0195	-0.30690	-27.626	53.769	2.3629
0.68603	43.726	-3.1992	-0.30257	-23.835	42.268	2.0337
0.73313	33.592	-2.9168	-0.29738	-24.308	34.059	1.8696
0.78022	30.237	-2.0365	-0.29481	-20.225	34.038	1.7715
0.82732	23.836	-1.6714	-0.30603	-19.752	28.577	1.5762
0.87441	17.713	0.1368	-0.31915	-28.417	24.649	1.6142
0.92151	14.512	3.3612	-0.38193	-38.716	25.974	1.9486
0.92151	17.991	1.1775	-0.44667	-26.773	25.868	2.2219
0.96075	15.986	2.2802	-0.50072	-27.535	24.795	2.3059
1.0	14.609	4.1885	-0.59698	-30.429	25.086	2.5472

EQUATION in K"

ξ	F'	G'	K'	F	G	K
0.54474	20.302	3.0124	-11.014	-11.693	25.910	0.0
0.59184	18.180	3.1841	-10.138	-10.763	25.015	0.0
0.63893	16.371	3.3416	-9.3906	-9.9696	24.115	0.0
0.68603	14.810	3.4850	-8.7462	-8.9760	23.288	0.0
0.73313	13.476	3.6217	-8.1840	-8.3990	22.443	0.0
0.78022	12.279	3.7373	-7.6902	-8.1644	21.465	0.0
0.82732	11.217	3.8388	-7.2522	-8.9826	20.126	0.0
0.87441	10.155	3.8824	-6.8616	-10.199	18.301	0.0
0.92151	9.1590	3.8888	-6.5112	-10.139	16.795	0.0
0.92151	8.0086	3.4004	-6.5110	-10.784	13.871	0.0
0.96075	7.3068	3.3722	-6.2451	-10.343	12.776	0.0
1.0	6.60	3.30	-6.0	-9.9372	11.531	0.0

TABLE 4.3.

Model μ (core) c.g.s,	1 0	2 10^7	3 10^9	4 10^{11}	5 10^{13}
k	0.281	0.275	0.275	0.243	0.055
h	0.606	0.601	0.600	0.530	0.109
ℓ	0.082	0.081	0.081	0.083	0.092
γ	0.675	0.674	0.675	0.713	0.946
δ	1.185	1.188	1.187	1.165	1.026
Λ	1.199	1.194	1.194	1.160	0.963
ℓ/h	0.135	0.135	0.135	0.157	0.844
$k - h/2$	-0.022	-0.025	-0.025	-0.022	0.000

The experimental results obtained for the diurnal wave O_1 which obeys such a static theory and is the least disturbed by indirect oceanic effects (see Chapter 11) are

$$\gamma = 0.676 \qquad \delta = 1.164.$$

The table 4.3 allows an upper limit to be put on the rigidity of the core which is probably 10^{10} cgs.

CHAPTER 5

Reduction of the Problem of Elastic Deformations of a Sphere to a System of Six Differential Equations of the First Order

This method of providing differential equations in the Runge Kutta form was developed first by Molodensky (1953) and later on, with slight modifications, by Jarosch, Pekeris and Alterman (1959).

We follow here Molodensky's notation.

5.1. EXTERNAL POTENTIAL DISPLACEMENTS AND CUBIC EXPANSION

Let us put, with Molodensky,

$$\bar{\omega} = \frac{W_n a^{n-1}}{g_o}, \tag{5.1}$$

where

$$g_o = G\frac{M}{a^2}$$

is the acceleration of gravity at the surface and

$$V_p = W_n + k_n W_n = R\frac{\bar{\omega}}{r^n}. \tag{5.2}$$

A solution will be found in the form

$$u = H\frac{x}{r}\frac{\bar{\omega}}{r^n} + T\frac{\partial}{\partial x}\left(\frac{\bar{\omega}}{r^n}\right) \tag{5.3}$$

which immediately gives

$$\zeta = ux + vy + wz = H\frac{\bar{\omega}}{r^{n-1}}. \tag{5.4}$$

H characterizes the radial displacement ζ. T characterizes the tangential displacement.

Moreover one has,

$$\xi = \frac{x}{r}u + \frac{y}{r}v + \frac{z}{r}w = H_n\frac{\bar{\omega}}{r^n} = H_n\frac{a^{n-1}}{r^n}\frac{W_n}{g_o}. \tag{5.5}$$

The cubic expansion

$$\theta = \frac{\partial u}{\partial x} + \frac{\partial v}{\partial y} + \frac{\partial w}{\partial z}$$

can be developed by taking advantage of Euler's theorem for homogeneous functions (as spherical harmonics are) and from the fact that

$$\frac{\partial T}{\partial x} = \frac{\partial T}{\partial r}\frac{\partial r}{\partial x} = \frac{x}{r}\frac{\partial T}{\partial r}$$

one obtains

$$\theta = f\,\frac{\bar{\omega}}{r^n} = \left(H' + \frac{2}{r}\,H - \frac{n(n+1)}{r^2}\,T\right)\frac{\bar{\omega}}{r^n}. \qquad (5.6)$$

5.2. TRANSFORMATIONS OF THREE FUNDAMENTAL EQUATIONS OF EQUILIBRIUM

One easily obtains

$$\Delta u = (T''-H'+f)\,\frac{\partial}{\partial x}\left(\frac{\bar{\omega}}{r^n}\right) + \left[(n+1)T'-(n+1)H+\frac{r^2}{n}\,f'\right] n\,\frac{\bar{\omega}x}{r^{n+3}}, \qquad (5.7)$$

$$x\,\frac{\partial u}{\partial x} + y\,\frac{\partial u}{\partial y} + z\,\frac{\partial u}{\partial z} = (rT'-T)\,\frac{\partial}{\partial x}\left(\frac{\bar{\omega}}{r^n}\right) + H'\,\frac{\bar{\omega}x}{r^n}, \qquad (5.8)$$

$$\frac{\partial \xi}{\partial x} - u + x\,\frac{\partial u}{\partial x} + y\,\frac{\partial u}{\partial y} + z\,\frac{\partial u}{\partial z} = (rT'-2T+rH)\,\frac{\partial}{\partial x}\left(\frac{\bar{\omega}}{r^n}\right) + 2H'\,\frac{\bar{\omega}x}{r^n}. \qquad (5.9)$$

The fundamental equation (4.93) can be rewritten as follows:

$$\left.\begin{aligned}&(\lambda+\mu)\,\frac{\partial \theta}{\partial x} + \mu\Delta u + \frac{1}{r}\,\frac{d\mu}{dr}\left(\frac{\partial \xi}{\partial x} - u + r\,\frac{\partial u}{\partial r}\right) + \frac{x}{r}\,\theta\,\frac{d\lambda}{dr}\\ &+ \rho\left[\frac{\partial}{\partial x}\,(W_2 + kW_2) + \frac{\xi}{r}\,\frac{dV_0}{dr} - \frac{x}{r}\,\theta\,\frac{dV_0}{dr}\right] = 0,\end{aligned}\right\} \qquad (5.10)$$

(and two similar relations for y and z) i. e. omitting inertial term $\rho\,\frac{\partial^2 u}{\partial t^2}$ it becomes

$$\left.\begin{aligned}&\left[\begin{aligned}&\rho\,(R + HV') + (\lambda+2\mu)f - \mu H' + \mu'\,(T' + H - \tfrac{2}{r}\,T)\\ &+ \mu T''\end{aligned}\right]\frac{\partial}{\partial x}\left(\frac{\bar{\omega}}{r^n}\right)\\ &+\left[\begin{aligned}&(R' + V'H' + V''H - V'f)\rho + n(n+1)\mu\,\frac{T'-H}{r^2} + \lambda' f\\ &+ 2\mu' H' + (\lambda+2\mu)f'\end{aligned}\right]\frac{\bar{\omega}x}{r^{n+1}} = 0,\end{aligned}\right\} \qquad (5.11)$$

and two similar relations for y and z.

The term $\rho(R+HV')$ represents the variation of pressure calculated in the hydrostratic hypothesis:

$$-\,\frac{\partial}{\partial x}\left(\xi\,\frac{dp_0}{dr}\right) = +\,\frac{\partial}{\partial x}\left(\xi\rho\,\frac{dV_0}{dr}\right).$$

The three equations (5.11) will be satisfied if the expressions between brackets, which are the coefficients of harmonics of different order, are null. Then the initial system of three equations (5.10) is replaced by two equations:

$$\left[-\mu\,(T'+H-\frac{2}{r}\,T)\right]' = \rho\,(R+V'H) + \lambda f + 2\,\frac{\mu}{r}\left(2H+T' - \frac{n^2+n+1}{r}\,T\right), \qquad (5.12)$$

$$\left.\begin{aligned}&-(\lambda f+2\mu H')' = \rho\,(R+V'H)' - \rho V'f + 4\,\frac{\mu}{r}\left(H' - \frac{H}{r}\right) -\\ &\frac{n(n+1)}{r^2}\,\mu\left(T' + H - 4\,\frac{T}{r}\right).\end{aligned}\right\} \qquad (5.13)$$

To solve the problem which has six unknowns (the tensor of tensions is a symmetric tensor of the second order) we need a third differential equation of the second order which is Poisson's equation.

5.2.1. Equation in R. (Poisson's Equation)

Equation (4.89) can be written here

$$\Delta\left(R \frac{\bar{\omega}}{r^n}\right) = 4 \pi G (\rho_0 F + \rho' H) \frac{\bar{\omega}}{r^n} \qquad (5.14)$$

as

$$\Delta (ff_1) = f_1 \Delta f + f \left(\frac{d^2 f_1}{dr^2} + \frac{2}{r} \frac{df_1}{dr} \right) + 2 \frac{\partial f_1}{\partial r} \frac{\partial f}{\partial r}$$

if f_1 depends only on r, which is the case for R.

Thus

$$\Delta \left(R \frac{\bar{\omega}}{r^n} \right) = R \Delta \frac{\bar{\omega}}{r^n} + \frac{\bar{\omega}}{r^n} (R'' + \frac{2}{r} R') + 2R' \frac{\partial}{\partial r} \frac{\bar{\omega}}{r^n},$$

where $\frac{\bar{\omega}}{r^n}$ is a surface harmonic function.

Therefore

$$\Delta S_n = \frac{-n(n+1)}{r^2} S_n$$

and (5.14) becomes

$$R'' = - \frac{2}{r} R' + \frac{n(n+1)}{r^2} R + 4 \pi G (\rho f + \rho' H), \qquad (5.15)$$

which describes the variation of potential produced by the new distribution of masses within the deformed body f, H. The three equations of the second order (5.13), (5.14) and (5.15) are the equilibrium equations of the elastic gravitating sphere. Introducing boundary conditions we will be able to determine the the three auxiliary functions:

H radial displacement,

T tangential displacement,

R potential variation.

However, these equations contain the derivatives of experimental functions: λ, μ and ρ. To overcome this difficulty one introduces the six parameters H, N, T, M, R, L, which are tied by differential equations of the first order.

5.3. COMPONENTS OF THE TENSION

Boundary conditions will be obtained from the characteristics of the tension at the surface. Two of the new variables that we have to choose to reduce (5.13) and (5.14) to the first order will be defined in such a way that we can easily write the boundary conditions:

a) normal tension balanced by the hydrostatic pressure;
b) tangential tension vanishing at the surface.

The component of the tension parallel to Ox

$$N_x = \frac{x}{r} X_x + \frac{y}{r} X_y + \frac{z}{r} X_z$$

or

$$N_x = \lambda \theta \frac{x}{r} + \frac{\mu}{r} \left(\frac{\partial \xi}{\partial x} + r \frac{du}{dr} - u\right) \tag{5.16}$$

may now be written:

$$N_x = (\lambda f + 2\mu H') \frac{\bar{\omega} x}{r^{n+1}} + \mu (T' - \frac{2}{r} T + H) \frac{\partial}{\partial x} (\frac{\bar{\omega}}{r^n}). \tag{5.17}$$

The normal component of the tension is

$$P = \frac{x}{r} N_x + \frac{y}{r} N_y + \frac{z}{r} N_z = N \frac{\bar{\omega}}{r^n}, \tag{5.18}$$

where N is one of the new parameters to be introduced.

From (5.17) we get

$$N = \lambda f + 2\mu H' \tag{5.19}$$

because $r \frac{d}{dr} \frac{\bar{\omega}}{r^n} = 0,$

$\bar{\omega}/r^n$ being a surface harmonic.

Then (5.6) is transformed into a first-order differential equation:

$$N = (\lambda + 2\mu) H' + \lambda \left(\frac{2}{r} H - \frac{n(n+1)}{r^2} T\right). \tag{5.20}$$

Eliminating the normal component P from N_x we get the projection on Ox of the tangential component

$$M_x = N_x - \frac{x}{r} P = \mu (T' - \frac{2}{r} T + H) \frac{\partial}{\partial x} \left(\frac{\bar{\omega}}{r^n}\right) \tag{5.21}$$

and one puts

$$Mr^{-2} = \mu (T' - \frac{2}{r} T + H), \tag{5.22}$$

a second differential equation of the first order which defines the new parameter M.

Introducing

$$u_r = HS, \qquad u_\theta = T \frac{\partial S}{r \partial \theta}, \qquad u_\lambda = T \frac{\partial S}{r \sin \theta \partial \lambda}, \qquad \theta = \xi S,$$

where S is a surface harmonic function, we obtain the components of the stress tensor in spherical coordinates:

$$\left.\begin{aligned} \sigma_{rr} &= (\lambda f + 2\mu H') S = NS, \\ \sigma_{r\theta} &= \mu \left(T' r^{-1} - 2Tr^{-2} + H r^{-1}\right) \frac{\partial S}{\partial \theta} = Mr^{-3} \frac{\partial S}{\partial \theta}, \\ \sigma_{\theta\lambda} &= \mu T r^{-2} \frac{\partial^2 (S/\sin \theta)}{\partial \theta \partial \lambda}, \\ \sigma_{r\lambda} &= \mu M r^{-3} \frac{\partial S}{\sin \theta \partial \lambda}. \end{aligned}\right\} \tag{5.23}$$

The dimensions of these parameters are such that

$$(N) = (T \mu r^{-2}) = (H \mu r^{-1}) = (M r^{-3}).$$

5.4. TRANSFORMATION OF POISSON'S EQUATION INTO TWO EQUATIONS OF THE FIRST ORDER

To perform this transformation we evidently have to introduce a new parameter which will complete the set of six parameters and provide the final system of six linear differential equations to replace Herglotz's equation of the sixth order.

We define

$$L\, r^{-2} = R' - 4\,\pi\, G\, \rho\, H, \tag{5.24}$$

where $G\,\rho\,H$ is the gravitational attraction of the column of displaced matter and L represents the change of radial gravitational flux of density. This has the advantage of the elimination of the derivation of ρ which was a great inconvenience because it is a purely experimental function.

Deriving (5.24) with respect to r one has

$$L'\, r^{-2} - 2L\, r^{-3} = (R'' - 4\,\pi\, G\, \rho'\, H) - 4\,\pi\, G\, \rho\, H'. \tag{5.25}$$

Replacing L by its definition (5.24) and combining (5.15), (5.6) and (5.25), one has

$$L' = n(n+1)R - 4\,\pi\, G\, \rho\, n(n+1)T. \tag{5.26}$$

Poisson's equation is replaced by the system (5.24) and (5.26).

Now, introducing the new parameters in (5.12) and (5.13) we can transform them into two equations of the first order giving respectively M' and N' as a function of the six variables (for the details of calculations see Melchior, 1972).

The system of the six linear differential equations is, finally:

$$N = (\lambda+2\mu)H' + \lambda\left(\frac{2}{r}\,H - \frac{n(n+1)}{r^2}\,T\right), \tag{5.20}$$

$$Mr^{-2} = \mu(T' - \frac{2}{r}\,T + H), \tag{5.22}$$

$$Lr^{-2} = R' - 4\,\pi\, G\, \rho\, H, \tag{5.24}$$

$$L' = n(n+1)R - 4\,\pi\, G\, \rho\, n(n+1)T, \tag{5.26}$$

$$M' = -\left(\rho V' r^2 + \frac{2\mu(3\lambda+2\mu)}{\lambda+2\mu}\, r\right)H - \frac{\lambda}{\lambda+2\mu}\, Nr^2 - \rho R r^2 + \frac{2\mu}{\lambda+2\mu}\left[\lambda(2n^2 + 2n - 1) + 2\mu\,(n^2 + n - 1)\right]T, \tag{5.27}$$

$$N' = \frac{n(n+1)}{r^2}\, Mr^{-2} + 4\mu\,\frac{3\lambda+2\mu}{\lambda+2\mu}\, Hr^{-2} - 2\mu\,\frac{3\lambda+2\mu}{\lambda+2\mu}\, n(n+1)Tr^{-3} - \rho Lr^{-2} - \frac{4\mu}{\lambda+2\mu}\, Nr^{-1} + \rho V'\left[4\, Hr^{-1} - n(n+1)\, Tr^{-2}\right]. \tag{5.28}$$

In the case of a zero rigidity this system is reduced to the fourth order as $M=0$, $N=\lambda f$.

5.5. RESOLUTION OF THE SYSTEM

This system is a canonic system of the first order as it is formed with a number of equations which is equal to the number of unknowns, and it is resolved with respect to the highest order of derivatives of the unknown functions.

The system is linear because the boundary conditions are linear. It is not homogeneous because the condition at the surface is not homogeneous for the potential.

The solution of such a system of differential equations constitutes a six-dimensional vectorial space.

The integration will be performed six times for initial conditions which secure the independence of the solutions obtained. To realize that condition it is sufficient to secure the independency locally at the surface.

The general solution may be presented under the form of linear combinations of six particular solutions $(H_1\ T_1\ \ldots\ R_1)$, $(H_2\ T_2\ \ldots\ R_2)$, ... $(H_6\ T_6\ \ldots\ R_6)$ as follows:

$$\left.\begin{aligned} H &= C_1 H_1 + C_2 H_2 + \ldots + C_6 H_6,\\ T &= C_1 T_1 + C_2 T_2 + \ldots + C_6 T_6,\\ &\ldots\\ R &= C_1 R_1 + C_2 R_2 + \ldots + C_6 R_6. \end{aligned}\right\} \qquad (5.29)$$

The six arbitrary constants C_i will be determined in such a way that the boundary conditions will be satisfied at the surface of the Earth, at the surfaces of the core and inner core and at the centre of the Earth.

The system (5.29) gives

$$H_i = \frac{\partial H}{\partial C_i}. \qquad (5.30)$$

One can thus write for each of the six functions, represented here more generally by the function ϕ_i,

$$\phi_i(r) = h\,\frac{\partial \phi i}{\partial h} + \ell\,\frac{\partial \phi i}{\partial \ell} + (1+k)\,\frac{\partial \phi i}{\partial k} + N_o\,\frac{\partial \phi i}{\partial N_o} + M_o\,\frac{\partial \phi i}{\partial M_o} + L_o\,\frac{\partial \phi i}{\partial L_o}, \qquad (5.31)$$

where $1+k$, N_o, M_o, L_o are constants corresponding to the choice of the six series of six particular solutions $(H_1, T_1, \ldots, R_1)$, $(H_2, T_2, \ldots, R_2)$, ..., $(H_6, T_6, \ldots, R_6)$ which obey very simple boundary conditions:

$$H = h, \qquad T = \ell, \qquad R = 1 + k,$$

$$N = N_o, \qquad M = M_o, \qquad L = L_o.$$

Thus $\partial\phi/\partial h$ is a particular integral corresponding to the conditions

$$H_1\ (a) = 1, \qquad T_1 = M_1 = N_1 = L_1 = R_1 = 0$$

at the surface. This gives the solution $(H_1, T_1, \ldots, R_1)$ while $\partial\phi/\partial N_o$ correspond to the conditions

$$N(a) = N_o, \qquad H = T = M = L = R = 0$$

and gives the solution $(H_4, T_4, \ldots, R_4)$, etc...

Conditions for the Tensions at the External Surface (r=a)

Assuming that the Earth is placed in a vacuum, there is no resulting stress across the free external surface, i. e.

$$N\,(a) = 0, \qquad\qquad M\,(a) = 0.$$

As $M(a) = C_5\ M_5(a)$ and $N(a) = C_4\ N_4(a)$ it follows that

$$C_4 = C_5 = 0, \qquad\qquad N_o = M_o = 0,$$

and the general solution is reduced to

$$\phi_i\ (r) = h\,\frac{\partial\phi i}{\partial h} + \ell\,\frac{\partial\phi i}{\partial \ell} + (1+k)\,\frac{\partial\phi i}{\partial k} + L_o\,\frac{\partial\phi i}{\partial \ell_o}. \tag{5.32}$$

Discontinuity of the Potential at the Free Surface

This discontinuity is well known from Chasles's theorem: the discontinuity of the gradient is proportional to the surface density produced by the displacement

$$\left(\frac{\partial V_i}{\partial n}\right)_{ext} - \left(\frac{\partial V_i}{\partial n}\right)_{int} = -\,4\,\pi\,G\,\rho\,(a)\,H(a)\,S. \tag{5.33}$$

This formula is derived from Gauss's theorem

$$\int_S \frac{\partial V}{\partial n}\,d\,\sigma = -\,4\,\pi\,G\,M. \tag{5.34}$$

M is the masse included into S.

Gauss's theorem is applied to a volume where $d\sigma_2$ is obtained from $d\sigma_1$ by drawing the normals to S_1.

If

$$d\sigma = d\sigma_1 = d\sigma_2, \qquad\qquad m = \rho\,h\,d\,\sigma.$$

One has

$$\left(\frac{\partial V}{\partial n}\right)_{ext} d\sigma_1 - \left(\frac{\partial V}{\partial n}\right)_{int} d\sigma_2 = -\,4\,\pi\,G\,m$$

because the normal component is zero along the lateral surface. This gives

$$\left(\frac{\partial V}{\partial n}\right)_{ext} - \left(\frac{\partial V}{\partial n}\right)_{int} = -\,4\,\pi\,G\,\rho\,h. \tag{5.35}$$

Molodensky writes the perturbing potential as follows:

$$V_e + V_i = W_n + k\,W_n = R\,\frac{\bar{\omega}}{r^n} = \frac{K}{3}\,RS, \tag{5.36}$$

where

$$W_n = V_e = A_n^m\,r^n\,P_n^m\,(\theta)\,\cos\,(\sigma t - m\lambda) \tag{5.37}$$

and

$$S_n = P_n^m\,(\theta)\,\cos\,(\sigma t - m\lambda).$$

Thus

$$W_2 = \frac{\kappa}{3}\frac{g}{a} S r^2, \qquad A_2^{\ 1} = \frac{\kappa g}{3a}, \tag{5.38}$$

and

$$V_i = k W_n = \frac{\kappa}{3} RS - W_n = \frac{\kappa}{3}[R(r) - g \frac{r^2}{a}] S \tag{5.39}$$

gives at the free surface (r=a)

$$V_i = (\frac{\kappa}{3}) [R(a) - g\ a] S. \tag{5.40}$$

The external prolongation of this potential is

$$V_i\ (r>a) = (\frac{\kappa}{3}) [R(a) - g\ a] S \frac{a^3}{r^3}, \tag{5.41}$$

which gives

$$\left(\frac{\partial V_i}{\partial r}\right)_{ext} = - k(R - g\ a) S \frac{a^3}{r^4}, \tag{5.42}$$

while (5.39) gives

$$\left(\frac{\partial V_i}{\partial r}\right)_{int} = \frac{\kappa}{3} (R' - 2 \frac{gr}{a}) S. \tag{5.43}$$

Thus at the surface

$$\left.\begin{aligned} \left(\frac{\partial V_i}{\partial r}\right)_{int\ (r=a)} &= \frac{\kappa}{3} (R' - 2g) S, \\ \left(\frac{\partial V_i}{\partial r}\right)_{ext\ (r=a)} &= - \kappa (R - g\ a) \frac{S}{a}, \end{aligned}\right\} \tag{5.44}$$

and consequently we have the condition for R' at the free surface

$$\left(\frac{\partial V_i}{\partial r}\right)_{ext} - \left(\frac{\partial V_i}{\partial r}\right)_{int} = \frac{\kappa}{3}(R'-2g+3\frac{R}{a}-3g)S = -4\pi G\rho(\frac{\kappa}{3})\ HS. \tag{5.45}$$

The definition of L (5.24)

$$L\ r^{-2} = R' - 4\ \pi\ G\ \rho\ H$$

gives, taking condition (5.45) into account,

$$a^{-2}\ L + 4\ \pi\ G\ \rho\ H + \frac{3R}{a} - 5g = 4\ \pi\ G\ \rho\ H,$$

i. e.

$$L = 5\ g\ a^2 - 3\ a\ R \tag{5.46}$$

at the surface.

On the other hand, our definitions

$$\bar{\omega} = \frac{W_n\ a^{n-1}}{g} \qquad \text{and} \qquad W_n + kW_n = R\ \frac{\bar{\omega}}{r^n}$$

give

$$1 + k = R\left(\frac{a^{n-1}}{gr^n}\right) \tag{5.47}$$

which at the surface makes

$$R = (1+k)\ ag; \tag{5.48}$$

this provides an equation of condition which is

$$L = (2 - 3k)\ g\ a^2. \tag{5.49}$$

Using a as a unit of length, g as a unit for the force, ag as the unit of potential and ρ as unit of density, Molodensky writes

$$L = 2 - 3k. \tag{5.50}$$

The general solution is finally of the form

$$\phi = h\frac{\partial\phi}{\partial h} + \ell\frac{\partial\phi}{\partial\ell} + (1+k)\frac{\partial\phi}{\partial k} + (2 - 3k)\frac{\partial\phi}{\partial L}. \tag{5.51}$$

5.6. SOME ADDITIONAL CONSIDERATIONS CONCERNING SURFACE CONDITIONS

At the free surface the conditions N = 0 and M = 0 can be written respectively

$$\left\{(\lambda+2\mu)\frac{dH(r)}{dr} + \lambda\left[\frac{2}{r}H(r) - \frac{n(n+1)}{r^2}T(r)\right]\right\}_{r=a} = 0 \tag{5.52}$$

$$\left(\frac{dT(r)}{dr} - \frac{2}{r}T + H\right)_{r=a} = 0. \tag{5.53}$$

On the other hand, the definitions (3.1), (3.6) and (5.1), (5.4) give

$$\zeta = H_n(r)\frac{1}{g}\left(\frac{W_n}{r^n}\right)a^{n-1} = h_n(r)\frac{W_n}{g}, \tag{5.54}$$

i. e.

$$H_n(r) = \frac{r^n}{a^{n-1}}h_n(r), \tag{5.55}$$

while in the same way

$$T_n(r) = \frac{r^{n+1}}{a^{n-1}}\ell_n(r). \tag{5.56}$$

Therefore

$$\frac{dH(r)}{dr} = n\frac{r^{n-1}}{a^{n-1}}h(r) + \frac{r^n}{a^{n-1}}\frac{dh(r)}{dr}, \tag{5.57}$$

$$\frac{dT(r)}{dr} = (n+1)\frac{r^n}{a^{n-1}}\ell(r) + \frac{r^{n+1}}{a^{n-1}}\frac{d\ell(r)}{dr}, \tag{5.58}$$

which, at the surface r=a, gives

$$H(a) = ah(a), \tag{5.59}$$

$$T(a) = a^2\ell(a), \tag{5.60}$$

$$\left(\frac{dH(r)}{dr}\right)_{r=a} = n\ h(a) + a\left(\frac{dh(r)}{dr}\right)_{r=a}, \qquad (5.61)$$

$$\left(\frac{dT(r)}{dr}\right)_{r=a} = (n+1)\ a\ell(a) + a^2\left(\frac{d\ell(r)}{dr}\right)_{r=a}, \qquad (5.62)$$

Consequently the conditions at the surface become, for n=2:

$$(\lambda + 2\mu)\left[2h + a\left(\frac{dh(r)}{dr}\right)_{r=a}\right] + \lambda\ (2h - 6\ell) = 0, \qquad (5.63)$$

$$a\left(\frac{d\ell(r)}{dr}\right)_{r=a} + \ell + h = 0 \qquad (5.64)$$

(this corresponds to condition [(3.57): $e_{\lambda r} = e_{r\theta} = 0$] and (5.63) can be written

$$ah' = -2(1+m)h + 6m\ell, \qquad (5.65)$$

where

$$m = \frac{\lambda}{\lambda+2\mu} = 1 - 2\ (\frac{\beta}{\alpha})^2, \qquad (5.66)$$

λ, μ are the Lamé elastic moduli, α, β are the velocities of body P and S waves, respectively, and ν Poisson's coefficient:

$$m = \frac{\nu}{1-\nu},$$

then the first equ. (3.48) becomes (for n=2)

$$\eta = -\ mh + 6m\ell. \qquad (5.67)$$

When

$\lambda = \mu$ or $\nu = 0.25$, $m = 1/3$

one has simply

$$ah' = -\ \frac{8h-6\ell}{3}. \qquad (5.68)$$

Using "mean crustal values" $\alpha = 6.1$ km s^{-1}, $\beta = 3.54$ km s^{-1} (i. e. m=0.3264) we get for a classical Bullen B Earth model:

for semi-diurnal tides:	h=0.6199	ℓ=0.0880	
for diurnal tides:	h=0.6173	ℓ=0.0878	(5.69)
from observed data:	h=0.638	(ℓ=0.088)	

This yields:

for semi-diurnal tides:	ah'=-1.472	η=-0.232	aℓ'=-0.7079	
for diurnal tides:	ah'=-1.466	η=-0.231	aℓ'=-0.7051	(5.70)
from observed data:	ah'=-1.520	η=-0.244	aℓ'=-0.726	

Equation (5.67) gives from observed data, ah' = -1.525.
Molodensky (1953) found for several models

$$-0.287 > \eta > -0.311. \qquad (5.71)$$

Hence $-1.563 > ah' > -1.587.$ (5.72)

Takeuchi (1951) puts the solution into the form (4.98):

$$\zeta = u_r = \frac{2F(r)+r^2G(r)}{r} W_2 = h(r)\,\frac{W_2}{g},$$

and for the numerical computations

$$\xi = \frac{r}{a}.$$

Hence from

$$\left.\begin{aligned} F(r) &= a^2\,F(\xi), & G(r) &= G(\xi), \\ \frac{d\,F(r)}{dr} &= a\,\frac{d\,F(\xi)}{d\xi}, & \frac{d\,G(r)}{dr} &= \frac{1}{a}\,\frac{d\,G(\xi)}{d\xi}, \end{aligned}\right\} \qquad (5.73)$$

it follows that

$$\left.\begin{aligned} a\,\frac{dh(r)}{dr} &= \frac{ag}{r^2}\left[2r\,\frac{dF(r)}{dr} + 2r^2G(r) + r^3\,\frac{dG(r)}{dr} - 2F(r) - r^2G(r)\right] \\ &= \frac{ag}{r^2}\left[2ra\,\frac{dF(\xi)}{d\xi} + r^2G(\xi) + \frac{r^3}{a}\,\frac{dG(\xi)}{d\xi} - 2a^2F(\xi)\right] \end{aligned}\right\} (5.74)$$

and at the surface

$$a\left(\frac{dh(r)}{dr}\right)_{r=a} = ag\left[2\,\frac{dF(\xi)}{d\xi} + G(\xi) + \frac{dG(\xi)}{d\xi} - 2F(\xi)\right]_{(\xi=1)}. \qquad (5.75)$$

Numerical integration by Takeuchi yielded at the surface (ξ=1):

$$F(\xi)_1 = \frac{0.044557}{4\pi Ga^2}, \qquad G(\xi)_1 = \frac{0.24159}{4\pi Ga^2}, \qquad (5.76)$$

$$\left(\frac{dF(\xi)}{d\xi}\right)_1 = -\frac{0.33069}{4\pi Ga^2}, \qquad \left(\frac{dG(\xi)}{d\xi}\right)_1 = -\frac{0.29500}{4\pi Ga^2}. \qquad (5.77)$$

As $g = 4\pi G\bar{\rho}a/3,$

$\bar{\rho}$ being the mean density of the Earth equal to 5.51, one concludes:

$$ah' = -1.477, \qquad \eta = -0.25. \qquad (5.78)$$

Conditions at the Core-Mantle boundary (r=b)

The functions H, N, M, R, L have to be continuous at this boundary. On the contrary T, the tangential component of the displacement, need not to be continuous across a solid-liquid boundary, while because of the vanishing of shear stresses within the liquid core, the continuity of stress requires the vanishing of shear components on the solid side of the boundary, that is at the bottom of the mantle.

5.7. THE PROCEDURE FOR THE NUMERICAL INTEGRATION OF THE EQUATIONS-RESULTS

In general very little or no information is given by the different authors to explain the procedure they have followed, particularly for starting the integration. We follow here the explanations given by Takeuchi and Saito (Methods in computational Physics, vol 11, pp 217-302, 1972).

The most important point in the integration is that it should be carried out upward from below. If the equations were integrated downward from the surface, the error would grow exponentially with depth. However one cannot actually start integrations from $r = 0$ in the case of a sphere because the differential equations become singular at $r = 0$. Therefore one assumes the material inside a sphere $r = r_1$ to be homogeneous and isotropic. Such an assumption is permitted because r_1 may be taken small enough so that the properties inside this sphere do not affect the result. The solution in an homogeneous sphere is

$$H(r) = J_n(k_\beta r) \tag{5.79}$$

$$N(r) = \frac{\mu}{r}\{(n-1)\, J_n(k_\beta r) - (k_\beta r)\, J_{n+1}(k_\beta r)\} \tag{5.80}$$

with $$k_\beta^2 = \omega_i^2\, \rho/\mu \tag{5.81}$$

J_n being the spherical Bessel function of the first kind which may be evaluated at $r = r_1$ by using power series solution which provide the initial values to start the integration from there. The integration is continued up to the free surface $r = a$ where the boundary conditions $N = M = 0$ have to be met. The method of Runge - Kutta - Gill is generally adopted.

Various authors have performed the integration of the system of equations for different models of the Earth. Molodensky (1953) restricted his computations to the order $n = 2$ (Table 5.1). Longman, Kaula, Takeuchi Kobayashi and Saito, Farrell, have given evaluations of h, k, ℓ up to the order $n = 29$ (Tables 5.2 and 5.3).

Alsop and Kuo (1964), by integration of the equations with many different models, showed that second-order Love numbers are practically independent (a) of the presence of the solid inner core, and (b) of the presence or absence of a low-velocity channel in the upper mantle.

They also showed that the continental or oceanic crustal model (change of properties in the upper 50 km) is only of second order with respect to the effects of the oceans. They found for a zero rigidity of the core:

Model	h	k	ℓ	k/h	γ	δ	ℓ/h
Gutenberg Bullen	0.619	0.305	0.088	0.493	0.686	1.162	0.142
Jeffreys Bullen B	0.618	0.304	0.088	0.492	0.686	1.162	0.142
Jeffreys Bullen A	0.607	0.299	0.081	0.493	0.692	1.158	0.133
Gutenberg Bullen A	0.608	0.300	0.082	0.493	0.692	1.158	0.135

Moreover, they solved the equations for ten different values of the core rigidity, starting with zero and going to 10^{13}. The results obtained for γ and δ put a limit between 10^{10} and 5×10^{10} c.g.s., results obtained from 5×10^{11} and over being surely unacceptable.

TABLE 5.1. Second-order Love numbers obtained with Molodensky's Models (1953)

Model	μ(core) 10^{12}	k	h	ℓ	γ	δ	k - h/2	ℓ/h
1	0	0.327	0.662	0.107	0.665	1.172	-0.004	0.162
2	0.0367	320	649	106	671	1.217	-0.004	0.163
3	0.3675	276	559	103	717	1.145	-0.004	0.184
4	1.04	221	446	100	775	1.116	-0.002	0.224
5	∞	069	136	089	933	1.033	+0.001	0.654
6	0	310	619	091	691	1.159	0	0.147
7	0.073	298	596	091	702	1.149	0	0.153
8	0.73	229	458	090	771	1.115	0	0.197
9	1.4	191	382	089	809	1.096	0	0.233
10	∞	060	124	086	936	1.034	-0.002	0.694
11	1.4	192	386	088	806	1.098	-0.001	0.228
12	0	306	617	090	689	1.158	-0.002	0.146
13	0	344	683	096	661	1.167	+0.002	0.141
14	0	314	622	095	692	1.151	+0.003	0.153
15	0	315	628	088	687	1.156	+0.001	0.140
16	0	287	567	074	720	1.137	+0.003	0.131

Characteristics of Molodensky's Models (1953)

Model	ρ crust	ρ core	μ core	λ core
1	4.199	12.128	0	
2	4.199	12.128	0.037 X 10^{12}	"
3	4.199	12.128	0.368 X 10^{12}	"
4	4.199	12.128	1.04 X 10^{12}	"
5	4.199	12.128	∞	"
6	From 3.344 to 5.657	12.128	0	
7	From 3.344 to 5.657	12.128	0.073	"
8	From 3.344 to 5.657	12.128	0.73	"
9	From 3.344 to 5.657	12.128	0.73	"
10	From 3.344 to 5.657	12.128	∞	"
11	= Mod. 9 but $\rho = \rho(r)$ to 0.30 $r < 0.30: \rho$ = cte		1.4 X 10^{12}	$\lambda = \lambda(r)$ $r < 0.30: \lambda = \infty$
12	= Mod. 6 but heterogeneous core			
13	= Mod.12 with core radius = 0.575			
14	= Mod.12 but with flattening 1/293			
15	= Mod.14 with a change in the crustal thickness of 1/80			
16	= Legendre's law of density, fluid core (theoretical boundary case)			

In 1963 Molodensky has given

$h_2 = 0.6168, k_2 = 0.3015, \ell_2 = 0.0808, \gamma_2 = 0.6847, \delta_2 = 1.1646, k_2/h_2 = 0.4888$

while Saito (1974) obtained

$h_2 = 0.6085, k_2 = 0.3003 \qquad \gamma_2 = 0.6918, \delta_2 = 1.1581, k_2/h_2 = 0.4935$

$h_3 = 0.2915, k_3 = 0.0931 \qquad \gamma_3 = 0.8017, \delta_3 = 1.0702, k_3/h_3 = 0.3194$

TABLE 5.2. Love numbers for orders 2 to 29

n	h_L	h_K	h_T	k_L	k_K	k_T	ℓ_L	ℓ_K	ℓ_T
2	0.612	0.624	0.592	0.302	0.317	0.280	0.083	0.085	0.076
3	0.290	0.293	0.274	0.093	0.095	0.083	0.014	0.014	0.010
4	0.175	0.177	0.161	0.042	0.042	0.035	0.010	0.010	0.007
5	0.129	0.130	0.116	0.025	0.025	0.020	0.008	0.008	0.006
6	0.107	0.107	0.094	0.017	0.017	0.013	0.007	0.007	0.005
7	0.095	0.095	0.081	0.013	0.013	0.009	0.005	0.005	0.004
8	0.087	0.087	0.073	0.010	0.010	0.007	0.004	0.004	0.003
9	0.081	0.081	—	0.008	0.008	—	0.004	0.004	—
10	0.076	0.076	0.063	0.007	0.007	0.005	0.003	0.003	0.002
11	0.072	—	—	0.006	—	—	0.002	—	—
12	0.069	0.069	0.055	0.005	0.005	0.003	0.002	0.002	0.002
13	0.066	—	—	0.005	—	—	0.002	—	—
14	0.064	0.064	0.051	0.004	0.004	0.003	0.001	0.001	0.001
15	0.062	—	—	0.004	—	—	0.001	—	—
16	0.060	0.060	0.048	0.003	0.003	0.002	0.001	0.001	0.001
17	0.058	—	—	0.003	—	—	0.001	—	—
18	0.056	—	—	0.003	—	—	0.001	—	—
19	0.055	—	—	0.003	—	—	0.001	—	—
20	0.053	—	—	0.002	—	—	0.001	—	—
21	0.052	—	—	0.002	—	—	0.001	—	—
22	0.051	—	—	0.002	—	—	0.000	—	—
23	0.050	—	—	0.002	—	—	0.000	—	—
24	0.048	—	—	0.002	—	—	0.000	—	—
25	0.047	—	—	0.002	—	—	0.000	—	—
26	—	—	—	—	—	—	—	—	—
27	—	—	—	—	—	—	—	—	—
28	—	—	—	—	—	—	—	—	—
29	—	0.043	—	—	0.001	—	—	0.000	—

L = Longman, K = Kaula. T = Takeuchi et al.

Kakuta (1970) has obtained

$h_2 = 0.613, \; k_2 = 0.284, \; \ell_2 = 0.082$ thus

$\gamma_2 = 0.671, \; \delta_2 = 1.187, \; k_2/h_2 = 0.463, \; \ell_2/h_2 = 0.134.$

TABLE 5.3. Love numbers for a stress-free surface according to Farrell

n	Model	h	k	ℓ	k/h	γ	δ	ℓ/h
2	1	0.6114	0.3040	0.0832	0.4972	0.6926	1.1554	0.136
	2	0.6149	0.3055	0.0840	0.4968	0.6906	1.1556	0.137
	3	0.6169	0.3062	0.0842	0.4964	0.6893	1.1576	0.136
3	1	0.2891	0.0942	0.0145	0.3258	0.8051	1.0671	0.050
	2	0.2913	0.0943	0.0145	0.3237	0.8030	1.0685	0.050
	3	0.2923	0.0946	0.0147	0.3236	0.8023	1.0688	0.050
4	1	0.1749	0.0429	0.0103	0.2453	0.8680	1.0338	0.059
	2	0.1761	0.0424	0.0103	0.2408	0.8663	1.0351	0.058
	3	0.1771	0.0427	0.0104	0.2411	0.8656	1.0352	0.059

1, Gutenberg Bullen model. 2, Oceanic mantle. 3, shield mantle.

TABLE 5.4. Love numbers according to Pekeris and Accad

α - model

T/hour		h	k	ℓ	γ	δ	ℓ/h
2	—	0.8766	0.4463	0.1320			
4	—	0.7296	0.3728	0.1182			
6	—	0.7076	0.3619	0.1161			
12	—	0.6948	0.3556	0.1149	0.6543	1.1704	0.164
18	—	0.6924	0.3544	0.1147	0.6608	1.1648	0.165
24	—	0.6915	0.3540	0.1146	0.6620	1.1614	0.166
∞	$h^{(0)}$, $k^{(0)}$, $\ell^{(0)}$	0.6903	0.3535	0.1145	0.6625	1.1608	0.166
	$h^{(2)}$, $k^{(2)}$, $\ell^{(2)}$	2.403×10^5	1.014×10^5	1.65×10^4			
	static	0.5844	0.2983	0.1118	0.7139	1.1370	0.191

Polytropic models

T	h (β = -0.2)	h (β = 0.0)	h (β = 0.2)	k (β = -0.2)	k (β = 0.0)	k (β = 0.2)	ℓ (β = -0.2)	ℓ (β = 0.0)	ℓ (β = 0.2)
6	0.6243	0.6253	0.6256	0.3068	0.3076	0.3084	0.0847	0.0847	0.0848
12	0.6162	0.6152	0.6154	0.3022	0.3028	0.3035	0.0840	0.0843	0.0843
24	0.6120	0.6127	0.6128	0.3009	0.3016	0.3023	0.0841	0.0842	0.0842
∞	0.6118	0.6119	0.6119	0.3005	0.3012	0.3019	0.0840	0.0841	0.0842
	γ			δ			ℓ/h		
6	0.6825	0.6823	0.6828	1.1641	1.1639	1.1630	0.1357	0.1355	0.1355
12	0.6860	0.6876	0.6881	1.1629	1.1610	1.1602	0.1363	0.1370	0.1370
24	0.6889	0.6889	0.6895	1.1607	1.1603	1.1594	0.1374	0.1374	0.1374
∞	0.6887	0.6893	0.6900	1.1611	1.1601	1.1591	0.1372	0.1374	0.1376

As k and h are both decreasing when the rigidity in the core is increasing, the observed coefficients δ and γ vary slowly, and a limit for μ is difficult to determine with precision.

It is interesting to point out in this respect that the Love number ℓ is totally insensitive to that effect. Therefore the ratio ℓ/h easily obtained from extensometers (Chapter 10) could give a more efficient check for core rigidity.

The Alsop-Kuo programme involves integration of the six variables in steps throughout the entire sphere, so that the values of the particle motions may be saved for each of the three linearly independent solutions and then appropriately combined to obtain the actual particle motion of the forced oscillation as a function of depth.

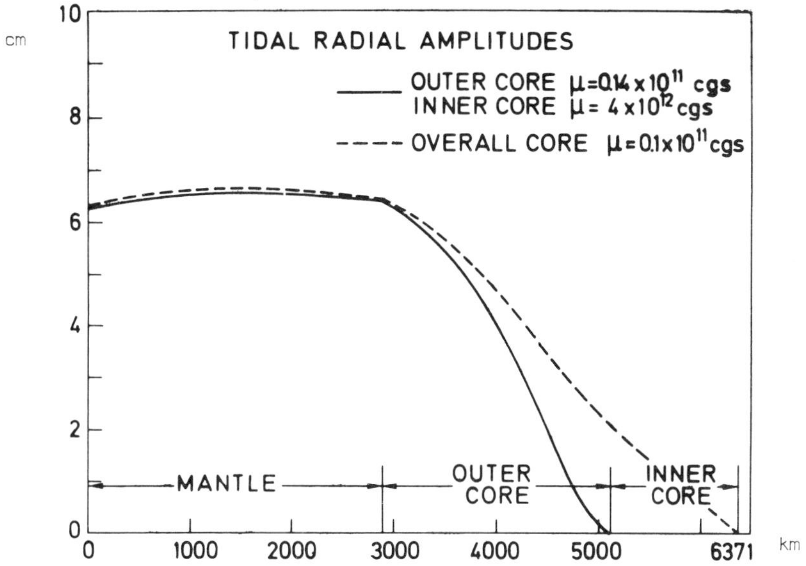

Fig.5.1. Comparison of relative radial amplitudes of semi-diurnal tide between models with and without a solid inner core; the liquid part of the core in both models is assumed to be slightly rigid; μ equals about 1×10^{11} c.g.s. (Alsop and Kuo, 1964).

TABLE 5.5. Correspondence of the notations used by different authors

Love Takeuchi	Longman	Alterman Jarosch Pekeris	Molodensky and this book
h_n	U	y_1	H
$F_n = \ell_n$	V	y_3	T
$K_n = 1+k_n$	P	y_5	R
		y_2	N
		y_4	M
		y_6	L
f_n	X		δ, θ

Figure 5.1 presents the radial particle motion of the forced oscillations for two different values of the rigidity of the outer core: an overall core rigidity of 0.1×10^{11} c.g.s. units is compared with that of a model with an outer core rigidity of 0.14×10^{11} c.g.s. units and an inner core rigidity of 4.0×10^{12} c.g.s. units. This shows that the particle motion in the mantle and in the top part of the outer core are similar. However, the motion in the bottom part of the outer core and in the inner core are dissimilar. The motion falls rapidly to zero in the bottom of the outer core and is completely excluded from the inner core in the model with a solid inner core. The net result is that the presence of the inner core has virtually no effect on characteristic numbers.

The effect of the ocean upon the observed Earth tides is complicated. As a first approximation to the effect of the ocean, Alsop and Kuo replace the top 5 km of the model Bullen B (modified with oceanic crust) with 5 km of ocean water (compressional velocity 1.52 km s^{-1}; shear velocity 0.0 km s^{-1}; density 1.03 g cm^{-3}).

Replacing a continental crust by an oceanic crust raises the values of h, k and ℓ slightly, but since all are raised by approximately the same ratio, the factors γ and δ do not differ appreciably between the two models. The effect of adding an ocean layer is more striking. The value of h is larger than for the model with the oceanic crust, but the value of k is much less even than for the model with the continental crust. Therefore there is a large effect upon the values of γ and δ (γ = 0.647 and δ = 1.214 for the core rigidity 0.0367×10^{11}), but the authors stressed that the values obtained, which assumes an idealized ocean, can give only an approximation to the effect of the ocean on Love numbers. Moreover these values are for the top of the crust and not for the top of the ocean.

TABLE 5.6. Love numbers according to Bodri (1975)

	n	h	k	ℓ	γ	δ	k/h	ℓ/h
B1	2	0.6341	0.3219	0.0886	0.688	1.151	0.508	0.140
	3	0.2993	0.0972	0.0152	0.798	1.070	0.325	0.051
B2	2	0.6335	0.3228	0.0882	0.689	1.149	0.510	0.139
	3	0.2983	0.0970	0.0150	0.799	1.070	0.325	0.050

5.8. THE PROBLEM RAISED BY THE ADAMS WILLIAMSON CONDITION IN A LIQUID CORE

In a liquid core the system of equations is reduced to a system of the fourth order because

$$\left.\begin{aligned} M &= 0 \\ \rho R &= \rho g H - N \end{aligned}\right\} \qquad (5.82)$$

This last equation being equation (5.27) when μ = o, which implies that

$$N = \lambda f.$$

Then the fundamental equations become, reintroducing the inertial terms

$$(5.12): \rho R - \rho g H + \lambda f + \omega_i^2 \rho T r = 0 \qquad (5.83)$$

$$(5.13): \rho \dot{R} + \rho g f - \rho \frac{d}{dr}(gH) + \frac{d}{dr}(\lambda f) + \omega_i^2 \rho H = 0. \qquad (5.84)$$

Deriving (5.83) with respect to r and subtracting (5.84) from this derivation we obtain

$$\omega_i^2 \left[\frac{d}{dr}(\rho T r) - \rho H\right] + \dot{\rho}(R - gH) - g\rho f = 0$$

and by taking (R-gH) again from (5.83)

$$(g + \lambda\dot{\rho}/\rho^2) f = \omega_i^2 (T + r\dot{T} - H) \qquad (5.85)$$

In the static case ($\omega_i = 0$) this requires either that f = 0 (that is incompressibility) or that

$$g + \lambda\dot{\rho}/\rho^2 = 0 \qquad (5.86)$$

Which is the Adams-Williamson relation. This relation must be satisfied or the core must simultaneously be homogeneous ($\dot{\rho} = 0$) and incompressible ($\lambda = \infty$). Now if we were to have f = 0 throughout the core we should also have

$$R = gH \qquad (5.87)$$

because of (5.83) when $\omega_i = 0$, and because of this additional algebraic relation between two of the variables of the problem, the number of arbitrary constants in the core would be reduced to one, which is insufficient to satisfy the boundary conditions (Longman 1963, Jeffreys-Vicente 1966).

The complexity of the problem at the core-mantle boundary results from this difference in the order of the differential equations which govern a solid and those which govern a fluid.

The tangential component of the displacement T need not to be continuous accross a solid-liquid boundary while, because the fluid is unable to support shear stresses, the continuity of stress requires the vanishing of shear components on the solid side of the boundary, that is at the bottom of the mantle.

The functions H, N, M, R, L have to be continuous at this boundary.

To avoid this difficulty it has been customary in tidal theories to assume that the relation of Adams Williamson is satisfied inside the core. Longman considered it as being implicit in the assumptions of elastic equilibrium under gravity.

Takeuchi (1950) even seems to have considered the Adams Williamson relation as a property of the static equilibrium of a fluid. But as he utilized for his numerical calculations a Roche quadratic law of density based upon the Bullen 1936 model ($\rho = 12.284\,(1 - 0.64014\,(r/a)^2)$) which was obtained on the basis of this Adams Williamson equation, there is no contradiction in his developments.

The density normally depends on the pressure, the temperature and an indefinite number of parameters describing the chemical composition. If one assumes that it only depends on the pressure one can write:

$$\frac{1}{k}\frac{dp_0}{dr} = -\frac{1}{v}\frac{dv}{dr} = \frac{1}{\rho_o}\frac{d\rho_o}{dr} \qquad (5.88)$$

where

$$k = \lambda + \frac{2}{3}\mu$$

and k = λ in the case of a liquid.

If the density stratification does not obey this distribution Pekeris and Accad (1972) define a function $\beta(r)$ by putting

$$\frac{dp_o}{dr} + \beta(r)\frac{dp_o}{dr} = \frac{\lambda}{\rho_o}\frac{d\rho_o}{dr} \qquad (5.89)$$

and as the initial configuration is in hydrostatic equilibrium ($dp_0/dr = -\rho_0 g_0$):

$$\beta(r) g_o = g_o + \lambda\dot{\rho}_o/\rho_o^2 \qquad (5.90)$$

When an infinitesimal fluid element is displaced radially, its density decreases because of the adiabatic expansion by an amount $\rho_0(1/\lambda)(dp_0/dr)dr$ while the neighbouring elements in its new position have a density $\rho_o - (d\rho_o/dr)dr$.

The restoring force applied to this element is proportional to the excess density or to

$$\left(\frac{\lambda}{\rho_o}\frac{d\rho_o}{dr} - \frac{dp_o}{dr}\right)$$

that is to $\beta(r)$ which, as defined by (5.89), is thus a dimensionless stability parameter. When $\beta(r) = 0$, the stratification is in neutral equilibrium; it is stable if $\beta(r) < 0$ but unstable when $\beta(r) > 0$.

Note that the restoring force can also be written as:

$$\left(\zeta\frac{d\rho}{dr} - \rho\frac{dp_o}{\lambda}\right)g(r) = \left(\frac{d\rho}{dr} + \rho^2\frac{g}{\lambda}\right)g(r)\,\zeta = -\rho N^2(r)\,\zeta$$

if we define

$$N^2(r) = -g(r)\rho^{-1}(r)\frac{d\rho}{dr} - g^2(r)\rho(r)\lambda^{-1}(r).$$

The equation of motion is then

$$\frac{d^2\zeta}{dt^2} + N^2(r)\zeta = 0$$

$N(r)$ is called the Brunt Väisälä frequency.

The stratification is stable when $N^2 > 0$, unstable when $N^2 < 0$ and neutral if $N^2 = 0$.

Of course

$$\beta(r) = -\frac{\lambda(r)}{g(r)\rho(r)}N^2(r)$$

When $\beta(r) = 0$, the Adams Williamson relation is satisfied which means that the core is chemically homogeneous and has adiabatic temperature gradients. Such conditions are unlikely to be met exactly everywhere in the actual core (see J.A. Jacobs, The Earth's Core, Academic Press 1975).

Moreover when $\beta(r) = 0$, one has the condition

$$H = r\dot{T} + T \qquad (5.91)$$

when $\omega_i \neq 0$. This makes the yielding in the core determinate and the tidal displacement *irrotational* (see (3.91)). For the other density stratifications ($\beta(r) \neq 0$) the vorticity does not vanish.

Let us now observe that the frequency ω_i enters the fundamental equations through a factor ω_i^2 coming from the inertial effect $\rho(\partial^2 u/\partial t^2)$ and that the "static solution" was obtained by dropping ω_i^2 terms. Therefore the complete equations suggest a solution by a perturbation expansion in the small parameter ω_i^2 with the static solution as the leading term.

We should therefore obtain a quadratic variation of $h, k, \ell, \gamma, \delta, \ldots,$ in function of the frequency ω_i of the form

$$\left.\begin{aligned} h(\omega_i) &= h^{(0)} + \omega_i^2\, h^{(2)} + \dots \\ k(\omega_i) &= k^{(0)} + \omega_i^2\, k^{(2)} + \dots \\ &\dots \\ \gamma(\omega_i) &= \gamma^{(0)} + \omega_i^2\, \gamma^{(2)} \end{aligned}\right\} \qquad (5.92)$$

which was proposed by Jeffreys and Vicente.

Pekeris and Accad (1972) have discussed at length this suggestion and their results for a spherically symmetric Earth can be summarized as follows:

"(a) In the case of uniformly unstable as well as uniformly neutral models. the ω_i^2 conjecture applies to Love numbers (Fig. 5.3).

"(b) Uniformly unstable and uniformly neutral models have no free oscillations with periods greater than about 53.7 min, which is the period of the fundamental spheroidal oscillation for n=2.

"(c) Uniformly stable models have an unlimited number of core oscillations with periods ranging from the fundamental of about 53.7 min to ∞, (Fig. 5.4).

"(d) In the case of uniformly stable models, the Love numbers show a nearly linear variation with ω_i^2, if we exclude the regions near the resonances."

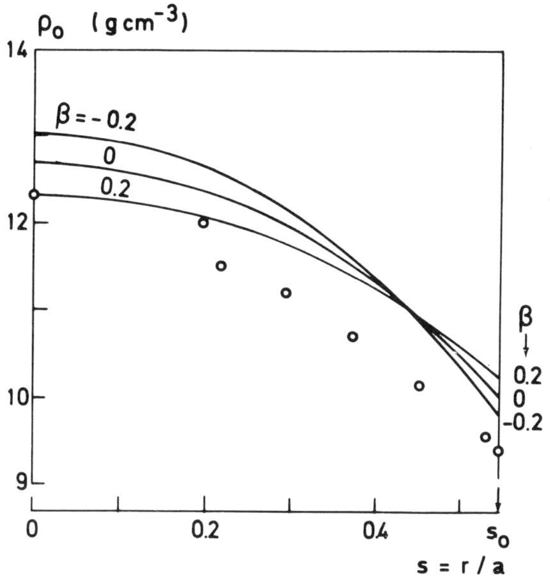

Fig. 5.2. Distribution of density $\rho_0(s)$ in the liquid core for uniform polytropic models β=-0.2, 0, 0.2 Gutenberg model as given by Pekeris and Accad (1972).

It is to be noted that the curves in Fig. 5.3 are close to those in Fig. 5.4 except of course near the resonances.

The free oscillations for periods greater than the fundamental spheroidal mode of about 53.7 min are "core oscillations" (Pekeris, Alterman and Jarosch 1963) in the sense that their amplitude is confined primarly to the interior of the core. This is shown by the plot of the function H(r) given in Fig. 5.5. Pekeris and Accad emphasize the interest of exploring the mechanical effects of the rapid drop in stress within the boundary layer at the top of the core and also to obtain an accurate solution of the passage of an earthquake pulse in the boundary layer.

In conclusion, Pekeris and Accad's resolution of the difficulty with the static limit in the case of a liquid core is as follows:

"1- The dynamics of the liquid core of the Earth do not impose any restriction on its density stratification.

"2- In the case of a uniformly unstable stratification ($\beta>0$) the stress, as well as the divergence of the displacement f, tend to zero with diminishing ω_i throughout the liquid core, except for a boundary layer of diminishing thickness near the core surface. Within the boundary layer stress rises steeply from a near-zero value to a finite value.
(The stress is here the radial one, since in a liquid the transverse components are zero anyhow).
This result implies that in the case $\beta>0$, the condition f=0 is approached through most of the core as $\omega_i \to 0$. The variation of stress within the boundary layer has a dynamic effect, so that, in the static limit, the jump in stress at the core boundary is not as arbitrary as is the discontinuity in tangential displacement T.

"3- In the case of a uniformly stable stratification ($\beta<0$) the stress has a term which oscillates with depth below the core boundary, the depth scale varying like ω_i. The stress and the divergence of displacement do not tend to zero because of the excitation of the free core oscillations.

"4- In all cases, including one of neutral equilibrium, the yielding inside the core in the static limit is determinate and not arbitrary."

Remark

It may be observed from the results of many different numerical integrations that the static values $h^{(0)}$, $k^{(0)}$, ... obtained by putting ω_i = 0 in the equations (5.92) systematically differ from those obtained from the static solution (5.29).

If Pekeris and Accad are right with their second conclusion the $h^{(0)}$, $k^{(0)}$, ... should be the correct values. However Denis (1974) has pointed out that a boundary layer do not exist for a stable stratified core and that in this case $h^{(0)}$, $k^{(0)}$, ... and h(static), k(static) ... should coïncide which is not the case in the Pekeris Accad results.

According to Denis (1974), when the core has a zero rigidity, the Pekeris Accad finite dynamic effect due to a boundary layer of zero width is not valid and the only correct way to calculate the static values of the Love numbers is by using the static equations.

If the core has a finite rigidity even extremely small, there is no more resonance, the lagrangian displacements are perfectly determined and one may extrapolate the dynamical values to obtain the static values.
Wunsch (1974) also notes that for a stratified perfect fluid, the solution for an arbitrarily small, but non-zero frequency, differs qualitatively from the solution when ω_i is exactly zero.

Therefore, according to Wunsch

"no matter how small ω_i becomes, one never obtains the solution that one finds from putting ω_i=0 ab initio in the equations. The limit ω_i=0 is a singular one. If one wants to deal with the limit in this case, one must use the equations of a real fluid of arbitrarily slight viscosity and diffusion. In a perfect fluid, the limit is meaningless. An arbitrarily small rigidity in the fluid can also serve to remove the zero frequency singularity".

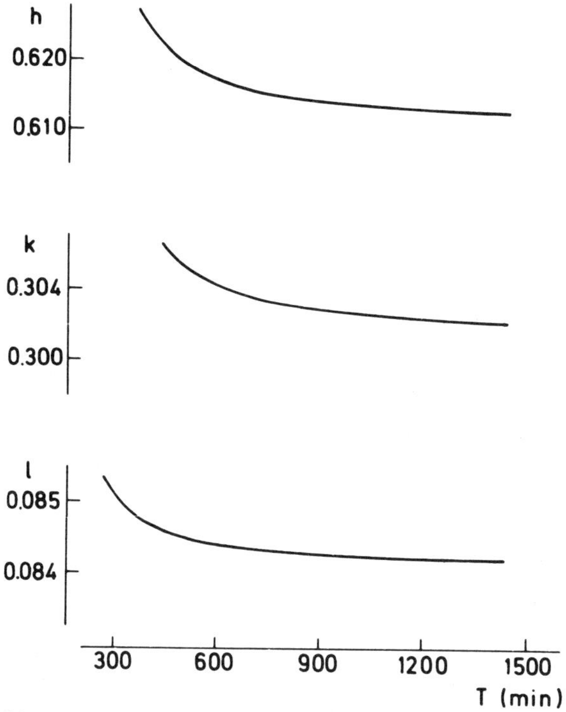

Fig. 5.3. Love numbers h, k, ℓ, as functions of the period T for a uniformly unstable liquid core model with β=0.2 n=2 (Pekeris a

We conclude with another quotation from Wunsch:

" That the difficulties are important can be seen from the fact that they led Longman (1963) to conclude that the core had to obey the Adams-Williamson condition, Smylie and Mansinha to permitting a discontinuity in radial displacement at the core-mantle boundary, and Dahlen (1974) and Chinnery (1974) to conclude that only an Eulerian and not a Lagrangian description of the core motions could be formulated.
One's intuition rebels at all of these conclusions".

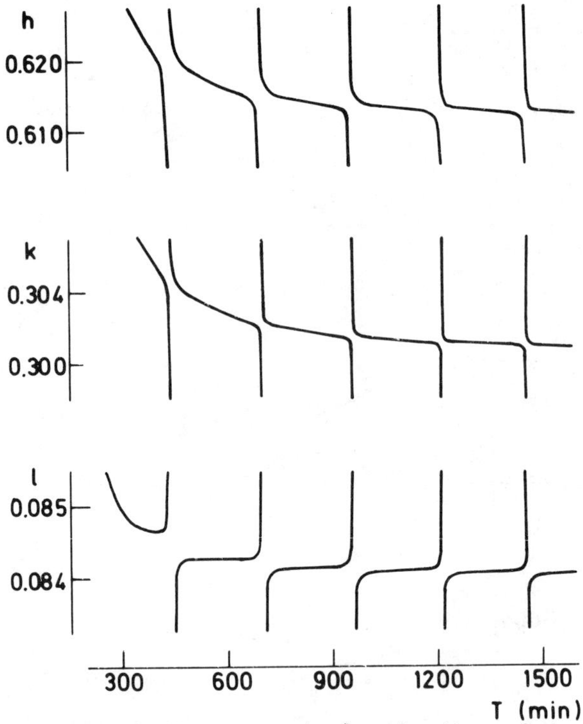

Fig. 5.4. Love numbers h, k, ℓ as functions of the period T for a uniformly stable liquid core model with β=0.2 n=2 as obtained by Pekeris and Accad (resonances are separated by about 4 h).

5.9. AN ATTEMPT OF MOLODENSKY TO ESTIMATE THE EFFECT OF VISCOSITY ON THE EARTH TIDE DEFORMATION

Molodensky (1963) also introduced the viscosity effect in this problem by the method of the variations of constants when the solution is known for an ideally elastic globe.

The modulus of rigidity (μ) and the six functions Φ_i (H, T, R, L, M, N) are replaced by complex expressions like

$$\left.\begin{aligned} \bar{\mu} &= \mu + im \\ \bar{\Phi}_i &= \Phi_i + i\,\Psi_i \end{aligned}\right\} \tag{5.86}$$

As the six differential equations are linear in μ and ϕ and homogeneous with respect to ϕ_i one obtains a new system

$$\left.\begin{aligned} \Phi'_k + \sum_{i=1}^{6} (A_{ik} + B_{ik}\mu)\,\Phi_i &= m_i \sum_{i=1}^{6} B_{ik}\,\Psi_i \\ \Psi'_k + \sum_{i=1}^{6} (A_{ik} + B_{ik}\,\mu)\,\Psi_i &= -m \sum_{i=1}^{6} B_{ik}\,\Phi_i \end{aligned}\right\} \tag{5.87}$$

(k=1, ... 6).

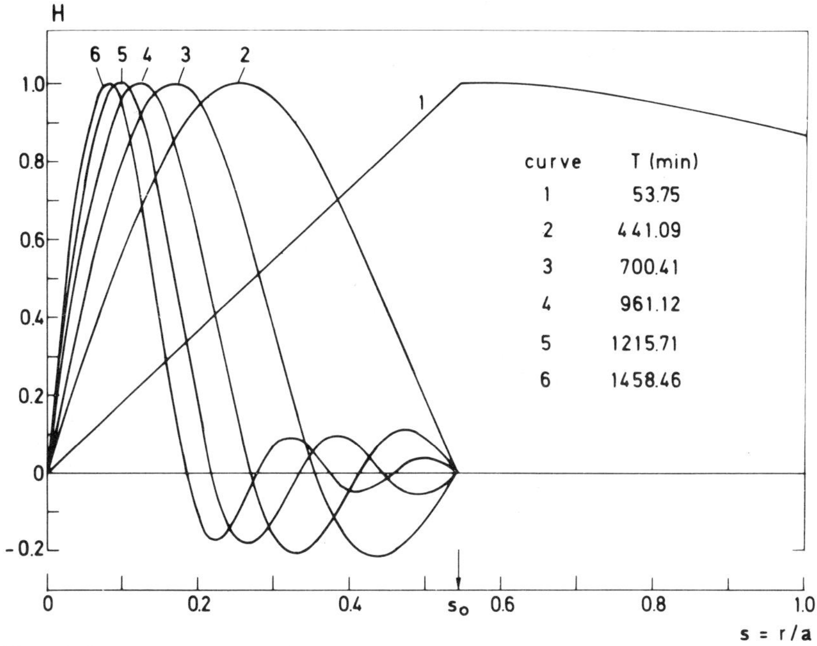

Fig. 5.5. The radial displacement H calculated by Pekeris and Accad (1972) for free spheroidal oscillations for n=2 in a uniformly stable liquid core model with β=-0.2 n=2

The solution of the homogeneous equations (ideally elastic globe) are used to calculate the second members.

Then some terms linear in m are added in the second members of the equ. (5.20) (5.22) (5.27) and (5.28) respectively

$$\left.\begin{aligned} m\chi_1 &= \frac{4m}{3}\left[\frac{N}{\lambda+2\mu} - \frac{3\lambda+2\mu}{\lambda+2\mu}\left(\frac{H}{r} - \frac{3}{r^2}T\right)\right] \\ m\chi_2 &= \left(\frac{M}{\mu}\right)_o \\ m\chi_5 &= \frac{3}{2}r^2\chi_1 + 4T_o \\ m\chi_6 &= -3\chi_1 \end{aligned}\right\} \qquad (5.88)$$

To determine the constants of integration one takes complex boundary conditions by introducing the complex value of the modulus of rigidity in the ordinary boundary conditions.

In this way one obtains complex Love numbers and consequently the phase lag of the tidal deformations. Molodensky obtains a phase lag of 4 minutes of time that is 1° for the diurnal waves and 2° for the semi-diurnals which is in quite good agreement with the most recent experimental results as given by eq. (3.133).

CHAPTER 6

The Liquid-Core Dynamic Theory

The existence of an earth liquid core with a radius of about 3480 km and this is well established by seismology, has important consequence for the behaviour of the Earth when tesseral tidal forces are applied to it.

The coupling mechanism between core and mantle is the key of the problem of some Earth rotation anomalies which depend from internal processes. Three mechanisms of coupling are proposed: inertial and topographic, viscous, electromagnetic. The only information we presently have to investigate the inertial and topographic couplings is from geodetic origin.

It is clear that if the core mantle boundary was perfectly spherical and the core material inviscid, no movement could be transmitted from the mantle to the core and vice versa. The two media could rotate independently. However the Clairaut-Radau theory allows us to calculate, by numerical integration, the ellipticities of the equipotential surfaces inside the Earth on the condition that the outside surface flattening as well as the density distribution are known. One finds in this way a flattening 1/392.15 (Denis, 1979) for the core-mantle interface a determination which rests on the hypothesis that the Earth is in hydrostatic equilibrium - which is not strictly true.

Such a flattening corresponds to a difference of 9 Km between the equatorial and polar radius of the core, a difference which the seismological technique has not yet allowed to measure. Evidently if a fluid is contained inside an elliptical cavity the motion of the mantle may be transmitted to it by pressure effects. This is what we call inertial coupling.

The hydrostatic equilibrium figure however is only a first approximation as has been established by the experimental description of the Earth's gravity field through an infinite development in spherical harmonics. This indeed contains large odd zonal terms and significative tesseral terms which demonstrate the non-hydrostatic state of matter inside the Earth. Hide suggested that the low order terms of this development (n=3,4,5) which necessarily correspond to deeply seated density anomalies could be related to some topography of the Core-mantle boundary like bumps of about 1 km or less height and hundreds km wide superposed to its ellipsoïdal figure. This creates what is called topographic coupling between core and mantle and very few is known about it.

It is moreover evident that this fluid which is a mixture of iron and some lighter element (probably sulfur) near to its eutectic point, exhibits some viscosity and we should thus consider the existence of a boundary layer. In this layer there is a strong velocity gradient which allows the adjustment of the internal fluid flow to the movement of the lower mantle elastic boundary.

Backus has shown (Philosophical Transactions, Royal Soc. London A 263, n°1141, pp 239-266, 1968) that on short periods it is reasonable to consider core motions to consist of a thin boundary layer of Ekman-Hartmann type close to the mantle-core boundary and an interior free-stream motion where the viscosity and resistivity are zero. Resistivity and viscosity modify the motion of the free stream in the boundary layer to fit the boundary conditions which the free stream cannot meet. Backus finds that for periods shorter than a century the Ekman layer is probably thinner than 120 km. More recently Gans

(Journal of Geophysical Research 77, p 360, 1972) obtained very low values for the dynamic viscosity in the core ($3.7 < \eta < 18.5$ centipoises) and Hide (Quarterly Journ. R. Meteor. Soc. 96, p 579, 1970) found accordingly that an extreme upper limit for the thickness of the boundary layer is 3km but "in any event very small indeed compared with the radius of the core".

Dissipation takes place inside this boundary layer, the thickness of which is proportional to the square root of the viscosity while *resonance* may happen when the oscillations of the boundary have a period very close to the period of free oscillations of the fluid in its container.

This possibility of resonance has been suspected indeed since the time of the discovery of the polar motion and a number of beautiful mathematical analyses were constructed by Hough, Sludsky and Poincaré. The liquid core effects are therefore often called Poincaré effects. There is a narrow range of frequencies in which core-mantle pressure coupling efficiently connects the core rotation to the mantle rotation by the nearly resonant excitation of the fluid core's tilt-over mode. This range of frequencies falls inside the diurnal part of Earth tide spectrum.

This problem was considered again under the impulse given by Jeffreys and numerous papers by many distinguished authors have been published since the Geophysical Year which lead to the conclusion that such a resonance effect is indeed observable in the form of perturbations of the amplitudes of some Earth tide waves having their period close to the resonance period (23h 56 min) and of the amplitudes of the associated astronomical nutations (cf equ. 2.56).

The existence of a solid inner core with a radius of 1210 km has only a very minor influence.

The stratification inside the liquid core, as described by the Brunt Väisälä frequency (see § 5.8) has a small effect but not totally negligeable as we will see later, which should allow to derive some informations about the core from special tidal measurements in the future with high precision instruments.

6.1. HISTORICAL BACKGROUND

The discovery of the polar motion at the end of the last century and the difficulties encountered to give a simple theoretical interpretation of its observed characteristics led several theoreticians to investigate the possible role of a liquid core in the perturbations of the rotation of the Earth.

The oscillations of a rotating ellipsoïdal rigid shell containing homogeneous fluid were simultaneously investigated by Hough (Philosophical Transactions A, CLXXXVI, 1895) and by Sloudsky (Bull. Soc. Imperiale des Naturalistes, Moscou IX, 1895).

We follow and summarize here the elegant procedure adopted by Poincaré to solve this problem (Bulletin Astronomique, vol. 27, 1910). Poincaré defines a motion as "simple" when the components of the velocity of any particle are linear functions of its coordinates and demonstrates, using Helmholtz's vortex theory, that if a motion is simple at the origin of time it will remain simple on the condition that the liquid completely fills its shell. Then, taking the axis of the ellipsoïdal shell as the axis of reference we can write the equation of the boundary surface as

$$\frac{x^2}{a^2} + \frac{y^2}{b^2} + \frac{z^2}{c^2} = 1 \qquad (6.1)$$

or

$$x'^2 + y'^2 + z'^2 = 1 \quad (6.2)$$

with the classical change of variables

$$x' = \frac{x}{a}, \qquad y' = \frac{y}{b}, \qquad z' = \frac{z}{c} \quad (6.3)$$

Let ($\dot{u}'$ $\dot{v}'$ $\dot{w}'$) be the components of the velocity of a particle inside the cavity (6.2). It is clear that because the cavity is closed, $\dot{u}'$ cannot be a linear function of x' but only of y' and z'. Similarly, $\dot{v}'$ cannot contain y' and $\dot{w}'$ cannot contain z'. Thus the only possible motion is a rotation $\vec{\omega}_1$ with respect to the shell, around an instantaneous axis of components p_1 q_1 r_1:

$$\dot{u}' = q_1 z' - r_1 y', \quad \dot{v}' = r_1 x' - p_1 z', \quad \dot{w}' = p_1 y' - q_1 x' \quad (6.4)$$

or, inside the ellipsoïd (6.1):

$$\dot{u} = \frac{a}{c} q_1 z - \frac{a}{b} r_1 y, \quad \dot{v} = \frac{b}{a} r_1 x - \frac{b}{c} p_1 z, \quad \dot{w} = \frac{c}{b} p_1 y - \frac{c}{a} q_1 x. \quad (6.5)$$

The absolute motion in space will be obtained by adding the instantaneous rotation $\vec{\omega}$ (p, q, r) of the shell itself:

$$\left.\begin{aligned} \dot{u} &= \frac{a}{c} q_1 z - \frac{a}{b} r_1 y + qz - ry, \\ \dot{v} &= \frac{b}{a} r_1 x - \frac{b}{c} p_1 z + rx - pz, \\ \dot{w} &= \frac{c}{b} p_1 y - \frac{c}{a} q_1 x + py - qx. \end{aligned}\right\} \quad (6.6)$$

The kinetic energy of the entire body is

$$2T = \sum m\,(\dot{u}^2 + \dot{v}^2 + \dot{w}^2),$$

i. e.

$$2T = Ap^2 + Bq^2 + Cr^2 + A_1 p_1^2 + B_1 q_1^2 + C_1 r_1^2 + 2Fpp_1 + 2Gqq_1 + 2Hrr_1 \quad (6.7)$$

where

$$A = \sum mz^2 + \sum my^2, \quad B = \sum mx^2 + \sum mz^2, \quad C = \sum my^2 + \sum mx^2, \quad (6.8)$$

are the moments of inertia of the whole system. The products of inertia are null because of the choice of the axes,

$$A_1 = \frac{b^2}{c^2}\sum mz^2 + \frac{c^2}{b^2}\sum my^2, \quad B_1 = \frac{c^2}{a^2}\sum mx^2 + \frac{a^2}{c^2}\sum mz^2, \quad C_1 = \frac{a^2}{b^2}\sum my^2 + \frac{b^2}{a^2}\sum mx^2, \quad (6.9)$$

are the moments of inertia of the core while

$$F = \frac{b}{c}\sum mz^2 + \frac{c}{b}\sum my^2, \quad G = \frac{c}{a}\sum mx^2 + \frac{a}{c}\sum mz^2, \quad H = \frac{a}{b}\sum my^2 + \frac{b}{a}\sum mx^2. \quad (6.10)$$

As the core is homogeneous one has

$$\left.\begin{aligned} A_1 &= \frac{1}{5}\left(\sum m\right)(b^2 + c^2), & F &= \frac{2}{5}\left(\sum m\right) bc,\\ B_1 &= \frac{1}{5}\left(\sum m\right)(c^2 + a^2), & G &= \frac{2}{5}\left(\sum m\right) ca,\\ C_1 &= \frac{1}{5}\left(\sum m\right)(a^2 + b^2), & H &= \frac{2}{5}\left(\sum m\right) ab. \end{aligned}\right\} \tag{6.11}$$

We shall write

$$A = A_1 + A_0, \qquad B = B_1 + B_0 = A_1 + A_0, \qquad C = C_1 + C_0 \tag{6.12}$$

A_0, B_0, C_0 are the moments of inertia of the shell.

The fundamental equation of hydrodynamics

$$\rho \frac{d\mathbf{v}}{dt} = - \mathbf{grad}\ p + \rho\ \mathbf{grad}\ V \tag{6.13}$$

will be written

$$\frac{\partial \mathbf{v}}{\partial t} + (\mathbf{v}\ \mathbf{grad})\mathbf{v} = -\frac{1}{\rho}\mathbf{grad}\ p + \mathbf{grad}\ V \tag{6.14}$$

which, using the formula

$$\frac{1}{2}\mathbf{grad}\ v^2 = \mathbf{v} \wedge \mathbf{rot}\ \mathbf{v} + (\mathbf{v}\ \mathbf{grad})\ \mathbf{v}, \tag{6.15}$$

becomes

$$\frac{\partial \mathbf{v}}{\partial t} - \mathbf{v} \wedge \mathbf{rot}\ \mathbf{v} = -\frac{1}{\rho}\mathbf{grad}\ p - \frac{1}{2}\mathbf{grad}\ v^2 - \mathbf{grad}\ V. \tag{6.16}$$

We put

$$\psi = \frac{p}{\rho} + \frac{1}{2}v^2 + V \tag{6.17}$$

to obtain

$$\frac{\partial \mathbf{v}}{\partial t} - \mathbf{v} \wedge \mathbf{rot}\ \mathbf{v} = - \mathbf{grad}\ \psi \tag{6.18}$$

of which we take the curl

$$\frac{\partial}{\partial t}\mathbf{rot}\ \mathbf{v} = \mathbf{rot}\ (\mathbf{v} \wedge \mathbf{rot}\ \mathbf{v}) \tag{6.19}$$

or

$$\frac{\partial}{\partial t}\mathbf{rot}\ \mathbf{v} + (\mathbf{v}\ \mathbf{grad})\ \mathbf{rot}\ \mathbf{v} = \mathbf{rot}\ \mathbf{v} \times \mathbf{grad}\ \mathbf{v}, \tag{6.20}$$

i. e.

$$\frac{d}{dt}\mathbf{rot}\ \mathbf{v} = \mathbf{rot}\ \mathbf{v} \times \mathbf{grad}\ \mathbf{v}. \tag{6.21}$$

This is Helmholtz's classical equation usually written as follows:

$$\left.\begin{aligned} \frac{D}{Dt}\left(\frac{\xi}{\rho}\right) &= \frac{\xi}{\rho}\frac{\partial u}{\partial x} + \frac{\eta}{\rho}\frac{\partial u}{\partial y} + \frac{\zeta}{\rho}\frac{\partial u}{\partial z},\\ \frac{D}{Dt}\left(\frac{\eta}{\rho}\right) &= \frac{\xi}{\rho}\frac{\partial v}{\partial x} + \frac{\eta}{\rho}\frac{\partial v}{\partial y} + \frac{\zeta}{\rho}\frac{\partial v}{\partial z},\\ \frac{D}{Dt}\left(\frac{\zeta}{\rho}\right) &= \frac{\xi}{\rho}\frac{\partial w}{\partial x} + \frac{\eta}{\rho}\frac{\partial w}{\partial y} + \frac{\zeta}{\rho}\frac{\partial w}{\partial z}. \end{aligned}\right\} \tag{6.22}$$

In the chosen rotating axes Helmholtz's equation is written as follows:

$$\frac{d}{dt}\,\mathbf{rot\ v} + \boldsymbol{\omega} \wedge \mathbf{rot\ v} = \mathbf{rot\ v} \times \mathbf{grad\ v}. \tag{6.23}$$

From (6.6) we derive the components of the curl:

$$\left.\begin{aligned}
\xi &= \frac{\partial \dot{w}}{\partial y} - \frac{\partial \dot{v}}{\partial z} = 2p + \left(\frac{c}{b} + \frac{b}{c}\right) p_1 ,\\
\eta &= \frac{\partial \dot{u}}{\partial z} - \frac{\partial \dot{w}}{\partial x} = 2q + \left(\frac{a}{c} + \frac{c}{a}\right) q_1 ,\\
\zeta &= \frac{\partial \dot{v}}{\partial x} - \frac{\partial \dot{u}}{\partial y} = 2r + \left(\frac{b}{a} + \frac{a}{b}\right) r_1 ,
\end{aligned}\right\} \tag{6.24}$$

and

$$\begin{aligned}
&+ \xi \frac{\partial \dot{u}}{\partial x} = 0,\\
&+ \eta \frac{\partial \dot{u}}{\partial y} = - \eta \frac{a}{b} r_1 - \eta\, r,\\
&+ \zeta \frac{\partial \dot{u}}{\partial z} = + \zeta \frac{a}{c} q_1 + \zeta\, q.
\end{aligned}$$

Equation (6.23) then becomes

$$\left.\begin{aligned}
bc \frac{d\xi}{dt} &= ab\, q_1 \zeta - ca\, r_1 \eta,\\
ca \frac{d\eta}{dt} &= bc\, r_1 \xi - ab\, p_1 \zeta,\\
ab \frac{d\zeta}{dt} &= ca\, p_1 \eta - bc\, q_1 \xi,
\end{aligned}\right\} \tag{6.25}$$

On the other hand, equations (6.24) give

$$\frac{1}{5}\left(\sum m\right) bc\, \xi = \frac{2}{5}\left(\sum m\right) bc\, p + \frac{1}{5}\left(\sum m\right)(b^2 + c^2)\, p_1 = Fp + A_1 p_1 = \frac{\partial T}{\partial p_1}$$

and, accordingly,

$$\frac{1}{5}\left(\sum m\right) ab\, \zeta = \frac{\partial T}{\partial r_1},$$

$$\frac{1}{5}\left(\sum m\right) ca\, \eta = \frac{\partial T}{\partial q_1}.$$

Finally, Helmholtz's equations are:

$$\left.\begin{aligned}
\frac{d}{dt}\frac{\partial T}{\partial p_1} + r_1 \frac{\partial T}{\partial q_1} - q_1 \frac{\partial T}{\partial r_1} &= 0,\\
\frac{d}{dt}\frac{\partial T}{\partial q_1} + p_1 \frac{\partial T}{\partial r_1} - r_1 \frac{\partial T}{\partial p_1} &= 0,\\
\frac{d}{dt}\frac{\partial T}{\partial r_1} + q_1 \frac{\partial T}{\partial p_1} - p_1 \frac{\partial T}{\partial q_1} &= 0,
\end{aligned}\right\} \tag{6.26}$$

while Lagrange's equations describing the rotation of the whole planet are:

$$\left.\begin{aligned}\frac{d}{dt}\frac{\partial T}{\partial p} - r\frac{\partial T}{\partial q} + q\frac{\partial T}{\partial r} &= L,\\ \frac{d}{dt}\frac{\partial T}{\partial q} - p\frac{\partial T}{\partial r} + r\frac{\partial T}{\partial p} &= M,\\ \frac{d}{dt}\frac{\partial T}{\partial r} - q\frac{\partial T}{\partial p} + p\frac{\partial T}{\partial q} &= N.\end{aligned}\right\} \quad (6.27)$$

Both systems of equations are explicitly written as follows:

$$\left.\begin{aligned}\frac{d}{dt}(Fp + A_1 p_1) + r_1 (Gq + B_1 q_1) - q_1 (Hr + C_1 r_1) &= 0,\\ \frac{d}{dt}(Gq + B_1 q_1) + p_1 (Hr + C_1 r_1) - r_1 (Fp + A_1 p_1) &= 0,\\ \frac{d}{dt}(Hr + C_1 r_1) + q_1 (Fp + A_1 p_1) - p_1 (Gq + B_1 q_1) &= 0,\end{aligned}\right\} \quad (6.28)$$

$$\left.\begin{aligned}\frac{d}{dt}(Ap + Fp_1) - r(Bq + Gq_1) + q(Cr + Hr_1) &= L,\\ \frac{d}{dt}(Bq + Gq_1) - p(Cr + Hr_1) + r(Ap + Fp_1) &= M,\\ \frac{d}{dt}(Cr + Hr_1) - q(Ap + Fp_1) + p(Bq + Gq_1) &= N.\end{aligned}\right\} \quad (6.29)$$

Now if we consider the body as a rotating ellipsoid such that

$$a = b, \qquad A = B, \qquad A_1 = B_1, \qquad F = G$$

and

$$H = C_1,$$

then the third equations of both systems give

$$\left.\begin{aligned}\frac{d}{dt}(Cr + C_1 r_1) + F(pq_1 - p_1 q) &= N,\\ \frac{d}{dt}(C_1 r + C_1 r_1) + F(pq_1 - p_1 q) &= 0.\end{aligned}\right\} \quad (6.30)$$

The component N of the torque exerted by the external forces is zero for a non viscous body (see Chapter 14). Then by taking the difference of equations (6.30) one has

$$\frac{dr}{dt} = 0,$$

i. e.

$$r = \text{constant} = \omega$$

and

$$C_1 \frac{dr_1}{dt} + F\,(pq_1 - p_1 q) = 0,$$

where the second term is of the third order with respect to the first one. Then

$$r_1 = \text{constant} = \omega_1 . \tag{6.31}$$

We can suppose that ω_1 is extremely small and can be neglected with respect to ω.

Then the other equations become:

$$\left.\begin{aligned} A\dot{p} + F\dot{p}_1 + (C-A)\,\omega q - F\,\omega q_1 &= L, \\ A\dot{q} + F\dot{q}_1 - (C-A)\,\omega p + F\,\omega p_1 &= M, \\ F\dot{p} + A_1\dot{p}_1 - C_1\,\omega q_1 &= 0, \\ F\dot{q} + A_1\dot{q}_1 + C_1\,\omega p_1 &= 0. \end{aligned}\right\} \tag{6.32}$$

If we put

$$\bar{\omega} = p + iq, \quad \bar{\omega}_1 = p_1 + iq_1, \quad L + iM = \kappa\, e^{-i\omega_i t}, \tag{6.33}$$

where ω_i are the tidal frequencies defined in (2.15) and (2.17), we obtain

$$\left.\begin{aligned} A\,\dot{\bar{\omega}} - i\,(C-A)\,\omega\bar{\omega} + F\,\dot{\bar{\omega}}_1 + i\,F\,\omega\bar{\omega}_1 &= \kappa\, e^{-i\omega_i t}, \\ F\,\dot{\bar{\omega}} + A_1\,\dot{\bar{\omega}}_1 + i\,C_1\,\omega\bar{\omega}_1 &= 0. \end{aligned}\right\} \tag{6.34}$$

The external forces which are susceptible to make the core rotate with respect to the shell are obviously tesseral forces like the diurnal tides and the Chandlerian polar motion.

Therefore in (6.34) we put

$$\bar{\omega} = \bar{\omega}\, e^{-i\omega_i t}, \qquad \bar{\omega}_1 = \bar{\omega}_1\, e^{-i\omega_i t}, \tag{6.35}$$

which gives

$$\left.\begin{aligned} \bar{\omega}\left\{-A\,\omega_i - (C-A)\,\omega\right\} + \bar{\omega}_1\, F\left\{(\omega-\omega_i)\right\} &= \kappa, \\ \bar{\omega}\left\{-F\,\omega_i\right\} + \bar{\omega}_1\left\{-A_i\,\omega_i + C_1\,\omega\right\} &= 0, \end{aligned}\right\} \tag{6.36}$$

and, remembering (2.17),

$$\omega - \omega_i = \Delta\,\omega_i, \tag{6.37}$$

we obtain

$$\left.\begin{aligned} \bar{\omega}_1 &= \frac{F\omega_i}{\Delta(\omega_i)}\,\kappa, \\ \bar{\omega} &= \frac{-A_1\,\omega_i + C_1\omega}{\Delta(\omega_i)}\,\kappa = + \frac{A_1\Delta\omega_i + (C_1-A_1)\omega}{\Delta(\omega_i)}\,\kappa, \end{aligned}\right\} \tag{6.38}$$

$$\Delta(\omega_i) = \begin{vmatrix} A\,\Delta\,\omega_i - C\,\omega & F\,\Delta\,\omega_i \\ F\,\Delta\,\omega_i - F\,\omega & A_1\,\Delta\,\omega_i + (C_1 - A_1)\omega \end{vmatrix} \tag{6.39}$$

i. e.

$$\left.\begin{aligned} \Delta(\omega_i) = -(C_1-A_1)\,C\omega^2 + A(C_1-A_1)\,\omega\Delta\omega_i + AA_1\,\Delta\omega_i^2 - A_1\,C\,\omega\Delta\omega_i + \\ A_1^2\,\omega\Delta\omega_i - A_1^2\,\Delta\omega_i^2 - A_1^2\,\varepsilon_i^2\,\omega\Delta\omega_i + A_1^2\,\varepsilon_1^2\,\Delta\omega_i^2. \end{aligned}\right\} \tag{6.40}$$

Now if the core is homogeneous:

$$\left.\begin{aligned} A_1 &= \frac{1}{5}\,M\,(a^2 + c^2), \\ C_1 &= \frac{2}{5}\,M\,a^2, \\ F &= \frac{2}{5}\,M\,a\,c, \end{aligned}\right\}$$

$$F^2 = 2\,A_1\,C_1 - C_1^2,$$

$$\left.\begin{aligned} \varepsilon_1 &= \frac{C_1-A_1}{A_1} = \frac{C_1^2-C_1A_1}{C_1A_1} = \frac{C_1A_1-F^2}{C_1A_1}, \\ \varepsilon_1^2 &= \frac{C_1^2 + A_1^2 - 2C_1A_1}{A_1^2} = \frac{A_1^2 - F^2}{A_1^2}, \end{aligned}\right\}$$

and neglecting the terms of the order of the square of the flattening we obtain

$$\bar{\omega} = \frac{(\Delta\omega_i + \varepsilon_1\omega)\,\kappa}{-\varepsilon_1 C\omega^2 + A\varepsilon_1\omega\Delta\omega + A\Delta\omega_i^2 - C\omega\Delta\omega_i + A_1\omega\Delta\omega_i - A_1\Delta\omega_i^2}. \tag{6.41}$$

If there was no fluid core ($A_1 \to 0$) we should have

$$\bar{\omega}_0 = \frac{(\Delta\omega_i + \varepsilon_1\,\omega)\,\kappa}{-\varepsilon_1 C\omega^2 - (C-A\varepsilon_1)\,\omega\Delta\omega_i + A\,\Delta\omega_i^2}. \tag{6.42}$$

Therefore we can write

$$\frac{\bar{\omega}}{\bar{\omega}_0} = \left(1 - \frac{A_1\,(\omega\Delta\omega_i - \Delta\omega_i^2)}{(C\omega - A\Delta\omega_i)\,(\varepsilon_1\omega + \Delta\omega_i)}\right)^{-1}, \tag{6.43}$$

a formula which gives the ratio of the amplitude of a nutation of frequency $\Delta\omega_i$ affected by the liquid-core movements with respect to the amplitude of the same nutation for a completely rigid body.

With the numerical values corresponding to the planet Earth,

$$A \sim C \sim 80{,}6 \times 10^{43}\ \text{g cm}^2,$$

$$A_1 \sim C_1 \sim 9.07 \times 10^{43} \text{ g cm}^2,$$

$$A_1/A = 0.112,$$

$$\varepsilon_1 = 0.002\,60 \sim 1/400,$$

and putting

$$\Delta\omega_i = \frac{\omega}{x}$$

(6.43) becomes

$$\bar{\omega} = \bar{\omega}_0 \left(1 - \frac{44.8}{400 + x}\right)^{-1} = a^{-1}\,\bar{\omega}_0. \qquad (6.44)$$

A resonance takes place for

$$x = -400,$$

that is

$$\Delta\omega_i = -\varepsilon_1\omega, \qquad T_i = 24\text{h}\,(1 - \frac{1}{400})$$

This resonance frequency essentially depends from the flattening of the core. The corresponding period should be 23h 56m 24s and appears as a new diurnal nutation. From (6.44) we can easily derive the values given in Table 6.1 for the main nutations.

TABLE 6.1. Liquid-core resonance effects on tesseral tides and nutations according to Poincaré's model

Tide		$\Delta\omega_i$	x	$\bar{\omega}/\bar{\omega}_0$
Precession	K_1	0	∞	1
Principal nutation	165.565	$-\frac{\omega}{6800}$	-6798	0.994
	(165.545)	$+\frac{\omega}{6800}$	+6798	1.007
Annual nutation	ψ_1	$-\frac{\omega}{365}$	-365	3.458
	S_1	$+\frac{\omega}{365}$	+365	1.062
Semi-annual nutation	ϕ_1	$-\frac{\omega}{183}$	-183	1.260
	P_1	$+\frac{\omega}{183}$	+183	1.083
Fortnightly nutation	(001)	$-\frac{\omega}{13.7}$	-13.7	1.131
	O_1	$+\frac{\omega}{13.7}$	+13.7	1.121

Such a correction should reduce the nutation constant calculated for a rigid Earth as 9".2272 to a value of 9".17 while the observed value is 9".20. The correction has indeed the correct sense, but is twice too large, which is due to the fact that we did not take account of the elasticity of the shell.

6.2. MOLODENSKY'S THEORY

6.2.1. Fundamental equation in a rotating system

In geodynamics we cannot use a system Σ_0 of absolute axes fixed with respect to space because the coordinates of the points in the Earth would change so much that very important non-linear terms would appear. To avoid these terms we have thus to introduce another system of axes with its origin at the centre of mass O of the Earth and rotating with respect to the fixed axis with the same speed ω as the Earth itself.

Moreover, to investigate the hydrodynamic effects of a liquid core in the Earth we obviously have to choose a system of axes of reference with respect to which the core-mantle boundary is at rest. This allows the boundary conditions to be expressed more or less easily.

This system Σ is supposed to be rigidly connected to the mantle, and therefore displacements of molecules in the mantle can be considered as small quantities of the first order with respect to the Earth's radius. This system rotates with the mantle at the angular speed ω.

Now the general equation of the motion of a molecule of an elastic deformable body is

$$\mathbf{\Gamma}_0 = \mathbf{F} + \frac{1}{\rho}\,\mathbf{div}\ \overline{\overline{T}} \tag{6.45}$$

where $\mathbf{\Gamma}_0$ is the absolute acceleration of the molecule, $\mathbf{F}$ is the body force acting on the mass unit, and $\mathbf{T}$ is the stress tensor acting on the surface of this molecule.

Our first task is to develop the expression of the acceleration in the system of axes Σ. In the fixed system Σ_0 one has

$$\mathbf{\Gamma}_0 = \frac{\partial \mathbf{v}_0}{\partial t} + (\mathbf{v}_0\ \mathbf{grad})\ \mathbf{v}_0\ , \tag{6.46}$$

while in the rotating system Σ

$$\mathbf{v}_0\ (P) = \mathbf{v}_0\ (O) + \mathbf{v}\ (P) + \boldsymbol{\omega} \wedge \mathbf{OP}\,. \tag{6.47}$$

We will suppose that in the system Σ the product of the speed of displacement by the deformation is negligible because the displacements are very small. If this is acceptable for the mantle, it is not certain if it is also acceptable for the core.

Nevertheless, we then put

$$\widetilde{\left(\frac{d}{dt}\right)_\Sigma} = \widetilde{\left(\frac{\partial}{\partial t}\right)_\Sigma} \quad \text{and} \quad \widetilde{\left(\frac{d}{dt}\right)_{\Sigma_0}} = \widetilde{\left(\frac{\partial}{\partial t}\right)_\Sigma} + \boldsymbol{\omega} \wedge \tag{6.48}$$

($\tilde{\cdot}$ represents a time derivative in the rotating system Σ)
then

$$\mathbf{\Gamma}_0(P) = \mathbf{\Gamma}(P) + \mathbf{\Gamma}_0(O) + 2\boldsymbol{\omega} \wedge \mathbf{v}\ (P) + \tilde{\dot{\boldsymbol{\omega}}} \wedge \mathbf{OP} + \boldsymbol{\omega} \wedge (\boldsymbol{\omega} \wedge \mathbf{OP}), \tag{6.49}$$

where

$2\boldsymbol{\omega} \wedge \mathbf{v}$ (P) is Coriolis's acceleration, $\dot{\tilde{\omega}} \wedge \mathbf{OP}$ is the acceleration of the system and $\boldsymbol{\omega} \wedge (\boldsymbol{\omega} \wedge \mathbf{OP}) = (\boldsymbol{\omega}\cdot\mathbf{r})\,\boldsymbol{\omega} - \omega^2 \mathbf{r}$ is the centrifugal acceleration.

Potential of the Centrifugal Force

In the rotating system Σ the instantaneous vector of rotation can be represented as follows:

$$\boldsymbol{\omega} = \omega(\mathbf{i}\,\varepsilon \cos \omega_i t + \mathbf{j}\ \varepsilon \sin \omega_i t + \mathbf{k}) = \omega\mathbf{k} + \varepsilon\omega\boldsymbol{\omega}^*, \tag{6.50}$$

where $\mathbf{i}, \mathbf{j}, \mathbf{k}$ are the unit vectors on Σ, ω is the undisturbed speed of rotation and $\boldsymbol{\omega}^* = (\mathbf{i} \cos \omega_i t + \mathbf{j}\ \sin \omega_i t)$ is the diurnal nutation associated with the diurnal Earth tide.

This nutation is retrograde and the sign of the ω_i is opposite to the sign of ω. We therefore have to put here

$$\Delta\omega_i = \omega + \omega_i \tag{6.51}$$

which is a small quantity.

The centrifugal acceleration becomes

$$\left.\begin{aligned}\boldsymbol{\omega} \wedge (\boldsymbol{\omega} \wedge \mathbf{OP}) = \omega^2\, \mathbf{k} \wedge (\mathbf{k} \wedge \mathbf{OP}) + \varepsilon\omega^2[\mathbf{k} \wedge (\boldsymbol{\omega}^* \wedge \mathbf{OP}) +\\ \boldsymbol{\omega}^* \wedge (\mathbf{k} \wedge \mathbf{OP})].\end{aligned}\right\} \tag{6.52}$$

As

$$\left.\begin{aligned}&\mathbf{OP} = x\mathbf{i} + y\mathbf{j} + z\mathbf{k},\\ &\ell^2 = x^2 + y^2 = r^2 \cos^2\theta = r^2 \left(\frac{2P_2 + P_0}{3}\right),\end{aligned}\right\} \tag{6.53}$$

one gets

$$\mathbf{k} \wedge (\mathbf{k} \wedge \mathbf{OP}) = -\,\mathbf{grad}\left(\frac{x^2+y^2}{2}\right) = -\,\mathbf{grad}\ \frac{\ell^2}{2} \tag{6.54}$$

$$\left.\begin{aligned}&\mathbf{k} \wedge (\boldsymbol{\omega}^* \wedge \mathbf{OP}) = z\boldsymbol{\omega}^*\\ &\boldsymbol{\omega}^* \wedge (\mathbf{k} \wedge \mathbf{OP}) = \mathbf{k}\ (\boldsymbol{\omega}^*\cdot\mathbf{OP})\end{aligned}\right\} = \mathbf{grad}\ z\ (\boldsymbol{\omega}^*\cdot\mathbf{OP}), \tag{6.55}$$

$$\boldsymbol{\omega}^*\cdot\mathbf{OP} = x \cos \omega_i t + y \sin \omega_i t = \ell \cos(\omega_i t - \lambda), \tag{6.56}$$

and, finally,

$$-\boldsymbol{\omega} \wedge (\boldsymbol{\omega} \wedge \mathbf{OP}) = \mathbf{grad}\left[\frac{\omega^2\ell^2}{2} - \varepsilon\omega^2 z\ell \cos(\omega_i t - \lambda)\right]. \tag{6.57}$$

Potential of the Diurnal Nutation

It is evidently

$$\phi = -\varepsilon\omega^2\, z\,\ell \cos(\omega_i t - \lambda) \tag{6.58}$$

as it results from a perturbation in the centrifugal force

$$\phi = d\left(\frac{\omega^2 r^2}{2}\cos^2\theta\right) = -\omega^2 r^2 \sin\theta\cos\theta\, d\theta \qquad (6.59)$$

with

$$d\theta = -(x_1\cos\lambda + y_1\sin\lambda), \qquad (6.60)$$

where $(x_1 y_1)$ are the pole coordinates, i. e.

$$x_1 = \frac{p}{r} = \varepsilon\cos\omega_i t, \qquad y_1 = \frac{q}{r} = \varepsilon\sin\omega_i t. \qquad (6.61)$$

Acceleration of the System

From (6.50) we write

$$\dot{\tilde{\omega}} = \omega_i\,\omega\,\varepsilon\,(-\mathbf{i}\sin\omega_i t + \mathbf{j}\cos\omega_i t) \qquad (6.62)$$

and

$$\dot{\tilde{\omega}}\wedge\mathbf{r} = \omega_i\,\omega\,\varepsilon\,(-\mathbf{k}\,y\sin\omega_i t - \mathbf{k}\,x\cos\omega_i t + \mathbf{j}\,z\sin\omega_i t + \mathbf{i}z\cos\omega_i t), \qquad (6.63)$$

where the same terms as those in (6.55) and (6.56) appear with a coefficient $\omega_i\ \omega\ \varepsilon$ instead of $\varepsilon\ \omega^2$ as shown by (6.52). Therefore we just have to introduce such terms in the gradient (6.57) as follows:

$$\mathbf{grad}\ \left[\phi_o + \left(1 + \frac{\omega i}{\omega}\,\phi\right]\right. = \mathbf{grad}\ \left[\phi_o + \frac{\Delta\omega i}{\omega}\,\phi\right], \qquad (6.64)$$

where

$$\phi_o = \frac{\omega^2\,\ell^2}{2}, \qquad \phi = -\varepsilon\,\omega^2\,z\,\ell\,\cos(\omega_i t - \lambda), \qquad (6.65)$$

taking care that the projection on z(**k**) is of opposite sign and that we then have to add twice this term in the third equation. (This is because the diurnal nutation potential is tesseral with Oz as a pole). Then (6.49) is written as follows:

$$\left.\begin{aligned}
\frac{d\dot{x}}{dt} &= \ddot{u} - 2\omega\,\dot{v} - \frac{\partial}{\partial x}\left(\phi_o + \frac{\Delta\omega_i}{\omega}\,\phi\right):\ \mathbf{i}\,,\\
\frac{d\dot{y}}{dt} &= \ddot{v} + 2\omega\,\dot{u} - \frac{\partial}{\partial y}\left(\phi_o + \frac{\Delta\omega_i}{\omega}\,\phi\right):\ \mathbf{j}\,,\\
\frac{d\dot{z}}{dt} &= \ddot{w} + \frac{2\omega_i}{\omega}\frac{\partial\phi}{\partial z} - \frac{\partial}{\partial z}\left(\phi_o + \frac{\Delta\omega_i}{\omega}\,\phi\right):\ \mathbf{k}\,,
\end{aligned}\right\} \qquad (6.66)$$

u (u v w) being the displacement components in the rotating system Σ.

The Newtonian gravitational potential being

$$V_o = G\iiint \rho\,\frac{dm}{r} \qquad (6.67)$$

and the tidal tesseral potential being

$$W_2 = D\,P_2^1\,r^2\cos(\omega_i t - \lambda) = Dz\,(x\cos\omega_i t + y\sin\omega_i t) \qquad (6.68)$$

we now can write the fundamental equation (6.45) as follows:

$$\frac{\partial \mathbf{v}}{\partial t} + 2\omega\ (\mathbf{k} \wedge \mathbf{v}) = \mathbf{grad}\ (W_2 + kW_2 + V_0 + \phi_0 + \frac{\Delta\omega_i}{\omega}\ \phi) - \frac{2\omega_i}{\omega}\frac{\partial\phi}{\partial z}\ \mathbf{k} + \frac{1}{\rho}\ \mathbf{div}\ \overline{\overline{T}}, \quad (6.69)$$

taking care that the second Love number k **is** not to be confused with the unit vector **k**.

Calculation of the Stress at the Point P

This is not so immediate as the particle, which was at P_0 and moved to P, has transported the stress initially supported by it at P_0, and is now submitted to an additional stress which results from the deformation.

Let us put

$$\mathbf{u}\ (u\ v\ w) = \mathbf{P_0P} \quad (6.70)$$

and

$$U = \mathbf{u}\ \mathbf{grad}, \quad (6.71)$$

then the initial undisturbed pressure at P is

$$p_0(P) = p_0(P_0) + U\ p_0\ (P_0) \quad (6.72)$$

and if the hydrostatic equilibrium was initially realized

$$\left.\begin{aligned} \mathbf{grad}\ p_0\ (P_0) &= \rho_0\ (P_0)\ \mathbf{grad}\ V\ (P_o), \\ \mathbf{grad}\ p_0\ (P) &= \rho_0\ (P)\ \mathbf{grad}\ V\ (P). \end{aligned}\right\} \quad (6.73)$$

We can write

$$p_0\ (P_0) = p_0\ (P) - \rho_0\eta. \quad (6.74)$$

If we put with Molodensky

$$\eta = \mathbf{u}\ .\ \mathbf{grad}\ V\ (P_0), \quad (6.75)$$

then

$$\mathbf{grad}\ p_0\ (P_0) = \mathbf{grad}\ p_0\ (P) - \rho_0\ \mathbf{grad}\ \eta - \eta\ \mathbf{grad}\ \rho_0, \quad (6.76)$$

η is proportional to the vertical displacement and to the local gravity. It is a small quantity, and in its expression we can therefore neglect the difference between P and P_0 coordinates. Then

$$\frac{\rho'\ (P_0)}{V'\ (P_0)}\ \eta = \frac{\rho'\ (M)}{V'\ (M)}\ \eta, \quad (6.77)$$

which means that according to the order of magnitude of the displacements the equipotential surfaces and the equal density surfaces keep the same normals.

Moreover

$$\mathbf{grad}\ V(P) = \mathbf{grad}\ V(P_0) + \mathbf{grad}\ \eta \quad (6.78)$$

gives, with consideration to (6.73) and (6.76),

$$\mathbf{grad}\ p_0(P_0) = \rho_0(P)\ \mathbf{grad}\ V(P) - (\mathbf{u}.\mathbf{grad}\ \rho_0)\ \mathbf{grad}\ V(P_0) - \rho_0\ \mathbf{grad}\ \eta, \quad (6.79)$$

where we can exchange **grad** ρ_0 and **grad** $V(P_0)$ because the equipotential and equal density surfaces are coïncident.

Calculation of the Variation of Density

We have

$$\rho_0(P) = \rho(P)\ (1 + \theta) + \mathbf{u}\ \mathbf{grad}\ \rho_0,$$

and because of (6.75)

$$\rho_0(P) = \rho(P)\ (1 + \theta) + \frac{\rho'}{V'}\ \eta. \quad (6.80)$$

In the calculation of div $\bar{\bar{T}}$ we have to consider a term

$$\mathbf{div}\ [p(P_o)\ .\ \bar{\bar{E}}],$$

where $\bar{\bar{E}}$ is the unit tensor.
Because of (6.76) we can write

$$\left.\begin{aligned} \mathbf{div}_P\ [p(P_o)\ .\ \bar{\bar{E}}] &= \mathbf{grad}_P\ p\ (P_o) \\ &= \rho_o(P)\ \mathbf{grad}_P\ V(P) - \rho_0\ \mathbf{grad}\ \eta - \eta\,\mathbf{grad}\ \rho_o \\ &= \rho_o(P)\ [1+\theta]\ \mathbf{grad}\ V - \rho\ \mathbf{grad}\ \eta, \end{aligned}\right\} (6.81)$$

neglecting $(\rho-\rho_0)$ when multiplied by a little quantity as **grad** η.

6.2.2. Fundamental Equation for a liquid Core

We put, of course,

$$\mu = 0,$$

then

$$\frac{1}{\rho}\ \mathbf{div}\ \bar{\bar{T}} = \frac{1}{\rho}\ \mathbf{grad}\ \lambda\theta - (1+\theta)\ \mathbf{grad}\ V + \mathbf{grad}\ \eta \quad (6.82)$$

and

$$\frac{1}{\rho}\ \mathbf{grad}\ \lambda\theta = \mathbf{grad}\ \frac{\lambda\theta}{\rho} + \theta\ \mathbf{grad}\ V \quad (6.83)$$

on the condition that the Adams-Williamson relation (with $\mu=0$) is satisfied (cf 5.84):

$$\lambda = \rho_0 \frac{dp}{d\rho_0} = \rho_0^2 \frac{dV}{d\rho_0} = \rho_0^2 \frac{dV}{dr}\frac{dr}{d\rho_0}$$

Let us now put

$$-\psi = W_2 + kW_2 + \frac{\Delta\omega_i}{\omega}\ \phi + \frac{\lambda\theta}{\rho} + \eta. \quad (6.84)$$

The fundamental equation (6.69) becomes

$$\frac{\partial \mathbf{v}}{\partial t} + 2\omega\,(\mathbf{k} \wedge \mathbf{v}) = - \mathbf{grad}\ \psi - \frac{2\omega_i}{\omega}\frac{\partial \phi}{\partial z}\,\mathbf{k} \tag{6.85}$$

or, explicitly,

$$\ddot{u} - 2\omega\dot{v} = -\frac{\partial\psi}{\partial x}, \quad \ddot{v} + 2\omega\dot{u} = -\frac{\partial\psi}{\partial y}, \quad \ddot{w} + \frac{2\omega_i}{\omega}\frac{\partial\phi}{\partial z} = -\frac{\partial\psi}{\partial z} \tag{6.86}$$

Considering that

$$\ddot{u} = -\omega_i^2\,u, \qquad \ddot{v} = -\omega_i^2\,v, \qquad \ddot{w} = -\omega_i^2\,w,$$

one easily recombines equations (6.86) according to the following form:

$$\left.\begin{aligned}
(\omega_i^2 - 4\omega^2)\,u &= \frac{\partial\psi}{\partial x} - \frac{2\omega}{\omega_i^2}\frac{\partial\dot{\psi}}{\partial y},\\
(\omega_i^2 - 4\omega^2)\,v &= \frac{\partial\psi}{\partial y} + \frac{2\omega}{\omega_i^2}\frac{\partial\dot{\psi}}{\partial x},\\
\omega_i^2\,w &= \frac{\partial\psi}{\partial z} + \frac{2\omega_i}{\omega}\frac{\partial\phi}{\partial z},
\end{aligned}\right\} \tag{6.87}$$

or

$$\mathbf{u} = (\omega_i^2 - 4\omega^2)^{-1}\left\{\mathbf{grad}\ \psi + \frac{2\omega}{\omega_i^2}\,\mathbf{k} \wedge \mathbf{grad}\ \dot{\psi} + \mathbf{k}\left[\frac{2}{\omega\omega_i}(\omega_i^2 - 4\omega^2)\frac{\partial\phi}{\partial z} - \frac{4\omega^2}{\omega_i^2}\frac{\partial\psi}{\partial z}\right]\right\}. \tag{6.88}$$

Taking the divergence of (6.88) and multiplying it by **grad** V gives the following two equations:

$$\left.\begin{aligned}
(\omega_i^2 - 4\omega^2)\,\theta &= \Delta\psi - \frac{4\omega^2}{\omega_i^2}\frac{\partial^2\psi}{\partial z^2},\\
(\omega_i^2 - 4\omega^2)\,\eta &= \psi' \mid V' + \frac{\partial V}{\partial z}\left(\frac{2(\omega_i^2 - 4\omega^2)}{\omega\omega_i}\frac{\partial\phi}{\partial z} - \frac{4\omega^2}{\omega_i^2}\frac{\partial\psi}{\partial z}\right) + \\
&\quad \frac{2\omega}{\omega_i^2}\left(\frac{\partial\dot{\psi}}{\partial x}\frac{\partial V}{\partial y} - \frac{\partial\dot{\psi}}{\partial y}\frac{\partial V}{\partial x}\right).
\end{aligned}\right\} \tag{6.89}$$

In the case of incompressibility the first one of these equations gives

$$\omega_i^2\left(\frac{\partial^2\psi}{\partial x^2} + \frac{\partial^2\psi}{\partial y^2}\right) + (\omega_i^2 - 4\omega^2)\frac{\partial^2\psi}{\partial z^2} = 0. \tag{6.90}$$

The last term of the second equation can be transformed if we use $\ell = \sqrt{x^2+y^2}$ as a variable:

$$\frac{\partial V}{\partial x} = \frac{x}{\ell}\frac{\partial V}{\partial \ell}, \qquad \frac{\partial V}{\partial y} = \frac{y}{\ell}\frac{\partial V}{\partial \ell},$$

and as

$$x = \ell \cos \lambda, \qquad y = \ell \sin \lambda,$$

(for the tidal potential of second order: m=1)

$$\frac{\partial \psi}{\partial \lambda} = x \frac{\partial \psi}{\partial y} - y \frac{\partial \psi}{\partial x}, \quad \frac{\partial \psi}{\omega_i \partial t} = - \frac{\partial \psi}{m \partial \lambda}, \quad \frac{\partial^2 \psi}{\partial t^2} = - \omega_i^2 \psi.$$

Finally, we have the system

$$\left.\begin{aligned}
&(\omega_i^2 - 4\omega^2)\, \theta = \Delta\psi - \frac{4\omega^2}{\omega_i^2} \frac{\partial^2 \psi}{\partial z^2}, \\
&(\omega_i^2 - 4\omega^2)\eta = \psi'|V' + \frac{\partial V}{\partial z} \left(\frac{2(\omega_i^2 - 4\omega^2)}{\omega\omega_i} \frac{\partial \phi}{\partial z} - \frac{4\omega^2}{\omega_i^2} \frac{\partial \psi}{\partial z} \right) - 2 \frac{m\omega}{\ell\omega_i} \frac{\partial V}{\partial \ell} \psi, \\
&\frac{\Delta(kW_2)}{4\pi G\rho'} \cdot V' = \eta + \rho \frac{V'}{\rho'} \theta = \eta + \lambda \frac{\theta}{\rho}, \\
&\psi + W_2 + k\, W_2 + \frac{\Delta\omega_i}{\omega} \phi = -\eta - \frac{\lambda\theta}{\rho}.
\end{aligned}\right\} \quad (6.91)$$

Using the first two equations for θ and η, the others become

$$\left.\begin{aligned}
\psi + W_2 + kW_2 + \frac{\Delta\omega_i}{\omega} \phi &= \omega^{-2} F(\psi) - \frac{2}{\omega\omega_i} \frac{\partial \phi}{\partial z} \frac{\partial V}{\partial z} \\
&= - \frac{V'}{4\pi G\rho'} \Delta (kW_2)
\end{aligned}\right\} \quad (6.92)$$

with

$$\left.\begin{aligned}
&F(\psi) = \frac{\omega^2}{4\omega^2 - \omega_i^2} \left(\rho \frac{V'}{\rho'} \left(\Delta\psi - \frac{4\omega^2}{\omega_i^2} \frac{\partial^2 \psi}{\partial z^2}\right) + \psi'|V' - \frac{4\omega^2}{\omega_i^2} \frac{\partial \psi}{\partial z} \frac{\partial V}{\partial z} - \frac{2m\omega}{\ell\omega_i} \psi \frac{\partial V}{\partial \ell} \right), \\
&\mu = 0.
\end{aligned}\right\} \quad (6.93)$$

6.2.3. <u>The Solution in the Core</u>

Molodensky tries to find a solution of the system (6.92), (6.93) in the form

$$\left.\begin{aligned}
&\psi = \kappa(2\omega + \omega_i)\, \omega_i\, \beta\, \ell\, z \cos(\omega_i t - \lambda) + \omega^2 \psi_1, \\
&V + \psi = \kappa[\, \alpha\, \Phi + \omega^2 (V_1 + \psi_1)],
\end{aligned}\right\} \quad (6.94)$$

by chosing for Φ the solution of

$$\Delta\Phi = - \frac{4\pi G\tilde{\rho}'}{\tilde{V}'} \Phi \quad (6.95)$$

with

$$V = V_e + \frac{\Delta\omega_i}{\omega} \phi + V_i. \quad (6.96)$$

β is the resonance parameter which considerably increases when ω_i is near to ω. On the other hand, the parameter α and the functions Φ, V_1 and ψ_1 change very little with ω_i. V_1 and ψ_1 represent small correction terms, but Molodensky has not given any estimates for them.

The solution of (6.95) has the form

$$\Phi = K\,(r)\;S(\theta,\lambda) \tag{6.97}$$

with

$$S(\theta,\lambda) = \frac{\ell z}{r^2}\cos(\omega_i t - \lambda);$$

then replacing into (6.95)

$$K'' + 2\,\frac{K'}{r} + \left(\frac{4\pi G\tilde{\rho}'}{\tilde{V}'} - \frac{6}{r^2}\right)K = 0, \tag{6.98}$$

which is similar to Clairaut's famous equation

$$\varepsilon'' + \frac{6\rho r^2}{3\int_0^r \rho r^2\,dr}\,\varepsilon' + \left(\frac{6\rho r}{3\int \rho r^2 dr} - \frac{6}{r^2}\right)\varepsilon = 0,$$

or, if we put (theorem of the mean),

$$3\int_0^r \rho r^2 dr = \bar{\rho}\;r^3$$

we can transform it as

$$(r^3\,\bar{\rho}\,\varepsilon)'' = 6\,\varepsilon\,r\,\bar{\rho} + 3\,\varepsilon\,r^2\,\rho'.$$

Then

$$\Delta\left(r^3\,\bar{\rho}\,\varepsilon\,\frac{S}{r}\right) = [\,(\varepsilon\,\rho\,r^3)'' - 6\,\varepsilon\,\rho\,r]\frac{S}{r},$$

which suggests identification of

$$\Phi \equiv r^3\,\bar{\rho}\,\varepsilon\,\frac{S}{r}$$

and takes as a particular solution of (6.95)

$$\Phi_1 = \varepsilon\,\bar{\rho}\,r^2\,S(\theta,\lambda) = K_1\,S(\theta,\lambda) \tag{6.99}$$

which is zero at the centre of the Earth.

The general solution is

$$K = \alpha_1\,K_1 + \alpha_2\,K_2, \tag{6.100}$$

but the K_1 and K_2 solutions must be independant which is satisfied if the Wronskian

$$\begin{vmatrix} K_1(r) & K_2(r) \\ K_1'(r) & K_2'(r) \end{vmatrix}$$

is different from zero, i. e.

$$K_1 K_2' - K_1' K_2 \neq 0. \tag{6.101}$$

From (6.98) one has

$$(K_2'' K_1 - K_2 K_1'') + \frac{2}{r}(K_2' K_1 - K_2 K_1') = 0. \tag{6.102}$$

Therefore

$$r^2(K_2' K_1 - K_2 K_1') = \text{constant}$$

and consequently

$$K_2 = CK_1 \int_b^r K_1^{-2} u^{-2}\, du \tag{6.103}$$

(b is the radius of the core - if one took another limit of integration it should correspond to the addition of a constant which we take here as null at the core boundary).

From (6.99)

$$K_1 = \varepsilon \rho r^2$$

and one easily obtains

$$\frac{K_1'}{K_1} = \frac{\eta-1}{r} - \frac{4\pi G\rho}{V'},$$

η being Radau's parameter $\left(\frac{a}{\varepsilon}\frac{d\varepsilon}{da}\right)$ with

$$\frac{3}{2}\frac{C}{Mr^2} = 1 - \frac{2}{5}\sqrt{1+\eta}.$$

Equation (6.94) gives

$$\Delta V = \kappa\alpha\, \Delta\phi + \kappa\omega^2\, \Delta V_1 \tag{6.104}$$

while

$$\frac{\partial\phi}{\partial z} = -\varepsilon\, \omega^2\, \ell \cos(\omega_i t - \lambda) \tag{6.105}$$

and

$$\frac{\partial V}{\partial\theta} = z\frac{\partial V}{\partial\ell} - \ell\frac{\partial V}{\partial z}. \tag{6.106}$$

Then, putting

$$\nu = 2\frac{\omega}{\omega_i}\left(\frac{\Delta\omega_i}{\omega}\beta - \frac{\varepsilon}{\kappa}\right), \tag{6.107}$$

Molodensky obtains (*)

(*) Assuming that $V'\tilde{\rho}' = \tilde{V}'\rho'$ and that V_1 is a solution of (6.95), which is not very clear to us.

$$F(\psi_1) = \alpha\Phi + \beta \frac{\partial V}{\partial \theta} \cos(\omega_i t - \lambda) + \nu\ell \frac{\partial V}{\partial z} \cos(\omega_i t - \lambda) \qquad (6.108)$$

where $\beta(\partial V/\partial\theta)$ represents the variation of V along a meridian: this term should vanish if the core boundary was spherical (ε=0).

On the equipotential s,

$$V(r\theta\lambda) = V(s) = V\,\{r[1-e(r)\;S(\theta,\lambda)]\} \qquad (6.109)$$

one has

$$\frac{\partial V}{\partial \theta} = \frac{\partial V(s)}{\partial s}\frac{\partial s}{\partial \theta} = \frac{\partial V(s)}{\partial s}\, re\, \frac{\partial S}{\partial \theta} = -\,4\pi G\bar{\bar{\rho}}\varepsilon\ell z = 3\,\tilde{V}'\,\bar{\varepsilon}\,r\,S = -\,\tilde{V}'\,r\,e^2\,S \qquad (6.110)$$

$$\left(e^2 = \frac{a^2-b^2}{a^2} = -\,3\bar{\varepsilon} \text{ in (6.109) is the eccentricity}\right).$$

$\partial V/\partial\theta$ is proportional to Φ_1 and therefore Molodensky *chooses* it to define K_1 on the core-mantle boundary (r=b), putting

$$K_1\,(b) = (V'\,r\,e^2)_{r=b}.$$

Thus

$$\frac{\partial V}{\partial \theta} = -K_1\,S,$$

$$\Phi = K\,\frac{\ell z}{r^2}\cos(\omega_i t - \lambda),$$

$$F(\psi_1) = \left[\,(\alpha K - \beta K_1)r^{-2} + \nu\,\frac{\partial V}{z\partial z}\right]\ell z \cos(\omega_i t - \lambda).$$

Accepting the spherical approximation for the coefficient of $\nu(\frac{\partial V}{\partial z} = \frac{z}{r}\frac{\partial V}{\partial r})$ one has, finally, the equation

$$\left.\begin{aligned} &[\,(\alpha K-\beta K_1)r^{-2} + \nu\,\frac{V'}{r}]\,\ell z\,\cos(\omega_i t-\lambda) = \\ &\frac{\omega^2}{4\omega^2-\omega_i^2}\,\frac{V'}{\rho'}\,[\,\rho(\Delta\psi_1 - \frac{4\omega^2}{\omega_i^2}\,\frac{\partial^2\psi_1}{\partial z^2}) + \frac{\rho'}{r}\,(r\psi_1' - \frac{4\omega^2}{\omega_i^2}\,z\,\frac{\partial\psi_1}{\partial z} - \frac{2\omega}{\omega_i}\,\psi_1)]\,. \end{aligned}\right\} \qquad (6.111)$$

Molodensky searches for a solution in the form

$$\psi_1 = \cos(\omega_i t-\lambda)\sum_{n=2}^{\infty} X_n\,(r)\,r^n\,P_n^1\,(\theta), \qquad (6.112)$$

a development in tesseral polynomials where the unknown are functions of r only as the rheological parameters ρ, μ, λ are supposed to be functions of r only.

The development of (6.111) with the use of Legendre polynomials is not difficult in principle but is a fairly long procedure. It is based upon several formulas demonstrated by Catalan in the last century (Acad. Roy. Belgique, 31, 1886, 47, 1889):

$$(1-\mu^2)\,P_n' = (n+1)\,(\mu\,P_n - P_{n+1}),$$

$$(1-\mu^2)\,P_n' = n\,(P_{n-1} - \mu\,P_n),$$

$$(2n+1)\ \mu\ P'_n = n\ P'_{n+1} + (n+1)\ P'_{n-1},$$
$$(1-\mu^2)\ P''_n = (n+2)\ \mu\ P'_n - n\ P'_{n+1},$$
$$(1-\mu^2)\ (P^1_n)' = -n\ P^1_{n+1} + (n+1)\ \mu\ P^1_n,$$
$$(2n+1)\ \mu\ P^1_n = n\ P^1_{n+1} + (n+1)\ P^1_{n-1},$$

with

$$\mu = \cos\theta, \qquad (P^1_n)' = \frac{\partial P^1_n}{\partial\mu}$$

then after some twelve pages of algebraic manipulations one obtains for n=2 the following equation:

$$\left.\begin{aligned}&\left[\left(\frac{12\omega^2}{7\omega_i^2} - 1\right)\rho\, X'_2\, r^6 + \frac{2(\Delta\omega_i)\ (2\omega-\omega_i)}{\omega_i^2}\rho\, X_2\, r^5 + \frac{80}{63}\frac{\omega^2}{\omega_i^2}(X_4\, r^9)'\,\rho\, r^{-1} + \right.\\ &\frac{4\omega^2-\omega_i^2}{3\omega^2}\nu\rho r^5 - \frac{4\omega^2-\omega_i^2}{12\pi G\omega^2}\left(\frac{\alpha K-\beta K_1}{r^2}\right)'\, r^6 = \\ &= \frac{2(\Delta\omega_i)\ (2\omega-\omega_i)}{\omega_i^2}\rho\,(X_2\, r^5)' + \frac{4\omega^2-\omega_i^2}{3\omega^2}5\nu\,\rho\, r^4.\end{aligned}\right\} \quad (6.113)$$

This can be simplified by the introduction of the work η as defined by

$$\eta = -\ (W_2 + kW_2 + \frac{\Delta\omega_i}{\omega}\phi) - \psi - \lambda\frac{\theta}{\rho}$$

which becomes

$$\eta = \frac{\kappa\omega^2 V'}{\omega_i^2 - 4\omega^2}\left(\psi'_1 - \frac{4\omega^2}{\omega_i^2}\frac{z}{r}\frac{\partial\psi_1}{\partial z} - \frac{2\omega}{r\omega_i}\psi_1\right) - \kappa\left(\beta\frac{K_1}{V'} - r\nu\right)V'\ \frac{\ell z}{r^2}\cos(\omega_i t-\lambda) \quad (6.114)$$

and therefore can be written

$$\eta = \sum_{n=2}^{\infty}\eta_n\, r^n\, P^1_n(\theta)\ \cos(\omega_i t-\lambda). \quad (6.115)$$

Terms containing P^1_2 are present in n=2 and n=4 terms of $z\ \partial\psi_1/\partial z$ (*); then all the first terms in (6.113) are found again and can be replaced as follows:

(*) Noting that
$$z\frac{\partial}{\partial z} = r\cos^2\theta\frac{\partial}{\partial r} + \sin^2\theta\frac{\partial}{\partial(\cos\theta)}$$
one has
$$(2n+1)\frac{1}{z}\frac{\partial}{\partial z}(X_n\, r^n\, P^1_n) = \frac{n(n+1)}{2n+3}\frac{1}{\cos 2\theta}P^1_{n+2}\, r^{n-1}\, X'_n + \frac{n(n+2)}{2n+3}\frac{1}{\cos^2\theta}P^1_n\, r^{n-1}\, X'_n + \frac{(n+1)(n-1)}{2n-1}\frac{1}{\cos^2\theta}P^1_n\, r^{-n-2}\,(X_n\, r^{2n+1})' + \frac{(n+1)n}{2n-1}\frac{1}{\cos 2\theta}P^1_{n-2}\, r^{-n-2}\,(X_n\, r^{2n+1})'.$$

$$\left\{ \frac{r^6}{4\pi G} [(\alpha K - \beta K_1) r^{-2}]' - 3\rho\eta_2 \frac{r^6}{\tilde{\kappa V'}} - \beta \frac{K_1}{\tilde{V}'} \rho r^4 \right\}' = \tag{6.116}$$

$$-5\nu\rho r^4 - \frac{6\omega^2 (\Delta\omega_i)}{(2\omega+\omega_i)\omega_i^2} \rho (X_2 r^5)'.$$

As

$$\psi = [(2\omega+\omega_i) \omega_i \beta \ell z + \omega^2 \sum_2^\infty X_n r^n P_n^1] \cos(\omega_i t - \lambda),$$

the addition of a constant to X_2 corresponds to a modification of the β parameter.

Therefore Molodensky defines this parameter by the additional condition

$$\int_c^b \rho (X_2 r^5)' \, dr = 0, \tag{6.117}$$

where c and b are the radii of the core boundaries.

This condition, by the way, allows elimination of the last term of (6.116) when we integrate this equation in the core:

$$\left| \frac{3}{\tilde{\kappa V'}} \rho r^6 \eta_2 + \beta \frac{K_1}{\tilde{V}'} \rho r^4 - \frac{1}{4\pi G} r^6 [(\alpha K - \beta K_1) r^{-2}]' \right|_b^c = 5\nu \int_c^b \rho r^4 \, dr. \tag{6.118}$$

If we consider the core as completely fluid, the expression (6.118) may be transformed by using

$$c = 0, \quad K = K_1, \quad \alpha K' = R' + 2(2\omega+\omega_i)\omega_i \beta r,$$

and becomes

$$\left| L r^2 - 2 R r^3 - \beta r^4 (K_1' - 2 \frac{K_1}{r} + \frac{4\pi G\rho}{V'} K_1) \right|_b^0 = -20\pi G\nu \int_0^b \rho r^4 \, dr. \tag{6.119a}$$

This relation gives a condition equation between β and

$$\nu = 2 \frac{\omega}{\omega_i} (\frac{\Delta\omega_i}{\omega} \beta - \frac{\varepsilon}{\kappa})$$

depending on the radial displacement $\eta_2/\tilde{V}'$ at each of the two boundaries (r=b), (r=c) of the core, avoiding in this way the necessity of computing the movements inside the core itself.

A second relation is necessary to eliminate ε and will therefore be obtained from the kinetic momentum equation in 6.2.5.

6.2.4. Boundary Conditions

We have shown in Chapter 5 that the general solution of the system of differential equations is

$$\phi(r) = h \frac{\partial\phi}{\partial h} + \ell \frac{\partial\phi}{\partial \ell} + (1+k) \frac{\partial\phi}{\partial k} + (2-3k) \frac{\partial\phi}{\partial L}$$

with ϕ = H, L, T, M, N, R successively.

At the liquid core-solid mantle boundary, only the tangential displacement T can have a discontinuity while the other five parameters give five continuity conditions which are as follows.

1. Parameter M

It is zero within the core. Therefore at r=b

$$M(b) = h\left(\frac{\partial M}{\partial h}\right)_b + \ell\left(\frac{\partial M}{\partial \ell}\right)_b + (1+k)\left(\frac{\partial M}{\partial k}\right)_b + (2-3k)\left(\frac{\partial M}{\partial L}\right)_b = 0$$

or

$$\left[\left(\frac{\partial M}{\partial k}\right)_b + 2\left(\frac{\partial M}{\partial L}\right)_b\right] + k\left[\left(\frac{\partial M}{\partial k}\right)_b - 3\left(\frac{\partial M}{\partial L}\right)_b\right] + h\left[\left(\frac{\partial M}{\partial h}\right)_b\right] + \ell\left[\left(\frac{\partial M}{\partial \ell}\right)_b\right] = 0. \quad (6.119b)$$

The coefficients in this relation depend only on the mantle structure. For his two models, Molodensky has obtained the following relations:

$$\left.\begin{array}{l}\text{(I)} \; 0.3873 + 0.6993\,k - 0.4754\,h - 3.7084\,\ell = 0 \\ \text{(II)} \; 0.3833 + 0.7053\,k - 0.4793\,h - 3.7155\,\ell = 0\end{array}\right\} \quad (6.120)$$

from which we can derive Table 6.2.

TABLE 6.2. Love numbers as related by boundary conditions (Molodensky's dynamical theory)

Model I $\ell = 0.10444 + 0.1886\,k - 0.1282\,h$

k	0.28	0.29	0.30	0.31	0.32	
h						
0.56	0.0854	0.0873	0.0892	0.0911	0.0929	ℓ
0.58	0.0828	0.0847	0.0866	0.0885	0.0903	
0.60	0.0803	0.0822	0.0841	0.0860	0.0878	
0.62	0.0777	0.0796	0.0815	0.0834	0.0852	
h						
0.56	0.153	0.156	0.159	0.162	0.166	ℓ/h
0.58	0.145	0.148	0.151	0.154	0.157	
0.60	0.134	0.137	0.140	0.143	0.146	
0.62	0.125	0.128	0.131	0.134	0.137	

Model II $\ell = 0.1032 + 0.1898\,k - 0.1290\,h$

k	0.28	0.29	0.30	0.31	0.32	
h						
0.56	0.0841	0.0860	0.0879	0.0898	0.0917	ℓ
0.58	0.0815	0.0834	0.0853	0.0872	0.0891	ℓ
0.60	0.0789	0.0808	0.0827	0.0846	0.0865	ℓ
0.62	0.0763	0.0783	0.0802	0.0821	0.0840	ℓ
h						
0.56	0.150	0.154	0.157	0.160	0.164	ℓ/h
0.58	0.140	0.144	0.147	0.150	0.153	ℓ/h
0.60	0.132	0.136	0.140	0.141	0.144	ℓ/h
0.62	0.123	0.126	0.129	0.132	0.135	ℓ/h

2. Parameter H

In the mantle, according to (5.5),

$$\zeta = H \frac{a}{r^2} \frac{W_2}{g_o}$$

with

$$\eta = \zeta g, \quad W_2 = V_e + \frac{\Delta\omega_i}{\omega}\,\phi, \quad V_e = \frac{\kappa}{3}\frac{g}{a}\, r^2\, P_2^1 \cos(\omega_i t - \lambda).$$

In the core

$$\eta = \eta_2\, r^2\, P_2^1 \cos(\omega_i t - \lambda).$$

Thus, neglecting the deviation of the vertical with respect to the radius vector as well as the little term $\Delta\omega_i/\omega\ \phi$, one has

$$H = \frac{3\eta_2 b^2}{\kappa V'} \qquad \text{at } r=b=0.55. \qquad (6.121)$$

3. Parameter R

In the mantle,

$$V = \kappa\, R\, \frac{\ell z}{r^2} \cos(\omega_i t - \lambda).$$

In the core,

$$V = \kappa\left[\alpha_1 K_1 + \alpha_2 K_2 + (2\omega+\omega_i)\omega_i\, \beta\, r^2\right] \frac{\ell z}{r^2} \cos(\omega_i t - \lambda).$$

Therefore at $r=b$,

$$R(b) = \alpha_1 K_1(b) + \alpha_2 K_2(b) - (2\omega+\omega_i)\omega_i \beta b^2. \quad (6.122)$$

When the core is completely fluid the term $\alpha_2 K_2$ disappears.

4. Parameter L

In the core

$$V'_C = \kappa[\alpha_1 K'_1 + \alpha_2 K'_2 - 2(2\omega+\omega_i)\omega_i \beta r] \frac{\ell z}{r^2} \cos(\omega_i t-\lambda),$$

but

$$V'_n = \kappa R'_N \frac{\ell z}{r^2} \cos(\omega_i t-\lambda),$$

while

$$\left(\frac{\partial V}{\partial n}\right)_{\text{mantle}} - \left(\frac{\partial V}{\partial n}\right)_{\text{core}} = R'_M - R'_C = 4\pi G(\rho_M-\rho_C)H,$$

and from the definition of L

$$L(b).b^{-2} = \alpha_1 K'_1(b) + \alpha_2 K'_2(b) - 2\omega_i \beta (2\omega+\omega_i)b - 4\pi G\rho_C(b) H(b). \quad (6.123)$$

The term $\alpha_2 K'_2$ disappears when the core is completely fluid.

5. Parameter N

The definition of ψ where $\lambda\theta = N$ in the case of a liquid gives

$$[R + V'H + \frac{N}{\rho} + (2\omega+\omega_i)\omega_i \beta r^2]_b = 0. \quad (6.124)$$

6.2.5. The Kinetic Momentum Equation

This equation is, in the rotating system of axes,

$$\left|\frac{d}{dt} \boldsymbol{\sigma}_a\right|_\Sigma + \boldsymbol{\omega} \wedge \boldsymbol{\omega}_i = \boldsymbol{m}_{ext} \quad (6.125)$$

with

$$\boldsymbol{\omega} = \omega(\varepsilon \cos \omega_i t.\mathbf{i} + \varepsilon \sin \omega_i t\, \mathbf{j} + \mathbf{k}) \quad (6.126)$$

$$\boldsymbol{\sigma}_a = \int_{\text{Earth}} \mathbf{OM} \wedge [\mathbf{V}_a(O) + \mathbf{V}_r(M) + \boldsymbol{\omega} \wedge \mathbf{OM}]\, \rho d\tau = \boldsymbol{\sigma}_r\, \mathbf{I}\, \boldsymbol{\omega}, \quad (6.127)$$

$\mathbf{I}$ being the inertia tensor:

$$\mathbf{I} = \begin{pmatrix} A & -I_{xy} & -I_{xz} \\ -I_{xy} & B & -I_{yz} \\ -I_{xz} & -I_{yz} & C \end{pmatrix} \qquad \bar{\omega} = \begin{vmatrix} \omega\varepsilon \cos \sigma t \\ \omega\varepsilon \sin \sigma t \\ \omega \end{vmatrix}$$

and

$$\mathbf{I}\,\boldsymbol{\omega} = \begin{vmatrix} A & \omega\varepsilon \cos\sigma t - I_{xy} & \omega\varepsilon \sin\sigma t - I_{xz} & \omega \\ -I_{xy} & \omega\varepsilon \cos\sigma t + B & \omega\varepsilon \sin\sigma t - I_{yz} & \omega \\ -I_{xz} & \omega\varepsilon \cos\sigma t - I_{yz} & \omega\varepsilon \sin\sigma t + C\omega & \end{vmatrix}$$

Putting the kinetic momentum relative to the rotating axes as

$$\sigma_r = (\Delta M_x, \Delta M_y, \Delta M_z), \tag{6.128}$$

one obtains the projections of the kinetic momentum on these axes as follows:

$$\left.\begin{aligned} M_x &= \omega(A\ \varepsilon \cos \omega_i t - I_{xz}) + \Delta M_x , \\ M_y &= \omega(A\ \varepsilon \sin \omega_i t - I_{yz}) + \Delta M_y , \\ M_z &= \omega C + \Delta M_z , \end{aligned}\right\} \tag{6.129}$$

neglecting the small quantities of higher order. M_x and M_y are responsible for the nutations.

As the axes are fixed with respect to the mantle, $\boldsymbol{\sigma}_r$ is determined by the movements inside the liquid core only, and we can write, for example,

$$\Delta M_x = \int_{\text{core}} \rho\ (y\ \dot{w} - z\ \dot{v})\ d\ \tau. \tag{6.130}$$

From

$$-\frac{1}{\omega_i}\frac{\partial\psi}{\partial t} = x\frac{\partial\psi}{\partial y} - y\frac{\partial\psi}{\partial x} = \frac{\partial\psi}{\partial\lambda}$$

we derive

$$\frac{\partial\dot{\psi}}{\partial y} - \omega_i\frac{\partial\psi}{\partial x} = -\omega_i\frac{\partial}{\partial\lambda}\frac{\partial\psi}{\partial y}$$

and a new form for the fundamental equations (6.87):

$$\left.\begin{aligned} (\omega_i^2 - 4\omega^2)\ u &= \frac{\omega_i - 2\omega}{\omega_i}\frac{\partial\psi}{\partial x} + \frac{2\omega}{\omega_i}\frac{\partial}{\partial\lambda}\frac{\partial\psi}{\partial y}, \\ (\omega_i^2 - 4\omega^2)\ v &= \frac{\omega_i - 2\omega}{\omega_i}\frac{\partial\psi}{\partial y} - \frac{2\omega}{\omega_i}\frac{\partial}{\partial\lambda}\frac{\partial\psi}{\partial x}, \\ \omega_i^2\ w &= \frac{\partial\psi}{\partial z} + \frac{2\omega_i}{\omega}\frac{\partial\phi}{\partial z}, \end{aligned}\right\} \tag{6.131}$$

when one calculates

$$\int_{\text{core}} \rho\ (yw - zv)\ d\tau$$

by introducing v and w as given by (6.131), the terms containing the derivatives with respect to λ are vanishing in the integration. One then applies the operator

$$\frac{d^2}{dt^2} = -\omega_i^2.$$

Considering that $\Delta\dot{M}_x = -\omega_i \Delta M_y$ and $\Delta\dot{M}_y = \omega_i \Delta M_x$ (because of the trigonometric functions in ψ and ϕ) one obtains:

$$\left.\begin{aligned}\Delta M_x = \frac{2\omega}{\omega_i(2\omega+\omega_i)} \int \rho\, x \frac{\partial\psi}{\partial z}\, d\tau + \frac{2}{\omega} \int \rho\, x \frac{\partial\phi}{\partial z}\, d\tau + \\ \frac{1}{2\omega+\omega_i} \int \rho\, (x \frac{\partial\psi}{\partial z} - z \frac{\partial\psi}{\partial x})\, d\tau.\end{aligned}\right\} \quad (6.132)$$

The different terms have to be evaluated which is a long but not difficult procedure. One has only to calculate them from

$$\phi = -\varepsilon\, \omega^2\, \ell\, z \cos(\omega_i t-\lambda),$$

$$\psi = \kappa(2\omega+\omega_i)\beta\, \sigma\, \ell\, z \cos(\omega_i t-\lambda),$$

$$\psi_1 = \sum_n X_n\, r^n\, P_n^1 \cos(\omega_i t-\lambda),$$

with

$$\int_c \rho\, x\, z\, d\tau = \int_c \rho\, x\, y\, d\tau = \int_c \rho\, y\, z\, d\tau = 0,$$

$$\int_c \rho\, x^2\, d\tau = \int_c \rho\, y^2\, d\tau = C_1/2,$$

$$\int_c \rho\, z^2\, d\tau = \int_c \rho\, (y^2+z^2) d\tau - \int_c \rho\, y^2\, d\tau = A_1 - C_1/2,$$

A_1 and C_1 being the moments of inertia of the core.

One obtains

$$\left.\begin{aligned}\frac{1}{\omega}\Delta M_y &= \kappa[(\beta - \frac{\varepsilon}{\kappa})A_1 + \frac{\omega_i \nu}{2\omega}(C_1-A_1)] \sin \omega_i t \\ \frac{1}{\omega}\Delta M_x &= \kappa[(\beta - \frac{\varepsilon}{\kappa})A_1 + \frac{\omega_i \nu}{2\omega}(C_1-A_1)] \cos \omega_i t.\end{aligned}\right\} \quad (6.133)$$

6.2.6. Products of Inertia

They appear because of:

a) changes of the internal density as given in Poisson's equation

$$\delta\rho = -\Delta V/4\pi G;$$

b) changes of the surfaces of discontinuity which correspond to the addition of a layer of density $\rho\eta/W'$ at each discontinuity.

Thus

$$I_{yz} = \int_c \delta\rho yz\, d\tau + \int_{S_+} \frac{\rho\eta}{W'}\, yz\, ds, \quad (6.134)$$

while Green's formula gives

$$\int_c \Delta V yz\, d\tau = \int_{S+} \left(\frac{\partial V}{\partial n}\, yz - \frac{\partial(yz)}{\partial n}\, V\right) ds.$$

We shall obtain

$$I_{yz} = -\frac{\kappa}{4\pi G}\left[a^{-2} L(a) - 2\frac{R}{a}\right]\iint a^{-2} y^2 z^2 \sin\omega_i t\,(a^2 \sin\theta\, d\theta\, d\lambda)$$

and, because of

$$L(a) = (2-3k)g\,a^2, \quad R(a) = (1+k)\,g\,a, \quad g = \frac{G\,\mathbf{m}}{a^2}$$

Finally

$$I_{yz} = \frac{\kappa k}{3}\,\mathbf{m}\,a^2 \sin\omega_i t$$

and

$$M_x = M\cos\omega_i t, \qquad M_y = M\sin\omega_i t$$

with

$$\frac{M}{\omega} = (A-A_1)\varepsilon + \kappa\beta A_1 + \frac{\kappa\omega_i\nu}{2\omega}(C_1-A_1) - \frac{\kappa}{3}\bar{m}\,a^2\,k.$$

Now the moment of the external forces is

$$\mathbf{L} = \int \rho(\mathbf{OM}\wedge\mathbf{grad}\,V_e)\,d\tau = \frac{g\kappa}{a}(C-A)\,(\mathbf{i}\,\sin\omega_i t - \mathbf{j}\,\cos\omega_i t).$$

The components of the three terms in (6.125) are thus:

$$\frac{d\boldsymbol{\sigma}a}{dt} = \dot{\mathbf{M}} \quad \left|\begin{array}{l} -\omega_i M_y \\ +\omega_i M_x \\ C\omega \end{array}\right|$$

$$\boldsymbol{\omega}\wedge\mathbf{M} \quad \left|\begin{array}{l} \omega\,\varepsilon\,\sin\omega_i t\,M_z - \omega\,M_y \\ -\omega\,\varepsilon\,\cos\omega_i t\,M_z + \omega\,M_x \\ \omega\,\varepsilon\,\cos\omega_i t\,M_y - \omega\,\varepsilon\,\sin\omega_i t\,M_x \end{array}\right|$$

$$\mathbf{L} \quad \left|\begin{array}{l} \frac{g\kappa}{a}(C-A)\sin\omega_i t \\ -\frac{g\kappa}{a}(C-A)\cos\omega_i t \\ 0 \end{array}\right|$$

and the equation in x gives the relation

$$\varepsilon - \frac{\Delta\omega_i}{\omega}\left(\frac{A-A_1}{C}\varepsilon + \kappa\beta\frac{A_1}{C} + \frac{\kappa\omega_i\nu}{2\omega}\,\frac{C_1-A_1}{C} - \frac{\kappa\,k\,\mathbf{m}\,a^2}{3C}\right) = \kappa\,\frac{C-A}{Cq}, \quad (6.135)$$

which is the additional equation needed to tie the nutation ε, the tidal frequency ω_i and the resonance parameter β.

For the nutation of a rigid Earth this should give

$$\varepsilon_0 - \frac{\Delta\omega_i}{\omega} - \frac{A}{C}\varepsilon_0 = \kappa\,\frac{C-A}{Cq}. \qquad (6.136)$$

The system of equations determining h, k, ℓ, β, α and $\varepsilon/\varepsilon_0$ are (6.118), (6.120), (6.122), (6.123), (6.124), (6.135), (6.136).

6.2.7. The Solution

The boundary conditions expressing R(b) and L(b) give

$$(\alpha K r^{-2})'_b = b^{-4}\,[L(b) + 4\pi G\rho b^2 H(b) - 2bR(b)] \qquad (6.137)$$

while Molodensky's equation gives for c=0

$$2Rb^3 - Lb^2 + \beta K_1 b^4 \left(\frac{4\pi G\rho}{W'} + \frac{K_1'}{K_1} - \frac{2}{b}\right) = 15\nu G \frac{C_1}{2} \qquad (6.138)$$

and Radau's formula gives

$$\frac{K_1'}{K_1} = \frac{\xi-1}{r} - \frac{4\pi G\rho}{W'}.$$

Then (6.138) becomes

$$2Rb - L + \beta K_1 b(\xi-3) = 15\nu G \frac{C_1}{2b^2}. \qquad (6.139)$$

The definition of ν (6.107) enables (6.135) to be transformed as follows:

$$\varepsilon = \kappa \frac{C-A}{qC}\left(1 + \frac{\Delta\omega_i}{\omega}\frac{A-C_1}{C}\right) + \frac{\Delta\omega_i}{\omega C}\left(\beta A_1 - k\frac{ma^2}{3}\right)$$

(neglecting $\frac{C_1-A_1}{A_1}\frac{\Delta\omega_i}{\omega}$ with respect to unity).

Then

$$\frac{\varepsilon}{\varepsilon_0} = 1 + \frac{\Delta\omega_i}{\omega}\frac{C_1 q}{C-A}\left[\beta - \frac{C-A}{Cq} + k\frac{ma^2}{3C}\left(\frac{C}{C_1}\right)\right]. \qquad (6.140)$$

With

$k = 0.3$, $\qquad \frac{C_1}{C} = 0.118$,

$\frac{C-A}{Cq} \sim 1$, $\qquad \frac{ma^2}{3C} \sim 1$,

Molodensky obtains for his two models

$$\left.\begin{array}{ll} \beta = \dfrac{41.87}{0.2136-100\,\dfrac{\Delta\omega_i}{\omega_i}} + 1.9 & \text{from (6.139)}, \\[2ex] \dfrac{\varepsilon}{\varepsilon_0} = 1 + 0.1250\,(\beta-4.3)\,\dfrac{\Delta\omega_i}{\omega} & \text{from (6.140)}, \\[2ex] \text{resonance for } \omega_i = -\dfrac{\omega}{0.997864} = -15°073\,265\,1 & \\ \qquad (1 \text{ sidereal day} - 3\text{m } 4\text{s}). & \end{array}\right\} \text{I}$$

$$\left.\begin{aligned}
&\beta = \frac{41.15}{0.2159-100\,\dfrac{\Delta\omega_i}{\omega_i}} + 1.7 \quad \text{from (6.139)},\\
&\frac{\varepsilon}{\varepsilon_0} = 1 + 0.1224\,[\beta-4.1]\,\frac{\Delta\omega_i}{\omega} \quad \text{from (6.140)},\\
&\text{resonance for } \omega_i = -\frac{\omega}{0.997841} = -\,15^{\circ}\!.0736\ 125\\
&\qquad\text{(1 sidereal day - 3m 6s)}.
\end{aligned}\right\}\quad \text{II}$$

6.2.8. The Inner Core

If there is an inner core the conditions at the centre introduced for the functions X_n and K are no longer valid and must be replaced by conditions at the inner-core boundary. The kinetic momentum equation is also modified.

Because of lack of information Molodensky assumes that the inner core is homogeneous ($\rho' = 0$, $\mu' = 0$, $\lambda' = 0$) and incompressible. As its moment of inertia is very weak this will not change our results very much.

The fundamental equations thus become:

$$\left.\begin{aligned}
&R'' + \frac{2}{r}R' - \frac{6}{r^2}R = 0,\\
&N = 2\,\mu\,H',\\
&H' + \frac{2}{r}H - \frac{6}{r^2}T = 0,\\
&M = \mu\,r^2\left(T' - \frac{2}{r}T + H\right).
\end{aligned}\right\} \qquad (6.141)$$

To solve the last two equations we have to introduce two arbitrary constants to express H and T:

$$H = C_1\,H_1 + C_2\,H_2, \qquad T = C_1\,T_1 + C_2\,T_2. \qquad (6.142)$$

(H_1, T_1) and (H_2, T_2) are two particular solutions of the H equation.

If we take a particular solution

$$H = r^k, \qquad T = \ell\,r^s,$$

it follows that

$$k\,r^{k-1} + 2r^{k-1} - 6\ell\,r^{s-2} = 0$$

and

$$k-1 = s-2, \qquad k+2 = 6\ell$$

or

$$s = k+1, \qquad \ell = (k+2)/6.$$

Thus one can choose the general solution, for example, in the form

$$H = C_1 r + C_2\,r^3, \qquad T = \frac{1}{2}\,C_1\,r^2 + \frac{5}{6}\,C_2\,r^4. \qquad (6.143)$$

Molodensky discarded a particular solution like $H = r^2$ to avoid H being expressed as a function of R which did not depend of it.

Consequently

$$T' = 7\frac{T}{r} - \frac{5}{2}H. \tag{6.144}$$

The R equation is a Euler equation having for general solution

$$R = C_1\, r^{s1} + C_2\, r^{s2}$$

(s_1, s_2) being the solutions of the equation

$$s(s-1) + 2s - 6 = 0$$

which are +2 and -3. But as r becomes null at the centre of the Earth the solution r^{-6} must be discarded and it will become

$$R = C_3\, r^2 \tag{6.145}$$

and

$$R' = \frac{2}{r}R.$$

It follows that

$$\left.\begin{aligned} M &= \mu\, r^2\left(\frac{5}{r}T - \frac{3}{2}H\right),\\ N + \rho\,(R+V'H) &= \mu\left(\frac{13}{2}\frac{H}{r} - 9\frac{T}{r^2}\right).\end{aligned}\right\} \tag{6.146}$$

The conditions at the surface of the inner core are:

1) Continuity of M:

$$\left(\frac{5}{r}T - \frac{3}{2}H\right)_{r=c} = 0. \tag{6.147}$$

2) Continuity of R:

$$R = [\alpha_1 K_1 + \alpha_2 K_2 - (2\omega+\omega_i)\omega_i\, \beta\, r^2]_{r=c}. \tag{6.148}$$

3) Continuity of L:

$$L = r^2(R' - 4\pi G\rho H) = 2rR - 4\pi G r^2 \rho H, \tag{6.149}$$

and from (6.123)

$$[\frac{2}{r}R + 4\pi G(\rho_i - \rho)H = \alpha_1 K_1' + \alpha_2 K_2' - 2\omega_i\beta(2\omega+\omega_i)r]_{r=c}. \tag{6.150}$$

4) Continuity of N:

$$[(\rho_i - \rho)(R+V'H) + \frac{19}{5}\mu\frac{H}{r} + (2\omega+\omega_i)\omega_i\,\beta\, r^2]_{r=c} = 0, \tag{6.151}$$

i. e. because of (6.147)

$$\frac{13}{2}\frac{H}{r} - 9\frac{T}{r^2} = \frac{19}{5}\frac{H}{r}.$$

5) Continuity of H: the inner core can precess with respect to the mantle and we can represent such a rotation by the vector

$$\boldsymbol{\omega}_1 = \omega\,\varepsilon_2\,(\cos \omega_i t\ \mathbf{i} + \sin \omega_i t\,\mathbf{j}\ ) \qquad (6.152)$$

and the *relative* displacements $(\boldsymbol{\omega}_1 \wedge \mathbf{r})$ have a component normal to the boundary:

$$\frac{1}{V'}\left[\left(\boldsymbol{\omega}_1 \wedge \mathbf{r}\right)\mathbf{grad}\ V\right]. \qquad (6.153)$$

One has

$$\left(\boldsymbol{\omega}_1 \wedge \mathbf{r}\right) = \omega\,\varepsilon_2\left[\cos \omega_i t\ (\mathbf{i} \wedge \mathbf{r}) + \sin \omega_i t\ (\mathbf{j} \wedge \mathbf{r})\right]$$

$$= -\,\omega\,\varepsilon_2\ r\ \sin(\omega_i t-\lambda)\,\mathbf{d}_\theta$$

$$\mathbf{grad}\ V = \left(\frac{\partial V}{\partial r}\,\mathbf{d}_r,\ \frac{\partial V}{r\partial\theta}\,\mathbf{d}_\theta,\ \frac{\partial V}{r\sin\theta\partial\lambda}\,\mathbf{d}_\lambda\right)$$

(see Figure 6.1).

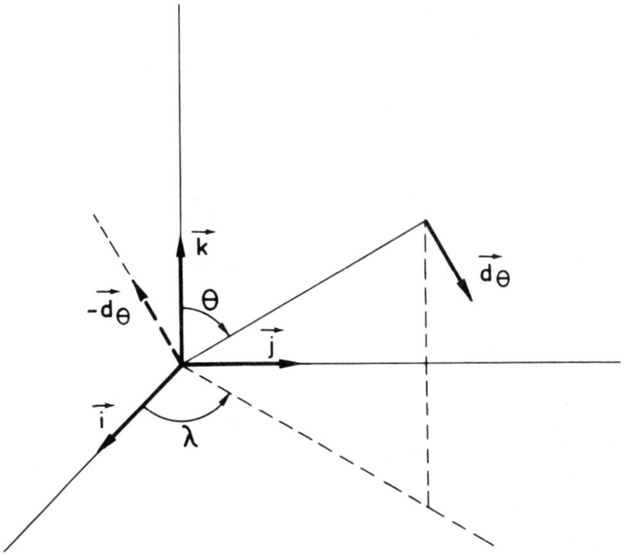

Fig. 6.1.

Therefore

$$\frac{1}{V'}\left[\left(\boldsymbol{\omega}_1 \wedge \mathbf{r}\right).\ \mathbf{grad}\ V\right] = -\,\frac{1}{V'}\,\omega\,\varepsilon_2\ \sin(\omega_i t-\lambda)\,\frac{\partial V}{\partial\theta}.$$

But we have shown that

$$-\,\frac{\partial V}{\partial\theta} = +\,V'\ r\ e^2\ S$$

with

$$S = \frac{\ell z}{r}, \qquad e^2 = -3\varepsilon$$

and the displacement speed (6.153) becomes

$$\omega\, \varepsilon_2 \sin(\omega_i t-\lambda)\; e^2\; r\, \frac{\ell z}{r^2}.$$

Then the displacement is obtained by integration

$$-\frac{\omega}{\omega_i}\, \varepsilon_2 \cos(\omega_i t-\lambda)\; e^2\; r\, \frac{\ell z}{r^2}. \tag{6.154}$$

Such a displacement is of the order of e^2 and therefore can be neglected with respect to the errors made in the boundary conditions and the continuity condition for H remains unchanged:

$$\frac{3\eta_2}{V'}\, r^2 = \kappa\, H. \tag{6.155}$$

However, the existence of the inner core modifies the kinetic moment equation where we have to exchange

$$C_1 - C_2 \text{ with } C_1$$

and

$$A_1 - A_2 \text{ with } A_1,$$

(A_2, C_2 being the moments of inertia of the inner core.

Moreover, we have to add the kinetic momentum due to the rotation of the inner core with respect to the axes fixed in the mantle, i. e. $A_2\, \omega\, \varepsilon_2$.

This gives

$$\left.\begin{aligned} \frac{M}{\omega} = A\,\varepsilon + (\kappa\beta-\varepsilon)\,(A_1-A_2) + \frac{\kappa\omega_i\nu}{\omega}\,[(C_1-C_2)-(A_1-A_2)] \;- \\ \frac{\kappa}{3}\, m\, a^2\, k + A_2\, \varepsilon_2. \end{aligned}\right\} \tag{6.156}$$

The determination of ε_2 is a real problem and one should have an additional equation similar to (6.135).

But Molodensky considers this as useless because

$$A_2 \sim 10^{-3}\, A \qquad \text{and} \qquad \varepsilon_2 < \varepsilon.$$

Indeed, the precession-nutation of the inner core must have an amplitude lower than that of the mantle because its dynamical flattening is inferior. The third term in (6.156) is also negligible with respect to the others.

Then

$$\varepsilon - \frac{\Delta\omega_i}{\omega}\left[\frac{A}{C}\,\varepsilon + (\kappa\beta-\varepsilon)\,\frac{A_1-A_2}{C} - \frac{\kappa k m a^2}{3C}\right] = \kappa\, \frac{C-A}{qC}.$$

6.2.9. The Numerical Results

It is not possible to reproduce the development of the numerical results of Molodensky but only the results.

The characteristics of the models are:

	Model I No inner core	Model II With inner core
Flattening	1/298.3	1/297
C_1/C	0.118	0.106
e^2	5092×10^{-6}	5072×10^{-6}
At the core boundary	$V' = -1.104$ g	$V' = -1.044$ g

And in the Mantle

Radial distance	ρ ($g\ cm^{-3}$)	μ/ρ ($km^2\ s^{-2}$)	$(\lambda+2\mu)\ \rho$ ($km^2\ s^{-2}$)
1.00	3.28	18.26	58.03
0.94	3.62	25.27	82.20
0.86	4.64	40.16	129.25
0.55	5.69	54.24	190.90

The computations results are:

MODEL I

$$h = 0.6206 + 0.4711 \times 10^{-3} \frac{\omega_i(\omega_i+2\omega)}{\omega^2} \beta,$$

$$k = 0.3070 + 0.2384 \times 10^{-3} \frac{\omega_i(\omega_i+2\omega)}{\omega^2} \beta,$$

$$\ell = 0.0904 - 0.0112 \times 10^{-3} \frac{\omega_i(\omega_i+2\omega)}{\omega^2} \beta,$$

$$\alpha_1 = 0.6092 + 0.3264 \times 10^{-3} \frac{\omega_i(\omega_i+2\omega)}{\omega^2} \beta.$$

MODEL II

$$h = 0.6168 + 0.4276 \times 10^{-3} \frac{\omega_i(\omega_i+2\omega)}{\omega^2}\beta - 0.0916\alpha_2,$$

$$k = 0.3015 + 0.2109 \times 10^{-3} \frac{\omega_i(\omega_i+2\omega)}{\omega^2}\beta - 0.1125\alpha_2,$$

$$\ell = 0.0808 - 0.0152 \times 10^{-3} \frac{\omega_i(\omega_i+2\omega)}{\omega^2}\beta - 0.0095\alpha_2,$$

$$\alpha_1 = 0.5625 + 0.2475 \times 10^{-3} \frac{\omega_i(\omega_i+2\omega)}{\omega^2}\beta - 0.4300\alpha_2.$$

We deduce from these relations:

$$\text{Model I: } k = 0.495\ h + 0.0054 \times 10^{-3} \frac{\omega_i(\omega_i+2\omega)}{\omega^2}\beta,$$

$$\text{Model II: } k = 0.489\ h + 0.0019 \times 10^{-3} \frac{\omega_i(\omega_i+2\omega)}{\omega^2}\beta - 0.0677\alpha_2$$

TABLE 6.3. Molodensky's Theory

		A	Frequency ω_i	$\Delta\omega_i = \omega_i - \omega$	$\omega/\Delta\omega_i$	β Mod.1	β Mod.2
Q_1	135.655	0.07216	-13.398 660 87	1.642 407 77	9.157 938	5.26	5.00
O_1	145.555	0.37689	-13.943 035 57	1.098 033 07	13.698 193	7.08	6.50
M_1	155.655	-0.02964	-14.496 693 93	0.544 374 71	27.629 992	12.45	12.06
π_1	162.556	0.01029	-14.917 864 69	0.123 203 95	122.082 681	42.23	41.20
P_1	163.555	0.17554	-14.958 931 36	0.082 137 28	183.121 117	56.86	55.50
S_1	164.556	-0.00423	-15.000 001 96	0.041 066 68	366.259 758	87.89	85.73
	165.545	0.01050	-15.038 862 23	0.002 206 41	6816.987 155	185.30	180.17
K_1	165.555	-0.53050	-15.041 068 64	0.000 000 00	∞	197.92	192.30
	165.565	-0.07182	-15.043 275 05	-0.002 206 41	-6816.987 155	212.41	206.20
	165.575	0.00154	-15.045 481 47	-0.004 412 83	-3408.485 856	229.21	222.94
ψ_1	166.554	-0.00423	-15.082 135 30	-0.041 066 66	-366.259 758	717.52	728.04
ϕ_1	167.555	-0.00756	-15.123 205 90	-0.082 137 26	-183.121 117	-125.17	-124.06
	168.554	-0.00044	-15.164 272 59	-0.123 203 95	-122.082 681	-61.91	-61.22
J_1	175.455	-0.02964	-15.585 443 33	-0.544 374 69	-27.629 992	-10.87	-10.86
OO_1	185.555	-0.01623	-16.139 101 69	-1.098 033 05	-13.698 193	-4.45	-4.54
ν_1	195.455	-0.00311	-16.683 476 39	-1.642 407 75	-9.157 938	-2.44	-2.57

Argument		Nutation period		Molodensky		η/η_0 Jeffreys	
		M. sid. day	M. sol. day	Mod.1	Mod.2	Mod.1	Mod.2
Q_1	135.655	9.157 938	9.132 933	1.0137	1.0121		
O_1	145.555	13.698 193	13.660 791	1.0254	1.0214	1.0269	1.0266
M_1	155.655	27.629 992	27.554 550	1.0369	1.0353		
π_1	162.556	122.082 692	121.749 353	1.0389	1.0372		
P_1	163.555	183.121 117	182.621 117	1.0359	1.0344	1.0350	0.9707
S_1	164.556	366.259 758	365.259 710	1.0286	1.0273		
	165.545	6816.987 155	6798.373 824	1.0037	1.0032	1.0036	1.0012
K_1	165.555	∞	∞				
	165.565	-6816.987 155	-6798.373 824	0.9962	0.9964	0.9964	0.9989
	165.575	-3408.485 856	-3399.179 212	0.9918	0.9921		
ψ_1	166.554	-366.259 758	-365.259 710	1.2472	1.2448		
ϕ_1	167.555	-183.121 117	-182.621 117	1.0884	1.0857	1.0895	1.1420
	168.554	-122.082 692	-121.749 353	1.0678	1.0814		
J_1	175.455	-27.629 992	-27.554 550	1.0687	1.0663		
OO_1	185.555	-13.698 193	-13.660 791	1.0798	1.0772	1.0768	1.0670
ν_1	195.455	-9.157 938	-9.132 933	1.0924	1.0897		

TABLE 6.4. Love numbers obtained by Molodensky's dynamical theory (see Fig. 6.2).

	Model I					Model II			
	h	k	γ	δ		h	k	γ	δ
Q_1	0.621	0.307	0.686	1.160		0.615	0.300	0.685	1.165
O_1	0.618	0.305	0.687	1.160		0.614	0.300	0.686	1.164
π_1	0.601	0.297	0.696	1.155		0.599	0.293	0.694	1.160
P_1	0.594	0.294	0.700	1.153		0.593	0.290	0.697	1.158
S_1	0.579	0.286	0.707	1.150		0.580	0.283	0.703	1.155
165.545	0.533	0.263	0.730	1.138		0.540	0.259	0.719	1.151
K_1	0.527	0.260	0.733	1.137		0.535	0.261	0.726	1.143
165.565	0.521	0.256	0.735	1.137		0.529	0.252	0.723	1.151
ψ_1	0.959	0.478	0.520	1.242		0.928	0.455	0.527	1.246
ϕ_1	0.680	0.337	0.657	1.174		0.670	0.328	0.658	1.178
J_1	0.626	0.310	0.684	1.161		0.621	0.304	0.683	1.166
OO_1	0.623	0.308	0.685	1.161		0.619	0.303	0.684	1.165
STAT.	0.621	0.307	0.686	1.160					

TABLE 6.5.

Model	RESONANCE Frequencies (°/hour)	Periods h m s
Jeffreys-Vicente II	14.938 789 4	24 05 54.0
Molodensky I	15.073 265 1	23 53 00.0
Molodensky II	15.073 612 5	23 52 58.0
Jeffreys-Vicente I	15.074 760 6	23 52 51.5
Jeffreys-Vicente II	15.101 684 1	23 50 18.2

Shen and Mansinha (1976) have constructed a general first-order theory describing those small oscillations of a rotating oblate earth that are affected by the presence of a liquid outer core. They include the effects of the nutation due to the rotation of the outer core relative to the solid earth and demonstrate that at a period close to the sidereal day the tesseral spheroidal mode S_2^1 is accompanied by a toroidal rigid rotation of the liquid outer core T_1 with respect to the solid earth. The resonance effects due to the existence of $S_2^1\ T_1$ are reflected in the diurnal tidal Love numbers and are responsible for the existence of a diurnal nutation. The numerical results obtained by Shen and Mansinha for neutral cores (Adams Williamson condition satisfied - see Section 5.8) agree with those of Molodensky.

Moreover, when using earth models with uniform polytropic cores as given by Pekeris and Accad they found practically no difference for β = -0.2, 0.0 and +0.2 (stability parameter, (5.87)). This shows that the $S_2^1\ T_1$ core mode is not sensitive to the density distribution in the core.

6.3. THE JEFFREYS-VICENTE THEORY

In the development of a theory of the motions in the liquid core and their effects in the Earth's dynamics, the attention of the early workers was attracted first by the nutations phenomena (free and forced) because of the precise astronomical observations which have been available since about 1870. At that time no experimental data existed on earth tides and it was not foreseen that they would be available in the near future. Moreover it was only much later that it was realized that the nutations are associated with the tesseral diurnal tides (Chapter 2).

Sir Harold Jeffreys (1949) was the first to stress the significance of the Poincaré theory for geodynamics and geophysics and he gave the first numerical estimate of the resonance effect on Love numbers in 1956. He showed for the first time that the discrepancy between the observed and calculated forced nutation amplitudes can only be explained by consideration of the fluidity of the outer core. He and Vicente later on made important extensions of this theory by using a variational method. This method however does not indicate clearly the degree of approximation and some of the steps are not particularly obvious.

For what concerns the density distribution Jeffreys and Vicente consider that through most of the core, the variation of density is mainly due to compression and that, provided the moments of inertia and the ellipticity e are correct, the detailed distribution of density is of secondary importance.

They adopt the Bullard's value for the core ellipticity:

$$e = 0.002\ 567 = 1/389.6 \qquad (6.157)$$

(Bullen's value, 1936, was 1/384.6)
and take two models:

a) homogeneous and incompressible, with an additional particle at the centre (representing the inner core) chosen to make the mass and moment of inertia of the core correct.

b) Roche density law:

$$\rho(r) = \rho_0 - \rho_1 \left(\frac{r}{b}\right)^2 \qquad (6.158)$$

(b being the radius of the core), the variation of density being wholly due to pressure variation (Adams Williamson condition - see Section 5.7).

They also estimate that the Radau approximation will always give solutions valid within one per cent. The numerical results of their two models are reproduced in Table 6.6 but they clearly are in conflict with the numerous observations now available principally for the δ factor but not for the γ factor and this gives a k/h ratio which looks very improbable.

In turn Jeffreys and Vicente (1966) have objected to Molodensky's solution for his second model on the basis that, for diurnal tides, the differential equation to be satisfied is hyperbolic, and "with the new boundary conditions imposed by an inner core, the complementary functions needed are of a totally different form".

Information about the liquid core which may be derived from Earth tides theory and observations

The static theory as developped in chapters 4 and 5 has essentially been used to fix tentatively an upper limit to the rigidity of the core.

The Love numbers h and k indeed depend very much upon this parameter when it exceeds 5×10^{10} cgs but the effect is more or less cancelled when we combine them to form the observed parameters γ and δ. This is not the case however for the ratio ℓ/h which can be easily determined with extensometers without any need for calibration.

Therefore when looking at the table 4.3 from Takeuchi, one can propose an upper limit of 10^{10} cgs while the Molodensky model 2 in the Table 5.1 shows that the value 3×10^{10} cgs is already too large but unfortunately does not give information between that value and zero.

The results obtained by Alsop and Kuo, on the other hand, should permit to fix an upper limit at 1.5×10^{11} cgs.

Jeffreys and Vicente (1966) emphazise the fact that the theory of the dynamical effects of a liquid core is the only known explanation for the experimentally observed alteration of the nutations amplitudes and their associated tesseral diurnal tidal waves amplitudes by a resonance effect.

This dynamical theory would not be applicable "if the rigidity was more than would make a transverse wave traverse the core's radius (3400 km) in a day. This critical value would make the velocity v_t about 0.04 km s^{-1} and the corresponding rigidity ($v_t^2\rho$) would be about 2.10^8 cgs" which is well below any limit suggested by the static models so far described.

TABLE 6.6. Love numbers obtained by Jeffreys and Vicente's dynamical theory

Central particle model								
$\Delta\omega_i/\omega$		h	k	ℓ	γ	δ	ℓ/h	k/h
-1/13.7	OO1	0.590	0.244	0.082	0.654	1.224	0.139	0.413
-1/183		0.523	0.218	0.084	0.695	1.196	0.161	0.417
-1/6800		0.490	0.205	0.086	0.715	1.182	0.176	0.418
0	K1	0.492	0.206	0.086	0.714	1.183	0.175	0.419
1/6800		0.494	0.207	0.086	0.713	1.184	0.174	0.419
1/183	P1	0.555	0.231	0.082	0.676	1.209	0.148	0.416
1/13.7	O1	0.584	0.242	0.082	0.658	1.221	0.140	0.414
static		0.585	0.289	0.082	0.704	1.152	0.140	0.494

Roche model								
$\Delta\omega_i/\omega$		h	k	ℓ	γ	δ	ℓ/h	k/h
-1/13.7	OO1	0.597	0.258	0.070	0.661	1.210	0.117	0.432
-1/183		0.710	0.298	0.072	0.588	1.263	0.101	0.420
0	K1	0.551	0.244	0.082	0.693	1.185	0.149	0.443
1.183	P1	0.568	0.264	0.084	0.696	1.172	0.148	0.465
1/13.7	O1	0.603	0.261	0.078	0.658	1.211	0.129	0.433
static		0.598	0.273	0.082	0.675	1.189	0.137	0.456

DISSIPATIVE CORE MANTLE COUPLING

A sharp resonance as shown on the figure 6.2 is in fact not possible because of friction effects which are acting essentially within the core-mantle boundary layer and are most prominent at the resonant frequency.

The consequences of a dissipative core mantle coupling have been investigated first by Toomre (1974) who used the failure to detect a phase lag in the nutation to set up an upper limit to the core viscosity ($\nu < 10^4$ poises). Then Rochester (1976) showed that the secular decrease of obliquity suggested by Aoki as another consequence of the core viscosity was extremely small, far below the level of detectability.

Sasao, Okamoto and Sakai (1977) have estimated the effect of damping on the sharpness of the resonance and obtained the classical representation as given on the figure 6.3 where α is the damping coefficient of the free core nutation.

TABLE 6.7. Resonance Effect on astronomical nutations according to the Molodensky's models

		Tidal arg.			Frequency	Model 1 Amplitude	Model 1 Period	Model 2 Amplitude	Model 2 Period
34	M	135.655	N	Q1	−13.39866087	0.07315	9.157938	0.07303	9.157938
149	M	195.455	N	NU1	−16.68347639	−0.00340	9.157938	−0.00339	9.157938
57	M	145.555	N	O1	−13.94303557	0.38646	13.698192	0.38496	13.698192
139	M	185.555	N	OO1	−16.13910169	−0.01753	13.698192	−0.01748	13.698192
82	M	155.655	N	M1	−14.49669393	−0.03073	27.629992	−0.03069	27.629992
122	M	175.455	N	J1	−15.58544333	−0.03168	27.629992	−0.03161	27.629992
112	S	168.554	N		−15.16427258	−0.00047	122.082681	−0.00048	122.082681
92	S	162.556	N	PI1	−14.91786469	0.01069	122.082681	0.01067	122.082681
95	SM	163.555	N	P1	−14.95893136	0.18215	183.121117	0.18189	183.121117
109	S	167.555	N	FI1	−15.12320590	−0.00823	183.121117	−0.00821	183.121117
99	S	164.556	N	S1	−15.00000196	−0.00435	366.259758	−0.00434	366.259758
106	S	166.554	N	PSI1	−15.08213530	−0.00528	366.259758	−0.00527	366.259758
101	M	165.545	N		−15.03886222	0.01054	6816.987155	0.01053	6816.987155
103	M	165.565	N		−15.04327504	−0.07155	6816.987155	−0.07156	6816.987155

$E_{\zeta} = 0.0164120$ Rigid Earth 9.2100

			Tidal arg.		Period	Model 1 $\sin\theta \cdot \Delta\psi$	Model 1 $\Delta\theta$	Model 1 Arg.	Model 2 $\sin\theta \cdot \Delta\psi$	Model 2 $\Delta\theta$	Model 2 Arg.
34	149	M	135.655	195.455	9.157938	−0.010483	0.011505	85.455	−0.010466	0.011485	85.455
57	139	M	145.555	185.555	13.698192	−0.082940	0.090822	75.555	−0.082614	0.090474	75.555
82	122	M	155.655	175.455	27.629992	0.028300	0.000430	65.455	0.028250	0.000417	65.455
112	92	S	168.554	162.556	122.082681	−0.020477	−0.022360	58.554	−0.020416	−0.022340	58.554
95	109	SM	163.555	167.555	183.121117	−0.522696	0.572164	57.555	−0.521975	0.571323	57.555
99	106	S	164.556	166.554	366.259758	0.057886	0.005590	56.554	0.057766	0.005590	56.554
101	103	M	165.545	165.565	6816.987155	6.825822	9.184261	55.565	6.828060	9.184261	55.565

$E_{\zeta} = 0.0164427$ Rigid Earth 9.2272

			Tidal arg.		Period	Model 1 $\sin\theta \cdot \Delta\psi$	Model 1 $\Delta\theta$	Model 1 Arg.	Model 2 $\sin\theta \cdot \Delta\psi$	Model 2 $\Delta\theta$	Model 2 Arg.
34	149	M	135.655	195.455	9.157938	−0.010503	0.011526	85.455	−0.010486	0.011507	85.455
57	139	M	145.555	185.555	13.698192	−0.083096	0.090992	75.555	−0.082769	0.090643	75.555
82	122	M	155.655	175.455	27.629992	0.028353	0.000431	65.455	0.028303	0.000417	65.455
112	92	S	168.554	162.556	122.082681	−0.020515	−0.022402	58.554	−0.020455	−0.022382	58.554
95	109	SM	163.555	167.555	183.121117	−0.523674	0.573235	57.555	−0.522951	0.572392	57.555
99	106	S	164.556	166.554	366.259758	0.057994	0.005600	56.554	0.057874	0.005600	56.554
101	103	M	165.545	165.565	6816.987155	6.838591	9.201441	55.565	6.840832	9.201441	55.565

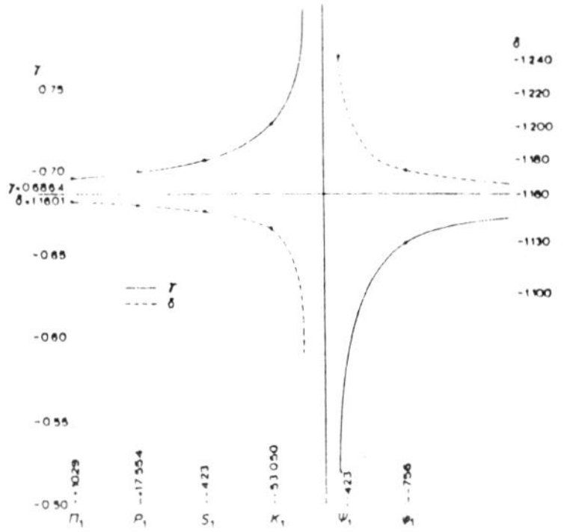

Fig. 6.2. The resonance effect on the observed tidal factors γ (solid line) and δ (dashed line) according to the Molodensky model 1.

Theoretical amplitude coefficients are indicated in correspondance of the tidal waves symbols which are put in abscissae as functions of their frequencies.

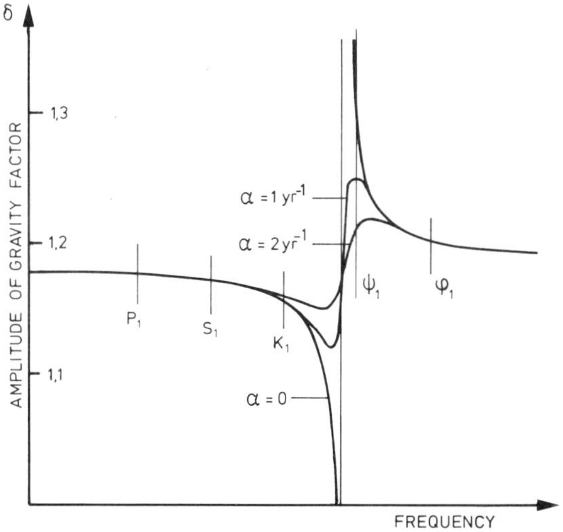

Fig. 6.3. The viscosity in the liquid earth core determines how strong will be the resonance (according to Sasao, Okamoto and Sakai).

The difficulty in the precise experimental determination of the amplitude of the very small wave ψ_1 has, until now, not allowed to evaluate the viscosity in the liquid core in this way.

6.4. RECENT THEORETICAL DEVELOPMENTS

6.4.1. Introduction

Until now the development of the spheroïdal tidal deformations of the Earth, which may be considered as infinitesimally small disturbances, was made for a SNREI model Earth: spherical, non rotating, elastic, isotropic. It is well known indeed that with such a model the dynamical equations of equilibrium become separable into two sets: a system of order 6 which has purely spheroïdal oscillations σ_n^m as solutions and a system of order 2 which has purely toroïdal τ_n^m oscillations as solutions (no radial displacements and consequently no cubic dilatations, no density changes, no gravity variations). In the toroïdal system the restoring force is proportional to the shear modulus μ and consequently such oscillations do notexist inside the liquid core.

The ellipticity and rotation were introduced only for some restricted fluid core motions by Poincaré and later in the Jeffreys theory and in the Molodensky theory but these theories have been specifically formulated for the diurnal tesseral earth tides and require an Adams Williamson core which makes them unsuitable for a generalized treatment.

On the other hand, the theory of astronomical precession-nutations that we developped in 1968 and reproduce here in the chapter 2 shows that, due to the flattening of the Earth, a net torque appears which results from the tesseral tidal forces (equations 2.8 or 2.22) and that, because the Earth is rotating this torque produces a classical gyroscopic effect which is the precession-nutations itself (equation 2.53).

This particular case showed thus that the σ_2^1 spheroïdal tidal deforming potential becomes coupled to a rigid rotation τ_2^1 around an axis rotating itself in the XOY plane (equation 2.37).

These considerations have been generalized by M.L. Smith (1973) which showed that there is a coupling of *any* spheroïdal displacements with toroïdal displacements as a direct result of ellipticity and rotation which excite the toroïdal mode.

The coupling series is an infinite sequence of terms which has to be truncated to allow to build a numerical solution.

This approach of the problem is extremely fruitful because it does not separate the gravitational modes of the rotating elliptic earth from all other elastic modes.

In 1911, already, in his book "Some Problems of Geodynamics" Love discussed the effects of inertia (Chapter V) and ellipticity of the earth (Chapter VI) on the tidal deformations. He investigated the perturbation on the M_2 wave and indicated the dependency of the correction from the fourth order function $(7z^2-r^2)/a^2$ which is equal to $(7 \sin^2\phi-1)$ at the surface of the Earth. His conclusion was that "without calculation we should expect the correction for inertia to be small of the order $\omega^2 a/g$ [which is 1/289 (0.0035)]and the correction for ellipticity to be small of the order ε [which is 1/297 (0.0033]".

Moreover, in the case of a homogeneous incompressible Earth Love found that the inertia effect is multiplied by a small coefficient while the coefficient by which ε is multiplied differs little from unity. The correction coefficient could thus be expected to be around 0.004.

6.4.2. Spheroïdal and Toroïdal displacements

The classical theory of Earth Tides involves only spheroïdal fields as the tides essentially cause radial deformations and cubic dilatations.

However a general field of displacements involves also the so called toroïdal fields. Considering the typical geometry of the planetary problems, it is clear that surface spherical harmonics will play a major role in the solutions of our equations. To any surface harmonic

$$Y_n^m(\theta,\lambda) = P_n^m(\cos\theta)\, e^{im\lambda} \tag{6.159}$$

we can associate two distinct varieties of vectorial fields:

- a spheroïdal field S_n^m which is defined by prescribing the form of two scalar functions of the radius

$$\bar{\sigma} = \sum_{n=0}^{\infty} \sum_{m=0}^{n} \{\hat{r}\, U_n^m(r)\, Y_n^m + V_n^m(r)\, \bar{\nabla}_1\, Y_n^m\} \tag{6.160}$$

$$\bar{\nabla}_1 = \hat{\theta}\,\frac{\partial}{\partial\theta} + \hat{\lambda}\,\frac{\partial}{\sin\theta\,\partial\lambda} \tag{6.161}$$

$\bar{\nabla}_1$ being the gradient operator on the unit sphere.

- a toroïdal vectorial field T_n^m which is defined by prescribing the form of one scalar function of the radius

$$\bar{\tau} = \sum_{n=0}^{\infty} \sum_{m=0}^{n} W_n^m(r)\, \{-\hat{r} \wedge \bar{\nabla}_1\, Y_n^m\} \tag{6.162}$$

which obviously has no radial component and no cubical dilatation

$$\bar{r} \wedge \bar{\nabla}_1\, Y_n^m = -\,\mathrm{rot}\,(\bar{r}\, Y_n^m)$$

Consequently it does not modify the density, nor the gravity field. Such displacements involve shear movements only and therefore are confined to the mantle. In the core

$$W(r) = 0 \qquad (0 \leqslant r \leqslant b)$$

However their resonant frequencies depend on the boundary conditions at the core mantle interface and on the nature of the core mantle coupling. In this way they can give informations on the nature of this interface.

The functions U(r), V(r) and W(r) define the positions and number of nodal surfaces associated to an infinitesimal deformation. U(r) is everywhere continuous but V(r) and W(r) can have discontinuities at interfaces.

The parameters n(degree) and m (order) determine through the distribution of their nodal lines the surface pattern of the movements during the deformation associated to a particular oscillation:

$$\left.\begin{aligned}
(\sigma_n^m)_r &= U_n^m(r)\, P_n^m(\cos\theta)\,\cos(\omega_i t - m\lambda)\\
(\sigma_n^m)_\theta &= V_n^m(r)\,\frac{\partial}{\partial\theta} P_n^m(\cos\theta)\,\cos(\omega_i t - m\lambda)\\
(\sigma_n^m)_\lambda &= m\, V_n^m(r)\,\frac{1}{\sin\theta}\, P_n^m(\cos\theta)\,\sin(\omega_i t - m\lambda)
\end{aligned}\right\} \tag{6.163}$$

$$\left(\text{as } \frac{\partial}{\partial\lambda} \to m\,\sin(\omega_i t - m\lambda)\right)$$

$$\left.\begin{aligned}
(\tau_n^m)_r &= 0 \\
(\tau_n^m)_\theta &= -m\, W_n^m(r) \frac{1}{\sin\theta} P_n^m(\cos\theta) \cos(\omega_i t - m\lambda) \\
(\tau_n^m)_\lambda &= -\, W_n^m(r) \frac{\partial}{\partial\theta} P_n^m(\cos\theta) \sin(\omega_i t - m\lambda)
\end{aligned}\right\} \qquad (6.163)$$

We shall thus introduce in the fundamental equations

$$\left.\begin{aligned}
u_r &= \sum_n \sum_m U_n^m(r)\, P_n^m(\cos\theta) \cos(\omega_i t - m\lambda) \\
u_\theta &= \sum_n \sum_m V_n^m(r) \frac{\partial}{\partial\theta} P_n^m(\cos\theta) \cos(\omega_i t - n\lambda) \\
&\quad - \sum_n \sum_m m W_n^m(r) \frac{1}{\sin\theta} P_n^m(\cos\theta) \cos(\omega_i t - n\lambda) \\
u_\lambda &= \sum_n \sum_m m\, V_n^m(r) \frac{1}{\sin\theta} P_n^m(\cos\theta) \sin(\omega_i t - m\lambda) \\
&\quad - \sum_n \sum_m W_n^m(r) \frac{\partial}{\partial\theta} P_n^m(\cos\theta) \sin(\omega_i t - m\lambda)
\end{aligned}\right\} \qquad (6.164)$$

the total field of displacements is thus to be written in the form

$$\bar{S}(r,\theta,\lambda) = \bar{\sigma} + \bar{\tau} = \sum_n \sum_m \{ \bar{\tau}_n^m(r,\theta,\lambda) + \bar{\sigma}_{n+1}^m (r,\theta,\lambda)\} \qquad (6.165)$$

an infinite series of terms.
However the symetries existing in the earth model considerably restrict the form of the displacements:

(1) there is invariance of the rotating elliptic earth model for a rigid rotation around OZ (there is indeed no difference between the initial earth and the earth after a rotation by an angle λ).
It results that the developments contain only those terms which share the same functional dependence in λ. All terms therefore must have the same m numerical value.

(2) there is invariance in a pointwise inversion through the center of mass of the earth: the mass particles associated in this way cannot be distinguished. Martin Smith has demonstrated that consequently

$\bar{\sigma}_n^m$ is even or odd according to n is odd or even

$\bar{\tau}_n^m$ is even or odd according to n is even or odd

and as the total field must be either even, either odd, the displacement field of the particles, associated to a specific normal mode must have the truncated form

$$\left.\begin{aligned}
&\bar{s}(r,\psi,\omega) = \bar{\tau}_n^m(r,\psi,\omega) + \bar{\sigma}_{n+1}^m(r,\psi,\omega) + \bar{\tau}_{n+2}^m(r,\psi,\omega) \\
\text{or}\quad &\bar{s}(r) = \bar{\sigma}_n^m(r,\psi,\omega) + \bar{\tau}_{n+1}^m(r,\psi,\omega) + \bar{\sigma}_{n+2}^m(r,\psi,\omega)
\end{aligned}\right\} \qquad (6.166)$$

which gives a system of order 10 for the mantle (6S + 2T + 2T) and of order 4 for the fluid core (2S + 1T + 1T).

6.4.3. The role of the Coriolis force as an example of coupling

As said before, the differential equations for a SNREI earth do not couple any σ_ℓ^m or τ_ℓ^m to any other σ_ℓ^m or τ_ℓ^m.

It is very easy to show how the introduction of the Coriolis force in the equations results in a coupling between σ_ℓ^m and $\tau_{\ell-1}^m$, $\tau_{\ell+1}^m$.

When we write indeed the first member of the first differential equation (in u_r) as

$$\frac{\partial^2 u_r}{\partial t^2} - 2\,\omega \sin\theta \frac{\partial u_\lambda}{\partial t} - \ldots \tag{6.167}$$

we immediately observe that the effect of the Coriolis term is to produce a coupling between the movements along the different axes of reference: here in (6.167) u_λ with u_r. The first term of (6.167) having no toroïdal part simply gives:

$$\frac{\partial^2 u_r}{\partial t^2} = -\,\omega_i^2\; U_n(r)\; P_n^m(\cos\theta)\,\cos(\omega_i t - m\lambda) \tag{6.168}$$

while the second is to be developped into two parts

spheroïdal

$$-\,2\,\omega \sin\theta \frac{\partial u_\lambda}{\partial t} = -2\,\omega\,\omega_i\, m\, V_n^m\, P_n^m(\cos\theta)\,\cos(\omega_i t - m\lambda) \tag{6.169}$$

toroïdal

$$-2\,\omega \sin\theta \frac{\partial u_\lambda}{\partial t} = +\,2\,\omega\,\omega_i\, W_n^m \sin\theta \frac{\partial}{\partial\theta} P_n^m(\cos\theta)\,\cos(\omega_i t - m\lambda) \tag{6.170}$$

where

$$\left.\begin{aligned} &\sin\theta \frac{\partial}{\partial\theta} P_n^m(\cos\theta) = -\sin^2\theta \frac{d\,P_n^m(\mu)}{d\mu} \\ &\mu = \cos\theta, \qquad\qquad 1-\mu^2 = \sin^2\theta \end{aligned}\right\} \tag{6.171}$$

the classical relation

$$(1-\mu^2)\frac{d\,P_n^m(\mu)}{d\mu} = -\frac{n(n-m+1)}{2n+1} P_{n+1}^m + \frac{(n+1)(n+m)}{2n+1} P_{n-1}^m \tag{6.172}$$

allows to write

$$(1-\mu^2)\frac{d\,P_{n-1}^m}{d\mu} = -\frac{(n-1)(n-m)}{2n-1} P_n^m + \frac{n(n-1+m)}{2n-1} P_{n-2}^m \tag{6.173}$$

as well as

$$(1-\mu^2)\frac{d\,P_{n+1}^m}{d\mu} = -\frac{(n+1)(n-m+2)}{2n+3} P_{n+2}^m + \frac{(n+2)(n+m+1)}{2n+3} P_n^m \tag{6.174}$$

There are thus P_n^m polynomials in the W_{n-1}^m term and in the W_{n+1}^m term as well but not in W_n^m. Thus the Coriolis term becomes

$$+2\,\omega\,\omega_i \left\{\frac{(n-1)(n-m)}{2n-1} W_{n-1}^m - m\,V_n^m - \frac{(n+2)(n+m+1)}{2n+3} W_{n+1}^m\right\} P_n^m(\cos\theta)\,\cos(\omega_i t - m\lambda) \tag{6.175}$$

This gives rise, in the core, to a coupling between the spheroïdal and toroïdal displacement fields which appear with different n degrees.
For n=2, m=1 (6.175) is written as

$$2\,\omega\,\omega_i\,\left(\frac{1}{3}W_1^1 - V_2^1 - \frac{16}{7}W_3^1\right) P_2^1\,(\cos\theta)\,\cos(\omega_i t - \lambda) \qquad (6.176)$$

By coupling the surface displacement u_λ with the radial displacement u_r the rotation transfers energy from the spheroïdal to a toroïdal oscillation. The amplitude of the coupling depends of the proximity of the spheroïdal and the toroïdal frequencies.

As W_1 is a linear function of the radius (degree is 1) the core motion is a rigid rotation with respect to the mantle and the inner core.

6.4.4. Results of J. Wahr

However the convergence of the series has not been demonstrated and there is no reason to believe that any normal mode of the rotating Earth is represented by the series (9.17) or (9.18) or by any finite linear combination of spheroïdal and toroïdal fields. One should therefore consider the results of calculations with caution unless one can check them by observations or by calculations made with a series containing more terms.

The coupling sequences for the order 2 are

$$S = \tau_1^1 + \sigma_2^1 + \tau_3^1 + \sigma_4^1 + \ldots \qquad (6.177)$$

and

$$S = \sigma_2^2 + \tau_3^2 + \sigma_4^2 + \ldots \qquad (6.178)$$

(as n cannot be lower than m).

The nutation is represented in this way by

$$\bar{S}(r,\psi,\omega) = \tau_1^1\,(r,\psi,\omega) + \sigma_2^1\,(r,\psi,\omega) \qquad (6.179)$$

As seen before τ_1^1 is a rigid rotation around an axis which himself slowly rotates in the equatorial plane while σ_2^1 is the elastic deformation corresponding to the centrifugal potential associated to a rigid rotation τ_1^1.

In a more general way one can write

$$\bar{S} = \tau_1^{\pm 1} + \sigma_2^{\pm 1} + \tau_3^{\pm 1} + \sigma_4^{\pm 1} + \ldots \qquad (6.180)$$

where + corresponds to a retrograde rotation and - to a prograde rotation. One could also consider a rigid rotation of one shell of radius a and b by writing that $\tau_1^1(\bar{r})$ and $\tau_1^{-1}(\bar{r})$ are associated to a form $W_1^{\pm 1}(r) = \alpha r$ inside $a \leqslant r \leqslant b$, α being a convenient constant.

It is clear that the truncated series (6.180) will give for the tidal variation of gravity a form like

$$\Delta g = -\frac{2g}{r_\emptyset}\left\{\delta\,Y_\ell^m + \delta_+\,Y_{\ell+2}^m + \delta_-\,Y_{\ell-2}^m\right\} A_{\ell m i}\,(\omega_i) \qquad (6.181)$$

where δ_+ and δ_- are the latitude dependent corrections to the classical δ amplitude factor. It is thus convenient to write now

$$\delta = \delta_o + \delta_+\,\frac{Y_{\ell+2}^m}{Y_\ell^m} + \delta_-\,\frac{Y_{\ell-2}^m}{Y_\ell^m} \qquad (6.182)$$

and to keep the usual formula

$$\Delta g = -\frac{2g}{r}\,\delta .\ A_{\ell mi}\ (\omega_i)\ Y_\ell^m\ (\theta,\lambda) \tag{6.183}$$

In the cases which interests us ℓ=2, m=1 and =2, m=2 we shall thus simply have

$$\left.\begin{aligned} \delta^{(2)}\ (\text{semi diurnals}) &= \delta_o^{(2)} + \delta_+^{(2)}\ \frac{Y_4^2}{Y_2^2} \\ \delta^{(1)}\ (\text{diurnals}) &= \delta_o^{(1)} + \delta_+^{(1)}\ \frac{Y_4^1}{Y_2^1} \end{aligned}\right\} \tag{6.184}$$

or

$$\left.\begin{aligned} \delta^{(2)} &= \delta_o^{(2)} + \delta_+^{(2)} \left\{\frac{\sqrt{3}}{2}\,(7\sin^2\phi - 1)\right\} \\ \delta^{(1)} &= \delta_o^{(1)} + \delta_+^{(1)} \left\{\frac{\sqrt{6}}{4}\,(7\sin^2\phi - 3)\right\} \end{aligned}\right\} \tag{6.185}$$

we evidently find in (6.185) the function indicated by Love in 1910.
J. Wahr (1981) has calculated the δ_o and δ_+ for different most recent earth models, neutral (β=0) or stable (β<0) fluid cores (see § 5.8).
These models are 1066 A from Gilbert and Dziewonski, PEM-C from Dziewonski, Hales and Lapwood and C 2 from Jordan and Anderson.

The results are so close to each other that there is presently no hope at all to make a choice on the basis of earth tide data.
However Wahr obtains coefficients which can be checked by observations:

$$\left.\begin{aligned} \delta_o^{(1)}\ (O_1) &= 1.152 & \gamma_o^{(1)}\ (O_1) &= 0.689 \\ \delta_o^{(1)}\ (P_1) &= 1.147 & \gamma_o^{(1)}\ (P_1) &= 0.700 \\ \delta_o^{(1)}\ (K_1) &= 1.132 & \gamma_o^{(1)}\ (K_1) &= 0.730 \\ \delta_o^{(1)}\ (\psi_1) &= 1.235 & \gamma_o^{(1)}\ (\psi_1) &= 0.523 \\ \delta_o^{(1)}\ (\phi_1) &= 1.167 & \gamma_o^{(1)}\ (\phi_1) &= 0.660 \\ \delta_o^{(1)}\ (OO_1) &= 1.154 & \gamma_o^{(1)}\ (OO_1) &= 0.687 \end{aligned}\right\} \tag{6.186}$$

with

$$\left.\begin{aligned} \delta_+^{(1)} &= -0.006 \text{ for the neutral 1066 A} \qquad \gamma_+^{(1)} = -0.001 \\ &= -0.007 \text{ for PEM - C} \\ &= -0.007 \text{ for C2} \\ \delta_o^{(2)} &= 1.160 \quad \gamma_o^{(2)} = 0.692 \quad \delta_o^{(0)} = 1.155 \quad \gamma_o^{(0)} = 0.689 \\ \delta_+^{(2)} &= -0.005 \quad \gamma_+^{(2)} = -0.001 \quad \delta_o^{(0)} = -0.007 \quad \gamma_+^{(0)} = -0.001 \end{aligned}\right\} \tag{6.187}$$

There exists thus, *in principle*, a possibility to check if the stratification is neutral or not from the coefficient $\delta_{+}^{(1)}$ if its determination was precise enough.

A comparison of the data provided by the Trans World Tidal Gravity Profiles of ICET which include many equatorial and tropical stations with the high latitude data collected mainly in Europe is given in the Chapter 13.

These experimental results confirm the latitude dependence indicated by Love and Wahr but give a different value for δ_o.

For what concerns the hydrodynamical resonance in the liquid core, Wahr obtains essentially results similar to those of Molodensky (model I) Po Yu Shen and Mansinha.

He has also demonstrated that the Love numbers k_o, h_o, ℓ_o depend on the diurnal frequency according to formulas like:

$$k_o(\omega) = k_o(\omega_{O_1}) + k_1 \frac{\omega - \omega_{O_1}}{\omega_1 - \omega} \qquad (6.188)$$

where ω_{O_1} is the frequency of O_1, ω_1 the frequency of the free core nutation and ω the frequency of the considered tidal diurnal component.

CHAPTER 7

Tidal Analysis

7.1. INTRODUCTORY REMARKS

Tidal analysis is a very old technique known for its applications to oceanic tides, which were the only tides to be regularly measured before the International Geophysical Year (1957). At this epoch continuous records of Earth tides began and raised different difficulties in their analysis.

In investigations of oceanic tides the theoretical static value of the amplitude and phase of each tidal wave separated by analysis of experimental data has no importance and no meaning. Even the problem of the mass of the Moon is without interest. The purpose of the analysis is an empirical but precise determination of the tidal constants aiming at the preparation of an ephemeris to be used for practical navigation problems as well as the construction of cotidal charts in the ocean concerned.

In Earth-tides studies, on the other hand, the computation of the theoretical amplitude and phase has to be made with the highest accuracy as their comparison with the observed corresponding constants furnishes the parameters related to the Earth's rheology. This method of interpretation is sufficiently explained by the fact that the Earth tide is the only geophysical phenomenon for which the acting forces (luni-solar attraction) are known beforehand, in frequency and amplitude.

This feature supplies the importance and the peculiarities of this problem. The aim of the analysis of the observations is to give the response of the terrestrial globe to each frequency in the force acting.

Consequently we have to solve clearly several problems: (1) Considering a given length of record, what tidal waves can be separated? (2) Reconstruct from the theoretical tidal spectrum the waves homologous to those groups separated by the analysis of experimental data. (3) Have the best astronomical fundamental constants (the mass of the Moon is presently known to four exact figures). (4) Be sure that the instrument is correctly and regularly calibrated during the recording period (the calibration of the best instruments is now achieved to four exact figures).

Earth-tide analysis differs from oceanic-tide analysis with respect to another important peculiarity, which is of instrumental character.

The effects to be measured are very minute and very high sensitivity must be obtained from the instruments. For example, one can calculate that the mean square error of one reading corresponds to a deflection of the beam of a gravimeter or of the foot of an horizontal pendulum of about one nanometer.

In such conditions it is no wonder that slow variations in the instrumental conditions of any origin (rheologic, thermic, molecular), acting either on the suspensions or inside the amplificators, produce a so-called "drift" of the instrumental zero. Such a drift is all but linear. For good instruments it is probably parabolic of higher order. But external effects acting on or inside the Earth's crust may more or less change such drift coefficients as, for example, underground water-level, barometric pressure, and unknown local effects.

For good instruments, correctly installed, the normal drift can be a few microgals per day in the vertical component and 1 to 3 ms of arc per day in horizontal component. But abnormal conditions can happen during some intervals, as in the case of a flood of a neighbouring river.

The analytical process of Earth-tide harmonic analysis must take into account the possibility of an irregular drift and separate it carefully from the semi-diurnal and diurnal waves. The noise which is present in all experimental measurements does not allow separation of waves having very close frequencies. In our case, the geophysical noise level, mainly due to atmospheric and seismic excitation, far exceeds the instrumental noise level. Therefore we should have records as long as possible but, looking at the tidal spectrum, one can clearly understand that it will always be impossible to separate completely all the waves from each other. If there were no noise at all, we could derive, for example, 100 different waves (their amplitudes and phases) from 200 tidal readings on the condition that these readings should extend on a sufficiently broad time interval (if not, the matrix should be ill conditionned).

Of course the mean square error on the tidal parameters to be derived increases without any limit if one tries to separate waves of too close frequencies. The fine structure of the spectrum can only be obtained from very long records. The rule generally adopted is that two waves can be separated within a given interval of time if, during this interval, their relative difference of phase reaches 360°. However, it may be possible to separate waves even if the interval is shorter. As an example, Godin (1970) takes as a rule for separation a minimum phase difference of 288°.

Nevertheless, if we accept as a safety rule that waves are separable only when their phase difference has reached 360°, we find a practical way by considering Doodson's argument number; one can separate waves having the first digit different with one-day registrations — diurnals from semi-diurnals.

Among the same species (same first digit) one can separate waves having the second digit differing by one unit into one month (M_2 from N_2 and L_2) and those having the second digit differing by two units into half a month ($M2$ from S_2).

Looking now at groups of waves having the same first two digits as, for example, the group 27 having as constituents T_2, S_2, R_2, K_2 and other minor components, one easily understands that the (T_2 + S_2 + R_2) waves can be separated from K_2 with 6 months'data while one needs one complete year to separate S_2 from T_2 and R_2. But the K_2 constituent itself is a combination of three waves: 275.555, 275.565, 275.575 which could be completely separated only with 18 years'observations.

We shall also consider group 16 which is the most important one among the diurnal species because it contains

π_1	162.556	K_1	165.555
P_1	163.555	ψ_1	166.554
S_1	164.554 + 164.556	ϕ_1	167.555

All these waves have deep geophysical meaning as they are related to the dynamical effects of the Earth's liquid core. With 6 months'observations, one will separate π_1, P_1, (S_1 + K_1 + ψ_1) and ϕ_1. One year will be necessary to separate S_1 from K_1 and ψ_1.

Moreover, we should consider that S_1, as a real tidal component, has a very minute amplitude but can in fact be much more important. Its period being one solar day and the thermic perturbations acting on the instruments combine their effects within a S_1 meteorological wave while the barometric effects rather produce a S_2 meteorological component. The effect is very troublesome with metallic instruments and in shallow stations. The meteorological S_1 wave is satisfactorily reduced with well-thermostatized and sealed gravimeters, with quartz horizontal pendulums installed in galleries without ventilation, at a minimum depth of 40 m. Nevertheless one cannot be sure that these spurious meteorological effects have been separated from the real tidal diurnal waves unless one has analysed a whole year of data.

7.2. HISTORICAL METHODS

7.2.1. Darwin's Method

The first step in the analysis of records of oceanic, gravimetric, or clinometric tides consists of measuring the ordinates of the curve registered at each exact mean solar hour. The beginning of each day is taken at 00 h 00 UT, regardless of the longitude of the station and of the local hour in use.

Readings are tabulated with 31 (or 30) lines with 24 columns numbered from 0 to 23. It becomes obvious that if we arrange these data in a table with 12 columns and 60 lines, and average each column, we shall obtain 12 points of a sinusoid representing fairly closely the wave S_2, whose period is exactly 12 solar hours. One can also determine the amplitude and the phase corresponding to the central day of the month being reduced.

One could also prepare a similar table for the wave M_2, but to do this it is necessary to read the values on the original curve for each lunar hour M_2, or calculate the values by interpolation in the original table. The same problem is involved for each wave to be investigated; N_2, L_2, K_1, O_1, etc. This would be a very long and tedious task.

To avoid such difficulties at a time when there were no calculating machines, Darwin (1883) introduced the double-point method. Because of the relative abundance of initial data (24 X 31 = 744 readings), the intrinsic precision of the hourly marks of the recorded ordinates, and the readings which are made, an approximate interpolation was sufficient. Darwin's method consists of recopying table S_2 but displacing progressively by one hour alternately every 28 or 29 h (60 min every 28 h 30 min corresponding to 50 min every 24 h, since we know that the period of M_2 is 12 h 25 min).

To carry out this displacement one has to copy two consecutive hours in each division of the table (prepared for M_2) marked with a point, hence the name "double-point" method. The distribution of the double points obviously differs for each wave, and is carried over to the appropriate standard tables. For O_1 the double points are separated by intervals of 12 and 13 h; for N_2 by intervals of 17 and 18 h. This is not a very precise method and has been abandoned for the more modern methods of Doodson, Doodson-Lennon, Lecolazet, Pertsev and Matvéev.

7.2.2. Labrouste's Theory

Let us first define the idea of selectivity based on the properties of the combination of ordinates which have been clearly set forth by Labrouste and Labrouste, in tables and an atlas, both rich in detail and easy to use.

(a) Elementary Combinations Y, Z-Selectivity

Let us consider a series of readings of equidistant ordinates

$$(y) = y_{-m}\ y_{-m+1}\ \cdots\ y_{-3}\ y_{-2}\ y_{-1}\ y_0\ y_1\ y_2\ y_3\ \cdots\ y_m$$

and the two elementary combinations

$$\text{first type:} \qquad Y_m = y_{-m} + y_m, \tag{7.1}$$

$$\text{second type:} \qquad Z_m = \pm (y_m - y_{-m}), \tag{7.2}$$

called, respectively, combination of addition and combination of subtraction.

By convention

$$Y_0 = 2y_0, \qquad Z_0 = 0. \tag{7.3}$$

Thus, for example, the application of the combination Y_2 to the series y consists in replacing the ordinate y_0 by the combination $(y_{-2} + y_2)$, the ordinate y_1 by $(y_{-1} + y_3)$, etc.

If one supposes that the series y is a series of ordinates read from a sinusoid of period $T = 2\pi/\omega$,

$$y \equiv y_i = a_n \sin (\omega x_i + \kappa), \tag{7.4}$$

x_i being the abscissae corresponding to the equidistant reading time intervals (T therefore is a multiple of this reading interval $x_{i+1} - x_i$), one has

$$\left.\begin{aligned} Y_m = y_m + y_{-m} &= a_n \sin[\omega(x_0 + m) + \kappa] + a_n \sin[\omega(x_0 - m) + \kappa] \\ &= 2a_n \cos m\omega \sin (\omega x_0 + \kappa) \end{aligned}\right\} \tag{7.5}$$

which shows that Y, transformed from y by the combination Y_m, is a sinusoid not dephased in relation to the original sinusoid, but whose amplitude is multiplied by the factor

$$\alpha_m = 2 \cos m\omega = 2 \cos m\,(360°/T). \tag{7.6}$$

The amplitude factor defines the selectivity of the combination Y_m with regard to the period T. If the initial series y is represented by the sum of a large number of sinusoids with different periods (this is true for tides) one can prepare a selectivity curve which gives the amplification obtained for each imaginable period. Y combinations uniformly amplify the long period components.

Let us now consider an application of the combination Z_m to the sinusoid

$$\left.\begin{aligned} Z_m = y_m - y_{-m} &= a_n \sin[\omega(x_0+m) + \kappa] - a_n \sin[\omega(x_0-m)+\kappa] \\ &= 2a_n \sin m\omega \cos (\omega x_0 + \kappa) \\ &= 2a_n \sin m\omega \sin(\omega x_0 + \kappa + 90°) \end{aligned}\right\} \tag{7.7}$$

This shows that Z transformed from y by the combination Z_m is a sinusoid which is in quadrature with the initial sinusoid (the combinations Z are therefore dephasing) and the amplitude of which is multiplied by the factor

$$\beta_m = 2 \sin m\omega = 2 \sin m\,(360°/T). \tag{7.8}$$

For $T = \infty$, $\beta = 0$, and we see that combinations of type Z filter the long periods. They amplify waves of medium period. One notes that the selectivity curves of the combinations Y and Z are essentially different. They are shown in Fig. 7.1.

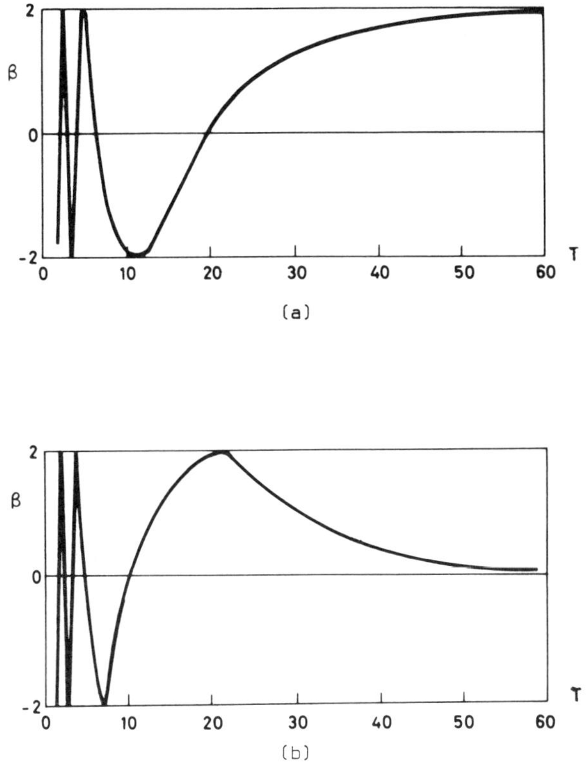

Fig. 7.1. (a) Filter Y; (b) Filter Z.

(b) Complex Combinations of Addition

$$\left.\begin{aligned}
s_m &= y_{-m} + \dots + y_{-1} + y_0 + y_1 + \dots + y_m = y_0 + Y_1 + \dots + Y_m,\\
S_m &= Y_0 + Y_1 + \dots + Y_m = S_m + y_0 ,\\
(s_m)_N &= y_0 + Y_N + Y_{2N} + \dots + Y_{mN} ,\\
(S_m)_N &= Y_0 + Y_N + Y_{2N} + \dots + Y_{mN} = (S_m)_N + y_0 ,\\
(s_m)_{N/2} &= y_0 - Y_{N/2} + Y_{2N/2} - Y_{3N/2} + \dots + (-1)^m\, Y_{mN/2} ,\\
(S_m)_{N/2} &= (S_m)_{N/2} + y_0 .
\end{aligned}\right\} \quad (7.9)$$

Their selectivity will be defined respectively by:

$$\left.\begin{aligned}
\sigma_m &= 1 + \alpha_1 + \ldots + \alpha_m = 1 + 2\sum_{i=1}^{m} \cos i\,\frac{2\pi}{T} = \frac{\sin(2m+1)\pi/T}{\sin \pi/T},\\
\Sigma_m &= 2 + \alpha_1 + \ldots + \alpha_m = 2\,\frac{\sin \pi(m+1)/T \cos \pi(m/T)}{\sin \pi/T},\\
(\sigma_m)_N &= 1 + \alpha_N + \ldots + \alpha_{mN},\\
(\Sigma_m)_N &= 2 + \alpha_N + \ldots + \alpha_{mN},\\
(\sigma_m)_{N/2} &= 1 - \alpha_{N/2} + \ldots + (-1)^m\ \alpha_{mN/2},\\
(\Sigma_m)_{N/2} &= 1 + (\sigma_m)_{N/2}.
\end{aligned}\right\} \tag{7.10}$$

The combinations of spaced ordinates $(s_m)_N$ and $(S_m)_N$ cause a large amplification of the period N, and are therefore very useful when one knows in advance the period to emphasize. It should be noted that they also amplify the harmonics of N and the long periods. The combinations $(s_m)_{N/2}$ and $(S_m)_{N/2}$ have the same property with respect to the period N, but amplify simultaneously only the uneven harmonics.

(c) Complex Combinations of Difference

These are defined as follows:

$$T_m = Z_1 + Z_2 + \ldots + Z_m \tag{7.11}$$

which have a selectivity

$$\tau_m = \beta_1 + \ldots + \beta_m = 2\,\frac{\sin \pi(m/T) \sin \pi(m+1)/T}{\sin \pi/T} \tag{7.12}$$

(since $Z_0 = 0$). One can also use

$$(T_m)_{N/4} = Z_{N/4} - Z_{3N/4} + Z_{5N/4} + \ldots \pm Z_{mN/4}. \tag{7.13}$$

(d) Multiple Combinations

This involves the successive application of Y, s or Z. In this case the final selectivity factor is the product of the successive selectivity factors. We represent the multiple combination by a symbolic product

$$\Pi\ (Y) = Y_m\ Y_p\ \ldots\ Y_r. \tag{7.14}$$

The selectivity factor is

$$\pi(\alpha) = \alpha_m\ \alpha_p\ \ldots\ \alpha_r. \tag{7.15}$$

We shall have therefore

$$\left.\begin{aligned}
\pi(s) &= s_m\ s_p\ \ldots\ s_r \quad \text{with} & \pi(\sigma) &= \sigma_m\ \sigma_p\ \ldots\ \sigma_r,\\
\pi(Z) &= Z_m\ Z_p\ \ldots\ Z_r, & \pi(\beta) &= \beta_m\ \beta_p\ \ldots\ \beta_r,\\
\pi(T) &= T_m\ T_p\ \ldots\ T_r, & \pi(\tau) &= \tau_m\ \tau_p\ \ldots\ \tau_r.
\end{aligned}\right\} \tag{7.16}$$

We should note here that the combinations Y should be very suitable for filtering out short periods and for amplifying long periods. The filtering in the zone of short periods will increase in efficiency as the number of zeros in α_i increases and are distributed as uniformly as possible in the part we wish to filter. The combinations $\pi(s)$ will be still more efficient.

(e) Fundamental Formulas

The principal formulas which simplify the associations of combinations by addition and multiplication are

$$\left.\begin{aligned} Y_m Y_r &= y_{m+r} + y_{-m+r} + y_{m-r} + y_{-m-r} = Y_{m+r} + Y_{m-r} \\ Y_m Y_r Y_p &= Y_{m+r+p} + Y_{m+r-p} + Y_{r-m+p} + Y_{r-m-p}\,, \\ Z_m Z_r &= Y_{r-m} - Y_{m+r} \qquad \text{non-dephasing,} \\ Y_m Z_r &= Z_{m+r} + Z_{r-m} \qquad \text{dephasing.} \end{aligned}\right\} \tag{7.17}$$

(f) Simple and Characteristic Examples

Combinations	Successive coefficients
$Y_1 = y_1 + y_{-1}$	1 0 1
$Y_2 = y_2 + y_{-2}$	1 0 0 0 1
$Y_1^2 = Y_2 + Y_0$	1 0 2 0 1
$Y_1^2 Y_2 = Y_2^2 + Y_2 Y_0 = Y_4 + 2Y_2 + Y_0$	1 0 2 0 2 0 2 0 1

But these elementary procedures are not applicable to combinations of high order. Labrouste and Labrouste (1943) have published, at the end of their treatise on the combinations of ordinates, tables for the calculation of the selectivities that are both extremely practical and easy to use. These tables give, for periods from T = 0.5 to T = 500, the factor of amplitude α_m, σ_m, β_m and τ_m.

(g) Property of the Combinations Z for the Elimination of the Drift

The drift can be considered as long-period waves superimposed on the tidal curve. One can assume that a parabolic arc $a + bt + ct^2$ fits quite correctly the drift on an interval of about 24 h.

If the combination of ordinates applied to isolate the waves of a certain species (diurnal or semi-diurnal) contains in factor a combination Z_i, its application will give

$$\begin{aligned} y_i - y_{-i} &= c(t_i^2 - t_{-i}^2) + b(t_i - t_{-i}) \\ &= (t_i - t_{-i})\,[c(t_i + t_{-i}) + b] \\ &= k(b + ct_j), \end{aligned}$$

where $t_j = t_i + t_{-i}$ and $(t_i - t_{-i})$ is a constant put equal to k. If the combination of ordinates contains a second combination Z_j, there will only remain a constant term k'c, where $k' = k(t_j - t_{-j})$.

As the coefficients a, b and c vary slightly during the month analysed, it is necessary to have a third combination Z in the product of the combinations to eliminate the variable coefficient c and thus all trace of such a drift in the results of the harmonic analysis.

Thus a combination Z_iZ_j completely eliminates the effects of a linear drift, a combination $Z_iZ_jZ_k$ completely eliminates the effects of a parabolic drift of the second order, etc.

7.2.3. Elimination of the Drift

It is obvious that the combination of ordinates used to isolate the drift should separate as perfectly as possible the components corresponding to diurnal and semi-diurnal tides.

In the methods of Lecolazet and of Doodson-Lennon the first combinations used for the separation of the various species of waves eliminate on account of their own structure a parabolic drift $(a + bt + ct^2)$ on each interval to which they apply. The combinations indeed contain $Z_iZ_jZ_k$ as a factor. Moreover, a special combination allows in each of these methods the calculation of the drift, whose study can be important for several reasons (geophysical or instrumental purposes).

A combination due to Doodson, simplified by Pertsev, had been chosen because it gives good selectivity and at the same time is fairly quickly applied. It is obvious that an infinite number of possible combinations exists. Table 7.1 gives a comparison of the selectivities of the combinations most frequently used.

(a) Doodson's Combination (X_0)

This combination relates to 39 consecutive ordinates of which only 30 are taken into consideration as shown in the following scheme:

1010010110201102112 0 2112011020110100101.

The combination is symetrical around the central ordinate, which is not included. It is thus written, according to Labrouste's notations:

$$(X_0) = 1/30(2Y_1+Y_2+Y_3+2Y_4+Y_6+Y_7+2Y_9+Y_{11}+Y_{12}+Y_{14}+Y_{17}+Y_{19})$$

with a convenient grouping of the combinations and using (7.17) we can write in a condensed form (Melchior, 1966):

$$(X_0) = [1/30(Y_5 + Y_5^2 - Y_0/2)](Y_0/2 + Y_8)Y_1 = 1/2(X_0')\ Y_1. \quad (7.18)$$

(b) Combination X_0' or Pertsev's Combination

Pertsev proposed another simpler combination in order to reduce the calculation involved:

$$(y_0 + Y_2 + Y_3 + Y_5 + Y_8 + Y_{10} + Y_{13} + Y_{18})/15.$$

Operating some groupings, the combination can then be written

$$\frac{1}{15}\left(Y_5 + Y_5^2 - \frac{Y_0}{2}\right)\left(\frac{Y_0}{2} + Y_8\right). \quad (7.19)$$

TABLE 7.1. Selectivity of the Combinations for Drift Computation

T (h)	$\rho[(X_0)]$	$\rho[(X_0')]$	$\rho[(L)]$
2	-0.200	0.200	0
4	0	-0.200	0
6	0	0	0
8	-0.058	-0.083	0
10	0.087	0.108	0.070
10.5	0.059	0.071	0.067
11	0.030	0.037	0.049
11.5	0.010	0.011	-0.024
12 S_2	0	0	0
M_2	0.0006	0.0007	-0.0150
12.5	0	0	-0.019
13	0.007	0.008	-0.031
13.5	0.019	0.022	-0.036
14	0.033	0.037	-0.034
14.5	0.047	0.051	-0.028
15	0.058	0.064	0.019
16	0.073	0.079	0
18	0.067	0.072	0.024
20	0.039	0.041	0.023
22	0.012	0.013	0.010
23	0.004	0.004	0.003
K_1	0	0.0002	-0.0005
24	0	0	0
25	0	0	-0.001
O_1	0.0029	0.0031	-0.0022
30	0.052	0.053	0.033
40	0.257	0.260	0.215
50	0.443	0.446	0.399
75	0.708	0.713	0.680
500	0.993	0.993	0.992
	1.000	1.000	1.000

This is the combination (X_0') in (7.18) and its selectivity factor is

$$\rho\ (X_0') = \frac{1}{15}\ (\alpha_5 + \alpha_5{}^2 - 1)(1 + \alpha_8). \qquad (7.20)$$

The essential role of Y_1 in Doodson's combination is to ensure a zero for T = 4 h because of the subharmonics M_6, $2M_6$, 2S M_6, S_0 of importance in oceanic tides. In theory these are absent from Earth tides, so it seems justifiable to delete this factor from the combination X_0.

One advantage of Pertsev's combination against Doodson's is that it is applied to a shorter interval of time and thus gives a better approximation for the actual drift.

(c) Summation of Lecolazet

Lecolazet's method of analysis (1956) was specially developed for the discussion of the observations of Earth tides taking into account the presence of a drift. This author did not eliminate the drift before the analysis as the combinations he uses to separate the diurnal from the semi-diurnal waves separate automatically a parabolic drift upon each interval of 21 h (they contain three combinations of Z-type).

Nevertheless, a special summation gives the mean position of the instrument every 21 h with very good precision:

$$(L) = \frac{1}{32}(Y_0+Y_1+Y_2+Y_3+Y_4+Y_5+Y_6+Y_7+Y_8+Y_9+Y_{10}+Y_{12}+Y_{13}+Y_{15}+Y_{18}+Y_{21})$$

which can easily be changed to

$$(L) = \frac{1}{32}[S_{21} - Y_{15}(Y_1+Y_4) - (Y_{17}+Y_{20})] \qquad (7.21)$$

and for which the selectivity factor is

$$\rho(L) = \frac{1}{32}[\Sigma_{21} - \alpha_{15}(\alpha_1+\alpha_4) - (\alpha_{17}+\alpha_{20})]. \qquad (7.22)$$

7.2.4. General Characteristics of the Methods of Analysis

The basis of the methods (Doodson, Pertsev and Lecolazet) rests on the characteristics properties of the combinations of ordinates described above. These methods resolve the problem of the harmonic analysis of one month's observations in two essential and distinct stages: (a) separation of species of waves — semi-diurnal, diurnal, long period, and (b) separation of different groups of waves contained in each of these types.

The methods can be distinguished during these operations by the choice of the combination of ordinates and their interval of application. Nevertheless, Pertsev's method is quite similar to that of Doodson's because it uses the same table of multipliers in operation (b).

Once the two operations (a) and (b) have been carried out, the methods of Doodson and Pertsev, on the one hand, and of Lecolazet, on the other, diverge. The first two continue the calculation in each group of waves until the separation of the constituents, using theoretical relations, while Lecolazet retains the groups of waves, as the actual analysis has given them, and compares them to the theoretical groups (which he calls "homologous waves") re-established for the central hour of the "month" of observations, from the theoretical waves of the development of Doodson, limited to 79 principal components (52 diurnal waves and 27 semi-diurnal waves whose amplitudes are at least 0.1 μgal in the vertical component at the latitude 45°).

7.2.5. Doodson's Method

(a) First Phase

The first operation consists in calculating for each day the contribution from each species of tide; these are obtained by linear combinations specially chosen to satisfy these requirements and denoted by Doodson by the symbols X_1, Y_1, for the diurnal tides and X_2, Y_2 for the semi-diurnal tides;

$$X_1 = Y_1^{\,2}\; Y_2(Y_2 - Y_0)\; Z_6^{\,2}. \qquad (7.23)$$

Y_1 has the same form but is simply displaced by 5 h.

Then Doodson operates the sums and differences of X_1 and Y_1. This is equivalent to multiplying the combination of type (7.23) by a combination of the order 5/2, of addition and subtraction, respectively:

$$X_1 + Y_1 = Y_1^{\ 2}\ Y_2(Y_2 - Y_0/2)\ Z_6^{\ 2}\ Y_{5/2}\ , \tag{7.24}$$

$$X_1 - Y_1 = -\ Y_1^{\ 2}\ Y_2(Y_2 - Y_0/2)\ Z_6^{\ 2}\ Z_{5/2}. \tag{7.25}$$

These combinations differ from the point of view of the elimination of the drift, because (7.24) has the same properties as (7.23) (elimination of a linear drift for 25 h), while (7.25) eliminates a parabolic drift for 30 h.

$$X_2 = -Y_1^{\ 2}\ Z_3^{\ 2}\ Y_6.$$

Y_2 has the same form and is simply displaced by 3 h.

Then

$$X_2 + Y_2 = -\ Y_1^{\ 2}\ Z_3^{\ 2}\ Y_6\ Y_{3/2}, \tag{7.26}$$

$$X_2 - Y_2 = Y_1^{\ 2}\ Z_3^{\ 2}\ Y_6\ Z_{3/2}. \tag{7.27}$$

Again, the combinations have different properties with regard to the elimination of drifts because (7.27) eliminates a parabolic drift for 16 h while (7.26) eliminates a linear drift for 13 h. These combinations can be applied 29 times within 30 days by successive shifts of 24 hours.

(b) Second Phase

One can, for example, write that

$$X_2 = A(S_2) + \sum_i A_i \cos \omega_i t + \sum_i B_i \sin \omega_i t \tag{7.28}$$

because the contribution from S_2 is necessarily constant during the month (time scale is indeed the mean solar time). On the other hand, the contribution from M_2 (as also for N_2, L_2, etc.) is variable and depends on the phase of M_2 with regard to the central day. It varies harmonically with a period of about 14 days.

It is simple in these conditions to isolate a selected wave of argument ω_m, that is to say, to calculate the values corresponding to A_m and B_m. The theory of least squares shows that we can obtain these values by multiplying, respectively, each of the successive values of X_2 by the corresponding value of $\cos \omega_m t$ and adding together the 29 partial products, and, in the same way, by the corresponding value of $\sin \omega_m t$ and adding together the partial products.

In practice Doodson replaces $2 \cos \omega_m t$ and $2 \sin \omega_m t$ by the nearest integer ± 2, ± 1, 0, and this ensured a "sufficient" selectivity for this level of the calculation. To isolate S_2 we have obviously coefficients uniformly equal to +1 which form a vector called d_0. For M_2, one multiplies the vectors X_2 and Y_2 by a vector d_2, formed according to the rule stated above.

Doodson used vectors d_0, d_1 ... d_7 (even) d_a, d_b ... d_g (odd) to separate seven groups. The scalar products by the vectors X_k Y_k give the even combinations X_{ki}, Y_{ki} and the odd ones X_{kj}, Y_{kj} which are then combined by addition and subtraction under the forms:

$$\left.\begin{aligned} C_{p1} &= X_{p1} + Y_{p1} + X_{pa} - Y_{pa}, \\ D_{p1} &= Y_{p1} - X_{p1} + X_{pa} + Y_{pa}, \text{ etc.} \end{aligned}\right\} \tag{7.29}$$

C_{20} and D_{20} therefore constitute the values of R cos r and R sin r for S_2 to which, at the end of the calculation one will add small corrections calculated, respectively, from C_{21} C_{22} C_{23} C_{24} D_{25} D_{12} and D_{21} D_{22} D_{23} D_{24} C_{25} C_{12}. It is the same for M_2 for which the principal part is contained in C_{22} and D_{22}. For K_1 it will be C_{10} and D_{10} and so on. Readers are referred to the publication of Doodson for details. It is obvious that S_2 so determined is in fact the resultant of the waves K_2, S_2, T_2, and K_1 is the resultant of K_1, S_1, P_1. Likewise L_2 and λ_2 are not separated, nor N_2 and ν_2, etc.

One can separate these various constituents if one has a year's observations available. The development of the calculation is exactly the same as the preceding one, just as one has replaced each "day" of observations by the functions X_1, Y_1, X_2, Y_2, which makes possible the separation of the species of waves, one will replace the "month" (of 29 "days" or of 29 functions X_1, 29 functions Y_1, 29 functions X_2, 29 functions Y_2) by 34 functions X_{10}, X_{11} ... Y_{24} Y_{2d} which achieves the separation of the groups.

Then from 12 month's observations one obtains a series of 12 values of X_{20} and Y_{20} which include a constant contribution from the wave S_2 and can be written

$$X_{20} = A(S_2) + A(K_2) \cos \omega(K_2)t + B(K_2) \sin \omega(K_2)t \qquad (7.30)$$

The application of a table of multiplicators m_i m_j constructed like d_i d_j, will make it possible to separate the amplitudes and phases of S_2 and of K_2. The resultant functions are given as X_{200}, X_{202}, X_{203}, etc. N_2 and ν_2, K_1 and P_1 can be separated in the same way.

7.2.6. Pertsev's Method (1957b)

The Pertsev's method is essentially a simplification of Doodson's method because the latter, originally designed for the study of oceanic tides, contains a part of the calculation involved by the presence of important subharmonics (shallow water tides). These subharmonics should not necessarily be calculated in the case of Earth tides, although they can be of great interest, particularly in coastal regions.

Pertsev strove to make the manual calculation simpler and quicker than in Doodson's method. This is a point that loses much of its interest when the work can be done with a computer.

7.2.7. Lecolazet's Method (1956)

One of the basic ideas of Lecolazet was to choose the combinations of ordinates that automatically eliminate the drift on the interval of application on condition that in this interval (always quite short) the drift is correctly represented by a polynomial of the second order. These combinations therefore contained the product of the three combinations Z.

As for the preceding methods, Lecolazet's method allows three phases corresponding to the same ways of separating the waves, daily linear combinations, monthly combinations, and resolution of the system of 7.28.

(a) First Phase

Lecolazet adopted an interval different from the fundamental shifting interv of 24 h to avoid a large part of the calculation tedious to carry out manually. He started from the following elementary considerations. When one wishes to isolate a wave of angular velocity ω, which can be written

$$A \cos \omega t + B \sin \omega t,$$

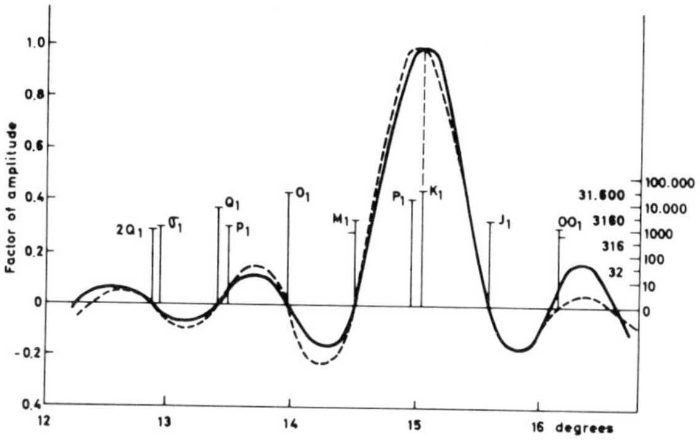

Fig. 7.2. Comparison of the selectivity of Lecolazet's monthly method (solid line) and Doodson's monthly method (dashed line) in separating the diurnal waves

one applies to the "month" a "daily" linear combination of the second type (the meaning of day here being an interval of n solar hours, where n is an undetermined integer). If one considers the result of this application to "day" p of the table, whose central hours is pn, one obtains by using a number proportional to

$$-A \sin(pn)\omega + B \cos(pn)\omega$$

and p will vary from $-m$ to $+m$ (p=c on the mean epoch of the observations, called "central day"). One thus obtains a series of $(2m+1)$ values which replace the $m(2m+1)$ hourly values of the month.

During the second stage of the calculations one immediately obtains a number proportional to B in applying to this series a combination of the first type; but one only obtains a number proportional to A by the application of a combination of the second type on condition that $n\omega \neq K\pi$ (K being any integer, n an integer to be determined). But, for all the tidal waves, ω differs only slightly from π/12 or π/6. One must therefore have

$$n \neq 12K \text{ and } n \neq 6K.$$

It follows that if, during the second stage, one wishes to obtain A and B from one and the same daily combination it is necessary that n is as different as possible from the most simple multiples of 6 and 12, which are 18 and 24. This led Lecolazet to choose the median value n = 21 h. It allowed him to use only one diurnal combination, and one semi-diurnal combination which he applied every 21 h, while Doodson, to obtain A and B, had to double his fundamental diurnal and semi-diurnal combinations in displacing them by 3 h and 5 h, respectively, because his method remains involved with the idea of a "complete day of 24 h" for the shifting procedure.

The two combinations of Lecolazet are written as follows:

$$\text{diurnal combination:}\quad Z_4\ Z_6(Z_5+Z_8)Y_3; \tag{7.31}$$

$$\text{semi-diurnal combination:}\quad Z_{3/2}^2\ Z_4\ Y_6(Y_5+Y_8). \tag{7.32}$$

Their application is done 33 times for a month of observations and therefore provides two series of 33 numbers, called diurnal series and semi-diurnal series.

(b) Second Phase

For the wave

$$A \cos \omega t + B \sin \omega t = R \cos(\omega t - r)$$

the daily diurnal combination provides a series of 33 values

$$\rho(\omega)\,[\,-A \sin(21\omega)p + B \cos(21\omega)p] \tag{7.33}$$

to which one should now apply a linear combination of the second type of order 16 to obtain

$$-\rho(\omega)\left[2\sum_{\mu=1}^{16} K_\mu (\sin 21\omega)\mu\right] A. \tag{7.34}$$

The coefficients K should be such that the bracket is as large as possible relative to the analogous terms coming from the waves of neighbouring periods. One could take

$$K\mu = (\sin 21\omega)\mu \tag{7.35}$$

exactly, and this is the solution of the least squares, but Cauchy-Tisserand's method has been applied which consists in replacing K by 0, -1, or +1 according to whether the sine is between -1/3 and +1/3, -1 and -1/3, or +1/3 and +1. This makes the manual calculations much easier.

Lecolazet's method can also be extended to treat the results of a series of monthly consecutive analyses. It can be applied to any interval sufficiently long composed of a group of equidistant monthly analyses, regardless of the displacement of time between consecutive analyses even if this is not constant (recovery of 10 1/2 days or 20 1/2 days were often used).

7.2.8. Calculation of Mean Square Error in the Results of a Harmonic Analysis

Let us suppose that the mean square error σ in the measurement of an ordinate is known. Under this condition it is theoretically easy to calculate the mean square error in the values R cos r and R sin r, relative to each tidal wave, using the combinations which made possible the calculation of these two values.

For example, for R cos r of the wave Q_1 in Doodson's method one has

$$10^6\ R \cos r = 583\ C_{13} + 31\ C_{10} + 11\ C_{11} + \ldots + 11\ D_{16}. \tag{7.36}$$

Thus the error in C_{13} practically determines the error in R cos r of Q_1. Now C_{13} is obtained from the 714 initial hourly data and one can calculate the coefficients a_i which affect each of the 714 hourly data and each of the waves. The m.s. error in R cos r of one wave will be given in the form

$$\sigma \left[\sum_{i=1}^{714} a_i^{\,2} \right]^{1/2} \tag{7.37}$$

The coefficient of σ is in general between 0.07 and 0.08.

Estimation of the Mean Square Error

It is clear that in carrying out a linear combination of the data, eliminating all the tidal waves and the instrumental drift, one obtains a result that is only dependent on observational errors. If one carries out this linear combination n times, each time using different data, one obtains n independent numbers from which one can derive the mean quadratic error of an observation. A combination chosen by Lecolazet to satisfy the average of the diverse conditions of the different methods of harmonic analysis is

$$(E_1) = Z^5_{1/2}\ Z^6(Y_3 - Y_0/2). \tag{7.38}$$

This applies to the 24 hourly observations of each day with the 24 coefficients symmetric about 11 h 30 min:

-1, 5, -10, 11, -10, 11, -11, 10, -11, 10, -5, 1/1, -5 ... -1.

This combination eliminates a polynomial of the 5th degree, its amplitude factor is exactly zero for the waves of periods 4, 6, 12 h and equal to 0.006 for waves with a period of 24 h. The sum of the squares of the coefficients is 1872. Thus from the residuals we obtain

$$\sigma = \frac{(e_1^{\,2} + e_2^{\,2} + \ldots + e_n^{\,2})^{1/2}}{1872n}. \tag{7.39}$$

It should be noted that there is a definite advantage in applying this combination of errors previously to all computation of harmonic analysis. Examination of the residuals e_i makes it possible to reveal immediately the eventual presence of a reading error in the diagrams and to locate it. As certain individual values are only affected by coefficient 1 whereas others are affected by coefficient 11, some of the eventual errors have a greater chance of being detected than others, and the surest process will be to apply the combination E_1 a second time, displacing it by 4 h.

A series of applications coming from different instruments has given monthly values that are always approximately the same:

σ ≈ 0.2 mm ≈ 2 μgals for gravimeters

σ ≈ 0.2 mm ≈ 0".0002 for pendulums.

7.3. METHODS PRESENTLY USED WITH COMPUTERS

Methods of analysis available in 1957 (Doodson, 1954; Lecolazet, 1956; Pertsev, 1957b; Matvéev, 1962) were developed when computers were not commonly available. Consequently everyone tried essentially to limit the calculation task. It was practically impossible to use a least-squares method, and the analysis had to be applied to a one-month or one-year standard length of recording. These methods did not accept any gap in the record. However, as Earth-tide measurements are very often made in mines in hard environmental conditions (usually 100% humidity), gaps of some days or even a week sometimes occur if an equipment failure happens between the weekly visits to the station. Interpolation cannot seriously be made for more than a few hours. It resulted that a "one-year" method like Doodson's could practically never be applied. Consequently the separation of P_1, S_1 and K_1 was generally not achieved until Lecolazet (1960) proposed a scheme based upon the results of successive monthly analyses, But even with these simplified methods, computations were time consuming until a necessary step was made by Melchior (1959) who programmed the Doodson and Lecolazet methods on the first, at that time, widely available computer, the IBM 650.

Now after those first years of tentative research on Earth tides, thanks to new instrumental developments, one had several years of better data with more accurate calibrations available, so that the simple repetition of monthly processes were no longer satisfactory. In fact Kramer (1964) as well as Saritscheva (1964) showed that the repetition of the monthly process with a shift of 1 hour, 1 day, or 10 days showed spurious periodic oscillations in the amplitudes and phases of the principal waves-periods of some 10 or 12 h to 6-8 days in M_2 amplitude of 12 days in O_1 amplitude. This demonstrated clearly that these methods were not selective enough. In any case the procedure consisting of taking the mean of many successive and non-independent monthly analyses obviously was wrong.

A new flexible scheme of computation was necessary which could be applied to records of any length with as many gaps as happen to occur. With new and faster computers available everywhere, the method of least squares had to be the basis of the procedure, giving the results with the highest possible accuracy and, at the same time, yielding an estimation of this precision. Several programs have successively been developed by Venedikov (1966b), Usandivaras and Ducarme (1969), Zetler and Cartwright (1970), Schuster (1970), Chojnicki (1972), Schüller (1976), Yaramanci (1978)

7.3.1. The Transfer function

The goal of tidal measurements is to determine the response of the Earth to the tidal forces F(t) through an instrument, using a modelling system. A first problem arises because we cannot separate in the instrumental output O(t) what is due to the instrument itself from what is the natural phenomenon.

We must therefore determine independently the transfer function of the instrument in amplitude as well as in phase, an operation which is nothing else than the "calibration" of the instrument.

The transfer function of the Earth s(t) is the unknown and, following Love's theory, we have supposed it to be linear. We can also reasonably suppose it as being stable and time invariant. The input having an harmonic form, we can consider the global system as non causal. In these conditions for a single input F(t) we can write the predicted output p(t) by a convolution integral like:

$$p(t) = \int_{-\infty}^{\infty} s(\tau)\, F(t-\tau)\, d\tau \qquad (7.40)$$

In the frequency domain the Fourier transform gives

$$p(\nu) = s(\nu)\, F(\nu) \qquad (7.41)$$

As p(t) and F(t) are here pure harmonic functions, their Fourier transforms exist only for a finite data set, often referred as a data window. In practice this is not a restriction as we cannot produce infinite series of observations.

The transfer function s( ) is a complex function:

$$s(\nu) = |s_{(\nu)}|\, e^{i\phi(\nu)} \qquad (7.42)$$

and we generally refer to $|s(\nu)|$ as the "amplitude factor" and to $\phi(\nu)$ as the "phase difference" between the output and the input. The introduction of a rheological model to describe the transfer function of the instrument

$$T(\nu) = |T(\nu)|\ e^{i\alpha(\nu)} \tag{7.43}$$

will give the true transfer function of the Earth:

$$|E_{(\nu)}| = |s_{(\nu)}| \ / \ |T(\nu)| \tag{7.44}$$

$$\varepsilon_{(\nu)} = \phi_{(\nu)} - \alpha_{(\nu)} \tag{7.45}$$

7.3.2. Direct solutions of the linear system

Considering our modelling system as perfectly fitting the physical system (which means that there are no errors in the observations), we put

$$O(t) = p(t) \tag{7.46}$$

The evaluation of the Fourier transforms then gives

$$s(\nu) = O(\nu) \ / \ F(\nu) \tag{7.47}$$

where $O(\nu)$ is obtained for a data window $\{-T/2,\ T/2\}$.

We also can use the autocorrelation functions $C_{OO}(\tau)$ and $C_{FF}(\tau)$ defined as follows:

$$C_{OO}(\tau) = \lim_{T\to\infty} \int_{-T/2}^{T/2} O(t)\ O(t-\tau)\ dt \tag{7.48}$$

$$C_{FF}(\tau) = \lim_{T\to\infty} \int_{-T/2}^{T/2} F(t)\ F(t-\tau)\ dt \tag{7.49}$$

The corresponding Fourier transforms $C_{OO}(\nu)$ and $C_{FF}(\nu)$ generally called power spectra allow an evaluation of $|s(\nu)|$:

$$|s(\nu)|^2 = C_{OO}(\nu) \ / \ C_{FF}(\nu) \tag{7.50}$$

With such spectral techniques error calculation is not possible as we have supposed errorless observations. However one can evaluate the mean noise amplitude ε around the tidal spectral lines and deduce signal to noise ratios:

$$\varepsilon_A = \varepsilon/A \tag{7.51}$$

where A is the amplitude of the tidal wave. The error on the phase difference α is then

$$\varepsilon_\alpha = (\varepsilon/A)_{rad} \tag{7.52}$$

In practice these spectral methods are very time consuming as one has to evaluate two Fourier transforms. Moreover they do not take advantage of the fact that we do know beforehand the spectral content of the input function $F(t)$.

7.3.3. Optimal linear system solutions

Considering the existence of observational errors one can try to optimize the representation by minimizing the residual power. Defining:

$$e(t) = O(t) - p(t) \tag{7.53}$$

the condition will be expressed by:

$$\bar{e}^2 = \lim_{T\to\infty} \frac{1}{T} \int_{-T/2}^{T/2} e^2(t)\, dt = \text{Min} \tag{7.54}$$

or, from (7.40):

$$\bar{e}^2 = \lim_{T\to\infty} \int_{-T/2}^{T/2} \left(\theta(t) - \int_{-\infty}^{\infty} s(\tau)\, F(t-\tau)\, d\tau\right)^2 dt. \tag{7.55}$$

The only unknown being $s(\tau)$, the minimizing conditions are

$$\frac{\delta\, \bar{e}^2}{\delta s(\tau)} = 0, \qquad \frac{\delta^2\, \bar{e}^2}{\delta s(\tau)^2} > 0$$

This directly yields

$$\lim_{T\to\infty} \frac{1}{\tau} \int_{-T/2}^{T/2} 2\left(\theta(t) - \int_{-\infty}^{\infty} s(\tau)\, F(t-\tau)\, d\tau\right) F(t-u)\, dt = 0. \tag{7.56}$$

By using again the definitions of the auto and cross covariance functions one writes

$$C_{OF}(u) = \int_{-\infty}^{\infty} s(\tau)\, C_{FF}(u-\tau)\, d\tau \tag{7.57}$$

a convolution integral which is known as the *Wiener-Hopf integral equation*.

Its Fourier transform gives, in the frequency domain:

$$C_{OF}(\nu) = s(\nu)\, C_{FF}(\nu). \tag{7.58}$$

Two classes of tidal analysis methods directly derive from the Wiener-Hopf integral equation:

in the time domain the so-called response method,

in the frequency domain the cross-spectral method.

Both of them start from the time domain representation of the tidal input ($F(t)$). In the response method (7.58) is written under its matrix form for a finite data length and one has to define the sampling interval of $s(\tau)$ and its length. Increasing the length allows more oscillations of the function $s(\nu)$ while increasing the sampling interval $\Delta\tau$ decreases the bandwidth for which $s(\nu)$ is defined.

In practice, equation (7.58) is solved independently for the different tidal harmonics (D, SD, TD). This method is used mainly at the Tidal Institute of Liverpool (Yaramanci, 1978).

The cross-spectral method is seldom applied but leads to very similar results.

7.3.4. Harmonic Analysis Methods

One considers the input function F(t) through its spectral representation F(ν) according to the general formula:

$$F(t) = \int_{-\infty}^{\infty} F(\nu)\, e^{2\pi i \nu t} \tag{7.59}$$

which gives for a discrete spectrum:

$$F(t) = \frac{1}{2} \sum_{n=-N}^{N} F(\nu_n)\, e^{2\pi i \nu_n t} \tag{7.60}$$

One can introduce this representation into equation (7.57) or directly into (7.55) and obtain for the real an imaginary parts of s(ν) a set of classical normal equations as in the least squares adjustment methods.

This harmonic formulation using the frequency domain representation of the tidal forces under the Cartwright-Tayler-Edden formulation is the most widely used.

Differences appear only in the formulation of the observation equations and in the prefiltering procedure. Some authors treat separately the different tidal bands (Venedikov 1966 and Venedikov 1975). Others try a global solution and differ mainly by the filter lengths: direct least squares solution with a minimum filter length (Usandivaras-Ducarme), Pertsev filter (Chojnicki).

In practice the results are equivalent but large differences can appear in the error evaluation: least squares solutions generally underestimate the errors as they suppose a white noise structure and uncorrelated observations i.e. a unit variance-covariance matrix.

The second hypothesis cannot be avoided as we do not known the real variance-covariance matrix.

We know however, from spectral analysis methods, that the noise is a so-called "coloured noise". Moreover there is noise accumulation in the main tidal bands. It is the reason why the methods dealing separately with the different tidal bands lead to more realistic error estimates as they approximate the noise level independently in each band.

The best error estimates are certainly given by the spectral analysis of the residuals. One can also study the variations of the real and imaginary parts of s(ν) in consecutive partial analysis of the data. That principle is widely applied in the HYCON method by Schüller (Hybrid Least Squares frequency domain convolution).

To describe briefly the principle of these methods, let us write the hourly ordinate reading at epoch $T_i + t$ as follows:

$$\ell_{i,t} = \sum_j H_j \cos\left[\omega_j t + \phi_j(T_i)\right] + C + \sum_{k=1}^{n} B_k P_k^i, \tag{7.61}$$

where H_j is the observed amplitude of wave of frequency ω_j, ϕ_j is the observed phase of the same wave at epoch T_0 and

$$\phi_j(T_i) = \phi_j + \omega_j(T_i - T_0), \tag{7.62}$$

where T_0 is a conventional fixed epoch and C is a constant depending upon the choice of the zero line. This choice is usually made to have only positive ordinates which avoids transcription errors. P_k^i are k polynomials fitting with constants B_k the instrumental drift as well as possible.

A preliminary numerical filtering eliminates the frequencies external to the model accepted for the description of tidal phenomena. Then the equations of observations are compensated. Lecolazet has shown that such a model is acceptable as a statistical analysis of the residuals establishes that their spectrum is aleatory.

7.3.5. Construction of Numerical Filters by the Least-squares Method

Since there are computers there is no practical advantage to using simple coefficients like +1 and -1 as in Labrouste's method. Thus more efficient filters can be built (either low-pass or high pass) like the filter devised by Jobert (1964) to eliminate drift and long periods. Venedikov (1966a) has developed a general method to construct, with a computer, filters with the desired properties.

If ℓ_{-n}, ℓ_{-n+1}, ..., ℓ_n are hourly readings,

$$n = (N - 1)/2 \text{ at epochs } -n, -n+1, \ldots, n,$$

and if f_{-n}, f_{-n+1}, ... f_n are the N coefficients of filter $\mathbf{F}$,

$$x = \sum_{t=-n}^{n} f_t \ell_t = \mathbf{F^*} \mathbf{L} \qquad (7.63)$$

is the application of $\mathbf{F}$ on $\mathbf{L}$ (convolution), and

$$\mathbf{F} = \begin{pmatrix} f_{-n} \\ \vdots \\ f_n \end{pmatrix}, \quad \mathbf{L} = \begin{pmatrix} \ell_{-n} \\ \vdots \\ \ell_n \end{pmatrix}, \quad \mathbf{F^*} = (f_{-n} \ldots f_n) \text{ transpose of } \mathbf{F}, \qquad (7.64)$$

x is the scalar product of vectors $\mathbf{F}$ and $\mathbf{L}$ in N-dimensional space. Even filters $\mathbf{F}$ or addition filters are such that $f_t = f_{-t}$. Odd filters $\mathbf{G}$ or difference filters are such that $f_t = -f_{-t}$.

Even and odd filters are orthogonal:

$$\mathbf{F^*G} = \sum_{t=-n}^{n} f_t g_t = 0. \qquad (7.65)$$

Now, if

$$v_t = e^{i\omega t} = \cos \omega t + i \sin \omega t, \qquad (7.66)$$

the product

$$r_v = \mathbf{F^*V} \qquad (7.67)$$

is the amplification factor, or response, of filter $\mathbf{F}$ for a component $\mathbf{V}$ of frequency ω and r_v is the Fourier transform of the filter $\mathbf{F}$.

Since we can represent Earth-tide readings by the form

$$\ell_t = \sum_{i=1}^{s} (\xi_i \cos \omega_i t + \eta_i \sin \omega_i t) + \sum_{j=0}^{k} B_j P_j(t) + \varepsilon_t, \qquad (7.68)$$

ε_t being the reading error, N successive readings give a system of N equations that can be written

$$\mathbf{L} = x_1 \mathbf{A}_1 + \dots + x_m \mathbf{A}_m + \varepsilon \tag{7.69}$$

or

$$\mathbf{L} = \mathbf{AX} + \varepsilon \tag{7.70}$$

with

$$\mathbf{X} = \begin{pmatrix} x_1 \\ \vdots \\ x_m \end{pmatrix}, \tag{7.71}$$

$(m = 2s + k + 1)$

where the unknowns are ξ_i, η_i, B_k and

$$\varepsilon = \begin{pmatrix} \varepsilon_{-n} \\ \vdots \\ \varepsilon_n \end{pmatrix} \tag{7.72}$$

$\mathbf{A}_1 \dots \mathbf{A}_m$, coefficients of the unknowns in (7.68), are the vectors in the matrix $\mathbf{A}$ (m columns, N rows).

If we wish to determine a particular component x_1 of frequency ω_1, we will construct a filter $\mathbf{F}$ which amplifies this component and completely eliminates the others, that is,

$$x_1 = \mathbf{F^*L}. \tag{7.73}$$

We should have

$$\mathbf{F^*A}_1 = 1, \tag{7.74}$$

$$\mathbf{F^*A}_i = 0 \quad (i=2, \dots, m). \tag{7.75}$$

The filter $\mathbf{F}$ has N coefficients since there are N successive observations. These conditions form a system of m equations with N unknowns which are the coefficients of the filter $\mathbf{F}$.

As evidently $N \gg m$, there is an infinite number of solutions.

When we take the obvious observational errors, we can see that the choice of the most convenient filter is important. If ε are random and independent errors, and if σ^2 = constant is their dispersion, then from (7.73) the dispersion on the desired component x_1 is

$$\sigma_1^2 = \Sigma f_1{}^2 \sigma^2 = \mathbf{F^*F}\, \sigma^2 = \frac{\mathbf{F^*F}}{(\mathbf{F^*A}_1)^2} \sigma^2, \tag{7.76}$$

$$= \frac{|\mathbf{F}|^2 . \sigma^2}{|\mathbf{F}|^2 \ |\mathbf{A}_1|^2 \cos^2\alpha} = \frac{\sigma^2}{|\mathbf{A}_1|^2 \cos^2\alpha} \tag{7.77}$$

where α is the angle between vectors $|\mathbf{F}|$ and $|\mathbf{A}_1|$.

As we have N - m degrees of freedom, it can happen that vector $|\mathbf{F}|$ will be nearly orthogonal to $|\mathbf{A}_1|$. Then it amplifies the observation errors so much that it is a very poor filter.

We must find for filter $|\mathbf{F}|$ a position such that α will be minimum. The solution is given by the least-squares method applied to (7.70)

$$\mathbf{X} = (\mathbf{A^*A})^{-1} \mathbf{A^*L}. \tag{7.78}$$

$(\mathbf{A^*A})^{-1}$ is a symmetric matrix of m^2 elements, its first row being

$$\mathbf{G} = (g_1\ g_2\ \dots\ g_m). \tag{7.79}$$

Thus

$$\mathbf{G(A^*A)} = (1\ 0\ 0\ \dots\ 0). \tag{7.80}$$

We obtain

$$x_1 = \mathbf{GA^*L} = \mathbf{F^*L}$$

with

$$\mathbf{F} = \mathbf{AG^*}, \tag{7.81}$$

a column vector of N elements, which is the desired filter.

The dispersion is minimum as x_1 has been obtained by least squares. This filter is obtained by operator $\mathbf{G}$ that transforms matrix $\mathbf{A}(m,N)$ into a vector column $\mathbf{F}(N)$.

Equation (7.81) shows that $\mathbf{F}$ is a linear combination of the m vectors $\mathbf{A}_i$ with coefficients $g_1\ \dots\ g_m$. It is sufficient to know $\mathbf{G}$ to estimate the qualities of filter $\mathbf{F}$ as the dispersion on the desired component x_1 given by

$$\sigma_1{}^2 = \mathbf{F^*F}\ \sigma^2 = \mathbf{GA^*AG^*}\ \sigma^2 = (1\ 0\ \dots\ 0)\mathbf{G^*}\ \sigma^2 = g_1\ \sigma^2. \tag{7.82}$$

Practically, the (m - 1) components that must be completely eliminated are restricted and we must know the response of our filter to the other components in the tidal readings. Let the matrix containing all components of interest be

$$\mathbf{B} = (\mathbf{A}_1\ \mathbf{A}_2\ \dots\ \mathbf{A}_m\ \mathbf{A}_{m+1}\ \dots\ \mathbf{A}_{m+p}) \tag{7.83}$$

and let us form

$$\mathbf{a} = \mathbf{A^*B} \qquad a_{ij} = \mathbf{A}_i{}^*\mathbf{A}_j, \tag{7.84}$$

The response of $\mathbf{F}$ is

$$\mathbf{r} = \mathbf{F^*B} = \mathbf{GA^*B} = \mathbf{Ga}. \tag{7.85}$$

Thus we can calculate the response using $\mathbf{G}$ and $\mathbf{a}$ without having the necessity to calculate $\mathbf{F}$.

Having decided which components we want to eliminate (e.g. S_2, N_2, K_2 ... when we want a filter for M_2), we calculate $\mathbf{G}$ by (7.79) and $\mathbf{a}$ by (7.84). Then (7.82) and (7.85) give us information about the qualities of the filter. If it is convenient, we calculate its coefficients by (7.81). One can show that if we augment the number of components to be strictly eliminated, the angle α is increasing as well as g_1 and σ_1. Thus the advantage of eliminating more components can be lost by the increase of σ_1. If we do not eliminate enoug components, then response $\mathbf{r}$ is not satisfactory. There is a compromise therefore which is a matter of personal choice.

7.3.6. Venedikov's Method (1966b)

The first step consists in the elimination of the instrumental drift superposed on long-period tides and at the same time in the separation of diurnal, semi-diurnal and ter-diurnal tides. The method for ter-diurnal waves was developed by Melchior and Venedikov (1968).

Unknowns have been chosen to make the computations easier and to be, at the same time, in agreement with the physical meaning of the phenomenon. For each species of tides, two filters are applied on a sequence of 48 hourly readings. One is an even filter (addition filter called C) and the other is an odd filter (difference filter called S).

The amplification factor of a filter is calculated according to

$$\left.\begin{aligned} c_j^{(\tau)} &= \sum_{t=-23.5}^{+23.5} C_t^{(\tau)} \cos \omega_j t, \\ s_j^{(\tau)} &= \sum_{t=-23.5}^{+23.5} S_t^{(\tau)} \sin \omega_j t, \qquad \tau = 1, 2, 3. \end{aligned}\right\} \tag{7.86}$$

The diurnal filters $C^{(1)}$, $S^{(1)}$ are constructed to amplify the diurnal waves and eliminate the semi-diurnal ones. Their amplification factor is between 3.64 and 4.50 for the diurnals while it is exactly 0 for N_2, M_2, S_2 and generally smaller than 0.01 for all other semi-diurnals. An opposite property is obtained with the semi-diurnal filters $C^{(2)}$, $S^{(2)}$ which have zero amplification for O_1, P_1, S_1 and 0.0005 for K_1.

The filters have been constructed in such a way that they eliminate the instrumental drift as an arbitrary combination of Chebichef orthogonal polynomials of order 3 for C filters and of order 2 for S filters.
Table 7.2 from Lennon (1961) and Venedikov (1966) shows the efficiency of this elimination. These six filters are applied to every continuous 48 h record and then shifted from 48 h to avoid any superposition and have independent results. When a gap occurs the filters must be reapplied when new data are available.

TABLE 7.2. Response of different filters with respect to long-period drift

Angular Velocity Period (days)	1° 15	2° 7.5	4° 3.8	6° 2.5	8° 1.9	10° 1.5	12° 1.25
Diurnal filters							
Doodson	0.0184	0.0724	0.2702	0.5404	0.8119	1.0162	1.1055
Doodson-Lennon	0.0012	0.0098	0.0728	0.2183	0.4365	0.6820	0.8880
Lecolazet	0.0015	0.0115	0.0852	0.2520	0.4944	0.7524	0.9483
$C^{(1)}$	0.0001	0.0016	0.0314	0.1066	0.2625	0.5208	0.7774
$S^{(1)}$	0.0014	0.0110	0.0985	0.2412	0.4620	0.7285	0.9271
Semi-diurnal filters							
Doodson	0.0036	0.0142	0.0524	0.1019	0.1447	0.1616	0.1362
Doodson-Lennon	0.0001	0.0010	0.0073	0.0213	0.0403	0.0561	0.0566
$C^{(2)}$	0.00001	0.0001	0.0017	0.0051	0.0099	0.0121	0.0073
$S^{(2)}$	0.00004	0.0003	0.0025	0.0047	0.0057	0.0037	0.0008

Since the filters eliminate a constant and the drift as well, it is of no importance that the successive parts of the records refer to different base lines. One only has to avoid any such jump within a 48h interval.

n applications of the filters to the hourly readings $\ell_{i,t}$ give six new series of n independent data:

$$M_i^{(\tau)} = \sum_{t=-23.5}^{+23.5} C_t^{(\tau)} \ell_{i,t} \tag{7.87}$$

$$N_i^{(\tau)} = \sum_{t=-23.5}^{+23.5} S_t^{(\tau)} \ell_{i,t}, \quad i = 1, 2 \ldots n; \; \tau = 1, 2, 3,$$

which can be written according to (7.61)

$$M_i = \sum_i c_j H_j \cos \phi_j(T_i), \quad N_i = -\sum_j s_j H_j \sin \phi_j(T_i) \tag{7.88}$$

The constant and the polynomials are eliminated by this operation; unknowns are H_j, ϕ_j.

One can presently accept from geophysical considerations that Love numbers and phases are not different for waves having very near frequencies within a same group, with the exception of K_1 group. A yearly analysis will in any case be necessary if one wants to derive correctly these waves. Venedikov distributes the tidal development into p groups of waves and, denoting by α_k and β_k the index number of the first and the last wave of one group (see Table 1.5) he writes:

$$\left.\begin{aligned} H_j &= \delta_k h_j \quad \text{or} \quad \gamma_k h_j, \\ \phi_j(T_i) &= \phi_j(T_i) + \kappa_k, \\ \xi_k &= \delta_k \cos \kappa_k, \\ \eta_k &= -\delta_k \sin \kappa_k, \\ \alpha_k &\leqslant j \leqslant \beta_k, \\ k &= 1, 2, \ldots, p \end{aligned}\right\} \tag{7.89}$$

Putting (7.89) into (7.88) one obtains a system of 2n equations (i=1, 2 ... n) with 2p unknowns (k=1, 2, ... p):

$$\left.\begin{aligned} M_i &= \sum_{k=1}^{p} \left[\xi_k \sum_{j=\alpha_k}^{\beta_k} c_j h_j \cos \phi_j(T_i) + \eta_k \sum_{j=\alpha_k}^{\beta_k} c_j h_j \sin \phi_j (T_i) \right], \\ N_i &= \sum_{k=1}^{p} \left[\xi_k \sum_{j=\alpha_k}^{\beta_k} s_j h_j \sin \phi_j(T_i) + \eta_k \sum_{j=\alpha_k}^{\beta_k} s_j h_j \cos \phi_j(T_i) \right], \end{aligned}\right\} \tag{7.90}$$

with 2n > 2p.

The classical technique of least squares is applied to this system, $\mathbf{M}$ is the column of 2n elements $M_i N_i$; $\mathbf{x}$, the column of 2p unknowns ξ_k, η_k, and $\mathbf{a}$, the matrix of coefficients of system (7.90). The system is written

$$\mathbf{M} = \mathbf{a}\mathbf{x} \tag{7.91}$$

and the system of normal equations

$$\mathbf{A}\mathbf{x} = \mathbf{B}, \tag{7.92}$$

with

$$\mathbf{A} = \mathbf{a}^*\mathbf{a} \qquad \mathbf{B} = \mathbf{a}^*\mathbf{M} \tag{7.93}$$

Then

$$\mathbf{x} = \mathbf{A}^{-1}\,\mathbf{B} \tag{7.94}$$

and

$$\delta_k^2 = \xi_k^2 + \eta_k^2, \qquad \kappa_k = \text{arc tg}(-\eta_k/\xi_k). \tag{7.95}$$

Then, the mean square error on one value of M_i or N_i is

$$\sigma = \left(\frac{\mathbf{M}^*\mathbf{M} - \mathbf{B}^*\mathbf{x}}{2n - 2p}\right)^{1/2}. \tag{7.96}$$

If $\mathbf{A}_{k,k}$ (k=1, 2, ..., 2p) are the coefficients of the diagonal of the inverse matrix $\mathbf{A}^{-1}$, one has for the errors on ξ and η:

$$\sigma_{\xi k}^2 = \sigma^2\,\mathbf{A}_{k,k}, \qquad \sigma_{\eta k}^2 = \sigma^2\,\mathbf{A}_{k+p,\,k+p}. \tag{7.97}$$

Since $\mathbf{A}_{k,k}$ and $\mathbf{A}_{k+p,\,k+p}$ are nearly equal and, because of the orthogonality properties, the error on δ_k is

$$\sigma_\delta^2 = \sigma_{\xi k}^2 + \sigma_{\eta k}^2 = 2\,\mathbf{A}_{kk}.\sigma^2$$

and on κ_k

$$\sigma_\kappa = \sigma_k/\delta_k. \tag{7.98}$$

A program written in FORTRAN for the IBM 360/44 by Venedikov and Pâquet (1967) was recently improved and adapted to Cartwright-Tayler development by Ducarme. The first step calculates M_i N_i functions and their corresponding epoch T_i. The second step takes into account the available number of $M_i N_i$ to decide about the wave separation according to Table 1.5, calculates the longitudes τ, s, h, p, N', p' and theoretical phases for epoch T_0; and forms (7.88) and (7.90).

The coefficients of equation M_i from line i in matrix **a** while those of N_i form line (p+i) in the same matrix. The program multiplies two by two the coefficients in each line to obtain the matrix **A**. It multiplies each coefficient of line i by M_i and each of line p+i by N_i to form the vector **B**. When all sets of values M_i N_i are considered, matrix **A** is inverted and (7.94-7.98) are resolved.

Results of this method are illustrated by Figs. 7.3, 7.4, 7.5, 7.6 showing how the precision increases with the length of the record. Table 7.3 shows that this method applied to a one-year data set, shifted by one-day steps, produces no periodic modulation in the results. This program has been widely used since 1966 at the International Centre of Earth Tides which applied it to more than 100,000 days'observations and published a large part of the results (Melchior, 1971).

7.3.7. The leakage effect

The main interest of the Venedikov procedure for tidal analysis when it was developped in 1965, was that it was very cheap in computer time and mainly that it allowed the presence of gaps in the data. However the method is sensitive to leakage effects.

One calls *leakage effect* the influence upon one given tidal frequency of the energy present in another spectral band. Due to unfavourable window functions coupled to the side lobes of the transfer functions of the Venedikov filters perturbation energies from even distant frequencies can leak into the tidal bands where they cause biases with respect to the tidal parameter estimates (Schuller, 1978).

For the standard filters on 48h, there is a maximum leakage from each frequency band situated at $\pm$ 7°.5/h in angular speed. For example the O_1 component gets a maximum effect from the frequency 21.44°/h.

To improve such a situation one can either smooth the window function or remove the side lobes by designing better filters. It is a reason why methods as Chojnicki (1973) or Schuller (1975) are less sensitive to leakage effects.

However data with gaps will always produce leakage effects by spoiling the window function.

7.3.8. Usandivaras-Ducarme Method (1969)

The program developed for this method is more elaborate than the Venedikov program as it gives in a single operation the diurnal, semi-diurnal, ter-diurnal and the long-period tides as well as the coefficients of a polynomial drift. Moreover it takes into account every pair of 2 hourly readings instead of intervals of 48h only. It is much more efficient for some observation stations where extremely difficult conditions produce many gaps in the records. The hourly readings are represented by (7.61) or (7.68).

TABLE 7.3. Station Dourbes, Belgium. Horizontal Pendulum Verbaandert-Melchior, N°8, East-West Component, Yearly Harmonic Analysis Displaced by 2 Days

	Amplitude factors						
	M_2	S_2	N_2	K_2	K_1	P_1	O_1
65 01 22 66 01 24	0.8568*	0.8074	0.8624*	0.8169	0.7335*	0.7072*	0.6548
65 01 24 66 01 26	8562	8071	8628	8142	7328	7043	6554
65 01 26 66 01 28	8562	8072	8635	8135*	7322	7040	6560
65 01 28 66 01 30	8559	8076	8630	8138	7321	7043	6556
65 01 30 66 02 01	8555	8079	8646	8156	7309	7030	6540
65 02 1 66 02 03	8553	8075	8659	8164	7302	6969	6521
65 02 3 66 02 05	8548	8066	8676*	8178	7299*	6974	6521
65 02 5 66 02 07	8550	8060	8661	8212*	7325	6942	6492
65 02 7 66 02 09	8548	8055*	8666	8204	7319	6933	6494
65 02 9 66 02 11	8545	8064	8665	8203	7312	6923	6503
65 02 11 66 02 13	8544*	8070	8644	8191	7316	6918*	6507
65 02 13 66 02 15	8544	8072	8636	8188	7317	6919	6509
65 02 15 66 02 17	8545	8079	8633	8192	7329	6938	6532
65 02 17 66 02 19	8546	8082*	8628	8202	7330	6960	6543
Mean	0.8552	0.8071	0.8645	0.8176	0.7319	0.6978	0.6514
Max. deviation	24	27	52	77	36	154	68

	Phase						
	M_2	S_2	N_2	K_2	K_1	P_1	O_1
65 01 22 66 01 24	3.192*	7.686	5.652	6.958	1.698*	5.091	1.143
65 01 24 66 01 26	3.170	7.776	5.758	6.727	1.653	5.200	1.060*
65 01 26 66 01 28	3.167	7.790*	5.806*	6.719	1.664	5.375	1.117
65 01 28 66 01 30	3.170	7.762	5.734	6.839	1.675	5.421	1.115
65 01 30 66 02 01	3.164	7.705	5.708	6.972	1.676	5.735	1.186
65 02 1 66 02 03	3.167	7.693	5.705	6.922	1.545	5.813	1.331
65 02 3 66 02 05	3.165	7.692	5.682	6.750	1.559	5.883*	1.275
65 02 5 66 02 07	3.131	7.625*	5.655	6.697	1.478	5.276	1.413*
65 02 7 66 02 09	3.113	7.650	5.666	6.547*	1.453	5.319	1.357
65 02 9 66 02 11	3.130	7.643	5.771	6.746	1.426	5.518	1.405
65 02 11 66 02 13	3.116	7.665	5.716	6.862	1.416*	5.427	1.368
65 02 13 66 02 15	3.108	7.669	5.730	6.906	1.422	5.402	1.338
65 02 15 66 02 17	3.085	7.663	5.647	7.075	1.501	5.010	1.290
65 02 17 66 02 19	3.072*	7.647	5.512*	7.128*	1.559	4.920*	1.291
Mean	3.139	7.690	5.695	6.846	1.552	5.385	1.264
Max. deviation	0.12	0.17	0.29	0.58	0.28	0.96	0.35

Introducing theoretical amplitude h_j and phase φ_j for each wave and putting

$$\left.\begin{array}{ll} H_j = \delta_j h_j^T, & a_j^i = h_j \cos\varphi_j, \\ \xi_j = \delta_j \cos\kappa_j, & b_j^i = h_j \sin\varphi_j, \\ \eta_j = -\delta_j \sin\kappa_j, & \psi_j = \phi_j - \varphi_j, \end{array}\right\} \qquad (7.99)$$

Equation (7.61) can be transformed into

$$\ell_i = \sum_j (a_j^i \xi_j + b_j^i \eta_j) + C + \sum_{k=1}^{4} B_k P_k^i. \qquad (7.100)$$

a_j^i, b_j^i can be calculated for any epoch of time, the unknowns being ξ_j, η_j, C, B_k.

To eliminate C, we consider as an observation equation the difference of two successive readings:

$$\ell_{i+1} - \ell_i = \sum_j [(a_j^{i+1} - a_j^i)\xi_j + (b_j^{i+1} - b_j^i)\eta_j] + \sum_{k=1}^{4} B_k(P_k^{i+1} - P_k^i). \qquad (7.101)$$

Groups of waves are formed according to the rule illustrated by Table 1.5. The treatment of (7.101) follows the same general rules of the least-squares method as in (7.91 - 7.95).

The numerical results are not significantly different from those obtained by the Venedikov procedure but the mean-square errors are slightly lower. One should recall that this method allows taking into account all lengths of recordings even those shorter than 48 h since it considers in fact every pair of two readings.

7.3.9. Chojnicki's Method

Chojnicki's method, very recently developed (1972), is based upon least squares with one iteration to realize an efficient elimination of the instrumental drift and all other long period effects. Chojnicki makes a first analysis on the raw data to derive with sufficient approximation the tidal constants of all main waves. Using these constants he subtracts a reconstructed "theoretical" tide from the raw data. This difference contains small tidal residuals, the drift and occasionally small discontinuities.

The Pertsev filter (7.19) is then applied which eliminates perfectly the tidal residuals owing to their minuteness. The result of this application is the real instrumental drift and all long period variations which are subtracted from the raw data. The corrected data are then analysed again to obtain the definitive tidal constants.

The use of the Venedikov's filters gives the same results as a direct application of the least square method on the hourly readings as demonstrated by Usandivaras and Ducarme (1970). These authors have recently shown (1975) that the results by the Chojnicki method are exactly the same as those by the Venedikov method even for the minor components.

They also show that the Chojnicki's estimate of the mean square error corresponds to its value in the ter-diurnal frequency band. In Geophysics, however, the noise is currently a "red noise" with an accumulation in the low frequencies. Moreover the meteorological noise has a diurnal period with subharmonics of decreasing amplitude. Therefore one usually observes an empirical relation of the following form:

m.s.e. (diurnal band) $\simeq$ 2 m.s.e. (semi-diurnal band)
$\simeq$ 3 m.s.e. (ter-diurnal band)

only a few high quality instruments, when a correction for atmospheric pressure has been applied, show no marked difference between the m.s.e. in these three bands.

Therefore we must consider the m.s.e. determined in the Venedikov method as the correct ones (Usandivaras, Ducarme, 1975). They fit very well indeed with the noise level exhibited by the Fourier transform method.

Schüller also showed (1975) that the Chojnicki iteration is equivalent to a direct application of the Pertsev filter to the raw data when its frequency characteristic (7.23) is taken into account in the theoretical amplitudes used for the following adjustment.

If one uses the Pertsev filter it is not possible to extract the long period tidal components from the data. Therefore Chojnicki uses the zero-point method proposed in 1956 by Lassovsky. This consists in calculating the epochs when the tidal component concerned is zero and in using the readings made at such epochs to reconstruct a drift curve which does not include the long period tidal components.

Chojnicki in this way succeeded in obtaining valuable constants for long period wave M_f which previously had not been obtained. However the determination of the true "zero points" is difficult and Chojnicki's procedure depends upon the assumptions made in this computation, particularly in regard to the phase lags. (see Wenzel, 1976).

7.3.10. Spectral Analysis

Only a brief account will be given here concerning spectral analysis since this technique is now widely used. We will mention here only those special aspects that are related to the specific Earth tide problem.

(1) The tidal function f(t) is experimentally known during a limited time interval (-L, +L). This means that one deals with a function

$$f_1(t) = f(t)\,a(t)$$

where

$$a(t) = +1 \quad \text{in interval } (-L, +L)$$
$$a(t) = 0 \quad \text{outside this interval.}$$

This is called a box car and produces in the spectrum

$$g(\omega_i) = \int_{-\infty}^{+\infty} f(t)\, e^{2\pi i \omega_i t}\, dt$$

secondary oscillations on both sides of each tidal peak which contaminate the other peaks. To avoid that phenomenon it is convenient to multiply the ordinates by

$$a(t) = [1 - t^2/L^2]^2$$

(apodization, hanning or hamming).

This unfortunately has the effect of diminishing the resolution of tidal lines which are very close in frequency. As remarked by Zetler (1964), the amplitude and phase of each line must be determined with a minimum of contamination from adjacent lines; thus, no smoothing can be tolerated unless, in the case of S_2 and K_2, the series to be analysed extends over a period of at least a few years.

Furthermore, even if contamination is avoided by a long enough series, the determination of the amplitude requires some devious technique of combining the amplitudes of several adjacent lines. The determination of the phase from the results spread over a few nearly Fourier lines appears to be even more unsatisfactory.

(2) f(t) is measured at discrete instants (usually every hour). This should be convenient only if the noise was negligible for frequencies higher than one cycle per hour. However, as the instrumental free period is around 50 cycles per hour or more, this is not true. The best way to overcome such a difficulty should be to diminish the reading interval. Attempts with 10 min or even 1 min intervals resulted in heavy computer work which can be solved only with powerful new computers.

(3) Spectral analysis usually is used for searching and determining unknown frequencies appearing as sharp peaks in the spectrum. In the case of Earth tides, all the frequencies are known, with exceptional precision. Time marks on tidal registrations are usually not precise enough to allow an experimental check of the astronomical frequencies.

In one Fourier transform of f(t) one can adjust the step in order to give the spectral density corresponding exactly to one pair of the desired frequencies but not the others because the periods of the principal tidal lines are incommensurable. Thus one has to make as many Fourier transforms as the number of pairs of frequencies desired.

(4) The presence of drift (called trend by statisticians) in the raw data to be analysed produces such important distortions in the spectrum that the real peaks can disappear. It is consequently necessary to eliminate any drift before performing the spectral analysis.

A perfect filter which does not produce any distortion of the tidal phenomenon and cuts all very short and very long periods will necessarily be very long and result in a considerable loss of information. However, a shorter filter gives a better approximation for the actual drift.

(5) Normalization of raw data is also to be done in order to reduce large variations of ordinates which are not essential and also mask periodic phenomena. Such an attenuation of the noise improves the sharpness of analysis. A normalization often used is given by

$$y'_i = \frac{y_i - \bar{y}_i}{(\Sigma\, y_i^2/n - 1)^{1/2}}.$$

The successive subtractions of ordinates used by Usandivaras-Ducarme (Section 7.33) meet this requirement.

7.4. FORMULAS TO BE USED IN THE PROGRAMS OF ANALYSIS

We shall explicitly describe here the theoretical tidal functions used in the new programs of computation written in Fortran at the International Centre of Earth Tides.

Let us write the tidal potential as

$$W = GE\mu \frac{r^2}{d^3} [P_2\ (z) + \frac{r}{d} P_3\ (z) + \ldots] \tag{7.102}$$

or

$$W = D(r)\ (W_2 + \frac{r}{a} W_3) \tag{7.103}$$

with

$$W_2 = \frac{2}{3}\,(3\cos^2 z-1)\,\left(\frac{c}{d}\right)^3,$$

$$W_3 = \frac{2}{3}\,(5\cos^3 z-3\cos z)\,\left(\frac{c}{d}\right)^4 \sin\pi:$$

To transform the expression (7.103) from the local to the equatorial coordinates system, and following Doodson's procedure, we separate the factors N_{ij}, functions of D and ψ (geocentric latitude), which are called geodetic coefficients, from the factors M_{ij} which are functions of δ, H and c/d. This allows development of the potential as a series of purely harmonic functions of the time. These functions are distributed according to the coefficient j of H, into the Laplace families of tides: long period (j=0), diurnal (j=1), semi-diurnal (j=2) and ter-diurnal (j=3).

$$W = \sum_{i=2}^{3}\sum_{j=0}^{3} W_{ij}, \qquad W_{ij} = M_{ij}\,N_{ij}. \tag{7.104}$$

The complete development was given by Doodson but as we are here concerned with the geodetic coefficients we give them explicitly:

$$\left.\begin{aligned}
N_{20} &= \frac{1}{2}\,D\,(1-3\sin^2\psi),\\
N_{21} &= D\sin 2\psi,\\
N_{22} &= D\cos^2\psi,\\
N_{30} &= 1.118\,03\,D'\sin\psi\,(3-5\sin^2\psi),\\
N_{31} &= 0.726\,18\,D'\cos\psi\,(1-5\sin^2\psi),\\
N_{32} &= 2.598\,08\,D'\sin\psi\cos^2\psi,\\
N_{33} &= D'\cos^3\psi,
\end{aligned}\right\} \tag{7.105}$$

with

$$D' = D\,\frac{r}{a}. \tag{7.106}$$

In the expressions derived from the chosen third-order potential, the numerical coefficients are such that the maximum value of the corresponding geodetic coefficient with respect to ψ is equal to 1. Hence the coefficients of amplitudes obtained in developing the M_{ij} functions will immediately give the order of magnitude of the harmonic constituents.

The three components of the tidal forces along the radial and the horizontal directions are, in geocentric coordinates,

$$\left.\begin{aligned}
-\frac{\partial W}{\partial r} &= C_v\,(r)\,\left(2\,W_2 + \frac{3r}{a}\,W_3\right),\\
-\frac{\partial W}{r\,\partial\psi} &= C_v\,(r)\,\frac{\partial}{\partial\psi}\left(W_2 + \frac{r}{a}\,W_3\right),\\
-\frac{\partial W}{r\cos\psi\,\partial\lambda} &= C_v\,(r)\,\frac{\partial}{\partial\lambda}\,\frac{W_2+\frac{r}{a}W_3}{\cos\psi},
\end{aligned}\right\} \tag{7.107}$$

with

$$C_v = \frac{D}{\bar{a}^2} = \frac{D(r)}{r^2}$$

For the third-order potential we also define

$$C'_V = \frac{r}{a} C_V. \qquad (7.108)$$

From the geodetic coefficients (7.105) we can derive the new geodetic coefficients for the components of the tidal force. However, the instruments physically measure with respect to the vertical direction given by the plumb line and we may consider that this is a good approximation of the normal to the ellipsoid. Now we have to rotate the axis in the meridian plane from the system (r, ψ) to the geographic coordinates system (n, ϕ).

If we define the components of the force along the three axis as follows:

Z: vertical pointing down along the normal n to the ellipsoid;
X: horizontal, positive to the South;
Y: horizontal, positive to the East;

then the deviations of the vertical $(\xi\eta)$ are positive to the North and positive to the West. Moreover, we define the longitudes positive to the West.

$$\left.\begin{aligned} Z &= -\frac{\partial W}{\partial n} = -\left(\frac{\partial W}{\partial r}\cos\varepsilon + \frac{\partial W}{r\partial\psi}\sin\varepsilon\right), \\ X &= -\frac{\partial W}{R_1\partial\phi} = -\left(\frac{\partial W}{r\partial\psi}\cos\varepsilon - \frac{\partial W}{\partial r}\sin\varepsilon\right), \\ Y &= \frac{-\partial W}{r\cos\psi\partial\lambda}. \end{aligned}\right\} \qquad (7.109)$$

The angle of rotation ε is the difference between the geographic and geocentric latitudes:

$$\left.\begin{aligned} \varepsilon &= \phi - \psi = e\sin 2\phi \\ \text{with} \quad \operatorname{tg}\psi &= (1-e)^2 \operatorname{tg}\phi \end{aligned}\right\} \qquad (7.110)$$

and R_1 is the radius of curvature of the meridian.

Then $\quad \xi = \frac{X}{g}, \qquad \eta = \frac{Y}{g} \qquad (7.111)$

Defining the coefficients

$$C_H = \frac{D(r)}{rg}, \qquad C'_H = C_H \frac{r}{a}, \qquad (7.112)$$

which appear in the development of the horizontal components, we obtain the new geodetic coefficients:

1° In the Vertical Component

$$\left.\begin{aligned} Z_{20} &= -C_V\,[(1-3\sin^2\psi)\cos\varepsilon - \tfrac{3}{2}\sin^2\psi\sin\varepsilon], \\ Z_{30} &= -3.354\,09\,C'_V\,[\sin\psi(3-5\sin^2\psi)\cos\varepsilon - \cos\psi(5\sin^2\psi-1)\sin\varepsilon], \\ Z_{21} &= -2C_V\,(\sin 2\psi\cos\varepsilon + \cos 2\psi\sin\varepsilon), \\ Z_{31} &= -0.726\,18\,C'_V\,[3\cos\psi(1-5\sin^2\psi)\cos\varepsilon - \sin\psi(15\cos^2\psi-4)\sin\varepsilon], \end{aligned}\right\} \qquad (7.113)$$

$$\left.\begin{aligned}
Z_{22} &= - C_V\ (2 \cos^2\psi \cos \varepsilon - \sin 2\psi \sin \varepsilon),\\
Z_{32} &= - 2.598\,08\ C'_V\ [3 \sin \psi \cos^2\psi \cos \varepsilon + (1-3 \sin^2\psi) \cos \psi \sin \varepsilon],\\
Z_{33} &= - 3C'_V\ (\cos^3\psi \cos \varepsilon - \cos^2\psi \sin \psi \sin \varepsilon).
\end{aligned}\right\} \quad (7.113)$$

2° In the North-South Component

$$\left.\begin{aligned}
\xi_{20} &= C_H\ [\tfrac{3}{2} \sin 2\psi \cos \varepsilon + (1-3 \sin^2\psi) \sin \varepsilon],\\
\xi_{30} &= 3.354\,09\ C'_H\ [\cos \psi\ (5 \sin^2\psi-1) \cos \varepsilon + \sin \psi(3-5 \sin^2\psi) \sin \varepsilon],\\
\xi_{21} &= - 2C_H\ (\cos 2\psi \cos \varepsilon - \sin 2\psi \sin \varepsilon),\\
\xi_{31} &= 0.726\,18\ C'_H\ [\sin \psi(15 \cos^2\psi-4) \cos \varepsilon + 3 \cos \psi(1-5 \sin^2\psi) \sin \varepsilon],\\
\xi_{22} &= C_H\ (\sin 2\psi \cos \varepsilon + 2 \cos^2\psi \sin \varepsilon),\\
\xi_{32} &= 2.598\,08\ C'_H\ [\cos \psi(2-3 \cos^2\psi) \cos \varepsilon + 3 \sin \psi \cos^2\psi \sin \varepsilon],\\
\xi_{33} &= 3C'_H\ (\sin \psi \cos^2\psi \cos \varepsilon + \cos^3\psi \sin \varepsilon).
\end{aligned}\right\} \quad (7.114)$$

3° In the East-West Component

$$\left.\begin{aligned}
\eta_{20} &= 0,\\
\eta_{21} &= 2C_H \sin \psi,\\
\eta_{31} &= 0.726\,18\ C'_H\ (1 - 5 \sin^2\psi),\\
\eta_{22} &= 2C_H \cos \psi,\\
\eta_{30} &= 0,\\
\eta_{32} &= 2.598\,08\ C'_H \sin 2\psi,\\
\eta_{33} &= 3C'_H \cos^2\psi.
\end{aligned}\right\} \quad (7.115)$$

These formulas were given by Wenzel (1974) and for the North-South and the East-West components do not differ from Lecolazet's calculation which is limited to the first approximation by putting

$$\cos \varepsilon = 1, \qquad \sin \varepsilon = e \sin 2\psi,$$

and neglects the ellipticity corrections for the terms coming from the third-order potential.
Lecolazet's constant F is our constant C_H where r is developed as

$$r = 1 - e \sin^2\phi + \frac{h}{a}, \qquad (7.116)$$

h being the height above the ellipsoid.

From the definition of Doodson's constant we derive the numerical expressions of C_V and C_H:

$$C_V = \frac{D(r)}{r} = \frac{D}{a^{-2}} r = \frac{3}{4}\frac{GE\mu \sin^3\pi}{a^3} r = 41.291\ 125\ (1 - e \sin^2\phi + \frac{h}{a}) \text{ microgals} \quad (7.117)$$

$$C_H = \frac{D}{ga^{-2}} r = \frac{3}{4}\frac{GE\mu \sin^3\pi}{ga^3} r \times 206265'' = 8.516\ 906 \times \frac{10^3}{g} (1-e \sin^2\phi+\frac{h}{a}) \text{ msec} \quad (7.118)$$

$$C'_V = C_V (1 - e \sin^2\phi + \frac{h}{a}), \quad (7.119)$$

$$C'_H = C_H (1 - e \sin^2\phi + \frac{h}{a}), \quad (7.120)$$

7.4.1. The Trigonometric Functions of the Arguments

In Doodson's original paper the arguments of the tidal constituents are associated with cosine or sine functions. A simple rule lets us know if we have to introduce a sine or cosine of the argument (ARG): if the sum of the numbers i (order of the potential) and j (defining the family of the tide) is an even number, one has to use a cosine function.

Thus:

for long-period terms: { M20 cosine; M30 sine

for diurnal terms: { M21 sine; M31 cosine

for semi-diurnal terms: { M22 cosine; M32 sine

for ter-diurnal terms: M33 cosine

In the computation program we usually express a sine function of ARG as cos (ARG - $\pi/2$) = cos ARG* which is valid for vertical and North-South components. For the East-West component, as we have a derivation with respect to the longitude (included in τ) we get:

$$d \cos \text{ARG} = - \sin \text{ARG} = \cos (\text{ARG} + \pi/2),$$

$$d \sin \text{ARG} = \cos \text{ARG} = \cos (\text{ARG}^* + \pi/2).$$

The addition of $\pi/2$ to all arguments gives the correct trigonometric function for the East-West component. One should remember that Lecolazet added π to the arguments in vertical and East-West components instead of changing the sign of the geodetic coefficient. To avoid the introduction of negative amplitudes he also added π to some arguments (Table 7.4).

7.4.2. Limits to the precision of our harmonic analyses

It is often difficult to obtain precise coordinates of stations in some parts of the world, so we may question the precision of the adopted values as a source of systematic error. It is easy to estimate the effect of an error in the longitude: 1' or 2 km in position corresponds to 4 s in time or 0.033° on the phase of the semi-diurnal waves and 0.016° on the phase of the diurnal waves. Therefore the error in longitude may be excluded from our considerations.

For what concerns the position in latitude we observe that in the equatorial zone the results are totally insensitive to a change of latitude of 1' for the semi-diurnal components but that for the diurnal components this change of latitude may give 1-1.5° change in the phase of the observed load signal while it has no sensible effect on its amplitude.

Comparing now the results obtained by the Venedikov method programmed at ICET (Bruxelles) upon a Univac computer with those given by the Chojnicky method as programmed at Bonn University upon an IBM computer we got only very small differences in the load signal for two different sets of data from stations in Norway as shown in the Table 7.4.
(The load signal is obtained by subtracting the calculated tidal effect for an elastic model earth from the corresponding observed effect (see 13.1).

TABLE 7.4. Comparison of observed load signals obtained by different methods of analysis and different computers

Station	Method/computer	M_2 wave		O_1 wave	
		B(µgal)	β(°)	B(µgal)	β(°)
Bergen	Venedikov/ICET	1.61	-133	1.59	155
	Chojnicki/Bonn	1.55	-130	1.86	153
Trondheim	Venedikov/ICET	4.87	-153	1.41	151
	Chojnicki/Bonn	4.95	-154	1.68	146

7.5. A GRAPHICAL TWO-DIMENSIONAL REPRESENTATION OF THE TIDAL VARIATIONS

A method devised by the Admiralty in the United Kingdom offers a nice visualization of the tidal variations. Having removed the drift one introduces the tidal heights (gravity, tilt or strain) in a two-dimensional rectangle where the days are put as abscissae and UT hours as ordinates. Then equal tidal heights lines are drawn accross the net to give a chart similar to those in Fig. 7.3.

Diagonal straight lines represent the Moon meridian transit at the observing place. The times of full moon, new moon and quarters are also indicated. This system was used a long time ago to make the analysis more easy when computers were not available. Now it could still be used to detect anomalous phenomena quickly and easily. The two drawings of Fig. 7.3 are obtained with two different Verbaandert-Melchior quartz pendulums operating simultaneously in the East-West component in two underground stations which are about 50 km distant. Some curious similar features are indicated by arrows.

TABLE 7.5. Corrections to Arguments

	Long period		Diurnal		Semi-diurnal		Ter-diurnal
	2,0	3,0	2,1	3,1	2,2	3,2	3,3
Vertical	0	$-\pi/2$	$-\pi/2$	0	0	$-\pi/2$	0
North-South	0	$-\pi/2$	$-\pi/2$	0	0	$-\pi/2$	0
East-West	-	-	0	$\pi/2$	$\pi/2$	0	$\pi/2$

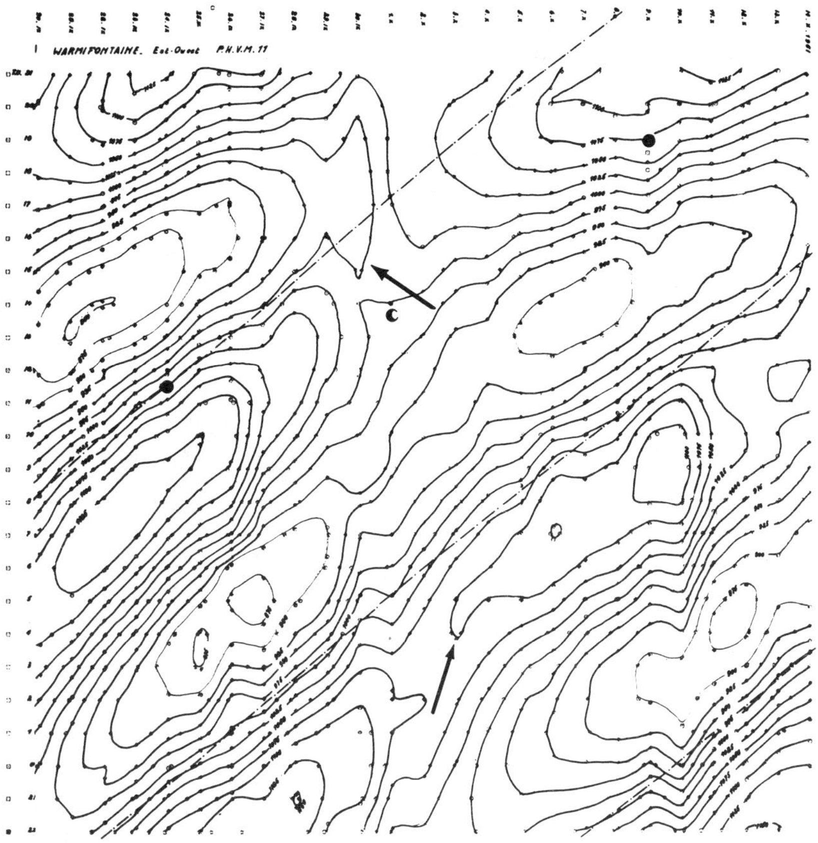

Fig. 7.3
(a) Station Warmifontaine (Belgium) East West Component
Horizontal Pendulum VM 11. 20/9 to 14/10 1961

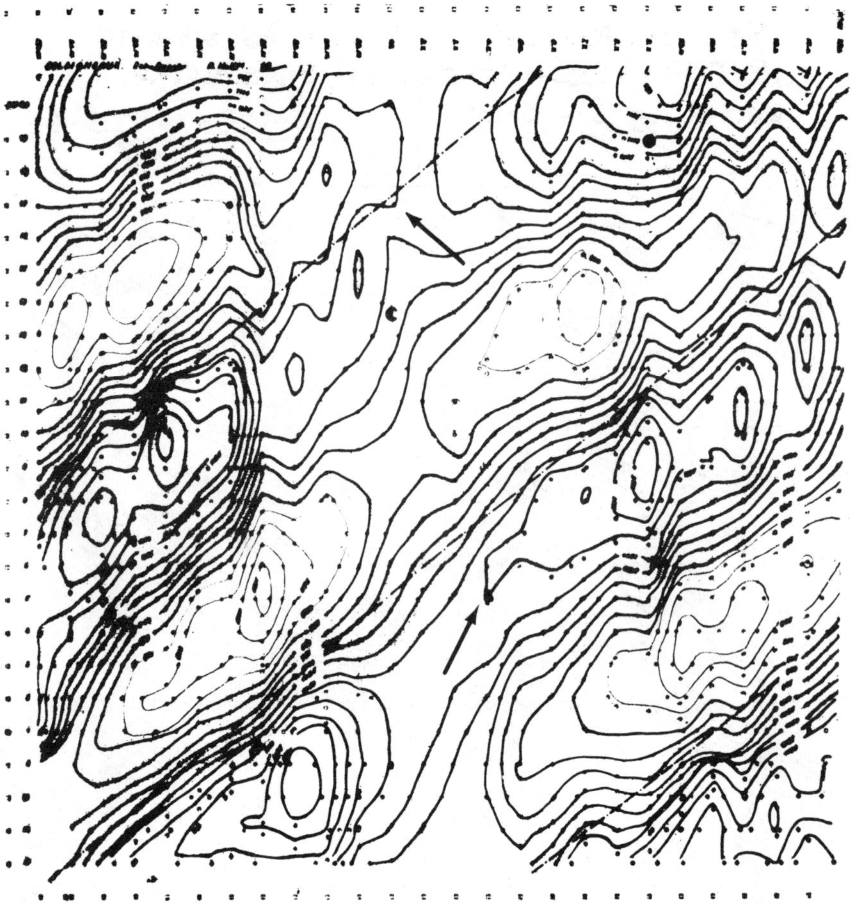

Fig. 7.3
(b) Station Sclaigneaux (Belgium) East West Component
Horizontal Pendulum VM 10. 20/9 to 15/10 1961

CHAPTER 8

Tilt: Deflexion of the Vertical with respect to the Earth's Crust, Instrument, Results (The Factor $\gamma = 1 + k - h$)

8.1. THE AMPLITUDE OF LONG-PERIOD OCEANIC TIDES

The study of the oceanic tides, as suggested by Kelvin, was the first means of measuring Earth tides. We know that equilibrium oceanic tides are due to periodic deviations of the vertical, so that the surface of the oceans continually adapts itself to the direction of the disturbed verticals. It is thus *a priori* evident that the observation of oceanic tides will lead to an estimation of the deviation of the vertical. The comparison of these observations with the theoretical values should enable us to deduce the factor (1+k-h) as shown by (3.35). But this comparison is only valid for long-period waves which, because of the slowness of the displacements and the presence of the continental barriers preventing the formation of currents, are little affected by the inertia, little damped and, therefore, should perhaps obey the static theory.

In 1881 Darwin analysed the records from 14 European and Indian ports, covering 33 years' observation; then Schweydar analysed records from 43 ports spread over the world and covering 194 years' observation.

Their results are as follows:

	Fortnightly tide, M_s:	Monthly tide, M_m:
Darwin	$\gamma = 0.675 \pm 0.084$	$\gamma = 0.680 \pm 0.387$
Schweydar	$\gamma = 0.6265 \pm 0.043$	$\gamma = 0.6053 \pm 0.102$.

Very serious objections were raised against this procedure for evaluating the factor γ, and the method is now only of historical interest: complex effects, which are impossible to eliminate by calculation, falsify the results.

As an example the gravitational forces exerted by the liquid layer raised by the tide itself was neglected in the theoretical calculations of the amplitude of the long-period tide. This secondary effect can be evaluated in the case of an ocean covering the whole Earth, which is assumed to be homogeneous.

Assuming that H is the height of the layer of water raised by the action of the potential W_2, the equation of the outer surface of the Earth becomes

$$r = a + H(\phi, \lambda) = a(1+H/a),$$

a being the mean radius of the Earth.

This superficial layer induces an additional potential given by (3.12) with $H = a\varepsilon$,

$$V = 4\,\pi\,G\,\rho_0\,aH/5,$$

while the gravitational acceleration is

$$g = \frac{4\pi G}{a^2}\int_0^a \rho\, r^2\, dr = 4\,\pi\,G\,\rho\, a/3,$$

ρ_0 and ρ being the mean densities of sea water and the Earth.

Thus

$$V = 3\,\rho_0\,g\,H/5\rho \tag{8.1}$$

and the height of the tide due to the additional attraction is

$$V/g = 3\,\rho_0\,H/5\rho \tag{8.2}$$

The height of the pure luni-solar tide ξ is the difference between the total observed height H and the height ζ due to the attraction of the layer itself, i. e.

$$\xi = H - \zeta = H\left(1 - \frac{3}{5}\frac{\rho_0}{\rho}\right), \tag{8.3}$$

From which

$$\frac{H}{\xi} = \left(1 - \frac{3}{5}\frac{\rho_0}{\rho}\right)^{-1}, \tag{8.4}$$

and assuming that $\rho_0/\rho \simeq 1/6$, we find that $H/\xi = 10/9$; therefore the observed heights should be reduced by 10% and, similarly, the numerical values of the coefficient which are deduced from it. Naturally this correction cannot be applied exactly in this form because the oceans only cover some two-thirds of the globe and are irregularly distributed. In addition the presence of the continental barriers modifies the amplitude and the phase of the theoretical tide.

Finally, static tides of the second kind (which result from the Coriolis force) can also exist, even in limited oceanic areas, as permanent currents which neither continental barriers nor friction seem to be able to impede: they are then imposed on ordinary static tides, and only observation will tell us to what extent they affect the phenomena.

It should be noted that in spite of all objections of a theoretical character, numerical values obtained by this method for the coefficient γ are in quite good agreement with those provided by other methods not liable to such objections.

In 1925 Proudman suggested trying to obtain a value of γ from the study of the tides in narrow seas or large lakes.

Grace tried this in the Red Sea but did not obtain a significant determination (the factor found was negative), and it is thought that this was due to the communication between this sea and the Indian Ocean.

8.2. LAKE TIDES

Evidence was available of the existence of tides in some large lakes - Huron, Superior, Michigan, Erie, Plattensee, Genfersee, Chiemsee, etc. - and several numerical values were available; unfortunately these resulted from too unselective and insufficiently explicit analyses unsuitable for discussion.

On the contrary, good results were obtained for lakes Baikal and Tanganyika. Records of the levels obtained at two ports on Lake Baikal have been analysed at different times by Sterneck, Grace, Ekimov and Krawetz, Parfianovitch and Aksentieva, and have provided the following numerical values:

Petchanaia	Wave M_2	Sterneck	γ = 0.52	α = -3°
		Grace	γ = 0.54	
		Aksentieva	γ = 0.72	α = +1°
	Wave K_1	Sterneck	γ = 0.73	α = +4°
Tankoĭ	Wave M_2	Aksentieva	γ = 0.55	α =+27°

The amplitudes are about 5-6 mm.

The study of Lake Tanganyika is quite simple in theory. The observations were made at Albertville, where the lake is about 75 km wide; a good series of observations were analysed by Melchior. The median position of Albertville is such that the level there is insensible to the uninodal longitudinal seiches whose period is about 4h (length of lake 637.5 km; mean depth 800 m); on the other hand, the binodal longitudinal seiche (period 2h) and the uninodal transverse seiche (period 40 min) should be observable. Taking into account the short period of the free oscillations, we can treat the problem of the luni-solar tides by the static theory.

At the latitude of Albertville (-5°54') the changes in the vertical in the sense East-West cause the periodic deviations whose amplitude is e = 0".015 64 for the wave M_2. The amplitude of the variations of level at the shores will be

$$\xi \doteq 1/2\ L \sin e, \tag{8.5}$$

the width L being 72.5 km; theoretically we shall have

$$\xi = 2.748 \text{ mm},$$

and the phase will obviously be 90° because the level is undisturbed when the East-West component of the deviation of the vertical is nil, i. e. when the Moon passes over the meridian.

The records of the tide-gauge make it possible to read the level with a precision of up to 0.48 mm, and we obtained for M_2:

amplitude 1.52 mm, phase 99°5'

giving γ = 0.55, α = 9°.

Unfortunately the degree of precision of the recordings was not known, and the value of the coefficient γ is liable to 10% error.

8.2.1. Experience in an Artificial Basin

Obviously the irregular form of the oceanic or lake basins complicates the interpretation. Therefore it is interesting to remind ourselves here of the measurements made in 1947 by Zerbe in one of the artificial basins situated at Carderock (near Washington) and designed for scale-model ship trials (David Taylor Model Basin). These long and narrow basins are deep enough for a long wave to travel from one end to the other within the required tidal period. They are enclosed under an arched roof and are ideal for making observations of the type required. However, the tide reaches a maximum of only a tenth of a millimetre. The basin used by Zerbe is 845 m long, 16 m wide and 6.7 m deep; it is pointed in the direction W16°N/E16°S. One gauge was fixed 17 m from the West extremity, the other at 15 m from the East extremity; they are therefore 813 m apart. The measurements thus made at the two extremities of the basin make it possible to eliminate the modification of level due to evaporation and reduce the observational errors. The gauges were read alternately every 15 min.

Unfortunately it has only been possible to make these measurements during one week, and it is therefore difficult to separate correctly the various components. Moreover, indirect effects due to the oceanic tides in the nearby Atlantic greatly disturb the measurements. Zerbe tried to evaluate them by a direct calculation on the masses of water involved, and finally obtained a coefficient γ = 0.74.

8.3. OBSERVATIONS CARRIED OUT ON LARGE WATER-LEVELS

In 1914 Michelson and Gale managed to observe the phenomenon in a specially built receptacle (of regular geometric shape) by microscopic then interferential measurements. They sank two tubes, 150 m long and with a diameter of 15 cm, to a depth of 1.8 m. These two tubes were about half-filled with water and oriented according to the meridian and the prime vertical. The first series of experiments were designed to study the variations of the level at the two ends of each tube. This was carried out by measuring, with a microscope, the distance separating the extremity of an immersed point from its image obtained by total reflection in the liquid. More precise results were obtained during a second series of observation using an interferential method with the help of the disposition depicted in Fig. 8.1, and which was placed at each end of the tubes.

A horizontal mirror was immersed in such a way that it was covered by 0.5 mm, which layer of water viscosity absorbed the minor shocks; any variation of the water level thus caused a change in the optical path of rays from the source S and a displacement of the interference fringes. The compensatory plate serves to seal up the tube. The displacement of the fringes were recorded on a film at a speed of 20 mm h^{-1}.

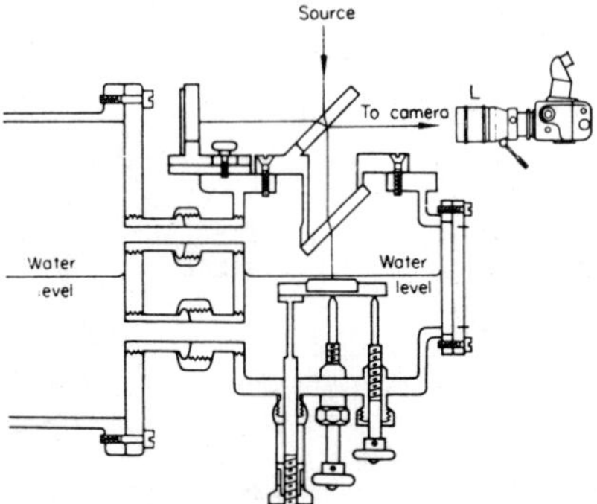

Fig. 8.1. Michelson-Gale interferometer for the measurement of variations of water-level.

These observations were carried out for a year (20 November 1916-20 November 1917). The films have been measured microscopically, the displacement of the fringes being measured to a tenth; the difference in the movement of the two ends of each tube give the observed tidal height.

The theory of tides in tubes of water has been developed by Moulton. The authors carried out a harmonic analysis not only on the experimental curve but also on the theoretically calculated variations in order to remove any distortion due to imperfect elimination of certain components. The result of the group of measurements is $\gamma = 0.69$.

In 1934 Egedal and Fjeldstad applied the same procedure in Europe by installing two water tubes in a tunnel 125 m long at Bergen (Norway). The apparatus (which they called a "level variometer") consisted of a cylindrical vase, 14 cm in diameter and 7 cm high, in communication with the water in the tube, which partially filled it. To avoid evaporation, the water was covered with a layer of Nujol (specially purified paraffin oil) about 1-2 cm thick on which floated a cylindrical glass float, 5 cm in diameter and 1-7 cm high. The vertical movements of this float were transmitted by a vertical rod to a gear pivoted to a mirror whose variations in inclination were observed by a small telescope.

Finally, one application of this method of very large levels occurred in a geodetic levelling operation in Denmark.

The difficulties encountered in carrying out a precise levelling between the various Danish islands has led Norlund to use since 1939 a new hydrostatic process for which the apparatus, similar to that of Michelson but much longer, should unfailingly lead to observations of Earth tides.

Two vertical glass tubes, marked off in millimetres, were set up in Fionie and Sjaelland and linked by a tube, 18 km long, filled with water, immersed at a depth of 60 m in the Great Belt. As the water surfaces in the two upright tubes belonged to the same surface level it was thus possible to relate the levels between the various islands. The tubes used were made of a lead-tin alloy (3% tin) with 10 mm bore and walls 3 mm thick, and weighed 1380 kg km^{-1}. The greatest difficulty was to remove all air-bubbles from the tubes.

In order that the two water surfaces belong to the same level it is necessary that

$$V(r + \Delta r, \phi + \Delta\phi, \lambda + \Delta\lambda) = V(r, \phi, \lambda),$$

which gives, with sufficient accuracy,

$$\frac{\partial V}{\partial r}\Delta r + \frac{\partial V}{\partial \phi}\Delta\phi + \frac{\partial V}{\partial \lambda}\Delta\lambda = 0. \tag{8.6}$$

Knowing the difference of latitude $\Delta\phi$ and of longitude $\Delta\lambda$ of the two stations, one can deduce the variations of r.

Between Fionie and Sjaelland, the levelling was carried out directly, and by using, as an intermediate, Sprogoe Island, the two results were practically the same. The differences between the readings of the upright tubes at the two terminal stations are obviously not constant and change regularly with time; the semi-diurnal wave has an amplitude of 2-3 mm.

The length of tube (18 km) considerably favoured the observation, which gave a coefficient

$$1 + k - h = 0.8.$$

More recently a water-level tiltmeter has been constructed by Bower (1973) which, having a length of 50 m, seems to have a sensitivity approximately equal to that of a horizontal pendulum. A dilatable crapaudine type calibrating device was constructed and used by Bower (see Section 8.8).

Another device, also 50 m long and based upon interferometric measurements, has been constructed and installed in a tunnel near Helsinki by Kääriäinen (1973).

No analysis of the data obtained with these instruments has so far been published.

8.4. THE INVENTION ABOUT 1830 OF THE "HORIZONTAL PENDULUM" BY HENGLER AND THE FIRST ZÖLLNER PENDULUM

At the beginning of the seventeenth century, with the discovery by Newton of the law of universal gravitation and its application in the interpretation of tidal phenomena, it became obvious that the luni-solar attractions could cause important movements at the surface of the Earth. The idea probably occurred to several experimenters to try to measure the influence of these attractions on the equilibrium of a freely suspended vertical pendulum. Although such an experiment was, even for that period, simple to carry out, no publication on this subject appeared before the nineteenth century.

In fact it was in 1828 that Gruithuisen of Munich published a paper which at that time was hardly noticed, and in which he proposed to suspend a long, vertical pendulum in a mine shaft so as to measure any deviation of the vertical with respect to the surrounding walls of rock. The author recalls in his report that as early as 1817 he published a note in which he described his experiments with a 3 m long pendulum to show the deviation which would be caused by changes in the apparent direction of gravity, by, among other things, the changing attractions of the Sun and Moon.

It is certain that Gruithuisen did not carry out any calculation of the theoretical movement of his pendulum due to luni-solar attraction before doing his experiments, since he would otherwise have realized the futility of his effort. As the combined attractions of the Sun and Moon are capable of producing an angular displacement of 0".05, this would be equivalent to a movement of the end of the pendulum through less than 1 micron, a quantity which was practically unmeasurable with the means at his disposal.

The importance of Gruithuisen is as a "promoter" since one of his students, Lorentz Hengler (born in 1806 at Reichenhofen, student of mathematics and astronomy at Munich, 1830-1), discovered a means of increasing by a very large factor the micro-displacements of the end of a pendulum in such a way as to make them usable in experiments. In contriving an ingenious bifilar suspension, Hengler invented the horizontal pendulum (Fig. 8.2), to which he gave the name Astronomische Pendelwaage.

He carried out numerous observations using this instrument in order to demonstrate the effect of luni-solar attraction. The movements of the tip of the pendulum were observed by means of a microscope, and the author declared that from these measurements he was able to show statistically a definite and significant effect in agreement with theoretical forecasts.

About 1872 Zöllner, taking advantage of new technical developments, constructed a horizontal pendulum based on the original idea of Hengler. The bifilar suspension since then has been called the "Zöllner suspension" because Zöllner's pendulum made it possible to carry out the first measurements equally applicable to seisms and to the slow movements caused by changes of the ground. Zöllner's pendulum, as shown on Fig. 8.3, consists of a very rigid bracket P fixed by a base plate to three levelling screws V_1, V_2 and V_3. At points A and B, situated practically on the same vertical, two metal wires are clamped and fixed at C and D to a horizontal metal beam.

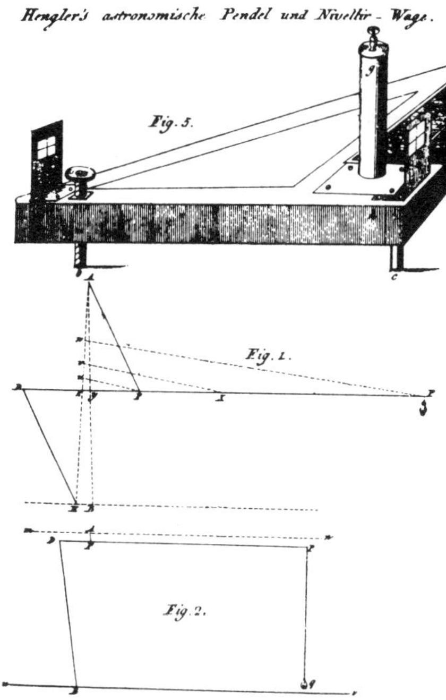

Fig.8.2. Figures taken from Polytechnisches Journal von Dingler 1832.
Fig. 1 is a schematic representation of the bifilar suspension by Hengler;
Fig. 2 a trial made by Hengler to demonstrate the rotation of the Earth with a bifilar;
Fig. 5 is the image of an horizontal pendulum constructed by Hengler himself.

The axis of rotation of the beam is therefore the line passing through A and B and which one can make almost vertical by suitable manipulation of the levelling screws.

To improve regulation in the modern pendulums, the levelling screws are arranged at the tips of a right-angled triangle. The screws V_1 and V_2, placed at the summits of the acute angles of the triangle, are, respectively, called the "sensitivity screw" and the "drift screw". By turning V_1 we modify the inclination of the pendulum in the sense of the moving beam, and this is done effectively by turning about a line passing through the points of contact of the two screws V_2 and V_3. Under these conditions, if the points of contact are on a parallel to the equilibrium plane of the pendulum, the moving beam is not subject to lateral deviation except for a possible change of 180° if the fictitious axis of rotation is reversed beyond the vertical.

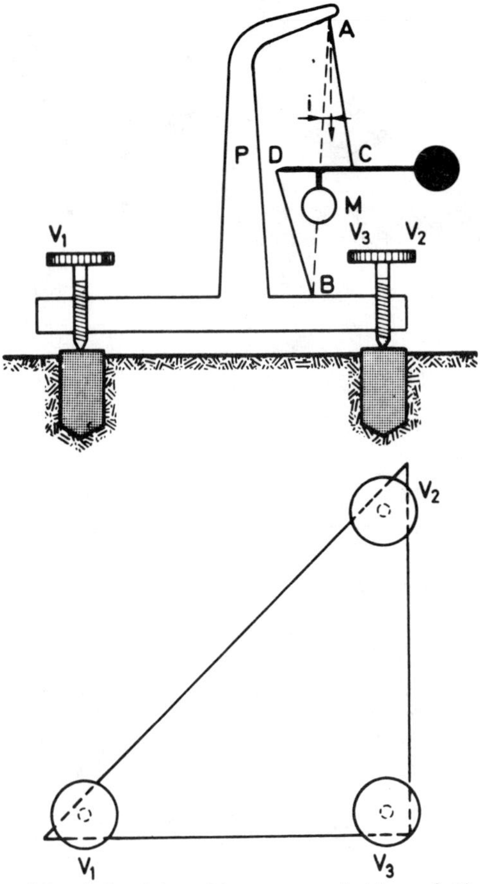

Fig. 8.3. Schematic representation of the Zöllner suspension.

On the other hand, by turning the screw V_2 the pendulum is turned about a line, obviously horizontal, passing through the points of contact of the screws V_1 and V_3, and the moving arm is subject to an important angular deviation.

Therefore with a single pendulum it will only be possible to measure the component of the deviation relative to the vertical, which takes place in a plane perpendicular to the moving beam. In order to determine completely the deviations of the vertical it will therefore be necessary to use two pendulums with the beam of one perpendicular to the beam of the other.

The mirror M fixed to the pendulum arm at the point O where the axis of rotation crosses the beam makes it possible, by using Poggendorf's method to record photographically and permanently the smallest movements of the

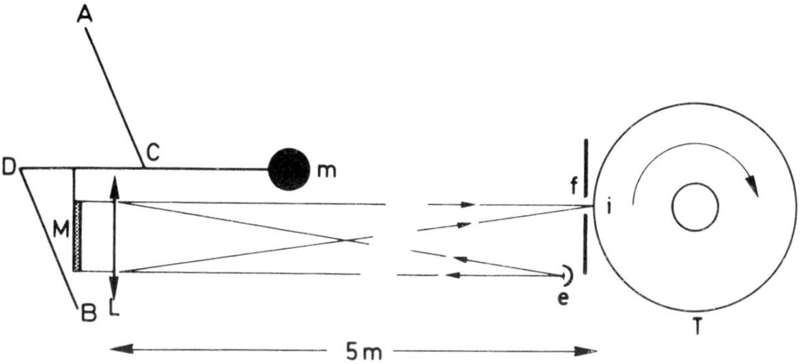

Fig. 8.4. Recording system of horizontal pendulum oscillations based on Poggendorf's method: M, plane mirror fixed to the pendulum arm; L, converging lens of 5 m focal length; e, electric lamp with rectilinear vertical filament; f, horizontal split (30 cm long and 0.1 mm aperture); I, image of the filament of lamp e reflected on M and projected on to the photographic paper born by the rotating drum (speed: 1 rotation per week).

pendulum (Fig. 8.4). A converging lens with a focal distance of about 5 m is placed in front of the mirror. A small electric lamp, with a rectilinear filament, is placed vertically at the focal point of the lens. The light from the filament is reflected by M, having first passed through the lens. The reflected light passes through the lens a second time and, behind a very fine horizontal slit, forms at I the image of a point of the filament on a drum, covered with sensitive paper: this drum turns regularly so as to make one complete rotation in a week. Every hour a small auxiliary lamp, whose illumination is governed by an accurate clock, lights up the slit for several seconds so that the time is marked on the sensitive paper by parallel and equidistant lines.

8.5. ELEMENTARY THEORY OF THE HORIZONTAL PENDULUM

Displacements relative to the vertical can arise not only from a real change in the direction of gravity but also from the instabilities of the ground on which the instrument rests.

Let us suppose that we have an ideal pendulum whose beam b has a mass which can be neglected *vis-à-vis* the terminal spherical weight of mass m (Fig. 8.5). Considering a sphere passing through the centre of this weight, and whose centre is located at the point O where the axis of rotation of the pendulum AB meets the beam b, the points Z (zenith) and N (nadir) are pierced points of the vertical in the spherical surface considered. The axis AB and the vertical form a small angle i which determines the sensitivity of the ideal pendulum. On the figure, the plane of oscillation of the pendulum is the plane PP' seen in profile. It forms an angle i with the horizontal plane HH'. The component mg sin i of the gravity along the beam direction maintains the pendulum in equilibrium. It is very small.

We choose as coordinate axes the axis of rotation of the pendulum (Oz), the direction of the beam in equilibrium (Oy) and an axis perpendicular to their plane at O (Ox).

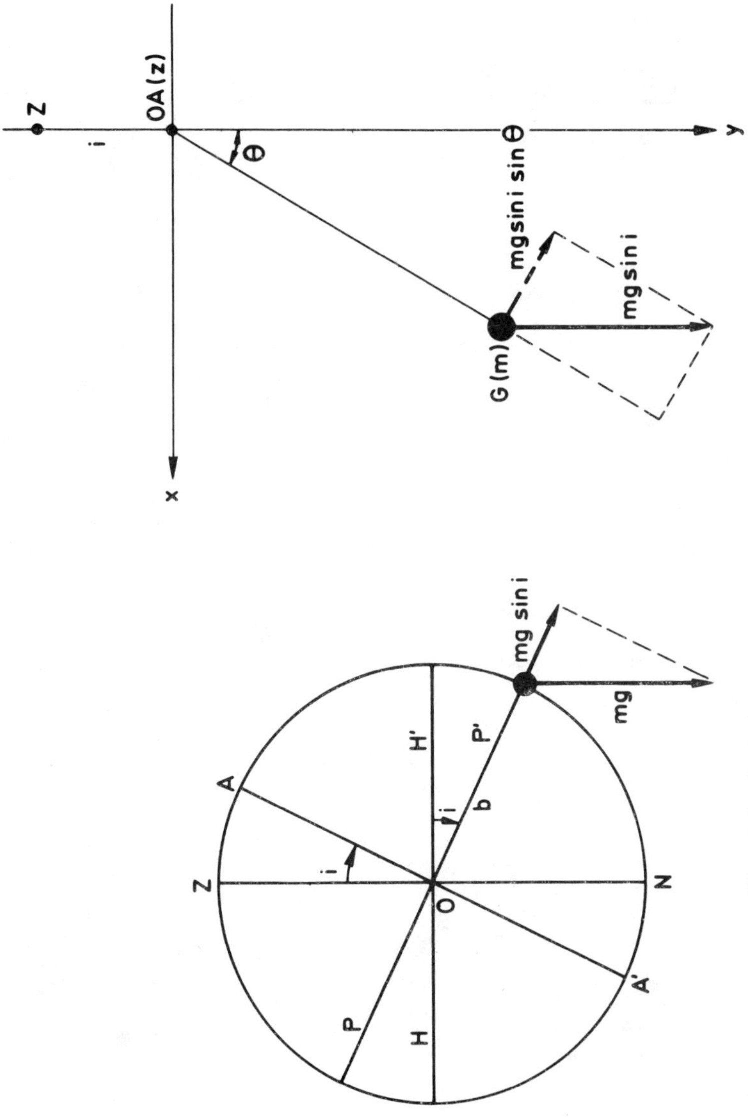

Fig. 8.5. Position of equilibrium of an horizontal pendulum

If we oscillate very lightly the beam about its axis of rotation and consider the position reached after a time t at an angle α to its original position, we find that the angular velocity perpendicular to the direction of the beam is

$$\mathbf{v} = \mathbf{OG} \cdot \frac{d\alpha}{dt}, \tag{8.7}$$

while the moment of the momentum with respect to the axis Oz is

$$\mathbf{M} = \mathbf{OG} \wedge m\mathbf{v} = m L^2 \frac{d\alpha}{dt} \tag{8.8}$$

with $|OG| = L$, the reduced length.

The coordinates of the centre of gravity are

$$x = L \sin \alpha, \qquad y = L \cos \alpha, \qquad z = 0, \tag{8.9}$$

and the components of the gravity applied to the pendulum are:

$$X = 0, \qquad Y = -m g \sin i, \qquad Z = -m g \cos i \tag{8.10}$$

The theorem of the projection of the kinetic moment on Oz thus gives the equation of the movement of the centre of gravity in the Oxy plane:

$$\frac{d}{dt}\mathcal{M}_0 (mv) = (xY - yX), \tag{8.11}$$

i. e.

$$\frac{d^2\alpha}{dt^2} + \frac{g \sin i}{L} \alpha = 0, \tag{8.12}$$

a differential equation whose general integral is

$$\alpha = \sin \left[\sqrt{\left(\frac{g \sin i}{L}\right)} t + \beta \right], \tag{8.13}$$

and the period of the free oscillation of pendular movement is thus

$$T = 2\pi \sqrt{\frac{L}{g \sin i}}. \tag{8.14}$$

The component Mg sin i only has an influence on the pendulum since Mg cos i remains perpendicular to the beam, and its effect is supported by the system of suspension. We could also consider that the oscillating mass is suspended at the end of a vertical pendulum whose virtual point of suspension is at the point at which the vertical of G meets the axis of rotation AB. The length of this fictitious pendulum is L cosec i; it is often called the equivalent length.

If it were possible to straighten the pendulum without breaking the suspensory filaments and without modifying the position of the point O on the arm b, the period of oscillation of the pendulum made vertical would be

$$T_0 = 2\pi \sqrt{\frac{L}{g}}. \tag{8.15}$$

The knowledge of T_i and T_0 thus makes possible the calculation of the angle i:

$$\sin i = T_0^2 / T_i^2. \tag{8.16}$$

Some makers fix a knife (whose edge passes through O) on the beam. By suspending the pendulum on this knife it is possible to measure T_0. This method is not very precise since it is almost impossible to make the knife-edge coincide perfectly with O. On the other hand, this increases the weight of the pendulum, when it is an advantage to keep it as light and as simple as possible.

This very elementary and very simplified theory of the horizontal pendulum only takes into account the movement of rotation of the pendulum about the axis AB since this is the only movement of interest in the study of Earth tides. The pendulum has other degrees of freedom and can carry out various other movements: rolling, pitching, and movement of the beam parallel to its direction of rest. These movements are rapidly damped in practice and they are only observed during a few minutes following the installation of the pendulum.

Let us now suppose that the vertical moves from OZ to OZ' (Fig. 8.6). The zenith describes a simple arc ZZ' = e along a great circle arc whose plane is normal to the pendulum's beam. Under the influence of this change in the vertical the pendulum becomes unbalanced, and in order to return to a point of stable equilibrium the weight moves from m to m' by an angle $\Delta\alpha$ so that its centre lies in the new vertical plane containing the axis AB. The arc of great circle Z'P defines, on the sphere, this new vertical plane; Z'P passes through the centre of m'.

In the elementary right-angled triangle ZZ'P,

$$\sin i = \tan e \cot \Delta\alpha, \tag{8.17}$$

or, considering the smallness of the angles,

$$\Delta\alpha = e/i = e\, T_i^2/T_o^2 \tag{8.18}$$

in which $\Delta\alpha$ is expressed in radians.

We define the sensitivity of the pendulum

$$s = e/\Delta\alpha = K/T_i^2. \tag{8.19}$$

The smaller i becomes the more important becomes the angular deviation $\Delta\alpha$ of the pendulum for a given deviation of the vertical.

The factor 1/i expressed in radians characterizes the amplifying power of the pendulum. If we could give i the value 1" the coefficient of amplification would become 206.265. In practice it is about 4".

However, one has to take into account the rigidity of the suspension wires because their resistance to torsion diminishes the sensitivity of the instrument. It is possible to avoid important resistance to torsion while maintaining sufficient resistance to breakage by a judicious selection of the dimensions and the material of the wires.

8.6. THE DIFFERENT MOUNTINGS FOR HORIZONTAL PENDULUMS

The inconveniences met with in the first constructions of Zöllner's system of suspension arose from the lack of quality in the wires used at that time which had a too high modulus of torsion: Hengler proposed single threads of silk or horse-hair; Zöllner used steel, and Orlov used platinum wires.

Several constructors thought of an analogous system of suspension, but without the use of suspensory filaments: von Rebeur Paschwitz developed such a system using two points acting in two different directions on the arm of an instrument whose weight was 42 g. Such instruments were designed mainly for seismic research.

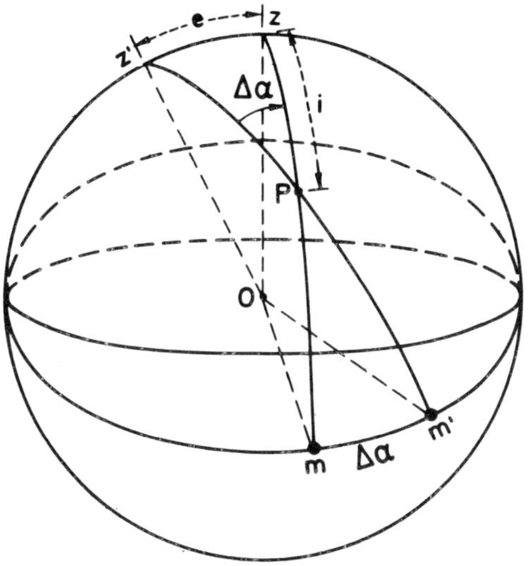

Fig. 8.6. Amplification of a deviation of the vertical by an horizontal pendulum.

Hecker perfected Von Rebeur Paschwitz's construction by arranging the points so that their axes and the direction of the gravity applied to the centre of gravity were concurrent.

The major problem in this abandoned type of suspension was to find suitable materials for the points and their planes of rest. The constructors generally used steel points resting on agate cups, but the wear of the points led to major inconveniences. In particular they introduced friction which greatly reduced the limits of sensitivity under which the apparatus worked. It seems that it was not possible to exceed periods of the order of 35 s, which gives a sensitivity of 0".008 per mm for 5 m of local length (Von Rebeur).

Milne and Shaw constructed an instrument in which only one of the suspension wires was replaced by a point. An apparatus of this type has been in use for many years at Bidston (Liverpool) where it recorded the tidal deformations of the ground.

The point, which was originally iridium, has been replaced by a sapphire of the type used in record players.

The period of the instrument in use is 17 s. A period of 24 s has been tried with success, but it is not certain that higher sensitivities can be obtained without throwing the apparatus out of balance.

The Zöllner suspension has several advantages over that of von Rebeur-Paschwitz. First, the simplicity of the construction, then the greater stability of the zero and of the sensitivity.

In Japan, Ishimoto was the first to introduce pendulums of fused quartz with quartz wires. We can criticize this apparatus because of a marked asymmetry and the introduction of a body like picein for fixing the quartz support to the base of the instrument. The apparatus is light (the wire has a diameter of 7 microns) and small in size. It is questionable whether the apparatus was installed with the necessary precautions, and in any case it does not seem to have given the results expected. Later, Nishimura returned to superinvar (wires of 30 microns), while Tomaschek and Schaffernicht used bands of phosphor bronze.

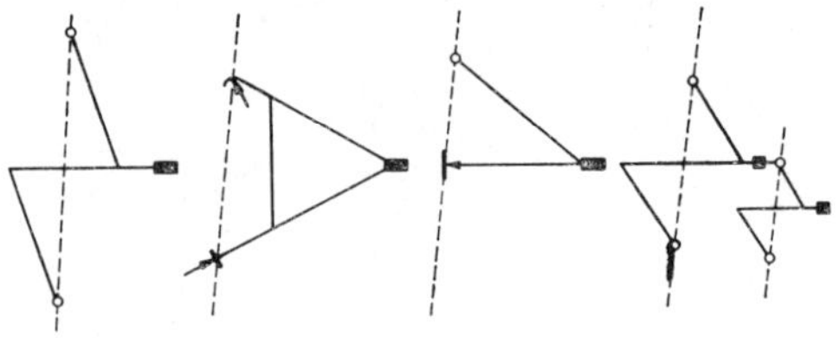

Fig. 8.7. Different mountings of horizontal pendulums.

Lettau obtained a greater sensitivity by coupling two horizontal pendulums (Fig. 8.7). This construction is quite delicate as it is almost possible to "couple" any two horizontal pendulums. To obtain good stability he used pendulums with very different masses and lengths.

8.6.1. Influence of the Rigidity of the Wires on the Value of the Amplification Coefficient

In the real pendulum the characteristics of the wires play a part, and it is necessary in computing the amplification coefficient to know the dimensions of the wires and the rheological properties of the substance from which they are made - the mass of the pendulum and the position of its centre of gravity relative to the fictitious axis of rotation AB.

To calculate the position of equilibrium of the pendulum we consider separately three elementary forces:

(1) A force derived from the deviation of the verticale to which the pendulum is supposed to react. The moment about O of this elementary force is

$$M = m\ L\ \Delta g = m\ L\ g\ \tan e \sim m\ L\ g\ e. \qquad (8.20)$$

(2) The restoring force of gravity which has a moment about O,

$$M_p = m\ L\ g\ i\ \Delta\alpha. \qquad (8.21)$$

(3) An additional restoring force which derives from the resistance to twisting of the two suspensory wires and has a moment about O given by

$$M_T = 2\ K_0 \Delta\alpha, \qquad (8.22)$$

where

$$K_0 = \pi\ \frac{\mu r^4}{2\ell}, \qquad (8.23)$$

r being the radius of the wires, ℓ their length and μ their modulus of rigidity.

The equilibrium is reached when

$$m\ L\ g\ e = m\ L\ g\ i\ \Delta\alpha + 2\ K_0 \Delta\alpha \qquad (8.24)$$

or

$$\Delta\alpha = \frac{m\ g\ L\ e}{m\ L\ g\ i + 2K} = \frac{e}{i + (2K/mgL)}. \qquad (8.25)$$

We put

$$\frac{2K_0}{m\ L\ g} = \varepsilon, \qquad (8.26)$$

ε being a certain angle whose value is expressed in radians and whose value is as low as the wire's quality is high. Then

$$\Delta\alpha = \frac{e}{i + \varepsilon}, \qquad (8.27)$$

where $1/i + \varepsilon$ is the amplification of the pendulum. This shows that the torsion of the wires slightly diminishes the sensitivity of the instrument.

In the case of Verbaandert-Melchior (VM) pendulums the angle ε expressed in seconds of arc has a value of 3", then one finds for T = 80 s, $i + \varepsilon$ = 10". The amplification coefficient is 20.626 under these conditions. But in practice it is necessary to take into account the amplification which results from the application of the optical method called the moving mirror of Poggendorf. The quartz beam, with a length of about 10 cm, is then replaced artificially by an optical beam 5 m long which, according to the laws of reflection, doubles the efficiency. The amplification of the optical method in these conditions is about 100 and the total amplification observed on the photographic record is of the order of 2×10^6. Thus the movements of the extremity of the light beam which affects the sensitive paper of the recorder has amplitudes equivalent to that of the movements of the extremity of a freely suspended vertical pendulum more than 200 km long.

8.7. PRINCIPLES OF CONSTRUCTION OF VERBAANDERT-MELCHIOR HORIZONTAL PENDULUMS

The deviations of the vertical relative to the ground have generally a *maximum* amplitude of less than 0".050. This angle, in the case of the VM pendulum, which has the short sides of the right-angled triangle base of 0.273 m, corresponds to relative variations in the height of the three points of the ground on which stand the three adjustable screws, which does not exceed 0.033 µm.

On the recording paper a variation of the ordinate of 0.2 mm, which can almost be estimated visually and corresponds to a deviation of the vertical not larger than 0".0002, also corresponds to changes in the level of the points of contact of about 0.3 nm. One nm is about the order of magnitude of the molecular unit of quartz. This will provide food for thought for any experimental worker wishing to construct tiltmeters. It shows that all parts of the apparatus should be such as to ensure a maximum rigidity. The materials used must have a thermal coefficient as low as possible, and should not be subject to internal variations such as slip or dislocations which occur in the crystalline structure of metals.

In constructing a metal pendulum it is very difficult to grip the filaments or suspension bands at the point required without producing spurious torsions. It is also necessary to introduce regulatory screws at the attachment points which diminish the stability of the apparatus.

For the bands, whose sections are 300 x 20 µm, it is impossible to define the points through which pass the axis of rotation of the pendulum. It is not even certain that such a point can be defined as such.

Because of their extreme stability, two materials seem to be particularly suitable for the construction of these extremely high precision instruments; these are fused quartz, or more precisely fused silica (SiO_2), and also a metal alloy discovered by Masumoto (63.5% iron, 31.5% nickel, 5.0% cobalt, with traces of manganese) called superinvar with a coefficient of expansion less than $10^{-7}/°C$.

Nevertheless, certain properties of silica, and in particular the facility with which the very fine wires of silica can be fixed by self-soldering to a quartz framework, make it preferable to superinvar. Fixing-screws and tightening-rings are thus eliminated.

Quartz is a crystalline variety of silica. By fusing it at a temperature of 2000°C the crystal structure is destroyed, and on cooling a vitreous mass is obtained which is isotropic and amorphous. This is called silica glass and it is almost insensitive to temperature changes to an even greater extent than quartz crystals.

The coefficient of linear expansion of fused silica is extremely small even over a wide temperature range, e. g. between 0 and 1000°C the average is 0.54×10^{-6}. This is the same order of magnitude as that of superinvar, which has a great disadvantage inasmuch as the minimum coefficient is obtained over a very small range of temperature.

Silica glass is very hard and scratches all other glasses; it has low specific weight, the density being 2.21.

One other particularly important quality is its extremely low viscosity, so that when a silica fibre is twisted there is practically no loss of energy through intermolecular friction such as is found in all metals. Mechanical hysteresis in fused quartz is the lowest of any substance, therefore using fine suspensory wires of silica it is possible to make an apparatus which returns to zero and which has a remarkable fidelity. Moreover, very fine wires of silica have a remarkable resistance to rupture, superior to the majority of most metals being inferior only to tungsten.

A remark concerning the free period T of the pendulum

A proof of these qualities is given by the horizontal pendulum itself. It is observed that for all metal pendulums the free period of oscillation is a function of the amplitude of the oscillation, and this is very troublesome for calibration when they are based upon measurements of the free period.

However, in a relation like (8.19) we suppose that T is a constant, i. e. independent from the amplitude of the beam oscillation and of the azimuth of the beam. This cannot be realized with metallic suspensions which use clamping devices to fix the metallic wires or bands to the support and to the beam: during the oscillation an anomalous flexure appears near the fixation, the rotation points are displaced producing a spurious tilt of the axis of rotation and consequently a variation of period and sensitivity. This effect does not exist with well-soldered quartz wires as shown by us (Melchior, 1964). When the soldering is carefully made, the rotation points do not change during the oscillations.

In many stations which still use metallic pendulums and where the modern crapaudine technique has not yet been introduced, the sensitivity determination still rests upon measurements of the period T (called the indirect method).

The dependency of T from the amplitude entails the making of complicated estimations and an uncertain "reduction to zero amplitude". This was a matter of concern and has been thoroughly and carefully investigated by several geophysicists: Schneider (1962), Goloubitsky (1970), and Skalsky and Picha (1969).

We quote the conclusion of these last two authors:

"The main source of inaccuracies of the indirect method of sensitivity determination is in the errors incurred in determining the period of oscillation T. With regard to the required accuracy, it would be necessary to determine the currently used period of oscillation T of 40 to 90 s with an error less than ± 0.05 to ± 0.11 s. This would not be a problem provided this quantity would remain constant during the measurement. This would enable us to increase the accuracy of its determination by increasing the number of directly measured oscillations. This assumption, according to our experience, is so far satisfied only with quartz VM pendula, apparently due to the use of quartz fibres. Tens of measurements, which we have carried out, showed that with these pendula a systematic change of the period of oscillation does not occur even with considerable amplitudes of the initial oscillation, exceeding 50'. On the other hand, with all tiltmeters having metal fibres, some change of the period of oscillation T occurs during its measurement. It has so far not been possible to explain the causes of all these changes. However, it is certain that the parameters of the fibres used, the methods of their clamping and the magnitude of the initial angle of oscillation have decisive influence."

This proves that the system of fixation by fusion is highly superior. As a consequence the calibration will be more exact for quartz pendulums than for metallic ones.

Fused silica is without doubt the most favourable material for the construction of excellent horizontal pendulums.

The VM pendulum (Fig. 8.9) is basically composed of a base in Duralinox (27.3 x 27.3 x 2 cm; weight 4.700 kg) with two screws (0.5 mm thread) and a support forming a right angle (the support being at the point of the right angle). This arrangement makes it possible to rapidly increase the sensitivity up to high values and ensure a simple and true correction of the drift during the regulation. The two screws have well-defined functions, as has been indicated. That which with the foot of the right angle forms a line parallel with the pendulum arm is used to put the pendulum "in period". That which forms the perpendicular is employed to regulate the azimuth of the arm and will be called the "drift screw". It can be seen that the foot of the right angle fulfils both of these functions at the same time, and that is the reason why it is not convenient to have a screw at that point.

The presence of a foot instead of a supplementary screw also serves to increase the stability of the apparatus.

A tetrahedron which forms the framework of the apparatus is constructed using tubes of quartz prepared by autogenous soldering. This tetrahedron is made integral with the base by the pressure of six springs; the three feet are, in addition, lightly blocked by small wedges, which prevent slipping during the mounting. They nevertheless allow the tetrahedron a certain amount of freedom so as not to cause stresses during the change in the thermal regime caused by the transfer from the laboratory, where the apparatus is set up, to the underground chamber.

The pendulum is set up in the interior of the tetrahedron using the Zöllner arrangement. It is made completely in quartz from a single piece of transparent quartz and consists of a hollow arm 90 mm long which ends in a mass of about 10 g. The suspension is carried out by means of two quartz wires with a diameter of 40 μm and 100 mm long fixed to the frame and to the pendulum by autogenous soldering.

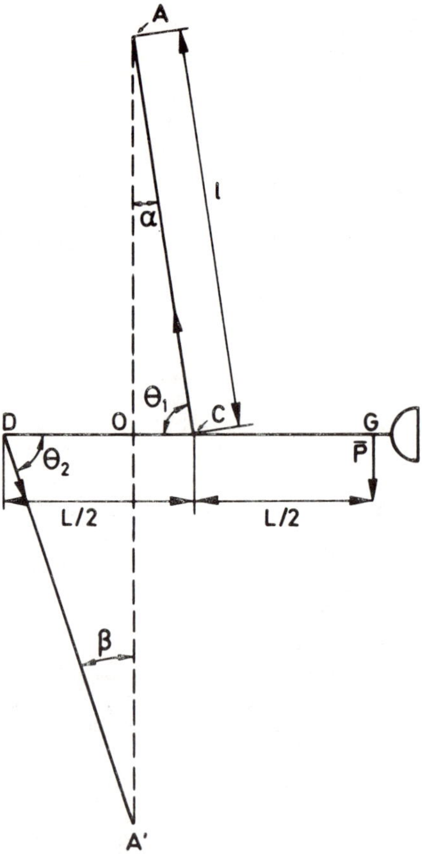

Fig. 8.8. Geometric characteristics of a simple pendulum.

Geometric Characteristics of a Simple Pendulum

According Fig. 8.8, we put

$L = DG$, T_1 is the tension of the upper wire and T_2 is the tension of the lower wire.

By construction,

$$CG = CD = \frac{L}{2}, \qquad OA = OB.$$

It follows that:

$$\left.\begin{aligned} T_1 \cos\theta_1 &= T_2 \cos\theta_2, \\ T_1 \sin\theta_1 &= T_2 \sin\theta_2 + m\,g, \\ T_2 \sin\theta_2 &= m\,g, \end{aligned}\right\} \qquad (8.28)$$

from which

$$T_1 \sin\theta_1 = 2\,m\,g = 2T_2 \sin\theta_2,$$

and

$$\tan\theta_1 = 2\tan\theta_2. \qquad (8.29)$$

Moreover

$$OA = OC\tan\theta_1 = OB = OD\tan\theta_2,$$

then

$$\left.\begin{aligned} 2\,OC &= OD \\ OC + OD &= L/2 \end{aligned}\right\}, \qquad \begin{aligned} OC &= L/6, \\ OD &= L/3, \\ OG &= 2L/3 \text{ is the reduced length} \end{aligned} \qquad (8.30)$$

The moment of inertia of the pendulum is

$$I = 4\,m\,L^2/9. \qquad (8.31)$$

For the evaluation of the tensions T_1 and T_2, because the angles α and β are small, we can take the sines as unity and $T_1 = 2T_2$.

The value of μ slightly varies with the diameter of the wire when this is very fine. For wires with a diameter of 40 µm the value is 3.3×10^{11} dynes cm^{-2}.

Thus for two wires of 100 mm length (8.26) gives

$$2K = \mu\,\frac{\pi r^4}{\ell} = 1.5 \text{ dyne cm}^{-1}, \qquad (8.32)$$

and with m = 10 g (8.29) gives $\varepsilon = 3''$.
With

$$OD = 30 \text{ mm}, \qquad OC = 15 \text{ mm} \qquad \text{and} \qquad OA = 100 \text{ mm},$$

equations (8.20) give

$$\theta_1 = 81.5°, \qquad \alpha = 8.5°,$$

$$\theta_2 = 73.3°, \qquad \beta = 16.7°.$$

We should stress the two fundamental characteristics of the tetrahedron-tripod frame: its shape is geometrically non-deformable and it is perfectly symmetric in relation to the suspension. After soldering, the tetrahedron is reheated and slowly cooled in an oven for 24 h. Before use, polarized light is used to check that no internal stresses exist.

The tetrahedron has a base of 185 mm, a height of 250 mm and the diameter of the tubes which form it is 6.5 mm. A plane mirror of 2 cm diameter is fixed to the pendulum in its axis of rotation: using an optical system of 5 m focal length, the image of a luminous spot is projected into the horizontal slit of a photographic recorder.

A device for automatic release makes it possible to raise and support the pendulum arm so as to remove the load from the suspensory wires. It is then possible to transport the apparatus over long distances without the risk of breaking the wires. This device also plays a very important role in the installation of the suspension or its replacement after an accidental breakage. For fixing the fine quartz wires by autogeneous soldering it is necessary, as experience has shown, to retain the pendulum arm firmly in the mean position which it will occupy later, when it has been freed for the installation of the apparatus.

The pendulums are actuated with a slow movement, very much reduced, acting on the drift screw, as well as a slow movement (with a reduction smaller than the former) acting on the period screw. These movements are operated by long transmission flexible rods which make a simple remote control possible.

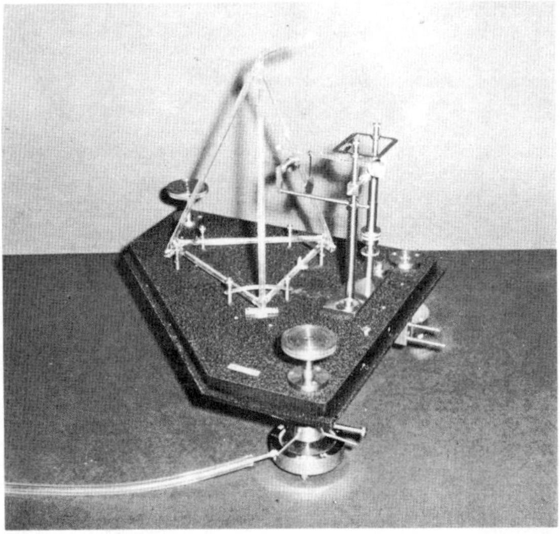

Fig. 8.9. The Verbaandert-Melchior quartz pendulum.

The soldering operation by the fusion of quartz at its fastening point eliminates all initial torsion of the fibre, a condition which cannot be realized when metallic wire is used. This circumstance, as well as the material utilized and the geometric figure chosen, are elements of the stability of the instrument which operates without any difficulty with periods of more than 80 s; this corresponds to a sensitivity of the order of 0."0007 per mm on the recording paper.

8.8. CALIBRATION WITH A DILATABLE CRAPAUDINE

The calibration of an instrument is unquestionably the most important operation to be done if one wishes to trust the informations derived from its performance. Precise calibrations can be performed by using mechanical deformations correctly kept under control. One may try to use for this purpose elastic deformations under pressure, piezoelectric deformations or possibly magnetostriction effects.

Verbaandert has invented the method of the dilatable bearing plate called in French "crapaudine" to produce an artificial and well-known tilt of the clinometer. A reservoir shaped like a manometric capsule with thick walls (Fig. 8.10) is placed under the drift screw of the instrument. Mercury is injected into the reservoir under a known pressure. As a result of the elastic deformation created by the increased internal pressure, the upper wall of the reservoir bulges and raises the drift screw of the pendulum which moves more or less according to its sensitivity.

The characteristics of the stainless-steel dilatable bearing plates constructed by the International Earth Tide Centre (ICET) are as follows:

Thickness of the upper wall	5 mm
External diameter	67 mm
Total thickness of the capsule	16 mm
Internal diameter	53 mm
Thickness of the manometric cavity	0.2 mm
Thickness of the lower closing plate	5 mm

The upper face has in its centre a small conic depression to accommodate the point of the drift screw.

An opening which is linked to the cavity makes it possible to inject the mercury under pressure by means of a very flexible plastic tube of small section which can be several metres long.

Of course the crapaudine ought to be calibrated itself before being used to calibrate pendulums. The calibration of a dilatable bearing plate is made by measuring the change of height ΔH of the mercury level which produces in the interferometer's ocular the substitution of one interference ring by the next one. The green line of mercury (λ = 0.546 074 micron) is used at the Brussels laboratory. Thus a dilatation of the crapaudine of $\lambda/2$ = 0.273 037 micron corresponds to ΔH. Such a calibration must be made when a pressure corresponding to the weight of the pendulum to be used is applied upon the centre of the bearing plate. For a VM pendulum this pressure is 3.55 kg.

In practice ΔH is comprised between 25 cm and 50 cm for the 140 bearing plates we have constructed. It is therefore possible to observe the transit of about eight interference rings and to check the linearity of the piece which in most of the cases is very good except, of course, in the vicinity of zero pressure. For exemple we have found:

Nr 132 $\quad \Delta H = 34.14 + 0.0000\ H,$

Nr 105 $\quad \Delta H = 40.75 - 0.0012\ H,$

Nr 158 $\quad \Delta H = 42.98 + 0.0001\ H,$

H being the mean mercury pressure represented by the height of the mercury level above the crapaudine, this height being given in centimetres.

It is convenient to install the level of mercury at a minimum height of 200 cm above the bearing plate when using it to calibrate a pendulum. Then if one uses a variation of height dh of 20 cm (as usual) one produces an artificial tilt of the pendulum

$$e = \text{arc sin} \frac{1}{\ell} \frac{\lambda}{2} \frac{dh}{\Delta H} = \alpha\ dh \qquad (8.33)$$

ℓ being the distance between the screws on the base of the pendulum.

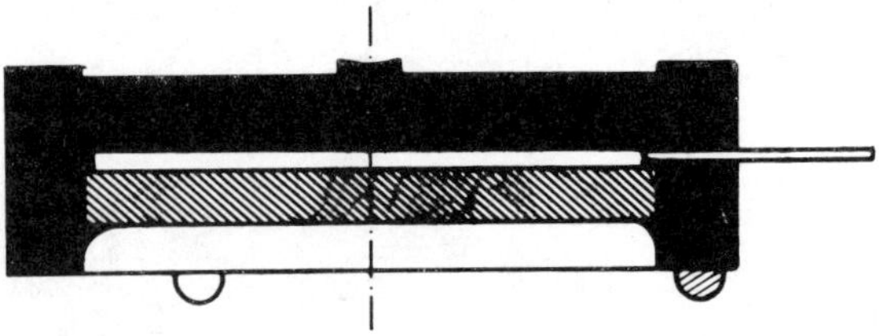

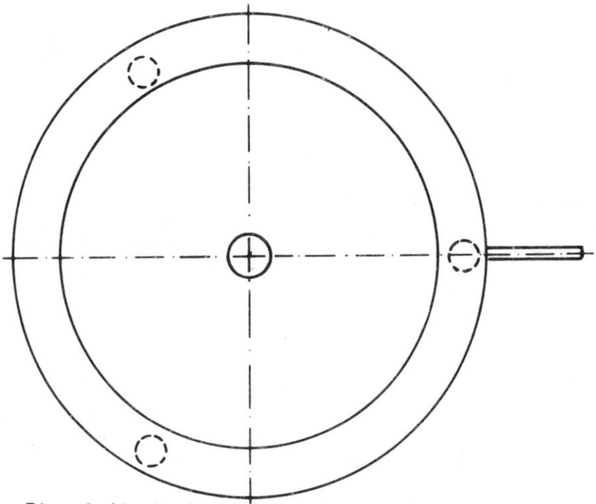

Fig. 8.10. Verbaandert dilatable bearing plate. (crapaudine)

By construction of the VM pendulums we have chosen $\ell = (\lambda/2)\ 10^6$ exactly; this gives for dh = 10 cm and ΔH = 40 cm

$$e = \text{arc sin}\ \frac{1}{4\times10^6} = 0''.05, \qquad (8.34)$$

which is just the size of the tidal tilts to be measured.

Many discussions took place concerning the time stability of the crapaudine coefficient because the first series, nos 1-76 were constructed in ordinary steel, chromed. These proved to suffer some evolution with time (1-2% within 2 or 3 years) mainly when maintained in hard ambient conditions of humidity (the water percolating in the mines often has a high degree of acidity).

Since 1968 we started with the construction of stainless-steel crapaudines (nos. 100-162) which have proved to have quite better behaviour. Our opinion in this respect is shared by Baker and Lennon (1976) who submitted at the Bidston Observatory different instruments and calibration procedures to detailed and precise experiences. Their comments are relevant:

(a) For any given crapaudine, individual determination of K at Bidston are mostly within ±1% of the mean value of K and generally appear to be randomly distributed about the mean, suggesting that between periods of 15 and 65 s there is no significant dependence of K on the period of swing.
(b) There are significant differences (from 3% to 7%) between the determinations of K using different crapaudines.
(c) Crapaudines 61 and 64 were subsequently recalibrated interferometrically. The corrected values of K are very close to the results from the stainless-steel crapaudines. This indicates that between 1967 and 1971 there was a very significant change in the response of the chromed-steel crapaudines.
(d) The remaining difference between the results from crapaudines 37 and 114 can possibly be attributed to changes occurring during transportation of the instrument.
(e) The indications are that the new stainless-steel crapaudines have shown a greater stability since their primary interferometric calibration in 1969. This is demonstrated by the fact that their determinations compare with the newly calibrated chromed steel crapaudines 61 and 64. The latter evidence is by inference only, so that further investigations using the new stainless-steel crapaudines are warranted.

Skalsky and Soukup at the Geophysical Institute, Prague report (1974) as follows

"The crapaudine and its transfer coefficient ($f = \lambda/2\ \Delta H$) have a decisive influence on the stability of the sensitivity and on the accuracy of observations with the compensating tiltmeter. The stability of the value of f was investigated with three types of crapaudines at the tidal station of Rimov near Ceské Budejovice between 16.3 and 8.6. 1972. The crapaudines were put directly under the levelling screws of the photo-electric tiltmeters and their deformations were measured by a contact vertical interferometer type IKPV 224, using a TKG 204 He-Ne laser as the source of light (λ = 0.6328 µm). On the whole 17 observational series were run (each comprised of 10-45 observations) under different observational conditions and with different crapaudine loads. The average mean error of one observation was ±0.50%, the average mean error in determining the transfer coefficient f from one observational series was ±0.11%. With the Belgian crapaudine nr 22, f varied roughly by ±1.4%. However, this was mainly due to the unsuitable construction of the "conical crater" in the upper part of the crapaudine which could not guarantee that the levelling screw and the crapaudine made contact at the same point each time the crapaudine was inserted under the levelling screw, rather than to real changes of the elastic properties of the material. *The mean value of the transfer coefficient of this crapaudine, $f=0.713 \times 10^{-6}$, agrees precisely with the values determined in 1964 (i. e. 8 years ago) during its first calibration in Brussels, as well as with the calibration repeated in 1967.* With the other two crapaudines of domestic production the observed "changes" of f did not exceed ±0.45 and 0.34%. No changes of the coefficient f were observed with loads varying from 3 to 9 kp. These results cannot be generalized, of course, because a larger number of crapaudines and a longer period for observations would be required for a detailed investigation of possible variations."

This large number of crapaudines is available at ICET Bruxelles which have been repeatedley calibrated. This allows indeed to generalize the Skalsky-Soukup conclusions:

1- there is no change of the transfer coefficient with time
2- temperature effects are not important (consider for exemple C136 which gives the same result at 5°C and at 17.5°C - Many such experiences were made before by Ducarme).

Moreover we should mention that a drift of annels during the interferometric calibration procedure is avoided by a very good thermic protection of the interferometric and a very quick operational procedure.

The dependence of the calibration from the absolute value of gravity can be easily taken into account.

The main danger, according to our experience, should come from air bubbles in the tubes and great care must be taken to eliminate such bubbles.

Finally we consider that the divergence observed in some cases between calibrations at Bruxelles - ICET and in situ calibrations cannot be attributed to the crapaudine but to some accident in the transportation of the pendulum. It must be emphasized in this respect that the calibration in situ is by its nature the correct one. The calibration made in the laboratory at Bruxelles is to be considered as a more or less precise indication and as a check for a correct behaviour of the instrument.

Some exemples

C 37

Epoch	1965.05	1965.10	1965.11	1966.08	1966.11	1969.04
ΔH	24.59	24.69	24.70	24.64	24.66	24.77

C 22

Epoch	1964.12	1967.05	Prague
ΔH	38.44	38.40	38.39?

C 112

Epoch	1969.01	1975.04
ΔH	36.502 ± 0.155	36.679 ± 0.171

C 113

Epoch	1969.01	1975.04
ΔH	37.449 ± 0.112	37.486 ± 0.140

8.8.1. Automatic Device for the Calibration of Pendulums in a Surface Laboratory

We have made the calibration automatic by using a rotating beam at the extremity of which the mercury reservoir is attached (Fig. 8.11). The beam is 48 cm long and the period of rotation is 1 h. The pressure inside the bearing plate therefore follows a sinusoïdal law in function of the time and causes amplified movements of the pendulum's beam which are recorded by the classical photographic method on a drum whose period of rotation is 4 h and which is placed 5 m from the pendulum. As this change of pressure is smooth and regular the recorded curves are sinusoïds with a quasi-constant amplitude.

But in a surface laboratory a permanent microseismic activity produces an excitation of the pendulum on its free period but of minor amplitude, which looks like a slight zigzag superposed on the sinusoïd (Fig. 8.12). This is, indeed, very useful as it enables us to determine with precision the free

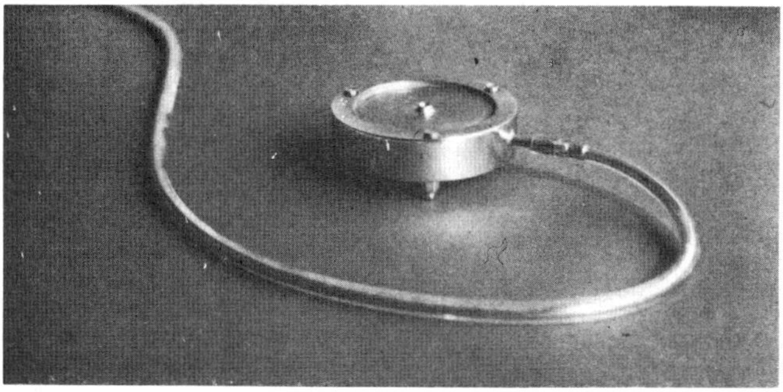

Fig. 8.11. The Calibration Devices for the Horizontal Pendulums
above: a "crapaudine" with its long plastic tube containing mercury
below: the clock for the weekly programme of calibrations. The mercury reservoir can be seen on the rotating arm. (Underground Laboratory of Geodynamics at Walferdange, Luxemburg).

period T of the pendulum using the time base provided on the registration by time marks produced every 15 m. It is thus possible to count a large number of free oscillations of the pendulum (generally about 60) and gives the photographic method of calibration a clear superiority over the manual one where we can hardly count more than 20 oscillations.

The sensitivity of an horizontal pendulum is related to its free period T by (8.19):

$$s\,T^2 = K, \tag{8.35}$$

s being the sensitivity defined as seconds of degree per millimetre on a photographic recorder installed at a standard focal length of 5 m:

$$s = \frac{i + \varepsilon}{2 \times 5000}, \tag{8.36}$$

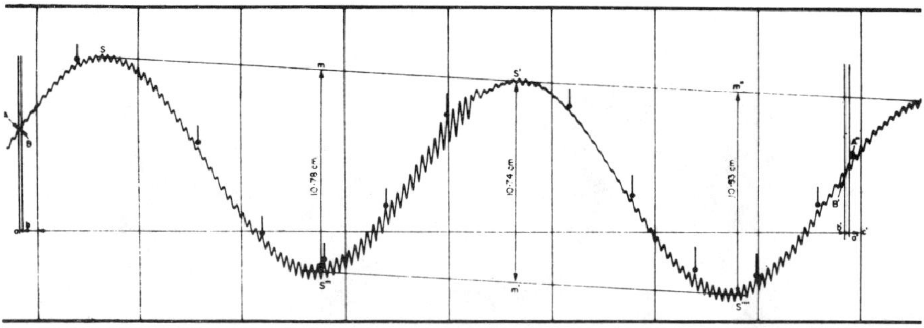

Fig. 8.12. Photographic registration of an automatic calibration (time marks are given each 15 min; the short oscillations of small amplitude are the free vibrations of the pendulum and allow the measurement of the period; large sinusoidal movement is the artificial tide produced by the dilatable bearing plate).

the factor 2 deriving from the fact that one reflection on a mirror (Poggendorf's method) duplicates the sensitivity.

K is the fundamental constant of the horizontal pendulum: when K is known, a measure of the free period T with a chronometer gives the sensitivity immediately.

Each calibration made by using the automatic device allows the derivation of one determination of K by measuring s and T on a registration like Fig. 8.16.

In practice we measure the total amplitude ΔL of the sinusoïd, and for a focal length D we have

$$\frac{\Delta L}{2D} = \frac{e}{i+\varepsilon}, \tag{8.37}$$

e being the inclination (8.33) produced by the crapaudine.

Thus

$$s = \frac{e}{10\Delta L} \times \frac{D}{500}. \tag{8.38}$$

The geometric characteristics of VM pendulums give as a mean of 264 independent determinations of 141 pendulums

$$K = 6.29. \tag{8.39}$$

The mean quadratic error in the determination of this constant is about 0.02, i. e. 0.3%.

8.8.2. A Programmed Device for the Automatic Calibration of Pendulums in the Underground Stations

The conditions of measurement of tilting of the ground necessitate the choice of deep sites which therefore are often far away from the institution involved in such research. In the case of horizontal pendulums non-specialized people may easily ensure the maintenance of the station, but the calibration remains a delicate procedure which, nevertheless, must be made regularly.

The classical method, which consists of producing a small excitation to the pendulum and measure its free period of oscillation T, disturbs the instrument.

We have therefore experimented successfully using the crapaudine in the underground stations. First of all we check that the introduction of a dilatable bearing plate permanently under the drift screw of the instrument does not modify the stability of the system since it does not produce any spurious drift. This is the result of 9 years' experience.

The programmed device represented on Fig. 8.13 consists of two parts (Melchior, 1966):

1) The "weekly clock": an electric motor drives a rotating disc in 7 days which may bear several contacts, a maximum of 7, but we use two to obtain two calibrations per week. Such a contact moves up a mercury inverter during 3 h.
2) When the mercury inverter is in the "up" position another motor drives the beam which bears the mercury reservoir bringing it within 15 min from the lower position to the upper position where it stops. When after 3 h the mercury inverter falls back to the "low" position, the mercury reservoir returns to its lower position within 15 min. This regular and smooth movement does not disturb at all the pendulum as shown on the piece of record reproduced on Fig. 8.15.

The difference of height of the mercury reservoir dh realized by this system is generally 20 cm. The calibration consists of measuring on the photographic paper the two displacements produced (up and down), take the mean and express in this way the millimetres on the photographic paper in terms of milliseconds of tilt.

A computer program makes such an interpretation from the dh value, the α constant of the crapaudine (form 8.33), and the two displacements d_1 and d_2 measured in millimetres on the photographic paper using the simple formula

$$s = \frac{2e}{d_1+d_2} \text{ per mm.} \qquad (8.40)$$

In practice $d_1=d_2 \approx 50$ mm and therefore 3 mm corresponds to 0ʺ001. This program produces the data in the form shown by the Fig. 8.20. These calibrations are introduced in the process of harmonic analysis (see Chapter 7).

8.8.3. Use of Piezoelectric Ceramics

Research has been performed at the International Centre of Earth Tides to take advantage of the piezoelectric effects in ceramics to calibrate pendulums (van Ruymbeke, 1972).

Ceramics very similar in size and dimensions to the mercury crapaudines can be found. The application of a tension of 1000 V produces a dilatation of 0.374 micron comparable to the dilatations produced by pressure variations in the crapaudines.

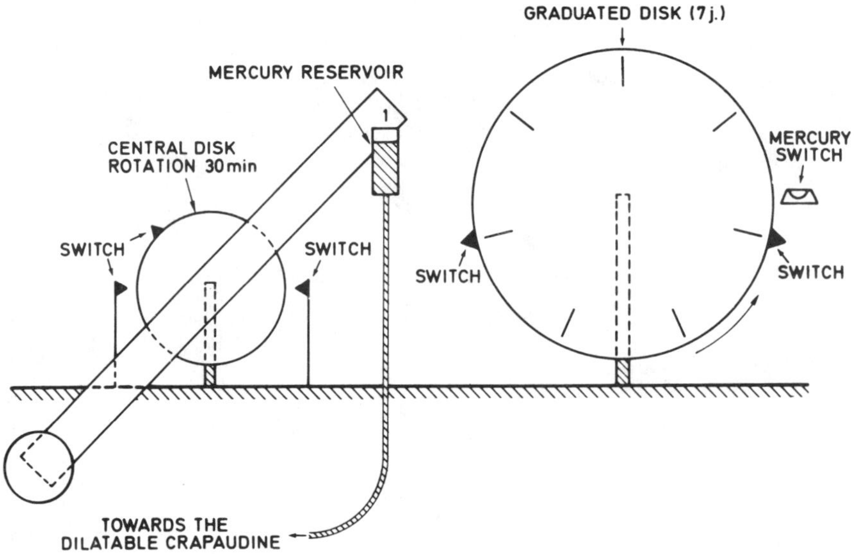

Fig. 8.13. Automatic calibration device.

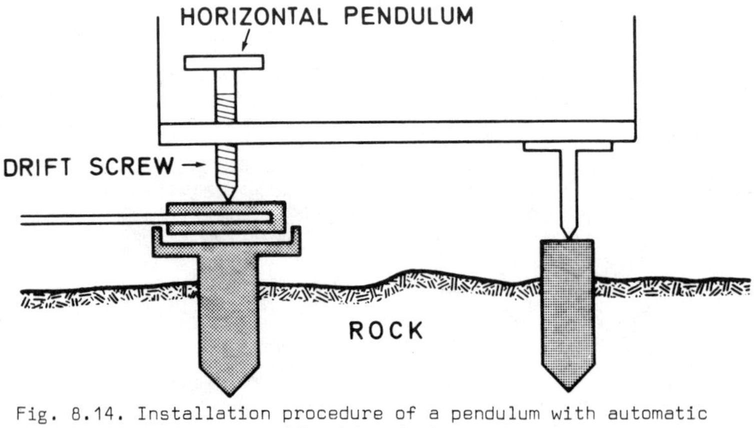

Fig. 8.14. Installation procedure of a pendulum with automatic calibration device.

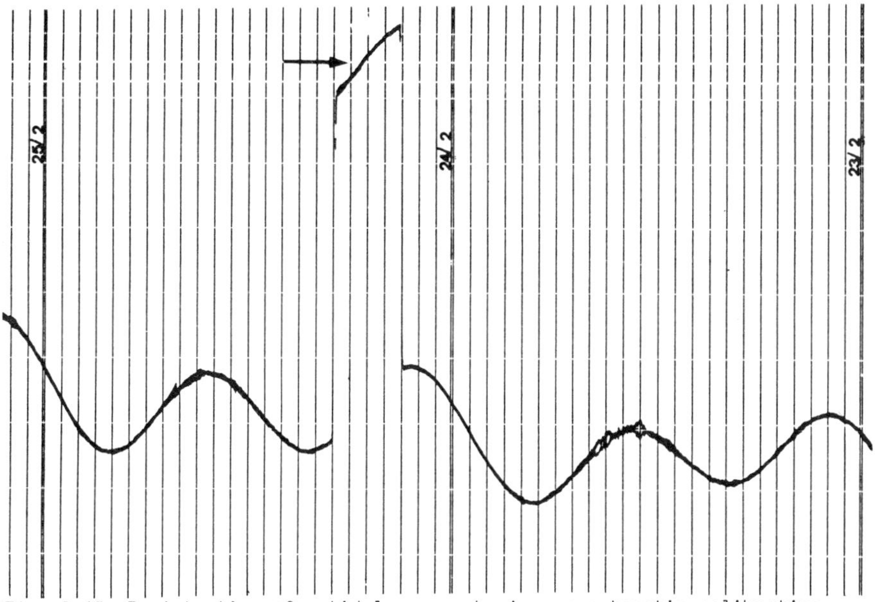

Fig. 8.15. Registration of a tidal curve showing an automatic calibration.

The advantage of the piezoelectricity effect is to allow large variations in the frequency of the artificial tilt produced. Consequently they can also be used under a gravity meter or a seismograph to determine its frequency response to vertical accelerations. Their disadvantage resides in an hysteresis which is not negligible at all.

8.9. THE INSTALLATION OF HORIZONTAL PENDULUMS IN UNDERGROUND STATIONS

It is obvious that the conditions under which a pendulum is installed will have as much bearing as its intrinsic qualities on the quality of the results. The instrument must rest on the rock with the minimum transition. The rock itself should be sufficiently homogeneous, and temperature at the point of installation should be as constant as possible without any diurnal variation. For this reason a depth of at least 30 m is needed to obtain reliable records. Most stations are installed at depths of more than 50 m.

We may point out by way of example that if a temperature difference of 0.003°C should accidentally occur between the drift screw and the other two adjusting screws of the apparatus, it would result in a parasitic deviation of approximately 5% of the maximum amplitude of the Earth tide. This is ample justification for all the precautions taken in installation.

Moreover, it has been shown by calibrations of VM pendulums that, with an operating period of 79 s and a focal distance of 5 m, 1 mm on the recording paper corresponds to 0"001, i. e. a difference of level of only 0.0012 µm. A precision of the order of 0.3 mm is effectively obtained in reading records, i. e. variations in the level of 0.4 nm, which is of the order of magnitude of molecular dimensions, are disclosed.

Such a result could be obtained only by a new installation method. The experience of others has also led to rejection of artificial pillars for the installation of apparatus of this type. In particular, concrete and cement should be rejected as foundation elements by virtue of their instability, their creep, and their sensitivity to humidity and humidity variations.

Areas where the rock is most homogeneous and strongest are selected in the rock walls, and niches for installation of the pendulums are chiseled out without the use of an explosive (Fig. 8.16). This procedure avoids all cracking of the rock that might adversely affect the stability of installation. Moreover, it has the effect of "embedding" the apparatus in the rock wall, which suppresses in large measure the sometimes disastrous effect of air currents capable of producing thermal changes.

Three stainless-steel cylinders (30 mm in diameter and about 50 mm in height) fitting into three precisely drilled holes in the rock are sealed in position to act as a bond between the pendulum and the rock. The bond is effected by a thin coating of thermosetting resin. The thickness of this adhesive laterally enveloping the steel cylinders is approximately 0.5 mm, and the lower part of the steel cylinders is directly in contact with the rock, which exactly fits its slightly conical end.

The upper part of these steel cylinders is cup-shaped and the cups are filled with inert liquid octyl sebacate of zero vapour tension which perfectly protects the supporting points of the apparatus. The plane upper surface of the cylinder, which projects several millimetres above the base of the niche (Fig. 8.16) acts as a classical bearing plate, distributed as follows, according to the role played by the three screws:

hole: drift screw
slit: fixed point (right angle)
flat: period screw

When the pendulum has been placed on the three supports firmly fixed to the Earth's crust, the niche is hermetically filled by insulated panels of polystyrene foam, 30 mm thick.

By placing up to six different pendulums for a total period of 5 years in the same niche we have found that all the instruments *drifted* with time in the same sense and at a regularly decreasing rate. This drift, which may rightly be attributed to an expansion effect of the rocks, is characteristic of the niche. For some niches it could initially be as much as 0".02 per day at the outset. At the present, after 9 years, it is of the order of 0".002 per day or less.

Once stabilized, niches possess the great advantage of affording guaranteed shelter for very long-term studies and three fixed reference points for the instrument. The geophysical aspect of the drift may be considered as soon as stabilization seems established, taking the precaution of duplicating the apparatus in each component.

It has been found that the sensitivity of the pendulums is very stable with time. For example, a pendulum installed at Warmifontaine (pendulum Nr23) maintained a self-period of 90.48 s (a sensitivity of 0".000 755 per mm) for almost a year with a total fluctuation of 2.9 s and a mean dispersion of ±0.25 s (see also Section 8.11).

The root-mean-square measurement error reduced progressively to 0".000 15. When the preparation of the niches, drilling and sealing of the bars has been terminated, it is our general rule to wait one week before setting the pendulums on the bars and starting the measurements. During that week the pendulums

Fig. 8.16. A view of a niche made in the slate mine at Warmifontaine (Belgium). The three bearing bars appear emerging from the base of the niche.

Fig.8.17. Photograph of an experimental cutting of bearing bar bonded in a slate block with phenolic resin.

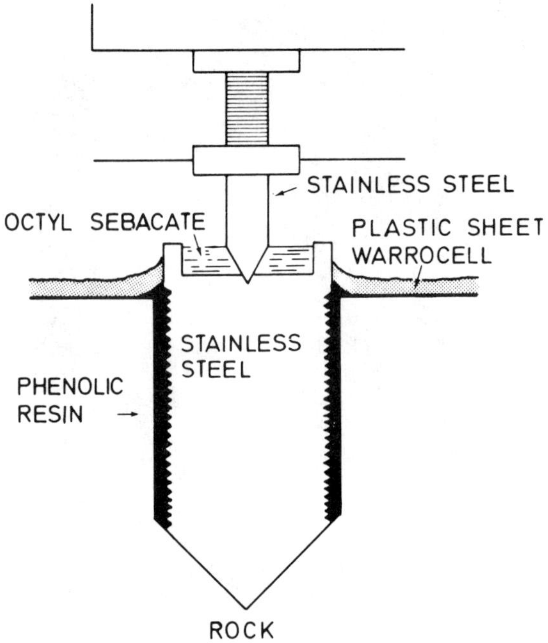

Fig. 8.17. Setting up the bearing bars for a horizontal pendulum in the rock.

are kept in the mine in order to stabilize their temperature as much as possible before putting them into operation. In these conditions their installation at a suitable sensitivity (free period of about 75 s) is an easy operation.

However, we have departed from our common policy when installing a station in Spitsbergen because of lack of time, and we installed the pendulums on the bars at a sensitivity of 0".001 per mm (period 75 s) immediately after sealing. A registration was started to record the behaviour of the instruments which is reproduced on Fig. 8.19.

The rock at that place was extremely hard flint - (we broke two drilling screws) - and this is probably an important factor of the observed stability. Nevertheless, the registration shows that after 12 h, the tidal curve appears very nice with, of course, a drift which is more important than in a station some months old.

Fix and Sherwin (1970) also observed that the concrete piers contribute to the noise in some way that was never completely isolated. To eliminate this effect they made all connections directly to the rock with expansion bolts set in holes drilled in the rock.

Subsidiary Apparatus

In caves or mining galleries, where these instruments have to function permanently, the hygrometric conditions are usually exceptionally bad: the air is supersaturated with water vapour and drops of water often fall from the ceiling. Therefore it is essential to visit the station once a week in order to

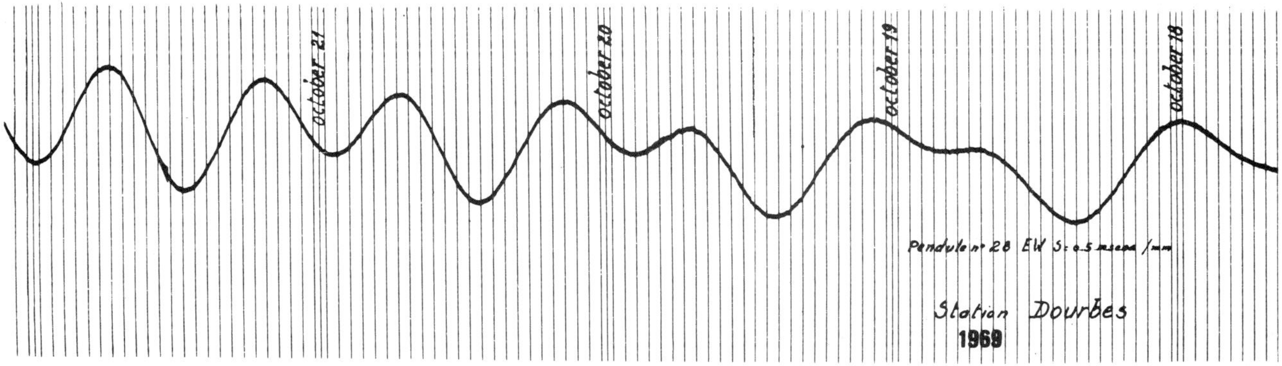

Fig. 8.19. A typical registration of a Verbaandert-Melchior quartz pendulum (East-West component, sensitivity 0"0005 per mm (scale reduction 1/2).

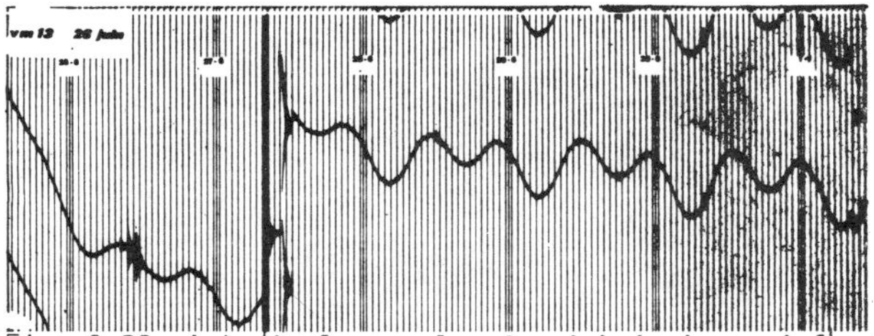

Fig. 8.20. A typical exemple of quick instrumental stabilization: the first tidal clinometric curve recorded in Spitsbergen (June 1969). June 25: Pendulum VM 13 is set in the niche. June 27: corrections for drift are made after one day registration; photographic paper was not removed. Reduction 1/7. Sensitivity: 1 mm = 0"0012.

be prepared to repair any defects in the operation of the equipment.

The recorder constructed at the Royal Observatory of Belgium is a drum recorder whose principal characteristics are as follows: diameter of the drum 30 cm, height 30 cm. The photographic paper used is of format 90 x 30 cm. The drum is turned by an electric motor with a large reduction gearing so as to make one complete turn per week. This corresponds to 6 mm of sensitive paper passing every hour before the slit of the recorder. The drum, which is made of bakelite, together with all associated parts of the recorder, is placed in a metal box which is kept under conditions which should be as nearly hermetic as possible.

A detailed description of all the above equipment can be found in Verbaandert and Melchior (1960).

8.10. AZIMUTH OF THE PENDULUM BEAM

When the beam is set very near to a North-South or East-West direction, a small error of azimuth practically does not affect the amplitude (0.16% for an azimuth of 5°, 0.64% for 10°) but is wholly reported in the phases.

We should determine this azimuth by auto-collimation on the mirror on the pendulum beam, but this also requires a knowledge of the angle the beam makes with the plane of the mirror, and this is very difficult to determine. In certain cases it is interesting to turn the plane of the mirror through a small angle in relation to the arm in order to simplify the installation in the plane of the meridian or of the prime vertical when local conditions do not lend themselves to this. The easiest procedure consists of determining by classical topographic methods the azimuth of the direction V_1 V_3 (Fig. 8.3) to which the beam is strictly parallel when we screw on completely the screw V_1 (sensitivity screw). When the pendulum is in that position one marks in the recorder's plane the position of the spot ℓ_0. Then one unscrews the screw V_1 until the correct working sensitivity (about 70 s) of the instrument is reached and one marks the new position of the spot ℓ_1. The correction to the azimuth V_1 V_3 is obviously arc tan $[(\ell_1-\ell_0)/d]$ if d is the distance between the recorder and the pendulum's lens.

In general the horizontal pendulums are oriented in the directions of the meridian and of the prime vertical, and practically all the stations conform to this arrangement, which has certain advantages. For stations situated near latitude 45° this arrangement makes it possible to separate completely the diurnal and semi-diurnal waves because the North-South component is almost exclusively semi-diurnal, while the East-West offers the most favourable conditions for the diurnal components, which are here at a maximum. In addition, the effect of an error on the determination of the azimuth is minimal for the East-West direction.

It is nevertheless necessary to consider the case of the instruments oriented in various directions because local conditions do not always make it possible to have a North-South or East-West orientation. Formerly, the custom was to place the apparatus in the bisecting directions in a way which made it possible to measure, with each apparatus, exactly identical tidal phenomena. It is then necessary to project the results on the principal directions in order to have comparable data for all the stations.

Let us consider a wave of angular velocity ω. The harmonic analysis of the observations provides us with the following expressions:

$$\left.\begin{array}{ll}\text{in the meridian} & \xi = H_N \cos(\omega t + \chi_N), \\ \text{in the prime vertical} & \eta = H_E \cos(\omega t + \frac{\pi}{2} + \chi_E),\end{array}\right\} \quad (8.41)$$

$$\left.\begin{aligned} &\text{in the azimuth } A_1 \;\; \xi_1 = H_1^* \cos(\omega t+\alpha_1+y_1) = \xi \cos A_1+\eta \sin A_1 \\ &\text{in the azimuth } A_2 \;\; \xi_2 = H_2^* \cos(\omega t+\alpha_2+y_2) = \xi \cos A_2+\eta \sin A_2 \end{aligned}\right\} \quad (8.42)$$

where X_N, X_E, y_1, y_2, are the phase-lags observed in each of the directions in relation to the corresponding theoretical phase, which is respectively 0, $\pi/2$, α_1, α_2.

The theoretical values α_1 and α_2 can be calculated from these formulas as a function of A_1 and A_2: it is sufficient to cancel out the observed phase-lags, and by using an elementary combination we obtain

$$\tan \alpha_i = \frac{H_E}{H_N} \tan A_i, \quad H_i^* = \sqrt{(H_N^2 \cos^2 A_i + H_E^2 \sin^2 A_i)}, \qquad (8.43)$$

where

$$\frac{H_E}{H_N} = \begin{cases} \dfrac{\sin \phi}{\cos 2\phi} & \text{for the diurnal waves,} \\[2ex] \dfrac{1}{\sin \phi} & \text{for the semi-diurnal waves.} \end{cases} \qquad (8.44)$$

We can thus state that the foot of the vertical passes in the azimuth A_i not at the time $\tau = (A_i/\omega)$, but later, at a time $\tau' = (\alpha_i/\omega)$. This deviation is sensitive. When the azimuth is 45° and the latitude 50° it amounts to:

52°33' - 45° = 7°14' for the semi-diurnal waves,
77°14' - 45° = 32°14' for the diurnal waves.

If we return to the complete formulas, which are suitable for the reduction of observed values with non-zero phase-lags, we obtain the following equations:

$$\left.\begin{aligned} H_N \cos \chi_N \cos A_1 + H_E \cos(\tfrac{\pi}{2} + \chi_E) \sin A_1 &= H_1^* \cos(\alpha_1+y_1) = X, \\ H_N \sin \chi_N \cos A_1 + H_E \sin(\tfrac{\pi}{2} + \chi_E) \sin A_1 &= H_1^* \sin(\alpha_1+y_1) = Y, \\ H_N \cos \chi_N \cos A_2 + H_E \cos(\tfrac{\pi}{2} + \chi_E) \sin A_2 &= H_2^* \cos(\alpha_2+y_2) = Z, \\ H_N \sin \chi_N \cos A_2 + H_E \sin(\tfrac{\pi}{2} + \chi_E) \sin A_2 &= H_2^* \sin(\alpha_2+y_2) = T; \end{aligned}\right\} \quad (8.45)$$

if the two pendulums are orthogonal, we naturally have

$$\sin(A_1 - A_2) = \sin A_1 \cos A_2 - \sin A_2 \cos A_1 = \sin \frac{\pi}{2} = 1,$$

and we deduce the simple expressions:

$$\left.\begin{aligned} H_N \cos \chi_N &= X \sin A_2 - Z \sin A_1, \\ H_N \sin \chi_N &= Y \sin A_2 - T \sin A_1, \\ H_E \cos(\tfrac{\pi}{2} + \chi_E) &= X \cos A_2 - Z \cos A_1, \\ H_E \sin(\tfrac{\pi}{2} + \chi_E) &= Y \cos A_2 - T \cos A_1. \end{aligned}\right\} \quad (8.46)$$

on the condition that there are no strong anomalies in one of the two azimuths.

The values of H_N, χ_N, H_E, χ_E can be easily calculated from the constants obtained in the azimuths of observations.

The example in which the two pendulums are not rectangular is seldom encountered but can be deduced from the same formulas by introducing into the first terms the coefficient $\sin(A_1 - A_2)$.

These formulas also make it possible to take account of the effect of an error in the determination of the azimuth. Schneider has published several interesting observations on this subject and has given a diagram making a very rapid calculation possible.

At latitude 50° we have an error of ±1° in the phases and 1% in the amplitude if we have made an error in the azimuth as shown in Table 8.1.

TABLE 8.1. Effect of an azimuth error

	Error of the phase ±1°		Error on amplitude of 1%	
	North-South	East-West	North-South	East-West
Semi-diurnal waves	±46'	±1°18'	±9°42'	±12°41'
Diurnal waves	±14'	±4°25'	±1°53'	±8°22'

It can be seen that the precise measurement of the diurnal components in the North-South direction in the middle latitudes, where most of the stations are located, requires a most exact determination of and, at the same time, such stability of the azimuth that it is difficult to attain it.

It should be noted indeed that when the spot reflected by a pendulum scans completely the field of the aperture of the recorder a progressive displacement of the azimuth of recording results. Under normal conditions, for a focus of 500 cm and an aperture of 30 cm, the scanning corresponds to a variation in the azimuth of 3°26'.

However the tidal variation of the azimuth of the beam which has the form

$$y_i = A\cos(\omega_i t + \alpha_i) \qquad (8.47)$$

produces a *small* variation of the phase itself, that is

$$\alpha_i = A\cos\omega_i t \qquad (8.48)$$

which introduces a non linearity effect in the registrations:

$$y_i = A\cos\omega_i t - \frac{A^2}{2}\sin 2\omega_i t \qquad (8.49)$$

the S_1 component, when important therefore spoils the S_2 tidal component. This is completely avoided with a zero-method.

Parallax of the Time Marks

The method of providing time marks by a brief illumination of the slit of the recorder requires that the filament of the lamp used for making the marks be parallel to, and at the same level as, the slit. This is difficult to achieve perfectly, and it is sufficient to determine carefully the small correction which must be applied to the phases determined by harmonic analysis.

To do this it is necessary, using a chronometer, to switch off one of the spots 5 min before the hourly mark and to switch it on again 5 min after.

It is then possible to measure if the hour mark is perfectly centred in relation to the interruption in the curve. Account must be taken of the thickness of the time mark (0.1-0.3 mm) when calculating this correction.

Timing

The correction to the clock should be determined each week using a good chronometer which has been compared with the broadcast time-signals.

The dispersion of the phases found for the phase of the wave M_2 is about one-tenth of a degree for long series which are now available, which corresponds to 12 s of time. It is sufficient therefore to have a precision of 10 s in the correction of the clock. It should also be noted here that the duration of contact (about 5-10 s) also limits precision, but this is yet unappreciable.

Presently all good stations are equipped with quartz clocks. On the other hand, as we can estimate the time mark position on the paper with a precision of ±0.2 mm, a chart speed of 6 cm h^{-1} should correspond to a precision of ±12 s of time upon each hourly reading.

With a much lower chart speed (6 mm h^{-1}), because of the very high number of hourly readings so far obtained (70.000 for one instrument at Dourbes), we nevertheless obtain a similar mean square error on the M_2 wave phase, i. e. 0.06° and 0.15° on the K_1 wave phase.

8.11. INSTRUMENTAL STABILITY

The stability of a horizontal pendulum must be examined from two points of view, namely period and drift. The latter forms the subject of many papers, but little attention has been given to stability in the period, although this is of the greatest importance since it directly determines the advantage of the apparatus and governs all quantitative interpretation of the recorded curves.

8.11.1. Stability of the Period

Many factors affect the constancy of the period for a horizontal pendulum. Firstly, it is essential that the two points which determine the axis of rotation should be well defined; these points must be fixed and must not vary in position with respect to the azimuth of the suspended beam, which would cause a very marked relation of period to azimuth, and, the more marked the greater the period used. In this case the angle between the axis of rotation and the vertical is so small that the slightest displacement of the fulcrum affects the angle (and hence the period) very greatly.

The effect is more pronounced with a metallic pendulum having wires gripped in chucks; it is even more if a strip is used. The first result of this is that it is often impossible to use periods above 30 to 35 s with a metal pendulum; moreover, the period becomes a function of amplitude.

For this reason it is essential, when fastening quartz wires, to produce a short joint by the capillary absorption of the primary material in order to ensure that the final form is a minute body of essentially hemispherical form.

Secondly, the period is affected by the stability of the anchoring of the pendulum in the crust. It would seem that our system of niches and reinforcement with scaled bars gives very good results.

Metal pendulums described in the literature show variations in period of the order of 3% per month, although the periods do not exceed 30 s. The stability of our quartz pendulums is much higher. The system of automatic calibration allows repetition of a test of sensitivity sufficiently often, usually twice a week. A good example is given by Fig. 8.20 representing yearly change of sensitivity of a VM quartz pendulum installed in Spitsbergen mines (depth 300 m; latitude 79°15').

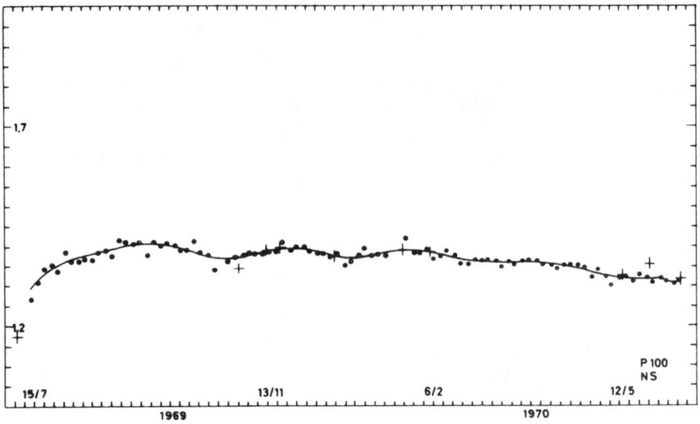

Fig. 8.20. Fluctuations of sensitivity with time of a Verbaandert-Melchior pendulum installed in Spitsbergen mines. The continuous fitting line has been adjusted according to the Vondrak-Pâquet procedure. Dots are the results of the automatic calibrations by using the device presented in the Fig. 8.15. Crosses are the results derived from chronometric measurements of the free period of oscillation of the beam (8.19).

Within about one year the sensitivity fluctuates between 0".001 25 and 0".001 45 per mm, which is a very satisfactory performance.

Two different approaches have been made to transform the hourly readings currently measured in tenths of a millimetre into milliseconds of arc (or microgals for gravimeters). (a) Linear interpolation between each pair of calibrations. This interpolation applies every 3 days and it is thus supposed that the sensitivity changes in a somewhat irregular way. One should remember, however, that changes of the order of a tenth of a millisecond of arc are of the size of 1.2 nm (as the basis of the instrument is 23.7 cm). (b) A curve fitting has been made by a method formerly devised by Whittaker and Robinson (1966), developed by Vondrak (1969), and recently improved by Pâquet (1972). This method consists in introducing an additional constraint by considering the sum of squares of third-order differences. If F is this sum and R the sum of squares of residuals, the best fitting is obtained through the minimum of

$$\Phi = R + \lambda^2 F, \tag{8.47}$$

where λ is a weighting parameter.

The problem is with a convenient choice of the parameter λ^2. Pâquet (1972) proposes to make a series of solutions for λ^2 values comprised between +1 and 10^9; his experiences have shown that $\lambda^2 F$ shows one or several minima with increasing λ^2. The best fit is realized with the λ^2 value corresponding to the minimum stationary value. The curve shown in Fig. 8.20 was obtained in this way.

Harmonic analysis has been performed using processes (a) and (b) for many different instruments. Amplitudes and phases are exactly the same, while mean-square errors show very little improvement for method (b). It seems thus that the sensitivity changes are smoothed phenomena.

8.11.2. Stability in Azimuth

Instability in azimuth commonly goes under the name of drift. Much work has been done on this in relation to geophysical interpretation and to the disturbing effects of drift in the determination of Earth-tide waves. Drift rates have been represented in polynomial form.

It has been found that drift rates drop considerably after a few months (Table 8.2). Some have asserted that there are variations of seasonal type, but it is difficult to be certain on this point. On the other hand, the drifts found with quartz instruments are generally much lower than those reported for stations with classical metal instruments.

TABLE 8.2. Time Decrease of the Drift at Sclaigneaux

North-South pendulum	Interval	Number of days	Daily drift unit: 0"001
4	1959.11.26-1960.01.04	40	+ 21.59
4	60.01.14- 60.03.07	54	+ 21.76
4	60.03.18- 60.04.09	23	+ 19.61
4	60.04.23- 60.07.11	80	+ 16.48
9	60.07.21- 61.02.10	205	+ 8.27
9	61.02.27- 62.03.25	392	+ 7.52
30	62.04.27- 62.08.23	119	+ 8.68
30	62.08.31- 62.10.15	46	+ 5.99
42	62.12.08- 63.03.05	88	+ 10.50
13	63.04.26- 63.07.18	85	+ 2.21

+ = drift in North direction)

Another example of a very long series of registration is provided by the East-West component at Sclaigneaux 1 where the measurements also started in 1960. During an uninterrupted registration over 900 days from April 1964 to September 1966 the daily drift was linear and amounted to 0".003. But since 1968 there is no longer any observable systematic drift, the trend having vanished.

Table 8.2 shows that the daily drift of five quartz instruments successively installed in the same niche is diminishing regularly to be finally of the order of 0".005; values from the literature are as follows: Winsford (England) 0".043 for one pendulum and 0".015 for the other: Kunrad (USSR) 0".03; Kondara (USSR) less than 0".2; Ashkabad (USSR) less than 0".2 for one pendulum, 0".05 for the other and from a graph for Bidston presented at the Munich Symposium, drift of 13" in 2 months for the North-South component (0".25 per day) and 0".04 per day for the East-West component.

The following are the probable causes of drift:

1. *Instrumental*

   Faults in the suspension.

   Unsymmetrical deformation of the support.

Deformation of the supporting parts in their screws.
Slip at the contact between support and props.
Change in the cement between prop and rock.
Deformation in the supporting rock.

2. *Geophysical*
Periodic deformation from variation in the flow in adjacent aquifers.
Instantaneous deformation in the crust consequent on zones of high or low pressure.
Deformation in the crust from loading by snow or surface water.

3. *Tectonic*
Flexure of the crust resulting from long-period tectonic causes and affecting large areas.

There are so many possible causes that no solution is to be expected from a single station with a single pair of pendulums.

However, one undoubted geophysical effect has been observed many times at Sclaigneaux (Belgium), a station which is not far from the River Meuse. The pendulums respond clearly to the rise and fall in the river's level.

8.11.3. An Electronic Version of the Verbaandert-Melchior Quartz Pendulums

Improvements of this pendulum should obviously consist in the realization of the zero method which fails as long as one keeps the optical signal device. To achieve such a goal van Ruymbeke (1976) has adapted with success a capacitative transducer on the beam of the pendulum.

The basic difficulty was that it is not possible to have an electric contact on the oscillating beam. To overcome it, van Ruymbeke designed the captor represented on Fig. 8.21 which can be described as follows: two pairs of parallel rectangular plates of surface S=hL (h being the height of the plate, L its width) are fixed on the bases, at a distance D from each other.

The capacity in the air is

$$C = \varepsilon_0 \frac{S}{D} = \varepsilon_0 \frac{hL}{D}$$

with

$$\varepsilon_0 = 8.85 \times 10^{-12}\ \mathrm{F\ m^{-1}}.$$

On the other hand, the hemispherical quartz mass at the extremity of the beam is replaced by a box made of two conductive shortcircuited plates P_1 and P_2 of height h.

This gives three different capacities:

$$C_1 = \varepsilon_0\, h\, (\frac{L-X}{D}), \qquad C_2 = \varepsilon_0\, h\, \frac{X}{D_1}, \qquad C_3 = \varepsilon_0\, h\, \frac{X}{D_2}$$

as shown on Fig. 8.21.

The equivalent circuit of one of the branches of the capacitative bridge shown on Fig. 8.22 is such that

$$C = C_1 + \frac{C_2 C_3}{C_2 + C_3} = \varepsilon_0\, h\, \{(\frac{L-X}{D}) + (\frac{X}{D_1 + D_2})\}.$$

If

$$p = \frac{D_1 + D_2}{D}$$

one writes

$$C = \frac{\varepsilon_0 h}{Dp}\, [\, pL + (1-p)\, X\,]$$

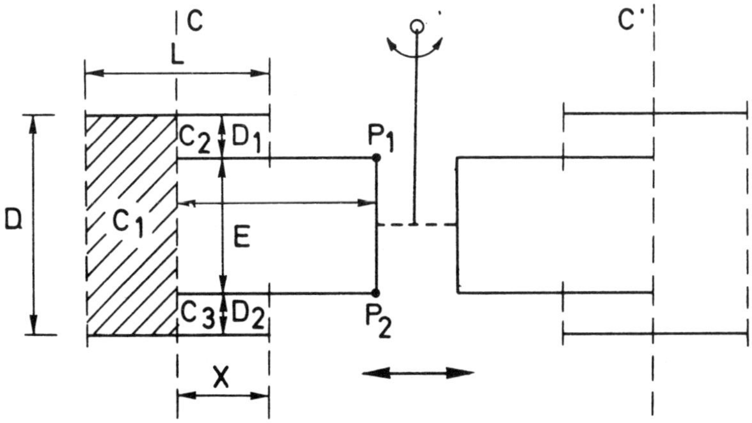

Fig. 8.21. Capacitative captor used by Van Ruymbeke for a VMR pendulum.

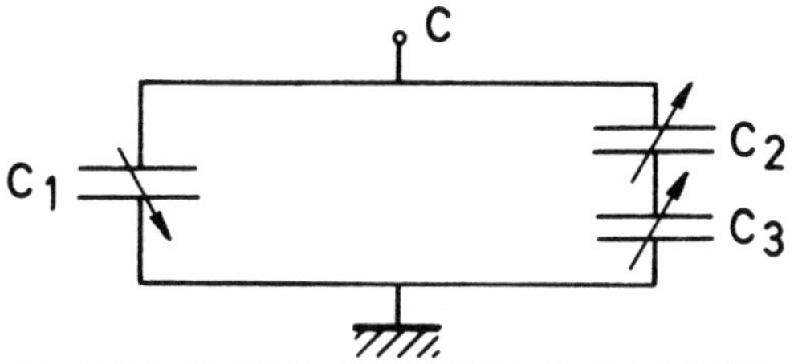

Fig.8.22. Equivalent circuit of one of the branches of the capacitative bridge shown on fig. 8.21.

and the sensitivity is given by

$$\frac{dC}{dX} = \frac{\varepsilon_0 h}{Dp}(1-p),$$

which ensures the linearity of this device. p being negligible by construction one has

$$\frac{dC}{dX} = \frac{\varepsilon_0 h}{Dp}.$$

The difficulties which van Ruymbeke solved in the realization of such a device are the light weight of the moving box (5 g), the parallelism of the plates and also in the quality of the electrodes P_1, P_2, which were not altered after 3 months' operation in the underground laboratory of Walferdange. Electric connections must also be of the highest quality.

In these new pendulums called VMR (and numbered from 500) the measurements are achieved by a classical Wheatstone bridge system comparing the two capacities

$$C = \frac{\varepsilon_0 h}{Dp}[pL + (1-p)X] \text{ and } \quad C' = \frac{\varepsilon_0 h}{Dp}[pL - (1-p)X],$$

which avoids any effect of a change of ε_0 due, for example, to humidity variations. The capacimeter used gives an output signal of one volt for a change of capacity in the bridge of 1.5 pF. The electrical noise is equivalent to a capacity change of 50 aF. It is interesting to note that this system of an oscillating box between parallel plates ensures some efficient damping which is helpful for the zero method procedure as well as for seismological investigations.

Several months experience with VMR pendulums at Walferdange show that they have the same behaviour and the same qualities as the original VM version.

The γ amplitude factors as well as the phases obtained by analysis remain the same.

The advantages of the instrument are:

a) no need of a 5 m focal length, any reduced space being sufficient;
b) remote control and registration;
c) future possibilities for the zero-method as well as for seismology;
d) the use of a free period of 40 s giving a better stability of sensitivity.

8.11.4. Other Quartz Clinometers

Copies of the VM instrument have been made in Sweden (Vogel) and Romania (Zugravescu); this instrument also greatly inspired a Chinese version realized at the Peking factory of seismographs. Other versions of a quartz pendulum have been realized by Bragard which operates at Kanne (Belgium) and by Blum which operate in France.

8.12. INSTRUMENTS OF VARIOUS TYPES

The Tsubokawa Electromagnetic Pendulum (TEM) (Fig. 8.23 , 8.26).

This instrument is a kind of electromagnetic level. The sensor is an aluminium plate supported by four metallic wires forming a cross-suspension as shown on Fig. 8.23. The system of filaments is fixed to the frame; the displacements due to the deviations of the vertical and ground inclinations are electrically detected by a differential transformer (transducer).

The container is filled with silicon oil for damping. Calibration is realized by turning the screws with a motor and the force depends upon the precision of the screw pitch and the precision with which it is known. It is unsatisfactory that the supporting screws are not rectangular to each other.

Comparisons of VM Horizontal Pendulums with the Tsubokawa Electromagnetic Tiltmeter

Very interesting comparisons of these instruments of completely different principle and design have been successively conducted at Akagane (north of Japan) and at the Walferdange underground laboratory (Luxembourg), i. e. in very different local and regional conditions. This has proven that the considerable anomalies measured at Akagane are not at all due to the instruments but to the regional indirect effects. At Akagane each set of instruments uses its own calibration device while at Walferdange all instruments were automatically calibrated by dilatable bearing plates.

It seems that the results obtained at Walferdange (Table 8.4) are more coherent due to this fact.

TABLE 8.3. A Comparison of TEM and VM Pendulums at Akagane (ϕ=39°10'N), Japan

		VM (486 days)		TEM (365 days)	
		γ	α	γ	α
North-South	M_2	0.381	-14.93°	0.386	-23.57°
	N_2	0.378	-15.00°	0.348	- 3.69°
	S_2	0.355	1.71°	0.435	-24.47°
East-West	K_1	1.156	-104.92	1.629	-106.23
	P_1	1.119	-107.67	1.641	-113.64
	O_1	1.058	- 82.74	1.392	- 82.14
	M_2	0.133	-135.53	0.235	-127.71
	N_2	0.315	-188.51	0.446	-174.63
	S_2	0.501	-115.22	0.492	-113.70

The agreement shown by such very strong phases in the East-West component with two quite different instruments is impressive.

At Walferdange three pairs of instruments installed at different places in the room were compared, two VM pairs with one TEM pair. The main results are as shown in Table 8.4.

TABLE 8.4. A Comparison of TEM and VM pendulums at Walferdange ϕ=49°40'N, Luxemburg.

		VM		VM		TEM	
North-South	M_2	0.625	-4.95	0.651	0.32	0.604	-2.96
	N_2	0.674	-5.80	0.694	-2.82	0.729	-5.78
	S_2	0.624	-1.84	0.640	7.81	0.764	5.04
East-West	K_1	0.759	-3.11	0.758	-5.61	0.729	-6.21
	P_1	0.741	-2.74	0.727	-6.86	0.691	-0.52
	O_1	0.680	0.76	0.644	-3.55	0.675	-6.71
	M_2	0.877	-8.78	0.920	-9.45	0.809	-8.75
	N_2	0.939	-5.27	0.943	-6.48	0.874	-4.87
	S_2	0.726	-14.19	0.781	-13.86	0.753	-13.61

It is to be observed that the agreement is better for the diurnal waves than for the semi-diurnals.

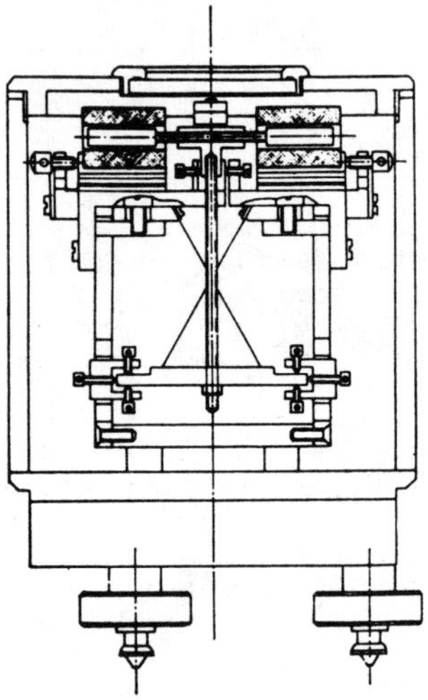

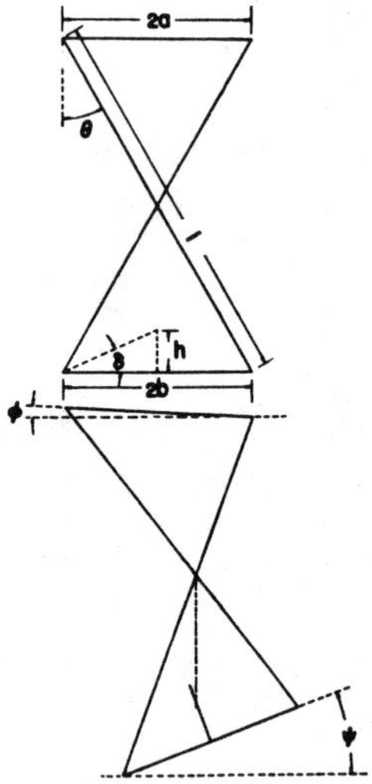

(a) (b)

Fig. 8.23. The Tsubokawa electromagnetic pendulum.

The Ostrovsky Pendulum (Figs. 8.24 and 8.25)

This instrument, largely used in the USSR, was devised at the epoch of the International Geophysical Year according to very conservative principles.

A metallic pendulum with a mass of 1 Kg is suspended horizontally by one vertical steel wire of 200 microns diameter and two horizontal tungsten wires of 100 microns diameter. Its free period is about 5 s. It has therefore a very low mechanical sensitivity and the amplification is provided by an electronic device borrowed from the old Askania GS 11 gravity meter: a light beam reflected from a mirror fixed on the pendulum is split by a prism into two photocells. The variable current from the photocells produces oscillations of a galvanometer which are registered photographically.

The base plate has three non-orthogonal levelling screws and the instrument is sealed against atmospheric pressure variations. This is, in our opinion, an erroneous concept because the atmospheric pressure variations have a negligible effect on an horizontal pendulum but produce sensible deformations of the supporting frame when it is sealed.

Electromagnetic impulses in solenoïds are used periodically to control the sensitivity of the instrument. These impulses are previously calibrated on a tilting platform by using tilt angles as great as 50" or even 500".

Recently more precise platforms have been designed which allow very small tilts to be produced.

Because of its short, free period this instrument is very sensitive to the microseismic activity and, like the Askania borehole pendulum, requires a strong electromagnetic damping. This introduces complications and uncertainties in the calibration procedure and instrumental phase-lags which should be determined.

As we have no personal experience of these instruments we quote some statements made by Skalsky and Picha (1969).

"The tilt of the instrument is in this case compensated by electric current, passed through coils, located on the pendulum in the field of a permanent magnet. The calibration of the tiltmeter is in the determination of its basic electromagnetic constant I_0, i. e. the current intensity necessary to compensate a tilt of one angular second. This "constant" depends on the magnitude of the gravity acceleration at the point of observation, on the mass and reduced length of the pendulum, on the total number of windings of the coil, on its distance from the rotational axis of the pendulum, and on the intensity of the magnetic field. As the effect of the different gravity accelerations at the calibration point and the tidal station can be neglected (or, if necessary, it can be introduced in a simple way), and all the parameters of the pendulum and coil may be considered as constants, the intensity of the magnetic field remains the only quantity capable of changing this basic "constant". By using the most up-to-date materials and suitable technological processing of the magnets, one may obtain magnets the intensity of which changes very little. Nevertheless, it is necessary to check this electrodynamic constant from time to time.

The disadvantage of this method of sensitivity determination is the complexity and the number of auxiliary devices, which, with regard to the small intensity of the impulses used (of the order of hundredths of a microampere), may become the source of a number of disturbing effects and perturbations. It is, therefore, necessary to devote special attention to all components, connections and sources, and to install these tiltmeters only under such conditions (especially climatic), which would guarantee the reliable function of the whole apparatus, with minimum disturbing effects".

Very many results obtained with this instrument have been published in the Russian literature and can be found in the French translation of the BIM series.

Most of these results concern the main wave M_2 only. The diurnal waves K_1 and O_1 are seldom given and exhibit strong variations from one month to the other, reaching 50% or more. The diurnal wave P_1 has never been isolated.

This may be a major criticism. The reason of this lack of determination of the diurnal components may be assigned to the metallic nature of the instrument and its high sensitivity to low thermic disturbances, to the low sensitivity of the instrument itself, its sensitivity to barometric changes and the number of sophisticated devices used for the amplification.

We also disagree upon the method of installation on concrete pillars in shallow shafts. It is evidently difficult to be sure if these arguments are really definitive. Nevertheless, the published results look rather poor.

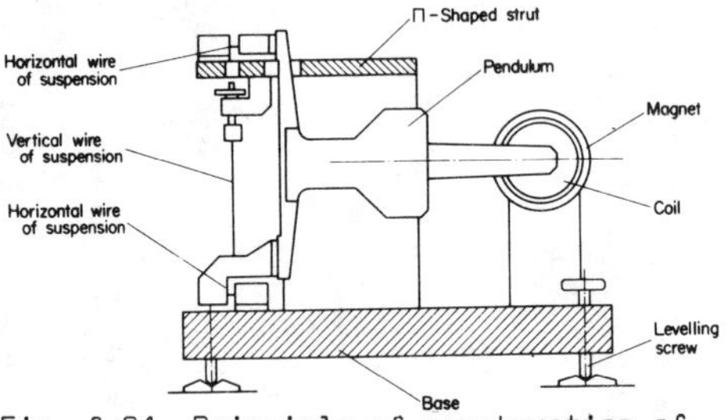

Fig. 8.24. Principle of construction of horizontal pendulum built by Ostrovsky.

Some Other Tiltmeters

A. D. Littles' biaxial tiltmeter employs a diamagnetic mass suspended in a potential well of a suitably shaped magnetic field (Simon, Enslie, Strong and McConnel, (1968)).

The depth of the potential minimum determines the restoring force acting upon the mass and consequently the pendulous frequency and sensitivity of the instrument. In this manner the elastic forces usually supplied by fine fibres are replaced by field forces and the mass is free to respond to tilts or accelerations without any trace of friction of external (Coulomb) or internal (anelastic) origin.

The magnetic field is derived from permanent magnets and no active circuit is necessary for operation of the suspension. At the same time the entire magnet system is small enough to permit packaging of the tiltmeter in an enclosure suitable for placement in boreholes.

The required amount of damping is obtained from eddy currents induced in the seismic mass by its motion in the field. The instrument requires no power to operate the suspension and it can be used over a wide range of temperatures. Under usual environmental conditions it requires no temperature control.

Hughes' tiltmeter consists of a 4 in. bubble trapped under a quartz optical flat (7.5 cm) which in turn is mounted on a slightly deformable fused quartz tripod. The bubble is sensed electrically through electrodes in the base of the bubble chamber which allows nullification of the system by a feedback mechanism. The tripod is deformed by torques applied about two perpendicular axes through a solenoid-permanent magnetic assembly. Currents in the solenoids

Fig. 8.25. The Ostrovsky pendulum as installed at the Obninsk station.

Fig. 8.26. A Tsubokawa electromagnetic pendulum.

are controlled by the amplified output of a bubble-position - sensing bridge in such a manner that the bubble is milled near the centre of the flat.

No permanent calibration device is provided with this equipment. This instrument appears to be extremely sensitive to air pressure changes (Schlemmer, 1976).

In a very recent paper Harrison (1976) describes the results obtained in the Poorman mine near Boulder, Colorado, with such devices. The same installation system as for VM pendulums has been used (quartz cylinders epoxied into vertical holes drilled into the rock to support the tripod) but on the floor of the tunnel. The drift of the instruments is similar to the drift of VM pendulums during the first year of installation.

The O_1 amplitude factors are extremely large and abnormal (γ = 0.82 - 1.02) which is strange to us where such a continental station is concerned, even if cavity effects may be suspected.

Vertical Borehole Pendulums

Schneider (1966) and Graf (1964) independently developed a vertical pendulum (2 m long) measuring the deviations of the vertical in two rectangular directions.

The Askania borehole tide pendulum which derives from these investigations, is a non-astatized vertical pendulum with remote control measuring plumbline variations for taking measurements in an encased borehole of 150 mm inside diameter and a depth of between 20 - 60 m.

The vertical measuring pendulum, which has an operative length of about 0.6 m, has a cross-spring joint suspension for measuring in the x- and y-direction by means of capacitative displacement transducers. The resolving power of these transducers is about 0.0002". It is provided with calibration devices for absolute calibration by means of definite displacements of a ball weight, one each for the x- and y-direction, and with devices for relative calibration by means of magnet arrangements and Helmholtz coils independent of the Earth and disturbance field.

An electrical remotely controlled fine adjustment device embodying a two-coordinate cross-sliding table adjusts the plumbline for each coordinate separately with an accuracy in the order of a hundreth of second of arc. The system is mounted in a pressure watertight tubular housing of 14 mm outside diameter and 1600 mm length. Its weight is about 75 Kg. A guided centring device is arranged in the bottom part while it is provided with forced fixing by spreading apart points arranged in the top part which penetrate into the borehole casing securing positive connection when the wire rope is slackened upon lowering the tubular housing into the borehole.

A grid plate fixed on the top part determines the location of the pendulum in azimuth with a telescope.

The sensitivity adjustment device allows an adjustment between about 2 min of arc on stage 1, and about 0.1 s of arc on stage 4 per full-scale deflexion of the recorder; then with a switch and control device one applies a compensation voltage for fine adjustment of the measured value on the recorder in the x- and y-directions.

A first version of this instrument has proved to be very satisfactory for Earth-tide measurements as the filtering was realized with a passive low-pass filter using large capacities which were depending on time, temperature and input voltage. They produced a time lag of about 40 min which made the calibration practically impossible and the tidal parameters uncertain.

This was needed by the fact that the free period of the beam being 1.36 s the noise amplitude can exceed the maximum Earth tide effect by a factor of 100. Recently Zschau (1976) in Kiel has reconstructed the calibration compensation and damping devices and obtaines a negligible phase and amplitude distortion. This was also corrected by Flach (1971), and Grosse-Brauckmann at Clausthal-Zellerfeld (1972) who replaced the original system by an active Butterworth-Bessel low-pass filter of fourth order (*) and improved the electronics. (Flach, 1975)

However some computations made at ICET were not giving very satisfactory results (confirmed by Flach, 1971).

Schmitz-Hübsch realized in upper Bavaria two North-South parallel profiles, 10 km apart, each of five 30 m boreholes at distances of 8 km. Measurements were made one year after boring to avoid mechanical disturbances. It is remarkable that the amplitude factors are as dispersed as in the stations equipped with classical horizontal pendulums but that the phases are far more abnormal (always around -35° to -40° for M_2 in both components) than in any other station except in the railway tunnel ones. Which kind of "cavity effect" can explain such a behaviour? (see Chapter 12).

There was considerable hope that such instruments could have been installed in *deep* boreholes drilled at any desired place and that the difficulty encountered in the selection of suitable sites for the horizontal pendulums could have been solved in this way.

One must admit that despite considerable and valuable efforts the situation is presently still disappointing as so few results of such observations have been published or made available except those from instruments installed in underground galleries just like horizontal pendulums (Freiberg, Erpel, Liverpool). This, of course, makes the determination of the azimuth easy. However, it is difficult to be convinced that the measurements azimuth is defined with the needed precision in the boreholes. Also the calibration is not free from systematic errors due to the strong dragging effect of the damping system, which makes a correct measurement of the produced artificial tilt quite hazardous.

Free Period and Damping

There is a clear advantage in having an instrument with a high free period of oscillation as it is then more or less insensitive to microseismic noise.

A pendulum like the VM quartz instrument, which is usually installed with a working free period of 60 - 80 s (sometimes even more) and is perfectly stable in these conditions is unaffected by the usual short-period microseismic activity. Only exceptional storms may disturb it by giving then a thick trace on the recorder.

Therefore no damping has been introduced in this instrument which avoids an important source of systematic instrumental error. This is not possible with short-period instruments such as the Askania vertical pendulum and the Ostrovsky pendulum which exhibit a noise several times bigger than the total tidal amplitude. For such devices an electromagnetic damping has been introduced (Tsubokawa has introduced an oil damping). This makes the registered phenomenon quite complex, and careful determinations of the effect of the damping on the amplitude and phase-lag *at different frequencies* must be made.

(*) Butterworth filters are "maximaly flat" filters calculated in order to have a flat response at the origin (derivatives of the transfer function are zero until a very high order). However they have not a fast step response which is a disadvantage in the calibration procedure. Therefore a Bessel filter is introduced which has a fast step response. In the passive filters the energy comes from the source signal itself. The active filters on the contrary introduce operational amplifiers which, when used before each filter separates them and isolate them, avoiding any influence between the different components of the filter.

With the Ostrovsky photoelectric tiltmeter also the instability of the electric current, keeping the pendulum "in the recording position", must be considered an "instrumental error".

8.12.1. Zero Method Tiltmeters

Great advantages should derive from the realization of a zero method that is a feedback method compensating in some way the deviation of the beam and maintaining it permanently in the same azimuth. The dilatable crapaudine is an obvious device to realize this operation.

However several preliminary attempts proved not to be very successful as the registered curve appears seriously distorted at the maxima and minima when the feedback force becomes extremely small.

Quite satisfactory results have recently been reported by Skalsky and Soukup (1974), associating an Ostrovsky photoelectric tiltmeter (modified by having right-angle levelling screws) with a Verbaandert crapaudine as a compensating device; the mercury height is recorded as the tidal signal.

Ostrovsky, Matveyev and Chliakhovoi (1975) propose using directly the output signal of the photocells of the Ostrovsky pendulum in a feedback system, sending its impulses in the solenoids to keep the beam at its zero position. The current needed to maintain the beam in this position is measured, providing a convenient tidal output. This is clearly the simplest way of obtaining the zero method.

The principle under development by van Ruymbeke on a VM pendulum is somewhat similar, but he uses the output of a capacitative transducer which seems to us a more convincing device than photocells.

8.13. RESULTS OF TILT REGISTRATIONS ANALYSIS

8.13.1. Internal Accuracy

This accuracy will evidently depend upon the length of the available record and the ability of the method to eliminate noise principally of a long period or polynomial form. It depends evidently also on the precision of readings on tidal curves and thus of the quality of the curve and the sharpness of time marks.

Figures 8.28 and 8.29 demonstrate how the internal mean-square error of the amplitude factor γ (horizontal component) diminishes as a function of increases of the recording duration.

If the improvement is slight for wave M_2, it is important for all the diurnal waves as the meteorological perturbations are better and better separated.

For a determination of waves of minor amplitudes such as P_1, Q_1, J_1, N_2, K_2 ..., several years records are clearly needed to obtain satisfactory results.

8.13.2. External Accuracy

External accuracy depends upon the existence of systematic errors such as calibration, azimuth, timing errors, insufficient compensation for barometric and thermic influences, non-linearity of recorders. Calibration should normally be correct to 1%. Exchange of crapaudines (stainless) do not produce any variation. Azimuth can be in error by 1° to 2° for horizontal pendulums, but this error might be much greater for vertical pendulums. Timing in the worse case should be exact to 30 s but usually is correct to 1 s.

A convenient check should be to look at the consistency of results obtained at different stations in a broad area for waves not suspected of having regional anomalies. Such anomalies are usually due to superficial deformations produced by the loading effect of oceanic tides (Chapter 11) and to cavity and topographic effects (Chapter 12). To avoid any doubt concerning the calibration or the azimuth one has advantage to look at ratios like $A = \gamma(O_1)/\gamma(K_1)$, $B = [1-\gamma(O_1)] / [1-\gamma(K_1)]$, $C = \gamma(M_2)/\gamma(O_1)$ given in Table 13.3 as well as to differences of phases like $\alpha(S_2) - \alpha(M_2)$ as given in Table 13.4.

A comprehensive discussion of the available results in tiltmetric stations will be given in Chapter 13 after consideration of the regional and local disturbances due to oceanic tides interaction, cavity and topography effects which can be large. However, it seems that we cannot conclude this chapter concerning the tiltmeter without first giving a comment about the results so far obtained.

Obviously a geophysical discussion of tilt registrations can only be done on data which offer a minimum of guarantees with respect to gross systematic errors. We therefore will restrict our discussion to those stations which satisfy the following criteria:

(1) Precise and correct calibrations have been repeated in a regular manner, demonstrating a quite stable sensitivity of the instrument (see Fig. 8.20).

(2) The harmonic analysis of data has been performed correctly: the results of the main diurnal components K_1, P_1, O_1 and the meteorological spurious component S_1 have been determined and published (see Table 8.5). This, of course, allows a proper judgement on the environmental disturbances to be made.

Experience shows that in most good stations the amplitude of S_1 is never higher than 0.3 ms (Table 8.5). Doubts can be expressed therefore when it reaches 1 ms. It is observed that in shallow stations (depths under 20 m) this factor commonly reaches 10 ms and even much more. Such stations have to be rejected for serious analysis. Therefore many ancient results will not be referred to here since they have been treated historically in our previous volume.

Before examining the tables of the most recent and precise data, a remark has to be made concerning the phase sign. For each tidal wave the direction of the vertical describes an ellipse having East-West and North-South directions as axes (see Fig. 1.10). The procedure of harmonic analysis applied separately on both components of the tilt registered by two different pendulums gives sinusoïdal waves which are more or less dephased with respect to the theory.

Combining such results obtained separately for the North-South and East-West components one obtains an ellipse which has its principal axes tilted with respect to the North-South and East-West directions as in the case represented on Fig. 8.27.

In this particular situation one can easily see that the maximum amplitude in the East-West component will be observed with a *lag* with respect to the expected time corresponding to the theoretical prediction. On the contrary, the maximum amplitude in the North-South component is observed in *advance* with respect to the theoretical prediction.

TABLE 8.5. The Observed S_1 component with horizontal pendulums

Theoretical amplitude 0"000 024 γ × 2 cos 2ϕ North South; × 2 sin ϕ East West

Station	ϕ	Nr	Observed amplitude in 0"001 North South	East West
Sclaigneaux	50°30'	1	0.231	0.406
		2	0.203	0.259
Dourbes	50° 6'	1	0.148	0.124
		2	0.048	0.062
Walferdange	49°40'	1	0.038	0.197
		2	0.225	0.160
Spitsbergen	78°12'		0.264	0.510
Dannemora	60°12'		0.070	0.028
Lohja	60°13'		0.328	0.406
Erpel	50°35'		0.367	0.114
Tiefenort	50°49'		0.106	0.420
Příbram	49°41'		0.036	0.359
Graz	47° 4'		0.051	0.621
Costozza	45°28'		0.131	0.109
Bari	40°52'	1	0.520	0.360
		2	0.610	0.439

Remarks 1. The S_1 component corresponds to the waves 121, 122, 123 in the Cartwright Taylor Edden Table.

2. a deflection of the vertical of 0"0002 that is 10^{-9} corresponds to an horizontal acceleration of one microgal only.

Therefore it is to be expected that the phases determined by analysis of the tidal curves will always be of opposite sign for the two principal directions. This is indeed what we observe in Tables 13.2 and 13.3.

8.13.3. First General Comments on the Results

Diurnal waves

1) In the East-West component the results for the four main waves K_1, P_1, O_1, Q_1 are quite satisfactory in the sense that they fit fairly well the theoretical predictions made by Jeffreys-Vicente and by Molodensky from models with a liquid core. This will be examined more in detail in Chapter 13. Results obtained by different instruments installed at different places in the underground "cavity" at Dourbes and at Walferdange are quite coherent as shown by the ratios A and B.
2) All stations being presently installed around the latitude 45°N, the amplitudes of the diurnal waves in the North-South component are much reduced. It is therefore difficult to estimate which is the part of environmental conditions on the strange and disappointing results obtained for the diurnal waves in this component.

Semi-diurnal Waves

1) Again for these waves the East-West component results are fairly regular and smooth all across Europe as shown on the map of Fig. 13.7. The γ factor is considerably increased with respect to the models, but this can be explained by indirect effects due to the oceanic tides (see Chapter 11).
2) The North-South component results are really puzzling. They seem to be completely meaningless as the numerical value of γ change by 100% on a distance of 50 km (ex Sclaigneaux-Dourbes).

This is difficult to understand as we observe that the same instruments in the same stations give contemporaneously a fairly homogeneous distribution of the γ factor for the M_3 wave.

Ter-diurnal waves

Although their amplitude is never more than 0ʺ.0001, nearly all instruments give surprisingly good results. This is due to the fact that we observe three cycles per day in a frequency band which is not much disturbed by noise (the period is 8 h 23 min).

As shown in Chapter 3, the higher harmonics are more dependent from structure of the external parts of the planet than the second harmonic. Unfortunately the numerical values obtained do not prove to be very sensitive to the different model structures, while the small amplitudes to be measured do not allow a sufficiently precise determination of γ_3 to enable us to make a choice between these models. Nevertheless, such a consistency between many different stations, components North-South or East-West, and different instruments show that the calibrations are correctly made.

The more detailed examination of all clinometric results is given together with the other components in Chapter 13.

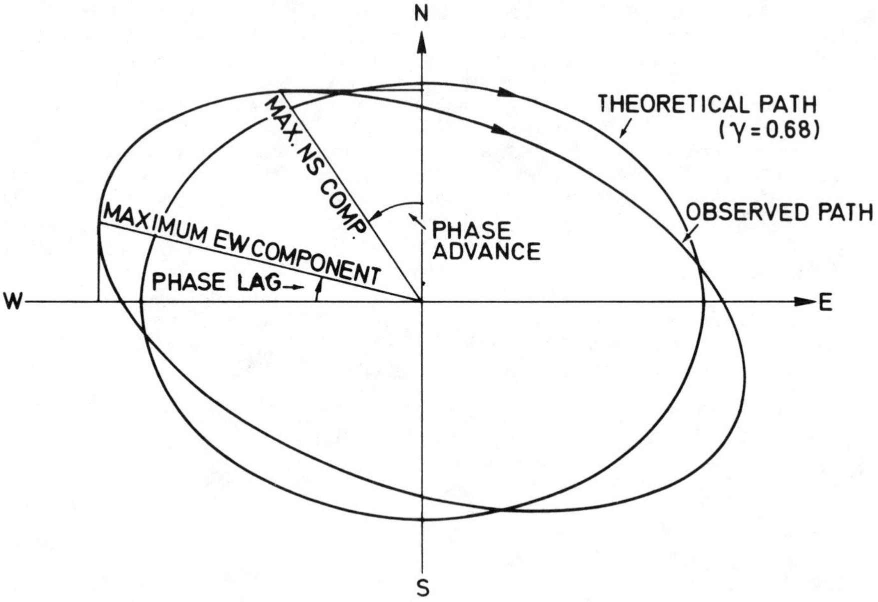

Fig. 8.27. Comparison of the observed path of the direction of the vertical for one specific tidal wave with its theoretical path. The phases differences in the North South and East West components have opposite signs.

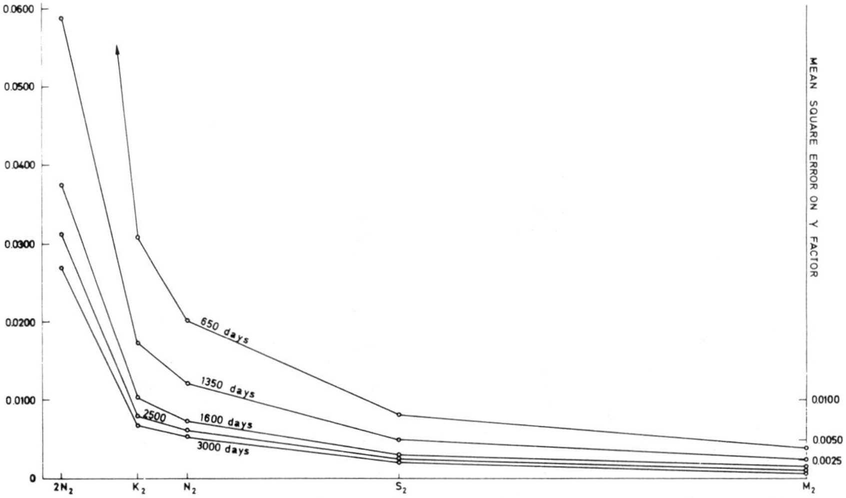

Fig. 8.28. Quartz horizontal pendulum VM8 - Station Dourbes (Belgium) - Semi-diurnal waves - the internal mean square error on the γ factor as a function of the number of registration days analysed. The lines approximate the equilater hyperbola y=m/x, y being the mean square error for one tidal wave of amplitude x.

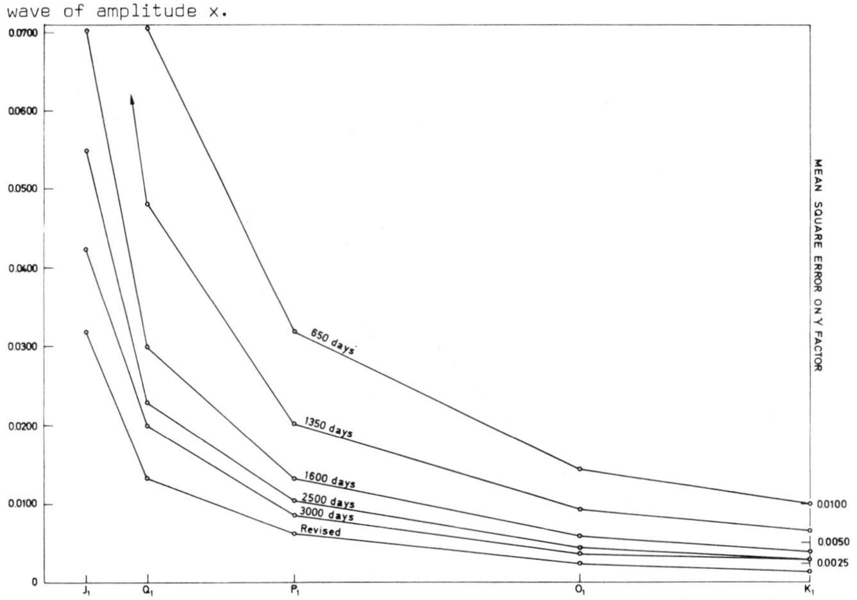

Fig. 8.29. Quartz horizontal pendulum VM8 - Station Dourbes (Belgium) -: same curves as in Fig. 8.29 but for the diurnal waves. The curve "Revised" corresponds to a careful revision of all data which concluded to a rejection of 126 days among the 3000 available.

Fig. 8.30. Installation of VM quartz horizontal pendulums (two components) in the Pribram station (Czechoslovakia) at a depth of 1300 metres. Mercury tube for calibration device is well visible.

Fig. 8.31 . Installation of VM quartz horizontal pendulums (two components) in the Longyearbyen station (Spitsbergen, latitude 78°12' North).

CHAPTER 9

Gravity Tide, Instruments, Results

9.1. GENERAL PRINCIPLES FOR GRAVIMETRIC MEASUREMENTS

We wish to register the time variations of gravity in order to determine the numerical value of the coefficient $\delta = 1 + h - 3/2\ k$ which is the ratio of the amplitude of the variations of g observed on the Earth's crust to the amplitude of the corresponding variations calculated in celestial mechanics. These are estimated to be of the order of 2×10^{-7} of g, and we cannot expect that the actual variations will be much different. Therefore, as a precision of 1% on the numerical value of δ is needed to be of practical value, the instrument used must be extremely stable and reach a sensitivity of 10^{-9}. If correctly measured the phase-lag should also be a parameter of major interest in the investigations on the Earth's viscosity.

The principle of all instruments devised to measure g and its variations (with space as well as with time) is to balance the mean value of g with a constant force equal to that mean value and to measure the possible slight differences. Until quite recently the only force being sufficiently stable with time was the elastic force exerted by a spring; an idea introduced by Herschel in 1849.

There now exists a supra-conductivity gravimeter which will be described in Section 9.9, but with this exception all instruments used in the world are elastic spring gravimeters.

It is quite evident that a spring is a convenient device to serve as a fundamental part in gravimeters as well as in seismographs. The primary function of a spring is to store energy, to apply a definite force or torque, to support moving or vibrating masses, or to indicate and control load or torque. They can be used either as compression springs or as tension springs, but only the second type is used in geophysical instruments.

A typical gravimeter setting is presented on Fig. 9.3 which shows how the torque $mgb \cos \gamma$ exerted by the gravity on a beam of length b is balanced by the torque exerted by the suspension spring $C(\gamma)$.

As seen on the different figures, helical tension springs are used, and consequently the primary stress is torsional. Therefore the spring acts just like a torsion wire which has been made compact in a very limited space by reeling. This has the advantage of considerably increasing the sensitivity and generally allows a more easy thermostatization which is essential to measure relative changes of 10^{-9} size.

Let

r = radius of the wire,
L = total length of the wire,
μ = modulus of rigidity of the material used, and
E its Young's modulus $= \mu(3\lambda + 2\mu)/(\lambda + \mu)$.

The torque exerted on this wire to produce an angular deflexion α is

$$N = \frac{\pi}{2}\frac{\mu r^4}{L}\alpha = K_o\,\alpha, \qquad (9.1)\ (*)$$

where

K_o is the constant of torsion of the wire.

Let us now reel this piece of wire to form an helical spring (Fig. 9.1) and let

R = the radius of the helix,

$e = \frac{R}{r}$ the spring index,

k = the pitch ($k/2R = \beta$ is the pitch angle),

$\ell = 2\pi R$ the length of one coil, (9.2)

n = the number of coils,

$H = nk$ the height of the spring, (9.3)

then $L = 2\pi nR$, (9.4)

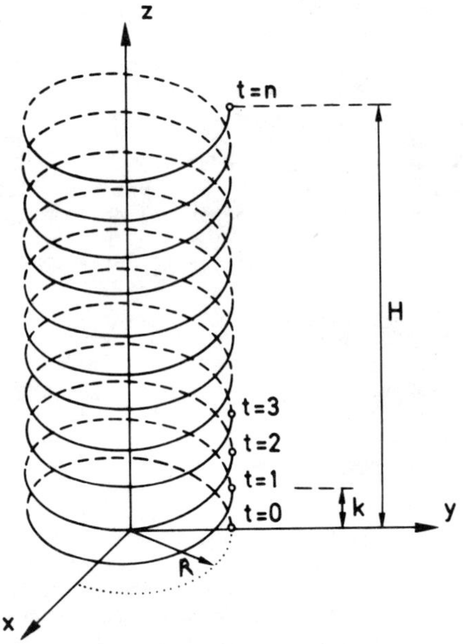

Fig. 9.1. The helicoïdal spring

(*) ($\pi/2\ r^4$) is the moment of inertia of a circle of radius r. If one uses a flat rectangular spring of section ($h \gg b$) the moment of inertia is

$$\frac{hb}{12}(b^2 + h^2) \approx \frac{h^3 b}{12}$$

which replaces ($\pi/2\ r^4$) in formulas (9.1), (9.13) and (9.14).

and the torsion is given by

$$\frac{1}{T} = \frac{2\pi k}{4\pi^2 R^2 + k^2} \tag{9.5}$$

while the curvature is

$$\mathscr{R}^{-1} = \frac{4\pi^2 R}{4\pi^2 R^2 + k^2} \tag{9.6}$$

(see Melchior, Physique et dynamique planétaire, vol. 2, Gravimétrie). When the pitch angle β can be neglected, the deformation is the same as the torsion of a wire. When it cannot be neglected the deformation produced by an axial stress P results from a torsion PR cos β and from a flexure PR sin β.

In practice, k is very small with respect to R (pitch angles are usually lower than 3°), so that

$$\mathscr{R}^{-1} = R^{-1} \tag{9.7}$$

and

$$\frac{1}{T} = \frac{1}{2\pi R}\,\frac{k}{R} = \frac{\alpha}{\ell} \tag{9.8}$$

for one coil of length $\ell = 2\pi R$.

A change of g results in an additional torque

$$N' = m\,\Delta g\,R \tag{9.9}$$

which produces a variation of the pitch Δk and therefore a change of the torsion of the helix wire (the coil radius also changes with this deflexion but this has only a very slight effect). Then

$$n\,\Delta k = \Delta H \tag{9.10}$$

and

$$\Delta\alpha = n\,\frac{\Delta k}{R} = \frac{\Delta H}{R} \tag{9.11}$$

so that

$$m\,\Delta g\,R = N' = \frac{\pi}{2}\,\frac{\mu r^4}{L}\,\Delta\alpha = \frac{\pi}{2}\,\frac{\mu r^4}{LR}\,\Delta H \tag{9.12}$$

and, finally,

$$m\,\Delta g = \frac{\pi}{2}\,\frac{\mu r^4}{LR^2}\,\Delta H = K'_0\,(H - H_0). \tag{9.13}$$

$\Delta H = H - H_0$ is the extension of the spring producing a torsion in the wire which is characterized by the constant

$$K'_0 = \frac{\pi}{2}\,\frac{\mu r^4}{LR^2} = \frac{K_0}{R^2}. \tag{9.14}$$

This is, of course, an approached value corresponding to infinitely small values of α and very high values of the spring index $e = R/r$.

The more exact formula is

$$K' = b^{-1}\,K'_0 \tag{9.15}$$

with

$$b = \frac{\cos \alpha}{1+\frac{3}{16}\frac{\cos^4 \alpha}{c^2 - 1}} + \frac{2\mu}{E} \sin \alpha \operatorname{tg} \alpha. \qquad (9.16)$$

The difference is under 1% if $\alpha < 10°$ and $c > 4$.

In conclusion, to reach a 10^{-9} sensitivity one has to measure ΔH changes of about 2×10^{-8} cm (H is about 20 cm), i. e. 0.2 nm.

If the mechanical device is so sensitive, the spring's course will be quite restricted with respect to gravity variations. To make such an instrument usable everywhere on the Earth an additional spring (zone spring) is needed to readjust the central "zero position" of the beam when the instrument is transferred to a place with a very different g value. Of course, a third very fine spring is also needed to measure with great precision the minute variations of g. It is called the "measuring spring" and in some instruments it is automatically governed to maintain the beam permanently at its zero position. In this case the method of registration is said to be a "zero method". Now such a beam is subject to a force which is equal and of opposite direction to the *variable* weight of the displaced air (Archimedes' push). To avoid the gravimeter reacting like a barometer it is absolutely necessary to compensate the mass m for buoyancy with an empty light box of convenient size fixed on the beam on the other side of the rotation point and at a correct distance from it. But such a compensation never being total, one has to enclose all the measuring device in an airtight case. This case will also be well isolated from thermic variations and the interior kept at a constant temperature (to about 0°.001) by a thermostatization. All these mechanical components are common to all gravimeters.

9.2. A STATIC SPRING GRAVIMETER: THE ASKANIA INSTRUMENT

The principle of this instrument is shown on Fig. 9.2. Two helical flat springs are used not as tension springs but as torsion springs in such a way that the mode of stressing is flexural. The initial curvature angle α of one coil is about 360°. The curvature is

$$C = \mathscr{R}^{-1} \sim R^{-1}.$$

A bending due to a change of gravity creates an angle of inclination $d\alpha$ which corresponds to a fraction $d\alpha/360°$ of a coil. The new radius of curvature R' is

$$\frac{R'}{R} = \frac{1}{1 + \frac{d\alpha}{360°}}. \qquad (9.17)$$

Then the curvature becomes

$$C' = \frac{1}{R'} = \frac{1}{R}\left(1 + \frac{d\alpha}{360°}\right) \qquad (9.18)$$

and we have

$$C'\alpha = C(\alpha+d\alpha). \qquad (9.19)$$

The equilibrium equation is

$$\text{curvature } C = \frac{\text{bending moment } K_1}{\text{Young's modulus } E \times \text{moment of inertia, } I}.$$

Putting

$$K_1 = E\,I\,C \tag{9.20}$$

one has for two springs,

$$Mg\ell = 2\,K_1\,\alpha, \tag{9.21}$$

$$M(g+dg)\ell \cos d\alpha = 2\,K_1'\alpha = 2\,K_1(\alpha+d\alpha). \tag{9.22}$$

Practically $\cos d\alpha \approx 1$, therefore one has

$$M\ell dg = 2\,K_1\,d\alpha, \tag{9.23}$$

and combining with (9.21)

$$d\alpha = \alpha\,\frac{dg}{g}. \tag{9.24}$$

The sensitivity is thus $s=\alpha$ and for a variation $dg=1$ µgal, one has

$$d\alpha = 360° \times 10^{-9} \approx 0''.0013.$$

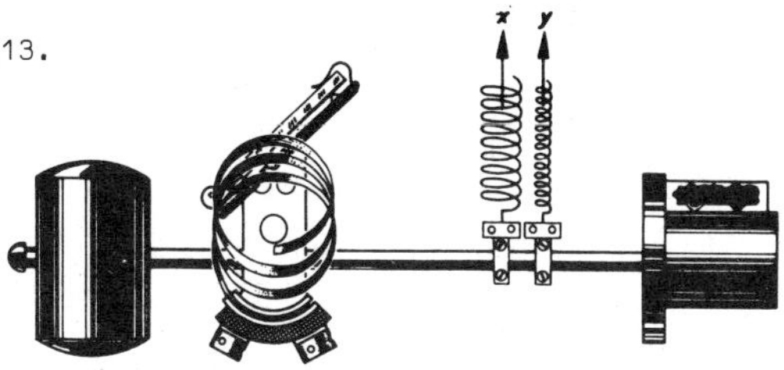

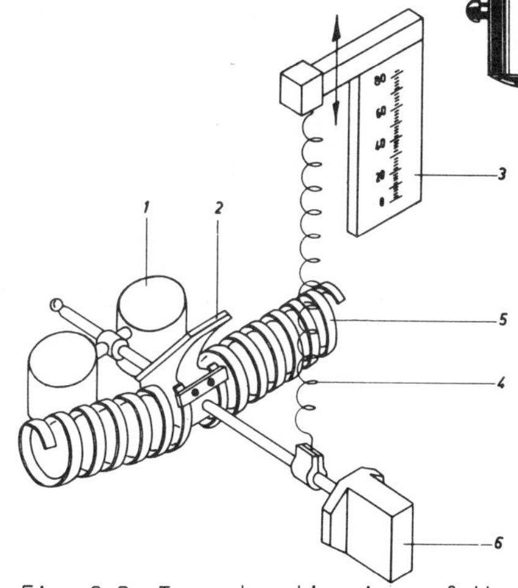

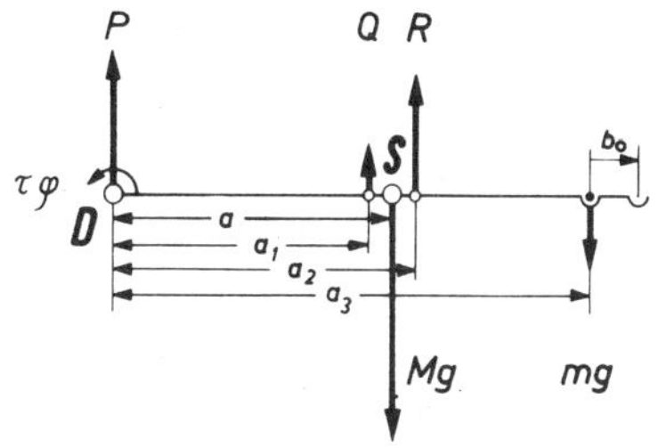

Fig. 9.2. Two schematic views of the spring suspension in an Askania static gravimeter with a diagram of forces.

1. Capsul for barometric compensation
2. Mirror
3. Micrometric scale
4. Measuring spring
5. Main springs
6. Mass

9.3. AN ASTATICIZED GRAVIMETER WITH ZERO HEIGHT SPRING: THE LACOSTE ROMBERG INSTRUMENT

The equation of equilibrium of an instrument similar to the one shown on Fig. 9.3 is

$$m\ g\ b = h\ F\ (H) \tag{9.25}$$

when the beam is strictly horizontal.

When the beam is displaced by an angle γ the resulting moment is

$$\mathcal{M}(\gamma) = m\ g\ b\cos\gamma - h(H + dH)\ F\ (H + dH) \tag{9.26}$$

which can be developed in a MacLaurin serie as follows:

$$\mathcal{M}(\gamma) = \mathcal{M}(o) + \gamma\ \mathcal{M}'\ (o) + \frac{\gamma^2}{2}\ \mathcal{M}''\ (o) + \ldots \tag{9.27}$$

with

$$\mathcal{M}(o) = m\ g\ b - h\ F = o, \tag{9.28}$$

$$\mathcal{M}'(\gamma) = -h\frac{dF}{d\gamma} - F\frac{dh}{d\gamma} - m\ g\ b\sin\gamma. \tag{9.29}$$

The geometry of the instrument gives

$$hH = ab\sin\beta, \tag{9.30}$$

$$H^2 = a^2 + b^2 - 2ab\cos\beta, \tag{9.31}$$

and one easily demonstrates that

$$\mathcal{M}''\ (\gamma) = \frac{3h}{H}\ (ab\cos\gamma - h^2)\ (\frac{F}{H} - \frac{dF}{dH}) - h^3\frac{d^2F}{dH^2}. \tag{9.32}$$

This term being multiplied by $\gamma^2/2$ in (9.27) does not change its sign with γ so that the moment is not symmetrical with respect to the zero position, and the free period of the beam is different on the zero position sides.

To eliminate such a dissymmetry one should have

$$\frac{dF}{dH} = \frac{F}{H}, \tag{9.33}$$

which means that the tension F is proportional to H and has as a consequence

$$\frac{d^2F}{dH^2} = 0 \tag{9.34}$$

and

$$H = 0 \tag{9.35}$$

when the spring is left free from any tension.

This means that when the spring is in this condition all the coils press on each other to try to fulfil the condition H = 0. When stress is applied the first result is merely to separate the coils. The importance of this property was discovered by LucienLaCoste.

It enables the property of astatization to be established.

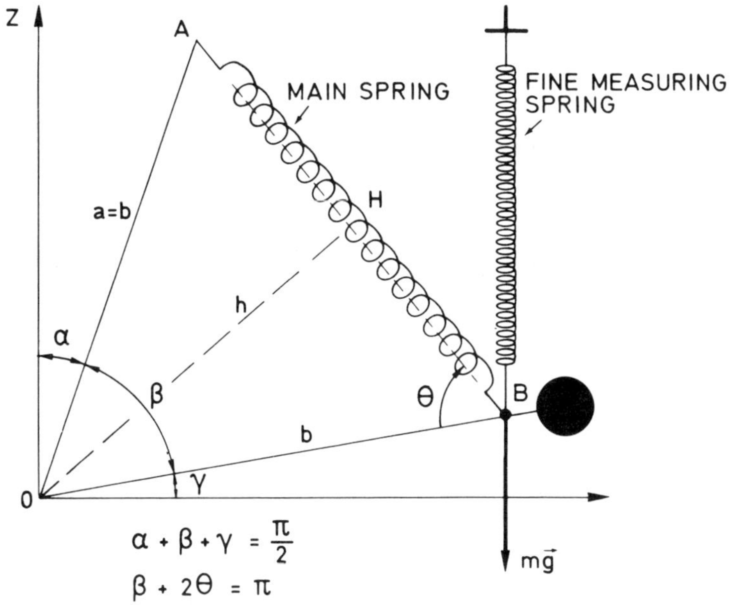

Fig. 9.3.

To construct a zero-height spring (*) it is necessary to obtain an helix with k = 0, i. e. 1/T = 0, which means that the torsion produced in the wire by reeling must be cancelled by an equal but opposite torsion given to it just before reeling the spring.

Now let us designate by H the height of the spring when no tension is applied to it and (H+x) its height when some tension is applied. Then, from (9.33),

$$F = Kx. \tag{9.36}$$

As

$$h = b \sin \theta, \tag{9.37}$$

$$K = \frac{\pi}{2} \frac{\mu r^4}{LR^2}, \tag{9.38}$$

from (9.25) and considering Fig. 9.3, the equations can be written

$$m\ g \cos \gamma = Kx \sin \theta, \tag{9.39}$$

(*) This is often called "zero-length spring" but we find it more correct to call it "zero-height spring".

$$(H + x) \sin \theta = a \sin \beta, \tag{9.40}$$

$$(H + x)^2 = a^2 + b^2 - 2ab \cos \beta, \tag{9.41}$$

and by differentiation

$$(H + x)\, dx = ab \sin \beta \, d\beta. \tag{9.42}$$

Combining these equations one obtains, if $\gamma + \beta = \pi/2$ (thus $\alpha = 0$),

$$d\beta = \frac{x(H+x)^2}{Hab \sin \beta} \frac{dg}{g}, \tag{9.43}$$

to be compared with (9.24) and which shows that the sensitivity of the instrument to variations of g

$$S = \frac{x(H + x)^2}{Hab \sin \beta} \tag{9.44}$$

becomes infinite when H = 0. This is the property of astatization. All United States made instruments are variations of the basic geometry of Fig. 9.3. Those used for continuous tidal registrations are the LaCoste Romberg and the North American (Geodynamics version) gravimeters. (Fig. 9.6).

9.4. LEVELLING OF THE GRAVIMETERS

When we look at the diagram principle of Fig. 9.3 of an astaticized gravimeter we clearly see that we have the possibility of independently changing two of the angles.

(1) The angle β, which obviously is related to a frame attached inside the box of the instrument, can be modified by turning the fine measuring spring. This operation enables the best mechanical astatization and consequently the best operational sensitivity for tidal recording to be obtained.

(2) The angle α can be modified by turning one of the three levelling screws of the gravimeter housing. By operating these screws one can find the best value to give to the α angle and adjust the meter to its minimum sensitivity to tilt.

This can be explained as follows. As we have a gravity torque

$$M_1 = m\, g\, b \cos \gamma = m\, g\, b \sin (\alpha+\beta) \tag{9.45}$$

and a spring torque

$$M_2 = Kx \sin \theta = Kx \sin \left(\frac{\pi}{2} - \frac{\beta}{2}\right) \tag{9.46}$$

which must remain in equilibrium (momentums with respect to O), we will observe in case (1)

$$\frac{\partial M_1}{\partial g} dg + \frac{\partial M_1}{\partial \gamma} d\gamma = \frac{\partial M_2}{\partial \theta} d\theta,. \tag{9.47}$$

and as

$$d\gamma = - d\beta = 2\, d\theta, \tag{9.48}$$

$$\frac{d\theta}{dg} = \frac{\partial M_1/\partial g}{\dfrac{\partial M_2}{\partial \theta} - \dfrac{\partial M_1}{\partial \theta}}. \tag{9.49}$$

In case (2)

$$\frac{\partial M_1}{\partial \alpha}\, d\alpha + \frac{\partial M_1}{\partial \beta}\, d\beta = \frac{\partial M_2}{\partial \theta}\, d\theta, \tag{9.50}$$

and as

$$d\alpha = -\, d\beta = 2\, d\theta, \tag{9.51}$$

$$\frac{d\theta}{d\alpha} = \frac{\partial M_1/\partial \alpha}{\dfrac{\partial M_2}{\partial \theta} - \dfrac{\partial M_1}{\partial \theta}} \tag{9.52}$$

Comparing (9.49) and (9.52) we obtain

$$\frac{dg}{d\alpha} = \frac{\partial M_1}{\partial \alpha} \Big/ \frac{\partial M_1}{\partial g} \tag{9.53}$$

or

$$\frac{dg}{g} = \cot g\, (\alpha+\beta)\, d\alpha. \tag{9.54}$$

This shows that any change $d\alpha$ in the levelling of the instrument has the same effect as a change of gravity dg and is amplified in the same way by the astatization property. Therefore, every month, the levelling has to be controlled carefully during the registrations. To minimize possible disturbance one has to settle the instrument at the minimum point of sensitivity to tilt (which obviously corresponds to $\alpha+\beta = \pi/2$). This position is easily determined by slightly turning the levelling screws and comparing the level-bubble readings with the position on the recorder.

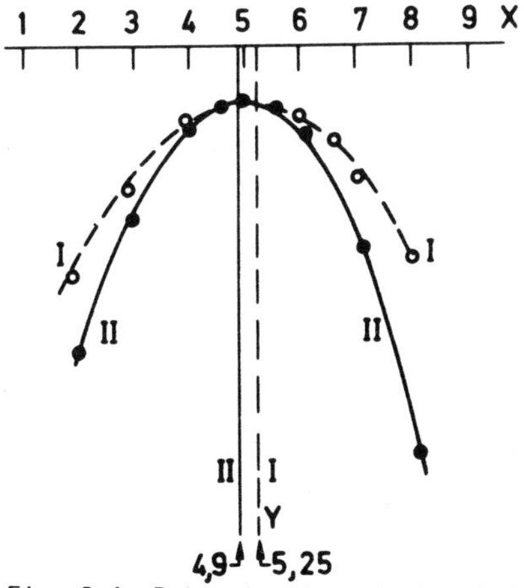

Fig. 9.4. Determination of the minimum point of sensitivity to tilt by observations of cross level (I) and longitudinal level (II).

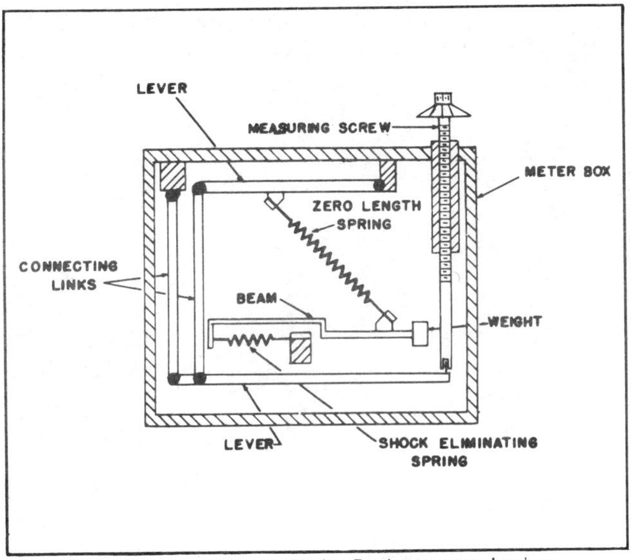

Fig. 9.5. Scheme of a LaCoste-Romberg gravimeter.

As the function has the form (9.54) one will obtain a graph similar to the one represented on Fig. 9.4. The position of the bubble, which has its centre corresponding to the summit of the curve, gives the correct setting of the gravimeter on its levelling screws because obviously it is in this position that a slight change of levelling has the minimal effect on the gravity readings.

In the practical adjustment, the angle β is determined by the position of the mobile "wire" index on one specific "reading line" in the micrometer. Then, maintaining this wire on this reading line, one adjusts the angle α by the tilting procedure described and illustrated by Fig. 9.4.

The role of this final operation is to restitute a horizontal position for the beam by changing α when β has modified (i. e. $\alpha+\beta=\pi/2$).

Tilting of the instrument can be used as a calibration procedure. Wenzel found that a tilt of 150" produces in an Askania instrument an apparent change of 300 μgals.

We can conclude by saying that much more care is needed when working with the complex modern gravity meters than with the quartz horizontal pendulums as they are so sensitive to the environmental conditions:

a) the temperature must be maintained constant with a very high precision, a variation of $0{\overset{\circ}{.}}0003$ producing a change of the order of 1 μgal;
b) the air-pressure variations produce buoyancy phenomena which are never completely compensated;
c) electrostatic effects can disturb the recording and must be eliminated by a good grounding;
d) magnetic effects may have some influence;
e) level instability disturbs the sensitivity control by tilting of the instrument.

Fig. 9.6. Photograph of the interior of a North American (Geodynamics) gravimeter.

9.5. REGISTRATION DEVICE

As the gravimeters were originally devised for field measurements they are provided with a microscope which controls optically the beam's position by observing an index fixed to it with respect to some scale. Therefore the first tentative attempt (which was a successful one) at making a recording instrument was by adapting a camera and registering on photographic film (Lecolazet, 1956).

The Askania (GS 11 and GS 12) instruments developed in 1957 for the IGY were equipped with two photocells, differentially mounted, which received a light reflected upon a mirror fixed to the beam. Any difference in the quantity of light received by these two elements causes a potential difference which produces a current passing through a galvanometer coil. A spot reflected on the galvanometer's mirror was used for the continuous registration. This system, although in use for some 10 years in many stations, was not very satisfactory because of the high noise in the photocells and mainly because of their instability with respect to calibration (change with power of light spot) (see Section 9.6). Therefore the many published results obtained with these instruments must be taken into account with caution and only when a great number of careful calibrations have been performed (which unfortunately is not always the case).

A much better system was devised some years ago by using a position-detection device by capacity variation called *capacitative transducer*. It is now used in the LaCoste Romberg, Geodynamics and Askania GS 15 gravimeters.(Fig. 9.7). A plate attached to the oscillating mass of the gravimeter moves between two fixed plates. The space X between the moving plate and the fixed one is about 1 mm, while the plates themselves are about 2cm x 2 cm, i. e. a surface $s \approx 4\ cm^2$.

The capacity of such a condensor is, in SI units,

$$C = 8.85\ \varepsilon\ \frac{S}{X} \approx 3.542\ \text{pF}, \tag{9.55}$$

ε being the dielectric constant of air (equal to 1.00059).

In the Geodynamics gravimeter this capacity is introduced in a resonant circuit (Fig. 9.7) the frequency of which is:

for the upper plate: $f_1 = \dfrac{A_1}{\sqrt{C_1}}$

and for the lower plate: $f_2 = \dfrac{A_2}{\sqrt{C_2}}$

A_1 and A_2 being adjustable constants which can be made equal.

The beating of the two frequencies gives

$$A(\sin 2\pi f_1 t + \sin 2\pi f_2 t) = 2A \sin 2\pi\left(\frac{f_1+f_2}{2}\right)t \cos 2\pi\left(\frac{f_1-f_2}{2}\right)t. \tag{9.56}$$

A demodulator eliminates the $(f_1+f_2)/2$ frequency and the low-frequency signal $(f_1-f_2)/2$ is transformed by a frequency-to-voltage converter into the voltage

$$\Delta U = B(f_1 - f_2) = AB\ \frac{1}{\sqrt{C_1}} - \frac{1}{\sqrt{C_2}}\ . \tag{9.57}$$

Any change of position ΔX of the moving plate results in a voltage variation producing an opposite variation ΔC of the capacities

$$\Delta U = \frac{AB}{C_o}\left(\frac{1}{\sqrt{1+\frac{\Delta C}{C_o}}} - \frac{1}{\sqrt{1-\frac{\Delta C}{C_o}}}\right) \tag{9.58}$$

as

$$C_1 = C_o + \Delta C, \qquad C_2 = C_o - \Delta C.$$

The development

$$\frac{1}{\sqrt{1+\frac{\Delta C}{C_o}}} = 1 - \frac{1}{2}\frac{\Delta C}{C_o} + \frac{3}{8}\left(\frac{\Delta C}{C_o}\right)^2 - \frac{15}{48}\left(\frac{\Delta C}{C_o}\right)^3 + \ldots$$

$$-\frac{1}{\sqrt{1-\frac{\Delta C}{C_o}}} = -\ 1 - \frac{1}{2}\frac{\Delta C}{C_o} - \frac{3}{8}\left(\frac{\Delta C}{C_o}\right)^2 - \frac{15}{48}\left(\frac{\Delta C}{C_o}\right)^3 - \ldots \tag{9.59}$$

gives by neglecting quantities of the third-order only:

$$\Delta U = \frac{-AB}{C_o}\frac{\Delta C}{C_o}. \tag{9.60}$$

From (9.55) we have

$$\frac{\Delta C}{C} = -\frac{\Delta X}{X}, \tag{9.61}$$

which is independent from any influence of time variations of the dielectric constant ε.

Around the mean position X, ΔU appears as a linear function of the displacement of the gravimeter's beam ΔX for $\Delta C/C$ is very small,

$$\Delta U = \frac{AB}{C_o}\frac{\Delta X}{X}. \tag{9.62}$$

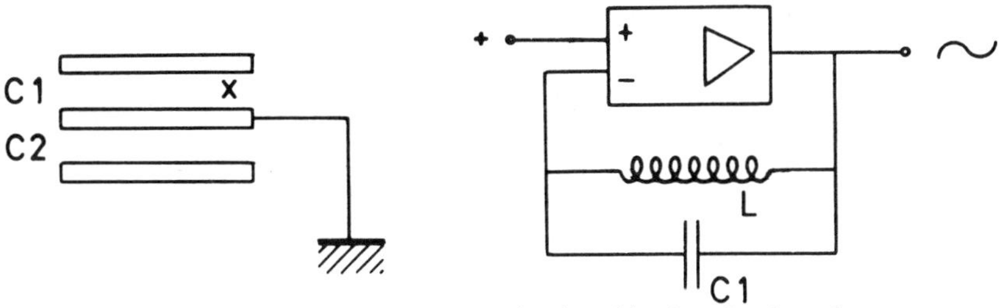

Fig. 9.7. The capacity in the resonant circuit of a Geodynamics gravimeter.

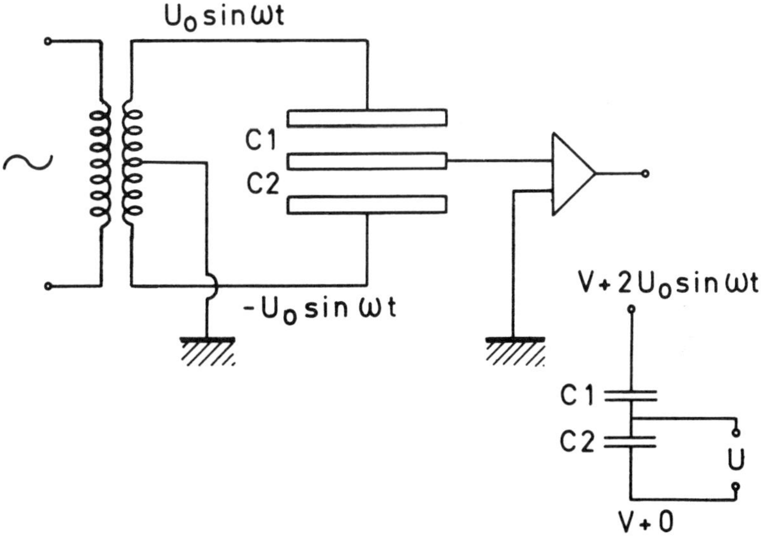

Fig. 9.8. Capacitive system as used in the Askania gravimeters.

In the system represented on Fig. 9.8, which is used in the Askania gravimeters, a modulation U_0 sin ωt opposite phase is induced on the fixed plates so that the output of the system is

$$U = V + 2U_0 \sin \omega t \frac{C_1}{C_1+C_2} \qquad (9.63)$$

with

$$V = - U_0 \sin \omega t.$$

As

$$C_1 = C_0 + \Delta C, \qquad C_2 = C_0 - \Delta C,$$

it follows that the bridge gives an amplitude $V + (\frac{C_1}{C_1+C_2})$ U with

$$U = U_0 \sin \omega t \frac{\Delta C}{C_0},$$

and again an amplitude modulated signal

$$U = - U_0 \frac{\Delta X}{X} \sin \omega t. \qquad (9.64)$$

It is a linear function of the displacement ΔX of the gravimeter's beam. In the first system the stability mainly depends upon the self (which therefore must be carefully thermostatized) but also upon the reference tension used in the demodulator electronics, while in the second system much depends upon the stability of the alimentation U_0 upon the transformer ratio and quality of the demodulator.

9.6. THE CALIBRATION OF GRAVIMETERS

The calibration of the recording paper is of primary importance in this research, and this seems not always to have been clearly understood, at least at the start of operations. The difficulty arises from the fact that the sensitivity of the system varies with time in a way which is by no means linear, and it is appropriate, therefore, to repeat the calibration operations practically every week while taking care not to perturb fundamentally the tide curve. One should be careful to choose the least disturbed periods (absence of micro-seisms) and also periods when the tide curve is the flattest, i. e. the phases of quadrature of the Moon and the Sun.

However, the only way to correctly calibrate an instrument is to produce *artificially* the phenomenon the instrument has to measure. Evidently one has to be able to evaluate with high precision the artificial effect.

Because we already have in the vertical component the very important gravity force - the corresponding instrument - the gravimeter is fundamentally different from the instruments used to measure the horizontal components. The calibration problems will also be very different.

The calibration of gravimeters in the range of fractions of a milligal is a difficult task because it is not easy to produce a precisely known artificial variation of g. Two methods are possible.

The first one is to change the height of the instrument by 1 m or so (that is nearly 0.3 mgal, i. e. about the size of the tidal effect). This has been done by different experimentalists using an ascending platform (Bonatz, 1964; Pariisky, unpublished). The objection against this procedure is that we do not accurately know the local gradient of gravity. It has to be measured independently (Bonatz, 1970).

The second method is to change the gravity acting on the instrument beam by the very near approach of a big mass (lead or mercury) (Groten, 1970). Unfortunately the moving weight produces at the same time a non-negligible flexion of the ground so that the correct operation would be to put the mass alternately over and beneath the instrument. This does not seem to have been worked out until now. Practically the most widely used method is to apply to the gravimeter a very stable force which has been precisely compared to the gravitational forces.

Tidal gravimeters are slightly modified field instruments still equipped with a fine measuring spring and a micrometer carefully calibrated along gravity bases. In the case of Askania gravimeters this can also be done by means of a ball-bearing device.

Once such a fundamental operation is carried out, this calibration will be used as a basis for the future calibration of recording paper in the following way: during some hours the spring will have to be operated about every 20 min. by turning the micrometric screw alternatively in the direction of increasing and decreasing values from 20 to 30 divisions in the case of LaCoste Romberg Model G instruments. Care is to be taken for the backlash effect in the micrometric screw when unscrewing it. Therefore one has to unscrew by about 5 divisions in excess to readjust the final reading to its desired value by a screwing operation. The micrometric measurement is then compared to the displacement of the pen on paper (Fig. 9.9) and the calibration coefficient, expressed in microgals per centimetre, can easily be deduced from each displacement. For LaCoste Romberg gravimeters this coefficient is usually about 1 5 μgal cm^{-1}.

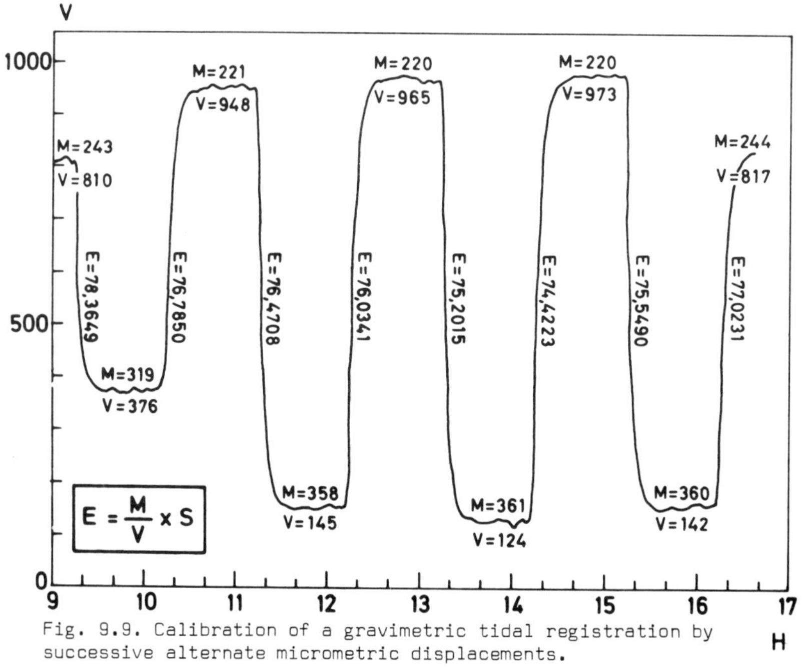

Fig. 9.9. Calibration of a gravimetric tidal registration by successive alternate micrometric displacements.

Two of the Askania GS 11 gravimeters used at the Brussels Observatory and the Walferdange Laboratory have been equipped with coils in order to enable automatic control of the sensitivity stability by electromagnetic attraction. A simple device with a weekly programme produces regular alternative displacements as shown on Fig. 9.9.

In the case of Geodynamics gravimeters (which are North American gravimeters conveniently transformed into recording instruments) the sensitivity can be checked regularly, and it is fairly constant versus time. The sensitivity is controlled by an artificial displacement of the gravimeter beam produced by the electrostatic deflexion of the mass when a very constant tension (50 V) is applied to a calibration plate mounted over the beam so as to produce an apparent variation of g of 100 μgal. A similar system had already been devised before by Lecolazet and Steinmetz in a North American gravimeter by using a condenser formed with two plates surrounding the mass and applying a potential difference of 550 V. However, this is not related in any way to the gravity unit, the milligal, and does not lead to a real "calibration" method but to a check of sensitivity stability with time.

To overcome these difficulties the easiest way is to install the instrument to be calibrated at a location where the tidal constants are already well known, having been derived with high accuracy from many different instruments.

Not only has the amplitude to be calibrated but also the instrumental phase-lag which is a function of the tidal frequency as we will now demonstrate.

Since the IGY (1957-8) the gravity tides have been recorded without interruption in the underground rooms of the Royal Observatory of Belgium at Brussels-Uccle. Three Askania gravimeters have recorded for more than 600 days each. The station now serves as the base station to link the tidal gravity measurements made in the United States, Europe (Kuo et al, 1972), Asia and Australia. Subsequently the station has been serviced by many instruments including Askania GS 15 and GS 11 rebuilt with capacitative transducers, Geodynamics, and LaCoste Romberg.

Therefore Brussels is also the fundamental station for the Trans-World Tidal Gravity profile undertaken in Asia and the Pacific in October 1973.

Comparing these observations enables us not only to determine the most probable tidal parameters at Brussels more accurately but also to evaluate the systematic instrumental characteristics.

The response of the instruments under the action of the tidal forces may be expressed in terms of their elastoviscous properties. When several springs or torsion wires are working the prediction of the global response becomes complicated.

One may determine the elastic system properties by measuring their response to a step function or to the input of sinusoidal waves from which Ducarme has constructed a rheological model for each of the instruments having registered at Brussels. For each gravimeter one must consider the complete system including the mechanical element, the filtering or the output by means of galvanometers or electronic output and the recording system.

Many authors (Ditchko, Korba and Wirth) pointed out that the galvanometric recording system of GS 11 and GS 12 instruments introduces a large phase-lag. Therefore, in determining the mechanical phase-lag of the gravimeter itself one has first to calculate the phase-lag of the galvanometer and thereafter determine the phase-lag of the complete system: gravimeter plus recording system.

The damping ratio β of the galvanometer depends upon the critical resistance Rc and the external resistance Rext according to

$$\beta = \frac{R_1 + R_c}{R_1 + R_e}, \qquad (9.65)$$

where R_1 is the internal resistance and R_e is the external resistance of built in galvanometer and photocells; it can vary between 2.4 and 1.3 kΩ.

From these characteristics the recording system phase-lag values for the tidal wave M_2 of period T=745 min are given by

$$b_1 = -\text{arc tan}\,\frac{2\beta T_o}{T} = -\text{ arc tan}\,\frac{2\pi\xi_1}{T}. \tag{9.66}$$

The global phase-lag is determined from the response of the whole system to a step function as well as to forced waves introduced through the electromagnetic calibration system (which exists on ASK 145 and 160).

The response of an overdamped electromechanical system to a step input function is

$$x(t) - x(0) = A(1 - e^{-t/\xi_2}). \tag{9.67}$$

The value of the retardation time ξ is thus obtained by fitting the best exponential function to the experimental decay curve

$$\xi = \beta/\pi. \tag{9.68}$$

On the other hand, the response of the system to a tidal wave of period T is characterized by a phase-lag and an amplitude reduction coefficient B (attenuation) as follows:

$$b_2 = -\text{ arc tan } 2\pi\,\xi_2/T, \tag{9.69}$$

$$B = 1/\sqrt{1 + (2\pi\xi_2/T)^2}. \tag{9.70}$$

The numerical results of the global phase-lag for various Askania gravimeters are given in Table 9.1. The agreement between formula (9.66) and Table 9.1 was found to be excellent, which means that the internal phase-lag of Askania gravimeters does not exceed 0.1° for the semi-diurnal waves and can be neglected in tidal analysis. The attenuation of the waves is thus consequently negligible, certainly less than 0.1%.

TABLE 9.1. Global phase-lag b_2 deduced from step function compared with the phase-lag b_1 deduced from (9.65) and (9.66).

Galvanometric type	Instrument	ξ_2 (min)	$b_1(M_2)$	$b_2(M_2)$
A	G 145 A	5.1	2.4°-2.7°	2.46°
	G 143	5.0	2.4°-2.7°	2.41°
B	G 160	3.3	1.4°-1.6°	1.61°
D	G 98	3.7	-	1.79°
	G 175	3.6	-	1.74°
	Global Phase-lag deduced from forced waves			
C	G 145		1.30°-1.35°	1.35°

Volkov and Zasimov (1976) have measured the instrumental phase-lags of the Askania GS 15 which, as we have seen, has a strong filtering, and found values as large as 1.85° for M_2 and 0.89° for O_1.

The Phase-Lag of Astaticized Gravimeters

Astaticized gravimeters exhibit very complex elastic characteristics. The retardation times of the visco-elastic elements of the system are considerably lengthened when the free period of the instrument is increasing. It is well known (Steinmetz, 1960) that the response of North American gravimeters to the application of a step function is a sum of exponential terms of the form

$$x(t) - x(0) = A(1+\varepsilon_1(1 - e^{-t/\theta_1}) + \varepsilon_2(1 - e^{-t/\theta_2}) + \dots \quad (9.71)$$

As an example Steinmetz found

	ε_1	ξ_1	ε_2	ξ_2
NA 167	0.0129	10 min	0.0191	170 min
NA 138	0.0140	12 min	0.0239	130 min

This is equivalent to a rheological model combining one Hooke body and two Kelvin bodies with respective contributions to the total displacement of A, $\varepsilon_1 A$ and $\varepsilon_2 A$.

The phase-lag of such a system is

$$\tan b = -\varepsilon_2 \frac{n(\xi_2 + \frac{\varepsilon_1}{\varepsilon_2}\xi_1)}{1 + \frac{1+\varepsilon_1}{\varepsilon_2}(1 + n^2\xi_2{}^2)}, \quad (9.72)$$

n being the pulsation of the tidal wave.

It is very important to point out here that the phase-lag of the instrument is no longer proportional to the angular speed of the tidal constituent.

For the first approximation, if we neglect the effect of the shorter retardation time ξ_1, the results of the amplitude factors are modified by less than 10%. Thus a realistic rheological model can be constructed for astaticized gravimeters combining one Hooke element with only the longer period Kelvin body associated with the retardation time θ_2.

9.7. DEFINITION OF THE BRUSSELS FUNDAMENTAL TIDAL STATION

From the previous consideration we decide to use only Askania gravimeters to define the mean tidal parameters at Brussels (Table 9.2). Furthermore, the diurnal wave O_1 was chosen to normalize the tidal amplitudes because its coherency is better than that for the M_2 wave. The K_1 component is not convenient for normalization purpose because one year of registration is needed to separate it from the S_1 and P_1 components. As all the errors nearly have the same magnitude, we take the arithmetic mean

$$\delta(O_1) = 1.161 \pm 0.003.$$

TABLE 9.2. Brussels Tidal Gravimetric Factors (O_1 wave).

Instrument	Number of days	Amplitude factor	Phase lag
Ask. 11 n°145A	1438	1.1622 ± 0.0089	0°.18 ± 0°.44
Ask. 11 n°145	322	1.1575 ± 0.0100	0°.09 ± 0°.49
Ask. 11 n°160	876	1.1579 ± 0.0095	0°.25 ± 0°.47
Ask. 11 n°191	518	1.1643 ± 0.0065	0°.32 ± 0°.32
Ask. 15 n°206	94	1.1628 ± 0.0079	-0°.59 ± 0°.39
Molodensky model 1:		1.160	0°

The scattering of the ratio $c = \delta(M_2)/\delta(O_1)$ is more important. Taking into account the results of other gravimeters a reasonable mean value seems to be

$$C = 1.027 \pm 0.04.$$

For the phase normalization the more suitable component is M_2 owing to its greater angular speed. Taking into account the instrumental phase-lag corrections we find the phase-lag values given in Table 9.1. The weighted mean gives for all Askania GS 11 instruments a phase advance

$$\kappa(M_2) = 2.90° \pm 0.25°$$

while for new Askania GS 15 and transformed GS 11 it is

$$\kappa(M_2) = 2.69° \pm 0.15°.$$

The phase-lag of the O_1 wave is small but not so well defined. As the indirect effects caused by oceanic tides on the coasts of Europe are very small for the diurnal components, we can expect very little phase shift. Old Askania gravimeters generally give a weak positive phase while the new ones giving slightly negative values. The arithmetic mean between the two series gives

$$\kappa(O_1) = -0.16°.$$

From comparisons performed in other stations of the European profile (Walferdange, Strasbourg) between static and astaticized instruments we believe that a slight negative value is realistic.
From these considerations the Brussels Fundamental Tidal Station is defined by the following parameters:

$$\delta(O_1) = 1.161$$
$$\alpha(O_1) = -0.20°$$
$$\alpha(M_2) = +2.80°$$

Intercomparisons made with different instruments at Hannover, Sèvres, Liverpool and New York have shown that this system was very consistent indeed (Ducarme, 1975). As an example, the instrumental phase-lag of the gravimeter Geodynamics 721 was determined at New York by Jachens as $+0\overset{\circ}{.}79$, then at Brussels by Ducarme as $+0\overset{\circ}{.}89$ and finally at Liverpool by Baker and Lennon as $+0\overset{\circ}{.}9$ (comparison with a zero-method instrument).

9.8. CHOICE OF A RHEOLOGICAL MODEL FOR ASTATICIZED GRAVIMETERS

The simplest model one can adopt for astaticized instruments of free period T_0 is a two-parameter one. It is completely determined if we know the value of ε which gives the contribution of the Kelvin body to the total displacement and the parameter $\tau = 2\pi\theta$ where θ is the retardation time. If the damping ratio β is large, we can neglect the terms containing $(T_0/T)^2$ with respect to $(\tau/T)^2$.

Then the phase-lag b and the attenuation factor B corresponding to each tidal component of period T are given by the formulas

$$\tan b = -\frac{\alpha}{1 + \dfrac{1+\alpha^2}{\varepsilon}} = -\frac{\alpha}{1+\varepsilon+\alpha^2}, \tag{9.73}$$

$$B = \sqrt{\frac{1 + \dfrac{\alpha}{(1+\varepsilon)^2}}{1+\alpha^2}} \tag{9.74}$$

where $\alpha = \tau/T = 2\beta(T_0/T)$.

The parameters ε and τ can be easily determined from the values adopted for $\kappa(O_1)$ and $\kappa(M_2)$ at the fundamental station.

It is important to point out that the ratio f of $\kappa(O_1)$ over $\kappa(M_2)$ cannot take any arbitrary value.

From the relation

$$\kappa^2(O_1) = (1+\varepsilon)\,\frac{f - 1/f1}{f_1 - f}, \tag{9.75}$$

where $f_1 = 2.078\ 75$ is the frequency ratio between M_2 and O_1, we immediately derive the condition

$$1/f_1 < f < f_1. \tag{9.76}$$

For a simple Kelvin body ($\varepsilon \to \infty$) we get

$$\left.\begin{aligned} f &= 1/f_1 = 0.481, \\ \tan b &= \alpha, \\ B &= 1/\sqrt{1+\alpha^2}. \end{aligned}\right\} \tag{9.77}$$

For a simple Hooke body (ε=0), without any phase-lag

$$\left.\begin{aligned} f &= f_1 = 2.078, \\ \tan b &= 0, \\ B &= 1. \end{aligned}\right\} \tag{9.78}$$

Ducarme (1975) has given the parameters of the rheological models computed for all the astaticized gravimeters from their phase characteristics measured at Brussels. He deduced the relative amplitude correction

$$B_1 = \frac{B(O_1)}{B(M_2)}$$

to be applied on the $\delta(M_2)$ factor. After having applied this correction (Table 9.3) he finds as the mean value of the $\delta(M_2)/\delta(O_1)$ ratio for astaticized gravimeters

$$C = 1.029 \pm 0.005,$$

which is in agreement with static gravimeters.

TABLE 9.3. $\delta(M_2)/\delta(O_1)$ Ratio for Astaticized Gravimeters at Brussels

	Raw	Correction	Corrected	
GEO 74/1	1.0160	1.0179	1.0342	1.0325
74/2	1.0093	1.0212	1.0308	
GEO 84	1.0197	1.0097	1.0297	
GEO 151	1.0048	1.0160	1.0209	
GEO 721	1.0138	1.0092	1.0231	
GEO 730	1.0123	1.0181	1.0307	
GEO 761	1.0093	1.0166	1.0262	
GEO 765	1.0213	1.0092	1.0307	
GEO 804/1	1.0155	1.0071	1.0227	1.0275
GEO 804/2	1.0266	1.0055	1.0322	
LCR 3	1.0116	1.0213	1.0332	
LCR 258	(1.0236)	1.0167	(1.0407)	
LCR 298	1.0201	1.0043	1.0245	
LCR 305	1.0154	1.0204	1.0361	
LCR 336	0.9814	1.0568	1.0362	

From this point of view the gravimeter LaCoste Romberg 336 is a very interesting instrument. Without phase-lag correction we obtained at Brussels

$$\kappa(M_2) = -1.10°, \quad \kappa(O_1) = -5.30°, \quad C = 0.9814.$$

Applying the instrumental phase-lag corrections in order to obtain

$$\kappa(M_2) = 2.80° \text{ and } \kappa(O_1) = -0.2°$$

we get

$$C = 1.036,$$

which is now in good agreement with all other astaticized as well as static gravimeters at Brussels.

9.9. A SUPERCONDUCTING GRAVIMETER (PROTHERO AND GOODKIND, 1968, 1972)

A new type of gravimeter known as a superconducting gravimeter has been developed in the United States by Prothero and Goodkind (1968). The instrument uses the near-perfect stability of superconducting persistent currents, making it a device of exceptional stability. It consists of a 2.54 cm superconducting sphere levitated in the magnetic field of a pair of current-carrying coils at liquid helium temperature. The sphere is an aluminium shell with a 1 mm wall thickness, plated with lead (about 0.025 mm). The coils produce a small force gradient in the vertical direction. Its position is detected by a capacitance displacement transducer and nullified by a feedback electronic system. The variations of the force on the sphere are observed by measuring the feedback voltage required to maintain the ball at the capacitance bridge null point.

Two techniques for measuring the position of the ball are used. The magnetic flux through a 1.57 cm diameter ring below the ball is measured, and the capacitance between the top plate and centre ring is compared to the capacitance between the bottom plate and centre ring in a bridge. The simultaneous use of these two devices makes it possible to distinguish between changes in the magnetic field which supports the ball and changes in other forces on it. This may be accomplished by operating the capacitance network with negative feedback on the ball so that its position relative to the flux detector remains fixed. Changes in the magnetic field will then produce a signal on both the capacitance network and the flux detector while changes in the gravitational force produce a signal only on the capacitance network. Thus a self-check on the stability of the magnetic "spring" is possible. Present measurements set an upper limit on magnetic support field changes of 1 part in 10^{10} per h.

The gravimeter, as shown in Fig. 9.10 is surrounded by a superconducting shield of lead-plate on a copper can. The entire system, including the superconducting ball, capacitor plates, superconducting magnet coils and superconducting shield are suspended in a vacuum and its temperature is regulated to a few μk. In this case the instrument noise is reduced to about 10^{-11} g while the instrument is insensitive to both temperature and pressure. No offsets occur during transfers. The signal-to-noise ratio is limited by variations in gravity resulting from the fluctuations in atmospheric pressure: strong correlations exist between 0.1-0.7 cycles per day and 4-7 cycles per day.

A strong correlation with the ocean-level fluctuations is also present at 3.86 cycles per day which is the M_4 frequency: this tidal wave is a result of the non-linear response of the shallow coastal waters.

An important source of systematic errors in the data is variations in the temperature of the helium bath. Care has been taken to eliminate from the instrument all materials that have a temperature-dependent magnetic susceptibility, since these materials could cause changes in the levitating force. However, there remains a temperature effect due to the temperature dependence of the slight penetration of magnetic field into the bulk of the superconducting wire of the support coils. This effect results in an apparent acceleration of 10 μgal for a helium bath temperature change of 0.001°K. Temperature variations are measured with a resolution of 10^{-6}°K by a germanium resistance thermometer mounted on the copper ball support coil form. These variations are reduced by a feedback system which controls the venting rate of the evaporating liquid helium by a servo-controlled needle valve. Thus the pressure and, consequently, the temperature of the helium bath are altered in response to the value of the germanium resistance thermometer, and regulation to better than 5×10^{-5}°K is obtained. The error signal is also measured to provide a further correction to the data.

To calibrate the instrument, a hollow steel sphere, mercury filled, is used (diameter 38 cm; weight 350 kg) being rolled at a distance 50 cm under the levitated gravimeter ball every 15 min producing a square wave of 10 μgal amplitude.

The instrument has been used at La Jolla and Pinon Flat, California (see results in Chapters 11 and 13). It has practically no drift and the quality of the curve is impressive.

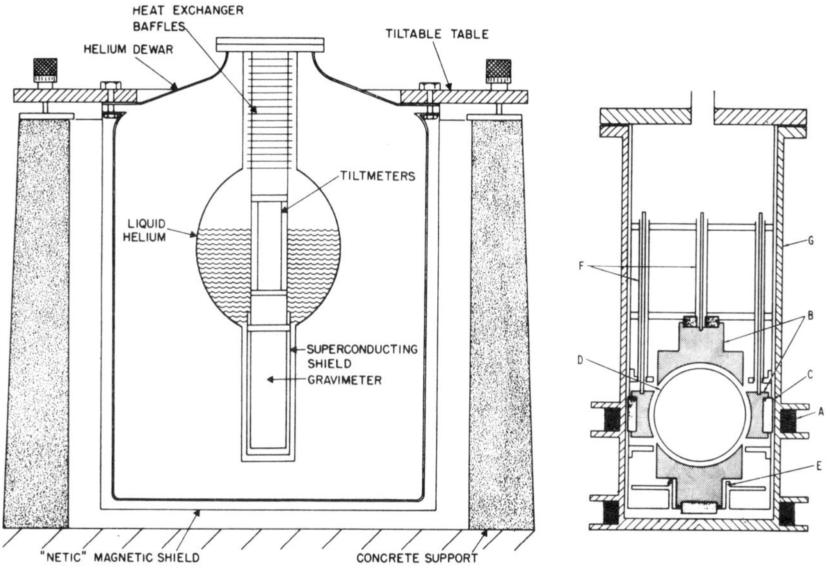

Fig. 9.10. Diagrams of the superconducting gravimeter
left: Diagram of the major components. The tiltable table is approximately 2 m above the floor.
right: Diagram of the portion of the gravimeter immersed in liquid helium: A, support coil; B, capacitance sensing plates; C, quartz insulator; D, ball; E, flux detector loop; F, capacitance plate leads; G, vacuum can.

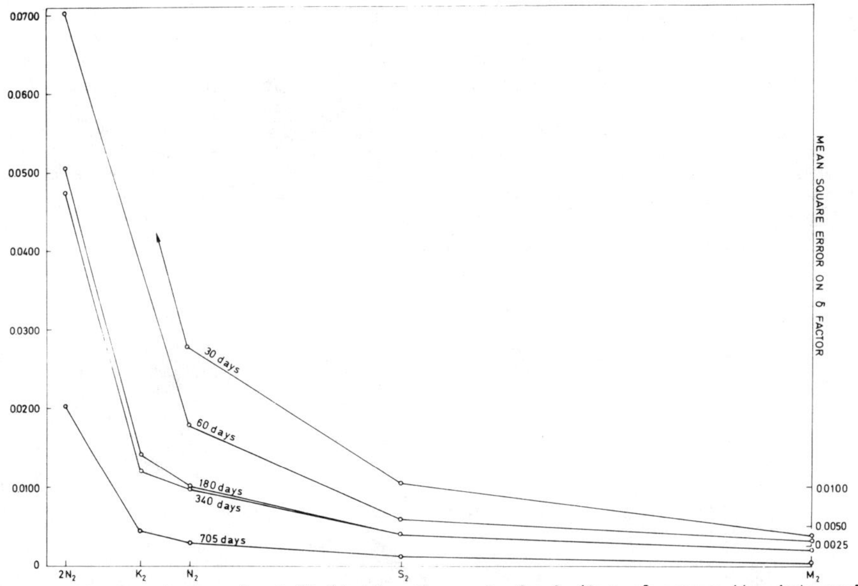

Fig. 9.11. Gravimeter Geod.84-Station Brussels-Semi diurnal waves-the internal mean square error on the δ factor as a function of the number of registration days analysed. The lines approximate the equilater hyperbola y=m/x, y being the mean square error for one tidal wave of amplitude x.

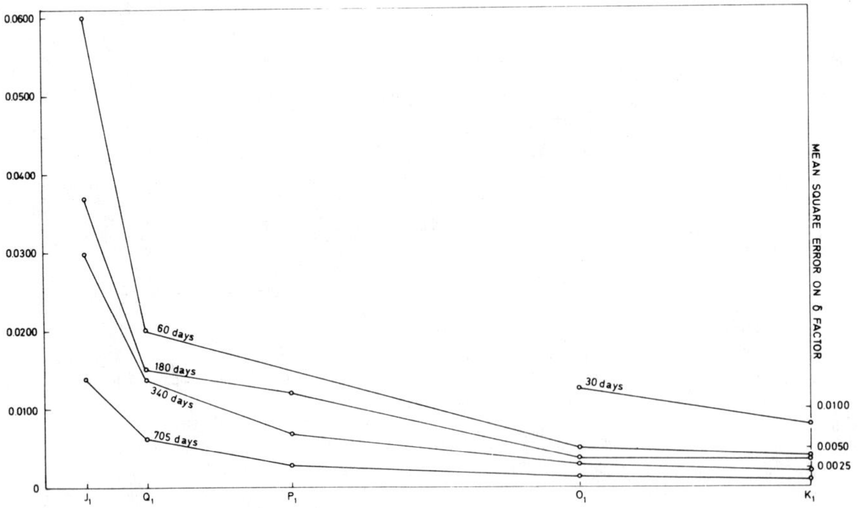

Fig. 9.12. Gravimeter Geod.84-Station Brussels-: same curve as in fig.9.11 but for the diurnal waves.

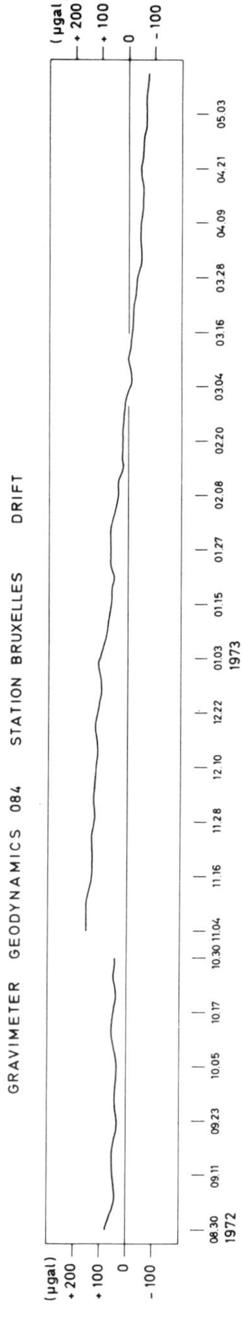

Fig. 9.13. Drift of the gravimeter Geodynamics 084 at the Brussels station.

9.10. RESULTS OF GRAVITY REGISTRATIONS ANALYSIS

9.10.1. Internal Accuracy

This accuracy will evidently depend upon the length of the available record and the ability of the method to eliminate noise principally of a long period or polynomial form. It depends evidently also on the precision of readings on tidal curves and thus of the quality of the curve and the sharpness of time marks.

Figures 9.11 and 9.12 demonstrate how the internal mean-square error of the amplitude factor δ (vertical component) diminishes as a function of increases of the recording duration.

If the improvement is slight for wave M_2, it is important for all the diurnal waves as the meteorological perturbations are better and better separated.

For a determination of waves of minor amplitudes such as P_1, Q_1, J_1, N_2, K_2 ..., several years records are clearly needed to obtain satisfactory results.

9.10.2. External Accuracy

External accuracy depends upon the existence of systematic errors in calibration, timing, compensation for barometric and temperature influences and non-linearity of recorders. Obviously most of the remarks made in the case of tilt measurements (§ 8.13.2) should be repeated here and we refer to these as well as to Chapter 13 where a comprehensive discussion of the results is given.

As we have seen before, the calibration of the instruments in amplitude and in phase is a really difficult operation. As an example of the present performances of the best instruments presently available we reproduce in the table 9.4 the results obtained at Liverpool by Baker and Lennon (1976) who obtained seven months simultaneous observations with an Earth Tides LaCoste Romberg gravimeter and with a Geodynamics gravimeter.

TABLE 9.4. Simultaneous observations of tidal variations of gravity covering seven months

The amplitudes of the main waves are given in microgals (Baker and Lennon 1976)

Waves	M_2		S_2		O_1		K_1	
Geodynamics 721	31.01	6.7°	14.64	6.2°	33.91	3.8°	49.01	4.3°
LaCoste Romberg ET 13	30.95	5.8°	14.59	5.7°	33.72	2.9°	48.45	3.3°
Difference	+0.2%	0.9°	+0.3%	0.5°	+0.6%	0.9°	+1.1%	1.0°

The Geodynamics 721 is one of the instruments used by Kuo, Melchior and Ducarme (1976) for the European profiles.

In some cases we have observed, after transportation of the gravimeter, unexpected jumps in the sensitivity which are difficult to explain. The relative positions of the plates of the capacitative transducers are suspected of having changed.

Another important check concerns the amplitude of the spurious S_1 component. Table 9.5 gives a number of examples which obviously show that the old version of the Askania gravimeters was rather unsatisfactory.

During recent years Bonatz has transformed quite a number of old Askania gravimeters (GS 11 and GS 12) by substituting a capacitative transducer system for the old photocells equipment. The results so far obtained are very encouraging, and many of these instruments have begun a new career. One of these modified instruments (BN09) installed in the underground laboratory for Geodynamics at Walferdange has been working since 1976 *without any heating*, taking advantage of the very constant temperature of the site (11°C). The first series of 180 days of registration has been analysed and shows (Table 9.5) a very low sensitivity indeed to the local meteorological perturbing effects. More data of course are needed before any definitive conclusion can be drawn in this respect.

9.10.3. First General Comments on the results

As for the clinometers, we will postpone any detailed geophysical discussion until (Chapter 13) after having discussed the possible regional and local systematic effects. However it is interesting to look now at the main characteristics of the results.

Diurnal Waves

The K_1 and O_1 main waves generally exhibit the behaviour due to the liquid core dynamical effect as predicted by Molodensky. The P_1 wave is not so often well determined as the gravimeters are practically always installed in shallow sites, and being more easily transportable than the clinometers they are transferred quite often from one station to the other for performing profiles. Of course the P_1 wave cannot be obtained when the duration of the observations does not reach 6 months.

By correcting the results in function of the parameters of an adjusted rheological model (Section 9.8) one obtains for the amplitude factors δ quite homogeneous results upon large continental areas while the phases are usually very near to zero.

Semi-Diurnal Waves

The results obtained with the best available instruments in Europe and the United States show quite homogeneous results with a marked indirect effect near the sea-shores which diminishes but never completely vanishes in the middle of the continents.

Ter-Diurnal Wave

As in the case of the quartz pendulums the M_3 wave is quite easily obtained.

TABLE 9.5. The Observed S_1 component with gravimeters

Theoretical amplitude: 0.35 δ sin 2ϕ µgal

A: Askania G: Geodynamics L: LaCoste Romberg

BN: Bonatz NA: North American ET: Earth Tide LaCoste Romberg

Station	Instrument	N	Observed amplitude (µg)	Phase (degrees)
Brussels	A 145	1438	0.66	30
	A 160	876	1.47	30
	A 191	518	0.96	31
	G 721	404	0.24	4
	G 84	782	0.63	-55
Cointe (Liège)	A 175	342	1.64	-43
Luxembourg	A 160	2086	5.09	21
Walferdange	BN 1	524	0.75	0
Walferdange	*BN 9**	*180*	*0.22*	*-8*
Sèvres	Sakuma	582	0.38	31
Frankfurt	L 98	1268	0.25	31
	NA 140	1176	0.30	47
Bonn	A 116	530	1.15	119
Hannover	A 130	312	0.88	115
	L 260	216	0.16	-21
	L 298	438	0.27	-47
Genova	A 97	702	1.21	118
Resina	A 141	874	3.87	-35
Trieste	A 108	338	1.49	-32
Stockholm	A 187	1254	2.22	37
Helsinki	A 168	1154	0.92	60
Godhavn (Greenland)	G 730	510	0.69	-27
Kerguelen	A 206	372	0.25	238
Fairfax/USA	G 804	364	0.41	278
Devils Lake/USA	G 718	174	2.00	290
Pittsburgh 1	L 8	140	1.24	-20
Pittsburgh 2	L 8	208	0.40	2
Ottawa 1	ET 12	1002	0.39	-35
Ottawa 2	G 786	566	0.60	20

**This instrument is working without thermostatization, being kept at the temperature of the mine (about 11°C).*

N is the number of days of observation

9.11. A CORRECTION FOR INERTIA

Parysky indicated the need of such a correction in 1961. In the case of gravimetric variations one has to take into account an additional spurious acceleration due to the up-and-down tidal movement of the ground supporting the instrument.

This movement was already defined as

$$\zeta = h\frac{W_2}{g} = h\frac{D}{g}\sum A_i \cos(\omega_i t + \alpha_i) \qquad (9.79)$$

and therefore the acceleration

$$\frac{d^2\zeta}{dt^2} = \sum \omega_i^2\, h\, \frac{D}{g} A_i \cos(\omega_i t + \alpha_i) \qquad (9.80)$$

combines with the tidal variation of gravity

$$\Delta g = -\frac{2\delta}{r} D \sum A_i \cos(\omega_i t + \alpha_i), \qquad (9.81)$$

and what we really measure is

$$\Delta g_{obs} = \Delta g - \frac{d^2\zeta}{dt^2}. \qquad (9.82)$$

Thus, for any tidal wave, the observed amplitude factor is

$$\delta + \frac{\omega_i^2 hr}{2g}, \qquad (9.83)$$

and to obtain the true δ value we must subtract the "inertia correction" $\omega_i^2 hr/2g$ which, for h=0.60 has the following numerical value:

$$\left.\begin{array}{llll} M_2 & 0{,}0038, & O_1 & 0{,}0009 \\ S_2 & 0{,}0041 & K_1 & 0{,}0010 \end{array}\right\} \qquad (9.84)$$

This correction is automatically applied in the computer's programs prepared at the International Centre of Earth Tides.

CHAPTER 10

Extensometry, Instruments, Results

10.1. HORIZONTAL COMPONENTS OF THE STRAIN TENSOR

The experimental investigations on strain-tensor components started in 1951 when Sassa, Ozawa and Yoshikawa published the first numerical results obtained with a superinvar wire, 20 m long. This instrument was extremely simple (Fig. 10.1). A wire (diameter 1.6 mm) is fixed at both its extremities to artificial pillars. A weight of 350 g suspended in the middle moves up and down when the deformations of the crust produce variations in the tension of the wire. This up-and-down movement of the weight is transformed into the rotation of a mirror through a bifilar suspension consisting of two superinvar wires 0.05 mm in diameter, under torque (Fig. 10.2).

Calibration is carried out by displacing one of the attachment points by means of a micrometer screw. For a focal distance of 2 m the sensitivity is 10^{-8} of the length per millimetre.

Other more recent instruments of this type (King-Bilham, Sydenham) use a capacitative transducer, the moving plate being connected to one of the wire's extremities through a mechanical device.

Another class of horizontal strainmeters uses rigid bars (superinvar*, zerodur**) or tubes (quartz) fixed at one extremity in the rock, and suspended with catenaries over a length of 10 - 30 m. In that case the sensor placed at the other "free" extremity can be either a displacement transducer or a horizontal pendulum. In that case there is a contribution from the pure tidal tilt of the beam, but this can be estimated correctly and taken into account in the analysis of the data.

The problem of the harmonic analysis of the recordings from extensometers raises special difficulties. Of course, no theoretical deformations exist for a rigid indeformable Earth, and consequently an "amplitude ratio" cannot be obtained in the same way as for the deviations of the vertical or the variations of gravity.

A convenient procedure is to put a fictitious theoretical amplitude equal to unity. Then the ratio taken to it directly provides the suitable combinations of Love numbers.

The Ozawa Extensometers

Since 1960, Ozawa, at the Geophysical Institute of the Kyoto University, successfully has devised many kinds of mechanical extensometers which are cheap, easy to install, easy for maintenance and which give elegant and large-

---

(*) Alloy: Co 4.8%, Ni 31.6%, Mn 0.5%, Fe 63.1% - temperature coefficient 10^{-6} per °C.

(**) Zerodur is a transparent glass-ceramic material, an inorganic non-porous material containing both glass and crystalline phases (Schott, Mainz, Germany).

Fig. 10.1. Observation room at Makimine (Japan: one Sassa extensometer and two Nishimura horizontal pendulums. Left of centre one can see the device carrying the bifilar suspension, under which the weight and the mirror can be seen (this part of the instrument is shown on Fig. 10.2).

Fig. 10.2. Bifilar suspension for a Sassa extensometer.

size tidal curves (see Fig. 10.5). In principle Ozawa's instruments are made with superinvar rods, 1 m long, interconnected to form a long bar of 10-20 m. One extremity A is firmly fixed to the bedrock while the other extremity B is "free" (see Fig. 10.3). The free-end position is compared with a reference C also fixed to the bedrock at a corresponding distance from A.

A very efficient device is obtained by fixing the upper suspension wire of a horizontal pendulum to the free end B while the lower suspension wire is fixed to some bolt attached at C to the ground, the neutral plane of the pendulum being perpendicular to the extensometer bar. Then the tidal variations of the distance AB produce oscillations of the axis of rotation BC of the pendulum which results in highly amplified oscillations of the beam (Fig. 10.5).

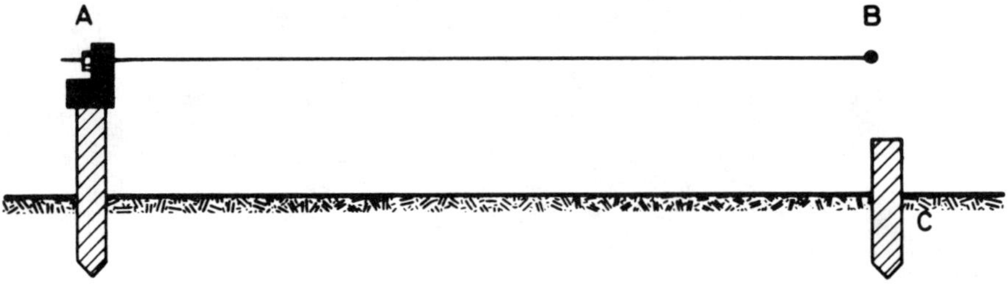

Fig. 10.3. Schematic representation of a bar extensometer.

Catenaries are used to maintain the bar horizontally: one suspension made of a long wire is used at every metre.

Ozawa has also combined two extensometers in opposite direction to evolve a "rotationmeter".

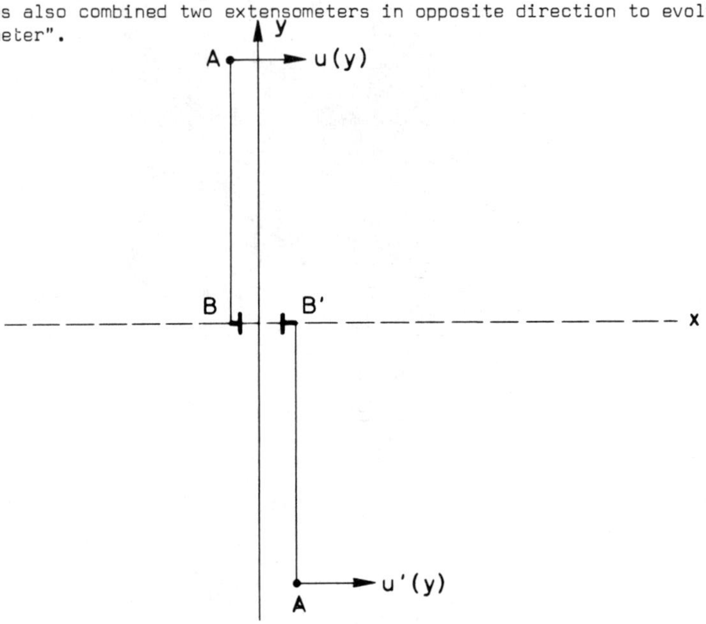

Fig. 10.4 Schematic representation of a rotationmeter.

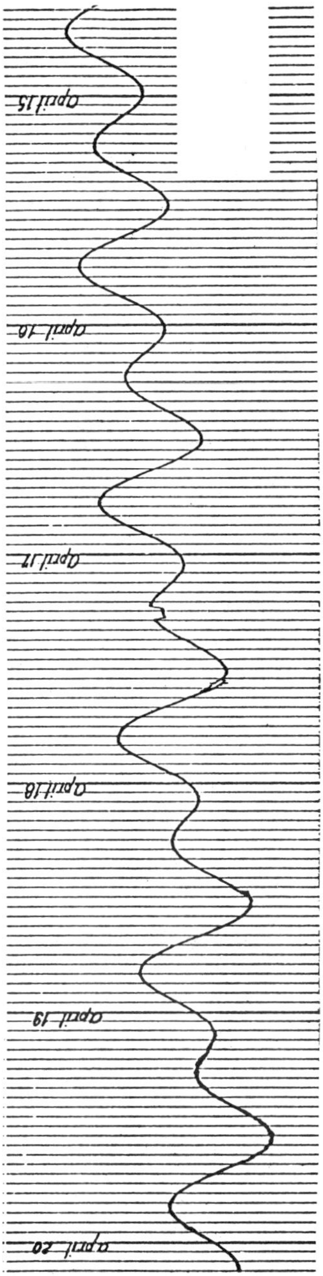

Fig. 10.5. Registration of the Ozawa extensometer at Walferdange underground laboratory (one week). One tentative calibration appears on the curve (17 April).

The system represented on Fig. 10.4 should measure $\partial u/\partial y$ = BB'. For such registrations two pendulums are used, being respectively suspended at B and B'. The light spot is successively reflected by the two pendulum mirrors and gives, therefore, their relative displacement. The pendulums must, of course, have the same sensitivity.

The installation of such instruments is not very difficult but must be made carefully as the relative displacements $\Delta L/L$ to be measured are of the order of 10^{-8}. It can be made according to a similar procedure as for horizontal pendulums by connecting them to bars strongly fixed in the rock. However, some authors prefer to put them simply without any fixation except for a heavy stone deposed on the ground (King Bilham).

Several research groups are presently working intensively in the development of strain measurements: King and Bilham at Cambridge (UK), Sydenham at Armidale (Australia), Gerard in New Zealand, Berger at UCLA (USA), Latinina and Boulatsen in the USSR.

Very detailed discussions about tensioning of the wires, pivots and transducer mounting to be used are given in the Sydenham papers (1972).

All systems currently use transducers which are now able to sense displacements of nanometre order over a small range. In that respect inductive solenoidal gauging units (LVDT: linear variable differential transformer) are preferable to capacitative methods as they are less sensitive to humidity.

The calibration of the extensometers is a difficult problem. The authors generally claim calibration accuracies of only 5-10%. If a displacement transducer is used the absolute value of the motion can be derived immediately.

When a superinvar bar is used, as in the Ozawa instrument, one can take advantage of the magnetic properties of ferronickel by producing magnetostriction effects. For cobalt alloys the magnetostriction constants are linearly proportional to the cobalt concentration. However this can not be used for a absolute calibration of the system but only to compare calibrated displacements of the LVDT sensor with corresponding oscillations of the suspended horizontal pendulum (for a same applied voltage).

Figure 10.6 shows the result of tests made at the Walferdange laboratory where a superinvar Ozawa bar extensometer 12 m long was installed in 1971. A coil has be threaded through the bar so that when applying a sinusoïdal voltage of frequency ω with a function generator one produces in the coil a current which produces a reduction of length of the bar whatever the sense of the current. Therefore the registration obtained either with a pendulum or a transducer has the form given by Fig. 10.6, i. e. a frequency 2ω.

On Fig. 10.6 the magnetostriction of the same bar has been observed:

a) With the amplifying pendulum device on a photographic recorder. The magnetostriction has a period of 30 min and is superposed to the tidal effects. Its amplitude on the recorder is 51.5 mm, and is of the same size as the Earth tide.

b) With a displacement transducer (LVDT) which has a sensitivity of 0.278 micron mV^{-1}. Its amplitude on the recorder is 61.4 mm, 1 mm corresponding to 0.004 mV.

Therefore the sensitivity on the photographic recorder was

$$s = \frac{61.4 \times 0.004 \times 0.278}{51.5 \times 12\ 10^6} = 0.011 \times 10^{-8}\ mm^{-1}.$$

At Ogdensburg Kuo uses a laser interferometry calibration device (1969): the transducer end of the extensometer is displaced longitudinally by means of an electromagnetic driving unit excited by a variable low-frequency oscillator (Fig. 10.7). The induced displacement of the tube is monitored by a Michelson interferometer which utilizes a neon-helium laser as its monochromatic

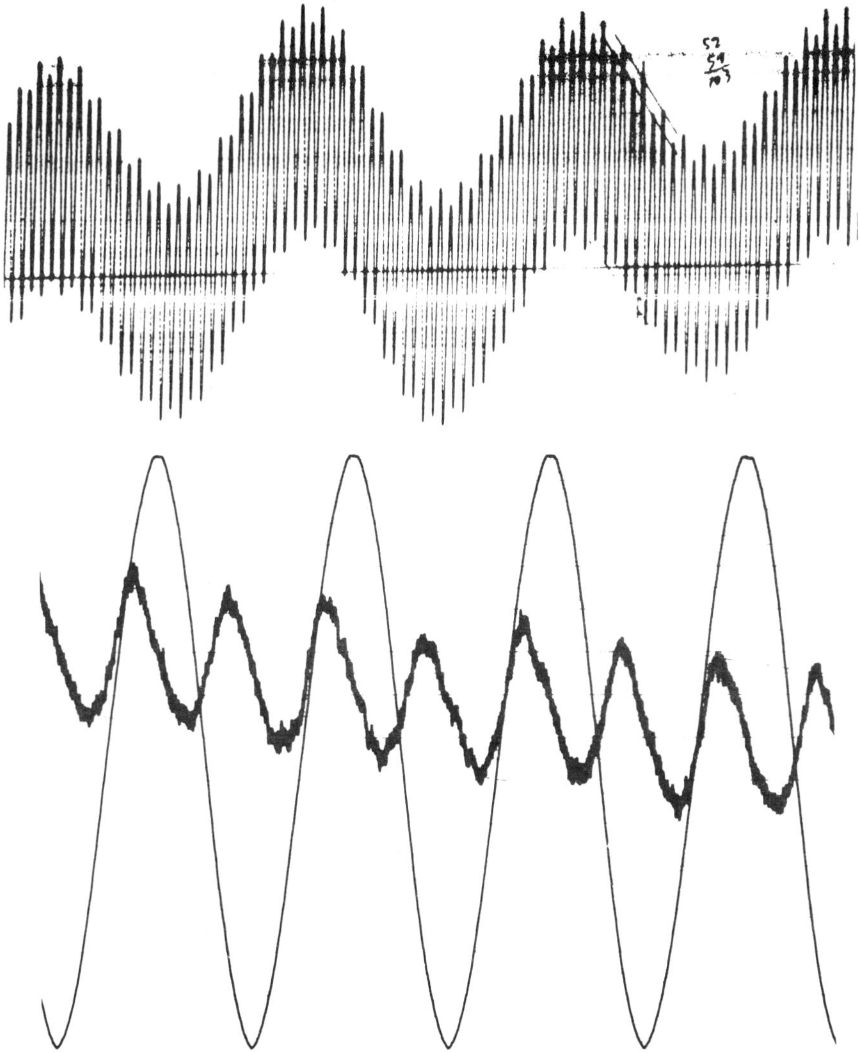

Fig. 10.6. Calibration by magnetostriction of a 12 m long superinvar bar (Ozawa extensometer) installed at the Walferdange laboratory.

a) above: photographic registration from the suspended horizontal pendulum; the magnetostriction oscillations are superposed to the tidal curve.

b) below: the small size sinusoïd is the registration from a displacement transducer (LVDT)-the large sinusoïd is the registration of the voltage variations applied to the coil (±50 volts) with a 30 min period.

light source. This system determines motion as small as 0.03 μm with good repeatability and errors of less than 5%.

The first paper reporting on a laser interferometer strainmeter was published in 1964 and it was expected that such instruments would be superior to the mechanical instruments. The development of the laser technique has made possible the use of optical interferometry on large distances, of the order of 1 km.

Laser strainmeters changes in distance in terms of the wavelength of light and therefore need no calibration. However, as the wavelength of a simple laser is not a good enough length standard (it is determined only to about 10^{-6}) it is necessary to stabilize it by reference to a better external length standard.

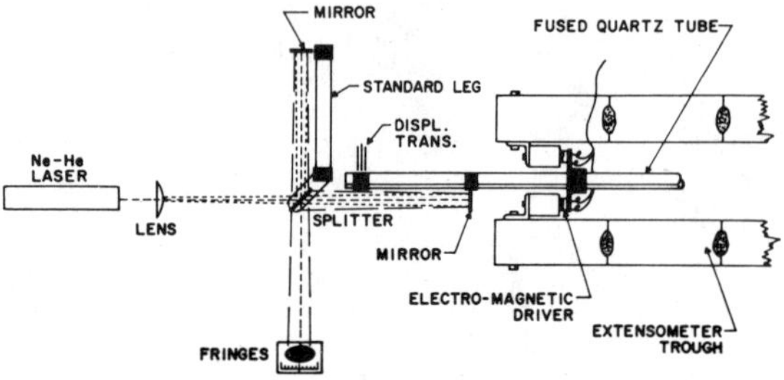

Fig. 10.7. Schematic diagram showing the actual setup of the Michelson-type interferometer for extensometer calibration (after Hade et al, 1968).

Moreover, because of atmospheric refractive index fluctuations, the path must be performed inside an evacuated tube. This makes this instrument very expensive.

Such an instrument (800 m long) devised by Berger and Lovberg is installed at the University of California, San Diego, at a distance of only 15 km from the Pacific Ocean, which makes tidal loading effects easily observable on the registrations.

At Queensbury the laser strainmeter system uses two helium-neon lasers which operate at 633 nm. One illuminates a Michelson interferometer with a 54 m long arm in which strain is measured. Corner cubes are used as references. By using a piezoelectric cylinder, the laser frequency is controlled to keep a particular fringe at the outputs of the interferometer as the path difference in the interferometer alters.

The other laser is stabilized by saturated absorption in iodine vapour and has high stability over long periods. The frequency difference between the lasers is measured to give a signal directly proportional to the strain. The registrations obtained during 12 weeks look good but no analysis or numerical results are given in the paper by Goulty, King and Wallard (1974).

In the Cooney Observatory near Armidale (New South Wales), Sydenham (1974) has extensively compared extensometers of different sizes and construction: quartz tubes, tensioned wires, laser. He concludes that, when coupled with data regarding temperature effects (the thermic influence on quartz is 5.10^{-7} per °K), the mechanical instruments are not inferior to laser interferometers but that using the data from only one instrument of any kind could lead to incorrect conclusions.

Quartz tube (which was introduced for the first time by Benioff) seems to be capable of the best performance. Ultra-low-expansion materials such as Schott Zerodur may even be better.

Quartz instruments are usually made of tubes, the wall thickness being about one-tenth of the tube diameter. The tube is supported at intervals with fine invar wire slings. Sydenham considers that quartz appears to be superior to invar on the grounds of better stability. This is also the conclusion reached at the Walferdange laboratory where a 26 m quartz tube extensometer has proved to be drift-free. Also in the USSR many quartz tube extensometers have been installed and used for various purposes: tides, secular effects, tectonics (Latinina).

10.2. THE VERTICAL COMPONENT OF THE STRAIN TENSOR

A simple vertical strainmeter has been installed with some success in the Walferdange laboratory. A hollow tube of circular section has an external diameter of 6 cm and a thickness of 6 mm. The total height available for installation was about 3 m. A cylindrical hole having a diameter of 64 mm and a depth of 20 cm has been drilled into the ceiling of the gallery. A 50 cm long piece of quartz tube was inserted into the hole and fixed with phenolic resins. A few days later, after consolidation of the resins, two more pieces of quartz tube were added, yielding a total length of 3 m. The lower end of the tube is provided with a well-planed quartz plate. Three stainless-steel rods, stuck vertically into the floor of the gallery to a depth of about 20 cm, support a heavy seismometer base-plate. The latter being almost 30 years old, it seems to be fairly free from internal stresses.

Strain is measured via two displacement transducers of the LVDT type. The electronic noise level is lower than 0.01 μm, and the linearity is better than 0.25% in a range of 3 mm about the central zero point. Although the calibration coefficients have been furnished by the factory itself, some have been checked at the Royal Observatory of Belgium by means of an expansible bearing plate. These checks confirmed to the nearest percentage the calibration coefficients published by the factory.

The base-plate itself is provided with a micrometric system bearing the two displacement transducers. This system enables one to press the transducer rods smoothly but firmly against the end-piece of the fused quartz tube; it allows correction of the drift of the instrument as well, but usually this is now done with a potentiometer in the recording system.

The whole system is heat-protected by a styrofoam casing inside of which a thermistance records the temperature variations. The records show very regular tidal curves the double-amplitude of which is about 4.5 cm under favourable conditions. Unfortunately, a noticeable drift necessitated one or more corrections a week. Earthquakes appear rarely on the records.

The radial strain at the surface is given by (3.48):

$$\varepsilon_{aa} = \frac{W_2}{ag}\left[a\left(\frac{dH(r)}{dr}\right)_{r=a} + 2H(a)\right] da = \eta\,\frac{da}{a}\,\frac{W_2}{g}. \qquad (10.1)$$

At Walferdange:

da = 3.40 m, taking into account one half of the length of tube and rods fixed within the rock.

$da/a = 5.3367 \times 10^{-7}$.

$\phi = +\,49°39'53''$.

for the transducer used: 1 mV = 0.2174 micron.

Taking account of the main diurnal and semi-diurnal tides, we get for Walferdange, $W_2/g = 72.39$ cm.

Assuming $\eta = -0.25$, this would produce a differential vertical displacement $\delta\eta \simeq 0.097$ µm (or 0.45 mV) (Table 10.1).

The actual observed double-amplitude is rather higher, about 4.5 cm at its maximum. Therefore the actual vertical strain factor measured at Walferdange is

$$\eta \approx \frac{4.5 \times (-0.25)}{2.21} = -0.509. \qquad (10.2)$$

A detailed harmonic analysis (Table 10.2) confirms this result. Expressing η in terms of Love numbers h and ℓ and Poisson's ratio ν we have at the surface of the Earth (3.60)

$$\eta = \frac{2\nu}{1-\nu}(3\ell - h). \qquad (10.3)$$

Assuming for Walferdange $h = 0.64$, $\ell = 0.09$, $\eta = -0.5$, we find $\nu = \eta/(\eta + 6\ell - 2h) \simeq 0.4$. This value seems fairly high, but is not inconsistent with the local geological structure. Indeed, the gallery is pierced into rather poorly consolidated gypsum-containing marls (Keuper strata). Notice that $\eta = -0.25$ yields $\nu \simeq 0.25$, which is an average value for well-compacted rocks. However, this amplification is most certainly due to a typical "cavity" effect (see Chapter 12).

TABLE 10.1. Expected amplitudes of the most important vertical strain tides at Walferdange (ϕ = 49°39'53") assuming $\eta = -0.25$

Tide	$\frac{W_2}{g}$ (cm)	$\delta\zeta$ (µm)
O_1	10.08	0.01345
P_1	4.69	0.00626
S_1	0.11	0.00015
K_1	14.18	0.01892
N_2	4.65	0.00620
M_2	24.30	0.03242
S_2	11.30	0.01508
K_2	3.08	0.00411
Sum	72.39	0.09659

10.3. RESULTS FROM AN ARRAY OF EXTENSOMETERS

The Japanese geophysicists have installed many extensometers in their country, including oblique and vertical instruments. At the Ozakayama tunnel station near Kyoto (ϕ = 34°59'6N), Ozawa has installed strainmeters in six directions - three horizontal components, two oblique, and one vertical. He can therefore reconstruct the factors:

$\eta = ah' + 2h$ from the vertical strainmeter,

$2(h-3\ell)$ from the sum of North-South and East-West components,

$ah' + 4h - 6\ell$ from the sum of vertical, North-South and East-West components,

as seen in Chapter 3 [(3.48), (3.54) and (3.80)] .

TABLE 10.2. Measured strain tidal values at Walferdange

Tide	η	Phase	ah'
O_1	-0.503 ± 0.020	$-2.18° \pm 2.23°$	-1.779
P_1	-0.389 ± 0.046	$+6.78° \pm 6.76°$	-1.537
K_1	-0.400 ± 0.015	$-2.06° \pm 2.13°$	-1.348
(S_1)	(-10.7 ± 2.7)	$(-24.31° \pm 14.83°$	
N_2	-0.425 ± 0.052	$+5.31° \pm 6.07°$	-1.701
M_2	-0.481 ± 0.009	$+2.06° \pm 0.98°$	-1.757
S_2	-0.523 ± 0.020	$+21.26° \pm 1.89°$	-1.799
K_2	-0.531 ± 0.064	$+18.96° \pm 6.05°$	-1.807

The numerical coefficient for the M_2 wave in the development of W_2/ag being simply

$$0.90812 \frac{D}{ag} \cos^2\phi = 3.8037 \cos^2\phi \times 10^{-8} = 2.559 \times 10^{-8} \text{ at Osakayama,}$$

Ozawa has obtained for the M_2 wave:

$$\left.\begin{aligned} ah' + 2h &= -0.280 \pm 0.036, \\ h - 3\ell &= 0.434 \pm 0.026. \\ ah' + 4h - 6\ell &= 0.588 \pm 0.027. \end{aligned}\right\} \tag{10.4}$$

Thus

The conditions at the free surface (3.57), (3.61) hold because the depth of the tidal stations is negligible with respect to the tidal wavelength. Equation (3.57) gives

$$a\ell' + h + \ell = 0. \tag{10.5}$$

Introducing the value h=0.62, which is well established by clinometric and gravimetric observations and fits fairly well the theoretical models, one can derive

$$\left.\begin{aligned} \ell &= 0.062 \pm 0.009, \\ ah' &= -1.520 \pm 0.036, \\ a\ell' &= -0.682 \pm 0.009. \end{aligned}\right\} \tag{10.6}$$

However, using observed values of six-components of extentions, the six-strain components of M_2 tide are calculated as follows by Ozawa:

$$\left.\begin{aligned}
e_{rr} &= 0.642\times10^{-8}\ \cos(2t-193.6°),\\
e_{\theta\theta} &= 0.957\times10^{-8}\ \cos(2t-29.6°),\\
e_{\lambda\lambda} &= 1.315\times10^{-8}\ \cos(2t-5.5°),\\
e_{\theta\lambda} &= 1.650\times10^{-8}\ \cos(2t-74.8°),\\
e_{r\theta} &= 0.463\times10^{-8}\ \cos(2t-226.4°),\\
e_{r\lambda} &= 0.539\times10^{-8}\ \cos(2t-183.3°),\\
\Sigma &= e_{\theta\theta} + e_{\lambda\lambda} = 2.223\times10^{-8}\ \cos(2t-12.3°),\\
\Delta &= 1.582\times10^{-8}\ \cos(2t-16.5°),
\end{aligned}\right\} \qquad (10.7)$$

which gives the representation of Fig. 10.8.

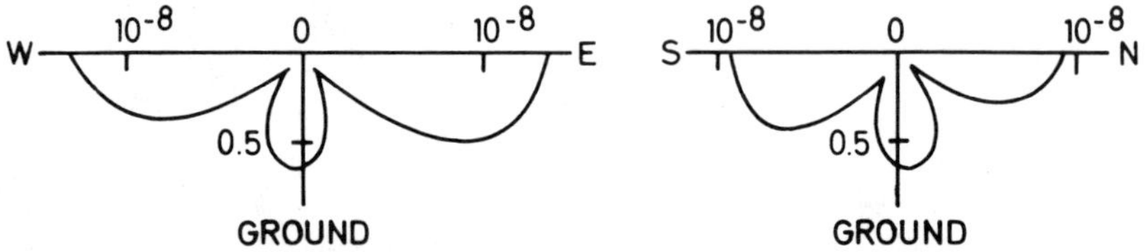

Fig. 10.8. Relation between the dip and the amplitude of the M_2 wave obtained by Ozawa at Osakayama.

We immediately observe from these data that

$$-\frac{1}{3}\,(e_{\theta\theta} + e_{\lambda\lambda}) = 0.76.10^{-8}\ \cos(2t-180°), \qquad (10.8)$$

which differs by 20% from e_{rr} in contradiction with (3.62) while $e_{r\theta}$ and $e_{r\lambda}$ are different from zero.

These anomalies are well emphasized by the distortion of the curves on Fig. 10.8 and 10.9 with respect to the theoretical curves given on Fig. 3.2 and 3.3.

A theoretical computation provides the following numbers for the Ozakayama latitude:

$$\left.\begin{aligned}
e_{rr} &= 0.539\times10^{-8}\ \cos(2t-180°),\\
e_{\theta\theta} &= 1.085\times10^{-8}\ \cos 2t,\\
e_{\lambda\lambda} &= 0.416\times10^{-8}\ \cos 2t,\\
e_{\theta\lambda} &= 0.872\times10^{-8}\ \cos 2t,\\
e_{r\lambda} &= e_{r\theta} = 0,\\
\Sigma &= e_{\theta\theta} + e_{\lambda\lambda} = 1.504\times10^{-8}\ \cos 2t,\\
\Delta &= 0.965\times10^{-8}\ \cos 2t,
\end{aligned}\right\} \qquad (10.9)$$

Few other numerical data concerning Love numbers have been published. At Ogdensburg (USA) Kuo obtained the following values:

	$h - 3\ell$	Lag
M_2	0.392	-1.2
O_1	0.331	-5.7

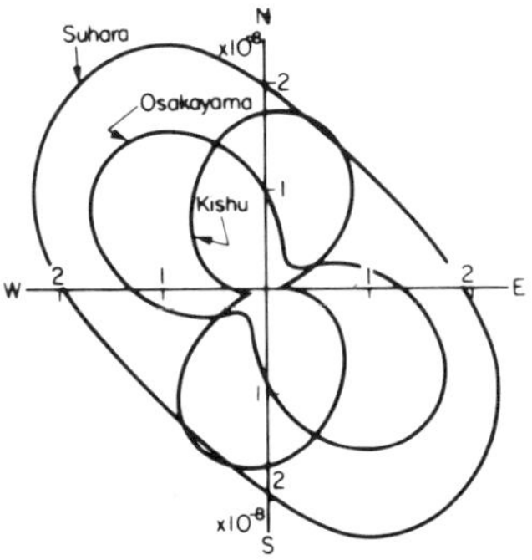

Fig. 10.9. Relation between the azimuth and amplitudes of the M_2 wave obtained by Ozawa from the observations at Osakayama, Kishu and Suhara.

At Isabella (California) Smith and Jungels (1970) obtained by analysis of the M_2 areal strain:

$$h - 3\ell = 0.475, \qquad \text{lag} \quad -6.0°.$$

A North-South invar wire strainmeter in Tiefenort (Harwardt and Simon), 21 m long, has a sensitivity of 6.10^{-10} mm^{-1}. This sensitivity is controlled by magnetostriction. However, real calibration control is not clear as when 18 months' data were analysed by the Venedikov method; only "normalized amplitudes" to the M_2 wave are given. They show, nevertheless, a good coherency.

Latinina and Karmaleyeva use quartz rods in a tunnel near Alma Ata (Talgar): along the tunnel axis (North-South) 26 m long; across the tunnel axis (East-West) 4 m long. The rods rest on rollers lying on intermediate platforms which are 1.5 m apart. The recording is made with a torsional device. Published results are in the North-South component:

$$e_{\theta\theta}\ (M_2) = 0.79 \times 10^{-8} \cos(2\tau + 6°)$$

$$e_{\theta\theta}\ (O_1) = 0.24 \times 10^{-8} \cos(\tau - 10°)$$

from which they derive (mixing diurnal and semi-diurnal components):

$$h = 0.46, \qquad 0.38.$$

$$\ell = 0.07, \qquad 0.07.$$

At Inkerman (Crimea) Boulatsen (1975) derives from waves O_1 and M_2:

$$\ell = 0.0752, \qquad h = 0.4425, \qquad \ell/h = 0.170.$$

He attributes the anomalous value of h to cavity effects (see Chapter 12). By a similar combination of O_1 and M_2 Kartvelishvili found at Tbilissi

$$\ell/h = 0.141.$$

However, the results obtained by mixing the observed results for M_2 and O_1 waves are not acceptable because of the influence of the oceanic indirect effects (Chapter 11) which are strongly frequency dependent.

Beaumont and Berger (1975) made a detailed and comparative analysis of the results obtained in seven sites in the United States with 17 instruments (Fig. 10.10).

The observed amplitudes are reproduced in Tables 10.3 and 10.4 in function of the azimuth. They are of about the predicted size (10^{-8} for M_2, 5×10^{-9} for O_1) but strongly dependent of the geographical location. The very precise discussion of Beaumont and Berger, in terms of oceanic influences and topography influence, is referred to in Chapters 11 and 12.

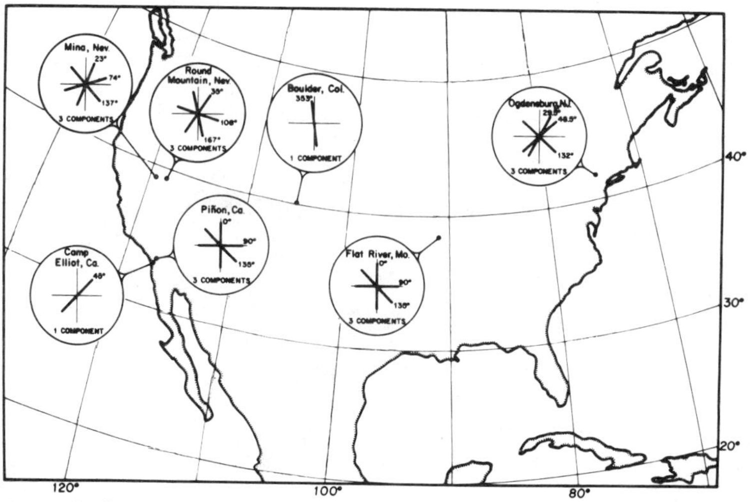

Fig. 10.10. Map showing seven location of strain observatories in the United States. The directions of the individual strainmeters at each observatory are indicated.

10.4. DETERMINATION OF THE RATIO ℓ/h

It has been shown in Chapter 3 that it is possible to obtain easily this ratio from the registrations of extensometers when they are installed in an azimuth which differs sufficiently from North-South or East-West directions.

Of course the calibration of the instrument is not needed in that case because it eliminates itself when we take the ratio of the cosine term to the sine term.

However, accurate timing is important because any error of the time marks should alter the phase and modify in opposite senses the amplitudes of the cosine and sine terms.

At the Walferdange laboratory there are three extensometers parallel to each other and installed in azimuth 35°, which is sufficient to make the sine term of the same order of magnitude as the cosine term.

Table 10.4 gives the numerical results obtained in this way by using formulas (3.77) and (3.78). This result, which looks quite interesting, may give the idea of making some kind of prospection with simple transportable strainmeters to be installed in azimuths comprised between 30° and 60° and which should not have to be calibrated. A quartz clock should be a part of such an equipment.

TABLE 10.3. United States strain measurements, analysis results according to Beaumont and Berger

Station	Azimuth (deg)	M_2 constituent Amplitude$\times 10^9$	M_2 constituent Local epoch	O_1 constituent Amplitude$\times 10^9$	O_1 constituent Local epoch
Camp Elliott	45	6.90 ± 0.15	−32.9°±1.0°	6.73 ± 0.15	+ 0.9°±5.0°
Piñon Flat	0	12.42 ± 0.12	− 2.2°±0.6°	3.72 ± 0.12	−11.8°±7.0°
	90	5.44 ± 0.15	−22.4°±1.0°	4.96 ± 0.26	− 0.6°±4.3°
	135	12.63 ± 0.08	− 3.0°±0.6°	4.85 ± 0.17	−11.0°±2.9°
Mina	23	4.52 ± 0.80	− 2.1°±8.0°	2.59 ± 0.20	−22.4°±6.0°
	74	3.36 ± 0.20	+18.9°±4.0°	6.00 ± 0.35	− 5.8°±4.5°
	137	7.79 ± 0.50	−19.8°±2.5°	4.43 ± 0.40	+17.6°±3.5°
Round	35.5	6.95 ± 0.25	+22.3°±3.0°	5.49 ± 0.30	−21.8°±3.5°
Mountain	108.0	5.09 ± 0.30	−12.6°±2.5°	6.60 ± 0.32	+ 6.0°±2.0°
	167.5	7.29 ± 0.20	− 3.5°±2.5°	2.97 ± 0.15	− 1.2°±2.5°
Poorman Mine	173	9.03 ± 0.06	−13.0°±3.0°	2.43 ± 0.04	−15.0°±3.0°
Flat River	0	14.09 ± 0.30	− 2.6°±1.4°	4.75 ± 0.30	− 2.6°±4.0°
	90	4.70 ± 0.15	+12.6°±2.0°	7.21 ± 0.35	− 6.0°±3.0°
	135	11.08 ± 0.50	−21.4°±3.0°	6.65 ± 0.60	+27.0°±4.0°
Ogdensburg	29.5	12.75 ± 0.30	+18.5°±1.6°	5.81 ± 0.40	−21.3°±1.5°
	48.5	13.05 ± 0.30	+21.2°±1.0°	7.25 ± 0.40	−18.4°±1.6°
	132	7.95 ± 0.40	−78.5°±3.0°	8.02 ± 0.50	+27.3°±3.0°

(Data intervals are given in the paper of Beaumont and Berger (1975)).

10.5. LOCAL EFFECTS

A very interesting experiment made by King and Bilham is described here. Several 10 m wire strainmeters aligned along the axis of the Queensbury Tunnel in Yorkshire show amplitude variations of the diurnal tides of up to a factor of three, with smaller variations in phase (Fig. 10.11). These observations are not easily explained because the semi-diurnal tides have different and smaller variations. This site had originally been selected for strain measure-

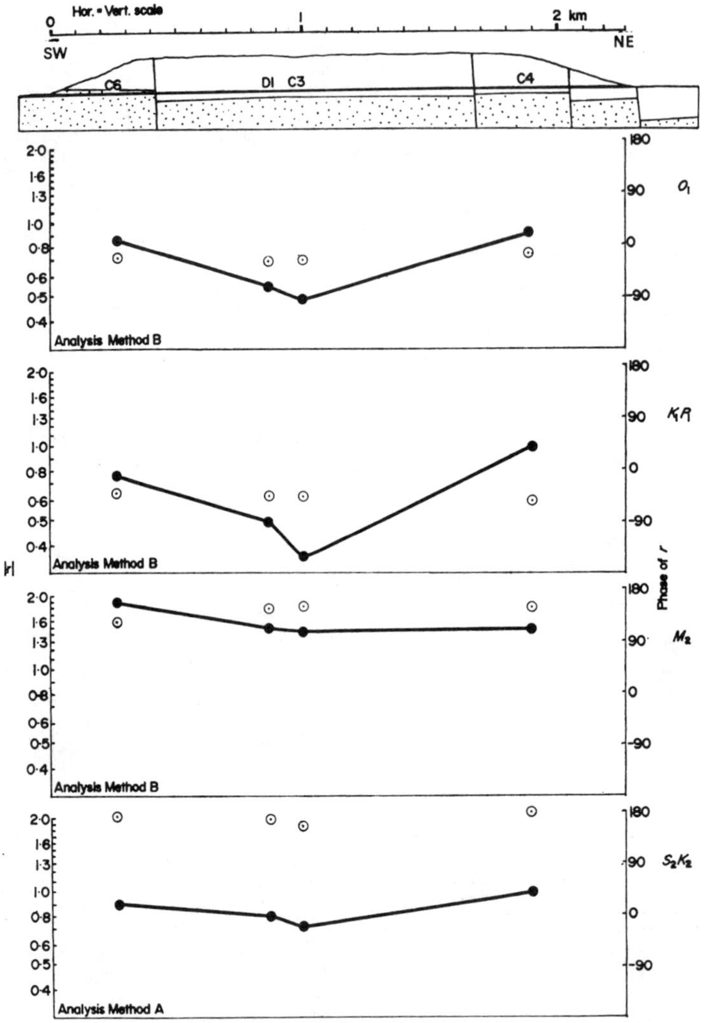

Fig. 10.11. Variation of |r| (solid dots) and phase of r (open circles) along the Queensbury Tunnel. The tidal r factor is suppressed at the centre of the tunnel for both diurnal and semi-diurnal tides. Note that the effect is approximately three times larger for the diurnal than for the semi-diurnal tides. (according to King and Bilham)

ments because the tunnel passes through weak coal-measure deposits with shallow dips. Such geological conditions would not be expected to produce large local variations in the tidal strain field (King, 1971), and the fact that they occur at this site suggests that they may be more widespread than has previously been believed.

One strainmeter was attached to the tunnel wall with bolts. Three model C and one model D wire strainmeters were also used, all of which were laid on the fill on the tunnel floor in a bed of sand.

TABLE 10.4. Walferdange parallel horizontal extensometers. Ratio ℓ/h.

Waves	Superinvar bar Ozawa	Invar wires C5	Invar wires D12	Models	
				Molodensky	Jeffreys
Q_1	0.176	0.217	0.191		
O_1	0.163	0.166	0.166	0.147	0.140
P_1	0.169	0.157	0.200	0.171	0.166
K_1	0.183	0.180	0.161	0.183	0.176
M_2	0.143	0.182	0.168	0.147	0.140
N_2	0.147	0.121	0.185		
S_2	0.142	0.180	0.165		
K_2	0.141	0.122			
M_3	0.027	0.101	0.092	0.050	
Number of measurements	22.464	9.888	2.592		

The intercalibration between the strainmeters is probably accurate to better than 10%. The only plausible explanation is in terms of local inhomogeneities, resulting from local geology, coal-mining operations near the tunnel, or poor coupling of the tunnel lining to the surrounding rock. More observations in this and other tunnels are planned to discover whether anomalies of the type found in Queensbury are widespread and to determine their origin.

Since several parallel measurements are necessary to reveal whether or not strain fields are affected by local conditions, it would seem hazardous to interpret Earth-strain data from single isolated strainmeters. Certainly it is not difficult to envisage that some of the anomalous results obtained by workers in the field of strain measurement may be attributed to interpreting data which has been effected in a similar way to data from Queensbury.

10.6. CONFINED FLUIDS AS CUBIC STRAINMETERS

Earth strains as small as those produced by Earth tides involving periodic cubic dilatations (θ) of about 2×10^{-8} induce easily observable water-level fluctuations in wells. When the wells are connected with sufficiently confined fluid bodies these fluctuations can amount to several centimetres.

Therefore the observation of luni-solar tides in underground beds is rather elementary and goes back to antiquity: the first written record is given by Pliny who observed tides in some wells in Italy (Como) and Spain.

Such observations, fully described in the first edition of *the Earth Tides* and listed in Table 10.5, are relative to confined aquifers (artesian wells) and unconfined aquifers (water-table system) as well.

As the extent of such aquifers is always extremely limited we can *a priori* dismiss the possibility of it being a real movement of the bed itself, and the cause must be sought in the deformations of the surrounding material. The liquid thus acts as a "deformation indicator" and is similar to a manometer. Thus the equation governing the variation of height dH in a well of cross-section $s=\pi r^2$ is a function of the cubic dilatation of the aquifer $\theta = dV/V$ and of the modulus of compressibility k_f of the fluid contained in the aquifer volume V:

$$- dV = s\, d\, H + g\rho\, dH \left(\frac{V}{k_f}\right) \qquad (10.10)$$

which gives

$$- dH = \frac{dV}{\pi r^2 + \rho g V / k_f} \approx \frac{k_f\, \theta}{\rho g} \ , \qquad (10.11)$$

and with the tidal dilatation

$$\theta = 2 \times 10^{-8}$$

one has in any case and everywhere

$$dH = 4 \text{ mm},$$

i.e. in contradiction with the observations recalled in Table 10.5.

The explanation proposed by Bredehoeft (1967) is that the amplitude depends on the specific storage of the aquifer. When the aquifer is compressed, the change in volume of the solid material due to deformation of the individual particles is small in comparison with the change in volume of the water. This assumption is apparently valid for granular aquifers, but, depending on the pore geometry and Poisson's ratio, it may not be valid for such aquifers as limestone and basalt. To the extent that the above assumption holds for a given aquifer, nearly all the volume change of a dilated aquifer goes into a volume change of the water in its pores. Hence the dilatation of the water is nearly equal to that of the aquifer divided by the volume porosity ϕ:

$$- dH = \frac{1}{\phi} \frac{k_f \theta}{\rho_g} \qquad (10.12)$$

It is convenient to consider first the dilatation that would occur if the fluid were not present and, then, to consider the dilatation produced by the change in pressure head. Neglecting in this way the saturating fluid, the tidal dilatation for an aquifer near the earth's surface, subjected to the earth tides is obtained by using (3.54) and (3.82):

$$\theta = \frac{1-2\nu}{1-\nu} (2h - 6\ell) \frac{W_2}{ag} \ . \qquad (10.13)$$

Then we must consider that the total dilatation in a saturated confined aquifer is the resultant of the tidal dilatation and the dilatation produced in the permeable rock of compressibility k_r by the change in the fluid pressure $p = \rho\, g\, dH$.

The "storage coefficient" of the formation or "hydraulic capacitativity" is defined as

$$S = \left(\frac{1-\phi}{k_r} + \frac{\phi}{k_f}\right). \qquad (10.14)$$

With this definition, the change in head produced by the tidal dilatation θ is given by:

$$-\,dH = \frac{\theta}{\rho g S} \tag{10.15}$$

By combining (10.13) and (10.15), we obtain an expression for the displacement of the water level in terms of the tidal disturbing potential and the specific storage $\rho g S$ of the aquifer:

$$-\,dH = \frac{1}{\rho g S}\left(\frac{1-2\nu}{1-\nu}\right)(2h-6\ell)\frac{W_2}{ag}\,. \tag{10.16}$$

Thus if Poisson's ratio is known, (10.16) enables the specific storage $\rho g S$ to be determined. (As an example, for sandstone of bulk modulus 2×10^{10} Nm^{-2}, thickness 60 m, and porosity 0.3, Jacob (1940) finds $\rho g S \sim 0.0001$ cm^{-1}. On the other hand, the barometric efficiency of the well is related to the specific storage as

$$B = \phi/k_f S, \tag{10.17}$$

and if we can determine S from tidal effects, the observation of barometric effects will give us the volume porosity ϕ of the aquifer volume in the vicinity of the well, a quantity which interests hydrologists and which is difficult to determine by other means.

Bredehoeft's equation (10.16) explains why the amplitude of the M_2 tide is greater than the predicted 4 mm. Moreover, the fact that the amplitude (corrected for latitude) increases with the depth of the well as shown in Table 10.6 probably depends on the porosity of the aquifer. In general, the porosity of geologic deposits decreases with depth, which would tend to increase the amplitude of the tidal fluctuation. The permeability of the confining layers also generally decreases with depth, so that deeper aquifers more closely approximate ideal artesian conditions. This, too, would tend to produce larger Earth tides in deeper wells. These are, of course, generalities; the geology of each site must be examined independently.

As the barometric effects have diurnal and semi-diurnal components it is worthwhile correcting the data for this before making the tidal analysis.

Bredehoeft applies (10.16) to the data of Turnhout, Iowa City and Carlsbad (New Mexico) for which the geologic information indicates as clearly artesian (Table 10.7).

The computations for Turnhout and Iowa City give good results. The various component waves give reasonable values for specific storage and porosity.

The results for the Carlsbad well are not as good. The specific storage is reasonably constant for the semi-diurnal waves; however, the diurnal components yield values almost twice as large as the semi-diurnal values. The values for the computed porosity seem unreasonably large which suggests that the tidal fluctuations should have been larger than observed. Either this well does not fit our assumptions for an ideal artesian system or something is operating to reduce the magnitude of the tidal fluctuations.

Dynamic Equations of the Aquifer Mechanics.

The percolation of fluids through slowly moving formations of isotropic permeable rock is a problem of transport like heat or viscous strain and is governed by an equation of the same kind:

$$\mathbf{q}_r = -C\,\mathbf{grad}\,p, \tag{10.18}$$

TABLE 10.5.

Stations		Kiabukwa Zaïre	Carlsbad (USA) New Mexico	Oak Ridge (USA) Tennessee	Iowa City (USA-IOWA)	Duchov (Czechoslovakia)	Turnhout (Belgium)	Basècles (Belgium)	Heibaart (Belgium)
Latitude		-7°47'	+32°18'	+35°55'	+41°39'	+50°37'	+51°19'	+50°32'	+51°23'
Length in days		62	60	15	28	93	62	201	448
epoch		(1954)	(1938)	(1952)	(1939)	(1879)	(1956)	(1962)	(1965 - 67)
Amplitudes (cm)	M_2	7.53	0.45	2.16	1.15	1.39	1.48	0.43	1.15
	S_2	3.95	0.25	0.93	0.50	0.72	0.68	0.29	0.53
	N_2	1.49	0.09	-	0.33	0.24	0.24	0.10	0.21
	K_1	-	0.16	1.13	1.41	1.35	1.41	0.63	1.38
	O_1	-	0.15	0.79	0.59	0.93	1.20	0.47	1.12
Phases (°) (theoretical values 180°)	M_2	183.7	174.0	180.4	181.4	169.0	185.1	155.5	188.4
	S_2	182.6	193.1	158.2	185.9	161.0	209.3	177.5	189.7
	N_2	170.8	178.3	-	192.2	164.0	176.6	140.7	182.0
	K_1	-	156.6	163.3	137.1	181.3	161.6	192.4	185.2
	O_1	-	207.4	145.6	187.2	193.6	151.2	191.7	185.6
Relative amplitudes									
Theoretical 0.465	S_2/M_2	0.524	0.555	0.431	0.435	0.518	0.459	0.674	0.461
Theoretical 0.194	N_2/M_2	0.198	0.200	-	0.287	0.172	0.162	0.233	0.183
Theoretical 0.710	O_1/K_1	-	0.937	0.699	0.418	0.689	0.851	0.746	0.811
Theoretical	O_1/M_2		0.33	0.37	0.51	0.67	0.81	0.81	0.81
Observed	O_1/M_2	0.11	0.52	0.60	0.74	1.01	1.04	1.07	0.97
Phases	S_2-M_2	-1.1°	+18.9°	-22.2°	+4.5°	-8.0°	+24.2°	+22.0°	+1.3°

TABLE 10.6.

Well	Depth (m)	$A(M_2)$ reduced to the equator (cm)
Basècles	40	1.05
Oak Ridge	75	3.29 *
Carlsbad	86	0.63
Milwaukee	120	2.00
Sontra	120	2.88
Iowa	252	2.06
Duchov	600	3.45
Heibaart	1660	2.93
Turnhout	2175	3.79
Kiabukwa	2400	7.67

*We had observations for only 15 days for this well.

TABLE 10.7. Comparison of data for the M_2, S_2, N_1, K_1 and O_1 waves for artesian wells at Turnhout, Iowa City and Carlsbad, according to Bredehoeft

	A	T	D	ρgS	B	ϕ
		Turnhout, Belgium, latitude + 51°19'				
M_2	1.48	10.2	0.83	0.6	0.77	0.10
S_2	0.68	4.7	0.38	0.6		0.10
N_2	0.24	1.9	0.16	0.7		0.11
K_1	1.41	14.8	1.2	0.9		0.15
O_1	1.20	10.5	0.86	0.7		0.12
		Iowa City, Iowa, latitude + 41°39'				
M_2	1.15	14.5	1.2	1	0.75	0.18
S_2	0.50	6.8	0.55	1		0.19
N_2	0.33	2.7	0.22	0.7		0.12
K_1	1.41	15.1	1.2	0.9		0.15
O_1	0.59	10.7	0.87	1		0.25
		Carlsbad, New Mexico, latitude + 32°18'				
M_2	0.45	18.6	1.5	0.3	0.65	0.42
S_2	0.25	8.6	0.70	0.3		0.36
N_2	0.09	3.5	0.29	0.3		0.40
K_1	0.16	13.7	1.1	0.7		0.78
O_1	0.15	9.7	0.79	0.5		0.65

A: Amplitude of Associated water-level from harmonic analyses (cm)
T: Theoretical amplitude of tidal potential $W_2 \times (1/g)$, (cm)
D: Theoretical amplitude of tidal dilatation $\Delta_t \times 10^{-8}$ (ν=0.25)
ρgS: Specific storage ($\times 10^{-8}$ cm^{-1}).
B: Barometric efficiency. ϕ: Porosity (ν = 0.25).

where $\mathbf{q}_r$ is a vector field which represents the relative mass flow per second across the unit area, p is the fluid pressure in the pores, C is the hydraulic conductivity which is assumed to be the ratio k_p/ν, k_p is the permeability of the rock and ν is the kinematic viscosity of the fluid.

Equation (10.18) is called Darcy's law by hydrologists. It is valid only for slow laminar flows of newtonian fluids when inertia may be neglected. This is acceptable for tidal effects but probably not for seismic phenomena. In this case, inertia effects must be included (Bodvarsson, 1970). Let $\mathbf{u}$ be the velocity field describing the movement of the permeable formation.

The absolute mass flow measured relatively to a stationary coordinate system is

$$\mathbf{q} = \mathbf{q}_r + \phi\,\rho\,\mathbf{u}, \tag{10.19}$$

ϕ being the volume porosity of the rock.

As the velocity is related to the mass flow by the expression

$$\mathbf{q} = \psi\,\rho\,\mathbf{v}, \tag{10.20}$$

where ψ is the area porosity, we have to introduce the inertia term

$$\rho\,\frac{\partial \mathbf{v}}{\partial t} = \frac{1}{\psi}\frac{\partial \mathbf{q}}{\partial t}, \tag{10.21}$$

and the Darcy law becomes

$$\frac{C}{\psi}\frac{\partial \mathbf{q}}{\partial t} + \mathbf{q} = -\,C\,\mathbf{grad}\;p + \phi\,\rho\,\mathbf{u} \tag{10.22}$$

(in most practical cases $\phi \approx \psi$).

The whole formation is subjected to small periodic dilatation. The volume porosity of the dilated rock becomes

$$\phi = \phi_o + \varepsilon(1-\phi_o)\theta, \tag{10.23}$$

where ε is an empirical factor ($\varepsilon \leqslant 1$) depending upon the rock cohesion. Then the specific fluid content of the porous formation under pressure p becomes

$$m = \phi\left(\rho + \frac{p}{k_f}\right) + \frac{p}{k_r}, \tag{10.24}$$

i. e.

$$m = \rho\,[\phi_o + \varepsilon(1-\phi_o)\theta] + \rho\,g\,S, \tag{10.25}$$

S being defined by (10.15).

On the other hand, the conservation of the fluid mass expressed in terms of the absolute flow $\mathbf{q}$ requires that

$$\operatorname{div}\mathbf{q} + \frac{\partial m}{\partial t} = 0. \tag{10.26}$$

The system (10.22), (10.25), (10.26) resumes the basic equations of the problem. Taking the divergence of (10.22) and substituting div $\mathbf{q}$ into m (10.26), which is eliminated by using (10.25), Bodvarsson easily obtains the resulting equation

$$\Delta p - \rho\,\frac{S}{C}\frac{\partial p}{\partial t} - \rho\,\frac{S}{\psi}\frac{\partial^2 p}{\partial t^2} = \rho\,\frac{\varepsilon}{C}\left(\frac{\partial \theta}{\partial t} + \frac{C}{\psi}\frac{\partial^2 \theta}{\partial t^2}\right), \tag{10.27}$$

second-order terms having been neglected.

The ratio $C/\rho S$ is called the hydraulic diffusity or transmissivity of the formation. When inertia is completely neglected, one finds the standard form of the transport properties equation (diffusion type)

$$\Delta p - \rho \frac{S}{C} \frac{\partial p}{\partial t} = \rho \frac{\varepsilon}{C} \frac{\partial \theta}{\partial t}. \qquad (10.28)$$

The dimensions of the aquifer itself are small with respect to the wave-length of the tidal phenomena, so that θ may be considered as spatially constant.

A very interesting experiment is being conducted at Heibaart (Belgium) by Sterling and Smets, by maintaining a constant pressure (greater than the atmospheric pressure) in a closed cylindrical bell-jar constructed on the tube of the boring. This eliminates the direct influence of the atmospheric pressure and considerably improves the quality of the tidal curves. Permanent registrations have been performed since 1964 in this well which is tubed up to a depth of 1196 m. All that we said before is not restricted to underground water, but similar observations also concern oil and molten lava.

Two essential criteria allow us to determine if the observed variations in the levels of wells have their origin in the alternating compression-expansion effects to which the underground bed is subjected by the Earth's crust. Firstly, the phase of the phenomena has to be 180° (opposition of phase), because we shall observe a high tide in the well at the moment when the calculation of the luni-solar attraction indicates a low tide. At low tide a compression takes place which causes a rise in the level of the underground water and vice versa.

The second criterion is also important: we shall be dealing with an Earth-tide phenomenon if the various tidal waves observed in the variations of the level are, between them, in the same ratios as forecast by celestial mechanics.

The levels of underground beds usually show a very great sensitivity to variations of atmospheric pressure. Figure 10.13 which shows the variations of the level at Turnhout (Belgium), is particularly suggestive in this respect. It is necessary to eliminate such effects before carrying out the harmonic analysis. But at Basècles, also in Belgium, this atmospheric pressure effect is not important. For Kiabukwa it is non-existent. For Turnhout we made a study in 1956 by a calculation of the correlation with variations in the atmospheric pressure. For the six other wells we have removed this "drift" by applying Pertsev's combination.

The tides at Kiabukwa, (Fig. 10.12) whose latitude is 7°S, are of an almost pure semi-diurnal type, whereas the tides in the six other wells (average latitude 45°N- show important diurnal components.

The results in Table 10.7 for each well clearly show that the proposed criteria in respect to the direct effects of Earth tides are confirmed. The ratios of amplitudes are taken in relation to M_2 in the semi-diurnal group, and in relation to K_1 in the diurnal group. The relation O_1/M_2 of the purely lunar terms, which have been chosen here for comparison in discarding carefully the meteorological effects, vary as a function of the tangent of the latitude ϕ.

Moreover, the amplitudes are not comparable from one station to another because the coefficients are functions of ϕ. Therefore in Table 10.7 we calculate the M_2 amplitudes "reduced to the equator" by dividing by $\cos^2\phi$. We can make two remarks: the amplitudes are always of the order of some centimetres, and with the exception of Oak Ridge, for which we have only 15 days observations, they increase with the depth of the layer.

Haubrich, of the University of Wisconsin, analysed the variations of level in two other wells in the United States: Nunn Busch, Milwaukee, with a depth of 12 cm (observations made in 1946), and Richland Centre (Observations made in 1958). In both cases the amplitudes are about 2-3 cm. Prof. Haubrich has analysed these records by the Fourier method and found for wave M_2 an amplitude of 1.07 cm which, reduced to the equator, gives 2 cm.

Brief Description of Different Observations

A list of 18 wells is given with a short description in the first edition. We should only repeat here those concerning not water-levels but oil or gas or have some very peculiar meaning.

Duchov (Czechoslovakia)

The measurement of the variations of level at Duchov is the first observation of the Earth tide in wells because it dates from 1879. It was made following a catastrophic flooding of a mine whose manager, Klönne, measured the variations in level.

Nienhagen (Germany)

This is an extremely important case because it concerns the variation in the flow of oil wells. The engineer (K. Sperling) demonstrated the effect of the tides M_2 and S_2; nevertheless, an exact harmonic analysis is not possible because the need for exploitation has made it impossible to obtain a recording.

Heibaart (Belgium)

A borehole whose depth is nearly 1660 m: amplitude ratios and phases are very consistent with Earth-tide theory.

Kiabukwa (Congo)

Hot-water spring flowing into a basin 2x4 m. Figure 10.12 shows the recorded curve and the curve of the cubic expansions calculated independently, and which is obviously traced with an arbitrary scale.

Vesuvius (Italy)

The observations made by Prof. Imbo during the 1944 eruption of Vesuvius are of great interest. During the course of each successive phase of eruption a variation was observed with a semi-diurnal lunar period (whose amplitude increases at the same time as the average index of eruptive intensity) in the frequency of the seismic shocks felt at the Vesuvius observatory. These shocks are associated with the process of degasification of the lava. Prof. Imbo interprets the phenomenon as follows: "This action is indirect in the sense that it is sustained by the magmatic masses following the tides of the crust." The result obtained is explained by a periodic variation of the state of compression of the magma from which variations in the excitation of the phenomena leading to seisms will be derived in the closed ducts, and in the open ducts a rise in the magmatic column with a tendency to accentuation in frequency and in the violence of the explosive phenomena when the depth decreases.

Tarka Bridge (Cradock, South Africa)

Young, at the same time of his observations made at the wells of Tarka Bridge stated that a release of methane was made obvious by the short movements of the float when each bubble of gas rose to the surface (Fig. 10.14). The frequency of these releases increases with the high tide in the well, i. e. when the compression is greatest. This is quite unexpected since the solubility of gas in water should increase with pressure.

Ernst noted a relation between the upper transit of the Moon and gaz production in South Germany: the content of methane (CH_4) increasing from 0.2 or 0.3% to 0.5-0.7%.

Tidal effects have even been observed along bedrock joints (Davis and Moore 1965) by using strain-gauge transducers.

Robinson and Bell (1971) have examined the level fluctuations of 30 wells in Virginia and identified tidal effects in 14 of these wells. They analysed the registrations of 10 wells by a least-squares analysis method. They obtain similar values as in Table 10.7 for the specific storage in the Appalachian Mountain Region (from 0.3 to 1.8×10^{-8} cm^{-1}).

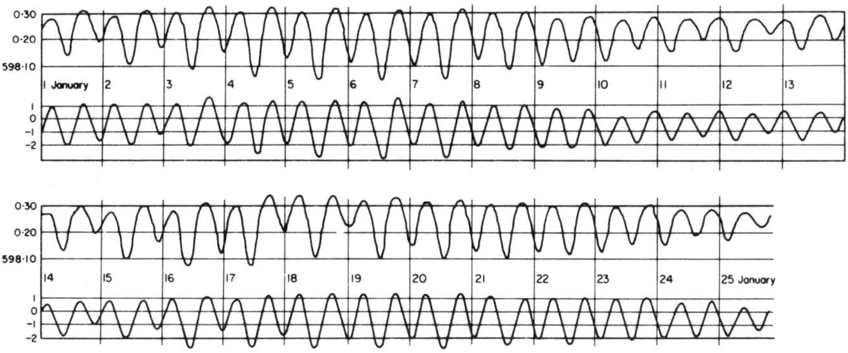

Fig. 10.12. Comparison of variations of level in the well at Kiabukwa (Congo) (upper curve) from 1 to 25 january 1954 with theoretical curve of cubic expansion (lower curve)

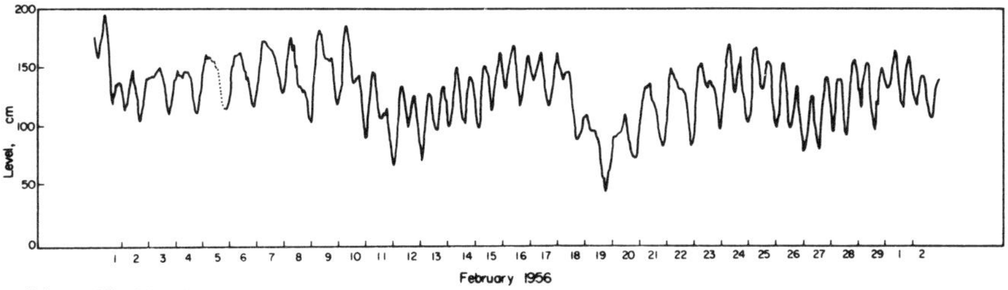

Fig. 10.13. Variations of level in the well at Turnhout (Belgium) after removal of atmospheric pressure effect.

Southern California

Teng and Mc Elrath (1976) report that an experiment performed in a southern California hotspring site shows that the groundwater radon emanation responds to subsurface stress variations, particularly to earth tides.

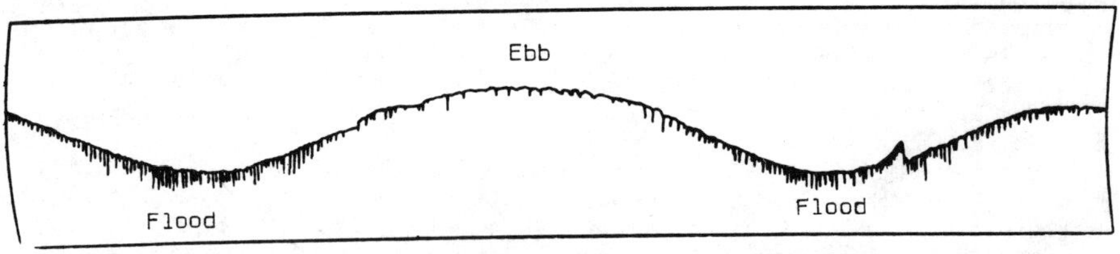

Fig. 10.14. Level oscillations of the Tarka Bridge well (South Africa) showing gas-bubble effects.

CHAPTER 11

The Indirect Effects, Deformations by Surface Loads of a Radially Symmetric Earth

11.1. HISTORICAL BACKGROUND

The first statement at the beginning of this century resulting from the examination of the results of early observations was of the systematic difference between the values of the factor γ provided by the meridian and by the prime vertical components of the tidal tilt, the latter generally leading to a larger value of γ.

The first to look for the reason of this disagreement appears to have been Hecker in 1907, who attributed it to "indirect effects" due to the complex influence exerted by the masses of water moving in the nearby seas. But even before that time D'Abbadie in France and Darwin in England pointed out the possible influence of the oceanic tides on the direction of the vertical near to sea-shores, and Darwin even tried to evaluate this effect (1882).

This influence takes three forms:

(a) the attraction of the masses of water causing the vertical to deviate and g intensity to vary;

(b) a variable flexure of the Earth's crust under the loading effect caused by the water; this is the most important part of the indirect effect;

(c) a variation of the potential of the Earth due to this deformation of the crust, an effect which is in the opposite sense to (a) and (b).

Shida calculated the (a) and (b) effects for the Kamigamo station (Kyoto). His result seemed to support Hecker's conjecture.

The problem, however, is not so simple. If these indirect effects are of major importance in the interpretation of Earth-tide observations, they do not explain all the observed peculiarities. For example, tilt measurements performed in the very centre of the largest continent at the Talgar station (near Alma Ata, Siberia) still exhibit a strong azimuth difference as

$$\gamma_{EW}(M_2) = 0.715, \qquad \gamma_{NS}(M_2) = 0.613,$$

in fair agreement with other Siberian results, a difference which can by no means be explained by indirect effects (Fandyushina, 1976).

11.2. COTIDAL CHARTS

The distribution of oceanic waters varying with the tides is represented by cotidal charts. This denomination is given to special maps of the oceans bearing a superposition of equal-height lines (co-amplitude lines) and equal phase lines (cotidal lines) describing for every tidal constituant the periodic movement of the oceanic waters under the influence of the tidal potential and the Coriolis force due to the rotation of the Earth.

At the beginning of this century Harris discovered that the cotidal lines were not progressing regularly across the oceans as was believed earlier but were turning around some points of zero tide height called amphidromic points.

As long as big computers were not available the drawing of cotidal maps was more or less empirically made according to the personal judgement of the author and using the amplitude and phase parameters derived from the oceanic tidal measurements made on sea-shores and islands. With computers, now available everywhere, it became possible to numerically solve the hydrodynamical equations of Laplace for the different tidal frequencies:

$$\left.\begin{aligned}
&\frac{\partial u}{\partial t} - 2\omega v \cos\theta = -g\,\frac{\partial(\xi-\xi)}{a \sin\theta\partial\lambda} \\
&\frac{\partial v}{\partial t} + 2\omega u \cos\theta = -g\,\frac{\partial(\xi-\bar{\xi})}{a\partial\theta} \\
&\frac{\partial \xi}{\partial t} + \frac{1}{a\sin\theta}\left[\frac{\partial}{\partial\lambda}(uD) + \frac{\partial}{\partial\theta}(vD\sin\theta)\right] = 0
\end{aligned}\right\}$$

(u, v are the velocity components; $D=D(\theta,\lambda)$ is the depth of the ocean; ξ is the height of the oceanic free surface and $\bar{\xi}$ the height of the static tide W_2/g above the undisturbed surface) with boundary conditions which differ according to the authors as shown in Table 11.1.

Methods and types of solution can be divided into two classes, according to whether the solution is derived solely from the tide generating potential and oceanic topography or whether it is constrained to agree with tidal measurements along the coastal boundaries.

Moreover, the different authors have used different grids and numerical schemes for their computations so that the results can be quite different. When integrated, these rather simple equations produce two paradoxes. First, the tides are subject to resonance catastrophes at certain depths (*) and second, on a stationary earth of modest depth all laplacian tides are inverted, exhibiting a 180° lag behind the corresponding newtonian static tides. But these difficulties can be overcome by introduction of the viscous forces: bottom friction and eddy dissipation. Even without these resonance problems, the necessity for introducing eddy dissipation forces follows from the enormous dimensions of the world oceans in which the tidal motions must be considered as entirely turbulent.

Thus one has to add two additional vector terms in the Laplace equations for modelling eddy dissipation and bottom friction. All the authors again differ in their treatment of these effects.

In their original form above, Laplace's equations do not include several non-negligible effects: ocean self-attraction, solid Earth tide and tidal loading of the Earth's crust by the oceanic tidal column itself.

These terms were introduced by Hendershott in 1972. The complete equations are now:

$$\left\{\begin{aligned}
&\frac{\partial u}{\partial t} - 2\,\omega\, v \cos\theta = (1+k-h)\,\frac{\partial W}{a\sin\theta\partial\lambda} + \frac{\partial W'}{a\sin\theta\partial\lambda} - g\,\frac{\partial\xi}{a\sin\theta\partial\lambda} - \\
&\qquad\qquad \nu\,\frac{\sqrt{u^2+v^2}}{D}\,u + \nu^*\,\Delta u \qquad\qquad (11.1)\\
&\frac{\partial v}{\partial t} + 2\,\omega\, u\cos\theta = (1+k-h)\,\frac{\partial W}{a\partial\theta} + \frac{\partial W'}{a\partial\theta} - g\,\frac{\partial\xi}{a\partial\theta} - \nu\,\frac{\sqrt{u^2+v^2}}{D}\,v + \nu^*\,\Delta v \\
&\frac{\partial\xi}{\partial t} - \frac{\partial\zeta}{\partial t} + \frac{1}{a\sin\theta}\left[\frac{\partial}{\partial\lambda}(uD) + \frac{\partial}{\partial\theta}(v\,D\sin\theta)\right] = 0
\end{aligned}\right.$$

(*) it results from this that relatively small changes in the mean oceanic depth cause significant changes in the tidal heights. The mean depth can also be intentionally altered just to avoid resonance.

where

W is the tidal potential

W' is the oceanic shelf gravitation and load potentials

ν is the kinematic viscosity (a quadratic law of bottom friction is introduced in (11.1))

ν^* is the lateral eddy (turbulent) viscosity

$$\Delta = \frac{\partial^2}{a^2 \cos^2\phi \, \partial \lambda^2} + \frac{\partial^2}{a^2 \, \partial \phi^2}$$

A numerical integration of these equations in realistically shaped oceans was performed for the first time by Hendershott (1972).

A solution by Munk, Snodgrass and Wimbush (1970) retaining the Earth-tide effect but neglecting the ocean self-attraction and tidal loading [i.e. assuming $\Gamma = (1+k_2)W_2$ and $\xi = h_2 \, W_2/g$] is used as a first approximation to calculate the neglected terms.

Hendershott's conclusion is that the Earth-tide effect is of the order of the astronomical potential itself while the other two effects are roughly one order of magnitude smaller. Also the existence of an appreciable Earth tide significantly modifies the usual expressions for stored tidal energy and for the rate of working on the oceans by tide-generating bodies which is estimated by Hendershott to be 3.04×10^{19} erg s^{-1} (which yields Q=34).

Omission of all effects due to solid earth tide makes an error of 30% (as $1+k-h \sim 0.7$).

In his more recent work, Schwiderski (1980) disagrees to some extent on such a large role from the oceanic loading effects. The extensive computer experiments on this author demonstrated that without any doubt there is some improvement with ocean loading effects, which should not be neglected. However, considerably more significant improvements were achieved through the inclusion of proper eddy dissipation and bottom friction terms in connection with a hydrodynamically defined ocean basin.
Schwiderski has discussed in details the analytical form of both terms. His conclusion was to adopt a linear law of bottom friction independent of ocean depth and he determined its coefficient b experimentally as depth independent:

$$B = b \, \mu \sin \theta$$

where μ is a longitudinal grading parameter chosen to achieve more uniform spherical mesh areas.

The eddy viscosity is a property of the flow itself and not a physical property of the fluid. Schwiderski proposes to define it as:

$$A = (a/r) \, L \, D(\lambda,\theta) \, (1 + \mu \sin \theta)$$

where a is a coefficient (in s^{-1}) to be chosen by trial and error computations.
L is the equatorial mesh length
μ is the longitudinal grading parameter.

As the mean lateral cross-section area of a local flow cell is

$$\Delta S = D \, L \, (1 + \mu \sin \theta)/2,$$

one sees that

$$A = a \Delta S$$

which is the essential feature introduced by Schwiderski, that the eddy viscosity depends strongly on the mesh area. Thus eddy viscosity is significant in deep ocean and negligible in shallow areas.

TABLE 11.1. Summary of Global Ocean Tidal Models since 1972.

Model	Constituent(s)	Mesh	n	Boundary Conditions	Form of Dissipation	Dissipation ($\times 10^{19}$ erg s^{-1})	Loading Terms
Hendershott (1972)	M2	6° mercator	1452	specified elevation	implicit, in shallow seas and shelves only	3.08	yielding to astronomical force only
Zahel (1973)	K1	4° spherical	1013	reflecting	bottom stress in shallow water	-	none
Zahel (1973)	M2	1° spherical	41068	reflecting	bottom stress in shallow water	3.77	none
Estes (1977)	M2,S2,N2,K2, K1,O1,P1	2° spherical		reflecting	bottom stress in shallow water	-	yielding to astronomical force only
Estes (1977)	M2	3° spherical		reflecting	bottom stress in shallow water	-	complete potential using Green's function
Schwiderski (1978)	M2	1° spherical	40855	reflecting	bottom friction $BL^2\ \mu \cos \phi$ eddy dissipation $a/2\ L\ H(\lambda,\phi)(1+\mu\cos\theta)$	-	estimate loading terms with $0.1\ \zeta_o$ $\gamma = 0.690$
Parke-Hendershott (1978)	M2,S2,K1	6° mercator	1452	specified elevation	implicit, in shallow seas and shelves only	M2 = 2.22 S2 = 0.208 K1 = 0.221	complete potential using Green's function $\gamma = 0.690$
Accad and Pekeris (1978)	M2,S2	2° mercator		modified Proudman condition using ramp shaped shelf edge	implicit, in shallow seas and shelved only	M2 = 2.55 S2 = 0.526	estimate loading terms with 0.085 $_0$
Bogdanov-Magarik ( )	M2,S2,K1,O1	5°	1770	specified elevation	in shallow seas and shelves		

Extensive exploratory computations have been performed by Schwiderski to select the best values for a and b such as to obtain the best fit with some thousands of selected islands and shore experimental measurements.
Schwiderski has incorporated into his models over 2000 empirical tide data collected around the world at continental and island stations. The grid he chooses is a 1°x1° one which means that the world ocean will be described by a net of 40855 polygones, each one being characterized for each tidal wave by a mean tidal amplitude, a mean phase, its coordinates and surface. Despite such a high number of polygones, some important regional seas still are not included in the Schwiderski maps: Baltic, Kattegat, Irish, Mediterranean seas, Red Sea, Seas of Japan, Sulu, Seram, Hudson, and Korean bays, Chihlian, Persian and Californian Gulfs.

Thus, for earth tide stations situated in the vicinity of these areas, special local cotidal maps will have to be reintroduced.

Schwiderski has produced cotidal maps for all important constituents of the tide: Q_1 O_1 P_1 K_1 N_2 M_2 S_2 K_2 M_f.

He has given an excellent comprehensive exposition of his method in "Reviews of Geophysics and Space Physics" vol. 18, 243-268, 1980.

However, even if these new maps may be considered as very precise (Schwiderski claims for a precision of 5cm in the heights for his M_2 cotidal map and of 10cm on the total prediction) they still have a serious defect for gravity attraction and loading computation as they do not perfectly conserve the tidal mass.

The distribution of non-conserved tidal mass is not known and therefore one uses an artifice such as an equal redistribution over all the oceans or corrections proportional to the tidal amplitude itself. Some authors (Pertsev, Groten) tried to represent the oceanic tides distribution by a Legendre polynomial development, a procedure which makes the computation of the indirect effects on the continents easier. However, to represent a tidal map by a Legendre polynomial development correctly, one must retain more than order n=30 (which means 960 terms).

Anyway, this raises the problem of a non-zero tide on the continents due to the truncation of the series and a problem of mass conservation. (see Table 15.3).

These remarks show that the calculation of the load tide is far more complicated than that of the body tide. Celestial mechanics provide us with an extremely precise description of the body force, while oceanography can only give us empirical observations or numerical integrations of the hydrodynamical equations.

For an exposure of the Oceanic Tides problem see:

M.C. Hendershott - Numerical models of Ocean Tides in "The Sea" vol.6, Wiley Int. Publ. 1977.

M.C. Hendershott - Ocean Tides in EOS, 54, 2, 76-86, 1973

E.W. Schwiderski - On Charting Global Ocean Tides Review of Geophysics and Space Physics 18, 1, 243-269, 1980.

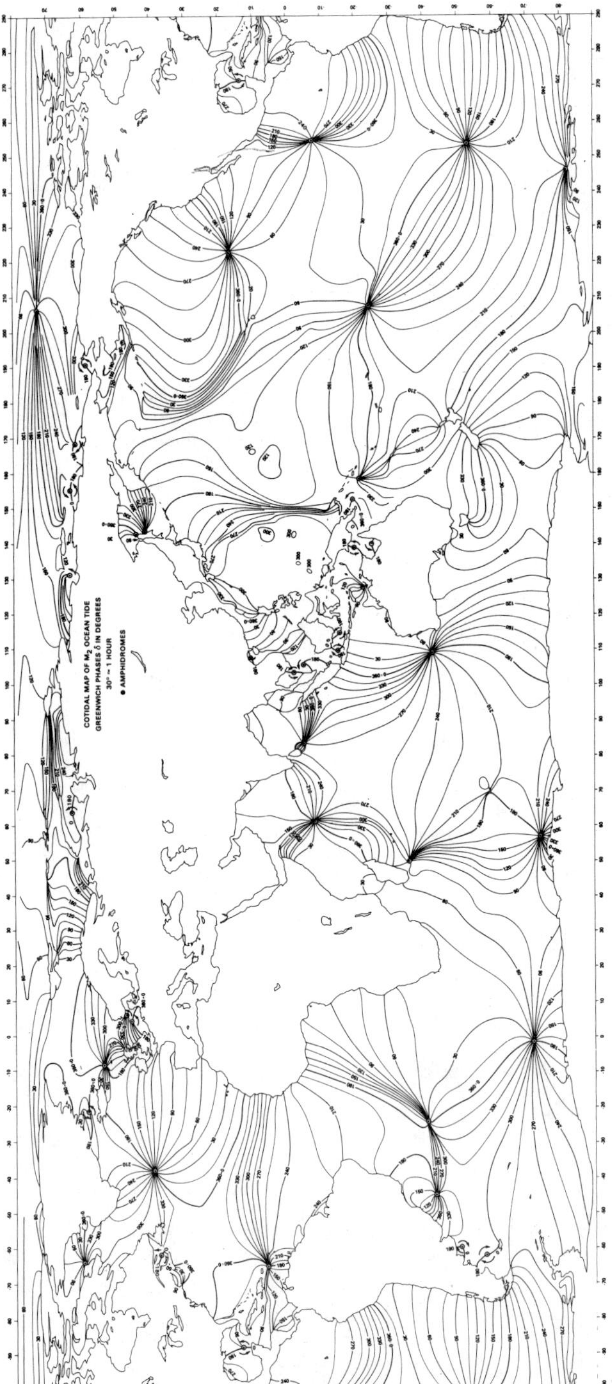

Fig. 11.1: Schwiderski M_2 cotidal map - Cotidal lines in degrees (30° ~ 1 hour).

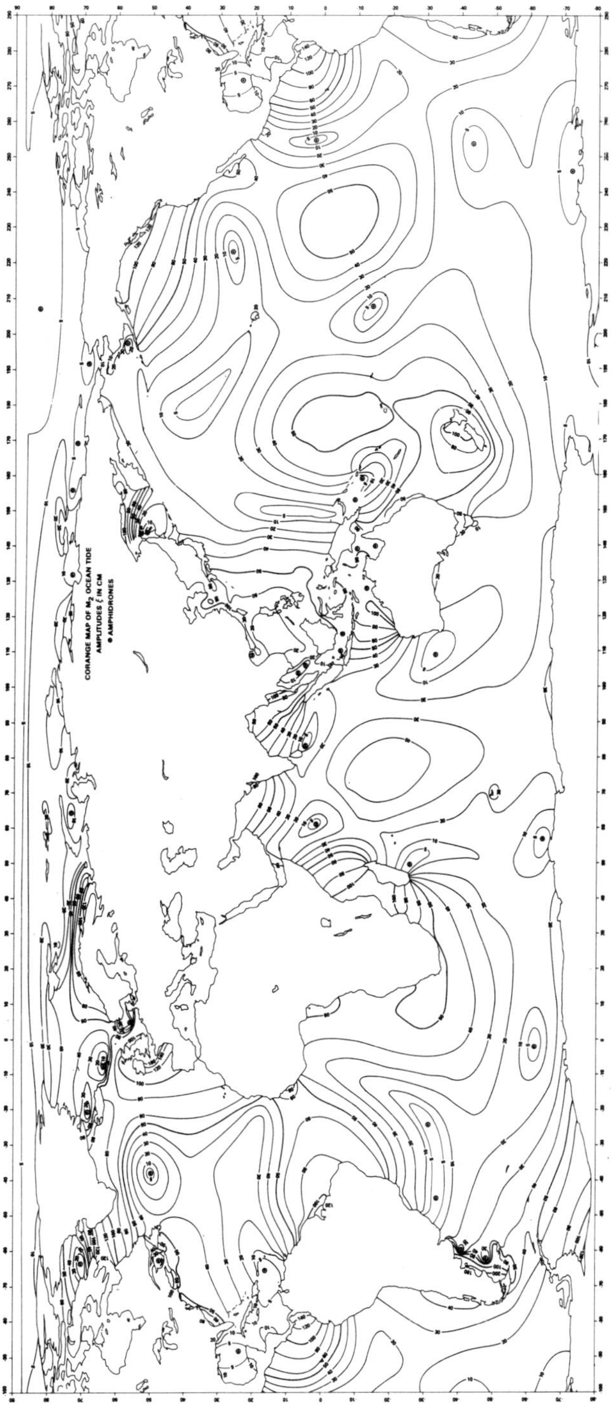

Fig. 11.2: Schwiderski M_2 corange map - amplitudes in cm.

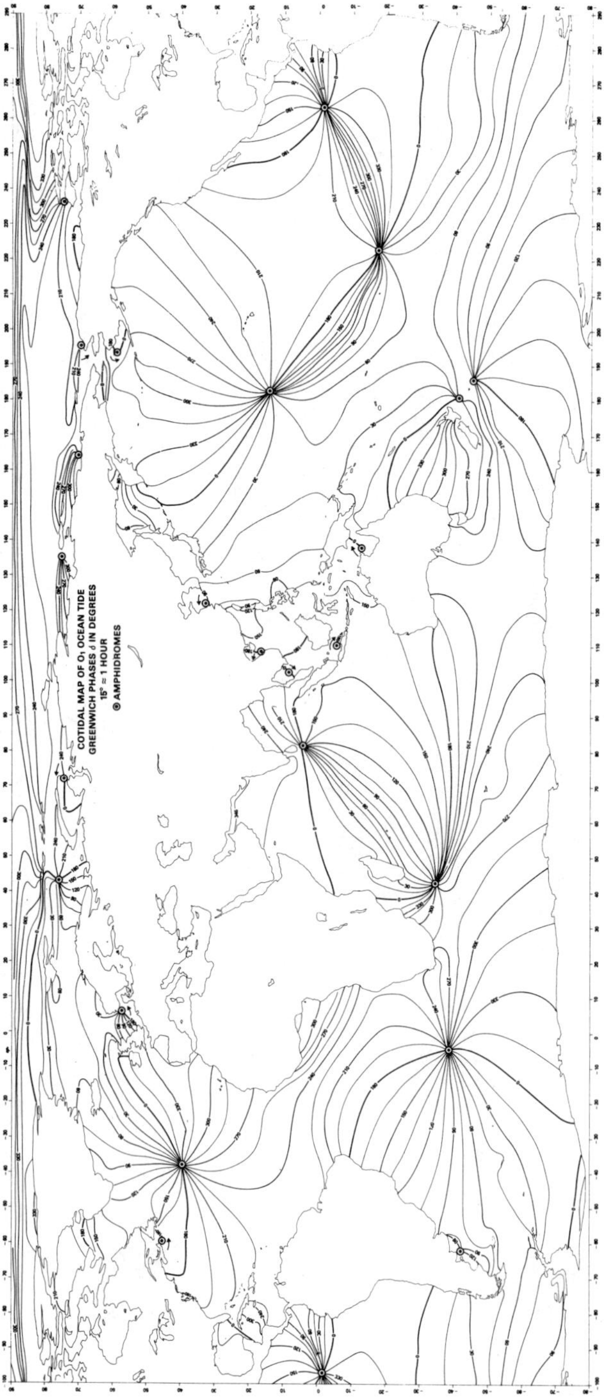

Fig. 11.3: Schwiderski O_1 cotidal map - cotidal lines in degrees (15° ~ 1 hour).

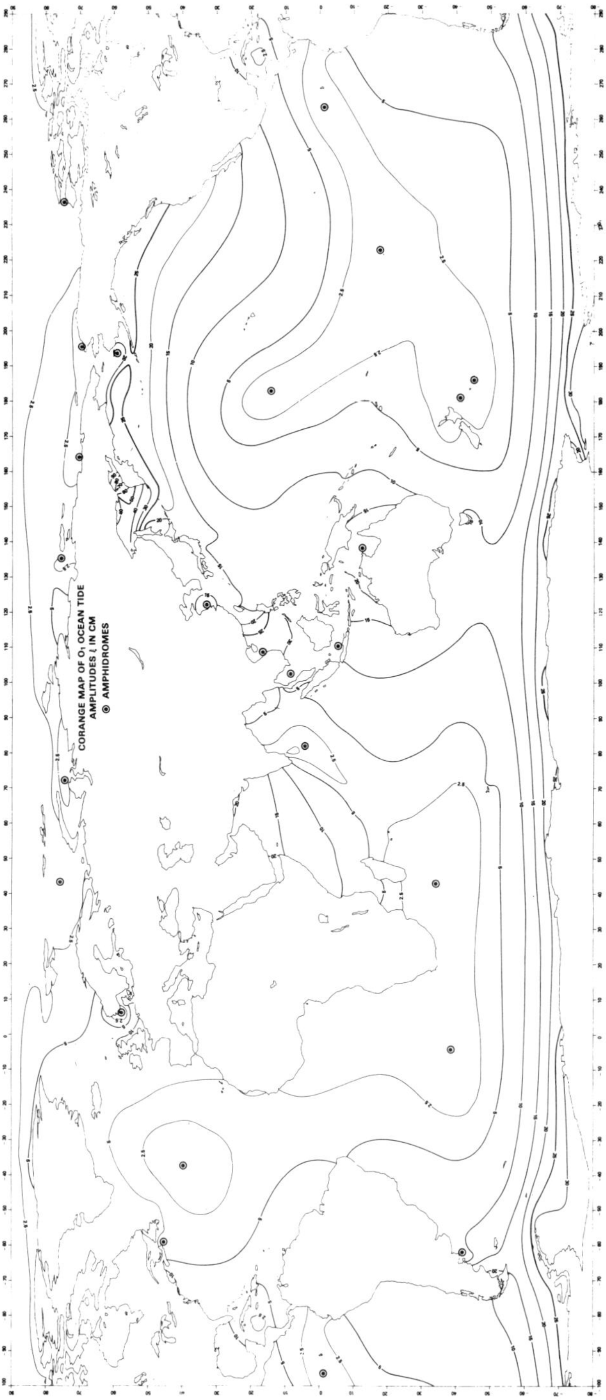

Fig. 11.4: Schwiderski O_1 corange map - amplitudes in cm.

The distribution of non-conserved tidal mass is not known and therefore one uses an artifice such as an equal redistribution over all the oceans or corrections proportional to the tidal amplitude itself. Some authors (Pertsev, Groten) tried to represent the oceanic tides distribution by a Legendre polynomial development, a procedure which makes the computation of the indirect effects on the continents easier. However, to represent a tidal map by a Legendre polynomial development correctly, one must retain more than order n=30 (which means 960 terms).

Anyway, this raises the problem of a non-zero tide on the continents due to the truncation of the series and a problem of mass conservation.

These remarks show that the calculation of the load tide is far more complicated than that of the body tide. Celestial mechanics provide us with an extremely precise description of the body force, while oceanography can only give us empirical observations or numerical integrations of the hydrodynamical equations which is still far from being perfect.

11.3. CALCULATION OF THE WATER ATTRACTION

Assuming here that such charts are precise enough we divide (Fig. 11.7) the oceanic surface into small sectors dS with constant amplitude h of the oceanic tide, for which the station occupies the summit and which are limited by the radii r_1, r_2 and the azimuths θ_1, θ_2 (reckoned counter clockwise).

The attraction caused by such a nearby element of volume $dV = h\ r\ dr\ d\theta$ and of mass ρdV (where ρ is the density of ocean water) is

$$d\alpha = G\ \rho\ h\ \frac{dr\ d\theta}{r} \qquad (11.2)$$

projecting on to Ox axis, for example,

$$\left.\begin{aligned} d\alpha_x &= G\ \rho\ h\ \frac{dr\ d\theta}{r}\cos\theta, \\ \alpha_x &= \int_{r_1}^{r_2}\int_{\theta_1}^{\theta_2} G\ \rho\ h\ \frac{dr\ d\theta}{r}\cos\theta, \end{aligned}\right\} \qquad (11.3)$$

which gives the components of the deviation of the nadir:

towards the South:

$$\xi'' = \frac{G\ \rho\ h}{g\ \sin 1''}\ (\log r_2 - \log r_1)\ (\sin\theta_2 - \sin\theta_1);$$

towards the East: (11.4)

$$\eta'' = -\frac{G\ \rho\ h}{g\ \sin 1''}\ (\log r_2 - \log r_1)\ (\cos\theta_2 - \cos\theta_1).$$

For distant masses the sphericity of the Earth must obviously be taken into account. One obtains in the vertical component (Fig. 11.7)

$$\Delta g = -\frac{Gh\rho}{4a^2}\iint\frac{ds}{\sin\kappa/2},$$

and in the horizontal components (11.5)

$$\begin{pmatrix}\xi\\ \eta\end{pmatrix} = -\frac{Gh\rho}{4a^2 g}\begin{pmatrix}\cos\alpha\\ \sin\alpha\end{pmatrix}\iint\frac{\cos\frac{\kappa}{2}\ ds}{\sin^2\frac{\kappa}{2}}.$$

Applications of this procedure require the digitization of the cotidal charts in a form suitable for computation, e. s. a very great number of spherical trapezia where three parameters are determined:

a) the percentage of sea in proportion to land;
b) the mean amplitude of each tidal wave;
c) the mean phase-lag of each tidal wave with respect to the maximum of the potential at the zero meridian.

In all these formulas we will have to replace h by h $\cos(\omega t-\psi)$ for each constituent of the tide.

11.4. CALCULATION OF THE FLEXURE OF A FLAT EARTH'S CRUST DUE TO THE LOAD OF THE WATER

In 1878 Boussinesq, using the logarithmic potential, calculated the distortion of an elastic and homogeneous semi-infinite body limited by a plane surface under the effect of an overload and showed that the vertical displacement of the free surface was proportional at each point to the gravitational potential of the load and was a function of the elastic constants of the medium (λ, μ, Lamé's constants).

In general the deviation of the vertical due to the attraction may be written similarly to (11.3),

$$\alpha = \frac{G\rho}{g}\frac{\partial}{\partial x}\int\frac{h\,d\sigma}{r}\,, \tag{11.6}$$

$d\sigma$ being the elementary surface (not necessarily a sector) r the distance and with the load per unit of area given by

$$P = g\rho h, \tag{11.7}$$

we write

$$dP = P\,d\sigma = g\,\rho h\,d\sigma. \tag{11.8}$$

Therefore, since P is a force acting normally to the surface of the body,the displacements to which it gives rise are given by Boussinesq's solution:

$$\left.\begin{aligned}
u &= \frac{-1}{4\pi(\lambda+\mu)}\frac{\partial\chi}{\partial x} - \frac{1}{4\pi\mu}\frac{\partial^2\psi}{\partial x\partial z},\\
v &= \frac{-1}{4\pi(\lambda+\mu)}\frac{\partial\chi}{\partial y} - \frac{1}{4\pi\mu}\frac{\partial^2\psi}{\partial y\partial z},\\
w &= \frac{-1}{4\pi(\lambda+\mu)}\frac{\partial\chi}{\partial z} - \frac{1}{4\pi\mu}\frac{\partial^2\psi}{\partial z^2} + \frac{\lambda+2\mu}{4\pi\mu(\lambda+\mu)}\Delta^2\psi,
\end{aligned}\right\} \tag{11.9}$$

where

$$\psi = \iint Pr\,dx\,dy = \int r\,dP, \tag{11.10}$$

$$\chi = \iint P\log(z+r)dx\,dy = \int\log(r+z)\,dP, \tag{11.11}$$

from which

$$u = \frac{-1}{4\pi\mu}\frac{\partial^2}{\partial x\,\partial z}\int r\,dP - \frac{1}{4\pi(\lambda+\mu)}\frac{\partial}{\partial x}\int\log(r+z)dP, \tag{11.12}$$

$$\left.\begin{aligned} v &= \frac{-1}{4\pi\mu}\frac{\partial^2}{\partial y\,\partial z}\int r\,dP - \frac{1}{4\pi(\lambda+\mu)}\frac{\partial}{\partial y}\int \log(r+z)dP, \\ w &= \frac{-1}{4\pi\mu}\frac{\partial^2}{\partial z^2}\int r\,dP - \frac{\mu}{4\pi\mu(\lambda+\mu)}\int\frac{dP}{r} + 2\,\frac{\lambda+2\mu}{4\pi\mu(\lambda+\mu)}\int\frac{dP}{r}, \end{aligned}\right\} \tag{11.12}$$

because $\Delta r = +2/r$.

Thus the vertical displacement is

$$w = -\frac{1}{4\pi\mu}\frac{\partial^2}{\partial z^2}\int r\,dP + \frac{2\lambda+3\mu}{4\pi\mu(\lambda+\mu)}\int\frac{dP}{r}$$

It is immediately evident that

$$\begin{aligned} -\frac{1}{4\pi\mu}\frac{\partial^2}{\partial z^2}\int r\,dP &= -\frac{1}{4\pi\mu}\frac{\partial}{\partial z}\int z\,\frac{dP}{r} \\ &= -\frac{1}{4\pi\mu}\int\frac{dP}{r} - \frac{1}{4\pi\mu}\int z\,dP\,\frac{z}{r^3}, \end{aligned}$$

and since our equipment is situated almost at ground level, we may put $z = 0$, so that

$$w = \frac{\lambda+2\mu}{4\pi\mu(\lambda+\mu)}\int\frac{dP}{r}. \tag{11.13}$$

This shows that the vertical displacement of the crust is proportional at each point to the gravitational potential of the load and is a function of the elastic constants (λ, μ) of the medium.

In the non-perturbed Earth's gravity field this produces a variation of gravity equal to

$$\Delta g = -2\,\frac{g}{a}\,\frac{\lambda+2\mu}{4\pi\mu(\lambda+\mu)}\int\frac{dP}{r}. \tag{11.14}$$

Integrating this expression on a small flat sector as represented on Fig. 11.7 gives

$$\left(\Delta g\right)_2^1 = -\frac{2g^2}{a}\,\frac{\lambda+2\mu}{4\pi\mu(\lambda+\mu)}\,(r_2-r_1)\,(\theta_2-\theta_1)\rho h \tag{11.15}$$

To obtain the slope of the ground in the x direction we differentiate (11.13) with respect to x and, taking relation (11.8) into consideration, we obtain

$$\bar{\omega} = \frac{\partial w}{\partial x} = \frac{\lambda+2\mu}{4\pi\mu(\lambda+\mu)}\,g\rho\,\frac{\partial}{\partial x}\int\frac{h\,d\sigma}{r}. \tag{11.16}$$

The deviation of the vertical given by (11.6) is similar in form to (11.16). Therefore the slope of the ground under the effect of the load may be written in the form

$$\bar{\omega} = \frac{\lambda+2\mu}{4\pi\mu(\lambda+\mu)}\,\frac{g^2}{G}\cdot\alpha = \bar{\omega}_1\,\frac{\lambda+2\mu}{\mu(\lambda+\mu)}\,\alpha = \beta\alpha \tag{11.17}$$

with

$$\bar{\omega}_1 = \frac{1}{4\pi}\,\frac{g^2}{G} = 11.48 \times 10^{11} \sim 12 \times 10^{11}\ \text{dynes cm}^{-2}.$$

We may consider two extreme cases:

(a)$\lambda = \infty$, an incompressible crust, for which we have

$$\bar{\omega}_A = \frac{1}{4\pi}\frac{g^2}{G}\frac{1}{\mu}\alpha = \bar{\omega}_1\frac{1}{\mu}\alpha$$

(b)$\lambda = \mu$, a highly compressible crust, which yields

$$\bar{\omega}_B = \frac{3}{4\pi}\frac{g^2}{G}\frac{1}{2\mu}\alpha = \frac{3}{2}\bar{\omega}_1\frac{1}{\mu}\alpha.$$

Numerical examples are given in Table 11.2.

TABLE 11.2.

Modulus of rigidity of the crust (dynes cm^{-2})	Incompressible crust	Highly compressible crust
$\mu = 6\times10^{11}$	$\bar{\omega}_A = 2\alpha$	$\bar{\omega}_B = 3\alpha$
$\mu = 3\times10^{11}$	$\bar{\omega}_A = 4\alpha$	$\bar{\omega}_B = 6\alpha$
$\mu = 1\times10^{11}$ (gneiss)	$\bar{\omega}_A = 12\alpha$	$\bar{\omega}_B = 18\alpha$
$\mu = 0.18\times10^{11}$	$\bar{\omega}_A = 67\alpha$	$\bar{\omega}_B = 100\alpha$
$\mu = 0.14\times10^{11}$ (Japanese tuff)	$\bar{\omega}_A = 86\alpha$	$\bar{\omega}_B = 129\alpha$

The response to ocean loads depends very much on the locally variable properties of the crust and mantle. A model composed of homogeneous, isotropic spherical layers is much more likely to be valid for the body tide, having significant displacements through most of the Earth's volume, than for the load tide whose displacements are appreciable only in the crust and upper mantle. Differences in crustal structure, as beneath ocean basins and continents, will therefore affect the load more than the body tide. For example, the surface deformation near the load is very sensitive to the properties of sediments.

Thus it is necessary to modify the solution by considering the half-space not uniform but layered in accordance with the upper strata of the crust. One has to take into account layers down to a depth of two or three times the horizontal distance between the point mass and the point of observation. Below it is sufficient to suppose the solid extends to infinite depth and to have the elastic constants of the lowest layer.

An interesting application of this formula had been made by Lambert (1970) for observations made at Rawdon, Nova Scotia (Canada). The loading effects on gravity and tilt caused by large amplitude oceanic tides in the Bay of Fundy (5 m for M_2) and by tides in the Gulf of St. Lawrence and the North Atlantic Ocean were measured with a pair of Verbaandert-Melchior quartz horizontal pendulums (8 months) and a LaCoste-Romberg gravity meter (3 short series). The results were derived mainly from pendulum data, those of the gravity meter being made principally as a check of the results of tilt measurements.

The tiltmeter site at Rawdon, Nova Scotia, is ideally situated for separating the tilting due to the Bay of Fundy tides from the St. Lawrence and Atlantic tides. The separation is carried out by taking advantage of the distinct differences in the distribution of phases and amplitudes of the main constituents (M_2, S_2, N_2) of the ocean tides in function of the azimuth around Nova Scotia.

Calculation of horizontal attraction is carried through for the M_2 and K_1 waves. The N_2 and S_2 waves and the O_1 wave are respectively deduced from M_2 and K_1 considering amplitude ratios and phase-lags with respect to the main waves.

Lambert computes the β factor by adjusting the attraction calculation to the observed tilt, corrected from the pure Earth-tide effect with γ factor tentative values of 0.5, 0.7 and 0.9.

This gave him for the North-South component

$$\beta = 5.8 \pm 0.4,$$

and assuming Poisson's ratio $\sigma = 0.26$ he derives

$$\mu = 3.5 \pm 0.3 \times 10^{11} \text{ c.g.s.},$$

in excellent agreement with seismic results for the crustal layer beneath Nova Scotia,

$$\mu = 3.4 \times 10^{11} \text{ c.g.s..}$$

For the East-West component, directed towards the Bay of Fundy, he finds a higher apparent rigidity which could result from the deformation of deeper, more rigid layers.

Several models with exponential distributions of rigidity in function of depth, as given by Takeuchi and Jobert, have been tested by Lambert.

A more detailed analysis of the Rawdon data has been made later on by Lambert and Beaumont by the finite elements method and Green's functions (see Section 11.8).

(c) Indirect Effects on Strain Measurements

Ozawa writes Boussinesq's equations (11.12) in the form:

$$\left.\begin{aligned} u &= \frac{P}{4\pi\mu}\frac{xz}{r^3} - \frac{P}{4\pi(\lambda+\mu)}\frac{x}{r(z+r)}, \\ v &= \frac{P}{4\pi\mu}\frac{yz}{r^3} - \frac{P}{4\pi(\lambda+\mu)}\frac{y}{r(z+r)}, \\ w &= \frac{P}{4\pi\mu}\frac{z^2}{r^3} + \frac{P(\lambda+2\mu)}{4\pi\mu(\lambda+\mu)}\frac{1}{r}, \end{aligned}\right\} \tag{11.18}$$

and derives the strains as follows:

$$\left.\begin{aligned} e_{xx} &= \frac{\partial u}{\partial x} = \frac{P}{4\pi\mu}\left(\frac{z}{r^3} - \frac{3x^2z}{r^5}\right) - \frac{P}{4\pi(\lambda+\mu)}\left[\frac{(r^2-x^2)(z+r)-rx^2}{r^3(z+r)^2}\right], \\ e_{yy} &= \frac{\partial v}{\partial y} = \frac{P}{4\pi\mu}\left(\frac{z}{r^3} - \frac{3y^2z}{r^5}\right) - \frac{P}{4\pi(\lambda+\mu)}\left[\frac{(r^2-y^2)(z+r)-ry^2}{r^3(z+r)^2}\right], \\ e_{zz} &= \frac{\partial w}{\partial z} = \frac{P}{4\pi\mu}\left(\frac{2z}{r^3} - \frac{3z^3}{r^5}\right) - \frac{P(\lambda+2\mu)}{4\pi\mu(\lambda+\mu)}\frac{z}{r^3}, \\ e_{xy} &= \frac{\partial u}{\partial y} + \frac{\partial v}{\partial x} = -\frac{P}{4\pi\mu}\frac{3xyz}{r^5} + \frac{P}{4\pi(\lambda+\mu)}\frac{2xy(z+2r)}{r^3(z+r)^2}, \end{aligned}\right\} \tag{11.19}$$

$$\left.\begin{aligned} e_{zx} &= \frac{\partial w}{\partial x} + \frac{\partial u}{\partial z} = -\frac{P}{4\pi\mu}\frac{6xz^2}{r^5}, \\ e_{yz} &= \frac{\partial v}{\partial z} + \frac{\partial w}{\partial y} = -\frac{P}{4\pi\mu}\frac{6yz^2}{r^5}, \\ \theta &= e_{xx} + e_{yy} + e_{zz} = \frac{P}{4\pi(\lambda+\mu)}\frac{2z}{r^3}. \end{aligned}\right\} \quad (11.19)$$

Then, very near to the surface ($z \sim 0$) one has:

$$\left.\begin{aligned} e_{xx} &= -e_{yy} = \frac{P}{4\pi(\lambda+\mu)}\frac{x^2-y^2}{r^4} = \frac{P}{4\pi(\lambda+\mu)}\frac{\cos 2\theta}{r^2}, \quad e_{zz} = 0, \\ e_{xy} &= \frac{P}{4\pi(\lambda+\mu)}\frac{4xy}{r^4} = \frac{P}{4\pi(\lambda+\mu)}\frac{2\sin 2\theta}{r^2}, \\ \Sigma &= 0, \\ \theta &= 0. \end{aligned}\right\} \quad (11.20)$$

The component e_{xy} is twice the e_{xx}, e_{yy} components when $2\theta = 45°$. The vertical and horizontal areal strain as well as the cubic dilatation are free from indirect effects in Boussinesq's solution while the rotational becomes:

$$\left.\begin{aligned} \omega_x &= -\frac{P}{4\pi}\left[\frac{\lambda+2\mu}{\mu(\lambda+\mu)}\right]\frac{y}{r^3}, \\ \omega_y &= \frac{P}{4\pi}\left[\frac{\lambda+2\mu}{\mu(\lambda+\mu)}\right]\frac{x}{r^3}, \\ \omega_z &= 0. \end{aligned}\right\} \quad (11.21)$$

For practical numerical evaluations equations (11.20) should be written:

$$\left.\begin{aligned} e_{xx} &= -e_{yy} = \frac{P}{4\pi(\lambda+\mu)}\int_{r_1}^{r_2}\int_{\theta_1}^{\theta_2}\frac{\cos 2\theta}{r^2}\, r\, dr\, d\theta \\ &= \frac{P}{4\pi(\lambda+\mu)}\frac{1}{2}(\log r_2 - \log r_1)(\sin 2\theta_2 - \sin 2\theta_1), \\ e_{xy} &= \frac{P}{4\pi(\lambda+\mu)}\int_{r_1}^{r_2}\int_{\theta_1}^{\theta_2}\frac{2\cdot\sin 2\theta}{r^2}\, r\, dr\, d\theta \\ &= \frac{-P}{4\pi(\lambda+\mu)}(\log r_2 - \log r_1)(\cos 2\theta_2 - \cos 2\theta_1). \end{aligned}\right\} \quad (11.22)$$

(d) Potential Variation

This effect is smaller than (a) and (b) and acts as if it were caused by a negative mass. Darwin tried to evaluate it in 1882 but Nishimura appears to be the first to have taken it into consideration; he suggests that for a first approximation it can be considered as proportional to α and puts the sum of the indirect effects

$$I = (1 + \beta - \varepsilon)\alpha. \quad (11.23)$$

TABLE 11.3. Correction of amplitude factors and phases of the M_2 wave at Akagane and Hosokura for indirect effects

North-South Component	Akagane			
	VM pendulum		TEM tiltmeter	
	γ 0.381 A cos α	α -14.9° A sin α	γ 0.386 A cos α	α -23.6° A sin α
Attraction	0.830	-0.144		
Flexure	1.308	-0.756		
Potential variation	-0.425	0.116		
Total	1.713	-0.784	1.713	-0.784
Observed value	2.844	-0.753	2.725	-1.189
Total	4.557	-1.537	4.435	-1.973
Corrected amplitude	4.809		4.857	
Theoretical amplitude	7.701		7.701	
Corrected factor γ	0.624		0.631	
Corrected phase-lag	-18°639		-23°969	

East-West Component	Akagane				Hosokura	
	VM pendulum		TEM tiltmeter		TEM tiltmeter	
	γ 0.133 A cos α	α -135.5° A sin α	γ 0.235 A cos α	α -127.7° A sin α	γ 0.29 A cos α	α -155.9° A sin α
Attraction	-0.251	-2.458			0.003	-1.582
Flexure	0.501	-4.423			0.006	-3.772
Potential variation	0.083	1.210			-0.001	0.791
Total	0.333	-5.671	0.333	-5.671	0.008	-4.563
Observed value	-1.157	-1.136	-2.247	-1.778	-1.451	-3.244
Total	-0.824	-6.807	-1.914	-7.449	-1.443	-7.807
Observed amplitude	6.857		7.691		7.801	
Theoretical amplitude	12.193		12.193		12.255	
Factor γ	0.562		0.631		0.648	
Phase-lag	-6°949		-14°410		-1°472	

VM = Verbaandert-Melchior

Tem = Tsubokawa electromagnetic pendulum.

The Japanese authors Shida, Terazawa, Sekiguchi, Takahasi and Nishimura have tried to take into account the variations of the elastic characteristics of the medium with depth by stating that the depth of the crust involved in the deformation was dependent on the distance r from the disturbing mass. They write

$$I = (1 + \beta (r) - \varepsilon)\alpha. \qquad (11.24)$$

To obtain empirically the numerical values of these coefficients Nishimura carried out simultaneous observations at two Japanese stations: Aso and Kamigamo. The difference between the components M_2 (or O_1) obtained for these two points is free from the direct effect, which is practically the same, and only includes the indirect effects which are dependent on the positions of the stations relative to the neighbouring seas. The interpretation of this difference allowed Nishimura to obtain the values

$$\beta(r) = \frac{12.6}{r + 3}, \qquad \varepsilon = 0.5, \qquad (11.25)$$

r being expressed in degrees (distance counted in degree between the observation station and point mass).

Melchior compared this experimental result with an older theoretical study carried out by Rosenhead who, preoccupied with the movements of the masses of air at the surface of the Earth and of their effect on the products of inertia of the Earth, tried in 1929 to estimate theoretically the importance of the compensation $\varepsilon\alpha$ due to the flexure of the Earth's crust under the weight of these masses of air.

Adapting Herglotz's theory of the elasticity of the Earth to the present example, he found

$$0.3 < \varepsilon < 0.4,$$

the same order of magnitude as that found by Nishimura.

In his calculation Rosenhead used a modulus of rigidity equal to zero for the core and 16.95×10^{11} dynes cm^{-2} for the shell.

Nishimura, estimating that the distant masses of water have the same influence on the observations of the neighbouring stations of Aso and Kamigamo, thinks that only the masses within a radius of 1000 km play a role in the difference of the results of the two stations for the indirect effects, and considering the depth of the layers affected by the phenomenon to be 1000 km, he obtained for the local rigidity $\mu = 6.17 \times 10^{11}$ dynes cm^{-2}, a value that agrees with that already obtained by Shida using the earlier observations from Kamigamo. This interpretation leads to a correct determination of the modulus of rigidity of the outer mantle (60-1200 km depth).

A similar approach has been used more recently with some success (Table 11.3) by Sugawa and Hosoyama to compare simultaneous observations made at Akagane and at Hosokura (north of Japan) which give amplitude factors and phases deeply affected by the loading effect and attraction in the nearby sea. Table 11.3 has been obtained by using the procedure described in the preceding pages and mainly (11.25).

11.5. LOAD NUMBERS

By analogy to Love numbers, Munk and MacDonald introduced in 1960 similar parameters - load numbers - to describe the deformation of the Earth under a variable surface load of potential U_n. If such a surface load is produced by a gravitating layer of density q_n, one has, of course, for a point internal to the Earth, at the distance r from the origin,

$$U_n = \frac{4\pi Ga}{2n+1} q_n \left(\frac{r}{a}\right)^n . \qquad (11.26)$$

Considering that such a load has two opposite effects, a depression of the surface under the normal stress ($p_n = -g\, q_n$) and an uplift due to the gravitational attraction of the loading mass they represent the combined effect by

$$\zeta'_n = h'_n \frac{U_n}{g},$$

while the gravitational potential arising from this distortion is

$$k'_n U_n .$$

As the depression is larger than the uplift these numbers are negative. A ℓ'_n load number is defined similarly for the lateral deformation.

11.6. THE LONGMAN AND FARRELL SOLUTIONS OF THE LOADING EFFECT ON A SPHERICAL EARTH

Longman introduced in 1962 the use of Green's functions for determining the deformation of a spherical Earth under a concentrated surface mass load instead of expressing the surface mass distribution by a summation of surface zonal harmonics.

The name Green is given to such functions which enable linear differential equations *with boundary conditions* at the external spherical surface to be solved. Green's functions are here the point-load response functions, i. e. the two displacement components and the gravitational potential perturbation caused by the point surface mass load as *a function of the angular distance from the load*. The proper weighted sums of the load numbers (for each n) must be added up to form these Green's functions.

A normalized Green's function is constructed by dividing for example the tilt Green's function by the apparent tilt that results from the newtonian attraction of the point mass on a perfectly rigid Earth. Then for a particular theoretical earth model the tidal loading tilt is computed by convolving the normalized Green's function with the newtonian attraction contribution of the surrounding ocean tide.

$$T(r_s) = \iint_{Oceans} \mathcal{G}(|\bar{r}-\bar{r}_s|)\, H(r)\, dS \qquad (11.27)$$

where $\mathcal{G}$ is the tilt Green's function, H(r) is the *newtonian tilt* due to a water mass distributed over the infinitesimal area dS, and $\bar{r}$ and $\bar{r}_s$ are vectors describing the positions of the water mass and the observation site respectively.

In the case of the homogeneous half space and neglecting the potential variation, the Green's function tilt is simply, as given by (11.17)

$$\mathcal{G} = \frac{\lambda+2\mu}{4\pi G\mu(\lambda+\mu)}\, g^2 ;$$

it is independent from r.

The strain tensor elements can also be represented by using Green's function. By convolving each of these functions with the ocean tidal amplitudes one obtains the corresponding load tide characteristics at any point on the Earth's surface. The convolution integral is the weighted sum of these functions, the weights being the amplitudes of the various point loads.

If we consider a surface density given by

$$S = \sigma, \qquad 0 < \kappa < \varepsilon,$$

$$S = 0, \qquad \varepsilon < \kappa < \pi,$$

i. e. a circular cap of semi-angle ε, we can represent it by

$$S = \sum_{n=0}^{\infty} K_n \, P_n \, (\cos \kappa). \tag{11.28}$$

Longman shows that when $\varepsilon \to 0$ we have

$$K_n = \frac{2n+1}{4\pi a^2} \tag{11.29}$$

then he solves the differential equations for free oscillations of Alterman, Pekeris and Jarosch adapted to the application of a point mass load to the surface, by setting the frequency equal to zero. Longman solves these equations subject to boundary conditions determined by the mass distribution (11.28) placed on the surface, and internal conditions at the centre r=0 and at interfaces.

At the deformed surface of the Earth we must have the normal and tangential stress equal to the gravitational stress acting on the surface distribution of mass, but all supplementary terms may be neglected in front of the zero order term $g_0(a)$ corresponding to the unperturbed density distribution of the Earth. Therefore the *boundary conditions* are simply

$$\left.\begin{aligned} N &= - g_0 \, (a) \, K_n \\ M &= 0 \end{aligned}\right\} \quad r = a \tag{11.30}$$

The presence of gravitational effects requires a third boundary condition on the potential gradient (L parameter). Therefore Longman introduces this boundary condition on the deformed surface for the perturbation ψ of the gravitational potential. We let ψ_e be the external (in air) perturbation in gravitational potential. Then at the external surface r=a we have the *boundary conditions*

$$\frac{\partial \psi}{\partial r} - \frac{\partial \psi_e}{\partial r} = 4\pi G \, [\rho_0(a) \; U(a) \, + \, K_n] \, P_n \tag{11.31}$$

and

$$\psi = \psi_e .$$

Since ψ_e is harmonic

$$\psi_e = D \, r^{-n-1} \, P_n \, (\cos \kappa) \tag{11.32}$$

where D is some constant dependent on n, and so we have

$$\frac{\partial \psi_e}{\partial r} = - \frac{n+1}{r} \, \psi_e ,$$

and the boundary condition on ψ at the surface can be written

$$\frac{\partial\psi}{\partial r} + \frac{n+1}{a}\psi = 4\pi G\,[\rho_o(a)\,U(a) + K_n\,]P_n \qquad (11.33)$$

or, with (5.24),

$$L + \frac{n+1}{a} R = 4\pi G K_n \qquad (11.34)$$

at r=a.

Equations (11.30) and (11.34) represent three boundary conditions to be satisfied at r=a. At an internal discontinuity in density all the parameters must be continuous, and at the core boundary all except T must be continuous. They are to be regular at r=0.

We will now refer to the procedure used by Farrell.
The transformed surface potential of the point mass load is

$$\Phi_{2,n} = \frac{4\pi Ga}{2n+1}\Gamma_n = \frac{ag}{M}, \qquad (11.35)$$

where $\Gamma_n = [P_{n-1}(\cos\kappa) - P_{n+1}(\cos\kappa)]/4\pi a^2(1-\cos\kappa)$, when $n > 0$ and $\Gamma_0 = 1/4\pi a^2$.

The surface vertical displacement at distance κ from the point mass load is

$$U_n(r) = \frac{h'_n(\kappa)}{g}\Phi_{2,n}(\kappa)$$

or

$$u(\kappa) = \frac{a}{M}\sum_{n=0}^{\infty} h'_n\,P_n(\cos\kappa) \qquad (11.36)$$

from the definitions of h'_n, and the surface potential of the point mass.

With large n, h'_n, $n\ell'_n$, nk'_n become constant and Farrell defines

$$\lim_{n\to\infty}\begin{Bmatrix} h'_n \\ n\ell'_n \\ nk'_n \end{Bmatrix} = \begin{Bmatrix} h'_\infty \\ \ell'_\infty \\ k'_\infty \end{Bmatrix} \qquad (11.37)$$

Approximate expressions for these limits, derived from Boussinesq's problem, have been written by Farrell, but as the exact values he takes those computed for the largest n. The approach to the limit is clearly shown on Fig. 11.8, while the differences between the approximate and computed limits are given in Farrell's original paper.

Using the asymptotic value for h'_n, (11.36) can be written (Kummer transformation)

$$u(\kappa) = \frac{ah'_\infty}{M}\sum_{n=0}^{\infty} P_n(\cos\kappa) + \frac{a}{M}\sum_{n=0}^{\infty}(h'_n - h'_\infty)\,P_n(\cos\kappa). \qquad (11.38)$$

TABLE 11.4. Load deformation coefficients

$-h'_n$						$-nk'_n$						$-n\ell'_n$			
n	Longman	Kaula	Takeu-chi	Farrell	Pertsev Ivanova	n	Longman	Kaula	Takeu-chi	Farrell	Pertsev Ivanova	n	Longman	Kaula	Far-rell
0	0.134				0.133										
1				0.290	0.297	1						1			0.113
2	1.007	0.981	1.034	1.001	1.000	2	0.620	0.606	0.624	0.615	0.614	2	0.060	0.054	0.059
3	1.059	1.050	1.078	1.052	1.053	3	0.591	0.591	0.573	0.585	0.588	3	0.225	0.219	0.223
4	1.059	1.058	1.083	1.053	1.054	4	0.532	0.536	0.504	0.528	0.533	4	0.248	0.248	0.247
5	1.093	1.093	1.121	1.088	1.087	5	0.520	0.525	0.480	0.516	0.522	5	0.245	0.245	0.243
6	1.152	1.153	1.185	1.147	1.145	6	0.540	0.540	0.426	0.535	0.540	6	0.246	0.246	0.245
7	1.223	1.223	1.260	1.291	1.215	7	0.574	0.574	0.504		0.573	7	0.259	0.259	
8	1.296	1.297	1.338		1.287	8	0.608	0.608	0.528	0.604	0.611	8	0.272	0.272	0.269
9	1.369	1.369			1.358	9	0.648	0.648			0.651	9	0.288	0.288	
10	1.439	1.439	1.486	1.433	1.427	10	0.690	0.690	0.580	0.682	0.690	10	0.300	0.300	0.303
11	1.506					11	0.726					11	0.319		
12	1.572	1.571	1.622			12	0.768	0.768	0.624			12	0.336	0.336	
13	1.631					13	0.806					13	0.351		
14	1.691	1.691	1.750			14	0.840	0.840	0.672			14	0.378	0.378	
15	1.747					15	0.870					15	0.390		
16	1.798	1.801	1.872			16	0.896	0.896	0.720			16	0.416	0.416	
17	1.852					17	0.935					17	0.425		
18	1.902			1.893	1.878	18	0.972			0.952	0.964	18	0.450		0.452
19	1.949					19	0.988					19	0.475		
20	1.994					20	1.020					20	0.500		
21	2.037					21	1.050					21	0.504		
22	2.078					22	1.078					22	0.528		
23	2.117					23	1.104					23	0.552		
24	2.156					24	1.128					24	0.552		
25	2.194					25	1.150					25	0.575		
26	2.223					26	1.170					26	0.598		
27	2.257					27	1.188					27	0.621		
28	2.291					28	1.204					28	0.616		
29	2.322	2.314				29	1.218	1.218				29	0.638	0.638	
30	2.351					30	1.230					30	0.660		

TABLE 11.4. (Part II)

$-h'_n$

n	Longman	Kaula	Takeuchi	Farrell	Pertsev Ivanova
31	2.380				
32	2.408			2.379	2.352
33	2.429				
34	2.455				
35	2.470				
36	2.497				
37	2.521				
38	2.535				
39	2.562				
40	2.581				
56				2.753	
100				3.058	2.969
180				3.474	3.309
325				4.107	
550				4.629	4.402
1000				4.906	4.783
1800				4.953	4.933
3000				4.954	4.980
10000				4.956	5.006
20000					5.008
30000					5.008
70000					5.010
∞				5.005	

$-nk'_n$

n	Longman	Kaula	Takeuchi	Farrell	Pertsev Ivanova
31	1.240				
32	1.280			1.240	1.255
33	1.287				
34	1.292				
35	1.295				
36	1.332				
37	1.332				
38	1.330				
39	1.326				
40	1.360				
56				1.402	
100				1.461	1.468
180				1.591	1.533
325				1.928	
550				2.249	2.074
1000				2.431	2.294
1800				2.465	2.380
3000				2.468	2.406
10000				2.469	2.418
20000					2.418
30000					2.418
70000					2.419
∞				2.482	

$-n\ell'_n$

n	Longman	Kaula	Farrell
31	0.651		
32	0.672		0.680
33	0.693		
34	0.714		
35	0.735		
36	0.720		
37	0.740		
38	0.760		
39	0.780		
40	0.760		
56			0.878
100			0.979
180			1.023
325			1.212
550			1.460
1000			1.623
1800			1.656
3000			1.657
10000			1.657
∞			1.673

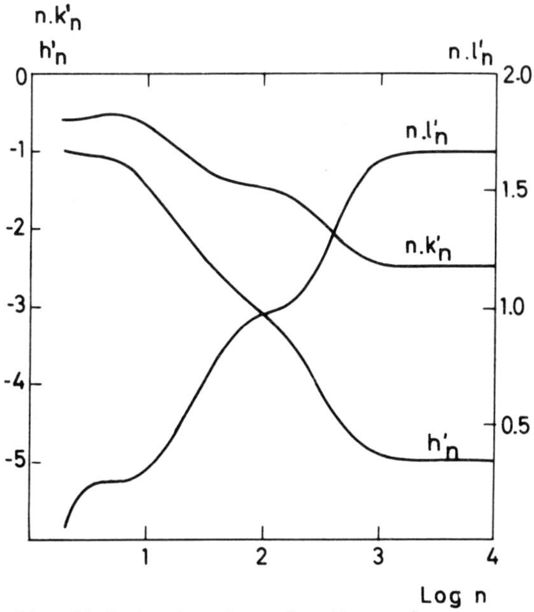

Fig. 11.8. Load numbers for the surface mass load on a Gutenberg-Bullen A Earth model. At n=10,000 the computed numbers agree with Boussinesq's approximation to within 1%. Load numbers for the other Earth models differ significantly from these numbers above n=20 to 30 (according to Farrell).

The first sum is

$$1/2 \sin \kappa/2 .$$

The second sum terminates after a finite number of terms, since $(h'_n - h'_\infty)$ is zero above n=N.

Then

$$u(\kappa) = \frac{ah'_\infty}{2M \sin \frac{\kappa}{2}} + \frac{a}{M} \sum_{n=0}^{N} (h'_n - h'_\infty) P_n (\cos \kappa). \qquad (11.39)$$

This finite sum, although better behaved than the infinite sum in (11.36), still converges rather slowly (mainly when κ is small) because the amplitude of P_n falls off only as $n^{-1/2}$. Farrell applies two further devices to speed the convergence. The first is to put a disc factor similar to (11.28) back into the transformed potential; the second is to use Euler's transformation on the series.

The maximum degree N, needed to evaluate the vertical displacement is given by Farrell as follows:

κ	180°	130°	90°	40°	20°
N	11	15	20	60	100

We refer the reader to the Farrell's paper for this procedure which involves the evaluation of a very large number of terms. But when n is large and κ is small, both h_n and P_n are slowly varying functions of n.

Of course, the load numbers must be known over a large range, typically $0 \leq n \leq 10{,}000$, but it is not necessary to integrate numerically the equations of motion for each integral n: interpolation can be used on a sparser table of the load numbers, and the spacing in degree between successive h_n', ℓ_n', k_n' can be a rapidly increasing function of n. Farrell calculated them until n = 10.000, Pertsev and Ivanova (1976) until n = 70.000 (Table 11.4).

The horizontal displacement at the surface is given by

$$v(\kappa) = \frac{a}{M}\sum_{n=1}^{\infty} \ell_n' \frac{\partial P_n(\cos\kappa)}{\partial\kappa}$$

$$= \frac{a\ell'_\infty}{M}\sum_{n=1}^{\infty}\frac{1}{n}\frac{\partial P_n(\cos\kappa)}{\partial\kappa} + \frac{a}{M}\sum_{n=1}^{\infty}(n\ell_n' - \ell_\infty')\frac{1}{n}\frac{\partial P_n(\cos\kappa)}{\partial\kappa}, \qquad (11.40)$$

TABLE 11.5. Mass-Loading Green's functions Gutenberg-Bullen Earth model; applied load 1 kg (according to Farrell)

κ, deg	Radial displacement, $\times 10^{12}(a\kappa)$	Tangential displacement $\times 10^{12}(a\kappa)$	g^E, $\times 10^{18}(a\kappa)$	t^E, $\times 10^{12}(a\kappa)^2$	$\varepsilon_{\kappa\kappa}$ $\times 10^{12}(a\kappa)^2$
0.0001	-33.64	-11.25	-77.87	33.64	11.248
0.0010	-33.56	-11.25	-77.69	33.64	11.248
0.0100	-32.75	-11.24	-75.92	33.64	11.253
0.0200	-31.86	-11.21	-73.96	33.62	11.278
0.0300	-30.98	-11.16	-72.02	33.58	11.322
0.0400	-30.12	-11.09	-70.11	33.52	11.382
0.0600	-28.44	-10.90	-66.40	33.30	11.537
0.0800	-26.87	-10.65	-62.90	32.92	11.709
0.1000	-25.41	-10.36	-59.64	32.35	11.866
0.1600	-21.80	-9.368	-51.47	29.82	11.988
0.2000	-20.02	-8.723	-47.33	27.78	11.714
0.2500	-18.36	-8.024	-43.36	25.29	11.083
0.3000	-17.18	-7.467	-40.44	23.09	11.303
0.4000	-15.71	-6.725	-36.61	19.84	8.779
0.5000	-14.91	-6.333	-34.32	17.85	7.538
0.6000	-14.41	-6.150	-32.78	16.81	6.682
0.8000	-13.69	-6.050	-30.59	16.25	6.019
1.0	-13.01	-5.997	-28.75	16.32	6.170
1.2	-12.31	-5.861	-27.03	16.33	6.535
1.6	-10.95	-5.475	-23.96	15.86	7.071

SI units.
$a = 6.371 \times 10^6$ m.
κ in radians in the normalization.

where again the first sum is known exactly to be

$$- \cos \frac{\kappa}{2} \, (1+2 \sin \frac{\kappa}{2})/2 \sin \frac{\kappa}{2} \, (1+\sin \frac{\kappa}{2})$$

and the second sum is evaluated just like the vertical displacement sum discussed previously.

For the strains, Farrell has obtained (always at the surface):

$$\varepsilon_{rr} = \frac{\partial u}{\partial r} = -\frac{2\lambda}{\sigma a} u + \frac{\lambda}{\sigma M} \sum_{n=0}^{\infty} n(n+1)\ell'_n P_n (\cos \kappa) \tag{11.41}$$

with $\eta = \lambda+\mu$, $\sigma = \lambda+2\mu$, then, along the tangent to the great circle joining the point mass to the observing point,

$$\varepsilon_{\kappa\kappa} = \frac{u}{a} + \frac{1}{a}\frac{\partial v}{\partial \kappa} = \frac{u}{a} + \frac{1}{M} \sum_{n=0}^{\infty} \ell'_n \frac{\partial^2 P_n (\cos \kappa)}{\partial \kappa^2}, \tag{11.42}$$

while

$$\varepsilon_{r\kappa} = 0, \tag{11.43}$$

and, according to (3.61),

$$\varepsilon_{rr} = -\frac{\lambda}{\lambda+2\mu} (\varepsilon_{\kappa\kappa} + \varepsilon_{\psi\psi}). \tag{11.44}$$

Of course, by combining the load effects of several point masses P_i at the observing site we will have to refer all of them to the same axis by choosing the East-West and North-South axis at the site. This means that a rotation matrix is to be applied to the strain tensor

$$\begin{vmatrix} e_{\kappa\kappa}^{(\kappa)} & 0 \\ 0 & e_{\psi\psi}^{(\kappa)} \end{vmatrix}.$$

For the higher values of n the deep internal structure of the Earth has less and less influence so that from n=9 the core can be taken as homogeneous.

Green's functions have been calculated for the Gutenberg-Bullen(GB) A Earth model. To explore the influence of the Earth's upper mantle, Farrell considers two additional models formed by replacing the top 1000 km of the GB Earth by the oceanic and continental shield structures of Harkrider (1970). Love numbers for n=2, 3, 4 are listed in Table 5.3, for these models, while Table 11.4 gives selected values of load numbers for the GB model, as well as Boussinesq's approximations at infinite wave number. Tabulations of Green's functions for all three models are given in Farrell's paper.

Normalized displacements for the GB model are plotted in Fig. 11.9. Near a point load Boussinesq's solution is the limiting value of the spherical solution; this will be convenient for normalizing spherical Green's functions. Therefore the normalizing functions chosen by Farrell,

$$\left.\begin{aligned} u^* (\kappa) &= -\frac{g\sigma}{4\pi\mu\eta(a\kappa)} = \frac{Gh'_\infty}{g(a\kappa)}, \\ v^* (\kappa) &= -\frac{g}{4\pi\eta(a\kappa)} = -\frac{G\ell'_\infty}{g(a\kappa)}, \\ (\eta &= \lambda + \mu), \end{aligned}\right\} \tag{11.45}$$

are taken from Boussinesq's problem. This normalization removes the basic r^{-1} dependence that characterizes Green's functions of a uniform half-space, and as $\kappa \to 0$ the normalized responses tend to 1.

The distant response is governed by low-degree Love numbers. On the contrary, when κ is small the terms with large n dominate the spherical harmonic expansion of the response.

The gravity anomaly and Green's functions tilt for the oceanic and continental shield Earth models are plotted in Fig. 11.10 and 11.11, the responses being normalized with respect to the direct attractions of the mass load.

The vertical acceleration when κ is small is approximately found by replacing h'_n by h'_∞, $(n+1)k'_n$ by k'_∞ in

$$g(\kappa) = \frac{g}{M} \sum_{n=0}^{\infty} [\, n+2h'_n-(n+1)k'_n \,]\, P_n\ (\cos\kappa), \tag{11.46}$$

and evaluating the sums

$$\sum_{n=0}^{\infty} n\, P_n\ (\cos\kappa) = -\ 1/4\ \sin\frac{\kappa}{2}\ ; \tag{11.47}$$

$$\sum_{n=0}^{\infty} P_n\ (\cos\kappa) = 1/2\ \sin\frac{\kappa}{2} \tag{11.48}$$

this gives

$$g(\kappa) = \frac{g}{4\ M\ \sin\frac{\kappa}{2}}\ (-1+4h'_\infty-2k'_\infty) \tag{11.49}$$

showing that near the load the ratio of the potential perturbation effect to the displacement effect is $-k'_\infty/2h'_\infty$.

At near and intermediate distances, the load number h'_n governs the elastic tilt: the horizontal acceleration from the perturbed density field is much less important. Thus the elastic tilt in this range is principally just the slope of the deformed surface. The ratio of the h'_n to k'_n contributions to the tilt is -100/1 at κ=1°, -10/1 at κ=10°, -3/1 at κ=30°.

A suitable normalizing function for the strain tensor components is the derivative of Boussinesq's tangential displacement, $\partial v/\partial r$, evaluated on the surface of the half-space

$$\varepsilon^* = \frac{g}{4\pi\eta(a\kappa)^2} = \frac{G\ell'_\infty}{g(a\kappa)^2}. \tag{11.50}$$

Figure 11.12 plots $\varepsilon_{\kappa\kappa}/\varepsilon^*$ for both mantle models; Fig. 11.13 shows the normalized areal strain $(\varepsilon_{\kappa\kappa} + \varepsilon_{\psi\psi})/\varepsilon^*$. We had seen that the surface areal strain vanishes in Boussinesq's problem.

11.7. PRACTICAL COMPUTATIONS OF OCEANIC LOADING EFFECTS

Following a procedure developed by Bower (1971) the co-range and equal phase lines of the cotidal charts of each constituent (when available) are used as a framework to divide the oceans into spherical polygons containing regions of approximately constant amplitude and phase. Each spherical polygon is then replaced by plane polygons which approximate the spherical surface. The basis of Bower's technique is the computation of the vertical, North, and East components of the gravitational attraction at an observation site due to each plane polygon. The appropriate Green's function is to be integrated over each polygon in the form of the first two terms of a Taylor series. The contributions of all the polygons covering the ocean are then added vectorially.

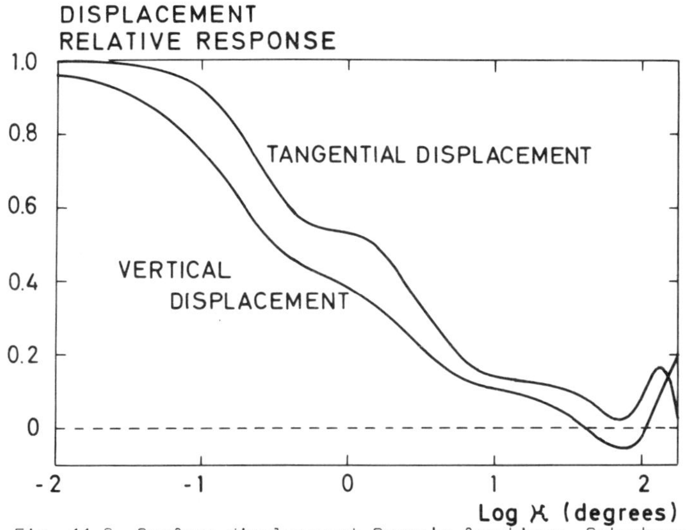

Fig. 11.9. Surface displacement Green's functions, Gutenberg-Bullen A Earth model. The displacements are normalized with respect to the response of a half-space with v_p=6.14 km s^{-1}, v_s=3.55 km s^{-1} or equivalently h'_∞=-4.96, ℓ'_∞=1.66. The displacements for the other Earth models differ significantly from these for κ<1 degree (according to Farrell).

To make such computations easy, the coordinates of all corner points of every polygon are stored on magnetic tape. This is to be done once for each set of cotidal charts.

11.8. AN APPLICATION TO THE BAY OF FUNDY TIDES

A most interesting application of this procedure has been made by Lambert and Beaumont for the famous Fundy Bay (Canada) where the oceanic tides have the largest amplitude in the world (5m for M_2) and are better known than the local earth structure. These authors had indeed, the possibility of checking the computations made for several crustal models with the observations performed with a pair of VM quartz pendulums installed in a mine (30 m deap) at Rawdon (Nova Scotia).

These authors chose the finite element method of solution because this method allows for the inclusion of laterally inhomogeneous models. The tilt response functions for layered elastic models predict significant variations among the tilts measured on different crustal models. Because of the fact, demonstrated by Alsop and Kuo, that the direct Earth-tide deformation appears not to be very sensitive to the local crustal structure, tilt measurements appear to be useful for determining crustal and upper mantle elastic properties providing that the areal distribution of the load is well known.

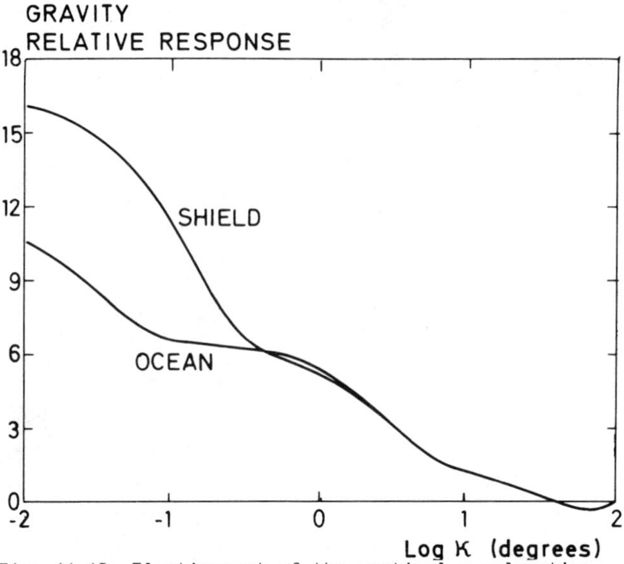

Fig. 11.10. Elastic part of the vertical acceleration Green's function. Responses normalized with respect to the vertical component of the newtonian acceleration of the mass load (according to Farrell).

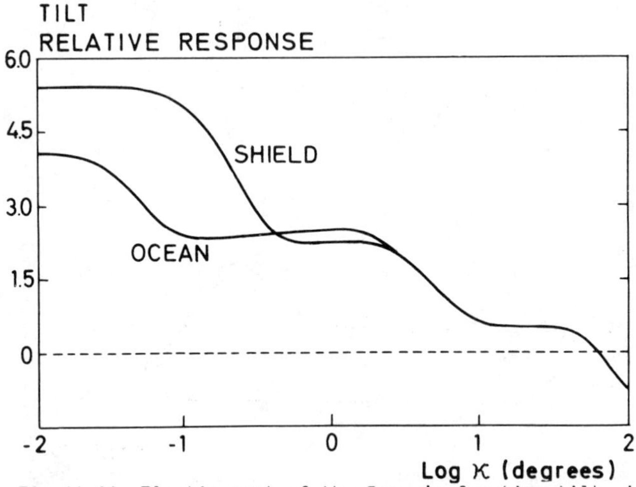

Fig.11.11. Elastic part of the Green's function tilt, including the horizontal acceleration of the perturbed density field. Response normalized with respect to the horizontal component of the newtonian acceleration of the mass load (according to Farrell).

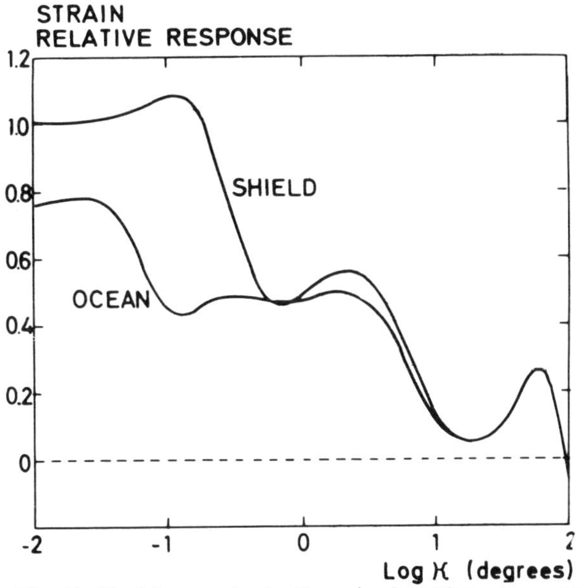

Fig.11.12. Linear strain Green's function, ε_{KK}, normalized to the response of a half-space with v_p=6.1 km s^{-1}, v_s=3.54 km s^{-1} or equivalently ℓ'_∞=1.83 (according to Farrell).

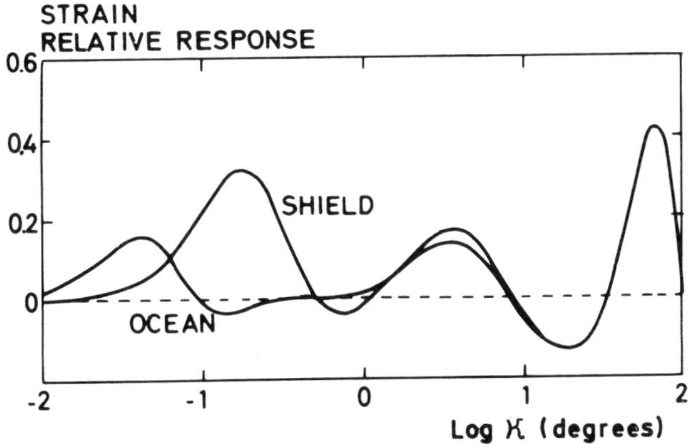

Fig.11.13. Areal strain Green's function, normalized to the linear strain on same half-space as in fig. 11.12 (acccrding to Farrell).

Moreover, the Green's function tilt has two distinct response sections, a near section where the tilt is sensitive to earth structure (0 km to approximately 150 km) and a distant section, where the tilts are very insensitive to earth structure (beyond approximately 150 km). Rawdon is a coastal station, but the near oceanic tides are well known while this is not the case for the deep ocean tides.

Therefore the measurements performed there enable the placing of strong constraints on both the local crustal structure and the far deep ocean tide distribution.

As stated by Beaumont and Lambert, the finite element method has been used extensively in civil engineering, soil mechanics, and rock mechanics to study the deformation of complicated structures under arbitrary loads. It is an approximate method of solution which relies on the division of a continuum into blocks that are interconnected only at nodal points. Forces are applied at the nodal points and the structure is solved for the equilibrium nodal point displacements by minimizing the total potential energy with respect to the nodal point displacements. The accuracy of the nodal point displacements depends on the size and configuration of the elements. A correct choice of a structure representing the body under stress is evidently critical. A large number of small elements will generally give accurate nodal-point displacements. In this paper only laterally homogeneous, isotropic, elastic models are used. However, the authors are able to assign different material properties to the elements; therefore, the finite element method does not restrict them to such simple models.

Beaumont and Lambert generated Green's functions for the various crustal and upper mantle models by using an axisymmetric, hemispherical, finite element model (Fig. 11.14).

A normal point load is applied on a single surface nodal point at the pole of the model. The elements are small near the pole where the displacements and tilts must be more accurate, and they increase in size in proportion to the distance from the pole. The nodal points along the bottom of the model are rigidly constrained in the vertical direction but are free to move laterally.

This model gives accurate tilts at distances closer than 1000 km from the load. But the tilts computed on a homogeneous version of the finite element model closely fit Boussinesq's solution. This result shows that within this region the effect of the Earth's sphericity on tilt is minimal, and that a plane-layered model would have sufficed for tilt calculations. However, a plane-layered model is no easier to solve by the finite element method than the hemispherical model.

Beaumont and Lambert have calculated the load tilt for six different models representative of certain prominent seismic structures (crustal thickness, upper mantle P_n velocity, tectonic characteristics, sediment thickness, and water depth).

Because of the diversity of surface sediments found throughout the world, the models have been simplified by omitting the upper sedimentary layers. For this reason the values of Green's functions within 10 km of the load are unrealistically low. The normalized tilt Green's functions corresponding to the six models are shown in Fig. 11.15.

Beaumont and Lambert have made very important remarks about these results as follows:

"Within 1000 km of the load the normalization is such that Green's functions would appear as horizontal lines for homogeneous Earth models. Thus the general decrease in the value of Green's function with distance from the load reflects the increase in elastic parameters with depth in the models.

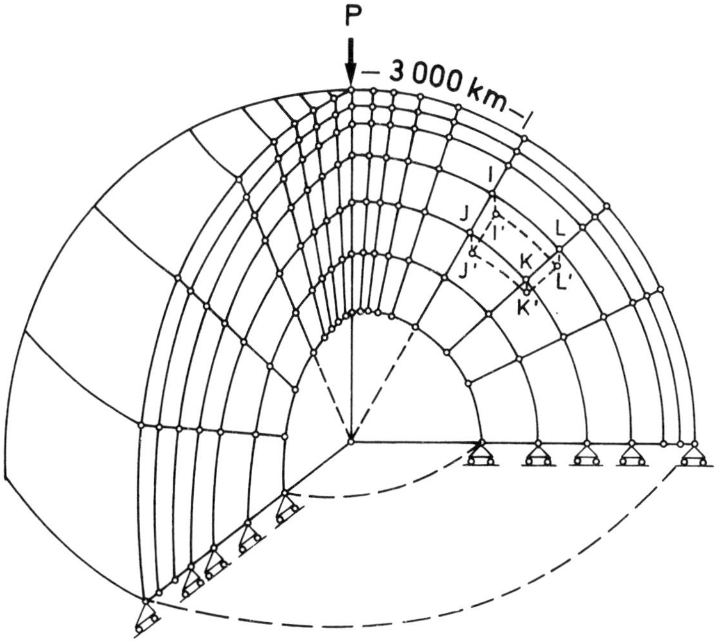

Fig. 11.14. Beaumont and Berger simplified diagram of a section through the finite element earth model. Under the point load P the nodal points I, J, K and L are displaced to I^1, J^1, K^1 and L^1. The axisymmetric finite element Earth model used in this study has 902 elements, 41 along the surface by 22 arranged in the radial direction. The elements are small near the origin, where the displacements must be accurate, and increase in size in proportion to the distance from the origin. A single nodal point load is applied at the surface on the axis. The solution gives the nodal point displacements throughout the model. The radial displacements of the surface nodal points give Green's function for radial displacement.

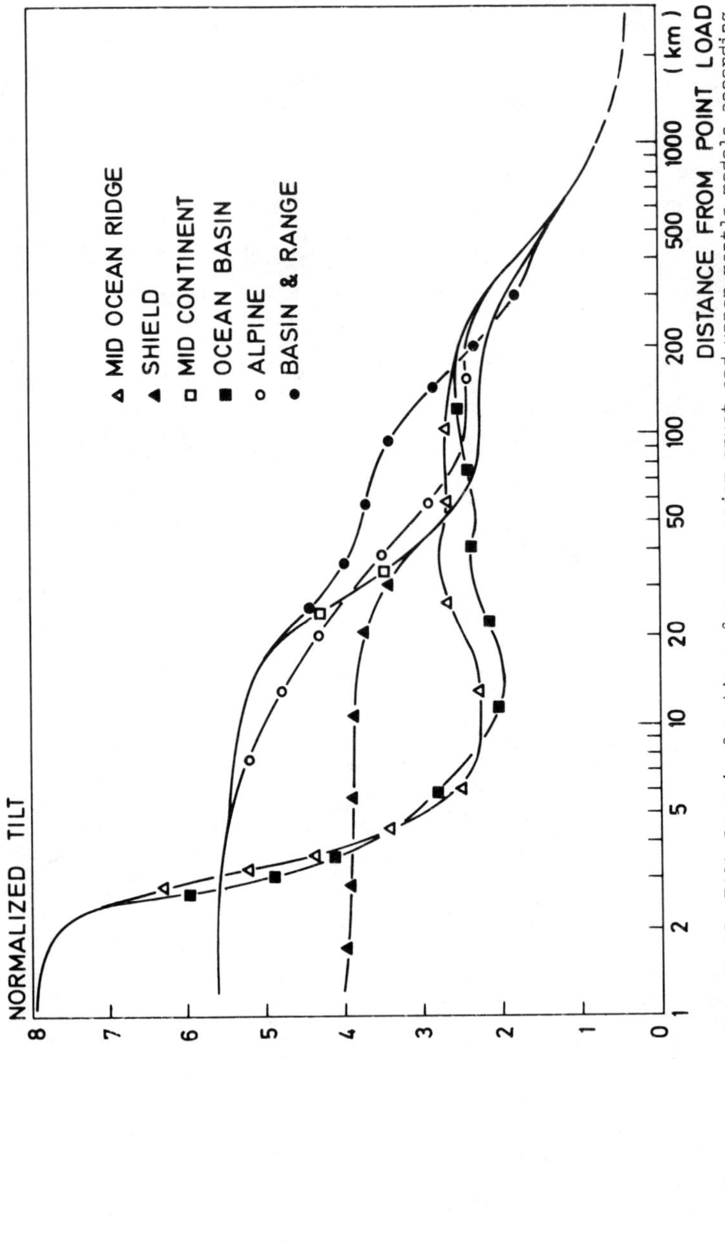

Fig. 11.15. Tilt Green's functions for some major crust and upper mantle models according to Beaumont and Lambert.

Beyond 1000 km from the load (dashed curve) where the effect of regional structure is negligible, Green's function is the same as that given by Bower (1971) for the Gutenberg Earth. The tilt responses of the various structures show very significant differences especially at distances from 10 km to 100 km from the load. A prominent feature in most of the curves is the steep drop in response which occurs due to the transition to a higher rigidity material at the Mohorovicic discontinuity (see Fig. 11.15). *The distance from the load at which this drop occurs is a crude measure of the thickness of the crust.* Hence there is an obvious division of Green's functions into continental and oceanic types, the distinguishing steepness of the oceanic models being due to the contrast between the thin low-velocity (rigidity) crust and the shallow higher-velocity (rigidity) mantle. It is interesting to note that the transition feature for the alpine structure is not so sharp even though the contrast in rigidity at the Moho is equivalent to that for the mid-continent structure. This result is due to the gradation in the elastic properties of the overlying crust. Equally interesting is the fact that the effect of the Moho for the basin and range model does not show up at all, even though there is a discontinuity in compressional velocity and density. Apparently Young's modulus shows a stronger dependence on shear velocity than on compressional velocity. The presence of a low-velocity channel (low shear velocity) in the mantle tends to cause a local maximum in the Green's function tilt or at least to flatten it at a distance approximately equal to the depth of the channel. However, as a result of the normalization, a bump in Green's function does not necessarily indicate the existence of a low-velocity channel. As can be seen from Green's function for the mid-ocean ridge structure, a thin, low rigidity upper layer over a thick more rigid layer also produces a bump."

Then the authors develop an application of these remarks for Nova Scotia structures (five models); they observe that the phases of the computed tilts are practically independent of the crustal structure, and *this places strong constraints on the tidal phase distribution in the deep ocean* provided that the tidal distribution in the near coastal zone is correct.

They conclude that the Zahel and Pekeris-Accad deep ocean cotidal charts fit the results better than the Tiron chart. On the other hand, the results "strongly suggest that the crust beneath central Nova Scotia is about 35 km thick and is underlain by normal mantle. The crust and upper mantle structure is not of the "Appalachian" type found beneath the Gulf of St. Lawrence."

Of course, more than one clinometric station is needed to obtain more precise informations on the crustal and upper-mantle structure as well as on the oceanic tides. Beaumont and Lambert consider also the lateral change in crustal structure pointing out that the transition from oceanic to continental structure that occurs at the continental margin (300 km South-East of Rawdon) has no effect on the tilts at Rawdon if this lateral change is limited to less than 35 km in depth. Their results show that beyond 200 km from the point load the tilts are insensitive to crustal structure.

11.9. THE LOADING EFFECTS ALONG THE COASTS OF WESTERN EUROPE

The oceanic tides exhibit quite important amplitudes along the coasts of Western Europe, particularly between the Gulf of Gascoigne and Jutland. Amongst the many investigations performed on this subject we will refer here to those which concern Britain and Schleswig-Holstein.

Bower (1970) has calculated the elastic depression of Great Britain and along the coasts of France, Belgium and Holland by using the Boussinesq procedure. His result (Fig. 11.16) shows that very strong flexure is produced in Cornwall which quickly increases to a maximum half amplitude of about 6 centimetres at Newlyn (N) at the extremity of the peninsula. This investigation has special importance since the high precision geodetic levelling network of the British Isles is based on mean sea level at Newlyn.

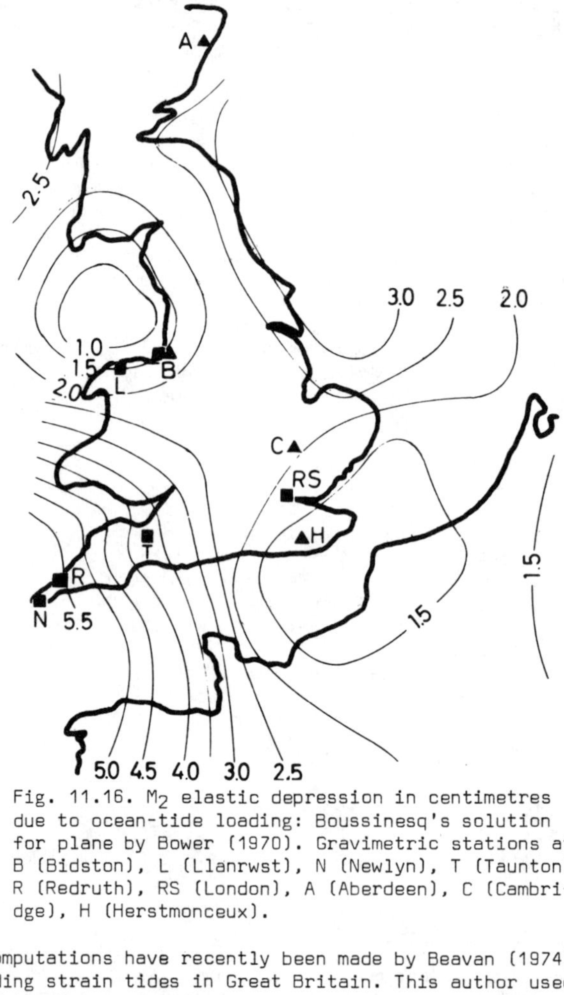

Fig. 11.16. M_2 elastic depression in centimetres due to ocean-tide loading: Boussinesq's solution for plane by Bower (1970). Gravimetric stations at: B (Bidston), L (Llanrwst), N (Newlyn), T (Taunton), R (Redruth), RS (London), A (Aberdeen), C (Cambridge), H (Herstmonceux).

Similar computations have recently been made by Beavan (1974) to estimate the ocean loading strain tides in Great Britain. This author used two different charts of the M_2 ocean tide around Great Britain and performed convolutions up to 400 km from each site. Differences between the two cases demonstrate the sensitivity of the load strain tide to variations in the tidal model. Perturbations in the strain field due to tides in more distant oceans are also calculated, but they are comparatively small and fairly constant over a region as small as Great Britain (Fig. 11.17).

An interesting feature of the M_2 load strain lobes is the large area of the country over which both the in-phase and the in-quadrature load tides are nearly zero in an East-West direction (Fig. 11.18). This coincides with an East-West zero of the theoretical in-phase strain tide at about 52°N, the theoretical in-quadrature component always being zero East-West (see Section 3.3). This implies that little M_2 strain should be measured by an East-West strainmeter in this area provided that effects due to distant tides, local inhomogeneities and geology are small.

Convolutions have been made for the O_1 tide in local seas, but the indications are that O_1 ocean load strain tides are strongly influenced by the tides in more distant oceans.

Zschau (1976), in Schleswig-Holstein (North Germany), conducted an extremely interesting investigation by comparing the indirect effects simultaneously recorded with three vertical pendulums installed in nearby boreholes at 30 m, 45 m and 60 m depth respectively. Zschau observes that the agreement between the three instruments is good in the East-West component but not in the North-South component. He attributes this to an influence of air-pressure variation which causes stronger tilts in the North-South than in the East-West component as shown by direct computations. According to Zschau a variation with depth would be explained by a combined action of air pressure on the surface and pore pressure in the highly porous subsoil of his station.

The drop in the experimental Green's function tilt gives, in the case of the North Sea region, a depth of about 30 km for the Mohorovicic discontinuity. Zschau attributes also a local maximum observed at distances greater than 100 km to a low shear velocity channel (the asthenosphere) in the mantle.

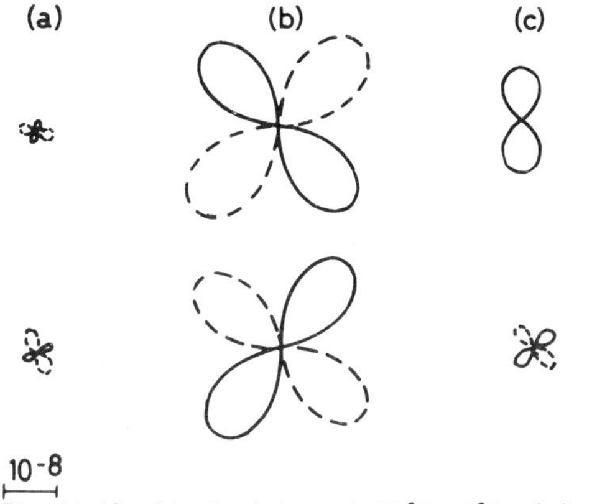

Fig.11.17. Strain lobes at 53°N, 2°W. (a) shows the load strain lobes due to distant tides; (b) shows those due to local tides (German data) on the same scale; and (c) shows the theoretical strain lobes in the absence of any tidal loading. The upper diagrams are in phase lobes, the lower ones are in quadrature. Phases are local phases. (From Beavan).

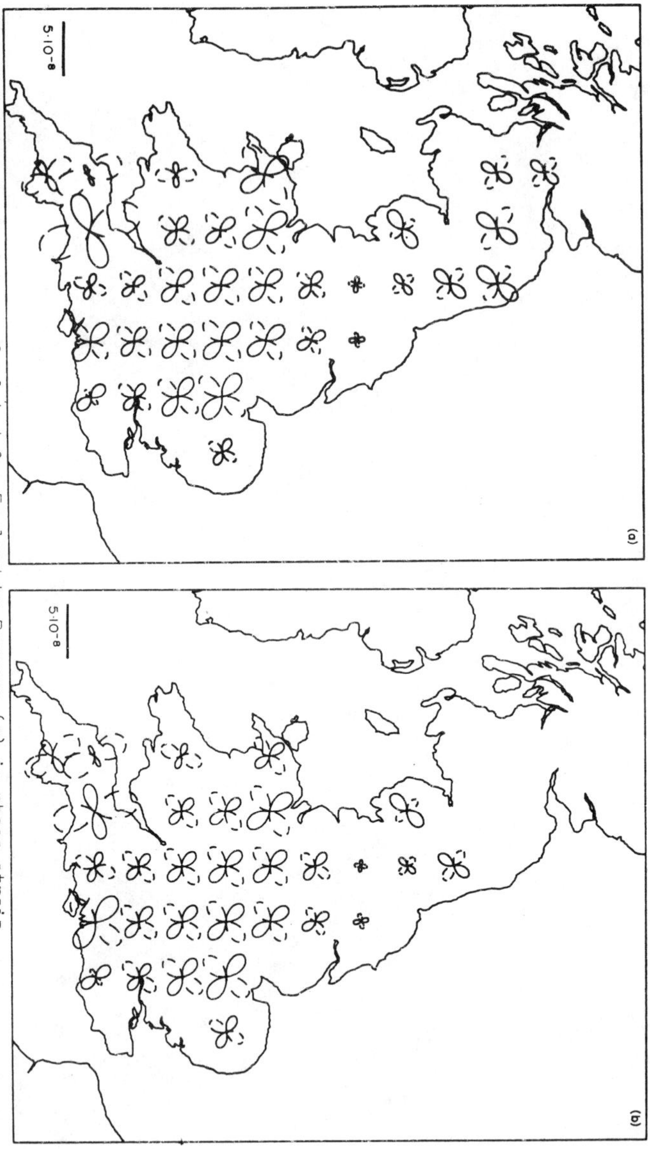

Fig.11.18. M_2 strain lobes calculated for England by Beavan: (a) in phase strain,
(b) in quadrature strain

11.10. THE U.S. TIDAL GRAVITY PROFILE

Another interesting contribution was made by Kuo (1969) who has given the general solutions to the problem of an inclined static load on a circular area of the surface of a multilayered medium. The inclined load is assumed to be a linear combination of a normal load and an asymmetrical shear load; each layer of the multilayered medium is assumed to be homogeneous and isotropic, and the interface of the layers are perfectly welded. Through the Thomson-Haskell matrix method, the generalized components of displacements and stresses which are expressed in the integral form, have been obtained on the surface, at the interfaces, and at any point in the multilayered medium. They are reducible to an axisymmetrical or asymmetrical surface load on an n-layered medium, including a half-space. This formulation has avoided the generally cumbersome nature of the problem and is adaptable to numerical evaluation. Through an elementary transformation, the general solutions for an inclined static load have been extended to evaluate the components of total displacements and total stresses due to a system of inclined loads of arbitrary orientation on the surface of a multilayered medium.

However, this can only be used for nearby loading effects as it does not take into account the sphericity of the Earth (flat Earth approximation).

In his paper of 1963 Longman concluded that:

"if sufficient accuracy is attainable, comparison of the effects of ocean tides distribution on a gravimeter and the observations of the earth tides may shed some light on the magnitude of ocean tides for from land where direct measurements have not so far been possible."

Following this idea, Kuo started a programme of trans-continental tidal gravity profiles in the United States with instruments ensuring an accuracy of 1% or better. His aim is, indeed, to consider the inverse problem of mapping ocean tides on the open oceans with extended Earth tidal gravity measurements on the adjacent lands supplemented by ocean-bottom stations (both tidal gravity and ocean-tide measurements) and the ocean tidal information derived from nearshore stations.

After completion of this profile, Kuo completed another along a meridian through the middle of the United States and two coastal profiles along the Pacific and Atlantic coasts. Then he conveyed three gravimeters to Brussels to ensure their calibration with respect to the very homogeneous tidal coefficient obtained in this station with three Askania gravimeters (3500 days measurements: $\delta(O_1) = 1.1610 \pm 0{,}0010$) (see Section 9.7) and undertook European transverse profiles with the co-operation of ICET and of several European scientific institutions.

Warburton, Beaumont and Goodkind have investigated the loading effects on the observations performed at two places (La Jolla and Piñon Flat) in southern California with the superconducting gravimeter (see Chapter 9). They apply Farrell's procedure upon modified cotidal charts of O_1 and M_2 waves. With this accurately calibrated instrument (which show a signal-to-noise ratio of 70 db) they obtain an agreement within 0.2%, which corroborates the placement of an M_2 amphidromic point at 1500 km south-west of San Diego.

The phase-lag of the solid Earth tide is found to have an upper limit of 0.1°. The Earth tide appears totally insensitive to crustal structure because the Moho discontinuity beneath Piñon Flat is at least deeper than it is at La Jolla, while no differential effect appears between the two stations.

These results give evidence for the overall validity of a δ=1.160.

11.11. OTHER INVESTIGATIONS

Pertsev (1976) having determined by numerical integration the loading Love numbers until n = 70.000 calculated the effect of the world oceanic tides on the Trans United States tidal gravity profile realized by Kuo. For the near zones defined by r < 1300 km Pertsev uses spherical segments 1°X1° while he uses segments 5°X5° for r > 1300 km. The result shows an increase of the δ values which contradicts the evaluation by Kuo himself. Groten and Brennecke reached similar conclusions as Pertsev by using the Zahel cotidal map.

Pertsev, like Groten-Brennecke, obtain, after correction, a phase of -2° for the O_1 wave as well as for the M_2 wave, which is difficult to accept. Tanaka (1972-73) investigated the tilts and strains (M_2 and O_1) on the basis of a group of stations in Japan (Akibasan, Oura, Kamigamo, Osakayama, Kishu, Rokko) and in China (Barim) by using Kuo's procedure (multilayered half-space) for the neighbouring sea (χ<30°) and Green's function method for the effect of the distant sea (χ>30°). Tanaka found that the surface tilt can be approximated by Boussinesq's solution multiplied by the Nishimura factor (11.25), and that according to these stations the best fit is obtained with ε=0.3, in fair agreement with Rosenhead's result.

Bozzi-Zadro has developed some analysis of the Trieste data in function of the frequency ω_i of the loading effect. She found different values for the modulus of rigidity which may be understood by the fact that the amphidromic point in the Adriatic Sea is nearer to Trieste for the semi-diurnal waves than for the diurnal ones. These latter have no amphidromic point indeed, but a nodal line at the latitude of Otranto. Therefore the semi-diurnal loading effect does not act as deep as the diurnal one according to Bozzi-Zadro who derives frequency dependent elastic parameters:

From the diurnals: $\mu = 0.58\times10^{12}$ dyne cm^{-2}.

From the semi-diurnals: $\mu = 0.45\times10^{12}$ dyne cm^{-2}.

From the seiches: $\mu = 0.35\times10^{12}$ dyne cm^{-2}.

Tilt and strain measurements are more difficult to organize than tidal gravity measurements but they are more sensitive to any kind of anomalies and therefore offer a great potential in the long term.

11.12. AN APPLICATION OF THE SPITSBERGEN OBSERVATIONS IN THE ARCTIC OCEAN

Theoretically the existence of an Earth-tide station at a very high latitude should allow a precise determination of the North-South and East-West components of the tesseral tidal waves (K_1, O_1, P_1 ...) because their amplitudes are maximum at the poles.

Access, problems of installation and maintenance are far more hard and difficult to solve at such a location than elsewhere. Prescriptions for horizontal pendulums drastically limit the choice of sites. From 1968 to 1970 a well-provided station (three Askania gravimeters and eight Verbaandert-Melchior horizontal pendulums) was under activity at Longyearbyen (78°12'N, 15°36'E, depth under free surface 350 m, distance to the sea about 5 km) in the frame of the "Astro-geo Project Spitsbergen" (Bonatz, Melchior, Blankenburgh, 1970) (Fig. 11.19).

Polar regions are also characterized by very small theoretical amplitudes of the other Earth-tide constituents: the diurnal vertical and the semi-diurnal horizontal and vertical components are null at the poles (Fig. 1.3, 1.4 and 1.5).

Oceanic semi-diurnal tide, on the other hand, has an appreciable amplitude (nearly 50 cm for M_2) along the Spitsbergen coast. Therefore it is clear that the indirect effect (deformations of the Earth's crust under periodical loading of the moving oceanic waters) will considerably modify the observed parameters.

However, the observations of tides in the Arctic Ocean have been episodic, of short duration and often inaccurate. Only one theoretical investigation (Goldsbrough, 1951) gives indications on the peculiarities to be expected in a polar ocean: because of the very small shape of the basin the tide is not generated in it and cannot co-oscillate with other oceans; the tide, thus, is necessarily derived from the North Atlantic ocean.

Moens (1969) collected all available data and tried to draw a cotidal chart for the principal constituent M_2 (Fig. 11.21) which indicated that the tide is propagating as a progressive wave from the North Atlantic Ocean, penetrating the Arctic Ocean between Greenland and Iceland for one part, and Iceland and Norway for the other part. This wave is gradually damped and deformed by the undersea topography of the European and Siberian shelves. Due to the accessibility of the west coast, the oceanic tide in Spitsbergen is fairly well determined except for the south-east region where uncertainty about the localization of an amphidromy between Hopen Island and Bjornoya remains.

The diurnal constituents K_1 and O_1 have an amplitude of a few centimetres only. Moens considers an attempt to draw precise cotidal charts for these waves as worthless, and in order to estimate the associated indirect effect he adopted an elementary distribution which, considering the very small amplitude of the indirect effects, gave sufficiently satisfactory results.

TABLE 11.5. Indirect gravimetric tide at Longyearbyen (Spitsbergen) according to Moens (1976)

Considering the oceanic tides effects up to 2000 km - an evaluation of the influence of the World Ocean on the basis of the Zahel's M_2 cotidal chart gives an amplitude of 0.5 µgals, only and on the basis of Bogdanov's and Magarik's O_1 chart, 0.05 µgals). Amplitudes expressed in microgals and phase-lags given with respect to the maximum of the potential at the zero meridian.

Elastic Model	M_2	K_1	O_1
Boussinesq: homogeneous half-space $\lambda = 1.50\ 10^{12}$ dynes/cm^2 $\mu = 1.22\ 10^{12}$ dynes/cm^2 (elastic properties at 500 km depth) Gutenberg-Bullen model	3.415 185.53°	1.256 31.26°	0.422 224.56°
Kuo: stratified flat model transposed from G.B. models	2.687 184.06°	0.969 30.99°	0.352 226.07°
Farrell: spherical Earth of G B type $a = 6.371\ 10^8$ m	2.323 184.26°	0.839 30.91°	0.311 226.43°

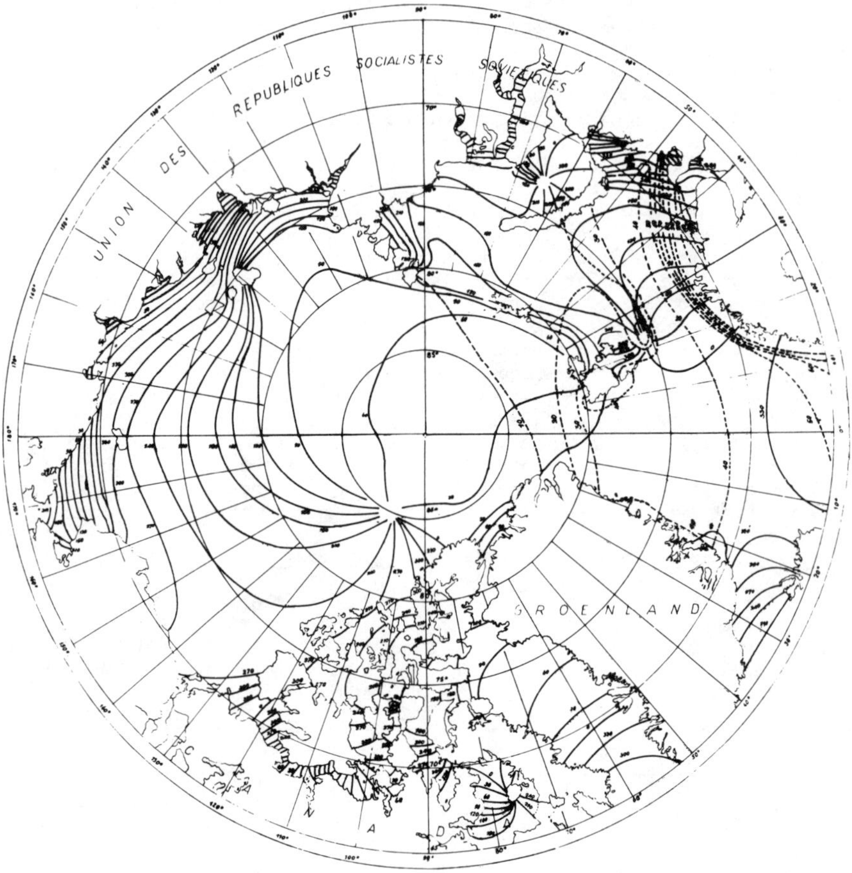

Fig. 11.21. Cotidal chart for arctic regions (M_2 tide) according to Moens.

This shows that Boussinesq's model is inadequate for the computation of the vertical component of the indirect effect (even for the near zones) because of the necessarily arbitrary choice of a mean value for λ and μ. With values of Lamé's parameters corresponding to a depth of 680 km the results would have been more acceptable. Discrepancy between Kuo's and Farrell's method is less than 15% on the amplitudes and probably reflects the difference between a flat and a spherical model. On the contrary, the choice of the model does not influence the phases of the indirect effect.

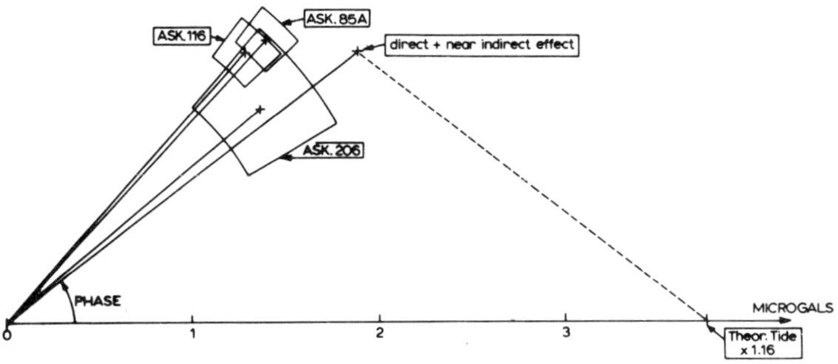

Fig.11.22. Polar diagram for the M_2 gravimetric wave according to Moens.

1. Elastic deformation and mass redistribution (A,C): 1.7599 μgals, 142.63° local.
2. Direct newtonian attraction (B): 0.5202 μgals, 141.96° local.
3. Total indirect effect (1+2): 2.3501 μgals, 142.46° local.
4. Direct effect (theor. tide x 1.16): 3.7649 μgals, 0° local.
5. Predicted effect (3+4): 2.3803 μgals, 36.98° local.

Instrument	Observed δ	Observed amplitude	Observed phase	Corrected δ	Corrected phase
Ask. 85A	0.6298±0.0374	2.0450	46.43°±3.41°	1.0081	0.87°
Ask. 116	0.5914±0.0403	1.9199	47.45°±3.90°	0.9739	-0.32°
Ask. 206	0.5531±0.0858	1.7867	39.21°±8.90°	1.0098	-5.32°

The amplitudes are in microgals. Thus the difference between δ=1.16 and δ=1.01 corresponds to only 0.4 μgal (a vertical displacement of about 1.5 mm).

The interpretation of the clinometric tide is critical. Moens used an elastic G B model but states that:

"a modified model with an oceanic crust would diminish the amplitude of the indirect effect, while a continental one would, at the contrary, increase it. Phase remains the same. In the M_2 East-West component, with the same phase, an increase of 30% of amplitude would better fit with pendulums VM 76 and 99, and an increase of 15% with VM 13. Geological maps attribute to Spitsbergen a continental-shelve structure, perhaps our model underestimates the amplitude by an amount of 2%."

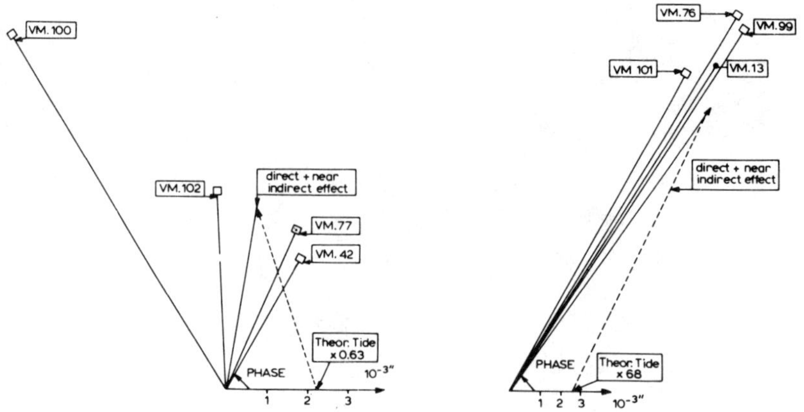

Fig.11.23. Polar diagram for the M_2 clinometric wave according to Moens.
left: North South component; right: East West component.

Units: 0.001"	N.S. Amplitude	N.S. Local phase	E.W. Amplitude	E.W. Local phase
1. Elastic deformation and mass redistribution (A, C)	3.9895	119.15°	10.5249	-29.81°
2. Direct newtonian attraction (B)	0.9889	59.15°	4.4108	-24.29°
3. Total indirect effect (1+2)	4.5650	108.33°	14.9213	-28.18° (+90°=61.82°)
4. Direct effect (theoretical tide x 0.68)	2.2440	0°	2.2916	-90°
5. Predicted effect (3+4)	4.4081	79.43°	16.1305	-35.37° (+90°=54.63°)

Ct.	Instrument	γ observed	Obs. Ampl.	Obs. phase	γ corr.	Corrected phase
NS	VM.42	1.0948±0.0154	3.6128	59.08°±0.81°	1.0654	-20.55°
NS	VM.102	1.4261±0.0173	4.7061	91.86°±0.70°	0.4046	16.10°
NS	VM.100	3.0313±0.0198	10.0033	121.73°±0.38°	1.7158	132.50°
NS	VM.77	1.2675±0.0315	4.1828	64.61°±1.42°	0.9928	-9.75°
EW	VM.13	5.4678±0.0188	18.4265	56.85°±0.20°	1.1242	36.90°
EW	VM.101	5.1713±0.0275	17.4273	60.02°±0.31°	0.7587	49.46°
EW	VM.99	6.1764±0.0313	20.8145	56.24°±0.29°	1.8213	42.57°
EW	VM.76	6.2702±0.0537	21.1306	57.90°±0.49°	1.8774	48.62°

Note: VM 99 and VM 100 are situated on the other side of the valley at 2 km from the other pendulums.

11.13 OBSERVATIONS AT THE SOUTH POLE

As shown in Chapter 1, fig. 1.4, on a symmetrical earth the vertical component of the diurnal or semi-diurnal gravity earth tide would vanish at the south pole. In fact, however 18 major constituents of these tides are observed there by B.V. Jackson and L.B. Slichter (1974) at amplitudes between 0.02 and 0.60 μgal, which are large relative to the ambient noise level (Fig. 11.24). Ocean tides are a primary cause of this observation and the amplitudes and phases of the major constituents observed provide constraints to be satisfied by inferred patterns of the world ocean tides.

The maximum amplitudes observed in the diurnal and semi-diurnal constituents were 0.61 μgal and 0.34 μgal, respectively. These amplitudes are much too large to be accounted for by any reasonable off-axis error in the location of the South Pole station, which is believed to lie within 2 km of the rotation axis.

11.14. SAITO EQUATION TYING THE LOVE NUMBERS WITH THE LOAD NUMBERS

On the grounds of a variational equation Saito demonstrates that

$$k'_n = k_n - h_n \tag{11.51}$$

a very important relation, the consistency of which is clearly evident from the tables 5.2, 5.3 and 11.4.

S.M. Molodensky (1977) also demonstrates the validity of equ (11.51) by using a different formulation based upon the auto-conjugate properties of the basic equations (5.20) to (5.28) which are emphasized by the symetry of the coefficients in these differential equations as established in another paper by the same author (1976).

Moens (1979) has shown how these relations can be obtained either from the symetry properties of the system of differential equations, either from a variational principal.

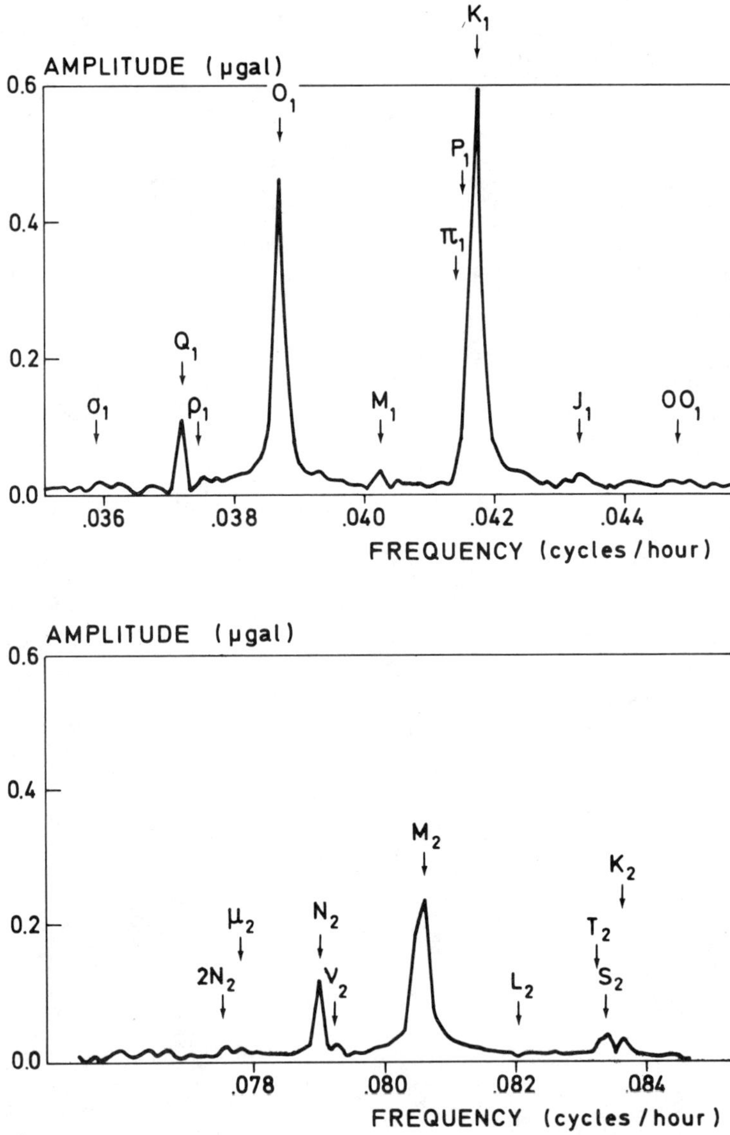

Fig. 11.24. Observed amplitude spectrum of diurnal and semi-diurnal constituents at the South Pole according to Jackson and Slichter (1974).

CHAPTER 12

Local Perturbations in Earth Tide Observations

We will discuss in this chapter some local perturbations in the observations which have now been clearly isolated and tentatively evaluated. We will deal with three major geophysical effects:

a) topo-geological
b) meteorological
c) hydrological

As a tentative conclusion of this discussion we will try to define the best conditions of installation of earth tide instruments.

12.1. TOPO-GEOLOGICAL FEATURES

Figures 11.22 and 11.23 give vectorial representations of the computed direct Earth tide and the oceanic loading effects. Many similar diagrams have been established as, for example, on Fig. 12.1, for strain measurements at Piñon Flat (USA).

The vectorial residual difference between such combinations of the calculated effects and the observed vector may represent the result of:

1. Uncertainties in the cotidal maps. The problem of the uncertainties of the cotidal maps was considered in Chapter 11, and possibilities of improving them from Earth-tide measurements themselves by an inverse procedure has been considered. This evidently cannot be done before clarifying the role of the outer topography and of the cavities and geological inhomogeneities on the measurements.
2. The departure of the Earth crustal constitution from the accepted radial stratification (lateral variations of the elastic parameters). From this point of view, one has to always remember that mining activity is associated with tectonic perturbations.
3. A distortion of the regional strain field produced by the topography of the free surface in the vicinity of the observatory (Fig. 12.2).
4. A distortion of the local strain field by the change of geometry of the underground cavity containing the instrument (mine, gallery or cave) (Fig. 12.3).

These effects alterate the phase and the amplitude of the tidal phenomenon. But the wavelengths of the tidal phenomenon are far much larger than the scale length of the local inhomogeneities which cannot therefore perturb the strain field at distances greater than their own dimension.

If the cavity where the Earth-tide station is located is deep enough with respect to the topographic features, then topographic effects and cavity deformations do not interact and can be treated separately. This means that an underground station must be in any case well below the nearby topographic features. Harrison (1976) remarks that "many underground strain installations are above the surrounding topography because of drainage considerations". This is, however, not the case for the underground installations like those at Walferdange, Warmifontaine, Sclaigneaux, Bad Grund, Grasse, Příbram, just to indicate

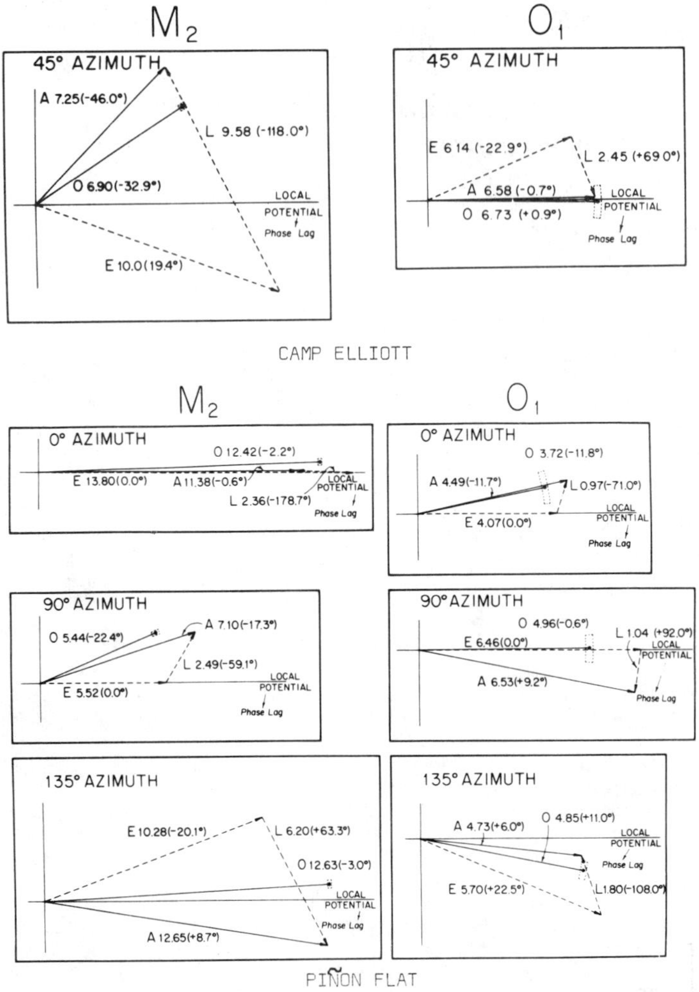

Fig. 12.1. E is the homogeneous solid Earth-strain tide; L the ocean load contribution; A the applied or the sum of E and L. The amplitudes of the vectors are in 10^{-9} strain units and the phases are in degrees of lag relative to the local potential. Dashed boxes around the end of the O vectors (observed tides) indicate the one-standard-deviation-error estimates. (According to Beaumont and Berger).

some typical examples.

In any case, deep burial in a small-size cavity considerably attenuates the effect as soon as the depth is greater than the height of the topographic feature.

On the contrary, cavities close to the free surface boundary produce very strong anomalies as one, indeed, verified it experimentally (such as the Chevron and Wiehl stations), principally when the topography is not smooth: shear strains generated by the topography are in this case magnified by the nearby cavity and, as shown by Harrison, produce rotations of the cavity which are observed as tilts.

12.1.1. The Topographic Effect

At the free surface of a uniform isotropic body one principal strain direction is always normal to the horizontal plane while the three principal directions of strain must be mutually perpendicular.

If the other two principal directions which lie in the horizontal plane make angles of ν and $\nu+90°$ to the East-West direction, one observes a rotation of the plane (xz), around Oy and

$$e'_{\alpha\beta} = g_{\alpha i}\, g_{\beta j}\, e_{ij} \tag{12.1}$$

gives

$$e'_{xz} = \frac{1}{2}\sin 2\nu\,(e_{zz} - e_{xx}) + e_{xz}\cos 2\nu. \tag{12.2}$$

If Ox', Oz' are the new principal axes, one has

$$e'_{xz} = 0 \tag{12.3}$$

and

$$\nu = \frac{1}{2}\,\mathrm{arc\,tan}\left\{2\,\frac{e_{xz}}{e_{xx} - e_{zz}}\right\}. \tag{12.4}$$

Then if the region has a gorge, the application of the stress e_{xx} causes a fixed rotation through an angle ν of the principal directions of strain in the (x/z) plane because at the sloping surface one principal direction must be normal to it.

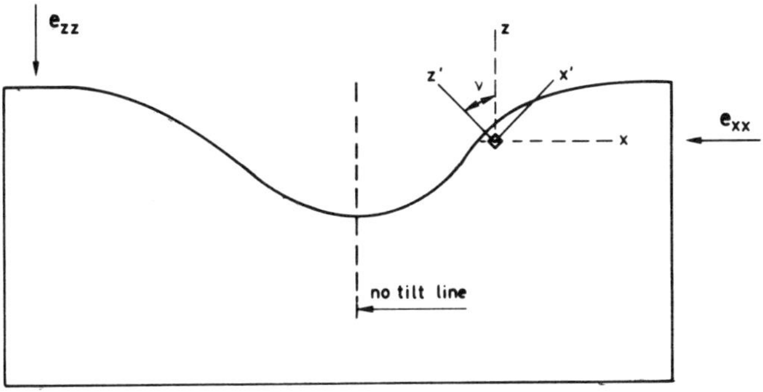

Fig. 12.2. Schematic representation of a region with a gorge.

The topography of the free surface is, of course, well known and therefore allows an easy modelling.

Harrison (1976) has calculated by the finite element techniques the effect of four different models of topography. He found that the strain is high in the valleys and low on the high ground at the valley margins; it disappears at a distance equal to about twice the size of the topographic accident.

Blair (1976) has constructed a reduced scale two dimensional model of the Armidale gorge (Australia) with resistive triaxial strain gauges (x, z and 45° directions) attached at various positions. The normalized strain distribution obtained in this way is similar to that obtained for the finite element solution. Blair moreover finds that the topography has a signifiant effect on the spectral nature of tidal strains.

12.1.2. Cavity Effect by Tilt-Strain Coupling

A cavity changes the boundary conditions and modifies the strain in its vicinity. King and Bilham (1973) show how tilts are induced in a tunnel wall due to a linear strain field imposed at a distance from the tunnel (Fig. 12.3).

The change of radius of a circular tunnel is given by

$$\delta_r = r_o \ (1 + 2 \cos 2z) \ e_{xx} \tag{12.5}$$

which produces a tilting at the tunnel wall

$$i = \frac{\partial r}{r \ \partial z} = - 4 \ e_{xx} \ \sin 2z; \tag{12.6}$$

this is maximum at $z = \pm 45°$.

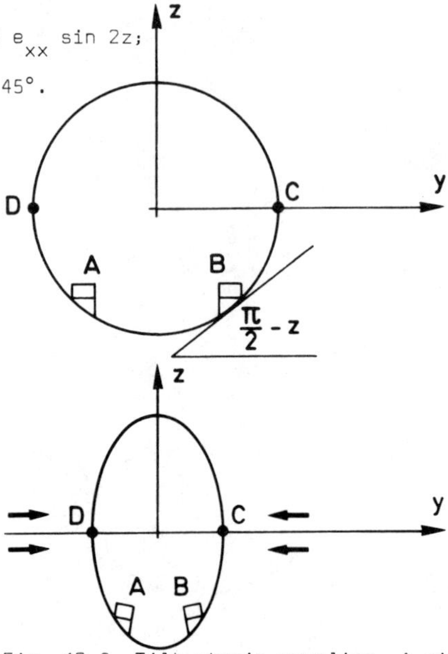

Fig. 12.3. Tilt-strain coupling. A circular tunnel is deformed by horizontal strain, and tiltmeters at A and B record anomalous tilts while C and D are not affected (Harrison, 1976).

As the strain tide is of the order of 10^{-8}, it could give rise to a local tilting of 4×10^{-8} radian, that is about 0"008.

It is to be noted that the East-West linear tidal strain is in quadrature with the East-West tidal tilt while in the North-South component strain and tilt are in phase. It is advisable to position the tiltmeters so as to minimize such effects because calculating the magnitude and phase of the strain-generated tilt is presently extremely difficult.

According to King and Bilham, provided that the tunnel is horizontal and beneath a uniform ground surface, tilts along the axis of the tunnel will be unaffected by tidal rock stresses. Thus when tiltmeters are sited in tunnels they should be arranged to measure tilt only along the tunnel axis. Separate tunnels should be used for the other components.

It is interesting to point out that when niches have been made to install pendulums they have often been established at points C and D of Fig. 12.3. This was precisely the case at Sclaigneaux. Nevertheless, the technique of niche excavations may have to be revised. Lecolazet and Wittlinger have given (1974) a very simple evaluation of what the contribution of cavity effect can be in clinometric measurements. Considering a small spherical cavity at the equator they obtain from formulas (3.48) and (3.58) the equations

$$\left.\begin{aligned} \frac{\sigma_{11}}{\mu} &= \frac{6(h-\ell)}{a}\frac{W_2}{g} + h'\frac{W_2}{g} + \frac{2\ell}{ag}\frac{\partial^2 W_2}{\partial\lambda^2} \\ \frac{\sigma_{22}}{\mu} &= \frac{6(h-\ell)}{a}\frac{W_2}{g} + h'\frac{W_2}{g} + \frac{2\ell}{ag}\frac{\partial^2 W_2}{\partial\theta^2}, \\ \frac{\sigma_{33}}{\mu} &= \frac{8h-6\ell}{a}\frac{W_2}{g} + 3h'\frac{W_2}{g}. \end{aligned}\right\} \quad (12.7)$$

At the Earth's surface $\sigma_{33} = 0$ gives

$$h' = -\frac{8h - 6\ell}{3a} \quad (12.8)$$

Then, when putting for convenience

$$W_2 = 12\ ga\ A\ \sin^2\theta\ \cos 2AH, \quad (12.9)$$

one has (at the equator)

$$\left.\begin{aligned} \sigma_{11} &= 4\mu\ A\ (10h-36\ell)\ \cos 2AH, \\ \sigma_{22} &= 4\mu\ A\ (10h-24\ell)\ \cos 2AH. \end{aligned}\right\} \quad (12.10)$$

The normal to the ellipse with the radius vector makes an angle which can be obtained from the Solomon formula:

$$e_2 = A\ (10h-36\ell)\ \cos 2AH\ \sin 2z, \quad (12.11)$$

z being the zenithal distance of the point considered.

The undisturbed East-West vertical deviation should be

$$e_1 = \gamma\frac{\partial W_2}{ga\partial\lambda} = 24\ \gamma\ A\ \sin 2AH. \quad (12.12)$$

At $z = 45°$, with $\gamma = 0.688$, one obtains

$$e_1 + e_2 = 24\ \gamma\ A \times 1.016\ \sin(2AH + 10°) \quad (12.13)$$

that is an amplitude amplification of about 2% and a phase lead of 10°. The effect, however, is sensibly greater for the North-South component as it reaches 13° on the amplitude and 28° on the phase. Equation (12.11) indicates that the effect should be minimum at $z = 0$.

One may point out that some parameters remain in perfect agreement as the γ_{EW} and even the γ_{NS} (M_2). On the contrary, the phase of O_1 in the East-West component exhibits a discrepancy of 32°, more than 2 h in time.

The geometry of the underground cavities is evidently well known in the case of mines but not natural caves. Nevertheless, it is always very difficult to model. A most important contribution to the solution of this problem has been recently given by J. C. Harrison (1976) who calculated the deformation of typical cavities by the finite elements method. Harrison first uses the Eshelby solution (1957) for the deformation of an ellipsoidal cavity in an infinite uniform isotropic medium in which the strain becomes uniform at large distances from the cavity.

Among the situations examined, a very interesting one is reproduced in Fig. 12.4: in some cases horizontal pendulums have been installed on a ledge, which proves now to have been a bad idea.

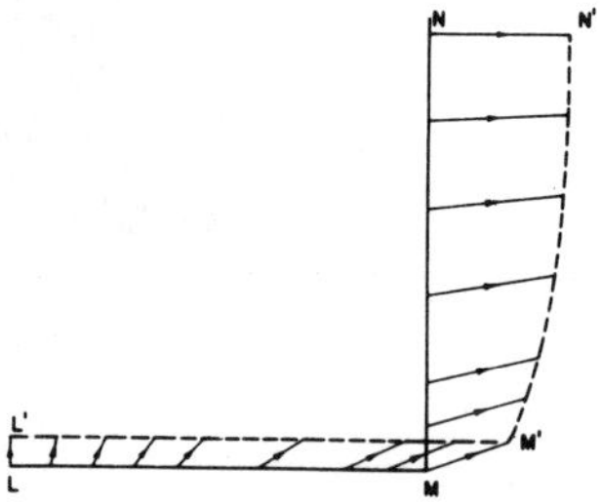

Deformation of a tunnel of square cross-section by horizontal tension. The undeformed corner is represented by a solid line and the deformed corner by a dashed line. Poisson's ratio is 0.25.

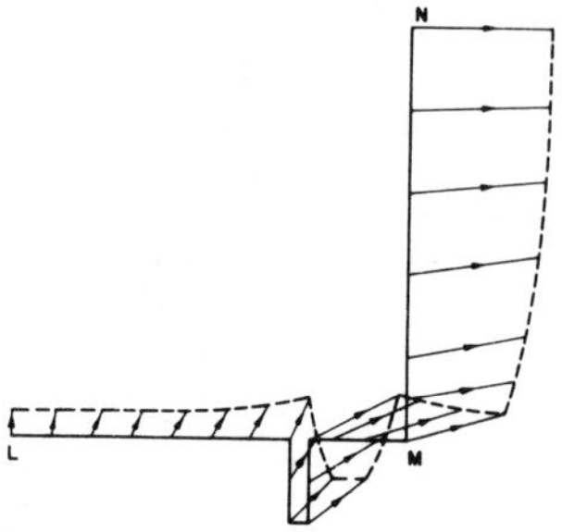

Effect of a crack in modifying the deformation of a tunnel of square cross-section. The undeformed corner is represented by a solid line, and the deformed corner by a dashed line. Poisson's ratio is 0.25.

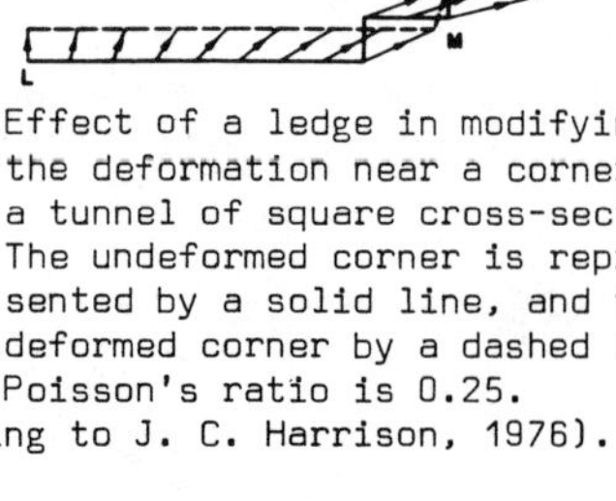

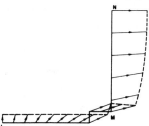

Effect of a ledge in modifying the deformation near a corner of a tunnel of square cross-section. The undeformed corner is represented by a solid line, and the deformed corner by a dashed line. Poisson's ratio is 0.25.

Fig. 12.4. (According to J. C. Harrison, 1976).

TABLE 12.1. Example of cavity effect: observed tidal parameters at the Erpel station

On the ledge	North-South γ	North-South α	East-West γ	East-West α
M_2	0.833	-12.0°	0.932	6.1°
O_1	1.700	11.7°	0.640	31.5°
On the wall, at 20 m from the ledge	γ	α	γ	α
1 M_2	0.774	-12.7°	0.955	-2.2°
1 O_1	0.941	25.1°	0.664	-0.1°
2 M_2	0.783	-8.3°	0.970	-2.4°
2 O_1	1.123	31.1°	0.629	-2.0°

TABLE 12.2. Comparison of amplitude factors (γ) and phases obtained in East-West component at Luxemburg and Walferdange

WAVE	LUXEMBURG Tunnel		WALFERDANGE 1 Mine		WALFERDANGE 2 Mine	
	γ	α	γ	α	γ	α
O1	0.646	-5.39°	0.680	0.15°	0.650	-3.82°
K1	0.642	-12.65°	0.763	-3.39°	0.753	-5.72°
P1	0.160	-28.41°	0.732	-2.24°	0.721	-5.82°
S1	382.44	—	5.425	—	4.374	—
M2	0.870	-4.75°	0.882	-8.84°	0.922	-9.16°
S2	0.740	-56.72°	0.727	-13.99°	0.789	-13.44°

A comparison of different tiltmeters at Llanrwst (Wales, UK) in a strong loading area

At Llanrwst, North Wales, in a gallery of a disused lead mine Baker and Lennon (1976) have installed three tilt equipments: a pair of quartz VM pendulums, a pair of metallic pendulums and one Askania borehole pendulum, this one being clamped rigidly to the vertical rock. The place is 18 km from the Irish Sea, at a depth of 41 m and distant 150 m from the entry of the mine.

The metallic pendulums were first installed for two years on a pillar and calibrated by measurements of their period of oscillation. They were later replaced by the VM pair which where calibrated by the crapaudine method.

The figures show the ellipses of oscillation of the vertical reconstructed from the results of analysis of the North South and the East West components. There is a fair agreement between the two kinds of horizontal pendulums but a strong anomaly for the Askania borehole instrument which mainly results from a difference in the component normal to the tunnel axis. In the tunnel azimuth however the difference is far less than the experimental uncertainties in determining the small marine effects on the gravimeter registrations. According to these authors:

"The excellent consistency of the measured phase lags at both Bidston and Llanrwst is worthy of note and will clearly be of importance in future theoretical comparisons."

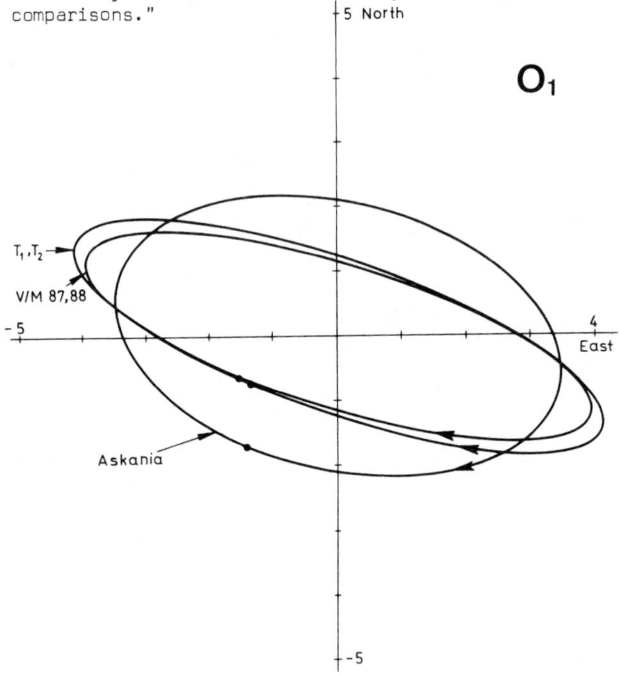

Fig.12.5. Llanrwst (North Wales). Wave O_1: path of the vertical as described by horizontal pendulums (T=Tomaschek, VM= Verbaandert Melchior) and one borehole vertical instrument (Askania)-scale 0"001-(Baker and Lennon 1976).

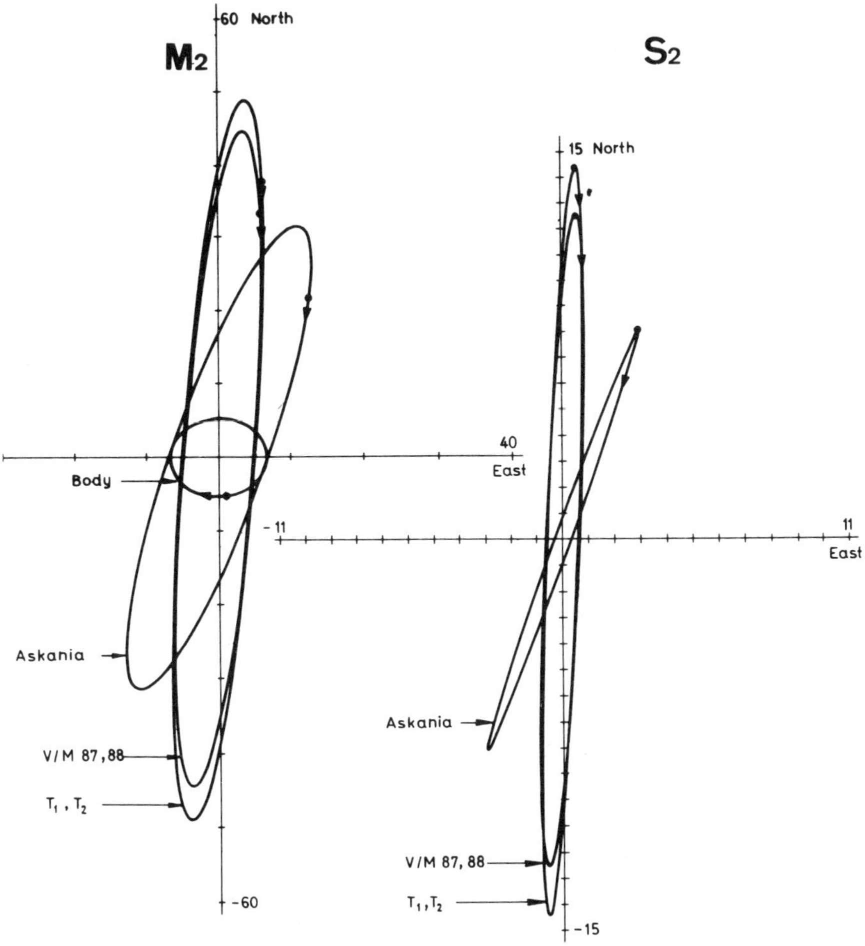

Fig.12.6. Llanrwst (North Wales). Waves M_2 and S_2: path of the vertical as described by horizontal pendulums (T = Tomaschek, VM = Verbaandert Melchior) and one borehole vertical instrument (Askania) - scale 0"001 - (Baker and Lennon 1976).

The tunnels

Old disused railway tunnels have been used quite often to install underground tidal stations because they seemed to offer many advantages: depth, length, easy access and logistic. Typical examples are Luxembourg (GD), Bussang (F), Queensbury (UK), Erpel (D) (see Table 12.1), Osakayama (Japan). However, we soon observed quite strong and unusual disturbances at Luxembourg which was therefore abandoned for Walferdange which looked more independent from such perturbations (Table 12.2).

These results justified abandoning the Luxemburg gallery which was open at both ends.

Itsueli, Bilham, Goulty and King report (1975) that extensometers installed in the Queensbury tunnel show observed effects in agreement with the computations made of cavity deformation. But the effective shape of the excavation may be different from the geometry of the tunnel because it is surrounded by a zone of low rigidity material, mainly beneath the tunnel floor. At Queensbury the agreement is better when using in the computations a cylindrical cavity rather than a cavity of the form of the inner walls.

Instruments installed *across* the tunnel show an amplification of slightly more than a factor of three, while the noise level is not increased to the same extent. The authors suggest that this cavity effect may prove a useful means of improving signal-to-noise ratios for studying low-level strain signals. Rock strain parallel to the tunnel is measured correctly. Cross-tunnel strain depends on the cross-and-along-tunnel strains at infinity (Harrison). The height and width of the tunnels being very often much less than the length of the strainmeter, Harrison (1975, fig. 9) calculates that the strain averaged over the length of the strainmeter is magnified by no more than 2%. Close to the end of the tunnel the strain is more enhanced because the dead-end of a tunnel is also a region of large strain-tilt coupling, but this disappears at a distance from the end equal to the tunnel's height.

Of course, mine galleries must produce the same kind of amplifications for the "across" strainmeters, and this is well observed at Walferdange with the vertical extensometer (Section 10.2).

Similarly, in the azimuth across the width of the Bidston vault (North-South component) there is considerable variation in tilt, of the order of 20%, depending upon position, while the East-West results are reasonably coherent (Baker and Lennon, 1976). A strainmeter far from the end of the tunnel (like Ogdensburg 29.5°) is unaffected by the cavity.

On strainmeters aligned along the gallery axis the effect is not larger than 2%. Most important is the fact that the shear strains induced by the topography are magnified by the cavity. Harrison states that

"In the case of a horizontal tunnel of circular cross-section, shear strains in a vertical plane perpendicular to the tunnel axis are tripled. Thus in this plane, tilt differences between vertical and horizontal elements calculated with the tunnel being neglected will be tripled by the cavity effect. In the vertical plane containing the axis of the cylinder the shear strains are doubled and there is a rigid body rotation equal to the distant strain. The net result is that there is no cavity effect for elements parallel to the axis of the cylinder, a result which is independent of Poisson's ratio and the shape of the cross-section. An analogous result holds for vertical boreholes, so that tiltmeters on the floor of a tunnel suffer no cavity effect along the length of the tunnel and tiltmeters attached to the vertical sides of a borehole are totally unaffected by the cavity. However, instruments attached to vertical bases in tunnels or horizontal bases across boreholes, experience cavity effects equal to twice the shear strain computed in the absence of the cavity."

This probably explains why the results until now obtained by borehole pendulums fixed on the bottom of boreholes of 15-30 m depth exhibit such strong phase anomalies as was shown by the experiences of parallel profiles conducted with great care by Schmitz-Hübsch in Bavaria (Table 12.3). Nevertheless, Harrison states that these perturbations, while extending to considerable depths, are fairly closely confined *laterally* to the generating topography especially in the case of tilt. This makes an installation near the end of a very long gallery like Walferdange (800 m from the entry in a hill which is only 90 m high) very favourable.

TABLE 12.3. Phase lags measured with borehole pendulums by Schmitz Hübsch for two parallel profiles in Bavaria

	Wave M_2			
	Profile 1		Profile 2	
	Zellsee		Wilzhofen	
North-South	0.793	-30.12°	0.828	-40.27°
East-West	0.688	-27.23°	0.575	-36.76°
	Hohenpeissenberg		Berg	
North-South	0.779	-41.73°	0.782	-35.64°
East-West	0.668	-34.79°	0.716	-43.63°
	Lettigenbichl		Murnau	
North-South	0.830	-38.73°	0.881	-42.10°
East-West	0.655	-35.08°	0.907	-31.48°

12.1.3. Lateral Inhomogeneities

In a second part of their analysis Berger and Beaumont (1976) analyse the effects of lateral inhomogeneities in structure. They first confirm that the cavity effects are less than 5% unless the strainmeters are installed transversely in tunnels. However, topographic effects may exceed 50% of the tide homogeneous strain in extreme cases.

Their conclusion is that, provided a very good knowledge of the ocean tides is ensured, a theoretical model including lateral inhomogeneities as well as the load and direct earth tides explains the observations.

Very important is the fact that these results demonstrate that secular stress, *in situ* stress and seismic strain, must all be corrected for site effects.

Unfortunately, geological effects are really difficult to model because the distribution of elastic parameters is not well known.

To perform such computations one has fortunately the possibility to take advantage of structural analysis programmes for static and dynamic response of linear systems, like prismatic solids. Such programmes have now been developed in several university departments of civil engineering.

Obviously geological inhomogeneities will play a role similar to that of the cavity, and Harrison even considers that a promising technique of geological modelling could derive from his analysis:

"The near-surface tilt and strain fields can then be calculated by the techniques described. Comparison with tidal observations would then allow the geological model to be refined. This provides the basis of an exploration technique, somewhat analogous to magnetotellurics, in which the local tilt or strain response to a large-scale periodic applied strain is studied. The information obtained would differ from that obtained seismically in that the response depends on the elastic constants but not the density of the materials involved; the periods of excitation are very different, and large responses are to be expected where there are rapid lateral variations of parameters, whereas seismic interpretations usually assume horizontal layering. Beaumont and Berger (1974) have pointed out that this technique is very sensitive in cases where the elastic parameters vary with time, that is, where a dilatant zone develops. In this case it is not necessary to determine the geological and topographic corrections absolutely but only to look for a variation in the tidal parameters with time. The tidal parameters can be accurately determined because the M_2 signal is large and at a frequency close to 2 cpd, where instrumental noise levels are low. The measurement of a secular tilt, developing over months as the dilatant zone expands, is much harder because of questions of instrumental stability and of distinguishing the dilatant expansion from seasonal effects. On the other hand, the secular tilts may be much larger, and a quantitative study is needed to decide which is the better technique for detecting the growth of a dilatant region."

12.2. THE INFLUENCE OF BAROMETRIC PRESSURE VARIATIONS

Attraction and load barometric effects on gravity.

Considering that the lateral extent of a typical barometric high or low is of the order of 1000 km one can simply calculate its gravitational attraction by the Bouguer formula:

$$\Delta g = -2\pi \; G \; \rho_a \; \Delta h$$

where ρ_a is the air density and h the thickness of the air sheet. If we express the pressure change in millibars we shall observe a variation of gravity

$$\Delta g_1 \text{ (microgals)} = - 0.43 \; \Delta p \text{ (millibars)}$$

due only to the attraction of the air masses. ($\Delta p = g \; \rho_a \; \Delta h$).

The load effect must also be considered and has an opposite effect to the attraction.

The surface depression due to a circular pressure cap is, according to Caputo (1962):

$$\Delta z = -12.5 \; 10^7 \; r \; \mu^{-1} \; \Delta p \text{ cm}$$

r being expressed in km; with $\mu \sim 10^{12}$ dyne cm^{-2};

$$\Delta z = -12.5 \; 10^{-5} \; r \; \Delta p \text{ cm}$$

which produces a free air anomaly:

$$\Delta g_2 \text{ (µgals)} = 39 \; 10^{-5} \; r \; \Delta p$$

If we take 350 km for r we have

$$\Delta g_2 \text{ (µgals)} = 0.14 \; \Delta p$$

and

$$\Delta g_1 + \Delta g_2 = -0.29 \; \Delta p$$

Warburton and Goodkind (1976) have used the superconducting gravimeter installed at La Jolla and at Piñon Flat to measure the influence of barometric pressure variations on gravity in the frequency range of 0.1 to 10 cycles/day. A condition for such an investigation is that the instrument has its intrinsec noise level below that of the pressure-induced gravity variations. Warburton and Goodkind have demonstrated that a major portion, if not all, of this noise is a response to barometric pressure fluctuations. To perform this analysis they first subtracted the pure Earth-tide effect calculated with $\delta_2 = 1.160$ and $\delta_3 = 1.067$. From the obtained residual they eliminate the remaining portion of the tidal signal which is due to the oceanic loading effects. This is done by a least-square fit at the frequencies of all Cartwright-Tayler terms.

The most important characteristics of atmospheric pressure variations is indeed the size of the region over which the variation takes place coherently.

The atmospheric tides, which are in strong correlation with gravity, are, of course, coherent on a worldwide scale: the aperiodic variations resulting from the motion of weather systems are coherent on a scale of hundreds of kilometres, but the fluctuations in the frequency range of a few cycles per day are coherent over distances of tens of kilometres only.
By such an evaluation for Piñon Flat, Warburton and Goodkind have found:

for 10 cycles/day	R = 15 km
for 8 cycles/day	R = 38 km
for low frequency	R = 600km.

The admittance is 0.30 μgal/mbar below 1 cycle/day and 0.33 μgal/mbar between 4 and 7 cycles/day. Warburton and Goodkind consider that these numbers may not vary much with location or time, but for 1 and 2 cycles/day they are strongly influenced by the oceans and therefore must be determined for each particular location. As a matter of fact, since the geographical distribution of the atmospheric tides is known up to 4 cycles/day (Chapman and Lindzen, 1970) it is possible to compute the response of gravity to them on a solid earth (Fredkin, 1976), but while the agreement is good at 3 and 4 cycles/day, Warburton and Goodkind find anomalously large admittances at S_1 and S_2. At S_1 frequency it reaches twice the continuum admittance, i. e. 0.68 ± 0.04 μgal/mbar (with a phase drift of 118°). They suggest that this amplification could be due to indirect effects of the atmosphere on gravity through its effect on S_1 and S_2 ocean tides.

The perturbation of barometric load on tidal gravity has been experimentally investigated by many other authors. Abours and Lecolazet also (1977) make an important distinction between coherent and incoherent barometric signals.

The incoherent signal, that is the signal which is not correlated with the tidal frequencies, is easy to identify and to eliminate in the data with an admittance of:

0.30 μgal/mbar atmospheric change of pressure below 1 cycle day^{-1};

0.33 μgal between 4 and 7 cycles day^{-1}, from a discussion of the superconducting gravimeter observations (Warburton and Goodkind, 1977);

0.28-0.43 μgal according to the size of the area affected by pressure changes (Varga, pers. comm., 1978);

0.29 μgal from a discussion of the Alice Springs data (Melchior et al., 1981).

It is impossible to separate the coherent signal, which has mainly P_1, K_1 and S_2 frequencies, from the purely tidal signal. Moreover, the admittance is dependent upon frequency.

The M_2 component in the lunar barometric tides has been extensively studied by Haurwitz and Cowley (1969), who demonstrated that it has a maximum amplitude of 80 μbar over Indonesia, decreasing as a regular function of latitude from the equator to higher latitudes (e.g. 24 μbar at Hobart latitude, as in Western Europe). Thus such a small barometric loading cannot contribute more than 0.02 μgal to the observed M_2 gravity tide.

On the other hand, the O_1 barometric tide reaches a maximum of 20 μbar only at a few places. This is surely too small to explain the observed residual amplitudes.

The most important barometric tide is S_2 which is about twenty times the corresponding M_2 tide. The explanation of this phenomenon is extensively discussed in Chapman and Lindzen (Atmospheric tides, Reidel 1970), the Kelvin resonance theory being finally excluded and the explanation found in ozone and water vapor absorption.

The amplitude is given to be

$$1250 \cos^3\phi \text{ microbar}$$

according to Chapman and Lindzen.
At some of our tidal gravity stations we have obtained:

	ϕ	Observed	Calculated
Alice Springs	-23°43'	913 254°	959
Bruxelles	50°48'	324 56°	316
Luxemburg Airport	49°40'	329 45°	338
Walferdange (underground)		348 63°	

As an example, when applying the corresponding corrections to the S_2 gravimetric results of Walferdange station we have obtained:

S2 observed gravimetric factors:	δ = 1.1950	α = 1°13
S2 corrected gravimetric factors:	δ = 1.1917	α = 0°84

Another example is at Alice Springs where we got:

S2 observed gravimetric factors:	δ = 1.1603	α = 0°70
S2 corrected gravimetric factors:	δ = 1.1568	α = 0°15

in both cases $\Delta\delta \sim -0.0035$.

In conclusion the gravimetric measurements could be corrected for pressure effects within 10% by assuming the admittances found by Warburton and Goodkind.

Unfortunately many current instruments are not fully compensated for pressure variations: by using an Askania gravimeter (old type GS 11) in 1964, Melchior observed a coefficient 5.72 μgal/mbar at the occasion of an exceptional storm accompanied with a sudden pressure variation of about 4 mbars, while a rough calculation gave him only 0.42 μgal/mbar which agrees with the Warburton -Goodkind very precise investigation.

Barometric effects on strain

The vertical extensometer operating at Walferdange has an S2 phase of 21°. It is strongly affected by the barometric effect and we have estimated that the admittance factor is approximately 0.005 if one writes:

$$10^9(\Delta\ell/\ell) \backsim 0.005\ \Delta p(\mu\text{bar}).$$

Then the correction for barometric pressure gives:

S2 observed vertical factors	η = -0.544	α = 20°9 (±1°2)
S2 corrected vertical factors	η = -0.403	α = -1°8
	±0.012	±3°0

The improvement is considerable.

The S_2 atmospheric tide excite oceanic tide at the same frequency but not in phase. This oceanic tide in turn produces Earth-strain tides. The generation of the S_2 tide is therefore more complicated than the other components.

Barometric effects on tilt.

Simon has made many experiments at the Tiefenort station, producing artificial variations in the air pressure in the mine by using a ventilator installed at a distance of 3 km (1974). He found a tilt effect of 0".001/mbar in the North-South component and of 0".0008/mbar in the East-West component. However, when he subtracts such barometric tilt-induced effects from the currently observed data, he obtains very abnormal amplitude factors (γ) for the O_1 and K_1 tides.

We are of the opinion that sealing horizontal pendulums covers is a wrong procedure because changes of pressure being no more balanced inside and outside the cover produce a deformation of the instrument's base and therefore additional spurious barometric disturbances. The installation of the instruments in air-tight chambers or niches can offer a better solution.

At Tiefenort niches were excavated in the ground itself in the middle of the rooms (Fig. 12.7) and hermetically sealed for temperature and pressure variations. The variations of temperature and pressure are controlled with recorders placed inside and outside the niches. This is clearly necessary if one uses sealed pendulums but not with the quartz pendulums which rest on plates which are freely submitted on all sides to air-pressure effects. In these niches Simon installed tiltmeters in completely air-tight steel boxes which have 10 cm thick baseplates. By loading such steel boxes with a weight of 4 kg these massive baseplates are deformed in such a manner that the horizontal pendulums react on the produced tilt changes to an order of magnitude equal to the tidal variations (0"020).

Simon uses this artificial deformation for controlling the sensitivity stability by filling a vessel with fluid with a system (1971) almost identical to the one introduced by Verbaandert and Melchior in 1961 (automatic calibration with mercury crapaudines).

Simon and Buchheim (1967) were of the opinion that changes in the atmospheric pressure can produce tilts confined to more or less small blocks in the

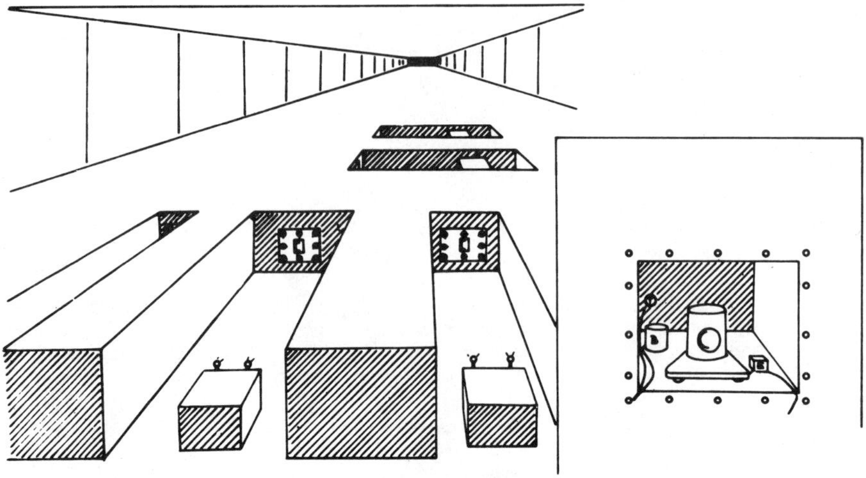

Fig. 12.7. New system of installation of the pendulums in the earth tide station at Tiefenort (German Dem. Rep.) (after Simon).

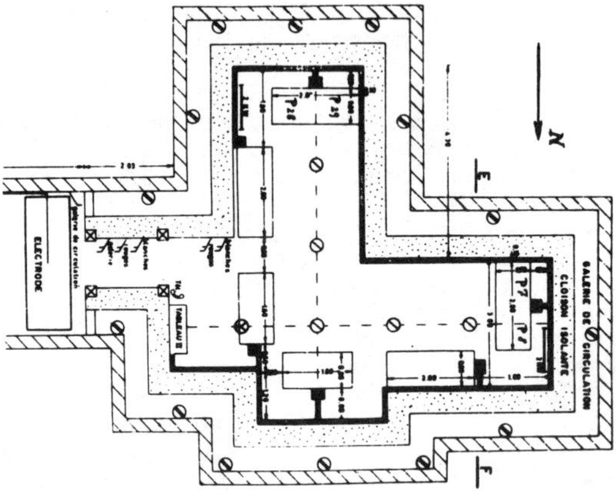

Fig. 12.8. The horizontal pendulums underground room at Dourbes (Belgium). One pair of VM pendulums (East West and North South component) is installed on each one of the big stone pillars situated at the two opposite extremities of the L room. Despite this different situation inside the cavity they give results in most excellent agreement.
Distance from the entry of the gallery: 50 m
Depth: 46 m, dimension of the cavity 6mX6m.

Earth's crust which therefore could react independently at distances as short as a few hundred kilometres. We consider this as still questionable because the results were derived from observations made with small precision metallic instruments which, moreover, are sealed.

Nevertheless, the experiments conducted by Simon at Tiefenort on barometric influences have demonstrated a dependence of the induced tilts from the installation site: installations in the walls being much more sensitive (up to 20 times) to pressure variations than the installations in the middle of the galleries or rooms.

In relation with these barometric effects, results showing extremely big fluctuations in the K_1 and O_1 parameters have been often presented by authors using metallic pendulums. We reproduce in Table 12.4 the results of the presently two longest available series of measurements from quartz VM pendulums.

Parallel Measurements with Exchange of Instruments

Such experiments have been performed in three places (at our present knowledge): Sclaigneaux (Belgium), Walferdange (Luxemburg) and Tiefenort (DDR).

One conclusion was clear using quartz VM pendulums: when exchanging one instrument with any other one placed on the same bolts in the rock, no change can be observed either in the tidal constants or in the drift.

At Tiefenort only metallic pendulums were used: some exhibited strong barometric dependence (Lettau, Tomaschek) while others did not show any relation (Hecker-Schweydar, Ostrovsky), which obviously demonstrates the instrumental origin of these effects.

12.3. THERMIC INFLUENCES

Temperature variations cannot be transmitted very deeply in the crust but, nevertheless, often appear due to artificial ventilation needed in operating mines. Considerable annual tilt changes have been observed with metallic instruments (Picha) but not with quartz VM pendulums. Such tilt is clearly of instrumental origin.

12.4. HYDROLOGICAL DISTURBANCES

These are probably the major long-period disturbances experienced in tidal measurements. Changes in the level of nearby rivers or of underground waters may produce strong deformations of the galleries where the instruments are installed.

Melchior and Brouet (1964) have published some typical qualitative examples of sudden drifts produced by river floods (Fig. 12.9 a, b) and also in the more unusual case of an important sudden drop in the river's level (Fig. 12.9 c). However, they concluded that a quantitative evaluation was probably impossible as the permeability of the surrounding rocks was not known and extremely difficult to introduce in the computations. They observe, for example, that when the river returns to its normal level the pendulums do not return exactly to their original position.

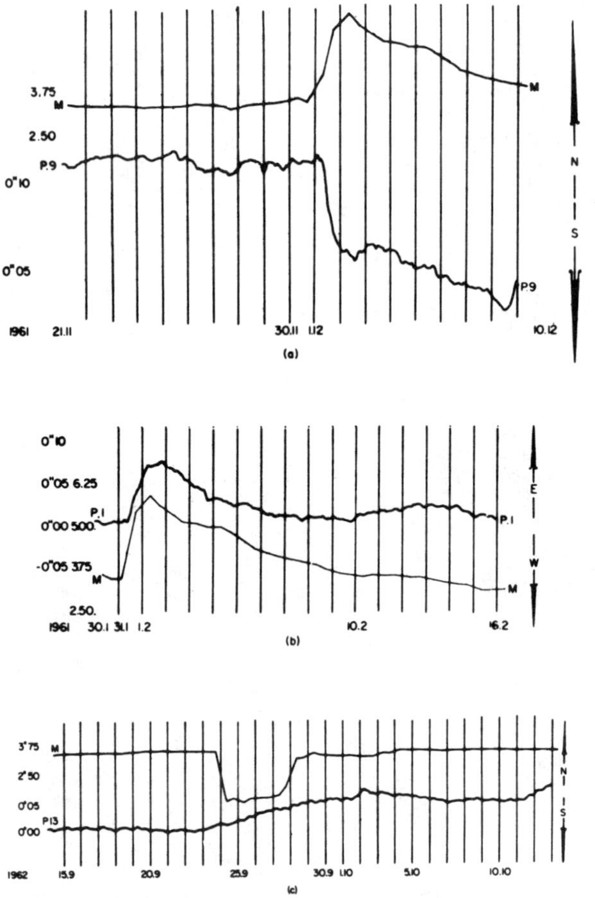

Fig. 12.9 Effect of the Meuse flood at Sclaigneaux in (a) December 1961, (b) February 1961, and (c) September 1962.

M: level of the Meuse in metres

P: drift of the pendulum.

12.5. THE CHOICE OF A SITE ON WHICH TO INSTALL A TIDAL OBSERVATORY

Clinometric and Extensometric Stations

The list of constraints which must be satisfied in order to obtain the best possible tidal observatory is a long one and we will try to classify them in order of importance.
The first constraint is obviously the noise consideration: the station should provide precise numerical information on the largest possible frequency band of the tidal spectrum, that is:

1) ter-diurnal waves (M_3);
2) semi-diurnal waves (M_2 S_2 N_2);
3) diurnal waves (K_1 O_1 P_1 ψ_1);
4) long-period waves (Mf, Mm, Ssa, Sa);

avoiding the introduction of any specific noise in each of these.

It is obviously not difficult to observe the semi-diurnal waves almost everywhere with the high-quality instruments now at our disposal, and consequently this will not be of any use as a criterion to be taken into account. The real criterion is the ability to derive correctly the diurnal components, that is to derive monthly determinations of amplitudes and phases of K_1 and O_1 which are not subject to excessive time variations, that is variations exceeding 10% of their mean value. Moreover, any annual analysis should give an acceptable result for P_1 and a very small S_1 component (see Table 8.5).

As a criterion we should propose that, for a tiltmeter installed in an East-West component at the latitude around 45°/50°, one should have

$$\gamma(S_1) \leq 5,$$

this limit being deduced from the experience obtained in many tidal stations. All stations situated at a depth greater than 40 m satisfy this condition.

Many shallow stations (5-15 m deep) have given extremely bad results for K_1 and O_1 amplitudes which vary by more than 100% month by month and, when an annual analysis could be made, an extremely high value for $\gamma(S_1)$, reaching 100 or more: this is obviously due to strong temperature effects. Such stations are not suitable for advanced tidal researches.

It should be added that in some deeper mine stations similar effects can be produced by artificial ventilation of the mine using surface air. In that case a non-ventilated gallery must be selected to install the station, but even this may be unsatisfactory because the whole system of galleries may be deformed by the ventilation which introduces a strong diurnal component. This has been observed in one of our stations when a strike of more than one month provoked the interruption of the artificial ventilation so that we could immediately observe a 10% change of the diurnal components parameters. This station has been closed for tidal investigations.

The stations satisfying condition (3) will normally give fairly good results for M_3 as the noise level is usually low in that frequency band.

For the last wave group, the long-period ones, there seems, however, to be no clear solution as the periods involved (half month, month, half year and year) are largely contaminated by other long-period effects mainly due to hydrological and barometric phenomena. The underground water-pressure pulsations cause slow crustal deformations which superpose to the tidal long-period deformations and the instrumental drift (which usually is parabolic and becomes very small with time increasing).

It has not been possible until now to separate correctly these effects, and they should not be considered as criteria for site selection unless strong hydrological changes should happen regularly because of pumping activities.

In the same category of criteria we should introduce the microseismic noise. It is evident that strong microseismic areas (sea shore or industrial area) must be avoided and that, when working in still active mines, only the abandoned galleries can be used. Usually a distance of about 500 m from the mining activity may be sufficient.

The second constraints are of a quite different kind. They are related to the existence of systematic effects:

1) oceanic indirect effects;
2) topography effects;
3) cavity effects.

It is clear that the real objective of the research must be defined when a selection has to be made according to these criteria. One may be interested precisely in some kind of prospecting of these effects and then search for a very special situation as, for example, when using extensometers to prospect the cavity effect on the ℓ/h ratio or when using gravimeters on islands or near sea shore to improve the cotidal charts.

But if the aim is to improve our knowledge of the pure body tides and particularly the liquid core hydrodynamics, all these spurious effects must be avoided, probably as follows:

1) The distance to the nearest sea is a basic but not sufficient parameter. By making a previous estimation of the indirect effects from the available cotidal maps one observes that some continental areas are still disturbed by world oceans. Very good sites are, in general, in the continental areas as, for example, Kathmandu (0.69 μgal for M_2 indirect effect) in Asia, Tucuman (0.61 μgal for the same effect) in South America. But this is not always true, as Alice Springs (in the centre of Australia) has a contribution of 2.36 μgals from these effects. In western Europe this contribution is about 2 μgals on the coasts and diminishes to 1.5 μgal in the centre.
2) A flat surrounding area will, of course, eliminate the topographic effects as the direction of principal strain, which must be perpendicular to the free surface, then remains vertical. Very deep stations will also be better in that respect as they will not be affected by smooth topographic undulations. In that sense Přibram and Dannemora are well-selected stations since they are very deep under a flat plateau. Sclaigneaux and Walferdange are in a moderate hilly zone, but the gallery in each case is very long (800-1000 m) and these stations also lie under a more or less flat plateau. Dourbes is not in the same situation because the gallery is too short (50 m), while Costozza is in a volcanic area with a complicated topography. Armidale is mainly in a very deep gorge (500 m depth) cutting a 1000 m high plateau, and this must obviously produce a very strong disturbance.
3) The cavity effect is avoided by installing the instruments where these effects are probably minimal. According to Lecolazet-Wittlinger and Harrison, the walls should be avoided as well as the extremities of the galleries. This rests upon theoretical calculations. In practice it is not so simple:

a) Diurnal waves (K_1 O_1 P_1) observed in East-West components in Europe in very different installation conditions do not show perceptible cavity affects (see Table 13.1). Indeed, Sclaigneaux instruments are in lateral walls; Dannemora and Příbram in end walls of the gallery; Dourbes and Sopron on stone pillars in the room; Walferdange and Bad Grund on the floor.
b) Ter-diurnal waves M_3 observed in East-West and North-South components in Europe do not show any sign of cavity effects (see Table 13.4).
c) Semi-diurnal waves (M_2 S_2 N_2 K_2) observed in the same stations show more geographic dispersion in East-West components (see Table 13.2), and very strong anomalies in North-South components (see Table 13.3). However, the strong anomaly observed at Dourbes (γ = 0.44) is exactly the same for two pendulums installed at two different edges of the observing room (see Fig. 13.4).

Other minor factors are the humidity percentage and the mean temperature. The VM quartz pendulums adjust quite well to any condition (humidity of 95% at Sclaigneaux, temperature of 41°C at Příbram, or -5°C at Longyearbyen/Spitsbergen).

Gravimetric Tidal Stations

Experience shows that the vertical component is much less sensitive to surface effects than tilt or strain. Therefore the constraints are not so severe for gravimeters as they are for tiltmeters or extensometers. Nevertheless, they must be taken into account carefully if one wants to obtain reliable results. We have put these constraints in the following order:

1) Stabilized power supply without interruptions to ensure that the heating of the instrument will not be interrupted as this creates a thermic shock in the sensor which may take many hours (up to 24 hours) to recover.
2) A firm pillar, isolated from the building construction on which to put the gravimeter, avoiding in this way any local industrial noise and mainly ensuring a fairly good stability with respect to tilt, which is an essential point for astatic instruments.
3) Thermic stability in the gravimeter's room to about 0.5°C a day.
4) A good maintenance ensuring the weekly calibrations, drift corrections when needed (which is seldom necessary with normal instruments) and checks after each strong earthquake for preventing the beam from sticking on the stops.

All the gravimeters used until now are thermostatized at temperature between 30° and 50°. They may not operate correctly when the room temperature is lower than 10° or higher than 30°.

An interesting experiment is being conducted in the Walferdange underground laboratory with two BN gravimeters (Askania GS 11 revised and transformed by Bonatz), one operating without any artificial thermostatization. The curves do not exhibit any drift and look perfectly smooth. However, more than a year's data has to be analysed before one can ascertain that no spurious thermic influence distorts the registered curve (see Table 9.5).

Timing Problem in Earth-Tide Stations

When analysing tidal curves of normal good quality the internal error on the M_2 phase angle is found to be as low as 0.06° (mean square error) in the 3000 days' observations with VM horizontal pendulums (East-West) at Dourbes. More usually it is 0.10° to 0.20° (Sclaigneaux, Walferdange, Příbram, Dannemora, etc ...). This corresponds to about 12 seconds of time. However, this is obtained with a recorder drum speed of about 6 mm/h or 0.1 mm/min of time. In the case of more recent gravimeters the recorder speed is 5 times higher, and therefore 0.1 mm corresponds to 12 seconds of time.

To reach such a precision, very fine time marks must be superposed on the tidal curve itself with a precision of about 5 s. This means that the time marks must cross the tidal curve at the correct time as shown on all the figures reproduced in this book (e. g. Fig. 8.19). Time marks put on one side of the recording paper, as is usual with multipoints recorders, are to be avoided as they are transported on to the curve with a real danger of parallax errors. Moreover, this operation represents a great waste of time.

Systematic clock errors are very dangerous, and therefore a good quartz clock is to be recommended. They are not much more expensive than good pendulum clocks. They are far easier to install as one has just to press a button to start them. Moreover, they seldom need a correction (once a year or so) and therefore very cheap to maintain.

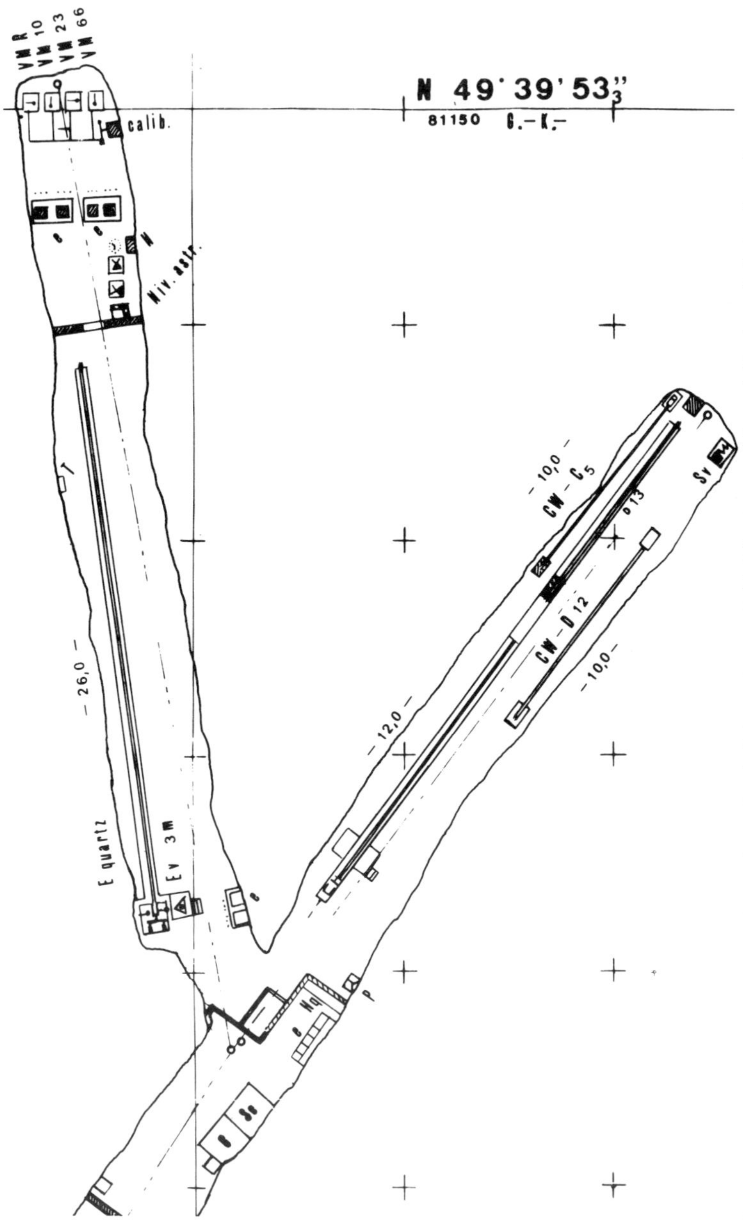

Fig. 12.10. The underground geodynamical laboratory of Walferdange (Grand Duchy of Luxemburg). VM: quartz horizontal pendulums ☒: Tsubokawa electromagnetic pendulums ▲ EV: vertical extensometer E quartz: 26 meters horizontal extensometer CW: horizontal wire extensometers Hq: quartz clock G: gravimeter

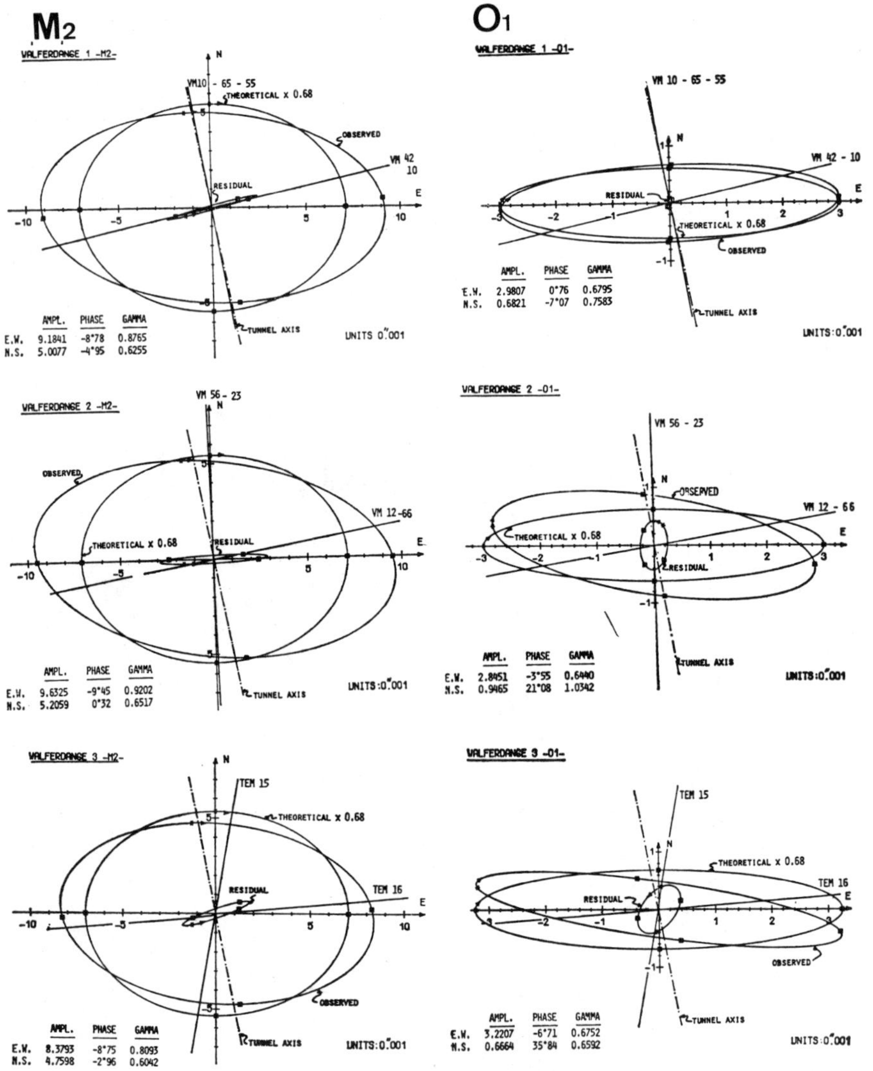

Fig. 12.11. Deflexion of the vertical at Walferdange 1-2-3, with two pairs of VM pendulums and one pair of Tsubokawa pendulums (below). Comparison of the observed tidal deflexion of the vertical with the theoretical deflexion (γ = 0.68). The small residual ellipses are obtained by subtraction of the theoretical from the observed ellipse. Graduated axis corresponds to NS and EW direction, units are 0"001. Azimuth of the pendulum and the tunnel axis are indicated.

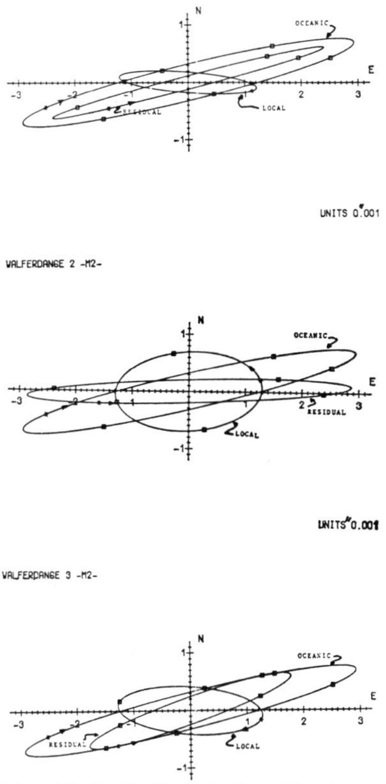

Fig. 12.12. M_2 Residual deflexion of the vertical at Walferdange with three pairs of pendulums (units 0"001). *Local* ellipses are obtained by subtracting the theoretical X0.68 and indirect effect from the observed tide.

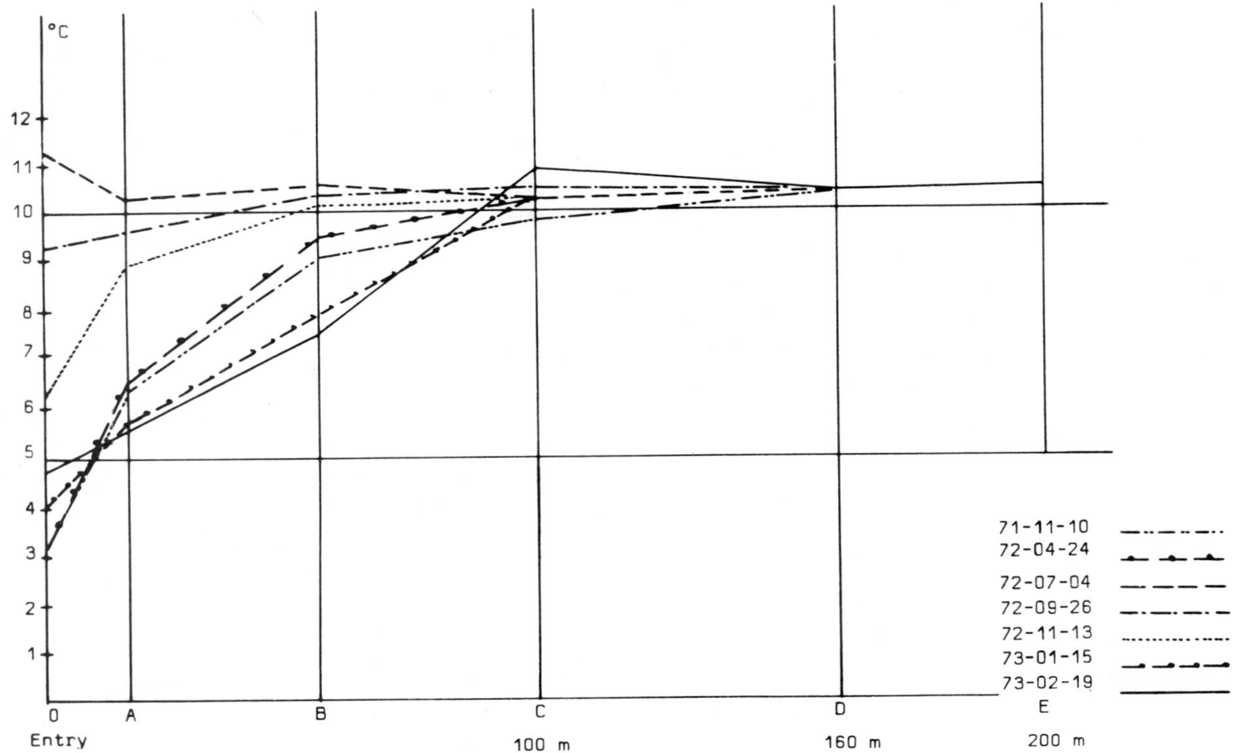

Fig. 13.13. Seasonal variations of temperature along the gallery at the underground geodynamical laboratory of Walferdange (measurements by Van Ruymbeke). The variations completely disappear at 200 m from the entry. The instruments are located at 800 m from the entry (Fig. 12.10).

CHAPTER 13

General Comparison of the Experimental and the Theoretical Results

13.1. TRANS WORLD TIDAL PROFILES

Because of the many different effects involved, comparison between experimental data obtained at different sites as well as the comparison of these results with the theoretical models is not an easy task. In this respect it is most unfortunate that observations are not made systematically all around the world as the first step towards a correct interpretation is the separation of the oceanic effects from the pure Earth Tide.

This needs a well distributed network of tidal stations. Since 1973 the International Centre of Earth Tides (ICET) has made a strong effort in this sense to fullfil the immense geographical gaps by developing two Trans World Tidal Gravity Profiles, one, extending over 17500 Km, through South Asia, South East Asia, Australia and the South Pacific, the other extending over 9500 Km in East Africa from Egypt to Madagascar. These Profiles have involved obviously quite a number of Developing Countries in this advanced research.

In 1970 a campaign of European profiles had been started at the initiative of J.T. Kuo (Kuo et al., 1972) and began with a comparison of calibrations of three Geodynamic gravimeters between New York (Columbia University) and Brussels (Royal Observatory). In this operation an agreement of 0.5% was achieved (Kuo et al., 1972). Thereafter 32 gravimeters from different makers (Geodynamics, La Coste-Romberg and Askania 15) were compared at the fundamental station of Brussels by recording the tidal variations during a minimum interval of 4 months each. These comparisons were simultaneous for groups of three to five instruments.

Using the tidal parameters derived from these registrations for the diurnal wave O_1 (amplitude and phase) and for the semi-diurnal wave M_2 (phase only), Ducarme (1975) adjusted a rheological model defined by a calibration factor and by a frequency dependent phase lag and attenuation factor to each instrument (see Sections 9.7 and 9.8). Comparisons of different instruments have been performed at Liverpool, Sèvres, Canberra and Wuhan (China) and provided a successful check of the models. These models are used for the reduction of the data obtained in the Trans World profiles: they ensure the homogeneity of the results.

The objectives of these measurements were defined as follows:

1. To determine to what extent measurements in continental stations such as Katmandu, Lanzhou, Urumqi and Alice Springs are free from oceanic tidal influence.
2. To compare coastal stations results with those from continental stations and to see if the tidal parameters (amplitude factor and phase lag) exhibit any regional behaviour and what the extension of such regions might be.
3. To check if any one of the existing cotidal charts allows the perfect correction of the observed data so that one will obtain identical tidal parameters at all the places which, moreover, fit the Love numbers obtained by the integration of the fundamental equations of an elliptical rotating

elastic isotropic planet when using the best models of the Earth's interior.

4. In the case that it would prove to be impossible, to see if any improvement or correction of the cotidal charts is to be done or if another geodynamical process or geophysical parameter has to be invoked to explain the observed anomalies, in particular lateral heterogeneities in the lithosphere.

For practical reasons it was decided that the first operation should extend in the East direction, starting from Brussels and, crossing South-East Asia, reach Australia, New Zealand and some of the Pacific islands. The second operation is developped all along East Africa. Moreover a network of nine stations has also been established in China as a cooperation between the State Bureau of Seismology of China, the Academia Sinica and the Royal Observatory of Belgium.

Considering the objectives of the project, the selected profiles were particularly interesting because, in 1973, the cotidal maps in the Indian Ocean, the China Sea and the South-West Pacific were still very uncertain and contradictory when considering the solutions proposed by different authors,

On the other hand, the crustal structure and plate tectonics areas of Himalaya, Indonesia, the Philippines, Papua, New Guinea, New Caledonia, Fiji and New Zealand, African rift, could offer excellent opportunities to observe new phenomena.

13.2. OCEANIC EFFECTS AND HETEROGENEITIES OF THE LITHOSPHERE IN TIDAL GRAVITY MEASUREMENTS

We have seen that precise Earth tide measurements carry much information about the tides of the ocean. This information (as stated for example by Hendershott in 1973) could be extracted in an *optimal manner* by solving an inverse problem for ocean tides using Earth tides as observables.

The earth tide measurements indeed provide areal means of ocean tides that are free of any local distortion resulting from shallow water effects but tidal loading effects depend much more upon the structure of the lithosphere than does the bodily Earth Tide: the tidal load displacements are appreciable only in the crust and upper mantle and are sensitive to differences in crustal structure as found beneath ocean basins and continents. Near the load the surface deformation is very sensitive to the properties of sediments.

At larger distances from the load, one has to take into account structures down to generally a depth two or three times the horizontal distance between the load and the point of observation. Lack of knowledge of these lithospheric features is why the loading effects are presently not predictable with a high precision.

To achieve such a goal it will be convenient to subtract from the observed tidal effects 1° the effect of the deformations of a convenient model earth, 2° the loading and attraction effects of a convenient model of the oceanic tide.

Let $\bar{A}$ (A_i, α_i) be the observed tidal vector for the wave i in the Cartwright-Tayler-Edden potential, A_i being its amplitude and α_i its phase. The corresponding theoretical vector for a standard elastic model earth, with liquid core, but oceanless, is $\bar{R}(\delta_i\ A_i, o)$ where A_i^T is the amplitude for a rigid body and $\delta_i = 1 + h_i - 3/2\ k_i$ (in the case of second order potential terms).

We define the *"residual vector"* $\bar{B}(B_i, \beta_i)$ as

$$\bar{B} = \bar{A} - \bar{R} \qquad (13.1)$$

or

$$B_i \cos(\omega_i t + \beta_i) = A_i \cos(\omega_i t + \alpha_i) - R_i \cos \omega_i t$$

The vector $\bar{B}$ is the indicator of the ocean-lithosphere tidal interactions. The corresponding calculated "load and attraction vector" $\bar{L}(L_i, \lambda_i)$ is obtained by the convolution procedure of Farrell applied to a given cotidal map.

We now define the *"end residue"* which is a vector $\bar{X}(X_i, \chi_i)$:

$$\bar{X} = \bar{B} - \bar{L} = \bar{A} - \bar{R} - \bar{L}$$

or (13.2)

$$X_i \cos(\omega_i t + \chi_i) = B_i \cos(\omega_i t + \beta_i) - L_i \cos(\omega_i t + \lambda_i)$$

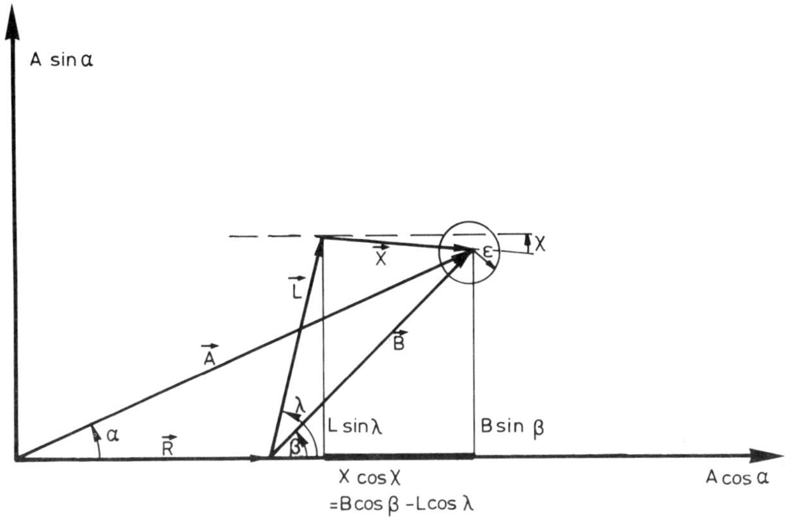

Fig. 13.1. Comparison of observed and calculated Ocean-Continent tidal interactions.

$\bar{A}$ observed tidal vector,
$\bar{R}$ calculated vector for an elastic model earth,
$\bar{B}$ Residual vector : $\bar{B} = \bar{A} - \bar{R}$
$\bar{L}$ Load (and attraction) vector calculated from the oceanic cotidal map.
$\bar{X}$ End residual vector: $\bar{X} = \bar{B} - \bar{L}$
ε instrumental error

For the semi diurnal wave M_2 a correct scale for this figure should be:

$R \sim A \sim 40$ (Europe) to 90 (Equator) μgals,
$\alpha \sim 0°$ to $\pm 5°$
$L \sim B \sim 2$ (Europe to 10 (South Pacific) μgals,
$X \sim 0.5$ to 5 μgal
$\varepsilon \sim 0.5$ μgal, (Europe) to 1 μgal (Equatorial zone)

$\vec{B} = \vec{A} - \vec{R}$; $\vec{B} - \vec{L} = \vec{X}$

These operations are represented by the figure 13.1 and illustrated in the Tables 13.1 to 13.14.

In the numerous evaluations (180 tidal gravity stations) made by Melchior and De Becker (1982) the Molodensky I model, the Wahr model and an experimental model have been used. They do not differ significantly in the tropical areas but the first two give rather larger values for δ at the mid latitudes (1.160 instead of 1.147 at 50°) and disagree from Wahr's even more at high latitudes.

To ensure homogeneity in the results the Schwiderski cotidal maps have been used as standards of the oceanic tide modelisation. Anyway, at the exception of those of Hendershott for the South Pacific and of Parke for Europe, all other maps are everywhere in contradiction with the land based tidal gravity measurements.

The vector $\vec{X}(X_i, \chi_i)$, which we have called the end residue, contains the unexplained part of the observed residual vector B. When $|X| > \varepsilon$ it is suspected to contain the following systematic effects:

Instrumental systematic errors:

Calibration (frequency-dependent)
Thermal influences
Barometric effects
Power supply, ground connection, dead band of the recorder
Drift

Geophysical effects:

Uncorrect cotidal maps
Lateral heterogeneity of the lithosphere
Load barometric effect

Computation errors:

Map digitization, computer processing
Errors in coordinates, imperfections of analysis method

However, from (13.2) we very simply derive for the semi-diurnal waves:

$$\begin{cases} B_i \cos \beta_i = A_i \cos^2\phi \,(\delta_i \cos \alpha_i - \delta_i^{th}) \\ B_i \sin \beta_i = A_i \cos^2\phi \,(\delta_i \sin \alpha_i) \end{cases} \qquad (13.3)$$

and for the diurnal waves:

$$\begin{cases} B_j \cos \beta_j = A_j \sin 2\phi(\delta_j \cos \alpha_j - \delta_j^{th}) \\ B_j \sin \beta_j = A_j \sin 2\phi(\delta_j \sin \alpha_j) \end{cases} \qquad (13.4)$$

A very important feature demonstrated by these formulas and clearly shown by the figure 1 is that the correlation between B sin β and L sin λ *is not affected by the choice of the Earth's model* ($\delta^{th}_{i,j}$) B sin β being independent of it on the condition that the viscous phase lag of the Earth is negligeable which has been demonstrated by Zschau (1978). This provides a check for the instrumental calibrations and for the oceanic cotidal maps.

On the contrary the correlation between B cos β and L cos λ is strongly affected by any imperfection in the "theoretical model" represented by the vector $\vec{R}$ so that the component:

$$X \cos \chi = B \cos \beta - L \cos \lambda$$

represents indeed the effects of the heterogeneity of the lithosphere while the component:

$$X \sin \chi = B \sin \beta - L \sin \lambda$$

is totally independent of this and contain only the instrumental errors and the imperfections of the modelisation of oceanic tides.

Melchior and De Becker (1982) have demonstrated from data of 180 world wide distributed tidal gravity stations that $X \sin \chi$ is of the order of accidental errors (less than 1 μgal) while $X \cos \chi$ has a very clear regional behaviour, reaching up to 5 μgals at several very characteristic places.

The figure 13.2 gives an example of their results for regression computations represented by the formulas:

$$\begin{pmatrix} B \cos \beta \\ B \sin \beta \end{pmatrix} = c_0 + c_1 \begin{pmatrix} L \cos \lambda \\ L \sin \lambda \end{pmatrix} \qquad (13.5)$$

A summary of the numerical results is given in the table 13.1 for the two waves M_2 and O_1.

TABLE 13.1. Numerical results of the regression between $\vec{B}$ and $\vec{L}$.

	M_2		O_1	
	sine terms	cosine terms	sine terms	cosine terms
correlation coefficient	0.857	0.815	0.778	0.516
c_0	-0.528	0.084	-0.061	0.010
c_1	1.403	1.725	1.415	2.708
mean square error				
on B	2.74	2.46	0.83	0.81
on L	2.05	1.56	0.63	0.46

The results for the other waves are similar and show a better correlation for the sine terms than for the cosine terms.

In view of the importance of this new result we shall examine separately each region where detailed informations about the measurements are now available successively for the semi-diurnal waves and the diurnal waves.

13.2.1. The semi-diurnal waves

Europe

There is a very high density of tidal stations in Europe, occupied with many different instruments from different institutions. The selection of stations presented here is taken from the ICET files. Most of them come from Bonatz and Richter (1975), Melchior et al. (1976), Gerstenecker and Groten.

The results, shown in Table 13.2 as "observed load vector", are very coherent, with phases between 40° and 65° but systematically lower in northern Italy and at Graz. Coastal stations in England, Portugal, Spain and Norway as well as Bordeaux in France of course have a large vector with a locally different phase but these features are very well explained by the cotidal maps. The best fit is undoubtedly given by the Schwiderski maps but the Parke M_2 also gives a very good fit on condition that the amplitude of the calculated B is reduced to 60%. The phase of the Schwiderski load vector for M_2 is often too large by a few degrees but if we had used the Wahr latitude-dependent factor δ (1.147 in Europe) this discrepancy would have been larger. The load vector (L,λ) calculated with the Parke map is correctly oriented but has too large amplitude: subtracting it vectorially from the residual vector (B,β) yields a vector (X,χ) with a phase χ of about -100 or -110°.

TABLE 13.2. Tidal wave M_2 (Europe). Residues with respect to Molodensky Model I and Schwiderski cotidal maps. Phases χ have been omitted for vectors of amplitude less than 0.3 μgal except when they agree for the two maps.

Station		Observed load vector		Load vector (Schwiderski map 1°x1°)		Vectorial difference	
		B(μgal)	β(°)	L(μgal)	λ(°)	X(μgal)	χ(°)
0011	Narssaq	5.93	11	4.45	6	1.55	25
0105	Cambridge	2.45	68	2.86	56	0.69	-171
0110	Herstmonceux	0.92	155	1.57	153	0.65	-29
0201	Brussels	2.01	64	1.88	63	0.14	
0212	Cointe	2.29	58	1.81	59	0.48	54
0220	Bruges	2.65	70	1.90	76	0.79	55
0252	Walferdange	1.84	57	1.84	61	0.13	
0271	Witteveen	1.90	41	1.48	35	0.46	60
0709	Hannover	1.28	51	1.31	48	0.07	
0710	Bad Grund	1.56	44	1.33	49	0.26	
0701	Bonn	1.51	58	1.66	55	0.17	
0706	Frankfurt	1.09	52	1.57	55	0.48	-118
0300	Strasbourg	1.49	62	1.71	60	0.23	
0303	Paris	2.66	61	2.47	72	0.53	-2
0736	Würzburg	1.43	52	1.46	54	0.06	
0737	München	1.61	54	1.38	53	0.23	
0711	Berchtesgaden	1.91	49	1.28	51	0.63	44
0310	Clermont	3.22	58	2.77	74	0.95	4
0311	Grasse	2.02	58	1.88	69	0.40	-6
0312	Bordeaux	5.79	74	5.55	83	0.92	3
0615	Zürich	1.98	52	1.68	60	0.39	15
0610	Chur	1.83	52	1.58	59	0.32	15
0516	Turin	1.69	37	1.80	65	0.85	-46
0508	Padua	2.84	19	1.64	64	2.04	-15
0509	Trieste	2.10	18	1.25	52	1.27	-15
0695	Graz	2.05	24	1.14	47	1.10	0
0480	Porto	5.10	110	7.57	114	2.51	-57
0822	Aarhus	1.48	38	1.05	51	0.51	10
0843	Bergen	1.61	-133	1.00	-109	0.81	-163
0844	Trondheim	5.00	206	0.78	185	4.28	-150
0846	Bodø	3.69	179	2.39	165	1.49	-158
0847	Tromsø	5.23	146	2.57	129	2.87	161
0882	Helsinki	0.41	30	0.54	31	0.13	
0887	Sodankyla	0.47	105	0.64	57	0.48	-170
0930	Pecny	1.06	32	1.13	46	0.28	
1117	Obninsk	0.95	-7	0.50	-4	0.45	-10

It is thus evident that a systematic distribution of the X phase over a sufficiently broad area is a valuable indication about the properties of the cotidal map from which it has been derived.

Obviously it would be difficult to consider residues lower than 0.3 μgal as significant, and even more so the phases attached to such small vectors; these phases therefore have been omitted from Table 13.1.

It is remarkable that the four fundamental European stations, Brussels, Walferdange, Hanover and Bonn (Melchior et al., 1976), where a large number of different instruments have been installed, compared and intercalibrated, offer a *perfect* fit with the Schwiderski M_2 map which indicates that the calibrations (and instrumental phase corrections) are correct to within 0.1 μgal, corresponding to 0.2% accuracy, the M_2 wave amplitude being about 40 μgal.

One should therefore pay some attention to the small but coherent end residues observed with the high-precision instruments at Strasbourg and Frankfurt: 0.3 -0.5 μgal with a -120° phase. However, this is not confirmed by the observations of Gerstenecker and Groten (1976) in the surrounding area, so this weak signal may be due to small calibrations errors.

A more important anomaly clearly appears for stations distributed around the Alpine belt (Table 13.3). At first sight one could have suspected a perturbation by the Mediterranean and Adriatic Seas but their effect does not reach more than 0.25 μgal as calculated by Chiaruttini (1976) for the Italian stations, who even concluded that "No help in the marine tide problem has to be expected from gravity measurements in the Italian peninsula".

When we consider the components X cos χ and X sin χ in the Table 13.3, we see that Chiaruttini is right: the anomaly does not come from marine tide. But we can add now that there is an anomaly and that it comes from lithosphere heterogeneities!

As shown by the figure 13.3 these stations are indeed situated just where the lithosphere reaches an exceptional thickness of about 130 km while it is in general of about 50-70 km elsewhere in continental Europe.

TABLE 13.3.

Wave M_2 residues in two specific areas in Europe (fig. 13.3)

		X cos χ	X sin χ
0310	Clermont Ferrand	0.95	0.07
0508	Padova	1.93	-0.24
0509	Trieste	1.24	-0.34
0695	Graz	1.10	0.00

An additional check of the validity of the Schwiderski maps in Europe can be offered by considering in the Table 13.4 those places where the N_2 load signal reaches more than 0.7 μgal. One can see that the agreement between observations and computations is outstanding.

Thus, as a conclusion we may say that even with a weak load signal (1 - 4 μgal for the M_2 tide) as recorded in Europe useful conclusions may be obtained concerning the cotidal maps and lithosphere heterogeneity. It must be emphasized however that, for a weak signal, the careful intercalibration of the instruments is essential.

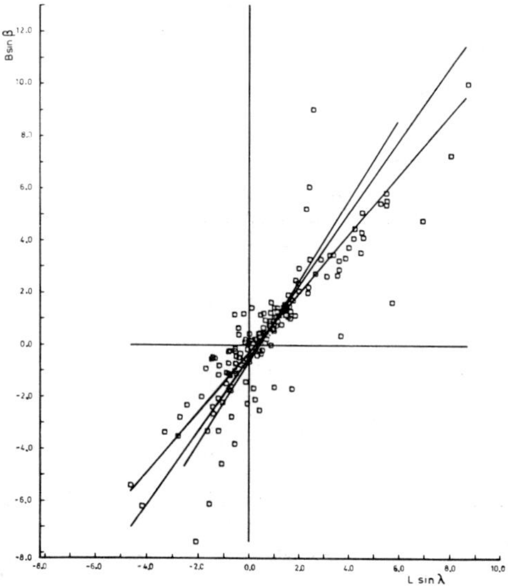

Fig. 13.2. Example of linear regression between the sine components of the observed and calculated oceanic load.

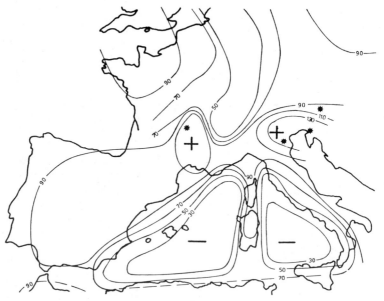

Fig. 13.3. Simplified map of Lithosphere thickness in Europe according to Panza et al (Pageoph 118, 1209, 1980) (interpretation by Mueller et al, Vermessung, Photogrammetrie, Kulturtechnik 12/80, 469, 1980).

* Tidal gravity stations exhibiting a 1 μgal anomaly on X cos χ.

TABLE 13.4.

Wave N_2, European stations where the residue has an amplitude larger than 0.7 μgal					
	Stations	Observed Residue		Schwiderski map	
0011	Narssaq	1.52	31	1.02	25
0312	Bordeaux	1.28	97	1.14	101
0404	Santander	2.31	107	1.88	111
0411	Santiago	1.92	117	1.96	127
0405	Burgos	1.05	103	1.18	114
0401	Valle L	0.94	103	0.92	120
0402	Madrid	0.79	112	0.86	119
0403	Granada	0.78	135	0.73	127
0406	S. Fernando	1.13	146	1.45	143
0408	Cuenca	0.77	119	0.73	116
040	Segovia	1.15	123	0.97	119
0409	Zamora	1.22	114	1.15	122

General considerations of the extra-European regions with large load signal

The regions of Africa, South and East Asia, Indonesia, Australia, South Pacific and Japan where many good observations are now available are characterized by large oceanic attraction and load signals: from 3 to 15 μgal for M_2. This makes the checking of cotidal maps much more clear so that, from a sampling of 70 stations, all outside Europe, we have observed at first sight a satisfactory fitting for the M_2 load in 45 cases for the Schwiderski map, 21 for the Parke map, 16 for the Hendershott map, 12 for the Zahel 4° x 4° map, 11 for the Zahel 1° x 1° map and 10 for the Bogdanov map. We also notice that the subtraction $\bar{B}$-$\bar{L}$ increases the amplitude of the residues (which is unsatisfactory) for 43 stations with the Hendershott map, 38 with the Zahel 4° x 4° map, 35 with the Bogdanov map, 33 with the Zahel 1° x 1° map, 24 with the Parke map and only 8 with the Schwiderski map.

We can clearly see that some large areas are poorly fitted: in the South China Sea, all maps fail except the Schwiderski map; the Bogdanov map is poor for Indonesia; the Hendershott map is poor for Africa; the Zahel 1° x 1° map is poor for Australia; and the Zahel 4° x 4° map is poor for Africa, Indonesia and Australia. Finally we may observe that the key station at Alice Springs in the very centre of Australia is correctly represented only by the Schwiderski and Parke maps.

East Africa and the Indian Ocean (Table 13.5).

We have performed, until now, observations in thirteen stations of East Africa but our profile in this region is still being measured. By including the UCLA station at Lwiro and the Bonatz stations at La Réunion and Kerguelen as well as some of our other stations around the Indian Ocean (Perth, Bandung, Colombo) we can offer a rather interesting picture of the situation around this ocean, as given in Table 13.5. We consider it as an exceptionally good fit, Helwan station being excluded (on experimental grounds) as well as Perth which is more astonishing as the observations there have been of the highest quality. The results obtained by UCLA at the Bunia station had to be rejected because they are in strong conflict with all neighbouring stations (Khartoum,

Nairobi, Lwiro) as well as with the load and attraction effects computed from Schwiderski maps for the three semi-diurnal waves M_2, N_2 and S_2. They are also in conflict with Nairobi and Lwiro for the diurnal waves O_1 and K_1.

The observed M_2, N_2 and S_2 phases, which are all positive in East Africa, are smoothly and systematically rotating anticlockwise from Aswan to La Réunion and Kerguelen. The M_2 oceanic tide has very large amplitudes (100 cm) all along the west coast of Madagascar, in the Mozambique Channel, while they are negligible on its east coast, due to the existence of an amphidromic point at the southeast. This makes very interesting the fact that the best fit is obtained for all three waves in Antananarivo. Kerguelen also gives a very good fit while the other stations are often quite satisfactory.

The stations Nosy Bé and Tolañaro respectively at the extreme north and extreme south of Madagascar are very close to the shore. This obviously explains their very large load signal. A detailed cotidal map around Madagascar will be needed to correctly fit these data.

The considerable discrepancies between the Indian Ocean cotidal maps proposed by the different authors are mainly due to the fact that this area is very near to resonance with tidal frequencies, so that slight changes or errors in the dimensions (which depend of course upon the grid dimensions) displace the eigen-frequencies and cause great differences in the results.

TABLE 13.5. Trans World Tidal Gravity Profiles

East Africa and Indian Ocean

STATION	M_2 OBSERVED RESIDUE		M_2 SCHWIDERSKI MAP		N_2 OBSERVED RESIDUE		N_2 SCHWIDERSKI MAP		S_2 OBSERVED RESIDUE		S_2 SCHWIDERSKI MAP	
	B	β	L	λ	B	β	L	λ	B	β	L	λ
3000 /Helwan/	2.47	-59	0.67	21	1.55	0	0.14	29	1.65	-112	0.28	-1
3005 Aswan	1.72	5	0.72	26	0.47	34	0.15	30	2.36	25	0.32	3
3010 Khartoum	1.55	38	0.94	43	0.29	66	0.18	46	2.76	61	0.46	17
3014 Addis Ab.	0.95	67	1.36	47	0.17	60	0.25	49	0.41	-3	0.74	14
3018 Arta/Djib	7.11	-4	1.72	-6	2.11	-11	0.41	-12	3.77	-4	0.94	-25
3020 Mogadis.	6.22	42	5.35	59	1.28	65	1.04	70	3.56	15	2.95	18
3030 Nairobi	3.75	45	3.32	69	0.71	55	0.58	81	2.29	27	1.68	32
3031 Voi	4.81	59	4.48	69	0.63	85	0.79	83	2.76	34	2.27	31
3420 Lwiro	2.08	87	2.11	71	0.92	58	0.38	78	1.37	63	1.01	40
3040 Dar Es Sa	7.94	71	8.47	71	1.46	83	1.50	88	4.05	43	4.28	32
3500 Maputo	8.71	60	6.26	64	1.32	60	1.14	70	4.66	30	3.33	31
3602 Nosy-Be	19.70	61	7.90	60	3.89	88	1.42	74	11.59	22	3.73	19
3601 Antananar	4.55	46	4.30	57	0.80	57	0.74	68	3.43	13	1.92	17
3604 Tolagnaro	5.16	92	3.69	41	1.03	289	0.63	46	3.45	52	1.51	2
3620 Reunion	5.15	140	2.60	111	1.48	136	0.55	130	2.52	148	1.18	121
9904 Kerguelen	5.30	196	4.27	192	1.10	214	1.07	225	2.29	140	2.14	136
4211 Perth	4.20	6	0.41	-77	0.73	24	0.16	108	1.66	-5	0.45	-148
4100 Bandung	6.29	-19	2.50	-48	1.19	-13	0.49	-26	2.89	62	0.84	-116
2460 Colombo	4.01	162	1.47	157	0.70	-171	0.18	-167	2.30	130	1.06	114

But probably the most interesting fact is the ($X \cos \chi$, $X \sin \chi$) result obtained at Arta/Djibouti (Table 13.8) in the famous Afar region as it clearly demonstrates a large tidal anomaly due to the heterogeneities in the lithosphere.

South and Southeast Asia (Table 13.6).

The 20 stations regularly distributed over this broad area are presented in Table 13.6. Seventeen have been established by ourselves since 1973, the other three being UCLA stations (New Delhi, Saigon, Baguio).
The phase β of the observed M_2 load exhibits a very regular trend all along this profile, from +10° in Istanbul and Ankara to -166° in Hyderabad. It also shows a very systematic distribution in five stations of Southeast Asia: -103 to -162° (except Kuala Lumpur) and another systematic picture at Hong Kong and in the Philippines (-19 to -66°).

However, the measurements made in the three stations in Pakistan (Peshawar, Quetta and Lahore, with two different instruments) and to a lesser extent in Teheran and Tabriz in Iran are in strong conflict with *all* the cotidal maps despite the fact that the Schwiderski M_2 map includes the Persian and Oman Gulfs where the amplitudes reach 70 cm.

An interesting feature is the strong difference between the observed loads at Penang and Kuala Lumpur (observations being made with the same instrument), unexplained by any world cotidal map and most probably due to the important semi-diurnal tides in the Malacca Strait.

TABLE 13.6. Trans World Tidal Gravity Profiles.

STATION	M_2 OBSERVED RESIDUE		M_2 SCHWIDERSKI MAP		N_2 OBSERVED RESIDUE		N_2 SCHWIDERSKI MAO		S_2 OBSERVED RESIDUE		S_2 SCHWIDERSKI MAP	
	B	β	L	λ	B	β	L	λ	B	β	L	λ
South Asia												
2000 Istanbul	1.32	10	0.69	19	0.19	13	0.14	32	0.68	8	0.24	-6
2005 Ankara	1.10	9	0.63	9	0.26	37	0.13	21	0.55	-1	0.23	-16
2202 Tabriz	2.41	-8	0.55	-37	0.53	-6	0.12	-27	1.08	-9	0.21	-64
2201 Teheran	2.37	-6	0.63	-58	0.43	-15	0.14	-47	1.12	5	0.24	-85
2352 Peshawar	2.20	-53	0.78	-113	0.38	-57	0.18	-99	1.15	61	0.31	-149
2350 Quetta	2.46	-27	1.26	-110	1.00	-45	0.29	-97	0.30	30	0.49	-145
2351 Lahore	1.69	-42	0.86	-118	0.36	-116	0.20	-105	0.97	21	0.35	-156
2450 Kathmandu	1.02	-166	0.87	-111	0.23	-121	0.20	-102	0.35	164	0.37	-156
2401 NewDelhi	1.76	-101	0.94	-121	0.80	-77	0.22	-107	1.00	-96	0.39	-160
2402 Hyderabad	0.59	-120	1.60	-117	0.36	-117	0.38	-103	0.27	125	0.64	-160
2460 Colombo	4.01	162	1.47	157	0.70	-171	0.18	-167	2.30	130	1.06	114
South East Asia												
2502 ChiangMai	1.00	-150	1.42	-110	0.20	-123	0.30	-106	0.99	160	0.59	-165
2501 Bangkok	1.19	-158	1.55	-114	0.39	180	0.33	-110	0.83	-175	0.72	-169
2551 Penang	2.56	-162	1.50	-129	0.42	179	0.39	-121	0.93	157	1.03	176
2550 K.Lumpur	0.66	81	0.50	-57	0.24	60	0.13	-77	0.30	58	0.37	176
2555 K.Kinabal	1.28	-128	0.77	-48	0.07	-176	0.12	-39	0.72	-165	0.17	-67
2701 Saigon	1.24	-103	0.67	-162	1.00	-128	0.18	-117	0.15	118	0.41	163
2601 HongKong	2.79	-19	1.86	-67	0.33	-77	0.43	-56	1.45	-6	0.59	-86
4010 Baguio	3.47	-53	2.55	-16	0.52	-28	0.44	-3	1.63	-47	0.94	-32
4011 Manila	4.17	-66	2.47	-13	0.56	-123	0.43	1	1.36	-71	0.93	-30

Indonesia - Australia and South Pacific (Table 13.7).

There are now 21 stations in this very extended area, 19 having been installed by ourselves and two at Taupo and Wellington, New Zealand by Dr. Dibble who joined us in the Lauder observations. The observed load vectors offer here an exceptional consistency over the whole area, a feature which is obvious when we consider the systematic behaviour of the phase β.

TABLE 13.7. Trans World Tidal Gravity Profiles.

	STATION	M_2 OBSERVED RESIDUE		M_2 SCHWIDERSKI MAP		N_2 OBSERVED RESIDUE		N_2 SCHWIDERSKI MAP		S_2 OBSERVED RESIDUE		S_2 SCHWIDERSKI MAP	
		B	β	L	λ	B	β	L	λ	B	β	L	λ
	Indonesia												
4105	Banjar B	3.05	-4	1.31	-37	0.53	-10	0.20	-26	1.48	-18	0.26	-84
4100	Bandung	6.29	-19	2.50	-48	1.19	-13	0.49	-26	2.89	62	0.84	-116
4110	Ujung Pa	4.41	-28	1.58	-130	0.82	-25	0.28	-122	1.73	-18	0.63	170
4111	Manado	6.08	-6	3.08	0	1.12	31	0.53	14	3.17	-33	1.78	-24
4120	Jaya Pura	5.24	-18	2.02	-21	1.14	-8	0.44	5	2.38	22	0.41	-13
4115	Kupang	4.76	-106	5.59	-130	1.09	-84	0.87	-121	1.74	145	3.11	170
4160	P.Moresby	4.92	-6	1.48	-93	1.47	-37	0.50	-61	3.76	9	1.14	-52
	Australia												
4210	Darwin	3.45	25	0.26	25	0.37	-11	0.11	49	2.92	-18	0.16	-9
4211	Perth	4.15	6	0.41	-77	0.72	31	0.16	108	1.54	-6	0.45	-148
4209	Alice Sp	0.72	-43	0.54	-79	0.16	-16	0.09	0	0.41	88	0.27	159
4208	Broken H	0.21	-87	0.87	-59	0.84	-8	0.21	-6	0.78	-19	0.19	159
4207	Charters	3.98	-56	1.46	-56	0.74	-31	0.37	-33	2.02	-42	0.63	-39
4205	Armidale	3.52	-52	3.26	-56	0.77	-38	0.64	-38	0.58	-104	0.74	-76
4206	Canberra	3.39	-43	2.69	-65	0.75	-41	0.55	-39	0.84	-2	0.43	-90
4220	Hobart	3.85	-65	2.77	-92	1.20	-63	0.53	-63	0.29	3	0.39	176
	New Zealand - South Pacific												
4420	Lauder	1.13	-76	1.16	-135	0.33	-82	0.29	-149	0.81	169	1.00	-155
4405	Wellingt.	5.10	-30	3.72	-29	1.10	-10	0.58	6	1.14	-100	1.21	-131
4401	Taupo	7.00	-29	5.17	-40	1.92	-1	0.92	-8	1.18	-78	1.39	-120
4400	Hamilton	8.00	-43	5.96	-51	1.14	-30	1.04	-21	1.75	-89	1.71	-119
4450	Noumea	12.74	-37	5.69	-22	2.32	-34	1.12	5	2.95	-45	1.25	-44
4460	Suva	11.23	-6	9.03	-8	2.33	-7	2.19	14	2.14	2	1.82	-10
4475	Apia	14.41	-25	8.03	-11	2.17	-17	2.04	4	4.43	-21	1.75	-10
4480	Papeete	2.33	14	1.00	97	0.61	66	0.27	111	0.21	104	1.05	149

This gives us much confidence in the experimental results which have been obtained with six different instruments.

The four stations Banjar Baru, Bandung, Ujung Pandang and Manado which are distributed over 2500 km in three large islands have *the same* χ *phase:* almost identical in the four stations. This means that the end residue $\bar{X}$ (about 3.6 μgal) is not due to local effects like the nearby coastal sea tides but probably to an extended regional tectonic effect.

The size of $\bar{X}$ is so large that no remote area in the cotidal maps could explain it with the exception perhaps of the Indian Ocean which has a peculiar tidal regime. However the maximum effect it can produce reaches 1.2 μgal according to our computations, which is much less than what we observe while its phase does not fit at all with χ.

The four stations along the east coast of Australia (Charters Towers, Armidale, Canberra and Hobart in Tasmania) (Table 13.7, where the Schwiderski M_2 tide reaches 50 - 60 cm, are extremely coherent as over a north-south distance of some 2700 km they keep exactly the same load vector: 3.7 μgal, -54°, which again means that nearby sea tides have very little effect. Four different instruments were used in these stations (all having been installed together for some months at Canberra). This again proves that instrumental errors are excluded at the size of the observed load vectors.

A most interesting feature is exhibited by the four stations established in New Zealand, where the observed M_2 loading effect is extremely large in the North Island corresponding to very high oceanic amplitudes (100 cm) and rather small in the South Island (1 μgal at Lauder). The end residue is still around 2 μgal *with a constant phase* (-12°). New Caledonia, Fiji and Samoa exhibit still larger load signals; they are the only places where it is larger than 10 μgal. Tahiti, in an area of smaller sea tide amplitudes, has a load reduced to 2.3 μgal.

There were serious problems with the Pacific Ocean. Quoting Hendershott (1973): "In the Pacific, the various numerical and normal mode models show the greatest divergence from one another and from Pacific Islands observations".

"Between Japan and New Guinea ... the various numerical maps either resolve this last region into two or three amphidromes [Hendershott 1972, Pekeris and Accad 1969] with realistically low amplitudes, or they fail entirely to be realistic [Pekeris and Accad 1969, Zahel 1970, Bogdanov and Magarik, 1967]."

"The Southeast Pacific is, tidally speaking, unknown territory. The only station between South America and Polynesia is Easter Island, and the Antarctic coast in this quadrant of the globe has no station".

Despite these difficulties however the Table 13.8 shows that, except for very coastal stations (Noumea, Apia) the total anomalies are concentrated within the $X \cos \chi$ component while $X \sin \chi$ remains less than one microgal and even often negligeable.

This demonstrates that we have now reached a precision such as to allow us to measure the effects of lateral heterogeneities in the tidal deformations. Such effects can reach up to 5 μgals on the M_2 amplitude.

TABLE 13.8. Effect of lateral heterogeneities in the lithosphere.

		$X \cos \chi$	$X \sin \chi$			$X \cos \chi$	$X \sin \chi$
Corner of Africa				*Iran/Pakistan*			
3018	Arta-Djibouti	5.39	-0.25	2201	Teheran	2.02	0.28
3014	Addis Ababa	-0.56	-0.13	2202	Tabriz	1.95	0.00
3020	Mogadishu	1.87	-0.40	2350	Quetta	2.62	0.05
				2351	Lahore	1.66	-0.35
				2352	Peshawar	1.64	-1.02
Indonesia/Torrès Strait				*New Zealand /South Pacific*			
4100	Bandung	4.28	-0.15	4220	Hobart	1.73	-0.70
4105	Banjar Baru	2.00	0.57	4400	Hamilton	2.11	-0.81
4110	Ujung Pandang	4.92	-0.78	4401	Taupo	2.16	-0.04
4111	Manado	2.96	-0.63	4405	Wellington	1.32	-0.51
4115	Kupang	2.28	-0.28	4420	Lauder	1.10	-0.27
4120	Jaya Pura	3.10	-0.89	4450	Noumea	4.94	-5.49
4160	Port Moresby	4.98	0.88	4460	Suva	2.23	0.08
4210	Darwin	2.91	1.30	4475	Apia	5.21	-4.53
				4480	Papeete	2.38	-0.42

China (Table 13.9)

The oceanic tides are very important all along the coasts of China and, in particular in the Gulf of Chihli which is not included in the Schwiderski maps. It is not astonishing therefore if the addition of the contributions from these areas strongly modifies the phase of the $\vec{L}$ vector. It is very encouraging to see how considerably it improves the agreement between observations and computations.

Figure 13.4 shows the M_2 load and attraction effect in China according to the Schwiderski map. It indicates the possible existence of an amphidromic point very near to Lanzhou but our results do not reconfirm this fact.

Japan and North Pacific (Table 13.10)

The six stations reported here are Aburatsubo from Shimada (1979); Memambetsu and Mizusawa (ICET); Wake and Honolulu from UCLA group; and Uwekahuna by R. Dibble who communicated his data to ICET. All stations are in fair agreement with the computed Schwiderski load signals for M_2 and even N_2 in the case of Mizusawa and Uwekahuna.

Remarks about the N_2 and S_2 waves

The N_2 load signal is often weak and was reported here only tentatively. When it reaches 0.5 μgal or more it is often in good agreement with the signal computed from the Schwiderski map (examples are Antananarivo, Kerguelen). The agreement for the S_2 component is less satisfactory and may be due to a meteorological component in the gravity measurements which could not be eliminated.

13.2.2. The diurnal tesseral waves

Our discussion will be mainly concerned with the O_1 wave which has a period of 25.819 h and can consequently be fairly well separated from the meteorological effects.

It is indeed difficult to separate rigorously the luni-solar wave K_1 from the solar wave P_1 and from the spurious meteorological component S_1 with only 4-6 months of observations even if the gravimeters used exhibit in general a very small S_1 instrumental component. This has been shown by the long records obtained at Brussels and Alice Springs. The difficulty of this separation is due to the fact that the K_1, P_1 and S_1 frequencies are very close to each other. On the other hand the P_1 wave has a slightly different δ factor than the K_1 wave because of the resonance effects of the liquid core of the Earth.

A reasonable agreement (maximum difference tolerated: 20°) between the phases β of the K_1 and P_1 load vectors has been taken as a criterion for their separation.

On the other hand all the diurnal waves vanish in the vertical component at the equator (being proportional to $\sin 2\phi$). As our stations in Sri Lanka, Malaysia and Indonesia are installed very near to the equator it will be extremely interesting to look at the diurnal observed amplitudes which, there, will be almost entirely due to the oceanic loading effects.

Europe

The calculated effects with the O_1 Schwiderski and Bogdanov maps are quite similar and uniform in non-coastal stations, giving an amplitude between 0.10 and 0.25 μgal with a constant phase very near to 160° (with both maps) along the line Brussels - Darmstadt with 166° at Kiel in the north and 190° at Florence in the south.

Such a small load vector is below the estimated level of significance of our measurements (about 0.5 μgal).

TABLE 13.9. Trans World Tidal Gravity Profiles.

STATION		M_2 OBSERVED RESIDUE		M_2 SCHWIDERSKI MAP		N_2 OBSERVED RESIDUE		N_2 SCHWIDERSKI MAP		S_2 OBSERVED RESIDUE		S_2 SCHWIDERSKI MAP	
		B	β	L	λ	B	β	L	λ	B	β	L	λ
	Central Asia												
1280	Frunze	0.58	-122	0.43	-112					1.03	-133	0.17	-151
1270	Tashkent	1.35	-125	0.51	-103					0.73	-136	0.20	-139
1293	Talgar	0.58	167	0.39	-117					0.38	167	0.16	-158
1289	Novosib	0.35	-94	0.22	-134					0.16	-99	0.08	175
2450	Kathmandu	1.02	-166	0.87	-111	0.23	-121	0.20	-102	0.35	164	0.37	-156
2620	Urumqi	0.71	-68	0.27	-137	0.21	-29	0.06	-113	0.49	32	0.12	173
2605	Lanzhou	0.72	-52	0.02	-27	0.15	-13	0.02	-36	0.24	49	0.07	85
2610	Shenyang	0.94	-18	0.66	46	0.17	-35	0.14	21	0.47	-7	0.34	18
2603	Beijing	0.89	14	0.42	49	0.22	9	0.10	18	0.56	13	0.25	27
2607	Wuhan	1.09	-40	0.56	-10	0.31	-19	0.15	-12	0.38	-9	0.23	-8
2604	Kunming	1.26	-45	0.39	-95	0.23	-48	0.09	-86	0.39	-7	0.13	-173
2613	Shanghai	2.39	-101	1.16	-64	0.46	-81	0.39	-61	0.60	-159	0.29	-73
2502	Chiangmai	1.00	-150	1.42	-110	0.20	-123	0.30	-106	0.99	160	0.59	-165
2600	Guangzhou	1.11	-75	1.09	-55	0.21	-33	0.25	-44	0.26	-35	0.30	-68
2601	HongKong	2.79	-19	1.86	-67	0.33	-77	0.43	-56	1.45	-6	0.59	-86

TABLE 13.10. Trans World Tidal Gravity Profiles.

STATION		M_2 OBSERVED RESIDUE		M_2 SCHWIDERSKI MAP		N_2 OBSERVED RESIDUE		N_2 SCHWIDERSKI MAP		S_2 OBSERVED RESIDUE		S_2 SCHWIDERSKI MAP	
		B	β	L	λ	B	β	L	λ	B	β	L	λ
	Japan-North Pacific												
2840	Aburats.	4.37	10	2.67	24			0.42	13	2.25	5	1.40	-4
2847	Mizusawa	2.51	31	2.10	48	0.36	18	0.23	40	1.57	11	1.11	10
2897	Memamb.	1.70	71	2.00	61	0.12	22	0.17	72	0.83	21	1.02	15
6001	Wake	1.79	67	5.89	76	0.19	-16	0.82	86	1.95	10	2.70	49
6003	Honolulu	5.33	80	2.41	72	1.04	-148	0.59	97	2.44	73	1.00	97
6004	Uwekahuna	4.59	46	3.10	68	1.09	72	0.74	92	1.07	14	1.18	87

East Africa and Indian Ocean (Table 13.11)

The amplitude of the observed load is between 0.15 and 1.26 µgal which makes its correct determination rather difficult. Therefore three stations only can be used for a check with computed load: Antananarivo (0.89 µgal), La Réunion (1.26 µgal) and Kerguelen (1.11 µgal). It is clear from Table 13.10 that the Schwiderski maps fit rather well as shown by the phase concordances.

The other stations around the Indian Ocean at Perth, Bandung and Colombo also fit the Schwiderski map remarkably well.

South Asia (Table 13.12)

There is a serious disagreement of the observed load vectors from Istanbul to Kathmandu with the vectors computed with either the Bogdanov map or the Schwiderski map. The computed vector is always very small while the observations have indicated a 1 µgal residual in Iran and 0.5 µgal in Pakistan. The only exception, amongst 11 stations, is Colombo which indicates an almost perfect agreement between observations and the Schwiderski map. The result is puzzling and this is surely an area where one should return to intensify the net we established.

In tropical countries, especially in arid regions and at altitude, strong changes of temperature occur between day and night, and also between summer and winter. Most of the stations of this area exhibited drifts at the time of a seasonal change, in particular Quetta and Peshawar.

Southeast Asia and Indonesia (Table 13.13).

We expected that this zone would be the most crucial test for three essential reasons.

(a) The South China Sea is a resonant diurnal system: the amplitudes of the K_1 and O_1 tidal waves reach more or less 0.5 m in the area.

(b) Because of obvious geographical reasons all our stations are more or less coastal.

(c) The direct Earth tide is extremely small as all our stations are very close to the equator.

This explains why our O_1, K_1 and occasionally P_1 load vectors amount to 1 - 3.5 µgal over all *this area*.

A first typical feature to be observed is that the phase difference in the K_1 and O_1 load signals fits everywhere with the corresponding difference in the neighbouring harbours, with only one exception in Manila. This shows evidently that the experimentally obtained load vector is mainly sensitive to the near-sea tides.

It is a great success for the O_1 and K_1 Schwiderski maps, that they reduce B to end residues X which are in general less than 0.5 µgal for O_1 and less than 1.5 µgal for K_1.

Success is normally better for O_1 than for K_1 which is very sensitive to barometric and thermal disturbances.

Australia, New Zealand and South Pacific (Table 13.14)

With diurnal load vectors which are of the order of 1 - 2 µgal in this area, we again observe a remarkable fit of our Australian stations at Perth, Armidale, Canberra and Hobart with the Schwiderski O_1 map.

In the South Pacific only Lauder (southern New Zealand) has a β phase compatible with the computed one. However the phase difference in the observed load vectors at Noumea and Suva, which is 129°, is reduced to 24° after subtraction of the Schwiderski map contribution. Similarly the Suva - Apia difference reduces from 93° to 33°.

More interesting is the fact that all end residues $\bar{X}$ have nearly the same phase, which is about +150°, and an amplitude of about 1.5 µgal in all of the South Pacific. This obviously reveals a distortion of the map which does not

TABLE 13.11. Trans World Tidal Gravity Profiles

STATION		O_1 OBSERVED RESIDUE B	O_1 OBSERVED RESIDUE β	O_1 SCHWIDERSKI MAP L	O_1 SCHWIDERSKI MAP λ	K_1 OBSERVED RESIDUE B	K_1 OBSERVED RESIDUE β	K_1 SCHWIDERSKI MAP L	K_1 SCHWIDERSKI MAP λ
	East Africa and Indian Ocean								
3000	Helwan	1.11	-97	0.20	157	1.34	-81	0.36	155
3005	Aswan	0.52	71	0.25	156	1.20	-122	0.48	154
3010	Khartoum	0.13	-153	0.30	155	1.99	-91	0.61	154
3014	Addis Ababa	0.43	178	0.52	145	1.29	-150	1.01	147
3018	Arta/Djibouti	0.75	116	1.06	142	1.75	130	2.11	146
3020	Mogadiscio	1.08	126	1.27	136	3.01	153	2.35	137
3030	Nairobi	1.30	9	0.55	144	1.69	36	1.01	143
3031	Voi	1.25	119	0.70	141	1.26	131	1.26	142
3420	Lwiro	0.37	15	0.27	159	0.10	-15	0.59	151
3040	Dar Es Salam	0.83	142	1.02	139	1.70	155	1.75	141
3500	Maputo	0.15	139	0.30	169	0.17	150	0.32	88
3602	Nosy Be	2.54	97	1.02	123	4.42	123	1.64	123
3601	Antananarivo	0.84	129	0.52	121	1.77	3	0.78	108
3604	Tolagnaro	0.69	3	0.31	119	5.42	27	0.50	67
3620	Reunion	1.26	47	0.53	82	1.44	177	0.91	69
9904	Kerguelen	1.11	-142	1.18	-130	1.35	-94	1.04	-108
4211	Perth	2.82	-139	1.77	-121	4.50	-154	2.15	-128
4100	Bandung	1.55	-89	1.54	-75	1.38	-123	1.91	-92
2460	Colombo	0.35	78	0.28	77	0.69	96	0.94	141

TABLE 13.12. Trans World Tidal Gravity Profiles

STATION		O_1 OBSERVED RESIDUE B	O_1 OBSERVED RESIDUE β	O_1 SCHWIDERSKI MAP L	O_1 SCHWIDERSKI MAP λ	K_1 OBSERVED RESIDUE B	K_1 OBSERVED RESIDUE β	K_1 SCHWIDERSKI MAP L	K_1 SCHWIDERSKI MAP λ
	South Asia								
2000	Istanbul	0.31	-33	0.14	155	0.48	-7	0.17	149
2005	Ankara	0.26	-37	0.16	151	0.31	19	0.22	151
2202	Tabriz	1.16	-11	0.23	138	1.77	-9	0.40	146
2201	Teheran	1.14	7	0.29	134	1.41	5	0.53	143
2352	Peshawar	0.29	-67	0.28	108	0.23	-87	0.57	127
2350	Quetta	1.43	-66	0.47	119	0.93	-11	0.94	131
2351	Lahore	0.52	-34	0.29	104	1.05	-12	0.60	126
2401	New Delhi	0.31	-120	0.27	100	0.76	-146	0.60	126
2402	Hyderabad	0.37	156	0.36	105	0.85	154	0.96	135
2450	Kathmandu	0.80	178	0.12	54	0.76	140	0.31	135
2460	Colombo	0.35	78	0.28	77	0.69	96	0.94	141

TABLE 13.13. Trans World Tidal Gravity Profiles

STATION		O_1 OBSERVED RESIDUE B	O_1 OBSERVED RESIDUE β	O_1 SCHWIDERSKI MAP L	O_1 SCHWIDERSKI MAP λ	K_1 OBSERVED RESIDUE B	K_1 OBSERVED RESIDUE β	K_1 SCHWIDERSKI MAP L	K_1 SCHWIDERSKI MAP λ
	South East Asia								
2503	ChiangMai	0.58	-114	0.45	-63	1.54	158	0.43	-121
2501	Bangkok	0.60	-52	0.59	-53	1.43	-108	0.51	-87
2551	Penang	1.10	-98	1.08	-84	1.43	-130	1.42	-127
2550	Kuala Lumpur	0.94	-88	1.11	-90	2.40	-134	1.34	-127
2555	Kota Kinabalu	3.25	-88	2.52	-79	2.70	-121	2.58	-115
2701	Saigon	1.51	-80	2.01	-80	3.78	-123	2.07	-119
2601	Hong Kong	2.26	-69	2.10	-66	2.60	-84	2.05	-100
4011	Manila	2.00	-76	1.80	-56	2.88	-79	1.65	-77
4010	Baguio	2.17	-70	1.96	-59	2.28	-90	1.75	-82
	Indonesia								
4105	Banjar Baru	2.27	-95	2.01	-84	2.43	-131	2.63	-117
4100	Bandung	1.55	-89	1.54	-75	1.38	-123	1.91	-92
4110	Ujung Pandang	2.85	-95	2.25	-90	6.02	-108	3.12	-110
4111	Manado	2.15	-125	1.79	-60	2.90	-140	2.26	-80
4115	Kupang	2.94	-124	2.40	-95	4.34	-129	3.31	-114
4120	Jaya Pura	1.81	-15	1.49	-21	3.52	-29	2.40	-32
4160	Port Moresby	0.96	5	1.57	-3	3.32	-31	3.10	-22

affect Alice Springs and consequently should be a coastal zone around New Zealand. The Alice Springs station has been established at the very centre of the Australian continent and three years of measurements have been completed. The O_1 observed load is exactly equal to the corresponding load calculated with the Schwiderski map.

This justifies a posteriori the choice of a model with

$$\delta(O_1) = 1.160$$

$$\delta(O_1) = 0°$$

China, Japan and North Pacific (Table 13.15).

The results given in this table exhibit a fine agreement for the wave O_1 when its amplitude is larger than $\varepsilon \sim 0.5$ µgal.

The results at Guangzhou (Canton) and Hong Kong are particularly impressive as they were obtained with two different instruments at very different epochs: LaCoste Romberg n°3 in 1974 at Hong Kong and Geodynamics n°783 in 1980 at Guangzhou. This is another proof of the quality of these carefull measurements.

The results for K_1 are practically as good as those for O_1.

The P_1 wave (Table 13.16).

Very few discussions have concerned the declinational solar P_1 wave up to now because of the difficulties in extracting it correctly from the data. Its period being equal to 23 h 53 m 57 s makes its frequency very close to the K_1 and S_1 frequencies so that only very good instruments, carefully protected again barometric as well as thermal disturbances, have been able to isolate it.

However, this tidal component, which is the third in amplitude in the tesseral family, is of major interest for investigations of the liquid core hydrodynamical oscillations. The load effects computed from the Schwiderski P_1 cotidal map are in close agreement with the observed P_1 loads in all the nine stations where we could separate it from K_1 and S_1 and where, of course, this P_1 signal was not too weak.

TABLE 13.14. Trans World Tidal Gravity Profiles

	STATION	O_1 OBSERVED RESIDUE		O_1 SCHWIDERSKI MAP		K_1 OBSERVED RESIDUE		K_1 SCHWIDERSKI MAP	
		B	β	L	λ	B	β	L	λ
	Australia								
4210	Darwin	1.64	161	1.15	-110	3.75	114	1.45	-124
4211	Perth	2.68	-137	1.77	-121	4.36	-154	2.15	-128
4209	Alice Springs	0.54	-123	0.53	-122	0.91	-124	0.55	-121
4208	Broken Hill	0.28	15	0.54	-174	0.42	52	0.35	166
4207	Charters	0.39	96	0.53	-2	0.71	4	1.31	-17
4205	Armidale	0.58	87	0.40	87	1.17	12	0.92	24
4206	Canberra	1.13	151	0.61	142	0.89	141	0.62	81
4220	Hobart	3.23	153	1.51	155	4.21	137	1.56	126
	New Zealand - South Pacific								
4420	Lauder	2.09	156	0.53	146	1.96	91	0.21	90
4405	Wellington	1.38	177	0.22	142	0.72	-168	0.33	-10
4401	Taupo	1.59	-166	0.04	109	2.13	-171	0.58	-21
4400	Hamilton	2.03	143	0.06	55	1.66	159	0.71	-20
4450	Noumea	1.15	104	1.10	0	1.82	-2	2.42	-22
4460	Suva	0.38	-127	0.96	-36	1.42	-76	1.84	-48
4475	Apia	1.39	140	0.61	-64	1.94	168	0.91	-68
4480	Papeete	1.17	158	0.30	-162	1.94	155	0.11	132

TABLE 13.15. Trans World Tidal Gravity Profiles

STATION		O_1 OBSERVED RESIDUE		O_1 SCHWIDERSKI MAP		K_1 OBSERVED RESIDUE		K_1 SCHWIDERSKI MAP	
		B	β	L	λ	B	β	L	λ
	Central Asia								
2450	Kathmandu	0.80	178	0.12	54	0.76	140	0.31	135
2650	Lhasa	0.10	3	0.15	7	0.56	-74	0.12	133
2606	Urumqi	0.28	0	0.17	57	0.38	-73	0.19	86
2605	Lanzhou	0.30	-24	0.30	2	0.48	-13	0.23	2
2610	Shenyang	0.58	-12	0.69	14	0.92	-10	0.78	1
2603	Beijing	0.75	12	0.51	10	0.92	22	0.54	0
2607	Wuhan M	0.83	-41	0.65	-20	1.11	-48	0.63	-32
2607	Wuhan B	0.79	-40	0.65	-20	0.87	-28	0.63	-32
2607	Wuhan C	0.63	-37	0.65	-20	1.31	-21	0.63	-32
2604	Kunming	0.96	-47	0.40	-45	1.03	-50	0.24	-82
2613	Shanghai	0.93	-22	1.24	-9	1.99	-33	1.49	-25
2502	ChiangMai	0.58	-114	0.45	-63	1.54	158	0.43	-121
2600	Guangzhou	1.21	-66	1.32	-60	1.40	-98	1.18	-90
2601	Hong Kong	2.26	-69	2.10	-66	2.60	-84	2.05	-100
	North Pacific								
2840	Aburatsubo	3.35	8	2.19	15	3.68	-9	2.73	-2
2823	Kyoto	0.71	90	1.69	11	1.09	-80	2.11	-6
2847	Mizusawa	2.31	14	2.06	22	3.69	-5	2.54	5
2847	Mizusawa B	2.59	3	2.06	22	3.73	-6	2.54	5
2897	Memambetsu	2.04	13	2.40	23	2.90	0	2.80	4
6001	Wake	1.43	-18	1.54	-1	1.67	-48	1.87	-28
6003	Honolulu	1.74	116	1.89	110	2.59	94	3.32	101
6004	Uwekahuna	1.82	93	1.97	112	3.43	82	3.45	102

TABLE 13.16. P_1 wave

STATION		OBSERVED RESIDUE		SCHWIDERSKI MAP		VECTORIAL DIFFERENCE	
		B(μgal)	β(°)	L(μgal)	λ(°)	X(μgal)	χ(°)
3020	Mogadiscio	0.57	104	0.78	137	0.43	2
9904	Kerguelen	0.40	-82	0.37	-106	0.16	-14
4105	Banjar Baru	1.57	-125	0.90	-115	0.70	-137
4115	Kupang	1.00	-116	1.10	-118	0.11	42
4160	Port Moresby	0.87	-16	0.91	-19	0.06	113
4209	Alice Springs	0.28	-106	0.18	-133	0.18	-27
4205	Armidale	0.46	7	0.26	38	0.27	-24
6004	Uwekahuna	1.17	83	1.03	102	0.39	23
2600	Guangzhou	0.33	-94	0.40	-92	0.07	97

13.2.3. Some additional remarks about the study of lateral heterogeneities in the Crust and Mantle

As stated in the US Geodynamics Committee report on Crustal Dynamics Part II Section 1 Oceanic Crustal Evolution, (January 1978) (Dr. J.R. Heitzler), "measurements of earth tides by recording gravimeters and other means across margins may be useful to study lateral variation of rheological properties in the lithosphere".

This implies of course a precise correction of the oceanic load and attraction effects on the measurements. It has been demonstrated here that this can be achieved by using the Schwiderski cotidal maps and by considering separately the cosine component $\bar{X} \cos \chi$ and the sine component $X \sin \chi$ of the final residue $\bar{X} = \bar{A} - \bar{R} - \bar{L}$ (see Tables 13.3 and 13.8).

S.M. Molodensky (1977) has pointed out that the tidal deformations, provided that they are observed with sufficient accuracy, have two advantages with respect to free oscillations:

1. they are sensitive to heterogeneity in the neighbourhood of the point of measurement while the free oscillations give informations about averaged heterogeneities.

2. the influence of lateral heterogeneity depends upon their orientation relative to the progression of the wave which, in the case of free oscillations, raises obvious difficulties. On the contrary the orientation of tidal waves is exactly known at any time.

With respect to ordinary seismic waves one can add that the flexure of the crust under the oceanic load depends upon elastic properties averaged over an area while the seismic wave samples the crust along the path of its ray. On the other hand, the tidal frequencies are several orders of magnitude lower than the seismic frequencies.

In 1980, S.M. Molodensky and M.V. Kramer have made calculations of body tide anomalies due to horizontal heterogeneities represented by differences of 5% in the speed of the seismic longitudinal waves under continents (lower speed) and oceans (higher speed), extending to a depth of 331 km while the transverse wave speeds are spherically symmetric. They represent these heterogeneities by a development in spherical functions up to the order 47.

In another model these authors extend the heterogeneity to the whole mantle, that is up to 2891 km depth. Their numerical results indicate for the δ factor a decrease of 0.20% over the continents in the first model and a possible decrease of 0.75% when the heterogeneity extends over the whole mantle. (fig. 13.4.). In the case the transverse seismic waves are slower under the oceans and the longitudinal wave speeds are spherically symmetric the decrease is replaced by an increase of the same size.

Tidal Tilt measurements are much more sensitive to lateral heterogeneities than the tidal gravity observations.

Beaumont and Boutilier (1978), using measurements from 5 tilt stations, have shown indeed, using different seismic models, that the most appropriate model for the crust beneath Nova Scotia is that given by a seismic refraction profile along the coast of Nova Scotia. The work shows the importance of using tilt differences to eliminate the uncertain tides from distances greater than 500 kilometres. The differential tilt signal then depends upon the crustal structure and the nearby shelf tide, which for M_2 is defined to an accuracy of 2% in amplitude and 2° in phase using ocean bottom tide gauges.

Baker (1977), using tilt observations in a mine at Llanwrst, North Wales, finds that a crustal model from seismic investigations in the Irish Sea with a three layered, 30 kilometre thick crust, fits the observations better than the standard Gutenberg Bullen Earth model with a 38 kilometre crustal thickness.

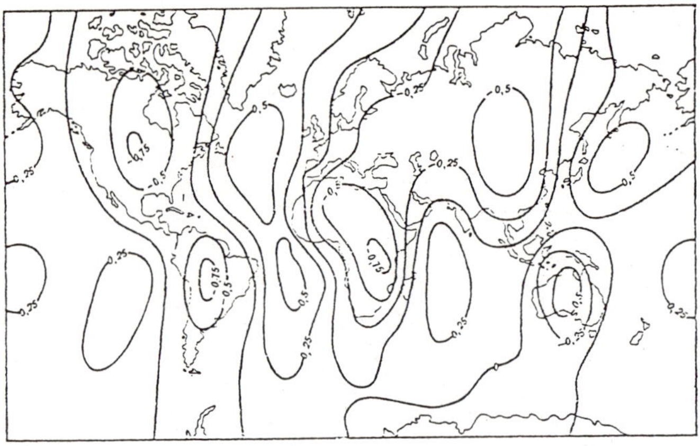

Fig. 13.4. δ variations in % according to Molodenskii when the heterogeneities extend over the whole Mantle.

Since 1974, M. Bonatz, C. Gerstenecker, R. Kistermann and J. Zschau have carried tilt measurements at six stations across a deep fault zone the Hunsrück-fault zone (FRG). The results of these six stations confirm a strong correlation between anomalies of the amplitudes and phases of the diurnal and half diurnal waves and the distances of the stations from the fault zone. Near the fault the amplitudes are amplified significantly. The phase lags change their signs across the fault.

Rosenbach and Grosse-Brauckmann have similarly undertaken a tilt profile in the Harz mountains.

13.3. LOADING AMPHIDROMIC POINTS

It is of interest to find points where the oceanic effects are minimized, even possibly zero. The question is whether such points exist. If they could be discovered it would be worthwhile to install there the best gravimeter to investigate the hydrodynamic effects of the Earth's liquid core. However the geographical position of such a point will be different for each tidal component.

At a certain distance from a sea the direct attraction of its water masses and their loading effect are equal and opposite, cancelling in such a way that the effect of the ocean is virtually zero. This is true only for the nearest sea and as the more distant oceans have a significant effect it is not easy to predict where the effects of all the oceans together will cancel. This should not necessarily happen just in the centre of each continent; nonetheless at such a place one might believe that the total effect would be minimized. One can even find a number of coastal or island points where all the world oceanic load effects cancel each other so that the place appears like an amphidromic point.

The Schwiderski O_1 map, for example, gives such points at Ostend (Belgium) as well as in some South Pacific islands (New Zealand, Tahiti), but these are artificial because the nearby coastal tides are not taken properly into account in all cases.

We should therefore restrict such a search to continental places and the only way to look for such places is empirical. We did this by computing for each

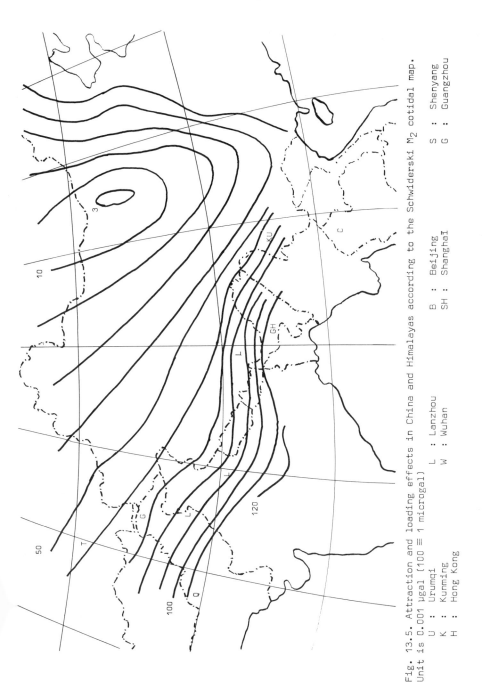

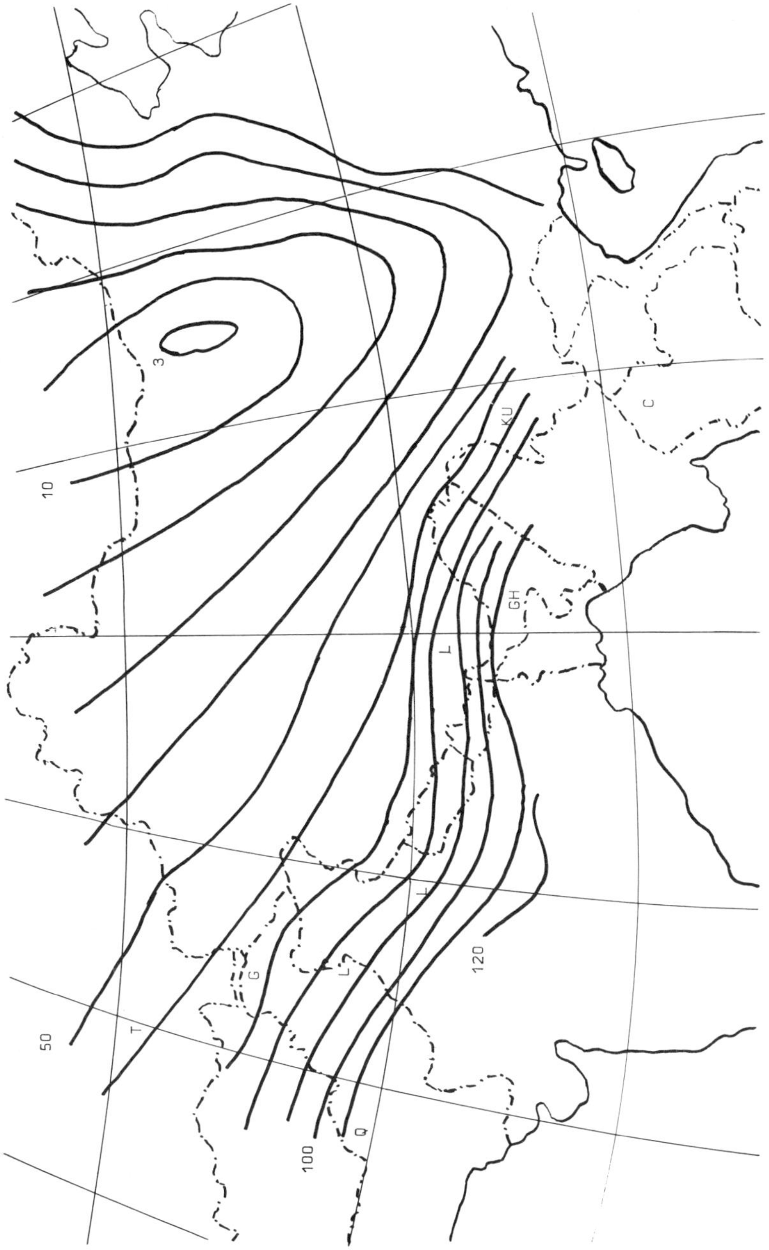

Fig. 13.5. Attraction and loading effects in China and Himalayas according to the Schwiderski M_2 cotidal map. Unit is 0.001 μgal (100 ≡ 1 microgal)

U : Urumqi	L : Lanzhou	B : Beijing	S : Shenyang
K : Kunming	W : Wuhan	SH : Shanghai	G : Guangzhou
H : Hong Kong			

cotidal map the oceanic effects on a 5° x 5° world grid, then by computing a finer grid for the regions which appear to be most interesting. On the other hand results of measurements can, just by chance, help to discover such a point.

It is clear from our Trans-World Tidal Gravity Profile that one such point must exist in Australia for the M_2 tidal wave, between Alice Springs and Broken Hill, which is indeed confirmed by the Parke map (at $\phi = -28°30'$, $\lambda = 137°30'$) but not by the Schwiderski or any other map. The M_2 Parke map indicates, moreover, that such points exist also in Argentina, near Tucuman ($\phi = -28°$, $\lambda = 297°$) and in Texas, west of Dallas ($\phi = +33°$, $\lambda = 260°$) while the M_2 Schwiderski map gives amphidromic points at $\phi = 33.5°$ N, $\lambda = 106.5°$ W in New Mexico and northwest of Lanzhou ($\phi = +37°$, $\lambda = 103°3°'$) in China.

With the O_1 Schwiderski map we have not found any amphidromic point in Australia even if the load is everywhere very small there. We also found that over a broad area in China (between 25° and 35°N, 84° and 92°E) the O_1 load is nearly uniformly small as it does not exceed 0.1 μgal. The situation is similar in Africa (northern Nigeria, Tamanrasset, Bangui).

13.4. EXPERIMENTAL DETERMINATION OF THE DYNAMICAL LIQUID-CORE EFFECTS

The first experimental determination of the three main waves K_1, P_1 and O_1 was obtained in 1966 when Melchior analysed some long series of records obtained in several different deep stations equipped with Verbaandert-Melchior (VM) quartz horizontal pendulums, correctly calibrated. These results were in significant agreement with the Molodensky models.

13.4.1. Vertical component

The success obtained with the Schwiderski cotidal maps in the interpretation of tidal ocean-continent interactions allows to correct with a rather fair precision the measurements of the main tesseral diurnal tides O_1, P_1 and K_1 in the vertical component.

Thanks to the ICET trans world tidal gravity profiles we have now a broad geographical coverage of gravimetric measurements with a duration sufficient to separate the wave P_1 from the wave K_1. We therefore consider only 48 series of which the results, corrected for oceanic effects according to the Schwiderski maps give by simple arithmetic means:

$$\begin{aligned} \delta(O_1) &= 1.1625 \pm 0.0088 \\ \delta(P_1) &= 1.1523 \pm 0.0140 \\ \delta(K_1) &= 1.1427 \pm 0.0076 \end{aligned} \qquad (13.6)$$

which fits with the predicted liquid core effect.

A discrimination between the different theoretical models proposed is however not possible until a precise determination of the very small wave ψ_1 wil be made with success.

13.4.4. Horizontal components

In the horizontal components only deep underground stations (depth of 40 meters as a minimum) allow a correct separation of the P_1 and K_1 waves from the spurious S_1 contributions. Such stations exist only in Europe (at the exception of Armidale, Australia) but the diurnal oceanic tides have very small amplitudes in the Atlantic Ocean and this fortunate circumstance allows the liquid-core effects to be correctly observed. Presently very long series of observations yield consistent results for these three waves despite the fact that the relevant stations are in completely different topographic environments and their instruments installed in different cavities and with different procedures (see Table 13.17).

Table 13.17. Short descriptions of the clinometric stations taken into account for the determination of the diurnal waves amplitudes

d: distance from entry p: depth from the free surface

SCLAIGNEAUX, Belgium	A single horizontal gallery inside a flat hill - old disused access gallery to a mine - 5 pendulums in niches at different distances. d = 600 - 800 m; p = 85 m.
DOURBES, Belgium	A single horizontal gallery inside a hilly region giving access to a L - shaped cavity - specially excavated for the purpose. Two pairs of pendulums installed at each extremity of the L (Fig. 12.7). d = 50 m; p = 46 m.
WALFERDANGE, Luxemburg	Horizontal gallery in a gypsum mine - very low mining activity - smooth hilly topography - pendulums installed on the floor. d = 800 m; p = 75 m.
BAD GRUND, Germany	Vertical shaft reaching a complex of mine galleries - mine in activity - smooth hilly region - pendulums installed on the floor. p = 380 m.
PRIBRAM, Czechoslovakia	Vertical shaft reaching an important complex of mine galleries - flat region - pendulums in niches. p = 1300 m.
TIEFENORT, DDR	Mine p = 295 m.
GRASSE, France	Natural cave under the Calern Plateau - pendulums in niches. d = 200 m; p = 75 m.
GRAZ, Austria	Horizontal gallery in the "Schlossberg", a 100 m dolomite hill in the centre of the town-pendulums in niches. d = 160 m; p = 85 m.
DANNEMORA, Sweden	Vertical shaft reaching a complex of mine galleries flat region - pendulums in niches. p = 350 m.
LOHJA, Finland	Vertical shaft reaching a complex of mine galleries - flat region - pendulums in niches. d = 200 m; p = 145 m.

However, extremely abrupt topography such as at Armidale/Australia, (a gorge 500m deep), Costozza, near Vicenza (a volcanic region), Roburent and Toirano in the western Italian Alps, Erpel (Germany) along the Rhine Valley, have given very different results and are not considered here.

In the other deep stations, equipped with quartz VM pendulums the spurious S_1 thermic component is always very small and the stability of the γ factors obtained by successive monthly analysis for K_1 P_1 and for O_1 is fairly good (see Fig. 13.4).

At each of these stations we always observe the same features which can be approximately described as follows:

$$\begin{gathered} \gamma(O_1) < \gamma(P_1) < \gamma(K_1), \\ 0.63 < \gamma(O_1) < 0.69, \\ 0.68 < \gamma(P_1) < 0.72, \\ 0.71 < \gamma(K_1) < 0.75. \end{gathered} \qquad (13.7)$$

These 14 stations shortly described in Table 13.17 are situated in Finland (Lohja), Sweden (Dannemora), Germany (Bad Grund), Belgium (Dourbes, Sclaigneaux), Luxemburg (Walferdange), France (Grasse), Austria (Graz), Czechoslovakia (Pribram) and Hungary (Sopron). Coastal stations (England and Italy) have not been considered in the group.

Because of the lack of any dependence from the geographical situation of the stations we have calculated mean values for the γ factors of the different waves which are given in Table 13.18 and obviously are in excellent agreement with the Molodensky models and the Jeffreys Vicente model I (central particle).

We have no tilt results outside of Europe except for coastal stations in Japan and Canada and a station in Australia (Armidale), which is a site exhibiting an important topographic effect related to its situation in a deep gorge.

The noise is much more important in the clinometric recordings, and only three very long series - 17 years - (two VM quartz horizontal pendulums, installe at two different places in the underground station Dourbes and one other at Walferdange), having been analysed at the International Centre, give for the ψ wave the following respective γ factors: 0.451, 0.505 and 0.395, which means a vanishing amplitude despite the high noise in that region of the spectrum, a fact that should be in agreement with Molodensky's models.

TABLE 13.18. Diurnal waves - Horizontal East-West component

Results from 14 European stations

Wave	Group	Amplitude 0".001	Number of series	$\gamma = 1 + k - h$
K_1	124-134	5.7	14	0.7429 ± 0.0045
P_1	114-120	1.7	14	0.7054 ± 0.0157
O_1	63-78	3.9	14	0.6788 ± 0.0056
		Minor components		
Q_1	33-52	0.7	12	0.6590 ± 0.042
NO_1	89-103	0.3	3	0.7150 ± 0.015
$\emptyset_1$	137-143	0.07	2	0.7410 ± 0.148
J_1	152-165	0.3	12	0.6690 ± 0.073
OO_1	173-183	0.2	11	0.6690 ± 0.075
ψ_1	135-136	0.02	3	0.5390(± 0.016)

For very small waves the series used were two from Dourbes and one from Walferdange.

TABLE 13.19. The smallest diurnal tidal waves as obtained in East-West component from the three longest available series.

Station	Instrument	Length of records in days	Number of hourly readings	Factor $\gamma = 1 + k - h$		
				NO_1	ψ_1	ϕ_1
Dourbes 1	VM 8	6443	107 952	0.729	0.548	0.748
Dourbes 2	VM 28	4392	88 080	0.699	0.520	0.590
Walferdange 2	VM 66	4060	59 376	0.718	0.549	0.886
Mean				0.715	0.539	(0.741)
				±0.015	±0.016	(±0.148)
Theoretical amplitude at the latitude of these stations:				0".000 32	0".000 025	0".000 066

TABLE 13.20. Derivation of the diurnal Love numbers from the experimental values of the γ and δ factors.

Wave	Doodson argument	$\gamma = 1 + k - h$	$\delta = 1+h-\frac{3}{2}k$	h	k	k/h
K_1	165.555	0.7429±0.0045	1.1427±0.0076	0.484±0.017	0.227±0.016	0.469±0.032
P_1	163.555	0.7054 0.0157	1.1523 0.0140	0.584 0.070	0.289 0.068	0.495 0.090
O_1	145.555	0.6788 0.0056	1.1625 0.0088	0.642 0.017	0.321 0.011	0.500 0.022
Q_1	135.655	0.6590 0.0420	1.1600 0.0180	0.70	0.36	

For γ: VM quartz horizontal pendulums
14 stations, 20160 days of registration, 483.840 hourly readings
For δ: Registrating gravimeters of different kinds
23 stations, 22986 days of registration, 551.654 hourly readings.

These results (Table 13.20) agree very well with relations of type

$$\gamma(\omega_i) = \gamma_0 + \omega_i^2 \gamma_2 + \ldots, \quad \delta(\omega_i) = \delta_0 + \omega_i^2 \delta_2 + \ldots \qquad (13.8)$$

and are in excellent agreement with the nearly hyperbolic curves of Molodensky, but they do not allow for determining the asymptote with the precision desirable. One has therefore to isolate the waves by having their frequency much nearer to this asymptote, that is ψ_1 and ϕ_1. This determination is extremely difficult because of their low amplitudes (Table 13.19). From our results one should conclude that $h(\phi_1) = 0.61$ and $k(\phi_1) = 0.29$, but the mean square errors are very high. On the contrary, we have no convincing results so far for $\delta(\psi_1)$ while $\gamma(\psi_1)$ looks quite coherent for the stations having extremely long records. Then, if we use $k/h = 0.498$ (Table 6.4, model I) as an additional theoretical relation, we should deduce $h(\psi_1) = 0.925$ and $k(\psi_1) = 0.461$.

Usandivaras and Ducarme (1976) attempted to determine the resonance frequency by determining the asymptote of an equilateral hyperbole adjusted to the results of all waves determined from seven of the nine fundamental gravimeters and the pendulums at Dourbes, Sclaigneaux and Walferdange. The resonance line is found displaced towards a somewhat higher frequency, between ϕ_1 and θ_1, where there is in fact no tidal line.

This results from the fact that the ratio $|1-\delta\ (O_1)|/|1-\delta\ (K_1)|$ is found lower than in Molodensky's models while the ratio $|1-\gamma\ (O_1)|/|1-\gamma\ (K_1)|$ is found higher.

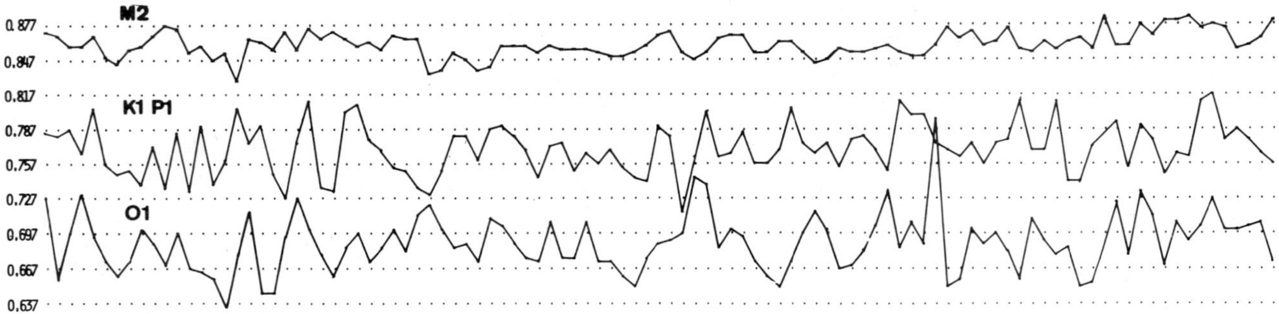

Fig. 13.14. Station Dourbes (Belgium) East West component Pendulum VM8. γ factor as deduced from successive *monthly* analyses 1963-1974.

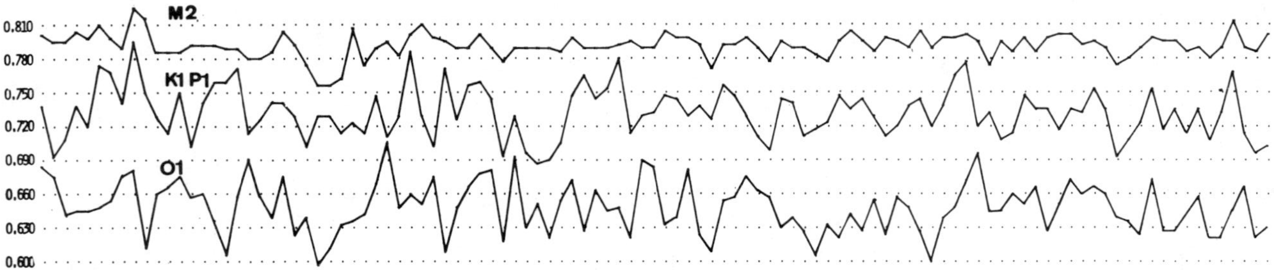

Fig. 13.6. Station Dourbes (Belgium)-East West Component-Pendulum VM28. γ factor as deduced from successive monthly analysis 1965-1974;

mean values are for VM8 $\gamma(O_1) = 0.6696$ for VM28 $\gamma(O_1) = 0.6556$

$\gamma(K_1P_1) = 0.7572$ $\gamma(K_1P_1) = 0.7499$

$\gamma(M_2) = 0.8581$ $\gamma(M_2) = 0.8157$

no correlation was found between amplitude or phase variations.

There does not seem to exist any abnormal behaviour on the side band components except perhaps for the phase lags.

13.4.3. Results obtained with extensometers

We have seen in 3.3.5 (equ. 3.77 and 3.78) that the ratio ℓ/h is very easy to obtain without, of course, any problem of calibration, on the condition that the azimut of the instrument is sufficiently different from 0° or 90°.

The only results obtained in this way are those of the Underground Laboratory at Walferdange, Grand Duchy of Luxemburg as given in the Table 13.21.

TABLE 13.21. Walferdange, horizontal extensometer, length 12m, Azimut 37°.

Ratio ℓ/h									
Series	y		O_1	P_1	K_1	N_2	M_2	S_2	K_2
1/	71-74	22464	0.159	0.163	0.180	0.150	0.148	0.152	0.151
3/	77-81	22176	0.163	0.169	0.183	0.147	0.143	0.142	0.141
Molodensky model I			0.147	0.171	0.183	0.146	→	→	→
Wahr model 1066A			0.139	0.146	0.167	0.140			

There is also at Walferdange a vertical extensometer which has given unique determinations of the factor $\eta = ah' + 2h = 2\nu(3\ell - h)$ (see equ. 3.65 and 3.72).

TABLE 13.22. Walferdange

Liquid core effects observed with the vertical extensometer					
	Observed		1971-78	1980-81	
Wave	$\Delta\ell$ nanometers	$(\Delta\ell/\ell).10^9$	n = 25344 LVDT	n = 7776 Capacitif	theoretical normalised η(*)
O_1	24.6	7.2	-0.446 ± 0.012 -4°48 ± 1°48	-0.463 ± 0.016 +6°55 ± 1.94	-0.463
P_1	11.4	3.3	-0.343 ± 0.023 +5°40 ± 3°84	-0.460 ± 0.029 +9°79 ± 3°62	-0.415
K_1	30.3	8.9	-0.356 ± 0.008 +0°44 ± 1°29	-0.406 ± 0.010 +4°20 ± 1°48	-0.318

(*) if $\nu = 0.326$ which should be the expected value, one should have, for O_1 $\eta = -0.225$. Here the values have been normalised to -0.463.

We may conclude from this review of recent experimental results that all kinds of measurements clearly indicate the influence of the liquid core effects as predicted by the theory. The very small amplitude of the crucial ψ_1 wave unfortunately has not yet permitted to use it as a criterion for the selection of the best model of the core. A reason is that the resonance is not very sensitive to the stratification inside the liquid core (see chapter 6).

13.4.4. Astronomical consequences of the liquid core effects

It was well known that the fundamental nutation constant associated to a spectral line very close to K_1 (see chapter 2), which cannot be presently separated in the tidal records, is not at all in agreement with the value calculated in celestial mechanics for a rigid body, that is 9".227. The very long available series of fundamental astronomical observations give values which do not exceed 9".20.

An elastic Earth model does not change the theoretical value but the models with a liquid core (Jeffreys Vicente, Molodensky, Po Yu Shen and Mansinha, Wahr) precisely reduce that value to 9".20. This is obviously a very strong argument in favour of these theories.

Moreover, with these "liquid-core models", the short period nutations like the semi-annual one (associated to P_1 and ϕ_1 waves) and the fortnightly one (associated to O_1 and OO_1 waves), are slightly corrected in such a way that their agreement with the astronomical experimental values is also improved as shown by the Table 13.23.

The astronomers therefore decided to change the constant of nutation in 1980. The previously adopted value (9".21) was, indeed, a "compromise". The same has to be done with the amplitudes of some of the short-period nutations, convincing arguments having been provided.

The formulas (2.56) obtained by Melchior and Georis show in a very simple way how the magnitudes of the axis of every elliptic nutation are directly related to the magnitude of one pair of diurnal tidal waves having their frequencies symmetric with respect to ω. They also show that the nutation frequencies $\Delta\omega_i$ can be immediately derived from the tidal frequencies ω_i by a simple subtraction $(\omega_i - \omega)$.

On the basis of these formulas, Melchior (1971) constructed a detailed table of nutations taking into account the experimental results he had derived from very precise Earth tide measurements.

This table, published in Celestial Mechanics, was used for the reduction of Lunar Laser Ranging by Williams (1976), Harris and Williams (1976), King et al. (1978) as well as for the reduction of VLBI measurements by Chopo Ma (Nasa,1978).

Chopo Ma gives some details about the improvement of the VLBI results showing that when using this table for the recent observations (since 1973) the root of weighted mean square delay residual always decrease, the improvement in the most recent data (1976-77) being *"quite startling, 37%"*.

Simultaneously "the diurnal polar motion scale factor is reduced in every data set when the nutation corrections are applied ... the phase angle is also changed by nutation corrections and the scatter in phase is reduced.

But even the classical techniques of Astrometry clearly indicate that the IAU nutation tables are not satisfactory. McCarthy et al. (1977) have shown that *"discrepancies with observations can accumulate to 0"1 in right ascension and significantly affect the determination of UT 1 and materially influence the derivation of the new fundamental catalogue of star positions and proper motions FK 5"*.

These authors suggested that in the absence of a non-rigid-Earth model which can satisfy all requirements, the coefficients found from the investigation of solid-Earth-tides should be adopted as a working standard and they concluded: *"Statistically, there is no significant difference in the fit of Melchior's coefficients, Molodensky I, Jeffreys Vicente I or Federsen models to the astronomical observational results, although the Melchior coefficients are the best overall. There is no obvious choice for the best Earth model"*.

The model II of Molodensky has to be discarded for an application to the precession-nutations problem because it has been calculated with an Earth flattening equal to 1/297 instead of 1/298.25.

For this reason the International Centre for Earth Tides (ICET) had chosen since several years the Molodensky model I as a provisional standard of comparison for Earth tide analysis. Moreover this model I slightly better fits the Earth tide measurements, typically for the fundamental wave O_1 (model I: $\delta = 1.160$; model II: $\delta = 1.164$).

It also appears that Molodensky model II nutations diverge from the more recent theories by up to 0"002 at six months period and 18.6 yr period. Such differences will be significant with the new astrometric techniques (VLBI, Lunar Laser).

More recent models as those presented by Mansinha and Po Yu-Shen (1976), Sasao et al. (1977) and principally Wahr (1979), are very suitable for astronomical applications.

As seen in the chapter 6, the Wahr model is based upon model 1066 A of Gilbert and Dziewonski (1975) and is the only one which considers a *non spherical rotating Earth* instead of the usual SNREI one.

Rotation and deformation are computed simultaneously. Elliptical and rotational effects are considered throughout both the core and mantle and at each internal boundary.

The results obtained by Melchior in 1971 and by Wahr in 1981 are extremely close and are compared to those derived from astronomical observations by detailed discussions published since 1976 in the Table 13.23.

TABLE 13.23. Nutation coefficients (units are arc s).

Term	Woolard	Astronomical observations	Rigid Earth with N=9.2050 (1)	Melchior (1971)(6)	Wahr
Principal					
Obliquity	9.2100	9.2050±0.0017 (1) 9.1990±0.0035 (2)	9.2050	9.2014	9.2035
Longitude	6.8584	6.8409±0.0025 (1) 6.8360±0.0035 (2)	6.8547	6.8386	6.8430
Annual					
Obliquity	0.0000	—	0.0000	0.0056	0.0055
Longitude	-0.0502	—	-0.0503	-0.0580	-0.0567
Semi-annual					
Obliquity	0.5522	0.578 ±0.002 (1)	0.5528	0.5724	0.5708
Longitude	0.5066	0.533 ±0.002 (1)	0.5072	0.5237	0.5215
Semi-monthly					
Obliquity	0.0884	0.0925±0.0014 (1) 0.0897±0.0007 (3) 0.0893±0.0022 (4) 0.0898±0.0016 (5)	0.0844	0.0910	0.0910
Longitude	0.0811	0.0853±0.0010 (1) 0.0818±0.0022 (4) 0.0824±0.0016 (5)	0.0811	0.0831	0.0834

(1) Mc Carthy et al. (1977)
(2) Yumi et al. (1978)
(3) Gubanov and Yagudin (1978)
(4) Iijima et al. (1978)
(5) Mc Carthy (1976): results of Washington and Herstmonceux PZTs
(6) Four models were given in the 1971 paper. One should have selected this one in 1979.

The same need for such a nutation table taking fluid core effects into account has also been experienced in the BIH reductions (Guinot, personal communication).

13.4.5. A direct determination of Love numbers from Very Long Base Interferometry on extra galactic sources.

Simultaneous observation of extra galactic sources with five radio-telescopes installed in the U.S.A. and Northern Europe permits the precise determination of the length of the chords joining the different instruments. Chords of about 5000 km are measured with a precision of 1 cm that is 2.10^{-9} results which has been discussed by a number of researchers from M.I.T., Haystack Observatory, Goddard Space Flight Center, Onsala Space Observatory, Bonn University and Rutherford-Appleton Laboratories (Herring at al. 9th New York Symposium on Earth Tides, 1981) who obtained

$$\begin{aligned} h &= 0.62 \pm 0.01 \\ \ell &= 0.11 \pm 0.03 \\ \phi &= -\ 1° \pm 1° \text{ (lag)} \end{aligned} \qquad (13.9)$$

13.5. EXPERIMENTAL DETERMINATION OF THE EFFECT OF THE EARTH'S FLATTENING AND ROTATION (INERTIAL EFFECTS) ON THE δ FACTOR

As we have seen in § 6.4.4 (equ. 6.184 and 6.185), these effects are of course functions of the latitude. Thanks again to the ICET profiles we have now a sufficient number of good observing stations in the tropical and equatorial areas as well as in the southern hemisphere. Moreover the Finnish Geodetic Institute has established a number of stations at high latitudes while there are also two very high latitude stations at Longyearbyen (Spitsbergen ϕ = 78°) and at Alert (Canada, ϕ = 82.5°).

Wahr results show that the effect on the tilt measurements (γ factor) is too small (coefficient -0.001) to be detectable presently. Even the poor distribution of tilt tidal stations in latitude would not permit an experimental search for this coefficient.

On the contrary the coefficient for tidal gravity measurements is larger and there is hope to determine it experimentally.

We have collected all data available at the International Center for Earth Tides, for which we could control the calibration procedure of the instruments and the analysis of the data. The observed δ factors and phases have been corrected for the attraction and loading oceanic effects by using the Schwiderski cotidal maps.

The results of some stations could be improved if we could introduce corrections from neighbouring seas not included in the Schwiderski maps. This is difficult to do for all concerned stations because of the lack of good *local* cotidal maps. We intend to make such corrections when detailed cotidal maps will be available for all concerned areas. But the use of the same world map for all stations has the advantage that it preserves the homogeneity of the results which could not be the case if the map was manipulated.

A first analysis was given by Melchior (1981) for the waves M_2 and O_1 by using respectively 120 and 73 stations. This difference in number comes from the fact that, on the equator, the tidal amplitude is zero for O_1 while it is maximum for M_2. Thus many equatorial and tropical stations from the ICET profile cannot be used in the analysis of O_1. On the contrary two very high latitude stations (Longyearbyen 78° and Alert 82°) can be used for O_1 but not for M_2 because the amplitude decreases as sin 2ϕ in the first case and as $\cos^2\phi$ in the second case.

The number of stations used now (Melchior and De Becker 1982) is greater even if we have eliminated some stations which are not properly corrected for the oceanic effects that is those stations where, after correction, we still get a phase lag or advance of more than one degree. This evidently very often happens when the amplitude is very low: for O_1, K_1, P_1 in the tropical areas, for M_2 in the high latitude area.

The waves S_2 and N_2 are not considered because M_2 is by far more representative of the tidal deformation in the semi-diurnal frequency band.

It is well known that a correct determination of K_1 and P_1 is more difficult than the determination of O_1 because of the interference with atmospheric noise represented by the spurious "wave" S_1. For some stations we have not been able to fix the K_1 amplitude factor and this is the reason why we include less stations in the K_1 computations than in the O_1 computations.

For P_1 there is this additional difficulty that one year data is a required minimum to separate it from K_1 and O_1. Thus only 37 stations can be used for that wave. The results of least squares adjustments are:

<u>for M_2</u>, with 139 stations instead of 120:

$$\delta = \underset{\pm 27}{1.1761} - \underset{\pm 13}{0.0032} \left\{ \frac{\sqrt{3}}{2} (7 \sin^2\phi - 1) \right\} \qquad (13.10)$$

and with 118 stations (eliminating particularly discordant results)

$$\delta = \underset{\pm 21}{1.1751} - \underset{\pm 10}{0.0046} \left\{ \frac{\sqrt{3}}{2} (7 \sin^2\phi - 1) \right\} \qquad (13.11)$$

thus practically no change with respect to Melchior's solution (1981);

<u>for O_1</u>, with 109 stations instead of 73:

$$\delta = \underset{\pm 16}{1.1618} - \underset{\pm 15}{0.0028} \left\{ \frac{\sqrt{6}}{4} (7 \sin^2\phi - 3) \right\} \qquad (13.12)$$

<u>for K_1</u>, with 81 stations:

$$\delta = \underset{\pm 12}{1.1458} - \underset{\pm 13}{0.0059} \left\{ \frac{\sqrt{6}}{4} (7 \sin^2\phi - 3) \right\} \qquad (13.13)$$

<u>for P_1</u>, with 33 stations:

$$\delta = \underset{\pm 29}{1.1522} - \underset{\pm 29}{0.0039} \left\{ \frac{\sqrt{6}}{4} (7 \sin^2\phi - 3) \right\} \qquad (13.14)$$

It is immediately observed that the latitude dependence fits fairly well in all cases with Love's conjecture and with Wahr's computations.

However there is a serious discrepancy in the independent term with Wahr's results and before to question these results one has to carefully discuss the calibrations of our gravimeters.

Calibration of instruments

There are essentially 3 families of instruments:

a) The Askania static instruments calibrated by the maker on a baseline near Berlin,

b) The LaCoste Romberg astatized instruments which were not really calibrated but have an automatic system of control of stability of their sensitivity.

For the many stations of our list, established with Geodynamics, a calibration of each of the seven instruments used has been derived from a long series of measurements made at the Royal Observatory of Belgium in Bruxelles. This has been obtained by constraining the δ factor for the O_1 wave to be: $\delta = 1.1610$, (Ducarme 1975). This numerical value was the result of previous very long series obtained with three Askania GS 11 instruments (102688 hourly readings). We have not found any systematic difference between this system and the calibrations made in Texas for the LaCoste Romberg gravimeters by the maker: from a comparison based upon the tidal wave O_1 as observed at Bruxelles, it appears that the possible correction for several instruments used is less than 1% when their astatization is not too high. Of particular interest in this respect are the measurements made in Spitsbergen (latitude 78°) with Askania gravimeters of the Bonn University (M. Bonatz) which give $\delta(O_1) = 1.1468$ after correction for oceanic tides and those made at Alert, Canada at the latitude 82° with an earth tide LaCoste and Romberg instrument (Bower et al.) which give $\delta(O_1) = 1.1431$ also after correction. At these latitudes the Wahr formula gives 1.138. Moreover the measurements made at the South pole by UCLA have finally given: $\delta(Mf) = 1.159 \pm 0.010$
while the Wahr formula for long period tides gives again a lower value: 1.149.

It is extremely difficult to accept the idea of an error of calibration on all kinds of gravimeters. These LaCoste Romberg calibrations were also found correct to $10^{-4}/10^{-5}$ on the absolute gravity base Paris-Bruxelles-Wiesbaden (Poitevin et al.). Moreover the amplitude factors measured at the fundamental station Bruxelles are in excellent agreement with those measured with a great number of instruments at Bonn and Hannover (Melchior 1976). Our measurements at Manila (Philippines) even give a lower amplitude factor than those made by Sato with other LaCoste Romberg instruments. Also the O_1 residue measured at the very centre of Australia, Alice Springs, rigorously fit the attraction and loading effects of the world Oceans as calculated from the Schwiderski map and Molodensky Earth model I.

There may be an alternative which is a global correction deriving from the M_2 oceanic cotidal map. It is to be observed in this respect that the M_2 Schwiderski map gives a slightly too small oceanic tidal effect on satellite orbits. When developped into spherical harmonics this map gives indeed: $C_{22} = 2.96$cm, $\varepsilon_{22} = 333°$ while six recent analysis of satellite orbits give, as a vectorial mean: $C_{22} = 3.22$ cm, $\varepsilon_{22} = 333°$. The correction for oceanic tides indeed has a systematic decreasing effect on the δ value. This effect could be underestimated by the Schwiderski maps. However we observe that, by subtracting from the observed M_2 and O_1 vectors (δ, α) a model (1, 160, 0°) we obtain residues which are in general in agreement with the cotidal maps (Melchior et al. 1981).

13.6. SEMI DIURNAL WAVES IN THE HORIZONTAL COMPONENTS

As already mentioned, a great number of stations only exists in Europe. Their results are quite coherent for that concerning the East-West component as shown on Fig. 13.12. On the contrary, the amplitude factors in the North-South component are extremely divergent, ranging from 0.38 to 0.90. This seems difficult to explain in terms of "cavity effects".

For horizontal pendulums, phase determination strongly depends upon the azimuth of the pendulum beam, the determination of which is not so easy when a precision better than one degree or half a degree is required. It is therefore worth while to take the differences between the observed phase lags, for example $\alpha(S_2) - \alpha(M_2)$ and $\alpha(N_2) - \alpha(M_2)$, which eliminates the azimuth problem. The results given in Table 13.10 for Europe give evidence of a clear regional pattern which must be associated to the distribution of oceanic tides. Most striking is the North-South differences for the Mediterranean stations Grasse, Roburent and Genova, which have the sea in their south direction.

On the other hand, along the Atlantic coast of Western Europe,

$$S_2 \text{ amplitude}/M_2 \text{ amplitude} = 0.3,$$

$$S_2 \text{ phase} - M_2 \text{ phase} = +50°.$$

From a computation of the attraction of oceanic tides in the Atlantic made by Groten on the basis of cotidal charts,

$$\alpha_{NS}(S_2) - \alpha_{NS}(M_2) = +3°30',$$

$$\alpha_{EW}(S_2) - \alpha_{EW}(M_2) = -3°0'.$$

These differences are in the same sense as the observed differences but half the amount of them. Of course, their opposite sign is in agreement with Fig. 8.27.

Assuming the loading effect to be proportional to the attraction effect, according to the Nishimura's formula for the total effect,

North-South difference +6°,

East-West difference -5°,

which more or less fits the experimental results in Table 13.10. The results obtained in Spitsbergen are even more striking. At the very high latitude (78° 15' N) of this station (Longyearbyen) the direct Earth-tide effect is practically zero for semi-diurnal waves and thus quite pure oceanic tidal loading effects are observed. The results given in Table 13.10 have been obtained from one-year observations with three quartz pendulums (VM type) in East-West component and three similar quartz pendulums in North-South component. Corresponding differences for oceanic tides in ten places on the main Spitsbergen island are in excellent agreement with both components of the crust tilting.

As the station is very near the coast (4 km) there is no difference observable between East-West and North-South components.

13.6.1. *A Remark Concerning the Solar Waves S_2, K_2 and T_2*

As the S_2 wave could be suspected of including a systematical atmospheric contribution of barometric of thermic origin, some interest was raised by the possibility of separating from it the K_2 declinational wave as well as the elliptic wave T_2 over intervals of one year of more. This has been done for 20 European stations equipped with quartz VM pendulums in both principal directions (North South, East and West) and for some 20 world stations equipped with new gravimeters.

The results demonstrate that the amplitude factors and phases of K_2 do not differ from those of S_2. Even for the longest series they are equal. This suggests that the oceanic cotidal charts of K_2 are identical to the S_2 charts. However, there seems to be a systematic phase lag for the wave T_2.

TABLE 13.24. Horizontal Components, Difference between the Phase Lags of the S_2 and M_2 waves

	North-South	East-West
Sclaigneaux 1	9.46	-4.63
Sclaigneaux 2	8.36	-10.61
Sclaigneaux 3		-8.51
Dourbes 1	10.79	-5.07
Dourbes 2	2.16	-5.41
Vielsalm	-9.0	-8.20
Chevron		-9.70
Remouchamps	-3.8	-6.60
Kanne	3.73	0.42
Warmifontaine 1	6.83	-5.02
Warmifontaine 2	13.06	-4.46
Walferdange 1	3.11	-5.41
Walferdange 2	7.49	-4.41
Walferdange 3	8.00	-4.86
Erpel	2.49	-0.68
Příbram	1.16	-3.14
Tiefenort	7.35	-1.45
Berchtesgaden	5.26	-1.04
Graz	4.05	0.79
Bov	1.07	-2.77
Bad Grund	11.77	1.24
Dannemora	8.91	2.00
Lohja	4.87	1.22
Sopron	8.89	-1.56
Grasse	-11.09	-2.19
Roburent	-13.54	-1.33
Genova	-9.81	3.18

13.7. SECTORIAL TER-DIURNAL WAVE M_3

The amplitude of the M_3 wave in the vertical component is $1.4646\delta_3\cos^3\phi$ µga which, at the latitude ϕ = 50° where most of the stations were established, gives an amplitude as small as $0.389\delta_3$ µgal.

The equatorial and tropical stations established by the ICET since 1973 (Tr World Tidal gravity profiles) allow a better experimental determination of the δ_3 parameter which, according to the theoretical models should be

$$1.067 < \delta_3 = 1 + \frac{2}{3} h_3 - \frac{4}{3} k_3 < 1.070$$

The highest observed value is 1.079, the lowest 0.972.

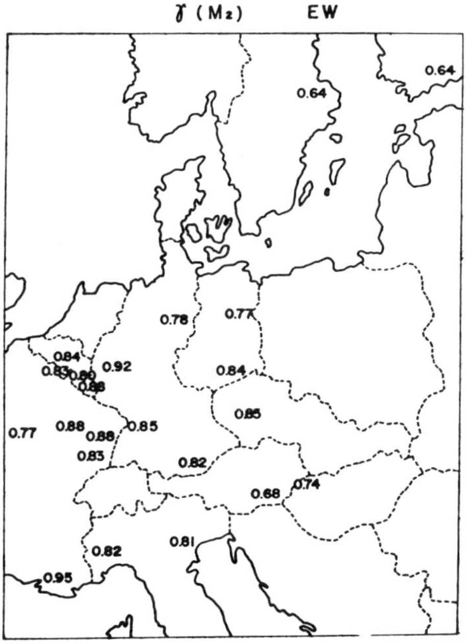

Fig. 13.7. Amplitude factors of M_2 of the vertical deflexion in Western Europe.

TABLE 13.25. Longyearbyen (Spitsbergen) phase differences

Total Observed Earth tide (direct + indirect effects)

Station	N	O_1-K_1	S_2-M_2	N_2-M_2	INSTR.
Longyearbyen 1 N EW	328	17.56	-44.24	23.78	VM 101
Longyearbyen 3 B EW	348	17.56	-43.52	26.60	VM 13
Longyearbyen 4 D EW	326	25.04	-39.86	26.01	VM 99
EQM			0.70	1.55	
Longyearbyen 1 N NS	312	5.63	-42.90	28.95	VM 102
Longyearbyen 3 B NS	324	6.65	-36.43	31.40	VM 42
Longyearbyen 4 D NS	320	15.25	-47.58	26.62	VM 100
EQM			1.20	4.10	
Oceanic tide					
Longyearbyen		139.2	-43.8	27.1	
Barentsburg		136.0	-42.0	25.0	
Kap Linne		80.0	-44.0	27.0	
Advent Bay		98.2	-37.7		
Ny Alesund		167.0	-46.0	30.0	
King's Bay		161.0	-48.0	21.0	
Gronf Jorden		136.0	-42.0	24.0	
Sorgf Jorden		200.0	-52.0	27.0	
Mossel Bay		171.0	-34.0	26.0	
Port Virgo		203.0	-32.0	25.0	

TABLE 13.26. Comparison of three solar semi-diurnal waves

	North-South component		East-West component	
$\gamma(S_2)/\gamma(K_2)$	0.991 ± 0.046	(21)	1.000 ± 0.055	(28)
$\alpha(S_2)-\alpha(K_2)$	+1.76° ± 3.09°	(21)	+0.45° ± 3.20°	(28)
$\gamma(S_2)/\gamma(T_2)$	1.004 ± 0.123	(5)	1.016 ± 0.038	(5)
$\alpha(S_2)-\alpha(T_2)$	+2.81° ± 2.63°	(5)	-4.35° ± 2.55°	(5)
	Vertical Component			
$\delta(S_2)/\delta(K_2)$	0.992 ± 0.025	(21)		
$\alpha(S_2)-\alpha(K_2)$	-0.12° ± 1.25°	(21)		
$\delta(S_2)/\delta(T_2)$	0.990 ± 0.037	(7)		
$\alpha(S_2)-\alpha(T_2)$	1.30° ± 1.61°	(7)		

The number of independent series is indicated between parenthesis. The mean values for S_2 are, respectively:

North-South	$\gamma = 0.697$	$\alpha = + 3.71°$
East-West	$\gamma = 0.711$	$\alpha = - 1.51°$
Vertical	$\delta = 1.199$	$\alpha = + 0.49°$

We have obtained the following results for four groups of 13 stations each:

latitude	n	δ_3
0° to 10°	13	1.083 ± 0.040
10° to 20°	13	1.092 ± 0.045
20° to 30°	13	1.074 ± 0.044
30° to 40°	13	1.087 ± 0.037
total	*52*	*1.084 ± 0.041*

The amplitude of the M_3 wave is extremely small in the horizontal components:

North-South $0''.000\ 308\gamma_3 \sin\phi \cos^2\phi$;

East-West $0''.000\ 308\gamma_3 \cos^2\phi$,

which, at the mean latitude of all the presently existing stations $\phi = 50°$, gives:

North-South $0''.000\ 097\gamma_3$;

East-West $0''.000\ 127\gamma_3$.

It is therefore quite astonishing that we can find coherent results from nineteen long series obtained with the quartz VM pendulums. The following mean result is derived from these series:

North-South component	$\gamma_3 = 0.812 \pm 0.040$ (8 series)
East-West component	$\gamma_3 = 0.803 \pm 0.041$ (11 series)
Theoretical models	$0.802 < \gamma_3 < 0.809$ (Chapter 5).

As an indication, the lowest value found experimentally is 0.741, the highest one being 0.870.

The only conclusion we may draw from these results is that our instruments seem to be well calibrated to obtain such a good comparison with the models derived from the seismological knowledge of the interior of the Earth. But the smallness of the amplitudes to be measured prevents the making of a choice between the different proposed models for which the third-order Love numbers, moreover, are unfortunately not very sensitive parameters.

13.8. THE ZONAL LONG-PERIOD TIDES

Because of the perturbations in tilt due to the underground waters it is not possible to derive information in this respect from the horizontal components. On the contrary, some good gravimetric series (but not all of those available) have been analysed for the long-period tides.

The best series is, as previously said, that of Brein at Frankfurt which gives the following results:

TABLE 13.27. (see Section 1.7.3. p. 32).

Wave	Period (days)	Amplitude at Frankfurt	δ	α
Ssa	182.6	2.8	1.258 ± 0.008	1.6° ± 0.3°
Mm	27.6	2.8	1.097 ± 0.007	0.3° ± 0.4°
Mf	13.7	6.0	1.233 ± 0.003	-0.3° ± 0.1°
Mtm	9.1	1.1	1.190 ± 0.014	1.0° ± 0.7°

Of course, the very high latitude (or equatorial) stations are in a more favourable position because the amplitude is increased. From the three gravimeters installed at Longyearbyen (Spitsbergen) by Bonatz, this author with Chojnicki derived the following mean results:

TABLE 13.28.

Wave	Period	Amplitude at Spitsbergen	δ	α
Ssa	182.6	6.1	1.094 ± 0.045	-5.5° ± 0.9°
Mm	27.6	7.4	1.180 ± 0.030	3.0° ± 0.3°
Mf	13.7	13.9	1.146 ± 0.031	-0.3° ± 0.6°
Mtm	9.1	2.7	1.028 ± 0.094	-1.8° ± 7.0°

This result for Mf seems more attendable than the high value obtained by Brein as the registrations made with the LaCoste Romberg tidal gravity-meter nr 4 at the South Pole itself during six years by the UCLA (University of California Los Angeles) team gives

$\delta(Mf) = 1.159 \pm 0.0015$ $\delta(Mm) = 1.164 \pm 0.0073$ $\delta(Mtm) = 1.164 \pm 0.0060$

(Rydelek and Knopoff, 9th Symp. Earth Tides, 1981).

13.9. TIME VARIATIONS OF TIDAL PARAMETERS

Shallow measurements of tilt (depth < 30m) are affected by severe perturbations related to thermic and barometric variations and ground water table variations related of course to rainfall. The instrumental drift therefore often contains an important seasonal component.

Baker and Alcock (1981) also demonstrated the reality of seasonal variations in 2 years tiltmeter data from Llanrwst (North Wales) of about 5% for the M_2 amplitude, 8% for the S_2 amplitude and 5° for the S_2 phase. They have shown that these variations are strikingly similar to those found in analyses of Irish Sea tide gauges and are therefore caused by seasonal variations in tidal loading.

Of course, seasonal variations of the oceanic tide parameters are most prominent in the shallow water areas like those around the British isles.

Also K_1 shows large seasonal variations - 30% amplitude and 15° in phase - which are remarkably coherent over a wide area. The O_1 amplitudes and phases also show significant seasonal variations but they are less coherent from place to place or from year to year.

These variations will affect Earth tide measurements through the ocean loading signal. They must be taken into account in short duration tidal gravity surveys and when looking for time variations of tidal tilt and strain prior to earthquakes in areas where tidal loading is of importance.

The most interesting observation is that of Lecolazet who derives time variations of relative tidal amplitudes (K_1/O_1, P_1/O_1, ψ_1/O_1) strongly correlated with the variations of the length of the day (see figure 13.4).

Since 1973, this author has performed digital recording of gravity tides in Strasbourg with an ancient LaCoste-Romberg gravimeter equipped with an electrostatic feedback.

The gravimeter itself is mounted in a sealed steel box located in a thermostatized room of an old buried small fort which protect it against the direct influence of barometric pressure variations.

However, it is well known that these variations have an indirect effect upon gravity. The observations have always been corrected for, the coefficient adopted being calculated for each yearly analysis (it is irregularly varying between 0.27 and 0.37 µGal/mb).

The gravity tide observations, carried out from August 1973 to June 1980 and analysed year by year, had shown a close correlation between the δ factors, normalized to $\delta(O_1)$ (the ratios $\delta/\delta(O_1)$ are free from changes in the sensitivity of the gravimeter), of the waves K_1 and ψ_1 (the most affected by the core resonance) and the mean length of day.

Trying to explain this phenomenon, the hypothesis had been put forward of the existence of variations in the nearly diurnal free wobble frequency, correlated with the Earth's rotation rate changes (Lecolazet, 1979) but the classical theory of the liquid core, improved by some new considerations, failed to justify this hypothesis (Hinderer, 1980).

Lecolazet concludes (1981) that:

1° There is an evident correlation between all the diurnal waves considered, which are purely solar (P_1, ψ_1, ϕ_1) or only partly, (K_1). Rather surprisingly, the variation of ϕ_1 is contrary to the others.

The best correlated waves are K_1 and P_1: the correlation confidence level is higher than 99%. This is deduced from the correlation coefficient value, 0.94, and the number of observations, 6.

2° There is also an evident correlation with the l.o.d. despite its last two values. The correlation confidence level of ψ_1 and l.o.d. remains slightly higher than 95%.

The correlation of the diurnal waves with the l.o.d. is partly confirmed by the results of Walferdange station.

3° The correlation extends to M_2, although this wave is purely lunar, but this is perhaps an artefact because the pressure correlation coefficient used is relative to the diurnal frequency band. However it must be noticed that a coherent M_2 wave has been found in the barometric pressure, whose amplitude is about 0.02 mb and phase approximately opposite to the one of the gravity tide M_2.
4° There is a rough relation between the amplitude of the $\delta/\delta(O_1)$ variation and the amplitude of the wave itself: the smaller is the wave, the larger is the variation.

Lecolazet considers as reasonable to attribute the gravity tide variations to the difference between the global effects of world-wide barometric pressure variations and the local effects. So, they would be correlated to the Earth's rotation rate variations by the evolution of the atmospheric general circulation. Dittfeld and al (1979) have already proposed a similar hypothesis for the annual variation they have observed at Potsdam.

However according to an analysis made by Usandivaras and Ducarme (1975) of the long and excellent series of tidal gravity measurements performed at Frankfurt a.m. (Germany) by R. Brein, there does not appear any significant variation of the amplitude factors and phases (see for example Fig. 13.9.).

An interesting representation of this possibility is also given by Fig. 13.10 constructed by these authors and associating the pairs of diurnal waves associated to each nutation.

This remains thus still controversial.

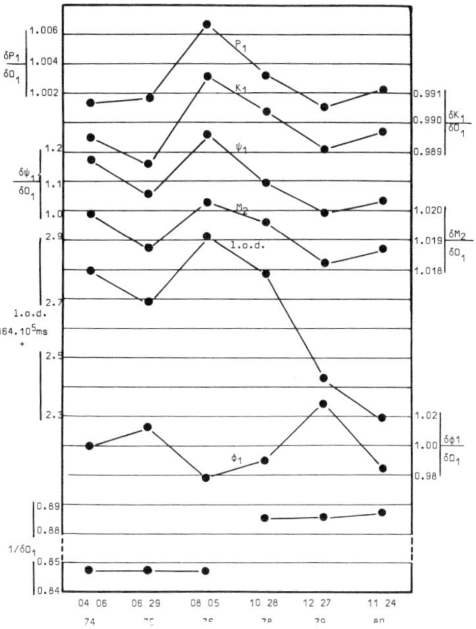

FiG. 13.8. Compared variations of $\delta/\delta(O_1)$ and mean length of day. In abscissa: middle dates of the yearly analyses; the time scale is not linear while the scale for the ordinates differs by a factor of 100 between $\delta K_1/\delta O_1$ and $\delta\psi_1/\delta O_1$.

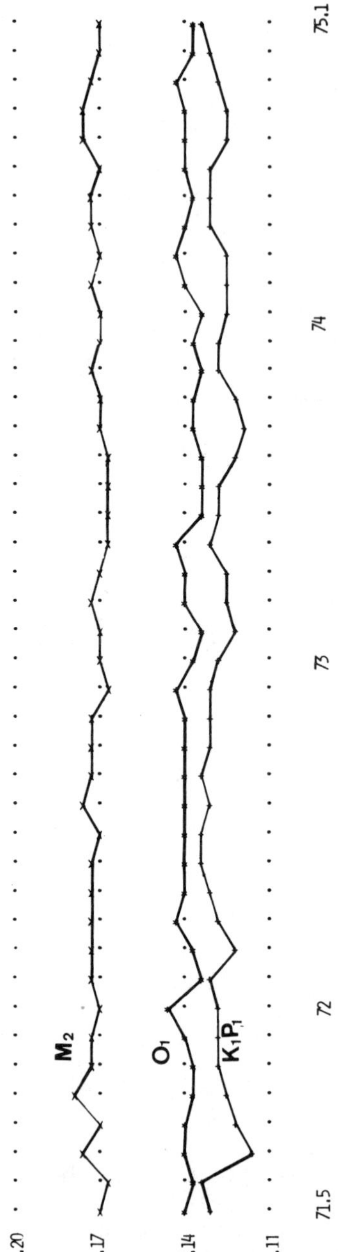

Fig. 13.9 Time variations of the δ factors observed at Frankfurt a. M. with the gravimeter LaCoste Romberg 98 by R. Brein - Analysis made by Usandivaras and Ducarme (1975).

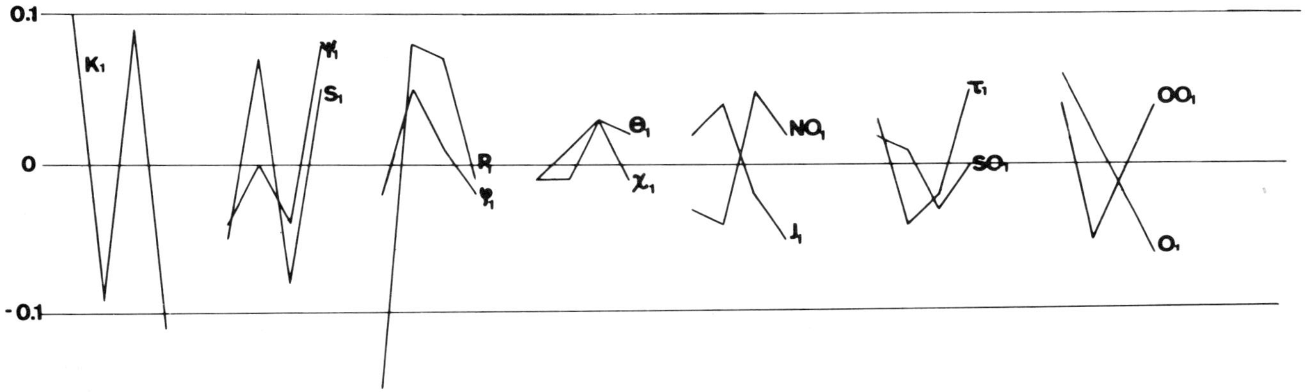

Fig. 13.10 Time variations of the δ factors for the associated pairs of diurnal waves corresponding to the precession (K_1) and the main nutations

Amplitudes are expressed in 0.1 microgal, for four consecutive years of observations (1970-71/1973-74) at Frankfurt a.M. with the gravimeter LaCoste Romberg 98 by R. Brein - Analysis made by Usandivaras and Ducarme (1975).

To obtain this representation the relative variation of the δ factors was multiplied by the theoretical amplitude of the corresponding wave. The fluctuations obviously do not depend from the amplitudes of the waves.

CHAPTER 14

Tidal Effects in Astronomy

14.1. THE EFFECTS OF THE DEVIATIONS OF THE VERTICAL IN THE FUNDAMENTAL ASTRONOMICAL OBSERVATIONS

At first sight it may appear doubtful that we can demonstrate a phenomenon of such a small amplitude in astronomic meridian observations whose precision reaches a maximum of a tenth of a second of arc. However, accurate observations carried out with the aid of instruments specially built for the precise determination of latitude and time followed methodically during many years have made it possible to demonstrate unquestionably the phenomenon and to provide an order of magnitude valid for the coefficient Λ defined by equation 3.47. It is obvious that we cannot hope to obtain here the degree of precision of the order procedures for the observation of Earth tides, but the result obtained is, on the whole, in as close agreement with theoretical studies as we should expect.

The observations concerned are time and instantaneous latitude. We suppose here that all the classical astronomical corrections have been taken into account (precession, nutation, aberration). However, a number of small effects which are not considered in the current practice of the astronomical reductions now clearly emerge from the noise level. Because of the sampling produced by the solar diurnal recurrence of the optical astronomical observations, the tidal components M_2 and O_1 appear with periods of respectively 14.765 and 14.192 days, while S_2 and K_1 have no resulting periodic effect (aliasing effect).

Indirect Effects

It is obvious that according to the principles of observation in meridian astronomy the flexure of the crust need not be considered here because it does not change the direction of the vertical to which the astronomical observations are referred with a spirit level or a mercury bath. We should have to consider here only the attraction of oceanic waters and the change of potential due to the deformation of the crust, this latter effect being a nearly 40% compensation of the first.

A very brief description of the instruments, of essentially different principles, having led to these results, will give an idea of the possibilities open to geophysics in the field of meridian astronomy.

1. Zenith Telescope

The mounting of this type of astronomical instrument, used since 1899 in the various stations of the International Latitude Service, is specially adapted to the precise determination of the latitude by the Horrebow-Talcott method. The latitude of a station can be obtained by measuring the zenithal distance z of a star whose declination is well known and, according to whether it is a star passing to the south or to the north of the zenith, one has

$$\phi = \delta_{south} + z_{south} ,$$

$$\phi = \delta_{north} - z_{north} .$$

The method consists in observing successively a star of each of these types passing at some minutes interval in the meridian and situated symmetrically to the zenith of the place so that $z_{south} \sim z_{north}$.

Under these conditions we obtain the latitude by the relation

$$\phi = \frac{\delta_{north} + \delta_{south}}{2} + \frac{z_{south} - z_{north}}{2}.$$

The half sum of the declinations is given in the catalogues (the methods of compensation used in the reduction of the results of all the stations makes it possible to improve them) while the difference in the zenithal distances is directly provided by the zenith telescope. The mounting called zenithal makes it possible, by a simple rotation of the telescope around the vertical axis placed in the centre of the frame, to find again the position exactly symmetrical in relation to the direction of the vertical: it follows that after the observation of the first star (south or north) of a "Talcott's pair" it is sufficient to carry out this rotation to find the second star (north or south) in the optical field of the instrument.

The relative position of the two stars in the field is measured with a high-precision micrometer screw, the difference between the readings giving the difference of the zenithal distances of the two stars. Two very sensitive and carefully selected spirit levels make it possible to control the stability of the instrument during the operations and to introduce any necessary corrections in the calculations of reduction.

The instrument does not include a divided circle, and this is its great advantage. The fact that the measurements are concentrated in a region near to the zenith and involve only very simple calculations of reduction, are other major advantages.

It is immediately obvious that the observations of the stars are related at each instant to the direction of the instantaneous vertical. They are thus affected by the luni-solar oscillations of the vertical. As the observations of the International Latitude Service are carried out each night using two groups of stars, one before and one after midnight, there is no effect from the solar tide in the latitude variations. The solar effect remains, however, present in the closing error defined as the sum, relative to a year's observations, of all the differences "morning group - evening group".

2. *The Floating Telescope of Cookson*

This is a photographic telescope (aperture 165 mm, focal length 166 cm) mounted on two floats immersed in an annular bath full of mercury. This bath of mercury provides the reference to the direction of the vertical. The principle of the observational method is that of Horrebow-Talcott, and experience has shown that during the rotation the small oscillations of the mercury are quickly decreased (in 2 min 30 s) so as not to affect the observations. Nevertheless, Kawasaki has shown that this instrument was particularly susceptible to wind influence, in which the direction plays the principal role as in the microseisms, the force seeming to play practically no role. This instrument was first employed at Greenwich. The analysis of the results has made it possible to determine the existence of a luni-solar effect, but attraction effects, due notably to the tides in the Thames, made the evaluation of the factor Λ

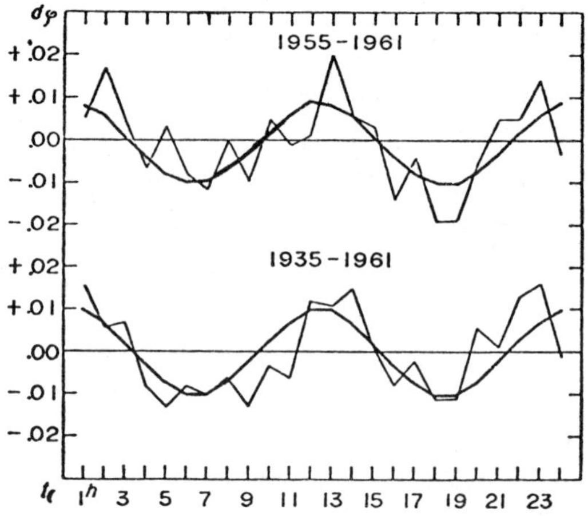

Fig. 14.1. Relation between the residual latitudes obtained at Mizusawa (Japan) with a visual zenith telescope and the Moon's hour-angle. Smoothed curves: M_2 terms (according to Sugawa (1965)).

difficult. The value given in Table 14.1 has been obtained after making a rough correction for the attraction of the water of the Thames only, but this cannot be considered as sufficient.

Another instrument of the same type, but more modern, has been put into service at Mizusawa, in parallel with a zenithal telescope. Sugawa made parallel determinations of the factor Λ for the visual zenith telescope and the floating zenith telescope from observations made simultaneously from 1943 to 1961 (Table 14.1).

3. *The Photographic Zenith Tube*

The first instrument of this type was built by Ross. It consisted of an objective (aperture 203 mm) with a large focal distance (5.16 m) with a bath of mercury placed below, at half the focal distance, which acts as a mirror and simultaneously defines the direction of the vertical. The objective is so constructed that the nodal point falls outside the objective, and can coincide, just below, with the focal point. As a result levelling errors do not affect the observations and the correction for "inclination" is always zero.

Besides this important property, the instrument makes it possible to eliminate the collimation error by reversing through 180° on each observed star and does not introduce an azimuthal error, being oriented towards the zenith. A photographic plate moves in the plane of the focus on a movable carriage synchronized by a quartz clock. The observation is therefore absolutely impersonal. Markowitz and Bestul have discussed the observations made in Washington with this instrument and took into account the indirect effect due to the attraction of the water but not that resulting from the variation of the potential due to the flexure of the Earth's crust; they obtained a coefficient Λ equal to 1.30. If we take into account a compensation of 40% in the attraction, caused by this variation in potential, the coefficient will be 1.27.

Some ten instruments of this type, of recent construction, are now in use, so we can now look forward to an improvement in the precision of the results.

4. The Danjon Impersonal Astrolabe

The astrolabe is an instrument observing according to the method of equal heights. The star light beam is duplicated by a reflection on a mercury bath acting simultaneously as a mirror and as a vertical direction indicator. The two light beams (1), (2) then enter the instrument through the two different faces of a 60° prism as shown on Fig. 14.2. Clearly the two images will be superposed when the zenith distance of the star is 30°.

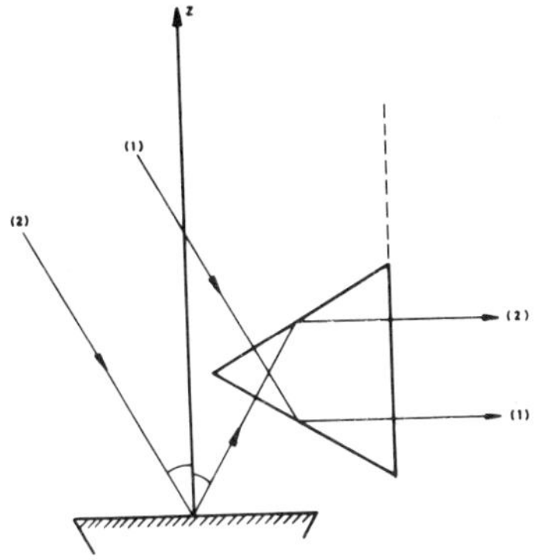

Fig. 14.2. Principle of the astrolabe.

The motor-controlled displacement of a Wollaston prism inside the instrument makes it possible to keep the two images together during the observation.

This displacement is registered on a chronograph to derive both coordinates of the place, that is time and latitude, which both depend on the vertical direction, a reference introduced in the measurements by the mercury bath.

14.2. THE PERIODIC VARIATIONS OF THE EARTH'S ROTATION SPEED

As demonstrated in Section 3.5, the zonàl deformations of the Earth corresponding to the long-period tides produce variations of the rotation speed which are proportional to the Love number k.

The main components can be calculated as shown in Table 14.3.

Many authors (Markowitz, Guinot, Pilnik, Djurovic and Melchior) have investigated the variations of the velocity of the Earth's rotation to discover these effects.

Spectral analysis of the differences (TU1-TUC) has given very convincing results recently obtained by Djurovic (Fig. 14.3), while the most recent results are given in Table 14.4.

TABLE 14.1. Latitude Observations

Theoretical formulas are:

$$\Delta v\ (M_2) = -0\overset{''}{.}00786\ \Lambda \sin 2\phi \cos (2\dot{s} - 2\dot{h})t$$

$$\Delta v\ (O_1) = 0\overset{''}{.}0065\ \Lambda \cos 2\phi \cos (2\dot{s} - \dot{h})t$$

Results of the analysis are obtained only from M_2 as the O_1 wave is vanishing at the mid-latitudes)

Station		Epoch	Phase	Λ	Author
Mizusawa	V	1900-12.9	+27°	1.00	Przybyllok
	V	1912-22.7	-100°	0.42	Nishimura
	V	1922-31.9	-11°	1.48	Kawasaki
	V	1900-31.9		0.58	Van Herk
	V	1900-34.9		0.38	Nishimura
	V	1900-34.9	-55°	0.64	Yevtushenko
	V	1935-54.9	-44°	1.30	
	V	1943-54	+15°	1.34	Sugawa
	F	1943-54	+23°	1.32	
	V	1955-61	+30°	1.22	Sugawa
	F	1955-61	-12°	0.84	Sugawa
Carloforte	V	1900-08.9	-4°	1.14	Shida
		1900-11.9	-2°	0.70	Przybyllok
		1900-12.0	-3°	0.85	Przybyllok
		1900-12.0	+82°	1.10	Nicolini
		1912-22.7	-26°	1.53	Nishimura
		1922-31.9	-3°	1.07	Kawasaki
		1922-34.9	+12°	1.42	Yevtushenko
		1900-31.9		0.79	Van Herk
		1900-34.9		1.07	Nishimura
		1900-34.9	-14°	1.17	Fedorov
Ukiah	V	1900-12.0	+7°	0.31	Przybyllok
		1912-22.7	+11°	1.36	Nishimura
		1912-22.7	+13°	1.17	Yevtushenko
		1922-31.9	-21°	1.60	Kawasaki
		1900-31.9		0.83	Van Herk
		1900-34.9		1.00	Nishimura
		1900-34.9	-1°	0.91	Fedorov

TABLE 14.1. Part II

Station		Epoch	Phase	Λ	Author
Gaithersburg	V	1900-12.0	+20°	0.99	Przybyllok
		1900-15.0		0.25	Nishimura
		1900-31.9		0.45	Van Herk
Tschardjui	V	1900-18.9		1.31	Nishimura
		1900-18.9		0.81	Fedorov
Cincinnati	V	1900-16.0		1.66	Nishimura
		1900-16.0		1.00	Van Herk
La Plata	V	1935-40.9	+16°	2.23	Carnera
Greenwich	P	1911-27	-9°	1.40	Kawasaki
Greenwich	P	1911-27	-3°	0.92	Spencer Jones
Washington	P		-10°	1.30	Markowitz
Poulkovo	V			1.05	Orlov
Poltava	V	1949-54.8	-1°	1.25	Filippov
Babelsberg	V	1914-17	+66°	1.60	Courvoisier
Pino-Torinese	V			1.37	Cerulli
Tokyo	P	1954	+65°	1.00	Torao
Paris	A	1956-63	+60°	0.81	Debarbat

V: visual zenith telescope. P: PZT. A: astrolabe.

Sasao and Sato (1981) have recently carefully calculated the Ocean Tide effects for the fundamental astronomical observations (latitude and time) at the five stations of the International Latitude Service, four of which being coastal stations. Such observations indeed use the local vertical as a reference while the purely tidal effects have more or less the same size as the unexplained residues in the irregularities of the polar motion or in the fluctuations of the velocity of rotation.

The tidal oscillations of the vertical must be corrected if the classical astronomical techniques have to be compared with the new techniques such as VLBI or Laser ranging to satellites which do not refer to the vertical.

Sasao and Sato have constructed the Green's function for the case of astronomical measurements with the combination $(1 + k'_n - \ell'_n)$ of the load Love numbers and show that at short distances the newtonian attraction is much more important than the pure loading effect. By using the Schwiderski cotidal maps they obtain amplitudes of one to two milliseconds of arc for the main waves (M_2, K_1, O_1) in latitude and greater in longitude, reaching up to 4.5 milliseconds at Ukiah for M_2.

TABLE 14.2. Time Observations

$$\Delta u\ (M_2) = 0^s.001\ 062\ \Lambda \sin (2\dot{s} - 2\dot{h})t$$

$$\Delta u\ (O_1) = 0^s.000\ 556\ \Lambda \tan \phi \cos (2\dot{s} - \dot{h})t \quad (\text{as } 0''015 \cos \phi = 0^s.001)$$

Results

Station	Phase	Λ	Author
Pulkovo V	-62°	1.06	Gubanov
Tokyo PZT	-11°	0.89	Gubanov
Greenwich PZT	+15°	1.51	Gubanov
Paris A	+43°	1.68	Débarbat
BIH O_1		0.806±0.246	Djurovic
BIH M_2		0.671±0.130	Djurovic

TABLE 14.3. Theoretical tidal variations of the Earth's rotation speed

Δu	Argument	Period	Tidal name
$+0^s.00032$ k	$3\dot{s} - \dot{p}$	9.133 days	Mtm
+ 13 k	$3\dot{s} - \dot{p} - \dot{N}$		
+ 249 k	$2\dot{s}$	13.661	Mf
+ 103 k	$2\dot{s} - \dot{N}$	13.633	
+ 10 k	$2\dot{s} - 2\dot{N}$		
+ 11 k	$2\dot{s} - 2\dot{p}$		
+ 23 k	$2\dot{s} - 2\dot{h}$		
+ 265 k	$\dot{s} - \dot{p}$	27.555	Mm
- 17 k	$\dot{s} - \dot{p} + \dot{N}$		
- 17 k	$\dot{s} - \dot{p} - \dot{N}$		
- 14 k	$\dot{s} + \dot{p}$		
- 6 k	$\dot{s} + \dot{p} - \dot{N}$		
+ 59 k	$\dot{s} + \dot{p} - 2\dot{h}$	31.812 (evection)	MSm
+ 60 k	$3\dot{h} - \dot{p}_s$	122	
+ 1529 k	$2\dot{h}$	183	
- 37 k	$2\dot{h} - \dot{N}$		
+ 488 k	$\dot{h} - \dot{p}_s$	365	
- 23 k	$\dot{h} + \dot{p}_s$	365	
+ 5150 k	$\dot{N}$	18.6 years	
- 27 k	$2\dot{N}$	9.3 years	

TABLE 14.4.

Period in days	Amplitude Fourier Transform	Amplitude Gibbs Transform	Theor. Ampl. for k = 0.317	Ampl. according to Guinot (*)	Ampl. according to Pilnik (*)
7.081	$0^s.00036$	$0^s.00033$	——	——	——
9.121	31	29	$0^s.00015$	——	——
13.661 (Mf)	121	120	109	$0^s.00065$	$0^s.00070$
14.192	54	59	51	74	94
14.765	49	52	7	66	48
27.555 (Mm)	45	60	73	66	80
31.664	35	34	19	——	27

(*) These authors separate Mf from the nodal term (2s-N). We have kept the two terms together here.

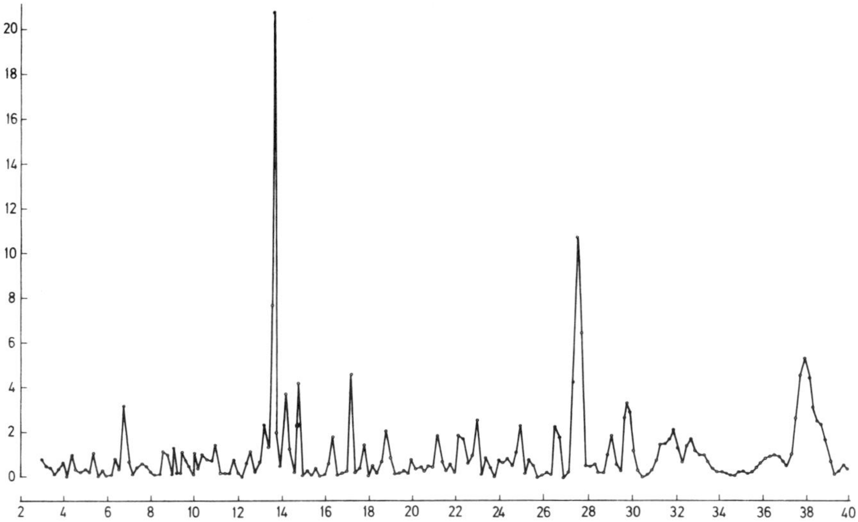

Fig. 14.3. Spectrogram of (TU1 - TUC) obtained by Djurovic from the BIH results of 1967 - 1974 (abscissa, the periods expressed in days).

TABLE 14.5. The Love numbers k as derived from the Earth's speed variations

Authors	Waves M_f	M_m
Guinot (1970)	0.331	0.265
Pilnik (1970)	0.300	0.282
Guinot (1974)	0.334	0.295
Djurovic (1975)	0.343	0.301

Mean square errors are, of course, given by these authors but are not comparable because the methods of analysis are different. Clearly one obtains a systematic difference which could be due to an effect of the corresponding oceanic and atmospheric tides.

14.3. THE ROLE OF VISCOUS TIDAL DEFORMATIONS IN THE SECULAR RETARDATION OF THE EARTH'S ROTATION

It is well known that the friction produced on the bottom of shallow seas by oceanic tidal currents fits fairly well the order of magnitude of the required power for explaining the secular retardation of the Earth's rotation. It is, however, surprising that the many calculations performed since 1923 on different data have always delivered the needed amount even when the astronomical conclusions were considerably changing.

The secular retardation is a phenomenon which has existed for more than a billion years as demonstrated by palaeontological discoveries. However, large climatic variations evidently have produced very important changes in oceanic tidal friction because of the disappearance of shallow seas and the creation of other shallow basins.

On the contrary, the contribution from internal Earth tides, which is the only one we will examine here, remained most probably constant even for such a long interval of time.

The total dissipation corresponding to astronomical and palaeontological evidence is 8.5×10^{26} erg yr^{-1}. In 1962 Jeffreys stated that

"it actually becomes rather difficult to see how so much energy gets into the shallow seas to be dissipated."

The most recent estimation made by Miller (1966) gave the dissipation in shallow seas as 4.4×10^{26} erg yr^{-1}, while Cox and Sandstrom (1962) estimate that the scattering into internal modes in open ocean is responsible for 1.2×10^{26} erg yr^{-1}. Miller also finds that in fact the flux from the deep seas is far too small to be considered. Kaula (1975) postulates that the dissipation takes place in shallow and shelf seas through processes unidentified. As Newton (1973) noted that the site of at least half the tidal dissipation has not been identified in the oceans, Bostrom (1976) considers that the lag in the bodily tides appears adequate to explain the missing part of the dissipation without recourse to an unidentified mechanism in the seas.

But Bostrom adds that the history of the tidal torque, if this is based on dissipation in the mantle, is likely to differ greatly from that of a torque based on dissipation in the seas because the dissipation in the mantle may be variable over a period of decades like the seismic energy release.

The internal tidal deformations cannot go forth without producing internal friction, the importance of which depends on the transport properties of matter. Viscous friction dissipates energy into heat and should be observed as a phase lag of the deformation. If we consider the Earth-Moon system as isolated from any external action, the law of conservation of energy

$$E_{kin} + E_{pot} + E_{diss} = \text{constant} \tag{14.1}$$

gives the rate of dissipated energy (power) as

$$\frac{dE_{diss}}{dt} = - \frac{dE_{kin}}{dt} - \frac{dE_{pot}}{dt}, \tag{14.2}$$

which implies that the speeds of rotation and of revolution of the Earth and the Moon will diminish with time (decreasing kinetic energy), while their distance will increase (decreasing potential energy).

The secular retardation of the Earth's rotation is a classical astronomical phenomenon deduced from the recorded longitude of the places of observations of the total eclipses in Antiquity. This is easily understood as follows. Let the length of the day be

$$J = 2\pi/\omega \tag{14.3}$$

with

$$\omega = \omega_o + \gamma t,$$

presently

$$\omega_{1970} = 7.29\times10^{-5} \text{ rad s}^{-1}$$

and let us take

$$\gamma = \frac{d\omega}{dt} = - 4.8\times10^{-22} \text{ rad s}^{-2} = -0.0031 \text{ rad century}^{-2} \tag{14.4}$$

which is, indeed, the value obtained from experimental measurements. One obtains

$$\frac{d\omega}{\omega} = -6.58\times10^{-18} \text{ s}^{-1} \text{ dt} = -1.83\times10^{-8} \text{ century}^{-1} \text{ dt}. \tag{14.5}$$

or

$$d\omega = -10^{-5} \omega (6.58\times10^{-13} \text{ s}^{-1}) \text{ dt}.$$

Thus for

$$dt = 3.2\times10^{12} \text{ s} \sim 100000 \text{ years},$$

$$d\omega \sim -2.1\times10^{-5} \omega, \tag{14.6}$$

and as

$$\frac{dJ}{J} = - \frac{d\omega}{\omega} \tag{14.7}$$

one obtains, after 100000 years,

$$dJ = + 2\times10^{-5} J = 2 \text{ s}.$$

After m rotations, i. e. m days, the accumulated lag will be

$$d\lambda = \frac{1}{2} \left(\frac{d\omega}{dt}\right) m^2 \text{ rad}, \qquad (14.9)$$

which allows the error in longitude for the calculated central line of the eclipses in Antiquity to be calculated 4000 years ago (that is, $m=-1.46\times10^6$ days ago), one has

$$\Delta\lambda \cong -4 \text{ rad} \cong -228° \cong -15 \text{ h},$$

while 2000 years ago $\Delta\lambda \cong 4$ h
and 100 years ago $\Delta\lambda \cong 26.4$ s.

As an example, a well-known eclipse is the one observed by Hipparchus in 123 BC which was observed 60° apart in longitude from the calculated spot.

This $\Delta\lambda$ obviously gives

$$\gamma = \frac{d\omega}{dt} \cong -4.8\times10^{-22}.$$

Some famous eclipses are listed in Table 14.6.

TABLE 14.6. Retardation of the Earth's rotation according to the eclipses of Antiquity

Author	Year	Eclipse	$\frac{d\omega}{\omega}$ in 10^{-5} per 100000 years
Dicke	-1889	Plutarchus (objected by Newton)	-1.77
	-1931	Plutarchus (objected by Newton)	-1.79
	-2088	Hipparchus	-1.79
	-2607	Archilochus	-1.76
	-3022	Babylon	-1.72
Curott	5 eclipses	Middle East	-1.612±0.076
Curott	15 eclipses	Middle East	-1.720
Curott	17 eclipses	China	-1.696
Newton	23 eclipses	Including 10 above	-1.661±0.062

Progress was made in this respect some 10 years ago when the palaeontologists discovered that fossil corals can be considered as fossil clocks. These animals develop themselves by secreting one ring every day, its width being a function of the quantity of light they have received. One can therefore expect and observe an annual modulation in their structure which allows to be counted more or less exactly the number of days (rings) contained in one year (wavelength).

An obvious check is that one finds 365 rings for the presently living corals. The fossil corals have been submitted to precise and delicate measurements conducted by Wells, Pannella and MacClintock. The results are represented by the numbers reproduced in Table 14.7.

TABLE 14.7. Retardation of the Earth's rotation according to Palaeontology

Epoch	Age in million years	Duration of one year in days	Duration of a day in hours	Relative duration of a day
Present	0	365.25	24	1
Cretaceous	72	370.33	23.67	0.986
Permian	270	384.10	22.82	0.951
Carboniferous	298	387.50	22.62	0.943
Devonian	380	398.75	21.98	0.916
Silurian	440	407.10	21.53	0.897

On the long geological time-scale one observes that the length of the day increases quite linearily. The duration of the day at the Devonian epoch was about 78 926 s which represents in 380 million years a loss of 7238 s with respect to its present value, that is 1.9 s in 100000 years, in very close agreement with the astronomical results for the last 3000 years.

Besides the eclipses, more recent observations of the Sun (declinations since 1760 and right ascensions since 1835), Venus and Mercury transits on the Sun's disk reconfirm the size of this phenomenon.

14.3.1. Calculation of the Torque Exerted by the East-West Horizontal Component of the Sectorial Tidal Force on the Sectorial Tidal Radial Deformation

The graphic representation of Fig. 14.5 is useful for understanding how the mechanism is acting. The figure aims to represent a section of the Earth by its equatorial plane wherein the Moon is supposed to be.

The Earth's rotation is counterclockwise (vector $\boldsymbol{\omega}$) as well as the Moon's revolution (vector $\mathbf{n}$). Moreover,

$$|n| \sim \frac{|\omega|}{27} < |\omega|$$

and therefore the tidal bulge (represented by the A and B protuberances) which is carried by the Earth in its rotation, has to rotate clockwise with the negative speed $(n-\omega)$ in order to remain on line with the attracting lunar body.

The internal friction causes a constant lag in this bulge rotation which is represented by the angle ε of the AB line with respect to EM. Then a schematic construction of the forces acting on A, B and O and of their differences with respect to the force acting on O (tides are, indeed, differential phenomena) shows that the resulting torque $\mathbf{f}$, $\mathbf{f}'$ acts opposite to the $\boldsymbol{\omega}$ direction and therefore acts as a drag for the Earth's rotation.
The potential of the sectorial semi-diurnal lunar tides at a point $P(r, \phi, \lambda)$ inside the Earth is, according to (1.9),

$$W_2 = \frac{3}{4} Gm \frac{r^2}{d^3} \cos^2\phi \cos^2\delta \cos 2H, \tag{14.10}$$

and the East-West horizontal component of the tidal force at P is

$$H_{EW} = - \frac{\partial W_2}{r \cos \phi \partial\lambda} = \frac{3}{2} Gm \frac{r}{d^3} \cos \phi \cos^2\delta \sin 2H. \tag{14.11}$$

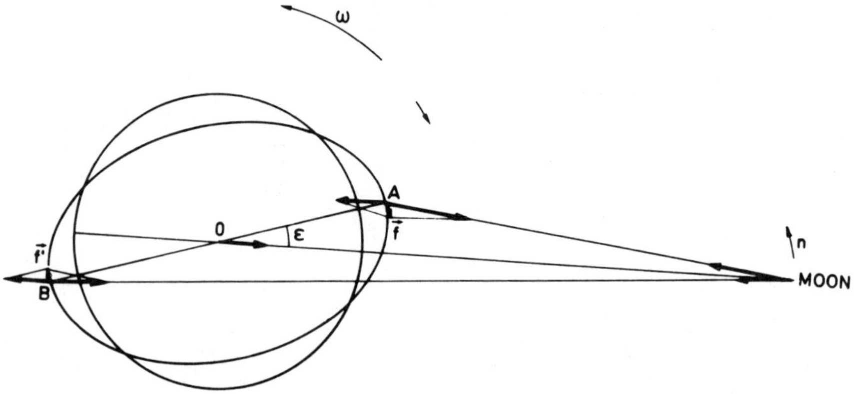

Fig. 14.5.

For brevity we put

$$A = \frac{3}{2}\frac{Gm}{d^3} s^{-2}. \tag{14.12}$$

Then the total torque exerted by the East-West forces applied to the whole mass of the Earth is

$$N = \int_{-\pi}^{\pi}\int_{-\frac{\pi}{2}}^{+\frac{\pi}{2}}\int_{0}^{a} (r \cos\phi)(A\, r \cos\phi \cos^2\delta \sin 2H)\, \rho r^2 \cos\phi \, d\phi\, d\lambda\, dr, \tag{14.13}$$

which is zero if there is spherical symmetry because

$$H = t_s - \alpha - \lambda$$

and

$$\int_{-\pi}^{\pi} \sin 2(t_s - \alpha - \lambda)\, d\lambda = 0. \tag{14.14}$$

Let us now introduce an elastico-viscous tidal deformation, the phase lag being ε:

$$\left.\begin{aligned} r &= r_0 + H(r)\frac{W_2(\varepsilon)}{g} = r_0 + \frac{H(r)}{g}\frac{Ar^2}{2}\cos^2\phi \cos^2\delta \cos 2(H-\varepsilon), \\ \rho &= \rho_0 - \rho_0\theta, \end{aligned}\right\} \tag{14.15}$$

$$\theta = f(r)\frac{W_2(\varepsilon)}{g} = r^2 f(r)\frac{S_2(\varepsilon)}{g}. \tag{14.16}$$

Then

$$\left.\begin{aligned} N = \int_{-\pi}^{\pi}\int_{-\frac{\pi}{2}}^{\frac{\pi}{2}}\int_{0}^{a} & \frac{A_2}{2}\cos^5\phi \cos^4\delta \sin 2H \cos 2(H-\varepsilon)\rho \\ & \left\{\frac{\partial}{\partial r}\left(\frac{r^6 H(r)}{g} - \frac{r^6 f(r)}{g}\right\}\right. dr\, d\phi\, d\lambda, \end{aligned}\right\} \tag{14.17}$$

where

$$\int_{-\frac{\pi}{2}}^{\frac{\pi}{2}} \cos^5\phi \; d\phi = \frac{16}{15}$$

and

$$\int_{-\pi}^{\pi} \sin 2H \cos 2(H-\varepsilon)d\lambda = \frac{1}{2}\int_{-\frac{\pi}{2}}^{\frac{\pi}{2}} \sin(4\lambda-2\varepsilon)d\lambda + \frac{1}{2}\int_{-\frac{\pi}{2}}^{\frac{\pi}{2}} \sin 2\varepsilon \; d\lambda = \pi \sin 2\varepsilon.$$

Thus

$$N = \frac{8\pi}{15} A^2 \cos^4\delta \int_0^a \rho \left\{ \frac{\partial}{\partial r} \left(\frac{r^6 H(r)}{g}\right) - \frac{r^6}{g} f(r) \right\} dr \sin 2\varepsilon, \qquad (14.18)$$

and using (3.102),

$$N = \frac{8\pi}{15} A^2 \cos^4\delta \frac{5\bar{\rho}a^6}{3g} k \sin 2\varepsilon. \qquad (14.19)$$

Evidently the retardation torque is proportional to the square of the Doodson constant because it results from the action of a *tidal* horizontal force upon a *tidal* radial deformation. Therefore the action of the Moon is about five times the action of the Sun.

It is also proportional to sin 2ε and vanishes when ε is zero. Replacing in (14.19), the well-known Earth constants give by their numerical values

$$N = 5.4 \sin 2\varepsilon \; 10^{24} \text{ ergs.} \qquad (14.20)$$

14.3.2. Dissipated Energy

A classical development of Darwin, based upon Kepler's third law,

$$n^2c^3 = G\,(M+m), \qquad (14.21)$$

allows to be calculated easily, the loss of mechanical energy in the Earth-Moon system by using a new variable ξ defined as

$$c = c_o\,\xi^2, \qquad n = n_o\,\xi^{-3}, \qquad (14.22)$$

the total angular momentum of the system being

$$\mathcal{M} = \frac{Mm}{M+m}\, n\, c^2. \qquad (14.23)$$

Its variation with time can be written

$$\frac{d\mathcal{M}}{dt} = \frac{Mm}{M+m}\, n_o\, c_o^2 \frac{d\xi}{dt} = N, \qquad (14.24)$$

while the third Euler equation gives for the Earth's rotation speed

$$C\frac{d\omega}{dt} = -N, \qquad (14.25)$$

then the corresponding kinetic energies

$$2\,E_{kin}^{(1)} = C\,\omega^2 \quad \text{(Earth rotation)}, \qquad (14.26)$$

$$2E_{kin}^{(2)} = \frac{Mm}{M+m} c^2 n^2$$ (planetary revolutions of the Earth and the Moon around their common centre of mass), (14.27)

vary as

$$\frac{d\,E_{kin}^{(1)}}{dt} = -\,N\omega, \qquad \frac{d\,E_{kin}^{(2)}}{dt} = -\,Nn, \tag{14.28}$$

while the newtonian gravitational potential energy

$$E_{pot} = -G\,\frac{Mm}{c} \tag{14.29}$$

varies as

$$\frac{d\,E_{pot}}{dt} = +\,2Nn, \tag{14.30}$$

so that we can conclude from (14.2) that

$$\frac{d\,E_{diss}}{dt} = N\,(\omega - n), \tag{14.31}$$

which shows that the rate of dissipated energy is equal to the torque exerted on the tidal bulge multiplied by the speed of the bulge itself.

Numerically,

$$\left.\begin{aligned}\frac{d\,E_{diss}}{dt} &= 3.8 \sin 2\varepsilon\; 10^{20} \text{ ergs s}^{-1}\\ &= 3.8 \sin 2\varepsilon\; 10^{10} \text{ kW.}\end{aligned}\right\} \tag{14.32}$$

A very rough estimate of the resulting heating could be made as follows:

$$\frac{dE_{diss}}{dt} = -\,J\sigma\,X\,M, \tag{14.33}$$

where J = 4.182 J (when dE/dt is expressed in watts), mechanical equivalent of heat; σ is the specific heat of the Earth material (heat quantity needed to increase the unity of mass by 1 degree); X the received heat of a unit of mass, expressed in degrees; and M the mass of the Earth.

Then

$$X = -\,\frac{N(\omega - n)}{J\sigma M} = 10^{-14} \sin 2\varepsilon \text{ deg s}^{-1}. \tag{14.34}$$

For example, if $2\varepsilon = 2°$, $\sin 2\varepsilon = 0.035$, X = 10 deg per 10^9 years, a value which could be acceptable.

Now if one supposes that all this energy will be transmitted outside of the Earth, this should produce a heat flow

$$H = \frac{N(\omega - n)}{JS} \text{ cal cm}^{-2}\text{ s}^{-1}, \tag{14.35}$$

S being the surface of the Earth (5×10^{18} cm^2).

Then

$$H = 1.8 \times 10^{-6} \sin 2\varepsilon \text{ cal cm}^{-2}\text{ s}^{-1}. \tag{14.36}$$

If $2\varepsilon=2°$ it should correspond to 5% of the observed heat flow which is about 1.2×10^{-6} cal cm^{-2} s^{-1}, which could be an acceptable value.
Additional remark: using (14.29), (14.30) and (14.20) one easily calculates that the distance of the Moon should increase by about 3 cm per year, a prevision which could be checked after some few years from the laser measurements to the reflectors recently installed on the Moon.

14.3.3. The Quality Factor Q

It is usual to define this parameter by its inverse Q^{-1} called the specific attenuation factor or factor of anelasticity:

$$Q^{-1} = \frac{\Delta E}{2\pi E} = \frac{1}{2\pi E}\oint \frac{dE}{dt}\, dt, \qquad (14.37)$$

where E represents the total energy which is circulated in the system or equivalently the peak energy attained by some part of it, whereas ΔE is that part of the total energy which is dissipated into heat in a complete cycle, the cycle being a single oscillation in the system in which energy is alternately stored and released. An example of such an oscillation would be the cycle of vibration of one of the frequencies in the tidal waves. (W. Kaula, An Introduction to Planetary Physics, 1968, p. 96).

Using the above definition of the specific attenuation factor one may quite easily express Q in terms of the phase lag ε for an harmonic motion of the frequency ω_i. Indeed, integrating the rate of loss of energy over one cycle,

$$\Delta E = \oint p\, de = -\oint p_o\, e_o\, \omega_i \cos \omega_i t \sin(\omega_i t+\varepsilon) dt, \qquad (14.38)$$

hence

$$\Delta E = \pi p_o\, e_o \sin \varepsilon, \qquad (14.39)$$

whereas

$$W = \text{surface OPM} = \frac{1}{2}\, p_o\, e_o \qquad (14.40)$$

(Fig. 14.5)

Then

$$Q^{-1} = \sin \varepsilon. \qquad (14.41)$$

If we consider that the best Earth-tide conclusion is according to (3.133):

$$\varepsilon = 1.45°$$

then $Q = 40,$ (14.42)

which is not in contradiction with the results of the Chandler notion of the pole ($Q \sim 60$).

14.4. ADVANCED RESEARCHES IN FUNDAMENTAL ASTRONOMY

In recent very high precision astronomical measurements like lunar laser ranging or VLBI (observation of quasars by very long base interferometry) the tidal effects appear so clearly that an elementary model is not sufficient to match the results properly.

As an example the nutation, Table 6.7, proposed by Melchior sensibly reduces the residuals in lunar laser ranging (Williams, personal communication,

1976) while VLBI already provides a direct determination of the h Love number (h_2 = 0.55 ± 0.05; Counselman, IAU Assembly, Grenoble, 1976).

The parameters describing the Earth's rotation were determined recently by lunar laser ranging (Erold program). A high accuracy reaching 0.″01 has already been obtained in a 194 single-day determination series (Bender et al, 1976) and will probably be improved in the near future. To reduce the data correctly, the authors have not only introduced consideration of the effect of zonal Earth tides on UT1 (with k = 0.29) but they also needed an improved nutation table taking the fluid core effects into account and therefore used Table 6.

The same need for such a nutation table, with fluid core effects, is now experienced in the BIH reductions (Guinot, personal communication), while new results obtained by McCarthy (1977) using photographic zenith tube data are supportive of Melchior's tidal results (Chapter 13). Lunar laser ranging and ratio interferometric reductions are aided by application of these empirical corrections as well (McCarthy, Seidelmann and Van Flandern, 1977).

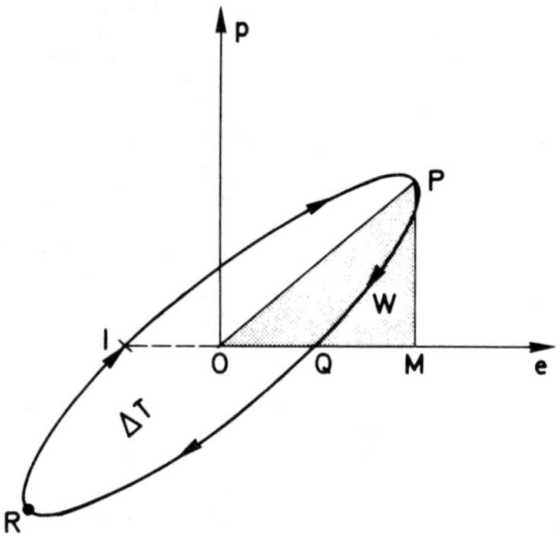

Fig.14.5. Comparison of the hysteresis loop with the peak energy represented by the surface OPM.

CHAPTER 15

Earth Tides, Satellite Orbits and Space Navigation

Kozai and Newton (1965) were the first to investigate these effects. Kaula gave an elegant mathematical treatment; Musen has published many valuable papers on the subject.

The effect of the tidal bulge on a satellite orbit depends upon the orientation of the bulge with respect to the orbit plane. Obviously the Earth's rotation does not enter into account and we have to refer to the orientation of the bulge in space. The K_1 bulge, as seen in Chapter 2, is directed towards the vernal equinox and therefore is practically fixed in space, producing a permanent distortion of the Earth's gravity field which is proportional to the Love number k. As the orbital plane is rotating with a period $2\pi/\dot{\Omega}$, which for the many satellites orbiting at nearly 1000 km height with an inclination of 45° is about 90 days, the K_1 dynamic effect on the satellite is a long-period period perturbation in about 90 days.

Since the satellite measures the integral effect of the tidal bulge potential (kW_2), the longer the period of the perturbation the larger will be its amplitude. K_1 effect, therefore, will very often be dominant. To investigate the role of other Earth-tide components we should subtract the tidal frequency ω_i from the Earth's rotational frequency, just as we did in the nutation theory and then subtract the node speed. However, when dealing with sectorial semidiurnal tides, we must realize that a satellite do not discriminate one bulge from the opposite one and therefore we have to subtract ω_i from 2ω and then subtract $2\dot{\Omega}$:

$$T_i(D) = 2\pi/(\omega-\omega_i-\dot{\Omega}) = 2\pi/\sigma_i; \qquad T_i(SD) = 2\pi/(2\omega-\omega_i-2\dot{\Omega}) = 2\pi/\sigma_i$$

TABLE 15.1.

	ω_i Tidal frequency	Frequency of bulge rotation in space (per day)	σ_i Frequency of orbital effect
K_1	$\dot{\tau}+\dot{s}$	0	$-\dot{\Omega}$
O_1	$\dot{\tau}-\dot{s}$	$2\dot{s}$ = 26°353	$2\dot{s} - \dot{\Omega}$
P_1	$\dot{t}-\dot{h}$	$2\dot{h}$ = 1°971	$2\dot{h} - \dot{\Omega}$
M_2	$2\dot{\tau}$	$2\dot{s}$ = 26°353	$2\dot{s} - 2\dot{\Omega}$
S_2	$2\dot{t}$	$2\dot{h}$ = 1°971	$2\dot{h} - 2\dot{\Omega}$

Table 15.2 gives a number of examples for satellites orbits of different inclinations

TABLE 15.2.

	a (km)	i	$\dot{\Omega}$/day	K_1		O_1		P_1		M_2		S_2	
				σ_i	T_i	σ_i	T_i	σ_i	T_i	σ_i	T_i	σ_i	T_i
BE-C	7 502 200	41°17	-4.2535°	-4.25°	85	30.60°	11.7	6.22°	57.9	34.85°	10.3	10.47°	34.4
Starlette	7 300 000	50°	-4.00°	-4.00°	90	30.35°	11.9	5.97°	60.3	34.35°	10.5	9.97°	36.1
Geos 1	8 072 900	59°38	-2.2465	-2.25°	160	28.60°	12.6	4.17°	86.3	30.85°	11.7	6.47°	55.6
Transit 1967.92A	7 453 770	89°3	-0°072	-0°072	5000	26.43°	13.62	2.04°	176	26.50°	13.58	2.12°	170
Geos 2	7 705 000	105°79	+1.40°	+1.40°	257	24.95°	14.4	0.57°	631	23.55°	15.3	0.83°	434
Geos C	7 220 000	115°	+2.73°	+2.73°	132	23.62°	15.2	0.76°	474	20.89°	17.2	3.49°	103

The frequencies σ_i are given in degrees per day.
The periods $T_i = 2\pi/\sigma_i$ are given in days.

Polar satellites like Transit 1962.92A (Table 15.2) having $\dot{\Omega} \sim 0$, do not allow to separate the main lunar waves O_1 and M_2 (13.6 days period) which however have a totally different behaviour in the solid earth tide and even more in the oceanic tides. Similarly P_1 cannot be separated from S_2 (170 days) period) for such satellites).

and shows that the periods are distinct from the other known perturbations and could be fairly well isolated except for the solar tide S_2 which evidently has practically the same period as the solar radiation pressure effect. This has to be accurately modelled in order to determine correctly the S_2 effect. As an example, Geos 2, because of its long S_2 period (434 days), exhibits a very large S_2 tidal effect which reaches 10", that is 300 m. A very long arc extending over 2 years, that is more than one cycle, was analysed at the GSFC (Douglas et al, 1972) and gave k=0.31, while 65 days arc of Geos 1 had given k=0.22 (for a tidal effect reaching 2" or 60 m). This effect, clearly dependent upon the inclination of the orbit, may be related to some latitudinal anomalies in the tidal phenomena. The Earth's rotation cancels the possible effects depending upon the longitude.

Two causes can be advocated to explain this phenomenon:

1) Lateral inhomogeneities in the upper mantle, but this is unlikely as demonstrated by the theoretical computations (Farrell, Alsop-Kuo).
2) The peculiar geographical distribution of the oceans, themselves tidally distorted, can most probably furnish the correct explanation. Newton (1965) considered the value of k he derived from polar satellites as "effective for the total earth and ocean tide", but as he found values all very near to k=0.31 he concluded that the ocean tide contributes little to the effect because being random in phase it was almost cancelling when averaged over the entire Earth.

This was proved to be wrong when non-polar satellites were considered. However the ocean tide perturbation is one order of magnitude smaller than the earth tide effect.

The tidal potential being given at the Earth's surface (r=a) by (1.1) as

$$W_2 = -\frac{GM}{d^3} a^2 P_2 (\cos z),$$

the tidal bulge potential is, at the satellite altitude d_s,

$$\mathcal{R} = \left(\frac{a}{d_s}\right)^3 k_2 W_2 \tag{15.1}$$

by Dirichlet's theorem or

$$\mathcal{R} = -\frac{GM}{d_s^3} \frac{a^5}{d^3} k_2 P_2 (\cos z). \tag{15.2}$$

The higher order terms in the potential have little influence on the satellites because their amplitude quickly decreases with $(a/d_s)^{n+1}$.

As for the Moon:

$$n^2 c^3 = G(E + M).$$

We have

$$\mathcal{R} = \frac{M}{E+M} \left(\frac{c}{d}\right)^3 \frac{n^2 a^5}{d_s^3} k_2 P_2 (\cos z). \tag{15.3}$$

On one orbital revolution one has

$$\frac{1}{2\pi} \int_0^{2\pi} \left(\frac{c_s}{d_s}\right)^3 dM = (1 - e^2)^{-3/2}. \tag{15.4}$$

The potential therefore becomes

$$\mathcal{R} = \frac{M}{E+M} n^2 \left(\frac{c}{d}\right)^3 \frac{a^5}{c_s^3} k_2 \, (1-e^2)^{-3/2} \, P_2 \, (\cos z). \tag{15.5}$$

The Lagrange equations have the form

$$\left.\begin{aligned}
\frac{di}{dt} &= \frac{1}{n_s c_s^2 (1-e^2)^{1/2} \sin i} \frac{\partial \mathcal{R}}{\partial \Omega} + \frac{\cos i}{n c_s^2 (1-e^2)^{1/2} \sin i} \frac{\partial \mathcal{R}}{\partial \omega}, \\
\frac{d\Omega}{dt} &= \frac{1}{n_s c_s^2 (1-e^2)^{1/2} \sin i} \frac{\partial \mathcal{R}}{\partial i}, \\
\frac{d\omega}{dt} &= \frac{-\cos i}{n_s c_s^2 (1-e^2)^{1/2} \sin i} \frac{\partial \mathcal{R}}{\partial i} + \frac{(1-e^2)^{1/2}}{n_s c_s^2 e} \frac{\partial \mathcal{R}}{\partial e}, \\
\frac{dM}{dt} &= n - \frac{(1-e^2)}{n_s c_s^2 e} \frac{\partial \mathcal{R}}{\partial e} - \frac{2}{n_s c_s} \frac{\partial \mathcal{R}}{\partial c}.
\end{aligned}\right\} \tag{15.6}$$

The variations of i and Ω are the most easily observed. The inclination of the orbit is the most favourable element for the investigations of the tidal perturbations (Fig. 15.1) because no drag or geopotential terms appreciably affect the analysis of the tidal perturbation in this element.

It is convenient to introduce the coefficient

$$A = \frac{M}{E+M} \frac{n^2}{n_s} \left(\frac{c}{d}\right)^3 \left(\frac{a}{c_s}\right)^5 k_2 \, (1-e^2)^{-2} \tag{15.7}$$

which has the dimension of an angular velocity:

$$\frac{M}{E+M} \sim 1/82.30, \qquad \left(\frac{c}{d}\right)^3 \sim 1,$$

$$\left(\frac{a}{c_s}\right)^5 \sim (0.85)^5 \sim 0.5, \qquad k_2 \sim 0.30,$$

$$\frac{n}{n_s} \sim \frac{2}{28 \times 24} \sim 0.003.$$

Thus

$$A \cong 0.0000\,054\; n.$$

The factor $(1-e^2)^{-2}$ shows that the tidal effects are considerably more important for eccentric orbits. Kaula (1964) has introduced a generalized form for the tidal potential:

$$\left.\begin{aligned}
\mathcal{R} = & \sum_{\ell=2}^{\infty} \sum_{m=0}^{\ell} \sum_{p=0}^{\ell} \sum_{q=-\infty}^{+\infty} \sum_{j=0}^{\ell} \sum_{g=-\infty}^{\infty} k_\ell \left(\frac{a}{c}\right)^\ell \frac{GM}{c} \\
& \frac{(\ell-m)!}{(\ell+m)!} (2-\delta_{om}) F_{\ell mp} (i_0) G_{\ell pq} (e_0) \left(\frac{R}{a_s}\right)^{\ell+1} \\
& F_{\ell mj} (i) G_{\ell jg} (e) \cos(\gamma^\circ_{\ell mpq} - \gamma_{\ell mjg} + \varepsilon_{\ell mpq}),
\end{aligned}\right\} \tag{15.8}$$

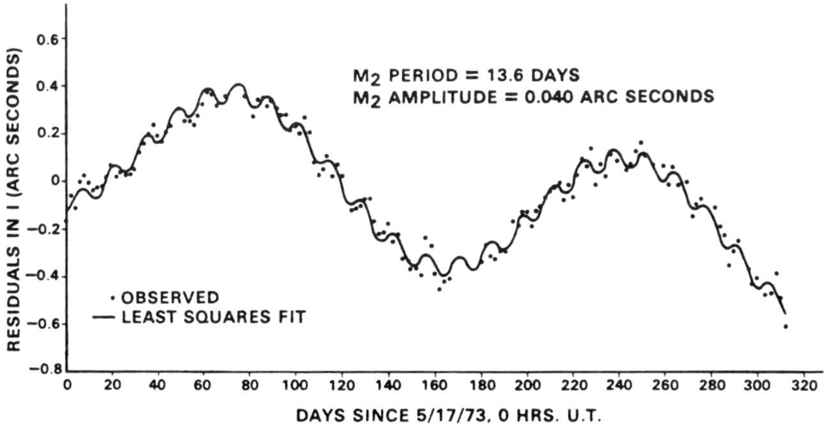

Fig. 15.1. Tidal perturbations on the inclination of the orbit of the polar satellite Transit 1967-92A. (Felsentreger, Marsh and Williamson, Journal Geoph. Res. 84, 4675, 1979).

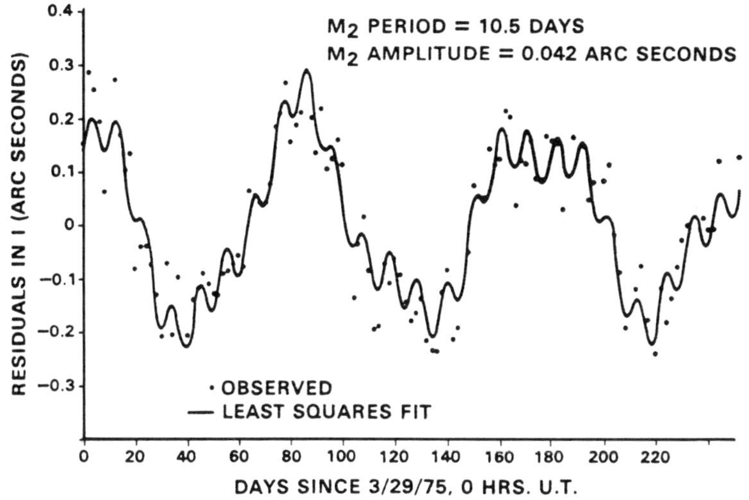

Fig. 15.2. Tidal perturbations on the inclination of the orbit of the satellite Starlette. (Felsentreger, Marsh and Williamson, Journal Geoph. Res. 84, 4675, 1979).

where

$$\left.\begin{aligned}\gamma &= (\ell-2j)\omega + (\ell-2j+g)M + m\Omega,\\ \gamma^\circ &= (\ell-2p)\omega + (\ell-2p+q)M_0 + m\Omega_o\end{aligned}\right\} \quad (15.9)$$

and

$$\varepsilon_{\ell mpq} = -(\ell-2p+q)\, n_0\Delta t + m\, n_e\, \Delta t, \quad (15.10)$$

the phase lag of the tidal bulge.

The polynomials F and G are given in most of the papers on celestial geodesy. The developments given by Kaula (1964), Newton and Kozai (1965) are in terms of instantaneous Kepler elements. The more recent developments by Musen and Estes (1972) and Musen and Felsentreger (1973) are based upon Brown variables like the Doodson expansion, and this facilitates the integration over long time periods due to their linear expression in function of time (Chapter 1).

Kaula was the first (1962) to indicate that ocean tidal effects could be perceptible. Therefore, to improve the artificial satellites orbits he introduced a modelling of the Love number k in function of the latitude:

$$\left.\begin{aligned}k_2 &= 0.351 - 0.055\, P_2^\circ\, (\sin\phi),\\ \varepsilon_2 &= 0.0195 - 0.0331\, P_2^\circ\, (\sin\phi) - 0.002\, P_4^\circ\, (\sin\phi).\end{aligned}\right\} \quad (15.11)$$

Admitting $k_2 = 1/2\, h_2$, this gives

$$\gamma_2 = 0.649 + 0.055\, P_2^\circ\, (\sin\phi),$$

$$\delta_2 = 1.175 - 0.027\, P_2^\circ\, (\sin\phi),$$

or, in function of the latitude,

	γ_2	δ_2
Equator	0.622	1.188
45°	0.663	1.168
Pole	0.704	1.148

In another paper Kaula has used the expression

$$k_2 = 0.372 - 0.240\, P_2^\circ - 0.013\, P_4^\circ. \quad (15.12)$$

These values do not fit with the results of the presently numerous stations for which we have results corrected for indirect effects (see chapter 13). They do not fit with the Wahr formula for an elliptic rotating Earth which gives δ_2=1.164 at the equator and shows, after Love, that the latitude dependence is in P_4^2 or P_4^1 but not in P_2^0, P_4^0 (see § 6.4).

Musen and Estes (1972) have replaced the Love numbers and phase lags by complex Love numbers corresponding to a Maxwell body, and because of the long period of tidal disturbances and because of the Earth's rotation these complex Love coefficients can be averaged along the parallels.

Balmino (1975), using the formalism of Kaula, tries to determine a longitude dependence of k and ε through possible resonances. He introduces, therefore, generalized Love numbers and phase lags defined as follows:

$$\left.\begin{aligned} k_{\ell mpq}(\phi,\lambda) &= \sum\sum k^{h\mu}_{\ell mpq} \cos\omega\,(\lambda-\lambda^{h\mu}_{\ell mpq})\, P_{h\mu}(\sin\phi), \\ \varepsilon_{\ell mpq}(\phi,\lambda) &\approx \varepsilon_{\ell mpq}(\phi) = \sum \varepsilon^{n}_{\ell mpq}\, P_n(\sin\phi). \end{aligned}\right\} \quad (15.13)$$

For a tentative evaluation of the perturbations on Geos B and Starlette he also uses

$$\left.\begin{aligned} k_2 &= 0.3 + 0.05\,P_{20} - 0.01\,P_{22}\cos 2(\lambda-70^\circ) + 0.01\,P_{40} \\ &\qquad + 0.05\,P_{42}\cos 2(\lambda+150^\circ), \\ \varepsilon_2 &= 2^\circ - 0.5^\circ\,P_{20} + 0.1^\circ\,P_{40}, \\ k_3 &= 0.1 + 0.02\,P_{20}, \\ \varepsilon_3 &= -\,0.2^\circ. \end{aligned}\right\} \quad (15.14)$$

Evidently oceanic tide effect superposes itself to Earth-tide effect, and this is responsible for the discrepancy between the "satellite" value of k_2 (0.25) and the "ground" value of it (0.30) and a great part of the latitude and longitude dependency of k and ε.

Lambeck and Cazenave (1973) have modelled the oceanic tide potential from the available cotidal charts by using Legendre polynomial developments like Pertsev introduced the use some years before.

$$U_n(a) = 4\pi G a \sum_{\ell}\sum_{m} \frac{1}{2\ell+1}\, C_{\ell m}\,\rho\, P_{\ell m}(\sin\phi)\,\sin(\omega_i t + m\lambda + \varepsilon_{\ell m}) \quad (15.15)$$

at the Earth's surface (ρ being the sea-water density) which for a non-rigid Earth gives at satellite height.

$$U_n(d_s) = \sum_{\ell}\sum_{m} (1+k'_\ell)\,\frac{4\pi G a^2}{(2\ell+1)c_s}\, C_{\ell m}\,\rho\left(\frac{a}{c_s}\right)^{\ell} \quad (15.16)$$

$$\sum_p F_{\ell mp}(i) \sum_q G_{\ell pq}(e) \begin{bmatrix} \sin\nu \\ -\cos\nu \end{bmatrix}^{\ell-m\ \text{even}}_{\ell-m\ \text{odd}}$$

with

$$\nu = [(\ell-2p)\omega + (\ell-2p+q)M + m(\Omega-t_s) + 2\pi n f T + \varepsilon_{\ell m}]. \quad (15.17)$$

They found that these oceanic perturbations introduce systematic errors diminishing the Love number k by 3% for high inclinations and by 17% for low inclinations while the phase is affected by several degrees.

Of course the authors have truncated this development to the order 4 and this may be justified by the fact that the perturbations on the satellite orbits being proportional to $(r/a)^{n+1}$ become negligible for $n>4$. This however produces an unpleasant effect for the geophysicist because such a truncation results in a coverage of the continents by important water masses and, of course, the total mass of the oceans is not kept constant. Interactions are badly represented and the result is not very interesting for oceanography and internal geophysics unless one uses the numerical values found for C_{22}, ε_{22}, C_{42}, ε_{42} from satellite orbital perturbations to constrain the modelisation of the cotidal map. The method however is powerful for the study of the long term evolution of the lunar orbit because only harmonics of order 2 have a role in the dissipation of energy.

Felsentreger, Marsh and Agreen (1976) then attempted to screen the solid Earth-tide effects from Geos 1 and Geos 2 satellite orbital data and to analyse the resulting data for ocean tidal parameters. They admitted in this approach that the value $k_2 = 0.30$ is now sufficiently well established. Their analyses have shown that long-period tidal effects are more easily determined from inclination data of close satellites. Thus, as shown by Table 15.2 only the K_1, P_1 and S_2 tidal waves could be detected. They derived, for the oceanic tide and the resulting loading effect, the following equations:

$$\left.\begin{aligned}
K_1:\ \delta I &= -\frac{9}{10}\, g\, \frac{\rho}{\rho e}\,(1+K_2')\left(\frac{a_e}{a}\right)^3 \frac{\cos i}{na^2(1-e^2)^2}\, C_{21}\, \frac{\cos(\Omega+\pi+\varepsilon_{21})}{\dot\Omega}\\
P_1:\ \delta I &= -\frac{9}{10}\, g\, \frac{\rho}{\rho e}\,(1+K_2')\left(\frac{a_e}{a}\right)^3 \frac{\cos i}{na^2(1-e^2)^2}\, C_{21}\, \frac{\cos(\Omega-2\lambda'+\pi+\varepsilon_{21})}{\dot\Omega-2\dot\lambda'}\\
S_2:\ \delta I &= -\frac{9}{5}\, g\, \frac{\rho}{\rho e}\,(1+K_2')\left(\frac{a_e}{a}\right)^3 \frac{\sin i}{n\,a^2(1-e^2)^2}\, C_{22}\, \frac{\sin(2\Omega-2\lambda'+\varepsilon_{22})}{2\dot\Omega-2\dot\lambda'}
\end{aligned}\right\}\quad (15.18)$$

λ' being the mean ecliptic longitude of the Sun. These forms can be easily understood on the basis of the preliminary remarks we made at the beginning of this chapter.

Geos 1 observations show an oscillation of about 1$''$.2 (36 m) in δI with the period of 160 days as given in Table 15.2. For Geos 2 the effect is 4$''$.5 (135 m).

A frequency analysis on the Geos 1 inclination residuals indicated the presence of perturbations having periods of 160, 85 and 56 days respectively induced by K_1, P_1 and S_2 as shown in Table 15.2. A least-squares fit gives the coefficients C_{21}, ε_{21} for K_1 and P_1, C_{22} and ε_{22} for S_2:

$C_{22}\ (S_2)$	=	1.7 ± 0.5	$\varepsilon_{22}\ (S_2)$	=	350° ± 21° (Geos 1)
		1.0 ± 0.2			62° ± 11° (Geos 2)
		2.25 ± 0.06			287° ± 1° (1978)
$C_{21}\ (K_1)$	=	8.8 ± 0.7	$\varepsilon_{21}\ (K_1)$	=	15° ± 4° (Geos 1)
		5.7 ± 1.6			334° ± 15° (Geos 2)
$C_{21}\ (P_1)$	=	5.0 ± 1.1	$\varepsilon_{21}\ (P_1)$	=	178° ± 15° (Geos 1)
		4.9 ± 1.1			127° ± 12° (Geos 2)

These results of Felsentreger, Marsh and Agreen are obviously very preliminary as they are based upon only two satellites, but their conclusions are very interesting and promising:

"We believe that these results indicate the feasibility of recovering global ocean-tide parameters from satellite data. This capability will increase as data from more satellites are studied and as the effects are seen in other orbital elements (in particular, the longitude of the ascending node, for which the effects are greater than for the orbital inclination but more difficult to isolate). An important reason for using satellite data for recovery of ocean-tide parameters lies in the fact that the perturbation due to any tidal constituent appears as a long-period effect with a frequency distinctly different from that of any other constituent, whether diurnal or semi-diurnal, so that separation of the constituents is thereby made quite easy. In addition, satellites sense the tidal effects all around the globe. Surface techniques provide information on local tidal variations and lead to separation of diurnal and semi-diurnal constituents. However, worldwide measurements are needed for the construction of a global model for any particular constituent."

TABLE 15.3. M_2 oceanic tide as developed in spherical harmonics

	C_{22} (cm)	ε_{22}	C_{42} (cm)	ε_{42}	
<u>Satellites</u>					
Geos 3	3.86	325°	1.26	102°	Cazenave et al. (1977)
Transit + Geos 1	3.07	337°			Lambeck (1977)
Geos 3 + 1967 - 92A	3.23	331°	0.87	113°	Goad and Douglas (1978)
Transit + Starlette + Geos 3	3.35	316°	0.97	109°	Felsentreger et al. (1978)
Transit + Starlette + Geos 3	3.42	325°	0.97	124°	Felsentreger et al. (1979)
Transit + Geos 3	3.21	352°	0.47		Daillet (1978)
<u>Cotidal Maps</u>					
Pekeris-Accad 1969	4.56	340°	1.49	170°	Daillet (1978)
Zahel 1970	4.89	335°	1.45	165°	Daillet (1978)
Hendershott I 1972	5.1	316°	1.2	115°	Lambeck et al. (1974)
Hendershott II	5.4	275°	1.1	75°	Lambeck et al. (1974)
Bogdanov-Magarik 1967	4.33	324°	1.7	116°	Daillet (1978)
Estes	3.31	333°	0.96	145°	Felsentreger et al. (1979)
Schwiderski 1979	*2.96*	*311°*	*1.01*	*125°*	*Schwiderski (1979)*
Equilibrium tide	4.00	270°	0.25	270°	Lambeck et al. (1974)

Rubincam (1976) derived the Love number k_2 from the perturbations in the orbital inclinations of three satellites (36 laser observations of BE-C, 142 Doppler observations of Geos 1, 113 camera and Doppler observations of Geos 2) (Table 15.3) for the main tidal waves.

All cotidal charts give a smaller k_2 than the solid earth k_2 (0.31) except the O_1 cotidal chart of Dietrich.

All three satellites indicate a remarkably small effective Love number for the P_1 tide which should imply a large ocean effect.

This is to be checked with the Schwiderski P_1 cotidal map which looks excellent (see § 13.2).

The solid tide phase lag has a non negligeable effect on the C_{22}, ε_{22} calculated values as shown by Goad and Douglas (JGR 83, 2306):

phase lag	C_{22}	ε_{22}
0°	3.23	331°
0.5°	2.76	325°
1.0°	2.32	318°

while C_{42}, ε_{42} do not suffer any change.

Rubincam estimates that the bulge phase lag should be 1.5° which fits quite perfectly our estimate 1.45° (3.133) derived from ground measurements.

However, the results of such corrections are still not very conclusive and even very poor concerning the phase lag. While of great importance for modelling the tidal effects on satellite orbits, these results are of little use if one wishes to describe the solid Earth-tide deformation. Here again the interaction ocean-continent appears so intricate that a *precise* separation of the effects is not really possible.

TABLE 15.4. (Rubincam, 1976)

			Ocean (assuming k_2=0.30 k=0°)	
			C (cm)	Phase
K_1	0.24175	2.407°	7.786	9.85°
K_2	0.30395	-0.417°	0.073	119.33°
P_1	0.19421	-3.551°	4.272	173.54°
S_2	0.30534	7.126°	1.831	4.50°

However, if we follow the same evolution as in gravity by mixing ground (or underground) measurements with satellite measurements, we would most probably derive new results and new ideas.

The reason is that on ground measurements the oceanic tides produce perturbations of local and regional extent while on the satellites we observe a global perturbation from the world oceanic tide.

A similar conclusion was derived by Musen (1975) that to make more progress it will be necessary to combine "the analysis of the perturbations of the satellites with the tidal observations on the Earth surface and obtain by this way a proper value for the exterior tidal potential. The present theory points out the existence of several tidal long-period and "cross effects" in the coefficients in the expansion of geopotential and in the motion of satellites. How long can we continue without including these effects and what are in fact the "average" elastic parameters which are being presently used to represent the observations of satellite? These questions constitute topics for a future research.

Gaposchkin (1976), by using precision satellite tracking with lasers, has significantly improved the knowledge of the geopotential. He considers that the effects of solid-Earth, ocean and atmospheric tides on both the gravity field and the station positions have now to be modelled and determined so that satellite data will become a source of information on tides and other deformations:

"The satellite acts as a filter, selecting certain combinations of spherical harmonics and transforming the spatial variations of the gravity field into a periodic temporal variation in satellite position. If temporal variations such as tides are properly modelled, centre-of-mass coordinates should be realized."

We should therefore have at our disposal

- a tidal stations network over the whole surface of the Earth
- a panoply of satellites with orbits well distributed in inclinations

as such geometries lead to different relative contributions from the different oceans.

CHAPTER 16

Solid Tides on the Surface of the Moon

The installation of a three-component seismograph on the Moon has raised the question of the tides of the Moon. Such a seismograph is designed primarily to record lunar seismic events, but it could have a sensitivity sufficient to record the tidal variations in lunar gravity and the tidal deviations of the vertical. These effects are relatively large on the Moon, and the dissipation produced by these tides is responsible for the frictional retardation of the rotation and the resulting fact that the Moon presents permanently the same side to the Earth, its rotation period being equal to its period of revolution.

The factors governing the amplitude of the tides on any planet are as follows:

a) distance of the perturbing body, the effect of which is proportional to the third power;

b) area, since the effect of the radius is proportional to the second power;

c) mass of the perturbing body.

We may deduce the amplitudes for the Moon by interchanging it with the Earth, in which case (a) is unaffected; but (b) reduces the amplitude greatly since the Moon's radius is 0.2725 of the Earth's, so the amplitude ratio is 0.074. As (c) is unaffected for the Sun, the solar tides are very small because of the factor 0.074, and the solar tide in gravity is only 5-6 μgal at maximum.

On the other hand, the Earth represents a perturbing mass 81 times larger, so that the tide is 81 X 0.074, or about 6 times larger than that produced by the Moon on the Earth, which is equivalent to a change of about 987 μgal.

The situation is best considered by using a treatment entirely analogous to that of Chapter 1. The geosolar perturbation potential, restricted to terms of the second order, is

$$W_2 = \frac{GE}{2}\frac{a^2}{r_☾^3}(3\cos^2 z_☾ - 1) + \frac{GS}{2}\frac{a^2}{r_\odot^3}(3\cos^2 z_\odot - 1) \qquad (16.1)$$

in which E and S are the masses of the Earth and the Sun respectively.

Taking the mass of the Earth as unity, we obtain Doodson's constant as

$$D = \frac{3}{4}\,GE\,\frac{a^2}{c^3} = 158\ 992\ \text{cm}^2\ \text{s}^{-2}. \qquad (16.2)$$

The parameters of the Moon are as follows:

Radius a = 0.2725 of the Earth's equatorial radius = 1.738 km.

Gravity = 0.166 (one-sixth) of the Earth's = 162.356 mgal.

Volume = 0.0202 of the Earth's volume = 218 830 X 10^6 km^3.

Mass = 0.01227 of the Earth's mass.

Density = 0.606 of the Earth's density = 3.3.

For the Sun

$$D' = 898\ \text{cm}^2\ \text{s}^{-2}, \qquad (16.3)$$

which is only 0.006 of the Earth term. Applying the formulas of Chapter 1 we have:

(a) Deviation of vertical

$$\frac{2}{\sin 1''} \frac{1}{a_{☾} g_{☾}} D = 2''324. \qquad (16.4)$$

So for the Earth $e = 2''324 \sin 2z_{♁}$;

for the Sun $e = 0''014 \sin 2z_{☉}$.

Now $z_{♁}$ varies by 40° at most, so the amplitude of the variation in the vertical may be up to 2"25.

(b) Variation in gravity

$$dg = 2 \frac{G}{a_{☾}} (\cos 2z + \frac{1}{3}) = 1.829 (\cos 2z + \frac{1}{3}) \text{ mgal.} \qquad (16.5)$$

The total variation is thus 2.44 mgal at maximum, the percentage variation being much larger on the Moon than on the Earth:

Moon: $\Delta g/g = 2.44/162,356 \approx 1.5 \times 10^{-5}$.

Earth: $\Delta g/g = 2 \times 10^{-7}$ (75 times smaller).

(c) Lunoid tide

$$\xi = \frac{W_2}{g_{☾}} = \frac{D}{g_{☾}} (\cos 2z + \frac{1}{3}) = \frac{158992}{162} = 979 (\cos 2z + \frac{1}{3}) \text{ cm.} \qquad (16.6)$$

These formulae allow direct calculation of tide variations in all three components for any point (ϕ, λ) on the surface. We merely introduce into each term the coordinates δ and $t = H + \lambda$ for the perturbing body relative to the Moon, as well as the distance r. Only the major solar term need be considered (this is a function only of the longitude), on account of the smallness of the solar contribution. The principal period of the variations is clearly the anomalistic month (27.55 days).

If we replace cos z by its expression (1.3), we observe that the three terms zonal, tesseral and sectorial are not decoupled in frequency (as in the case of Earth tides) because the speed of rotation is practically equal to that of revolution.

The eccentricity of the Moon's orbit produces a principal zonal wave of period 27.55 days whose pole is the point P that on average faces the Earth.

Librations of the Moon result from the ellipticity of its orbit about the Earth which produce changes in the speed of the Moon along its orbit (Kepler's second law) and from the inclination of its axis of rotation upon the normal to its orbit. The libration in longitude which reaches a maximum of ±8° gives rise to a principal sectorial wave a period 27.55 days, whose poles are the poles of rotation; whereas the libration in latitude which reaches ±7° gives rise to a principal tesseral wave with the same axis but with a period of 27.22 days. The secondary waves behave in the same way, having the same periods in each geometric type. The general distribution of the deformation is thus very complicated; the terms cannot be separated, and it is preferable to employ direct calculation for each point.

Harrison (1963) has suggested that a series expansion could be used to obtain a good approximation to the distribution of the various components; the first term should contain z_0, the value taken by z for zero libration and corresponding to a point P on the surface defining the mean direction to the Earth, the librations then being treated as small quantities.

Let us consider the local selenocentric sphere (Fig. 16.1) where T_0 is the mean position of the Earth (optical librations being zero); T is the position of the Earth as displaced by the optical librations ℓ and b; P is any point on the surface, of coordinates (ϕ, λ); and L is the pole of the lunar equator.

$$z = TP, \qquad z_0 = T_0P,$$

$$\lambda = T_0A, \qquad \phi = PA,$$

$$L = T_0B, \qquad b = TB.$$

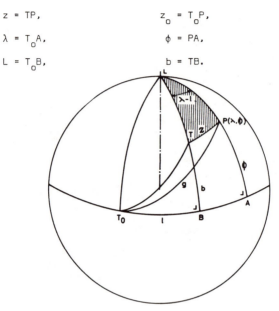

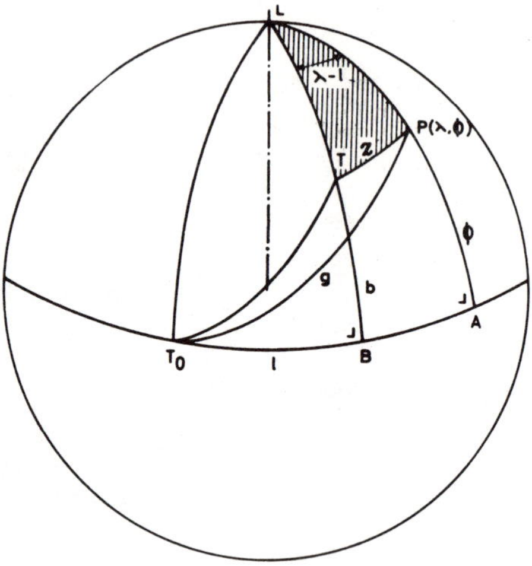

Fig. 16.1. The local selenocentric sphere. T_0BA is the lunar equator, L is the pole.

In the triangle PT_0A, where A is a right angle,

$$\cos z_0 = \cos \lambda \cos \phi.$$

Then, treating the librations ℓ and as small perturbations, cos z will be given in the spherical triangle LTP by:

$$\cos z = \sin \phi \sin b + \cos \phi \cos b \cos (\lambda - \ell)$$

This gives the following development:

$$\frac{4}{3}\left(\frac{3}{2}\cos^2 z - \frac{1}{2}\right) = \frac{4}{3}\left(\frac{3}{2}\cos^2 z_0 - \frac{1}{2}\right)$$
$$+ 4 \cos z_0 (\cos \phi \sin \lambda \sin \ell + \sin \phi \sin b)$$
$$+ 2 (\sin^2\phi - \cos^2 z_0) \sin^2 b - 2 \cos^2\phi \cos 2\lambda \sin^2\ell$$
$$+ 2 \sin \lambda \sin 2\phi \sin \ell \sin b + \ldots \qquad (16.7)$$

The orders of magnitude can be established by reference to the variations in gravity; the potential is then to be derived in terms of a, the common numerical factor being

$$L = (4/3)\ (2D/a_{☾}) = 2.440\ \mu gal. \qquad (16.8)$$

The product of the above expressions gives:

1. A term representing the fixed deformation of the Moon towards the Earth:

$$E_o = \frac{4}{3} \times 158\ 992\ (\frac{3}{2}\cos^2 z_o - \frac{1}{2}) \qquad (16.9)$$

for P ($z_O = 0$), the pole of the zone function and the point on average directed towards the Earth:

$$E_o = \frac{4}{3} D \times \frac{4}{3} \times 158\ 992 = 2.12 \times 10^5\ cm^2\ s^{-2},$$

$$\xi_o = \frac{4}{3}\frac{158\ 992}{162} = 13\ m,$$

and the maximum total variation in gravity is then (4D/3) (2/a) = 2.440 μgal. Here $\cos z_O = \cos\lambda \cos\phi$ is a constant of position.

2. Terms dependent solely on the variation in distance:

As $$(\frac{c}{d})^3 = 1 + 0.1647 \cos(\dot{s}-\dot{p})t + 0.030 \cos(\dot{s}-2\dot{h}+\dot{p})t$$
$$+ 0.024 \cos(2\dot{s}-2\dot{h})t + 0.013 \cos 2(\dot{s}-\dot{p})t$$

one obtains

$$L\ (\frac{3}{2}\cos^2 z_o - \frac{1}{2})\ \{0.1647 \cos(\dot{s}-\dot{p})t + 0.030 \cos(\dot{s}-2\dot{h}+\dot{p})t + 0.024 \cos(2\dot{s}-\dot{h})t + \ldots\} \qquad (16.10)$$

These tides are zonal tides about the pole P whose maxima lie where $\cos z_O = 1$, namely at P(0°, 0°) and P' (0°, 180°), and whose amplitude is proportional to ($3\cos^2 z - 1$), since this must be zero for the parallel of P(z_0 = 55°).

Employing Doodson's argument numbers, we get the components listed in Table 16.1, whose maximum amplitudes (at P) are readily found in μgal from (16.8) and (16.10).

3. Terms dependent on the librations

(a) in sin ℓ: here the coefficient $\cos z_O \cos\phi \sin\lambda = (1/2) \cos^2\phi \sin 2\lambda$ shows that we have a sectorial tide whose poles are those of rotation, as is obvious *a priori* since this tide is caused by libration in longitude.

(b) in sin b: here the coefficient is $\cos z_O \sin\phi = (1/2) \cos\lambda \sin 2\phi$, which shows that we have a tesseral tide caused by the libration in latitude, whose poles are those of rotation;

(c) in $\sin^2 b$, $\sin^2 \ell$, and sin ℓ sin b.

4. Cross terms, of which the main one is

$$\frac{3}{4} L\{0.1647 \cos(\dot{s}-\dot{p})t\}\ \{4 \cos z_O\ (\cos\phi \sin\lambda \sin\ell + \sin\phi \sin b)\},$$

which gives tides that are zero at P and at the edges of the visible disc. Table 16.1 summaries these results.

The optical libration in longitude (1) principally depends upon the angular distance of the Moon from its perigee (p). Its complete expression given in the astronomical tables is

$$\left.\begin{aligned}\sin \ell \approx \ell = {} & 22639''.5 \sin(\dot{s}-\dot{p})t + 4586''.4 \sin(\dot{s}-2\dot{h}+\dot{p})t \\ & + 2369''.9 \sin(2\dot{s}-2\dot{h})t + 769''.0 \sin(2\dot{s}-2\dot{p})t \\ & - 668''.1 \sin(\dot{h}-\dot{p}')t - 411''.6 \sin(2\dot{s}-2\dot{N})t \\ & - 211''.7 \sin(2\dot{h}-2\dot{p})t + 206''.0 \sin(\dot{s}-3\dot{h}+\dot{p}+\dot{p}')t \\ & + 190''.0 \sin(3\dot{s}-2\dot{h}-\dot{p})t + 165''.1 \sin(2\dot{s}-3\dot{h}+\dot{p}')t \\ & + 147''.7 \sin(\dot{s}-\dot{h}-\dot{p}+\dot{p}')t - 125''.2 \sin(\dot{s}-\dot{h})t \\ & - 109''.7 \sin(\dot{s}+\dot{h}-\dot{p}-\dot{p}')t - 55''.2 \sin(2\dot{h}-2\dot{N})t.\end{aligned}\right\} \quad (16.11)$$

The optical libration in latitude (b) principally depends upon the angular distance of the Moon from its node (N). Its complete expression is

$$\begin{aligned}& \sin b \approx b \\ -b = {} & 18461''.5 \sin(\dot{s}-\dot{N})t + 1010''.2 \sin(2\dot{s}-\dot{p}-\dot{N})t \\ & - 999''.7 \sin(\dot{p}-\dot{N})t + 623''.7 \sin(\dot{s}-2\dot{h}+\dot{N})t \\ & + 199''.5 \sin(2\dot{s}-2\dot{h}+\dot{p}-\dot{N})t - 166''.6 \sin(2\dot{h}-\dot{p}-\dot{N})t \\ & + 117''.3 \sin(3\dot{s}-2\dot{h}-\dot{N})t.\end{aligned} \quad (16.12)$$

The periodicity of the libration in longitude mainly depends upon the argument $(\dot{s} - \dot{p})t$ while the periodicity of the libration in latitude mainly depends upon the argument $(\dot{s} - \dot{N})t$.

The period of commensurability is thus $2\pi/(\dot{p} - \dot{N})$ which is practically six years (exactly 5.9968 years). Therefore the point $T(\ell, b)$ describes an ellipse around T_0, the major axis of it rotating by 180° within six years (Fig. 16.2).

On Fig. 16.2 the Earth T will cross the X axis when the Moon is at one of its nodes while it will cross the axis Y when it passes through its perigee.

ROLE OF ELASTICITY OF THE MOON

We now consider how the Moon may react to the perturbing potential of essentially terrestrial origin. We have to establish the possible Love numbers for the Moon. Here we are confined to speculations based on suppositions about the internal structure.

If we assume that the Moon is homogeneous (which is probably not far from being the case), we can use Kelvin's formulae (4.62), (4.65) and (4.75):

with $g = 162.356$ gal, $\rho = 3.33$, and $a = 1738 \times 10^5$ cm;

on the assumption that the rigidity of the Moon is (a) equal to that of the Earth's crust or (b) half of it, we have respectively μ as (a) 10^{12} or (b) 5×10^{11} cgs, so that $19\,\mu/2\,g\,\rho\,a$ is 100 or 50, whence

	h	k	ℓ
(a)	0.025	0.015	0.007
(b)	0.049	0.030	0.014

and so

(a) $\gamma = 0.990$ $\delta = 1.0025$ $\Lambda = 1.007$ $f = 0.055$

(b) 0.981 1.0040 1.016 0.108

All proposed models give numbers in between these two solutions.

This result is sufficient to show how difficult it could be to examine the properties of lunar matter with instruments designed for Earth tides, for the amplitudes are not sensitive to the internal structure, essentially on account of the Moon's small radius.

Kaula has recently calculated h and k for a series of models with mantle and core respectively homogeneous, as can be done from Herglotz formulae valid for two homogeneous media (see 1st edition, page 333). The core was taken as having a radius 0.5 to 1 relative to the Moon's radius: several different rigidities were used for the two media. In most cases δ did not exceed 1.01; the extreme values found were $\gamma = 0.856$ and $\delta = 1.05$, which relate to a fluid core and the following conditions: $\lambda_c = \lambda_m = \mu_m = 4 \times 10^{11}$, radius of core 0.75 $\rho_c = 3.81$, $\rho_m = 3.00$. A detailed knowledge of the interior is needed before further progress can be made. The observed tides will be so close to the theoretical ones that the instruments would have to be more accurate than any currently available in order to yield useful results; moreover, the periods of various waves are so long that drift will very much reduce the possibilities of analysis. Another effect that may offer better conditions for observation is that of cubic expansion. For the Moon

$$W_2/a\ g = \xi/a \approx 1.000/1.700 \times 10^5 = 0.6 \times 10^{-5}$$

and so for $f = 0.055$ we have $H = 3 \times 10^{-7}$, or for $f = 0.108$ $H = 6 \times 10^{-7}$, which represent an amplitude ten times that found on the Earth and hence proportionately more sensitive to the internal properties. This expansion effect has an important bearing on the formation of the lunar craters if any volcanic process is involved.

The tide effects may favour the expansion and recession of fusion waves; they may also act as a trigger to eruption. Here we may recall the variation in gas emission at Tarka Bridge and the effects observed on Vesuvius.

A strong correlation of the recorded lunar seismic events with consecutive perigees has been interpreted as tidal triggering of Moonquakes by many authors. The most active hypocentre is roughly halfway to the centre of the moon below an epicentre east of Mare Humorum. Special attention has been paid to the action of tidal forces at this hypocentre (Chapman, Middlehurst and Frisille, 1974). In particular the monthly times of minimum gravity at the epicentre correlates reasonably well with the Moonquake occurrences during 1970.

Difficulties in the interpretation of lunar measurements arise from the fact that because the rotation period is equal to the revolution period the several monthly and semi-monthly periods (sideral, synodic, anomalistic and draconitic) simultaneously appear in the orbital parameters in the coordinates of the lunar reflectors because of the librations and in tidal effects.

Therefore the three reflectors have been installed to form a nearly equilateral triangle and their distances to the Earth must be measured quasi-simultaneously by laser in order to be able to separate the different perturbations.

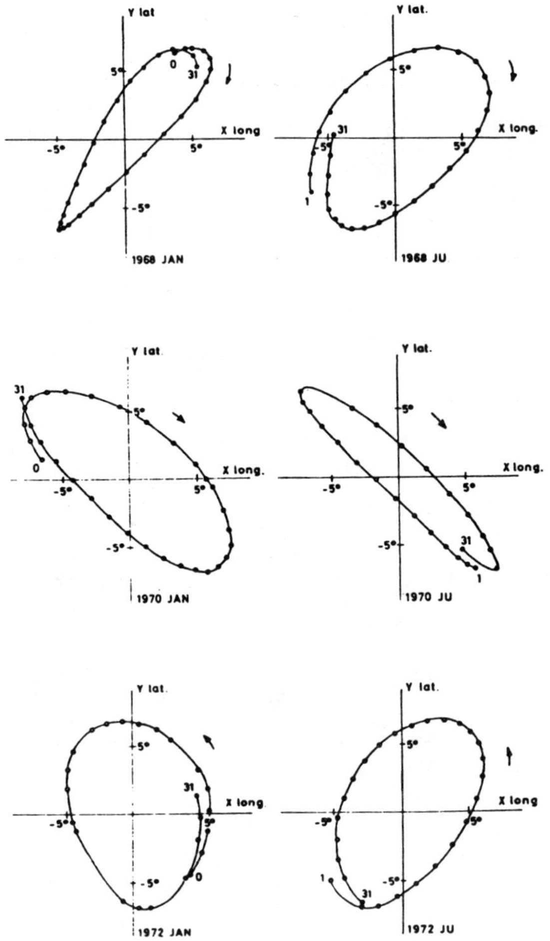

Fig. 16.2 a. Nearly elliptic curves described on the Moon's surface by the point T defining the direction of the Earth as seen from the Moon's centre. The curves are given for january and july of each of the six years 1968 to 1973.

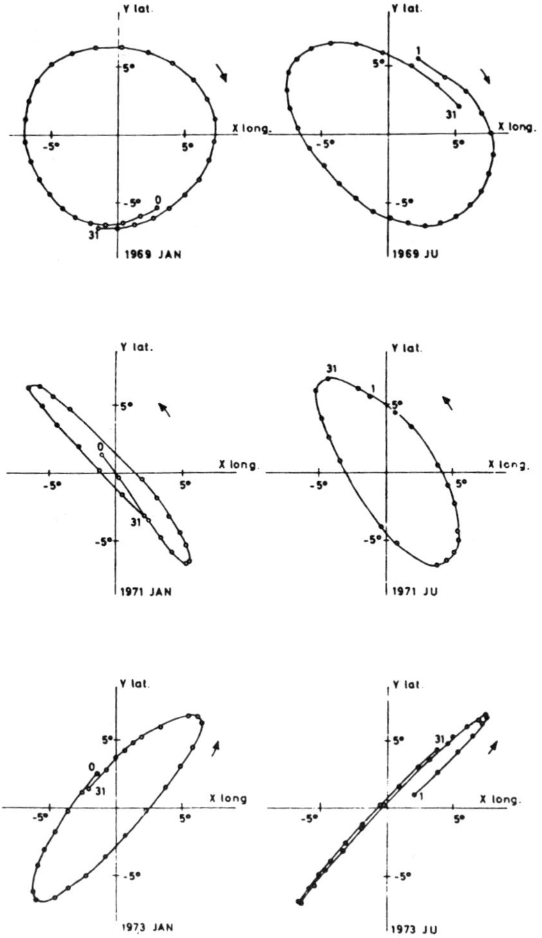

Fig. 16.2 b. Nearly elliptic curves described on the Moon's surface by the point T defining the direction of the Earth as seen from the Moon's centre. The curves are given for january and july of each of the six years 1968 to 1973.

TABLE 16.1. Principal tidal waves on the Moon

	Symbol	Argument number	Expression and amplitude in μgals	Period
	Eo	055.555		Permanent deformation
Zonal elliptic waves	Em	065.455	399 $\cos(\dot{s}-\dot{p})t$	27.55d(§)
$(3\cos^2 z_0 - 1)$	Esm	063.655	73 $\cos(\dot{s}-2\dot{h}-\dot{p})t$	31.81d(§§)
$(3\cos^2\lambda\cos^2\phi - 1)$	Esf	073.555	60 $\cos 2(\dot{s}-\dot{h})t$	14.76d(**)
		075.355	33 $\cos 2(\dot{s}-\dot{p})t$	13.78d(°)
Sectorial librational	Em	065.455	403 $\sin(\dot{s}-\dot{p})t$	27.55d(§)
waves	Esm	063.655	81 $\sin(\dot{s}-2\dot{h}+\dot{p})t$	31.81d(§§)
$\cos^2\phi\sin 2\lambda$	Esf	073.555	43 $\sin 2(\dot{s}-\dot{h})t$	14.76d(**)
Tesseral librational		065.565	424 $\sin(\dot{s}-\dot{N})t$	27.22d(×)
waves		075.465	18 $\sin(2\dot{s}-\dot{p}-\dot{N})t$	
$\sin 2\phi\cos\lambda$				
Second degree terms				
$\sin^2 g - \cos^2\lambda\cos^2\phi$		075.575	25 $\cos 2(\dot{s}-\dot{N})t$	13.61d(××)
Sectorial $(\cos^2\phi\cos 2\lambda)$		075.355	22 $\cos 2(\dot{s}-\dot{p})t$	13.78d(°)
Cross terms, $\cos\lambda\cos^2\phi$		075.355	33 $\sin 2(\dot{s}-\dot{p})t$	13.78d(°)
			35 $\sin(2\dot{s}-\dot{p}-\dot{N})t$	

(§) 27.55d = mean anomalistic month, these waves are elliptic waves of Eo.
(§§) 31.81d , these waves are evection waves of Eo.
(**) 14.76d = half synodic month, these waves are variation waves of Eo.
(°) 13.78d = half anomalistic month.
(×) 27.22d = mean draconitic month.
(××) 13.61d = half draconitic month.

CHAPTER 17

Miscellaneous

17.1. TIDAL TRIGGERING OF EARTHQUAKES

Many papers have been published on this argument. Knopoff considers the correlation as specious (1970) because one should expect that earthquakes would most likely occur when the tidal stresses are at their maximum. However, alignment of the Earth, Sun and Moon, which are advocated by authors claiming for a correlation, can in many cases produce a minimum tide instead of a maximum.

From a statistical point of view the problem seems to be still unresolved and even confused. But there is a strong objection to be raised concerning such a relation: diurnal and semi-diurnal tidal components should have far greater effect than the long period ones (fortnightly, 6 monthly) which are taken as only responsible of the triggering effect.

In a recent paper Heaton (1975) have reinvestigated the matter, taking into account the fault orientation parameters relative to 107 earthquakes grouped according to depth and focal mechanism. His conclusion is that shallow (<30 km) large magnitude (>5) oblique-slip and dip-slip earthquakes are triggered by tidal stresses while no correlation appears for shallow strike-slip earthquakes or for any type of intermediate or deep-focus earthquake.

17.2. TIDAL TRIGGERING ON VOLCANIC ERUPTIONS

Johnston and Mauk (1972) claim that 33 major primary Stromboli eruptions during the past 72 years are correlated with the phase of the fortnightly modulation envelope of the tidal curve (Fig. 17.1).

17.3. FLUCTUATIONS IN GEYSER ACTIVITY

In several papers Rinehart (1972) scrutinizes the variations in interval between the eruptions of three well-known geysers (including the Old Faithful Yellowstone) *vis-à-vis* the phases of the Moon. He concludes that the tidal forces exert a regulatory influence on geyser action. Again, here the long-period components, the fortnightly one and the 18.6-year period, are the only ones which seems evident, while semi-diurnal or diurnal components have no effect. To explain this Rinehart considers that a geothermal area is a very complex viscoelastic system which might filter the higher-frequency components like the semi-diurnal and diurnal tidal components.

This is not the case with aquifers (see Chapter 10). Moreover, each geyser should react in a different way under the solicitations of the tidal stresses.

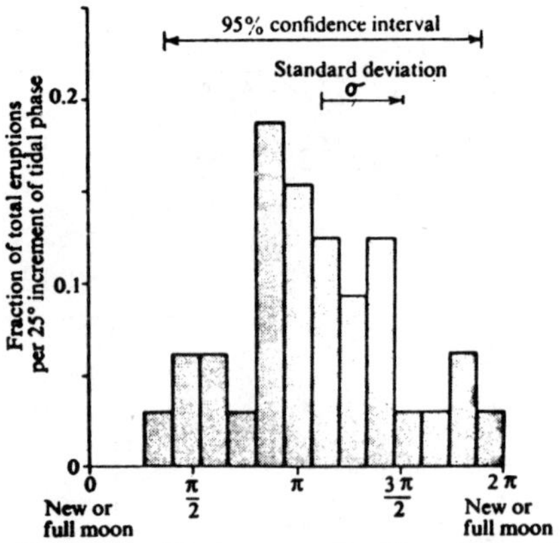

Fig.17.1. Histogram of all major primary eruptions from Stromboli since 1899 as a function 25° increments of fortnightly tidal phase. A 25° phase increment corresponds to approximately a day. The standard deviation from the mean is 2.9 days and the 95% confidence interval about the mean is 5.7 days. (Johnston and Mauk, 1972).

17.4. USING THE TIDAL INSTRUMENTS FOR FORECASTING DISASTER

17.4.1. Earthquake Prediction

According to Beaumont and Berger (1974) the observation of the M_2 tidal admittance of the Earth may offer a possibility of detection of dilatancy. Dilatancy is an inelastic volume increase that a rock undergoes near failure and may be observed by changes in the ratio of seismic velocities V_P/V_S prior to earthquakes.

Beaumont and Berger demonstrate that tilt and strain tidal amplitudes should exhibit large variations (50%) for moderate V_P/V_S changes (15%) and therefore provide sensitive earthquake predictors. Moreover, this technique has some potential advantages over the direct seismic observation of V_P/V_S, mainly because it is aseismic (no earthquake activity is needed to apply it) and continuous.

17.4.2. Forecasting Storm Surges

Zschau (1975) has found a significant cross-correlation between the tilts registered by several borehole pendulums installed near Kiel (Germany) and storm surges in the Atlantic. The important fact is that the tilt change is observed well in advance - 9-12 hours - with respect to the sea-level change. He interprets this in terms of changes of water masses in the sea which produces immediately a tilt while the water-level augments later on the coasts. This may provide a successful method of prediction.

17.5. THE EFFECT OF THE SOLID EARTH TIDE ON THE OCEANIC TIDES

The oceanic tides are governed by the famous Laplace tidal equations. However, in their original form several non-negligible effects are not included: ocean self-attraction, solid Earth tide and tidal loading by the oceanic tidal column itself.

Numerical integrations of these equations in realistically shaped oceans neglect these effects except the recent solution proposed by Hendershott (1972).

With consideration of the three mentionned effects, Hendershott writes the modified Laplace equations as follows:

$$\left.\begin{aligned}
&\frac{\partial u}{\partial t} - (2\omega \sin \phi)v = - g \frac{\partial}{a \cos \phi\, \partial\lambda} (\zeta - \frac{\Gamma}{g}) + \frac{\partial F}{\rho D \partial\lambda}, \\
&\frac{\partial v}{\partial t} + (2\omega \sin \phi)u = - g \frac{\partial}{a \partial\phi} (\zeta - \frac{\Gamma}{g}) + \frac{\partial F}{\rho D \partial\phi}, \\
&\frac{\partial \zeta}{\partial t} - \frac{\partial \xi}{\partial t} + \frac{1}{a \cos \phi} [\frac{\partial}{\partial\lambda} (uD) + \frac{\partial}{\partial\phi} (v.D \cos \phi)] = 0,
\end{aligned}\right\} \quad (17.1)$$

where (u v) are the components of the velocity, ζ and ξ are the geocentric ocean and solid earth tidal elevations, and Γ is the sum of all tidal potentials.

Thus

$$\xi_n = h_n \frac{W_n}{g} + h'_n (\alpha_n \zeta_{on}), \quad (17.2)$$

$$\Gamma_n = (1+k_n) W_n + (1+k'_n) g \alpha_n \zeta_{on}, \quad (17.3)$$

with

$$\zeta_0 = \zeta - \xi, \quad (17.4)$$

$$\alpha_n = \frac{3}{2n+1} \frac{\rho}{\rho_e}, \quad (17.5)$$

where F_λ and F_ϕ are the components of the bottom stress, ρ is the oceanic water density, ρ_e is the mean density of the Earth.

Then

$$\xi_0 = Z + \sum_n (1+k_n-h_n) \frac{W_n}{g} + \sum_n (1+k'_n-h'_n) \alpha_n \xi_{on}. \quad (17.6)$$

A solution by Munk, Snodgrass and Wimbush (1970) retaining the Earth-tide effect but neglecting the ocean self-attraction and tidal loading (that is assuming $\Gamma = (1+k_2)W_2$ and $\xi = h_2 W_2/g$) is used as a first approximation to calculate the neglected terms.

Hendershott's conclusion is that the Earth-tide effect is of the order of the astronomical potential itself while the other two effects are roughly one order of magnitude smaller. Also the existence of an appreciable Earth tide significantly modifies the usual expressions for stored tidal energy and for the rate of working on the oceans by tide-generating bodies which is estimated by Hendershott to be 3.04×10^{19} erg s^{-1} (which yields Q=34).

Omission of all effects due to solid Earth tide makes an error of 30% (as $1+k-h \sim 0.7$).

Finally, let us remark that some problems arise in the Indian Ocean: the first approximation solution gives amplitudes too great by a factor of 2 or 3.

17.5. OBSERVATIONS MADE ON THE OCEAN FLOOR

An instrumental package containing a Lamont feedback-controlled long-period triaxial seismometer, a three-component short-period seismometer, hydrophones and other instruments was installed in 1966 in the Pacific Ocean at a depth of 3903 m and at about 200 km West of San Francisco (Nowroozi, Kuo and Ewing, 1969). The oceanic tides and tidal variation of gravity were recorded from 10 April to 9 June 1967. The data were analysed by the different usual methods (harmonic analysis, convolution and least squares).

The amplitude of the main oceanic waves have been found to be:

O_1 (28 cm); K_1 (43 cm); M_2 (55 cm); S_2 (13 cm); N_2 (13 cm).

The attraction of the variable water mass has been estimated and subtracted from the observed variation of gravity.

Then the following δ factors were found:

	δ observed	δ corrected	Phase corrected
O_1	1.257	1.140	3.67
P_1	1.326	1.248	-1.50
K_1	1.126	1.114	-0.99
M_2	1.545	1.126	5.66

17.6. THE EFFECT OF OCEAN AND EARTH TIDES ON THE SEMI-DIURNAL LUNAR ATMOSPHERIC TIDE

Hollingsworth has shown (1971) that these effects, ignored until now in all studies of atmospheric tides are not trivial when considered as vertical oscillations of the lower boundary of the atmosphere. Using a linear theory he calculated the combined effects of the lunar potential, the Earth elastic tide potential, the oceanic tide potential as well as the lower boundary oscillations on a realistic model atmosphere. His result indicates that the ocean tide has a significant and probably dominant effect.

Forcing the model atmosphere with the lunar potential and the solid Earth tides produces a response which is uniform in longitude and has a maximum amplitude of about 28 μb at the equator. Thus the cotidal lines run North-South, the only amphidromes being at the poles.

The introduction of the additional forcing by the oceanic tides destroys the longitudinal uniformity, and the cotidal lines pattern becomes more complex, presenting a resemblance with the oceanic cotidal lines above the oceans. These results are in good agreement with the observations over Asia, East Africa and South America but not over Europe and North America.

Hollingsworth's calculations of tidal winds at the surface suggest that the hitherto unexplained sense of rotation of the observed tidal winds is an effect of the ocean tide. This work has the merit of showing that solid Earth and ocean tides must be taken into account for an explanation of the atmospheric lunar tide.

Miyahara (1972) also observes that a correction is to be introduced in the barometric lunar tides to take account of the up-and-down movement of the crust.

He estimates the amplitude to be 20-40% larger after correction, and the phase lag significantly reduced as well as its seasonal variation.

17.8. THE MAGNETIC FIELD

In 1971 Pekeris suggested that the magnetic field could be induced by the tide in the Earth's liquid core. The velocity required by the dynamo theory is estimated to be of the order of 0.1-0.01 cm s^{-1}. The M_2 component induces displacements of the order of 20 cm with a period of 12 h 25 m, that is 0.001 cm s^{-1}, which Pekeris considers to be close enough to the lower limit to merit examination.

Such motions could induce a periodic magnetic field having the frequency ω_i of the tide as well as multiple frequencies, including a steady term. The coupling coefficient for the steady term between the convectively inducing and induced fields is estimated by Pekeris to be of the order of $\omega_i H^2/\lambda$, H being the height of the equilibrium tide and $\lambda = 1/4\pi\kappa$ (κ: electrical conductivity of the core). But with $\omega_i = 1.4 \times 10^{-4}$ s^{-1}, H = 20 cm, $\kappa = 3\times10^{-6}$ e.m.u. Pekeris obtains a coupling coefficient of the order of 10^{-6} only as against unity in the case of the dynamo theory.

17.9. ABSORPTION OF GRAVITY

Majorana had inferred from the general relativity that some absorption of the gravity could appear during eclipses, for example. The estimated absorption coefficient could be of about 10^{-14} c.g.s. and therefore extremely difficult to measure.

Nevertheless, Tomaschek made some unsuccessful tentatives with horizontal pendulums. Later on registrations of high precision were made with a LaCoste Romberg tidal gravitimeter at Firenze (Italy) during the totality of the 1961 eclipse but no observable effect was detected.(Slichter et al. 1965).

It should be noted that if such an effect was existing within the limits of the present accuracy of our instruments, the semi-diurnal tidal should be modified because of the absorbing effect of the mass of the Earth itself and this should produce a M_1 component (as well as S_1, N_1, etc.).

17.10. THE INTERNATIONAL CENTRE FOR EARTH TIDES (ICET)

When it was decided to include within the framework of the IGY the study of Earth tides to the programme of discipline XIII, a World Data Centre (C) was established at the Royal Observatory of Belgium under the direction of P. Melchior. The activity of this Centre was not confined to the collection of data but developed in various directions: study of computation techniques, introduction of electronic computation (the use of computers for calculating the reduction of Earth tides was first introduced by the Brussels Centre in 1959), and publication of an *Information Bulletin*. In 1959 the International Association of Geodesy formed a Permanent Commission presided over by the late Prof. R. Tomaschek until 1963 and presently by Prof. R. Lecolazet. The activity of the Commission is supported by an International Centre established at Brussels.

Since 1958, ICET has performed an enormous amount of tidal analysis with different computers, starting with the IBM 650, continuing with IBM 1620, 7040, 360/44 and using presently Unidata 7440.

At the very beginning the manual methods of Doodson, Pertsev and Lecolazet were programmed, but in 1966 least-squares methods, with specially devised filters at ICET by Venedikov (see Section 7.3.2) and by Usandivaras-Ducarme (see Section 7.3.3), were employed for using all the possibilities of the computers. A chain of seven programmes, starting with the semi-automatic digitization of the curves and terminating with the edition of a report of results (Table 17.1), has been realized by Ducarme.

From the introduction of the code number of a station and the series number of the instrument considered to the automatic edition of the report as shown by Table 17.1, no manual intervention is possible which could alter any result. The computer takes all the information needed from hourly and calibration readings (which are on punched cards or magnetic tape) and from a library of stations stored on a disk pack for a station's description with position and constants.

A great number of stations from many different countries send their data to ICET to be analysed in this way. On the other hand, ICET has made its programmes available to any institution which requests them.

Other important activities of ICET are:

(a) The systematic translation and publication into French of all articles published in Russian (220 papers; 2298 pages).

(b) Editing of a *Bulletin of Informations* (BIM) which covers 4407 pages in 75 numbers since 1957 (700 copies printed).

(c) Keep a library of Earth-tide data (punched cards and magnetic tapes) from all stations which agree to make their measurements known and available to all interested scientists. This is a function of the World Data Centre.

(d) Keep in order a complete bibliography classified according to a special decimal coding. This being on punched cards is reproduced in the present book and contains more than 2000 references (Table 17.2).

(e) Instructs researchers in the utilization of instruments, calibration procedure and computation techniques. A great many of them have visited ICET since 1958.

(f) Organize international symposia on Earth tides: Brussels (1957), Munich (1958), Trieste (1959), Brussels (1961), Brussels (1964), Strasbourg (1969), Sopron (1973), Bonn (1977).

Table 17.1

Trans World Profile Asia Station Kathmandu
Station 2450 Kathmandu-Kirtipur Vertical Component Nepal
27 40 N 85 18 E H 1350 M P 0 M D 724 KM
Pillar at Ground level
Tribhuvan University, Inst. of Sc., Kirtipur Campus
Department of Physics Dr. K.L. Shrestha
Gravimeter Geodynamics 084 P. Melchior Trans World Profiles
Calibration Bruxelles - Fundamental Station
Installation B. Ducarme/P. Melchior
Maintenance L.P. Gewali
Grant AFOSR-73-2557 A Project-Task 8607-02
Method of least Squares/Venedikov Filters/ Hourly readings
Potential Cartwright Tayler Edden/ Full development
Correction for inertia proportional to the square of speeds
Computation International Centre for Earth Tides/FAGS/Brussels
Computer Siemens 7440/System BS 1000 Standard 77/2/23
Analysis Program B. Ducarme
Scale Factor 1.18255
Phase Lag O1 0.73 M2 0.76 O1/M2 0.96
Instrumental Delay 1028.60 Min
Differential Attenuation Correction M2/O1 1.00976 / Model 2 /

GEO	084	74 10 1	74 10 1	74 10 16	74 10 18	74 10 29	74 10 29
GEO	084	74 11 5	74 11 19	74 11 30	74 12 2	74 12 6	74 12 6
GEO	084	74 12 16	74 12 16	74 12 27	74 12 27	75 1 4	75 1 4
GEO	084	75 1 8	75 1 10	75 1 16	75 1 16	75 1 25	75 1 29
GEO	084	75 2 2	75 2 6	75 2 11	75 2 13	75 2 17	75 2 25
GEO	084	75 3 9	75 3 13				

Total number of days 74 1776 hourly readings

Group	Symbol	Amplitude Central	Phase Epoch	Ampl. Fact.	MSE	Phase Lag	MSE	Mean Amplitude
1- 62	Q1	4.9704	39.38	1.1225	0.0261	1.39	1.33	5.3459
63- 88	O1	27.8159	254.85	1.1278	0.0051	0.07	0.26	27.3157
89-110	M1	2.3542	246.86	1.1073	0.0494	-2.90	2.54	2.7470
111-143	P1S1K1	53.1935	92.65	1.1221	0.0033	0.68	0.17	41.4107
144-165	J1	1.9643	334.91	1.0499	0.0619	9.51	3.38	2.0678
166-197	OO1	0.8105	138.50	1.1844	0.1139	6.06	5.45	1.2838
M.S. Error		D	3.2051					
198-236	2N2	2.1324	46.56	1.1053	0.0521	0.50	2.70	2.4905
237-260	N2	12.9479	302.94	1.1461	0.0110	-0.56	0.54	13.6985
261-286	M2	68.2226	172.23	1.1431	0.0021	-0.22	0.11	68.2468
287-300	L2	1.2581	215.63	1.2011	0.0953	-0.25	4.63	1.5229
301-347	S2K2	25.5856	347.39	1.1479	0.0043	0.22	0.22	33.0228
M.S. Error		SD	3.3033					

O1/K1 1.0050 1-O1/1-K1 1.0464 M2/O1 1.0136

348-363	M3	1.0203	182.49	1.0387	0.0349	-0.09	1.91	1.1302
M.S. Error		TD	0.9557					

Reference Epoch TJJ = 2442403.0

Table 17.2. General Bibliography of Earth Tides

Number of published papers

Before the IGY		Since the IGY			
epoch	n	year	n	year	n
1800 - 49	3	1958	69	1970	188
1850 - 74	7	1959	39	1971	143
1875 - 99	64	1960	96	1972	114
1900 - 09	16	1961	91	1973	246
1910 - 19	46	1962	53	1974	114
1920 - 24	20	1963	91	1975	116
1925 - 29	33	1964	121	1976	195
1930 - 34	54	1965	68	1977	240
1935 - 39	30	1966	98	1978	195
1940 - 44	27	1967	78	1979	179
1945 - 49	27	1968	71	1980	109
1950 - 54	129	1969	86	1981	(95)
1955 - 57	116				
			961		1934
	572				
Total	3467				

Bibliography

HOW TO USE THE BIBLIOGRAPHY

This Bibliography probably covers all papers published on the subject of Earth Tides and related topics. In 1958 the International Commission for Earth Tides established a decimal classification which has been used to prepare this bibliography and is reproduced here below. This evidently does not correspond to the sequence of the chapters in this book, but we hope that the reader will have no difficulty in identifying any reference which is given as usual in date order under each heading.

Chapter	Bibliographic code
1	2
2	6.45
3	2.2
4	6.1/6.2
5	6.3
6	6.4
7	4
8	3.1
9	3.2
10	3.4/3.5
11	5.1
12	5.4/5.6
13	Part II: Results
14	3.3/6.7/6.8/7.1
15	6.82
16	9.0
17	1.2/7.4/8

ICET Decimal Classification

1. GENERAL

1.2 International organisations

1.21 General reports at the International Association of Geodesy (IAG). Regulations and Recommendations.
1.22 National reports at IAG
1.23 IAG Permanent Commission for Earth Tides
International Centre for Earth Tides (ICET)
1.24 Earth tide stations

1.3 Treatises, general articles

1.4 Documentation

1.5 Biographies

2. TIDAL THEORY

2.1 Tides for a rigid earth

- 2.11 Lunisolar potential
- 2.12 Astronomical constants
- 2.15 The three kinds of tides, geographical distribution
- 2.16 Development of the potential
- 2.17 Tables, nomograms

2.1 Tides for an elastic Earth

- 2.211 Love numbers
- 2.212 Variations of gravity
- 2.213 Variations of the direction of the vertical with respect to the crust
- 2.214 Variations of the direction of the vertical with respect to the stars
- 2.215 Deformation tensor
- 2.216 Cubic dilatations

3. OBSERVATION OF EARTH TIDES, INSTRUMENTS, RESULTS

3.1 Oscillations of the vertical with respect to the crust

- 3.11 Long period tides in the oceans
- 3.12 Lake tides
- 3.13 Tides in water tubes
  - 3.131 Interferential method
  - 3.132 Niveauvariometer
  - 3.133 Mercury tiltmeters
- 3.14 Horizontal pendulum
  - 3.141 Zöllner suspension
    - 3.141.1 Verbaandert-Melchior pendulums
    - 3.141.2 Ostrovsky pendulums
  - 3.142 Von Rebeur Paschwitz suspension
  - 3.143 Milne suspension
  - 3.144 Lettau suspension
- 3.15 Special pendulums, vertical pendulums
- 3.16 Calibration devices
- 3.17 Instrumental constants
  - 3.171 Measurements of constants
  - 3.175 Damping
- 3.19 Levelling

3.2 Variations of the gravity

- 3.21 Comparison of pendulum clocks with quartz or molecular clocks
- 3.22 Gravimeters
  - 3.221.1 Worden
  - 3.221.2 North American, Western
  - 3.221.3 LaCoste Romberg
  - 3.221.4 Geodynamics

3.221.5 Askania
3.221.6 Miscellaneous
3.221.7 Visual measurements
3.225 Tidal gravity profiles
3.23 Calibration of the gravimeters
3.24 Recording systems
3.245 Automatisation
3.25 Calibration of recorders, hysteresis, lags

3.3 Oscillation of the vertical with respect to the stars

3.31 Zenith telescopes, classical instruments
3.32 Photographic zenith tube
3.33 Astrolabes

3.4 Deformation tensor

3.41 Bar extensometers
3.42 Wire extensometers
3.43 Interferential method
3.44 Electronic methods
3.45 Laser
3.46 Vertical extensometers
3.47 Rotational

3.5 Cubic dilatations

3.51 Experimental dilatometers
3.52 Wells, borings
3.53 Volcanoes, geysers, gases

4. METHODS OF ANALYSIS OF THE TIDAL OBSERVATIONS

4.1 Elimination of the drift

4.15 Interpolation methods

4.2 Meteorological or hydrological perturbations

4.3 Harmonic analysis

4.31 Doodson's method
4.312 Doodson's-Lennon's method
4.32 Lecolazet's method
4.33 Pertsev's method
4.34 Least squares method
4.341 Venedikov's method
4.342 Matveev's method
4.343 Chojnicki's method
4.344 Usandivaras Ducarme's method
4.35 Older methods
4.38 Power Spectrum
4.39 Other methods

4.5 Methods of computation with electronic computers-programs

4.6 Effect of the accidental errors on the results of harmonic analysis

5. DISCUSSION OF RESULTS

5.1 Practical separation of direct and indirect effects

5.11 Boussinesq method
5.12 Longman's and Farrell's procedures
5.13 Corkan's empirical method
5.15 Cotidal maps
5.18 Atmospheric tides

5.4 Discussion of the local heterogeneities

5.40 Topographic effect
5.41 Cavity effect
5.42 Geological or tectonic effect
5.45 Hydrological effects

5.6 Secondary effects not having tidal frequencies

5.61 Meteorological effects
5.611 Temperature periodic variations
5.612 Pressure periodic variations
5.613 Non-periodic temperature variations
5.614 Non-periodic pressure variations

5.7 Instrumental drift, secular effects

6. RHEOLOGY

6.1 Deformations of an homogeneous Earth

6.2 Deformations of a non-homogeneous Earth

6.3 Earth's models

6.31 Ratio k/h

6.4 Dynamic effects of the liquid core

6.41 Core's dynamic
6.42 Results of observations
6.43 Core's constitution
6.45 Astronomical nutations
6.46 Free diurnal nutation

6.5 Free oscillations

6.7 Viscosity effects

6.8 Tidal friction

6.81 Secular retardation of the Earth's rotation
6.82 Effect of the satellite orbits
6.85 Oceanic friction
6.88 The orbit of the Moon

6.9 Heat Flow

* means a translation from the russian.

1.21
HECKER
SCHWEYDAR
LALLEMAND
HAID
RAPPORTS SUR LES MAREES TERRESTRES.
C.R.SEANCES 17E CONF.GEN.ASS.GEOD.INTER.PP52-54 HAMBOURG 1912

1.21
LAMBERT W.D.
RAPPORT PRELIMINAIRE SUR LES MAREES DE L ECORCE TERRESTRE.
TRAV.SECT.GEOD.UGGI,RAPP.GEN.2E ASS.GEN.TOME 4 MADRID 1924

1.21
FICHOT E.
RAPPORT DE LA COMMISSION MIXTE DES MAREES.
ASS.INT.GEOD.2E ASS.U.G.G.I. BULL.GEOD. 7 PP 323-326 1925

1.21
LAMBERT W.D.
RAPPORT SUR LES MAREES DE L ECORCE TERRESTRE.
TRAV.ASS.INT.GEOD.,RAPP.GEN.3E ASS.GEN.UGGI TOME 6 PRAG. 1926

1.21
LAMBERT W.D.
RAPPORT SUR LES MAREES DE L ECORCE TERRESTRE.
TRAV.ASS.INT.GEOD.RAPP.GEN.4E ASS.GEN.UGGI TOME 8 STOCK. 1930

1.21
LAMBERT W.D.
RAPPORT SUR LES MAREES DE L ECORCE TERRESTRE.
TRAV.ASS.INT.GEOD.UGGI RAPP.GEN.5E ASS.GEN.LISBONNE 1933

1.21
LAMBERT W.D.
RAPPORT SUR LES MAREES DE L ECORCE TERRESTRE.
TRAV.ASS.INT.GEOD.UGGI,RAPP.GEN.6E ASS.EDIMBOURG TOME 14 1936

1.21
LAMBERT W.D.
RAPPORT SUR LES MAREES DE L ECORCE TERRESTRE.
TRAV.ASS.INT.GEOD.UGGI,RAPP.GEN.7E ASS.GEN.WASH.1939
U.S.COAST GEODETIC SURV.,SP.PUBL.NO 223 PP 1-22 1940

1.21
N.
MAREES DE L ECORCE TERRESTRE
BULL.GEOD. NO 65-68 PP 140 1940

1. 21
LAMBERT W.D.
REPORT ON EARTH TIDES.
TRAV.ASS.INT.GEOD.UGGI,RAPP.GEN.8E ASS.OSLO PP 383-401 1948

1.21
LAMBERT W.D.
DARLING F.W.
RAPPORT GENERAL NO 10 SUR LES MAREES DE L ECORCE TERRESTRE.
TRAV.ASS.INT.GEOD.UGGI.RAPP.GEN.9E ASS.BXL.TOME 17 1951

1.21
LAMBERT W.D.
RAPPORT GENERAL NO 10 SUR LES MAREES DE L ECORCE TERRESTRE.
TRAV.ASS.INT.GEOD.UGGI.RAPP.GEN.10E ASS.ROME TOME 19 1954

1.21
MELCHIOR P.
RAPPORT SUR LES MAREES TERRESTRES.
BULL.GEODESIQUE NO 46 PP 28-52
OBS.ROY.BELG.COMM.NO 139 S.GEOPH. NO 46 1957

1.21
MELCHIOR P.J.
RAPPORT SUR LES MAREES TERRESTRES 1957-1960.
B.I.M.NO 20 PP 328-366 1960

1.21
MELCHIOR P.
RAPPORT SUR LES MAREES TERRESTRES. 1961-1963
B.I.M. N.33 PP 987-1012 1963

1.21
MELCHIOR P.
RESOLUTIONS PRISES A L ASSEMBLEE DE BERKELEY /AOUT 1963
B.I.M. N.34 P 1035 196

1.21
MELCHIOR P.J.
RAPPORT SUR LES MAREES TERRESTRES 1964-1967.
TRAV.ASSOC.INTERN.GEOD.23 PP 367-375 196

1.21
MELCHIOR P.
RAPPORT SUR LES MAREES TERRESTRES 1967-1971
XV ASS.GEN. UGGI - ASS.INT.DE GEODESIE MOSCOU AOUT 197

1.21
MELCHIOR P.
GENERAL REPORT ON EARTH TIDES /1972-1974/
XVI ASS.GEN.DE L UGGI GRENOBLE - AOUT 197

1.21
MORGAN P.
RESUME OF COMMISSION NO 5, MAREES TERRESTRES /EARTH TIDES/ SECTION V
XVI ASS.GEN.DE L UGGI GRENOBLE-AOUT 1975
B.I.M. NO 72 PP 4182-4185 19

1.22
N. COMMISSION POUR L ETUDE DES MAREES TERRESTRES /GROUPE XIII-GRAVIMETRIE DU COMITE SPECIAL POUR L ANNEE GEOPHYSIQUE INTERNATIONALE/.
RAPPORTS ET RECOMMANDATIONS / PARIS 7/9/1956
OBS.ROY.BELG.COMM.NO 100 S.GEOPH.NO 36 PP 32 1956

1.22
LAMBERT W.D. EARTH TIDES IN THE PROGRAM OF THE INTERNATIONAL GEOPHYSICAL YEAR.
OBS.ROY.BELG.COMM.NO 100 S.GEOPH. NO 36 1956

1.22
LAMBERT W.D.
BOULANGER J.D.
MELCHIOR P.J. PROGRAMME DES OBSERVATIONS DE MAREES TERRESTRES AU COURS DE L ANNEE GEOPHYSIQUE - PROGRAM OF OBSERVATIONS OF TERRESTRIAL TIDES DURING THE INTERNATIONAL GEOPHYSICAL YEAR.
OBS.ROY.BELG.COMM.NO 100 S.GEOPH. NO 36 1956

1.22
MELCHIOR P.J. QUELQUES COMMENTAIRES AU SUJET DU PROGRAMME D OBSERVATIONS.
OBS.ROY.BELG.COMM.NO 100 S.GEOPH. NO 36 1956

1.22
LENNON G.W. REPORT ON WORK AT BIDSTON DURING THE I.G.Y.
OBS.ROY.BELG.COMM.NO 142 S.GEOPH. NO 47 PP 73-81 1958

1.22
N. RAPPORT SUR LES TRAVAUX EFFECTUES EN TCHECOSLOVAQUIE PENDANT LES SIX PREMIERS MOIS DE L ANNEE GEOPHYSIQUE INTERNATIONALE.
B.I.M. NO 11 PP 190 1958

1.22
HARRISON J.C. A NOTE ON THE PAPER -EARTH TIDE OBSERVATIONS MADE DURING THE INTERNATIONAL GEOPHYSICAL YEAR-.
J.GEOPH.RES.68 5 PP 1517-1518 1963

1.22
LECOLAZET R. ETUDE DES MAREES TERRESTRES
RAPPORT NATIONAL SUR LES TRAVAUX FRANCAIS EXECUTES DE 1971 A 1974 - GEODESIE PP 10-11
COM.NAT.FRANCAIS DE GEODESIE ET GEOPHYSIQUE
XVI ASSEMBLEE GENERALE UGGI - GRENOBLE 1975

1.22
VALLIANT H.D. EARTH TIDES AND TILT METERS
NAT.REP.ON GRAV.RESEARCH IN CANADA 1971-1974 P 14
XVI ASSEMBLEE GENERALE UGGI - GRENOBLE 1975

1.22
X LES MAREES TERRESTRES
RAPP.TRAVAUX GEOD.EXECUTES EN POLOGNE,1971-1975 P 27-28
ACAD.POL.SC. - COMITE DE GEODESIE
XVI ASSEMBLEE GENERALE UGGI - GRENOBLE 1975

1.22
MARUSSI A. EARTH TIDES
ITALIAN NATIONAL REP.TO IAG PP 11-13
CONSIGLIO NAZIONALE DELLE RICERCHE - ROMA
XVI ASSEMBLEE GENERALE UGGI - GRENOBLE 1975

1.22
BONATZ M. ERDGEZEITEN
LANDESBERICHT DER BUNDESREP.DEUTSCHLAND UBER DIE IN DEN JAHREN 1971 BIS 1974 AUSGEFUHRTEN ARBEITEN PP 84-93
DEUTSCHE GEOD.KOM.REIHE B, H 212 MUNCHEN 1975
XVI ASSEMBLEE GENERALE UGGI - GRENOBLE 1975

RED DE ESTACIONES GEODINAMICAS DE MAREAS TERRESTRES
INFORME SOBRE TRABAJOS GEODESICOS 1971-1975 P 26
ESPANA - COM.NAC.GEODESIA.Y GEOFISICA
XVI ASSEMBLEE GENERALE UGGI - GRENOBLE 1975

1.22
X EARTH TIDES
GEODESY IN AUSTRALIA, NAT.REP.FOR 1971-74 PP 35-36
NAT.COM.GEOD.AND GEOPH. - AUSTRALIAN AC.OF SC.
XVI ASSEMBLEE GENERALE UGGI - GRENOBLE 1975

1.22
X — THE EARTH TIDE MEASUREMENT
REPORT ON THE GEODETIC WORK 1970-1974
THE ROYAL THAI SURVEY DEPT. PP 9-10
XVI ASSEMBLEE GENERALE UGGI - GRENOBLE 1975

1.22
X — MAREES TERRESTRES
GEODESIE - RAP.NAT.DE LA BELGIQUE 1972-1975 PP 16-21
INST.GEOGR.MIL.BRUXELLES
XVI ASSEMBLEE GENERALE UGGI - GRENOBLE 1975

1.22
X — EARTH TIDES
UNITED KINGDOM GEODESY REPORT 1971-1975 PP 42-44
THE ROYAL SOCIETY, LONDON 1975
XVI ASSEMBLEE GENERALE UGGI - GRENOBLE 1975

1.22
X — EARTH TIDES
REPORT GEODETIC WORK FOR THE PERIOD 1971-1974 P 9
NORGES GEOGRAFISKE OPPMALING
XVI ASSEMBLEE GENERALE UGGI - GRENOBLE 1975

1.22
X — EARTH-TIDE RESEARCH
GERMAN DEMOCRATIC REP.NAT.REPORT IAG PP 25-26
NAT.KOM.GEOD.GEOPH.AKAD.WISS. D.D.R.
XVI ASSEMBLEE GENERALE UGGI - GRENOBLE 1975

1.22
KUO J.T. — EARTH TIDES
UNITED STATES NAT.REPORT FOR GEODESY 1971-1974
REVIEWS OF GEOPH.& SPACE PHYSICS V.13 NO 3 PP 260-263
XVI ASSEMBLEE GENERALE UGGI - GRENOBLE 1975

1.22
HONKASALO T. — EARTH TIDES
GEODETIC OPERATIONS IN FINLAND 1971-1974 PP 17-19
XVI ASSEMBLEE GENERALE UGGI - GRENOBLE 1975

1.22
KUO J.T. — EARTH TIDES
U.S.NAT.REPORT 1971-1974 TO THE 16TH GEN.ASS.GRENOBLE
REVIEWS OF GEOPH.& SPACE PHYS.V.13 NO 3 PP 260-263 197

1.22
N. — 3EME RAPPORT DU SOUS-COMITE CANADIEN DE LA GEODYNAMIQUE
COM.NAT.CANADIEN DE U.I.G.G. 25 PAGES MARS 197

1.23
BAARS B.
JEFFREYS H.
TOMASCHEK R.
CORKAN — EARTH TIDES /GEOPHYSICAL DISCUSSION/.
THE OBSERVATORY 72 NO 866 PP 16-21 195

1.23
N. — COLLOQUE INTERNATIONAL SUR LES MAREES TERRESTRES,
PREPARATOIRE AUX TRAVAUX DE L ANNEE GEOPHYSIQUE
INTERNATIONALE /UCCLE 24-26 AVRIL 1957/
OBS.ROY.BELG.COMM.NO 114 S.GEOPH. NO 39 195

1.23
N. — DEUXIEME COLLOQUE INTERNATIONAL DE LA COMMISSION
DU C.S.A.G.I. POUR L ETUDE DES MAREES TERRESTRES
/MUNICH 21-26 JUILLET 1958/.
OBS.ROY.BELG.COMM.NO 142 S.GEOPH.NO 47 195

N. — CREATION D UN SERVICE PERMANENT DES MAREES TERRESTRES.
CHRON. U.G.G.I. NO 23 SEPTEMBRE PP 260-261 19

1.23
HIERSEMANN L. — 3 INTERNATIONALES SYMPOSIUM UBER ERDGEZEITEN IN
TRIEST/ITALIEN.
BERGAKADEMIE 9-10 PP 624-627 19

1.23
MELCHIOR P. — RAPPORT DU DIRECTEUR DU CENTRE INTERNATIONAL
DES MAREES TERRESTRES.
3E SYMP. MAREES TERRESTRES TRIESTE
BOLL.GEOFISICA TEORICA E APPL.VOL.2 NO 5 PP 14-16 19

1.23
MELCHIOR P.J. COMPTE RENDU DES REUNIONS DE LA COMMISSION PERMANENTE DES MAREES TERRESTRES / L ASSEMBLEE GENERALE D HELSINKI.
B.I.M.NO 21 PP 368-370 1960

1.23
MELCHIOR P.J. RAPPORT SUR L ACTIVITE DU CENTRE INTERNATIONAL DES MAREES TERRESTRES A UCCLE.
IVME SYMP.INTERNATIONAL SUR LES MAREES TERRESTRES.
OBS.ROY.BELG.COMM.NO 188 S.GEOPH.NO 58 PP 280-292 1961

1.23
SLICHTER L.B. EARTH TIDE RESEARCH IN THE UNITED STATES.
13 GEN.ASSEMBL.INT.UNION GEOD.GEOPH.BERKELEY PP 350-351 1963

1.23
MELCHIOR P. 5 EME SYMP.INT.SUR LES MAREES TERRESTRES.
BULL. GEOD. PARIS 73 PP 189-193 1964

1.23
N. NOUVELLES
B.I.M. N.38 P 1353 1964

1.23
BONATZ M. SYMPOSIUM UBER GEZEITEN DES FESTEN ERDE
XV GENERALVERSAMMLUNG I.U.G.G. JULI-AUGUST 1971
SONDERHEFTE ZEITSCH.F.VERMESSUNGSW. H.15 PP 36-39 1972

1.23
7E SYMP.INTERN.MAREES TERRESTRES
BULL.GEODESIQUE, PP 137-143, NO 112, JUIN 1974

1.23
SCHNEIDER M.M. INTERNATIONALES SYMPOSIUM UBER ERDGEZEITEN /SOPRON SEPTEMBER 1973/
VERMESSUNGSTECHNIK, 22.JG.,HEFT 3, PP 110-111 1974

1.23
MELCHIOR P. EARTH TIDES IN 1974 / A REPORT FROM THE INTERNATIONAL CENTRE FOR EARTH TIDES/
B.I.M. NO 70 PP 3982 - 4000 1975

1.24
CARNERA L. LA STAZIONE DEI PENDOLI ORIZZONTALI NELLE R.R.GROTTE DI POSTUMIA
RIC.SCI.4 T.2 NO 9 PP 344-353 1933

1.24
ELLENBERGER H. L ORGANISATION DE STATIONS DE MAREES TERRESTRES POUR LES MESURES DURANT L AGI
B.I.M.NO 3 PP 51-55 1957

1.24
N. L OBSERVATOIRE GRAVIMETRIQUE DE L ACADEMIE DES SCIENCES DE LA RSR D UKRAINE A POLTAVA 1957

1.24
SLICHTER L.B. INSTITUTE OF GEOPHYSICS EARTH-TIDE PROGRAM FOR THE INTERNATIONAL GEOPHYSICAL YEAR.
OBS.ROY.BELG.COMM.NO 142 S.GEOPH.NO 47 PP 68-69 1958

1.24
WITKOWSKI J. LE SERVICE DES MAREES TERRESTRES EN POLOGNE.
OBS.ROY.BELG.COMM.NO 142 S.GEOPH.NO 47 PP 94-95 1958

1.24
N. LISTE DES STATIONS DE MAREES TERRESTRES.
I. STATIONS GRAVIMETRIQUES
II. STATIONS CLINOMETRIQUES
CENTRE INTERN.MAREES TERR. UCCLE, 2FASC. JUILLET 1959

1.24
NORINELLI A. UNE NOUVELLE STATION POUR L ETUDE DE LA MAREE TERRESTRE.
IVME SYMPOSIUM INTERNATIONAL SUR LES MAREES TERRESTRES.
OBS.ROY.BELG.COMM.NO 188 S.GEOPH.NO 58 PP 163-164 1961

1.24
AMORIM FERREIRA E. INSTALLATION DE STATIONS DE MAREES TERRESTRES AUX ACORES ET A TIMOR. IVME SYMPOSIUM INTERNATIONAL SUR LES MAREES TERRESTRES.
OBS.ROY.BELG.COMM.NO 188 S.GEOPH.NO 58 PP 169-170 1961

1.24
AKSENTIEVA Z.N. SUR LE TRAVAIL DE L OBSERVATOIRE GRAVIMETRIQUE DE POLTAVA DE L ACADEMIE DES SCIENCES D U.R.S.S.

RESULT.PRELIM.RECH.VARIATIONS LATIT.ET MOUV.POLE TERR.
ACAD.SC.URSS.COMITE AGI MOSCOU PP 30-35 1961

1.24
BYL J. DIE NEUE EINRICHTUNG DER STATION FUR
ERDGEZEITENMESSUNGEN IN POTSDAM.
MONATSBER.DEUTSCHE AKAD.WISS.BERLIN 4.9 PP 577-585 1962

1.24
DOBROKHOTOV Y.S. STATIONS GRAVIMETRIQUES ET CLINOMETRIQUES POUR
OSTROVSKII A.E. L OBSERVATION DES MAREES TERRESTRES
PERTSEV B.P. AGI 1957-58-59 AKAD.NAOUK CCCP MOSCOU 1961
* B.I.M. NO 30 PP 743-747 1962

1.24
HARRISON J.C. EARTH-TIDE OBSERVATIONS MADE DURING THE INTERNA-
NESS N.F. TIONAL GEOPHYSICAL YEAR.
LONGMAN I.M. INST.GEOPH. UNIVERS. CALIFORNIA LOS ANGELES
FORBES R.F.S.
KRAUT E.A.
SLICHTER L.B. JOURN. GEOPH. RESEARCH V.68 N.35 PP 1497-1516 1963

1.24
JOBERT G. RAPPORT SUR L ACTIVITE DE L INSTITUT DE PHYSIQUE
DU GLOBE DE PARIS DANS LE DOMAINE DES MAREES TERRESTRES
ASS.BERKELEY UGGI B.I.M. N.34 PP 1036-1040 1963

1.24
PARIISKY N.N. EARTH-TIDES STUDIES IN THE USSR IN 1960-1963
ASS.BERKELEY UGGI B.I.M. N.34 PP 1043-1046 1963

1.24
OSTROVSKII A.E. INCLINAISONS DE MAREES D APRES LES OBSERVATIONS
PICHA J. AVEC LE CLINOMETRE PHOTO-ELECTRIQUE A PRIBRAM/PRAGUE/
SKALSKI L. RECH. MAREES TERR. ART.XIII SECT. PROGR. IGY.N.3 1963
MIRONOVA L.J.
WITMAN N.G. * B.I.M. N.35 PP 1167-1179 1964
1.24

1.24
SLICHTER L.B. REPORT OF EARTH TIDES RESULTS AND OF OTHER
MAC DONALD G.J.F. GRAVITY OBSERVATIONS AT UCLA.
HAGER C.L. 5 EME SYMP.INT.SUR LES MAREES TERRESTRES.
CAPUTO M. OBS. ROYAL BELGIQUE COMM. N.236 S.GEOPH.N.69 PP 124-130 1964

1.24
BLUM P.A. RAPPORT SUR L ACTIVITE DE L INSTITUT DE PHYSIQUE
CORPEL J. DU GLOBE DE PARIS DANS LE DOMAINE DES MAREES
GAULON R. TERRESTRES.
NIERES J.P. 5 EME SYMP.INT.SUR LES MAREES TERRESTRES.
JOBERT G. OBS. ROYAL BELGIQUE COMM. N.236 S.GEOPH.N.69 PP 172-177 1964

1.24
BYL J. BERICHT UBER DIE BISHER AM GEODATISCHEN INSTITUT
POTSDAM AUSGEFUHRTEN GEZEITENBEOBACHTUNGEN.
5 EME SYMP.INT.SUR LES MAREES TERRESTRES.
OBS.R.BELGIQUE COMM. N.236 S.GEOPH. N.69 PP 255-261 1964

1.24
BOSSOLASCO M. LA STAZIONE GRAVIMETRICA E GEOMAGNETICA DI ROBURENT
CANEVA A. /PROV.DI CUNEO/.
CICCONI G.
DAGNINO I
ELENA A.
EVA C. PUBL.IST.GEOF.GEOD.UNIV.GENOVA NO 152 196

1.24
N. INSTRUCTIONS FOR USE OBSERVATIONAL INSTRUMENTS.
DISASTER PREV.RES.INST.KYOTO UNIV.PP 1-48 196

1.24
BALAVADZE B.K. OBSERVATIONS DES PHENOMENES DE MAREES A TBILISSI.
KARTVELICHVILI K.Z. ROTATION ET DEFORM.MAREES DE LA TERRE,FASC.1
ACAD.SC.UKR.OBS.GRAV.POLTAVA KIEV PP 174-180 197

1.24
SIMON D. AUFBAU EINER NEUEN ERDGEZEITENSTATION IM SALZBERGWERK
TIEFENORT
VERMESSUNGSTECHNIK SOUS-PRESSE 197

1.24
BONATZ M. ERDGEZEITENREGISTRIERUNGEN IN DER ARKTIS

MELCHIOR P. INTERNATIONAL ASTRO-GEO-PROJECT SPITZBERGEN 1968/70
ZEITSCHRIFT FUR VERMESSUNGSWESEN HEFT 7 PP 305-309 1971
1.24
BALAVADZE B.K. OBSERVATIONS DES PHENOMENES DE MAREES A TBILISSI.
KARTVELICHVILI K.Z. ROTATION ET DEFORM.MAREES DE LA TERRE,FASC.1
ACAD.SC.UKR.OBS.GRAV.POLTAVA KIEV PP 174-180 1970
* B.I.M. N.62 PP 3232-3237 1971
1.24
SCHNEIDER M.M. BERICHT UBER DIE UBERWINTERUNG AN DER STATION WOSTOK
WAHREND DER 14 SOWJETISCHEN ANTARKTISEXPEDITION 1968-1970
GEOD GEOPH. VEROFF. R.III H.23 PP 5-32 1971
1.24
GREEN R. THE COONEY GEOPHYSICAL OBSERVATORY
SYDENHAM P.H. THE AUSTRALIAN PHYSICIST - V.8 N.11 PP 167-172 1971
1.24
KARTVELICHVILI K.Z. ETUDE DES MAREES A TBILISSI PAR OBSERVATIONS
GRAVIMETRIQUES ET EXTENSOMETRIQUES.
UNIVERS.TBILISSI PP 1-11 1965
* B.I.M. NO 63 PP 3293-3296 1972
1.24
GREEN R. FINE MEASUREMENT IN GEOPHYSICS
SYDENHAM P. ELECTRONICS TODAY - FEBRUARY PP 76-85 1972
1.24
BALAVADZE B.K. OBSERVATIONS DES MAREES TERRESTRES A TBILISSI
KARTVELICHVILI K.Z. AC.NAOUK GROUZINSKOI SSR INST.GEOPH.137 PAGES, TBILISSI 1972
1.24
SCHNEIDER M.M. RESULTS OF THE EARTH TIDE OBSERVATIONS AT THE
SIMON D. ANTARCTIC STATION VOSTOK 1969
2 INTERN.SYMP. GEODASIE UND PHYSIK DER ERDE MAI PP 1-8 1973
1.24
* EARTH TIDES SECTION
INST.OF COASTAL OCEANOGRAPHY AND TIDES PP 39-41 1973
1.24
SCHNEIDER M. OBSERVATIONS OF THE EARTH S CRUST TIDES AT
VOSTOK STATION
TROUDI SOV.ANTARKTITCHESKOI EXP.58, PP 174-183 1973
1.24
BILHAM R.G. THE LOCATION OF EARTH STRAIN INSTRUMENTATION
PHIL.TRANS.R.SOC.LONDON A 274 PP 429-433 1973
1.24
BRAGARD L. LE LABORATOIRE DE GEODYNAMIQUE DE L UNIVERSITE DE
LIEGE A KANNE /PROVINCE DU LIMBOURG/
REVUE DES QUESTIONS SCIENTIF.T.145 NO 1 PP 35-54 1974
1.24
UGRAVESCU D. STATIONS FOR EARTH TIDES RECORDINGS IN THE
SOCIALIST REPUBLIC OF ROMANIA
GEOPHYSIQUE, 18, PP 55-58, BUCAREST 1974
1.24
ORROJA J.NA. LA ESTACION DE MAREAS TERRESTRES DEL VALLE
IEIRA R. DE LOS CAIDOS
RTIZ R. ASAMBLEA NAC.DE GEOD.Y GEOF. IA 218 VOL 2 PP 171-175 1974
1.24
ATHER R.S. AN EXPERIMENT TO DETERMINE RADIAL DEFORMATION OF
RETREGER K. EARTH TIDES IN AUSTRALIA BY OCEAN TIDES
XVI ASS.GEN.DE L UGGI GRENOBLE - AOUT 1975
1.24
OVALEVSKY J. LE CENTRE D ETUDES ET DE RECHERCHES GEODYNAMIQUES
ET ASTRONOMIQUES /CERGA/
L ASTRONOMIE PP 1-16 1975
1.24
ARNOWIECKI WL. OBSERVATION STATIONS FOR EARTH TIDES IN POLAND
PUBLS.INST.GEOPH.POL.AC.SCI.VOL 94 PP 69-75 1975
1.24
ORROJA J.M. ESTUDIO DE MAREAS TERRESTRES EN ESPANA
IEIRA R. UNIV.COMPLUTENSE-FACULTAD CIENCIAS MADRID
RTIZ R.
EVILLA M.J. SEMINARIO ASTRON.GEODESIA NO 85 PP 101-120 1975
1.24
OSENBACH O. THE WORKING GROUP GEODESY AND GEOPHYSICS OF THE
FEDERAL REPUBLIC OF GERMANY

PROCEEDINGS 7TH INT.SYMP.EARTH TIDES, SOPRON 1973
AKADEMIAI KIADO BUDAPEST, P 127 1976

1.24
MALZER H. THE OBSERVATORY OF SCHILTACH
PROCEEDINGS 7TH INT.SYMP.EARTH TIDES, SOPRON 1973
AKADEMIAI KIADO BUDAPEST, PP 725-727 1976

1.3
DARWIN G.H. THE LUNAR DISTURBANCE OF GRAVITY.
SCIENT.PAPERS I, 14 PP 430-443 1882

1.3
MILNE J. EARTH PULSATIONS.
NATURE 28 PP 367-370 16 AOUT 1883

1.3
LALLEMAND MOUVEMENTS ET DEFORMATIONS DE LA CROUTE TERRESTRE.
ANN.BUR.LONG. 1909

1.3
FRANCOIS CH. SUR LES FORCES QUI AGISSENT SUR LE PENDULE.
CIEL ET TERRE XXXI NO 9 PP 358-362 NO 11 PP 455-459 1910

1.3
FRANCOIS CH. SUR LES FORCES QUI AGISSENT SUR LE PENDULE.
CIEL ET TERRE XXXII NO 3 PP 103-108
NO 10 PP 341-345 NO 11 PP 374-380 1911

1.3
SCHWEYDAR W. UBER DIE DEFORMATION DES ERDKORPERS.
PETERMANS GEOGR.MITTEIL.57.II. PP 74-75 1911

1.3
LALLEMAND CH. L ELASTICITE DU GLOBE TERRESTRE ET LES MAREES DE
L ECORCE.
BULL.ASTR.VOL.28 PP 368-388 1911

1.3
LAMBERT W.D. THE PROBLEM OF THE EARTH TIDES.
BULL.NAT.RESEARCH COUNCIL 3 PT 2 PP 18-26 192

1.3
LAMBERT W.D. LES MAREES DE L ECORCE TERRESTRE ET LEURS RELATIONS
AVEC LES AUTRES BRANCHES DE LA GEOPHYSIQUE.
BULL.GEOD. NO 5 PP 41-47 192

1.3
FIEDLER G. DEFORMACIONES PERIODICAS DE LA CORTEZA TERRESTRE.
BOL.ACAD.CIENCIAS FIS.MAT.NAT.TOMO XXIV NO 67 PP 57-63 192

1.3
LAMBERT W.D. EARTH TIDES.
PHYSICS OF THE EARTH-THE FIGURE OF THE EARTH.
BULL.NAT.RES.COUNCIL NO 78 1/5 193

1.3
HOPFNER F. DIE GEZEITEN DER FESTEN ERDKRUSTE.
HANDBUCH GEOPH.BAND I LIEF.II ABS.V 193

TOMASCHEK R. UBER DIE ZEITLICHEN SCHWANKUNGEN DER SCHWERKRAFT.
DAS WELTALL 32 PP 54-55 19

1.3
LARMOR J. THE TIDAL STRAIN OF THE EARTH.
NATURE LONDON 132 NO 3330 PP 313 19

1.3
MEROLA M. LE MAREE DELLA CROSTA SOLIDA E IL MODULO DI
RIGIDITA DELLA TERRA.
RIV.FIS.NAT.SCI.NAT.14 PP 289-304/357-368 19

1.3
VERBAANDERT J. LE PENDULE HORIZONTAL ET LA MESURE DE LA RIGIDITE
DU GLOBE TERRESTRE.
CIEL ET TERRE LIX NO 3-4 PP 132-134 1

1.3
DE CASTRO H. VARIACIONES TEMPORALES DE LA PESANTEZ POR INFLUJO
DE LA LUNA Y EL SOL.
CIENCIA X NO 1-2 PP 29-40 MEXICO 1

1.3
GOUGENHEIM A. LES MAREES TERRESTRES.
BULL.INF.COMITE CENTR.OCEANOGRAPHIE & ETUDES COTES III,8 1

1.3
DEFANT A. EBBE UND FLUT DES MEERES DER ATMOSPHARE UND DER ERDFESTE.
SPRINGER VERLAG BERLIN 1953

1.3
MELCHIOR P.J. LES MAREES TERRESTRES.
OBS.ROY.BELGIQUE-MONOGRAPHIES NO 4
CIEL ET TERRE LXX PP 1-22, 103-131, 191-217 1954

1.3
LOPEZ ARROYO A. MAREAS TERRESTRES.
REVISTA DE GEOFISICA XIII NO 49 PP 73 1954

1.3
ELLENBERGER H. DIE ERDGEZEITENFORSCHUNG UNTER BESONDER BERUCKSICHTIGUNG DES DEUTSCHEN BEITRAGES.
SONDERHEFTE ZEITSCHR.VERMESSUNGSWESEN H.2 PP 32 1954

1.3
TOMASCHEK R. THE TIDES OF THE SOLID EARTH AND THEIR GEOPHYSICAL AND GEOLOGICAL SIGNIFICANCE.
NATURE 173 NO 4395 PP 143-145 23 JANVIER 1954

1.3
TOMASCHEK R. PROBLEME DER ERDGEZEITENFORSCHUNG.
DEUTSCH.GEOD.KOMM.BAYER.AKAD.WISS.A NO 23 PP 16 1956

1.3
TOMASCHEK R. TIDES OF THE SOLID EARTH.
ENCYCLOP.PHYSICS VOL.XLVIII-GEOPH. II PP 775-845 1957

1.3
ELLENBERGER H. DIE ERDGEZEITENFORSCHUNG-UEBER-BLICK UND AUSBLICK.
FREIBERGER FORSCHUNGSHFFTE H.C.38 PP 19-35 1957

1.3
MELCHIOR P. LES MAREES TERRESTRES DANS LE CADRE DE L ANNEE GEOPHYSIQUE INTERNATIONALE.
OBS.ROY.BELG.COMM. NO 142 S.GEOPH. NO 47 PP 21-29 1958

1.3
KNEISSL M. DIE BEDEUTUNG DER ERDGEZEITEN FUR DIE GEODASIE.
OBS.ROY.BELG.COMM. NO 142 S.GEOPH. NO 47 PP 14-20 1958

1.3
MELCHIOR P.J. EARTH TIDES.
ADVANCES IN GEOPHYSICS VOL.4 PP 391-443 1958

1.3
GEZA T. A FOLDKEREG ARAPALYA.
MAGYAR TUDOMANY 8-9 1958

1.3
MUNK W.H.
MAC DONALD G.J.F. THE ROTATION OF THE EARTH
A GEOPHYSICAL DISCUSSION
CAMBRIDGE UNIV.PRESS 323PP 1960

1.3
MEDOUNINE A.E. L ETUDE, EN RUSSIE, DES DEFORMATIONS DE LA TERRE PROVOQUEES PAR L ATTRACTION LUNI-SOLAIRE /1892-1920/
ACAD.SC.URSS.INST.HIST.SC.NAT.ET TECH.PP 151-164 MOSCOU 1961

1.3
PARIISKII N.N. ETUDE DES MAREES TERRESTRES DEPUIS LE DEBUT DE L AGI
ACAD.SC.URSS GEOPH. BULL.NO 14 PP 179-187 1961

1.3
PARIISKII N.N. LES MAREES TERRESTRES ET LA STRUCTURE INTERNE DE LA TERRE.
NOUVELLES ACAD. SC. USSR. S.GEOPH. N.2 FEVR 1963
* B.I.M. N.33 PP 913-940 1963

1.3
MELCHIOR P.J. L EVOLUTION DES IDEES ET DES TECHNIQUES D OBSERVATION DANS L ETUDE DES MAREES TERRESTRES.
OBS.ROY.BELG.COMM.NO 235 S.GEOPH.NO 68 PP 3-32 1964

1.3
PARIISKII N.N. ETUDE DES MAREES TERRESTRES DEPUIS LE DEBUT DE L AGI.
BULL.GEOPH.ACAD.SC.URSS.NO 14 PP 179-187 1964

1.3
MELCHIOR P.J. — THE EARTH TIDES
PERGAMON PRESS,458 PAGES — 1965

1.3
MELCHIOR P.J. — LE DEVELOPPEMENT DES RECHERCHES EXPERIMENTALES SUR LES MAREES TERRESTRES.
ASKANIA-WARTE HEFT 66 JAHRGANG 22 PP 4-6 — 1965

1.3
LECOLAZET R. — LES MAREES TERRESTRES ET LA CONSTITUTION PHYSIQUE DE LA TERRE
ANNUAIRE BUR. LONGITUDES PP C1- C13 — 1966

1.3
MELCHIOR P.J. — CURRENT DEFORMATIONS OF THE EARTH S CRUST.
ADV.GEOPH.VOL.12 ACAD.PRESS INC.NEW-YORK PP 1-77 — 1967

1.3
MELCHIOR P.J. — PROGRES ACCOMPLIS DANS L ETUDE DES MAREES TERRESTRES /1957-1967/.
BULL.GEOD.NO 84 PP 159-185 — 1967

1.3
BUCHHEIM W. — POSITION AND OUTLOCK OF EARTH TIDE RESEARCH AFTER THE INTERNATIONAL GEOPHYSICAL COOPERATION.
MITTEIL.INST.THEOR.PHYS.GEOPH.BERGAK.FREIBERG NO 49 — 196

1.3
SOUSA MOREIRA — O FENOMENO DAS MARES TERRESTRES E AS SUAS CONSEQUENCIAS.
BOL.SERV.CARTOGR.EXERC.NO 17 VOL.III PP 12-19 — 196

1.3
SLICHTER L.B.
HAGER C.L. — EARTH TIDES - MAREA TERRESTRE.
GEOF.INT.INST.GEOF.UNIV.NAC. MEXICO VOL 8 N2-4 PP43-54 — 196

1.3
OKUDA T. — ANNUAL REPORT OF THE GEOPHYSICAL OBSERVATIONS MADE AT THE INTERNATIONAL LATITUDE OBSERVATORY OF MIZUSAWA FOR THE YEAR 1967.
MIZUSAWA — 196

1.3
SLICHTER L.B.
HAGER C.L. — EARTH TIDES
GEOFISICA INTERN.VOL.8 /2-4/ PP 43-54 — 196

1.3
MELCHIOR P.J. — THE EARTH TIDES. / IN RUSSIAN /
ED.MIR MOSKVA 482 PAGES — 196

1.3
SLICHTER L.B. — SUMMARY OF FIELD RESEARCH AUGUST-DECEMBER 1967
ANTARTIC J.UNIT.STATES, WASHINGTON N.3 H.2, P.45 — 196

1.3
N. — PROBLEMS OF RECENT CRUSTAL MOVEMENTS OF THE EARTH.
3D. INTERN. SYMPOSIUM LENINGRAD 1968
MOSCOW 564 PAGES — 196

1.3
BYL J. — EINE METHODE DER FEINNEIGUNGSMESSUNG IN DER ERDGEZEITENFORSCHUNG UND BEI DER UNTERSUCHUNG REZENTER KRUSTENBEWEGUNGEN.
FEINGERATETECHNIK 4 MITTEIL.GEOD.INST.POTSDAM N.106 — 19

1.3
BYL J. — DIE BESTIMMUNG DER VERTIKALKOMPONENTE DER ERDGEZEITEN.
Z.VERMESSUNGSTECHNIK 3 MITTEIL.GEOD.INST.POTSDAM N.113 — 19

1.3
MACHIN C. — MAREAS TERRESTRES.
UNIV. MADRID-SEMINARIO DE ASTR. Y GEODESIA PUB. N62 — 19

1.3
N. — TIDAL INSTITUTE AND OBSERVATORY REPORT FOR 1968-69 /EARTH TIDES PAGES 14-15/ THE UNIVERSITY OF LIVERPOOL. — 19

1.3
JEFFREYS H. — THE EARTH.
5E EDITION-CAMBRIDGE U.P. — 19

1.3
CHOJNICKI T. OBLICZANIE TEORETYCZNYCH PLYWOW SKORUPY ZIEMSKIEJ I ICH DOKLADNOSC
GEODEZJA I KARTOGRAFIA 19,3, PP 171-202 1970

1.3
BALAVADZE B.K. KARTVELICHVILE K.Z LABORATOIRE DES MAREES TERRESTRES A TBILISSI.
RAPP.SYMP.INTERN.LENINGRAD 1968 MESURES INCLINAISON
COMM.GEOPH.URSS.MOSCOU PP 214-216 1969
* B.I.M. N.57 MAI PP 2750-2752 1970

1.3
ZETLER B.D. INTERNATIONAL SYMPOSIUM ON EARTH TIDES.
EOS 51 1 PP 9-10 1970

1.3
BYL J. GEGENWARTIGE PROBLEME DER ERDGEZEITENFORSCHUNG
VERMESSUNGSTECHNIK SOUS-PRESSE 1971

1.3
SLICHTER L.B. EARTH TIDES
THE NATURE OF THE SOLID EARTH MC GRAW HILL 1971

1.3
MELCHIOR P.J. PHYSIQUE ET DYNAMIQUE PLANETAIRES - 4 VOLUMES
EDITEUR VANDER - RUE DEFACQZ, 21-1050 BRUXELLES 1971

1.3
BONATZ M. GRAVIMETRISCHE MESSUNGEN ZUR BESTIMMUNG DER GEZEITENDEFORMATIONEN DES ERDKORPERS
JEODEZI BULTENI - SAYI 5 - CILT 1 PP 225-240 1971

1.3
LENNON G.W. OCEANIC AND EARTH TIDES AND THEIR RELEVANCE TO GEODESY
OCEANIC AND EARTH TIDES CHARTERED SURVEYOR PP 467-468 1972

1.3
SLICHTER L.B. EARTH TIDES
THE NATURE OF THE SOLID EARTH, PP 285-320 1972

1.3
MELCHIOR P. MEASUREMENTS OF TILTS AND STRAINS
MEASUREM., INTERPRETAT., CHANGES OF STRAIN IN THE EARTH
PHIL.TRANS.ROY.SOC.SER.A, VOL.274, NO 1239 PP 203-208 1973

1.3
LENNON G.W. BAKER T.F. EARTH TIDES AND THEIR PLACE IN GEOPHYSICS
MEASUREM., INTERPRETAT., CHANGES OF STRAIN IN THE EARTH
PHIL.TRANS.ROY.SOC.SER.A, VOL.274, NO 1239 PP 199-202 1973

1.3
X. REPORT ON THE FIRST GEOP RESEARCH CONFERENCE
SOLID-EARTH AND OCEAN TIDES
EOS, VOL 54, 2, PP 96-100 1973

1.3
LENNON G.W. BAKER T.F. THE EARTH TIDE SIGNAL AND ITS COHERENCY
Q.JL.R.ASTR.SOC. NO 14 PP 161-182 1973

1.3
X SOLID EARTH AND OCEAN TIDES
REPORT ON THE FIRST GEOP RESEARCH CONFERENCE
E.O.S. VOL 54 NO 2 FEBRUARY PP 1973

1.3
BROSCHE P. DIE NATURWISSENSCHAFTEN
NATURWISSENSCHAFTEN 62 PP 1-9 1975

1.3
LECOLAZET R. PRESIDENTIAL ADDRESS
PROCEEDINGS 7TH INT.SYMP.EARTH TIDES, SOPRON 1973
AKADEMIAI KIADO BUDAPEST, PP 19-22 1976

1.3
ANDERSON A.J. GEODYNAMIC STUDIES USING EARTH-TIDE INSTRUMENTS
ACTA UNIV.UPSALIENSIS NO 396 PP 3-32 UPPSALA 1977

1.4
MELCHIOR P.J. DISPOSITIONS PRISES PAR LE CENTRE INTERNATIONAL DES MAREES TERRESTRES EN VUE D UNE CLASSIFICATION SYSTEMATIQUE DES RESULTATS DE MESURES.
B.I.M. NO 27 PP 607-611 1962

1.4
WITKOWSKI / LACOSTE-ROMBERG / VERBAANDERT / MELCHIOR P.J.
CATALOGUE GENERAL DES RESULTATS DES ANALYSES HARMONIQUES MENSUELLES D OBSERVATIONS DE MAREES TERRESTRES.
B.I.M.NO 32 PP 910-911 — 1963

1.4
MELCHIOR P.J.
CATALOGUE GENERAL DES RESULTATS DES ANALYSES HARMONIQUES MENSUELLES D OBSERVATIONS DE MAREES TERRESTRES AU 31 DECEMBRE 1962.
CENTRE INT.MAR.TERR.OBS.ROY.BELG.PP 148 — 1963

1.4
MELCHIOR P.J.
CATALOGUE DES DONNEES D OBSERVATIONS DE MAREES TERRESTRES FIGURANT AU 1 OCTOBRE 1964 DANS LA BIBLIOTHEQUE DU CENTRE INTERNATIONAL DES MAREES TERRESTRES.
B.I.M.NO 38 PP 1356@1361 — 1964

1.4
MELCHIOR P.
CATALOQUE DES STATIONS DE MAREES TERRESTRES
B.I.M. NO 56 PP 2688-2715 — 1969

1.5
JUNG K.
NOTICE NECROLOGIQUE DE WILHELM SCHWEYDAR 1877-1959.
ZEITSCH.GEOPH.JAHRGANG 26 PP 158-160 HEFT 3 — 1960

1.5
NAKAGAWA I.
NECROLOGIE - PROFESSOR E.NISHIMURA.
B.I.M.NO 38 PP 1354-1355 — 1964

1.5
N.
PRIX CH.LAGRANGE DE PHYSIQUE DU GLOBE.H.TACHEUCHI.
B.I.M.NO 42 PP 1609 — 1965

1.5
LECOLAZET R.
LE PROFESSEUR RUDOLF TOMASCHEK /1895-1966/.
B.I.M.NO 44 PP 1691 — 1966

1.5
MELCHIOR P.J.
LE PROFESSEUR RUDOLF TOMASCHEK /1895-1966/.
B.I.M.NO 44 PP 1692 — 1966

1.5
GROTEN E.
LE PROFESSEUR RUDOLF TOMASCHEK /1895-1966/.
B.I.M.NO 44 PP 1695 — 1966

1.5
LENNON G.W.
ARTHUR THOMAS DOODSON.
B.I.M.NO 51 PP 2317 — 1968

1.5
MELCHIOR P.J.
J.BROUET.
B.I.M. NO 52 PP 2396-2397 — 1968

1.5
N.
PRIX CH.LAGRANGE DE PHYSIQUE DU GLOBE. R.O.VICENTE
B.I.M.NO 52 PP 2486 — 196[illegible]

1.5
TARDI P.
WALTER DAVIS LAMBERT.
~~C.R.~~ACAD.~~SC~~.PARIS ~~268~~ PP 32-33 — 1969

1.5
AKSENTIEVA Z.N.
PHYSIQUE DE LA TERRE N.9- 1969
B.I.M. N.57 MAI PP 2717-2718 — 197[illegible]

1.5
N.
NECROLOGIE DE Z.N.AKSENTIEVA
ROTATION ET DEFORM.MAREES DE LA TERRE,FASC.1
ACAD.SC.UKR.OBS.GRAV.POLTAVA KIEV PP 337-339 — 197[illegible]

1.5
MELCHIOR P.
IN MEMORIAM JEAN VERBAANDERT
CIEL ET TERRE VOL 92, NO 1 PP 1-6 — 197[illegible]

2.1
BARTELS J.
GEZEITENKRAFTE.
HANDBUCH GEOPH.BAND I LIFF 2 ABSCH.IV — 193[illegible]

2.1
BARTELS J.
GEZEITENKRAFTE.
HANDBUCH PHYSIK-B.XLVIII-GEOPHYSIK II PP 734-774 — 195[illegible]

2.1
BOZZI-ZADRO M. MARUSSI A. — ON THE STATIC EFFECT OF MOON AND SUN ON THE SHAPE OF THE EARTH
BOLL.GEODESIA E SC.AFFINI-NO.4, PP 253-260 — 1973

2.11
PETERS C.A.F. — VON DEN KLEINEN ABLENKUNGEN DER LOTHLINIE UND DES NIVEAUS.
A.N. 22, 33 — 1845

2.11
HAGEN J.G. — ON THE DEFLECTION OF THE LEVEL DUE TO SOLAR AND LUNAR ATTRACTION.
A.N. 107 NO 2568 PP 371-380 — 1884

2.11
GAILLOT A. — INFLUENCE DE L ATTRACTION LUNAIRE SUR LA DIRECTION DE LA VERTICALE ET SUR L INTENSITE DE LA PESANTEUR.
BULL.ASTR.I PP 113-118, PP217-220 — 1884

2.12
DOODSON A.T. — THE HARMONIC DEVELOPPMENT OF THE TIDE-GENERATING POTENTIAL.
PROC.ROY.SOC.LONDON A.100 PP 305-329 — 1922

2.12
HORN WALTER — DIE ASTRONOMISCHEN GRUNDLAGEN DES HARMONISCHEN VERFAHRENS ZUR BERECHNUNG DER GEZEITEN
ARCHIV DEUT.SEEWARTE & MARINE OBS.61 BAND NO 8 28 P — 1941

2.12
DOODSON A.T. HORN W. GOUGENHEIM A. MARMER H.A. — STANDARD DEVELOPMENT OF TIDE-GENERATING POTENTIAL.
INTERN.HYDROGRAPHIC REVIEW MAY — 1954

2.12
PARIISKY N.N. — SOME REMARKS CONCERNING THE CALCULATION OF THEORETICAL AMPLITUDES OF TIDAL GRAVITY VARIATIONS.
B.I.M.NO 23 PP 466-478 — 1961

2.12
PALLAS W. — EINIGE BEMERKUNGEN UBER DIE EINARBEITUNG DER ASTRONOMISCHEN DATEN BEI DER HARMONISCHEN ANALYSE DER ERDGEZEITEN.
B.I.M. NO 26 PP 600-604 — 1961

2.12
MELCHIOR P.J. — LE NOUVEAU SYSTEME DES CONSTANTES ASTRONOMIQUES.
B.I.M.NO 38 PP 1362 — 1964

2.12
BURSA M. — TIDAL POTENTIAL DUE TO A NON-SPHERICAL LUNAR BODY
STUDIA GEOPH.ET GEOD. V.18 PP 1-7 — 1974

2.16
JUNG K. — UBER DIE DARSTELLUNG DER GEZEITENKRAFTE
-GERLANDS BEITRAGE ZUR GEOPHYSIK- 64 HEFT 4 PP 278-283 — 1955

2.16
CARTWRIGHT D.E. TAYLER R.J. — NEW COMPUTATIONS OF THE TIDE-GENERATING POTENTIAL
GEOPH.J.R.ASTR.SOC.23 PP 45-74 — 1971

2.16
WENZEL H.G. — ZUR GENAUIGKEIT DES THEORETISCHEN GEZEITENMODELLS
MITTEILUNGEN INST. THEOR.GEODASIE, HANNOVER, PP 11 — 1974

2.162
DRONKERS J.J. — TIDAL COMPUTATIONS IN RIVERS AND COASTAL WATERS
NORTH HOLLAND PUBL.CO.AMSTERDAM PP 518 — 1964

2.17
BRILLOUIN M. — OCEANS ET CONTINENTS. MAREES OCEANIQUES ET MAREES DU SOL. FORMULES NORMALISEES POUR LEUR CALCUL THEORIQUE.
C.RENDUS TOME 184 PP 849-853 4 AVRIL — 1927

2.17
ADLER J.L. — SIMPLIFICATION OF TIDAL CORRECTIONS FOR GRAVITY METER SURVEYS.
GEOPHYSICS VII NO 1 PP 35-44 — 1942

2.17
LAMBERT W.D. — NOTES ON EARTH TIDES.
GEOPHYSICS VIII NO 1 PP 51-56 — 1943

2.17
ELKINS TH.A. NOMOGRAMS FOR COMPUTING TIDAL GRAVITY.
GEOPHYSICS VIII NO 2 PP 134-145 AVRIL 1943

2.17
LOCKENVITZ A.E. THE PERIODIC VARIATIONS OF THE GRAVITATIONAL FORCE.
GEOPH.IX NO 1 PP 94-105 JANVIER 1944

2.17
DAMREL J.B. TIDAL GRAVITY EFFECT TABLES.
HOUSTON TECHN.LABO./WORDEN GRAVIMETER/.

2.17
MORELLI C. VARIAZIONE DIURNA DELLA GRAVITA IN EUROPA.
ANNALI GEOFISICA VI NO 1 PP 113-124, 295-307 1953

2.17
N. PREDICTIONS DE L EFFET LUNI-SOLAIRE DE LA GRAVITE.
C.R.COMITE NAT.FRANCAIS GEOD.GEOPH. PP 57-58 1953

2.17
N. VARIATION PERIODIQUE DE LA GRAVITE, PREDICTION.
C.R.COM.NAT.FRANCAIS GEOD.GEOPH. PP 72 1954

2.17
INGHILLERI G. CAMPO DI VALIDITA E CALCOLO GENERALIZZATO DELLE CURVE DI ATTRAZIONE LUNISOLARE.
POLITECNICO MILANO,PUBBL.IST.GEOD.TOPO. NO 81
RIVISTA GEOF.APPL. XV, NO 1 1954

2.17
NEUMANN R. ROLE JOUE PAR LA CORRECTION LUNI-SOLAIRE EN PROSPECTION GRAVIMETRIQUE.
GEOPH.PROSP. II NO 4 PP 290-305 1954

2.17
GOGUEL J. A UNIVERSAL TABLE FOR THE PREDICTION OF THE LUNI-SOLAR CORRECTION IN GRAVIMETRY.
GEOPH.PROSP. II SUPPL. PP 2-5 1954

2.17
GOGUEL J. TIDAL GRAVITY CORRECTIONS FOR 1954.
GEOPH.PROSP. II SUPPL. PP 6-31 1954

2.17
MORELLI C. TIDAL GRAVITY CORRECTIONS FOR 1954.
GEOPH.PROSP. II SUPPL. PP 32-42 1954

2.17
PETIT J.T. TABLES FOR THE COMPUTATION OF THE TIDAL ACCELERATIONS OF THE SUN AND MOON.
TRANS.AMER.GEOPH.UNION 35 NO 2 PP 193-202 1954

2.17
HAALCK H. DIE GEZEITENBEWEGUNG DES FESTEN ERDKOPERS UND DIE DADURCH FUR SEHR GENAUE GRAVIMETERMESSUNGEN NOTWENDIG WERDENDEN KORREKTIONEN.
GERL.BEITR.GEOPH. 64, 1, PP 1-15 1954

2.17
MORELLI C. TIDAL GRAVITY CORRECTIONS FOR 1955.
GEOPH.PROSP. III SUPPL. PP 30-42 1955

2.17
GOGUEL J. TIDAL GRAVITY CORRECTIONS FOR 1955.
GEOPH.PROSP. III SUPPL. PP 1-29 1955

2.17
N. TIDAL GRAVITY CORRECTIONS FOR 1957 /SERVICE HYDROGRAPHIQUE DE LA MARINE AND COMPAGNIE GENERALE DE GEOPHYSIQUE/.
GEOPH.PROSPECTING IV, SUPP.NO 1, DECEMBRE 1956

2.17
JONGOLOVITCH LES EPHEMERIDES APPROXIMATIVES DU SOLEIL ET DE LA LUNE POUR 1955-1960 ET LE NOMOGRAMME AUXILIAIRE POUR LE CALCUL DE L INFLUENCE LUNI-SOLAIRE SUR LA PESANTEUR
BULL.DE L INST.ASTRON.THEOR.T.6, NO 5, 78 PP 312-345 1956

2.17
HONKASALO T. ON THE TIDAL GRAVITY CORRECTION.
INTERN.GRAVITY COMM.-PARIS SEPTEMBER 1959

2.17
N. COURBES DE CORRECTIONS POUR LES VARIATIONS DE MAREES DE LA FORCE DE PESANTEUR POUR 1961-1961
INST.GEOD., PHOTOGRAMMETRIE ET CARTOGRAPHIE MOSCOU 1959

2.17
LONGMAN I.M. FORMULAS FOR COMPUTING THE TIDAL ACCELERATIONS DUE TO THE MOON AND THE SUN. J.GEOPH.RES.64 PP 2.351-2.355 1959

2.17
N. GRAPHIQUES DE CORRECTIONS DES VARIATIONS DE LA PESANTEUR DUES AUX MAREES POUR 1960 ET 1961. MOSKOV.INST.ING.GEOD.ASTROPHET.KARTOG 1959

2.17
N. COURBES DE CORRECTIONS POUR LES VARIATIONS DE MAREE DE LA FORCE DE PESANTEUR POUR 1962 INST.GEOD., PHOTOGRAMMETRIE ET CARTOGRAPHIE MOSCOU 1961

2.17
N. GRAPHIQUES DES CORRECTIONS POUR LES VARIATIONS DE MAREE DE LA FORCE DE PESANTEUR EN 1963-1964 INST.MOSCOU GEOD.PHOTOGRAPHIE AERIENNE ET CARTO.
* B.I.M. NO 32 PP 908-909 1963

2.17
N. GRAPHIQUES DES CORRECTIONS POUR LES VARIATIONS DE MAREE DE LA FORCE DE PESANTEUR EN 1963-1964 MINISTERE DE L ENSEIGN.SPEC.SUP.ET MOYEN DE LA R.S.F.S.R. INST.DE MOSCOU DE GEOD.,PHOTOG.AER.,CARTOG.MOSCOU 1962
* B.I.M. N.32 MAI PP 908-909 1963

2.17
N. TIDAL GRAVITY CORRECTIONS FOR 1965. GEOPH.PROSPECTING,THE HAGUE 12 SUPPL.1 1964

2.17
N. METHODS OF PREDICTION AND ANALYSIS ANNUAL REPORT 1964. TIDAL INST.AND OBS.UNIV.LIVERPOOL PP 8-10 1964

2.17
N. COURBES DE CORRECTIONS POUR LES VARIATIONS DE MAREE DE LA FORCE DE PESANTEUR POUR 1965 INST.MOSCOU GEOD., PHOTOGRAMMETRIE ET CARTOGR. 1965

2.17
DE ROP W. A TIDAL PERIOD OF 1800 YEARS OBS.ROY.BELG. TELLUS XXIII PP 261-262 1971

2.17
CHOJNICKI T. CALCULS DES MAREES TERRESTRES THEORIQUES ET LEUR PRECISION PUBL.INST.GEOPH.POLISH ACAD.SC.MAR.TERR.55 PP 3-42 1972

2.17
BROUCKE R.A.
ZUERN W.E.
SLICHTER L.B. LUNAR TIDAL ACCELERATION ON A RIGID EARTH AMER.GEOPH.UNION, GEOPHYS.MONOGRAPH J.G.R. VOL.16 PP 319-324 1972

2.17
POLLACK H.N. LONGMAN TIDAL FORMULAS RESOLUTION OF HORIZONTAL COMPONENTS J.G.R.VOL.78 NO 14, PP 2598-2600 1973

2.17
SEVILLA M.J. CORRECTION AUTOMATICA DE MEDIDAS GRAVIMETRICAS POR EFECTO DE MAREA 2EME ASAMB.NAC.DE GEOD.Y GEOF. P 48 MADRID 1976

2.2
PARIISKII N.N. CORRECTION FOR VERTICAL ACCELERATIONS DURING OBSERVATIONS OF TIDAL VARIATIONS OF GRAVITY GRAVIMETRICAL RES.1-PUBL.ACAD.SC.USSR, PP 39-40 1961
* B.I.M. NO 69 PP 3878-3880 1974

2.21
UO J.T.
UNKINS K.L. SOLID EARTH TIDES. LAMONT GEOL.OBS.COLUMBIA UNIV.PALISADES N-Y 1965

2.21
AKAGI S. EQUATIONS OF MOTION FOR THE EARTH TIDES. PUBL. INT. LAT. OBS. MIZUSAWA VOL VI N.2 PP205-223 1968

2.21
ZAWA I. PATTERN OF TIDAL DEFORMATION ON THE EARTH CONTRIB.GEOPH.INST. KYOTO UNIV. NO 14 PP 29-38 1974

2.211
OVE A.E.H. GENERAL THEORY OF EARTH TIDES. SOME PROBL.OF GEOD.CHAP.IV PP 49 1911

2.211
TOMASCHEK R. GROTEN E. — VORSCHLAGE ZUR EINHEITLICHEN BEZEICHNUNG DER GEZEITENQUOTIENTEN.
B.I.M.NO 33 PP 1029-1032 — 1963

2.212
SURVEY GROUP P.S. — GEODESIE SOUTERRAINE.
ORG.EUROP.RECH.NUCL.CERN,GENEVE AOUT — 1958

2.216
PEKERIS C. — NOTE ON TIDES IN WELLS.
TRAV.ASSOC.INT.DE GEOD.UGGI TOME 16 — 1941

3.0
WITKOWSKI J. — LATITUDE STATION OF THE POLISH ACADEMY OF SCIENCES AT BOROWIEC.
ACTA GEOPH.POLONICA-VOL.VII NO 1 NADBITKA PP 3-8 — 1959

3.0
WITKOWSKI J. — EARTH TIDES AND THEIR OBSERVATION AT THE LATITUDE STATION OF THE POLISH ACADEMY OF SCIENCES PP 54-64 — 1960

3.0
PARIISKY N.N. — OBSERVATION OF THE EARTH TIDES IN THE USSR FROM JUNE 1957 TO JUNE 1960.
B.I.M.NO 21 PP 371-386 — 1960

3.0
PARIISKY N.N. — STUDIES OF TIDAL VARIATIONS OF GRAVITY IN CENTRAL ASIA BY THE EARTH PHYSICS INSTITUTE OF THE USSR ACADEMY OF SCIENCES.
IVME SYMP.INTERNATIONAL SUR LES MAREES TERRESTRES.
OBS.ROY.BELG.COMM.NO 188 S.GEOPH. NO 58 PP 96-108 — 1961

3.0
BOSSOLASCO M. — SUR LES VARIATIONS DU VECTEUR GRAVITE A GENES.
IVME SYMP.INTERNATIONAL SUR LES MAREES TERRESTRES.
OBS.ROY.BELG.COMM.NO 188 S.GEOPH. NO 58 PP 109-121 — 1961

3.0
GERKE K. — RAPPORT SUR LES ETUDES DES MAREES TERRESTRES FAITES PENDANT LA PERIODE PASSEE PAR -L INSTITUT FUR ANGEWANDTE GEODASIE- A FRANKFURT AM MAIN /2E SECTION D.G.F.I./.
IVME SYMP.INTERNATIONAL SUR LES MAREES TERRESTRES.
OBS.ROY.BELG.COMM.NO 188 S.GEOPH. NO 58 PP 129-138 — 196

3.0
SIMON D. — METHODEN DER STORPEGELANALYSE UND IHRE BEDEUTUNG FUR DIE ERDGEZEITENFORSCHUNG.
RAPP.SYMP.INTERN.LENINGRAD 1968 MESURES INCLINAISON COMM.GEOPH.URSS.MOSCOU PP 95-131 — 196

3.1
DARWIN G.H. — ON AN INSTRUMENT FOR DETECTING AND MEASURING SMALL CHANGES IN THE DIRECTION OF THE FORCE OF GRAVITY
SCIENT.PAPERS CAMBRIDGE UNIV.PRESS V.I PP 389-429 1907
BRITISH ASSOCIATION REPORT PP 93-126 — 188

3.1
D ABBADIE — RECHERCHES SUR LA VERTICALE.
ANNALES DE LA SOC.DES SC.DE BXL. PP 37-51 — 188

3.1
DARWIN G.H. — THE LUNAR DISTURBANCE OF GRAVITY, VARIATIONS IN THE VERTICAL DUE TO ELASTICITY OF THE EARTH S SURFACE
SCIENT.PAPERS CAMBRIDGE UNIV.PRESS V.I PP 430-460 1907
BRITISH ASSOCIATION REPORT PP 95-119 — 188

3.1
EBLE L. — NOTE SUR LES DEVIATIONS DE LA VERTICALE ET LES MAREES DE L ECORCE TERRESTRE.
BULL.GEOD. NO 2 PP 144-147 — 192

3.1
EBLE L. — SUR LES VARIATIONS DE LA VERTICALE.
ANN.DE GEOPH.VOL.6 PP 128-129 — 19

3.1
NISHIMURA E. — ON EARTH TIDES.
TRANS.AMER.GEOPH.UNION 31 NO 3 PP 357-76 — 19

3.1
SCHNEIDER M. — ZUM INTERNATIONALEN STAND DER NEIGUNGSMESSTECHNIK IM HINBLICK AUF DIE ERFORDERNISSE BEI DER UNTERSUCHUNG DER REZENTEN ERDKRUSTENBEWEGUNGEN.
NAT.KOM.GEOD.GEOPH.DDR REIHE III HEFT 6 PP 31 — 19

3.1
BYL J. HORIZONTALPENDEL-NIVEAUVARIOMETER BETRACHTUNGEN UBER ZWEI VONEINANDER UNABHANGIGE METHODEN ZUR NEIGUNGSMESSUNG FEINGERATETECHNIK SOUS-PRESSE 1971

3.1
ALLEN R.V.
WOOD D.M.
MORTENSEN C.E.
SOME INSTRUMENTS AND TECHNIQUES FOR MEASUREMENT OF TIDAL TILT.
MEASUREM., INTERPRETAT., CHANGES OF STRAIN IN THE EARTH
PHIL.TRANS.ROY.SOC.SER.A, VOL.274, NO 1239 PP 219-222 1973

3.11
DARWIN G.H. ATTEMPTED EVALUATION OF THE RIGIDITY OF THE EARTH FROM THE TIDES OF LONG PERIOD
SCIENT.PAPERS CAMBRIDGE UNIV.PRESS V.I PP 340-346 1907
NATURAL PHILOSOPHY THOMSON-TAIT 1883

3.11
DARWIN G.H. ON THE DYNAMICAL THEORY OF THE TIDES OF LONG PERIOD
SCIENT.PAPERS CAMBRIDGE UNIV.PRESS V.I PP 366-371 1907
PROCEEDINGS OF THE ROY.SOC.LONDON XLI PP 337-342 1886

3.11
DARWIN G.H. DYNAMICAL THEORY OF THE TIDES OF LONG PERIOD.
PROC.R.S.LONDON 41 1886

3.11
RAYLEIGH NOTE ON THE THEORY OF THE FORTNIGHTLY TIDE.
PHIL.MAG./SER.6/ V. 1903

3.11
SCHWEYDAR W. HARMONISCHE KONSTANTEN DER VIERZEHNTATIGEN UND MONATLICHEN MONDFLUT.
PART.III-GERL.BEITRAGE GEOPH. 9 PP 41-77 1908

3.11
DAUBLESKY R. GEZEITENERSCHEINUNGEN IN DER ADRIA.
DENKSCHRIFT.AK.WISS.WIEN,MATH-NATURW.BD 96 PP 277-324 1919

3.11
STREET R.O. OCEANIC TIDE AS MODIFIED BY A YIELDING EARTH.
MTLY.NOT.R.ASTR.SOC.GEOPH.SUP.VOL.1 PP 292-306 1925

3.11
PROUDMAN J. TIDES IN NARROW SEAS.
THE OBSERVATORY XLVIII NO 619 PP 386-388 1925

3.11
GRACE S.F. THE SEMI-DIURNAL LUNAR TIDAL MOTION OF THE RED SEA.
MTLY.NOT.R.ASTR.S.GEOPH.SUP.VOL.II NO 6 PP 273-296 1930

3.11
GRACE S.F. THE SEMI-DIURNAL LUNAR TIDAL MOTION OF LAKE BAIKAL AND THE DERIVATION OF THE EARTH TIDES FROM THE WATER-TIDES.
MTLY.NOT.R.ASTR.S.GEOPH.SUP.VOL.II NO 7 PP 301-309 1931

3.11
DOHR G. STATISTISCHER NACHWEIS VON GEZEITEN IN BINNENSEEN, DARGESTELLT AM BEISPIEL DES BODENSEES.
ZEITSCHR.FUR GEOPH.23 H5 PP 256-271 1957

3.11
MAXIMOV I.V. MAREE LUNI-SOLAIRE A LONGUE PERIODE DANS L OCEAN MONDIAL.
RAPP.ACAD.SC.URSS.TOME 118 NO 5 PP 888-890 1958

3.11
VOROBIEV V.N. MAREES LUNI-SOLAIRES SEMI-MENSUELLES ET MENSUELLES DANS LES MERS DE L ARCTIQUE SOVIETIQUE.
RAPP.ACAD.SC.URSS.TOME 167 NO 5 PP 1039-1041 1966

3.11
WUNSCH C. THE LONG-PERIOD TIDES
REVIEWS OF GEOPHYSICS VOL.5 NO 4 PP 447-475 1967

3.12
KINOV A.P.
RAVIETS T.P.
EXAMEN PRELIMINAIRE DES MAREES DU LAC BAIKAL
TRAV.OBS.DE MAGNET.ET METEOR.D IRKHOUTSK NO 1 PP 54 1926

3.12
ARFIANOVITCH Y.A. ETUDE EXPERIMENTALE DES MAREES DU LAC BAIKAL
IZV.DE L INST.DE PHYS.MATH.STEKLOV.T.3 PP 189-200 1928

3.12
ROUDMAN J. NOTE ON FORCED TIDES IN A LAKE.
MTLY.NOT.R.ASTR.S.GEOPH.SUPPL.VOL.II NO 2 PP 96-97 1928

3.12
ENDROS A. GEZEITENBEOBACHTUNGEN IN BINNENSEEN.
ANNALEN HYDROGR.U.MARIT.METEOROL.58 H9 PP 305-314 1930

3.12
HALBFASS W. SEEN.
HANDBUCH DER GEOPH.VII,LIEF.1 BERLIN PAR.60 PP 152-153 1933

3.12
ZERBE W.B. THE TIDE IN THE DAVID TAYLOR MODEL BASIN.
TRANS.AMER.GEOPH.UNION VOL.30 PP 357-368 1949

3.12
MELCHIOR P.J. SUR L EFFET DES MAREES TERRESTRES DANS LES OSCILLATIONS DU NIVEAU DU LAC TANGANIKA A ALBERTVILLE.
BULL.ACAD.R.BELG.PP 368-371 3MARS
OBS.ROY.BELG.COMM. NO 95 S.GEOPH. NO 35 1956

3.12
AKSENTIEVA Z.N. SUR LES MAREES DU LAC BAIKAL
TROUDI POLTAVSKOI GRAV.OBS. T.II PP 106-117 1948
* B.I.M. N.34 NOVEMBRE PP 1104-1117 1963

3.12
MORTIMER C.H.
FEE E.J. FREE SURFACE OSCILLATIONS AND TIDES OF LAKES MICHIGAN AND SUPERIOR
PHIL.TRANS.R.SOC.LONDON A. VOL 281 NO 1299 PP 1-61 1976

3.12
RAO D.B.
SCHWAB D.J. TWO DIMENSIONAL NORMAL MODES IN ARBITRARY ENCLOSED BASINS ON A ROTATING EARTH - APPLICATION TO LAKES ONTARIO AND SUPERIOR
PHIL.TRANS.R.SOC.LONDON A. VOL 281 NO 1299 PP 63-96 1976

3.12
HAMBLIN P.F. A THEORY OF SHORT PERIOD TIDES IN A ROTATING BASIN
PHIL.TRANS.R.SOC.LONDON A.VOL 281 NO 1299 PP 97-111 1976

3.13
MOULTON F.R. THEORY OF TIDES IN PIPES ON A RIGID EARTH.
ASTROPH.JOURN.50 PP 346-355 1919

3.13
TSUMURA K. EARTH TIDAL OBSERVATION WITH A WATER TUBE TILTMETER.
3E SYMP. MAREES TERRESTRES TRIESTE
BOLL.GEOFISICA TEORICA E APPL.VOL.2 NO 5 PP 77-79 1960

3.13
OZAWA I. ON THE TIDAL OBSERVATION BY MEANS OF A RECORDING WATER-TUBE TILTMETER
JOURN.GEOD.SOC.JAPAN VOL.12 PP 151-156 1967

3.13
KAARIAINEN J. 50 M PITKA VESIVAAKA.
GEOFYSIIKAN PAIVAT MAI 1969

3.13
HONKASALO T. ERAITA TULOKSTA MAANKUOREN VUOKSITUTKIMUKSESTA.
GEOFYSIIKAN PAIVAT MAI 1969

3.13
HONKASALO T. MAANKUOREN VUOKSI-ILMION TUTKIMUKSISTA SUOMESSA.
GEOFYSIIKAN PAIVAT JUIN 1968 OULU 1969

3.13
BOWER D.R. ON WATER TUBE TILTMETER.
6 EME SYMPOSIUM MAREES TERRESTRES STRASBOURG
OBS.ROY.BELG.COMM. NO A9 S.GEOPH. NO 96 PP 219 1970

3.13
KAARIAINEN J. A LONG PIPE TILTMETER.
1970

3.13
BOWER D.R. A SENSITIVE WATER-LEVEL TILTMETER
MEASUREM., INTERPRETAT., CHANGES OF STRAIN IN THE EARTH
PHIL.TRANS.ROY.SOC.SER.A, VOL.274, NO 1239 PP 223-226 197[illegible]

3.131
MICHELSON A.A. PRELIMINARY RESULTS OF MEASUREMENTS OF THE RIGIDITY OF THE EARTH.
ASTROPH.JOURN.39 PP 105-138 191

3.131
MICHELSON A.A.
GALE H.G. THE RIGIDITY OF THE EARTH.
ASTROPH.JOURN.50 PP 330-345 191

3.131
KAARIAINEN J. UBER DIE 50-M LANGE ROHRLIBELLE ZUR UNTERSUCHUNG DER NEIGUNG DER ERDKRUSTE
GEODETISKA INST. - GEODEET.LAITOS PP 1-21 HELSINKI 1973

3.132
TAKAHASI R. PRELIMINARY REPORT ON THE OBSERVATION OF THE TILTING OF THE EARTH S CRUST WITH A PAIR OF WATER PIPES.
BULL.EARTHQU.RES.INST.TOKYO IMP.UNIV.VOL.8,2 PP 143-152 1930

3.132
HAALCK H. EIN QUECKSILBERNEIGUNGSMESSER VON HOHER EMPFINDLICHKEIT
ZEITSCHRIFT FUR GEOPH.8 PP 256-271 1932

3.132
EGEDAL J.
FJELDSTAD J.E. OBSERVATIONS OF TIDAL MOTIONS OF THE EARTH S CRUST MADE AT THE GEOPHYSICAL INSTITUTE, BERGEN.
GEOF.PUBL.VOL.XI NO 14 AKAD.I OSLO 1937

3.132
EGEDAL J. ON THE APPLICATION OF THE HYDROSTATIC METHOD TO LEVELLING AND TO DETERMINATION OF VERTICAL MOVEMENTS IN THE EARTH S CRUST.
PUBL.DET DANSKE METEOR.INST.MEDD.NO 10 1947

3.132
JARINOV N.A. PREMIERS RESULTATS DE L OBSERVATION DES MAREES TERRESTRES PAR DES VARIOMETRES A NIVEAU.
FISICA ZEMLI NO 10 PP 86-90 1970

3.132
DOBROKHOTOV IOU S. NOUVEAU DOMAINE D APPLICATION DES NIVEAUX DE HAUTE SENSIBILITE.
INST.PHYS.DE LA TERRE. ACAD.SC.URSS.MOSCOU
* B.I.M. N.62 PP 3190-3194 1971

3.133
EGEDAL J. OM HYDROSTATISKE MAALINGERS ANVENDELSE VED UNDERSOGELSER OVER JORDSKORPEBEVAEGELSER.
FESTSKRIFT TIL.N.E.NORLUND COPENHAGUE PP 113-116 1945

3.133
HOSOYAMA K. ON A MERCURY TILTMETER AND IST APPLICATION.
BULL.DISASTER PREVENT.RES.INST.KYOTO NO 6 PP 17-25 1953

3.133
STROOBANT A.D.
WALKER I.J. A HIGH SENSITIVITY DISPLACEMENT INDICATOR USING A CAPACITANCE BRIDGE
RADIO,ELECTRONICS AND COMM. - OCTOBER PP 13-16 1970

3.14
HENGLER L. ASTRONOMISCHE PENDELWAGE NEBST EINER NEUEN NIVELLIRWAGE ERFUNDEN UND DARGESTELLT.
DINGLER S POLYT.JOURN. BD.43 PP.81-92 1832

3.14
PERROT DEUX APPAREILS DESTINES A RENDRE MANIFESTES ET MESURABLES LES VARIATIONS OCCASIONNEES DANS L INTENSITE ET LA DIRECTION DE LA PESANTEUR A LA SURFACE DE LA TERRE, PAR LES DIVERS MOUVEMENTS DE NOTRE GLOBE ET L ATTRACTION DES CORPS CELESTES.
C.R. ACAD.SC. PARIS T.54 PP 728 1862

3.14
ZOLLNER F. ZUR GESCHISCHTE DES HORIZONTALPENDELS
BERICHTEN DER KGL.GESELLSCH.D.WISS. 1869 1871

3.14
ZOLLNER F. UEBER EINE NEUE METHODE ZUR MESSUNG ANZIEHERDER UND ABSTOSSENDER KRAFTE.
ANNALEN DER PHYSIK UND CHEMIE BD.CL P 131 1873

3.14
ZOLLNER F. BESCHRIEBUNG UND ANWENDUNG DES HORIZONTALPENDELS.
ANNALEN DER PHYSIK UND CHEMIE BD.CL P 134 1873

3.14
ZOLLNER F. ZUR GESCHICHTE DES HORIZONTALPENDELS.
ANNALEN DER PHYSIK UND CHEMIE BD.CL PP 140-150 1873

3.14
SAFARIK BEITRAG ZUR GESCHICHTE DES HORIZONTALPENDELS.
ANNALEN DER PHYSIK UND CHEMIE BD.CL PP 150-157 1873

3.14
ECH P. LORENZ HENGLER, ERFINDER DES HORIZONTALPENDELS.
ANNALEN DER PHYSIK UND CHEMIE CL N.11 P 496 1873

3.14
DAVISON C. NOTE ON THE HISTORY OF THE HORIZONTAL AND BIFILAR PENDULUMS. REP.65 MEETING BRIT.ASSOC.ADV.SC.HELD AD IPSWICH 1895

3.14
LEVITSKII G.V. INFORMATION SUR LES CONDITIONS DE COMMANDE DES PENDULES HORIZONTAUX. NELLE STE.RUSSE GEOGR.TOME 33 PUBL.5 PP 475-476 479-482 1897

3.14
VOZNESENSKII A.V. ORGANISATION D OBSERVATIONS A L AIDE DE PENDULES HORIZONTAUX A IRKOUTSK IZV.EMP.GEOGR.RUSSE T.34 PP 644-646 1898

3.14
ORLOV A.YA. SUR LA DETERMINATION DES CONSTANTES DU PENDULE HORIZONTAL PAR LE PROCEDE DE GALITZIN IZV.COMM.PERMAN.SEISM.ACAD.SC. T.3 PUBL.3 PP 101-104 1910

3.14
HECKER O. OBSERVATION A L AIDE DU PENDULE HORIZONTAL DES DEFORMATIONS DE LA TERRE DUES A L INFLUENCE LUNI-SOLAIRE RECUEIL ART.SUR QUESTIONS GEOD.S.P.B.TOPOGR.MIL.66 1911

3.14
ORLOV A.YA. RESULTATS DES OBSERVATIONS DE LA DEFORMATION LUNI-SOLAIRE DE LA TERRE A JURJEV, TOMSK ET POTSDAM TRAV.OBS.ASTRON.INST.NOVOROSSIJSK ODESSA NO 2 PP 1-281 1915

3.14
LAGRANGE E. SUR L HISTOIRE DU PENDULE HORIZONTAL EN ITALIE CIEL ET TERRE XLVIII N.10-11 PP 197-201 1932

3.14
BONTCHKOVSKII V.F. METHODE D OBSERVATION DES INCLINAISONS DE LA SURFACE TERRESTRE ET RESULTATS DE CES OBSERVATIONS TRAV.CONS.METH.ETUDE MOUV.DEFORM.ECORCE TERR.PP 252-262 1948

3.14
BONTCHKOVSKII V.F. METHODE D OBSERVATION DES INCLINAISONS DE LA SURFACE TERRESTRE TRAV.INST.GEOPH.ACAD.SC.URSS NO 5,132 PP 49-60 1949

3.14
HAMBOURTSEV G.A. DEUX REGIMES DE TRAVAIL DES PENDULES HORIZONTAUX IZV.ACAD.SC.URSS SER.GEOPH.NO 3 PP 201-208 1953

3.14
NAMSARAI S. ETUDE DES ERREURS INSTRUMENTALES DES CLINOMETRES THESE DE L AUTEUR DE CANDIDAT DES SC.PHYS.MATH. INST.MOSCOU M.L.LOMONOSSOV FAC.PHYS. 1953

3.14
BONTCHKOVSKII V.F. NAMSARAI S. SUR LA PRECISION DES INDICATIONS DES CLINOMETRES TRAV.INST.GEOPH.ACAD.SC.URSS NO 22/149/ PP 3-18 1954

3.14
TOMASCHEK R. ANLEITUNG ZUR MESSUNG MIT HORIZONTALPENDELN. OBS.ROY.BELG.COMM.NO 114 S.GEOPH. NO 39 1957

3.14
UHLIG G. EINFLUSS VON LUFTTURBULENZ UND LUFTTEMPERATURSCHWANKUNGEN AUF HORIZONTALPENDEL. OBS.ROY.BELG.COMM.NO 142 S.GEOPH. NO 47 PP 91 1958

3.14
UHLIG G. EINFLUSS VON LUFTTURBULENZ UND LUFTTEMPERATURSCHWANKUNGEN AUF HORIZONTALPENDEL. B.I.M. N.15 PP 235-244 30 MARS 195

3.14
MITTELSTRASS G. THE SUSPENSION OF SIMPLE HORIZONTAL PENDULUMS. B.I.M.NO 43 PP 1633-1641 196

3.14
MITTELSTRASS G. EIN VORSCHLAG FUR EINE NULLMETHODE BEI HORIZONTALPENDELREGISTRIERUNGEN. B.I.M.NO 44 PP 1714 196

3.14
MITTELSTRASS G. KONSTRUKTIONSPRINZIPIEN VON HORIZONTALEINFACHPENDELN. EIN BEITRAG ZUM BAU UND ZUR EICHUNG VON HORIZONTALPENDELN. WISSENSCH.ARB.GEOD.PHOTOGR.KARTOG.HANNOVER NO 26 PP 59 196

3.14
SCHNEIDER M. ON THE RESULTS OF TILT OBSERVATIONS AT FREIBERG/SAXONY. MITTEIL.INST.THEOR.PHYS.GEOPH.FREIBERG NO 49 196

3.14
SKALSKY L. / PICHA I.
SUR CERTAINS PROBLEMES DES OBSERVATIONS DE MAREES AVEC LES CLINOMETRES ET LEUR PRECISION. RAPP.SYMP.INTERN.LENINGRAD 1968 MESURES INCLINAISON COMM.GEOPH.URSS.MOSCOU PP 11-35 — 1969

3.14
VOGEL A.
A HORIZONTAL PENDULUM STATION WITH TILT COMPENSATION. 6 EMF SYMPOSIUM MAREES TERRESTRES STRASBOURG OBS.ROY.BELG.COMM. A9 S.GEOPH. NO 96 PP 213-215 — 1970

3.14
BAGMET A.L. / CHIROKOV I.A.
SUR LES PROCECES D INSTALLATION DES CLINOMETRES OBSERVATIONS GEOPHYSIQUES COMPLEXES A OBNINSK INST.PHYS.TERRE SCHMIDT,OBNINSK PP 109-116 — 1974

3.14
BAKER T.F. / LENNON G.W.
SPATIAL COHERENCY AND TIDAL TILTS PROCEEDINGS 7TH INT.SYMP.EARTH TIDES, SOPRON 1973 AKADEMIAI KIADO BUDAPEST, PP 479-493 — 1976

3.14
SHIROKOV I.A. / ANOKHINA K.M.
ON THE COMPARATIVE OBSERVATIONS OF TIDAL TILTS BY AS-KANIA BOREHOLE TILTMETER AND TILTMETER OF OSTROVSKY PROCEEDINGS 7TH INT.SYMP.EARTH TIDES, SOPRON 1973 AKADEMIAI KIADO BUDAPEST, PP 595-606 — 1976

3.14
CHIROKOV Y.A. / ANOKHINA K.M.
SUR LES OBSERVATIONS COMPARATIVES DES INCLINAISONS DE MAREES PAR LE CLINOMETRE VERTICAL ASKANIA ET PAR LES CLINOMETRES D OSTROVSKII ROT.ET DEFORM.DE MAREES DE LA TERRE VII PP 21-24 1975
* B.I.M. NO 73 PP 4219-4224 — 1976

3.141
ORLOV A.YA.
OBSERVATIONS SUR LA DEFORMATION DE LA TERRE SOUS L INFLUENCE DE L ATTRACTION DE LA LUNE, FAITES A JURJEV A L AIDE DES PENDULES HORIZONTAUX DE ZOLLNER IZV.ACAD.SC. T.4 NO 10 PP 775-784 — 1910

3.141
ORLOV A.YA.
PREMIERE SERIE D OBSERVATION AUX PENDULES HORIZONTAUX A JURJEV DES DEFORMATIONS DE LA TERRE DUES A L INFLUENCE DE L ATTRACTION LUNAIRE TRAV.OBS.ASTRON.DE JURJEV T.23 I PP 1-70 — 1911

3.141
ORLOFF A.
BEOBACHTUNGEN UBER DIE DEFORMATION DES ERDKORPER UNTER DEM ATTRAKTIONSEINFLUSS DES MONDES AN ZOLLNERSCHEN HORIZONTALPENDELN. ASTR.NACHRICHTEN BAND 186 N.4446 PP 81-88 — 1911

3.141
KOHLER F.
GEODETICKYM MERENIM DOKAZANE PRESINUTI KAMBRICKYCH VRSTEV NA BREZOVYCH HORACH, SBORNIK CES.SPOL.ZEMEVEDNE. PRAHA 84-89 — 1914

3.141
KOHLER F.
NOVE VYZKUMY V URCOVANI TBARU A VELIKOSTI ZEME, V STANOVENI HUTNOSTI ZEMSKE A JEJIHO VNITRNIHO SLOZENI, SBORNIK CES.SPOL.ZEMEVEDNE. PRAHA PP 183-187 — 1914

3.141
CECHURA F.
GEOPYSIKALNI A GEODETICKE MERENI V PRI BRAMSKYCH DOLECH. SBORNIK I SJEZDU SLOVANK.GEOGR. A ETNOGRAFU PRAZE 1924 PP16-18 PRAHA — 1926

CECHURA F.
PREDBEZNA ZPRAVA O POZOROVANI POHYBUVRSTEV ZEMSKYCH V DOLECH BREZOHORSKYCH. SBORNIK I.SJEZDU SLOVANSK.GEOGR. A ETNOGR. PRAZE 1924 PP 16-18 PRAHA — 1926

3.141
ABOLD V.K.
RESULTATS DES OBSERVATIONS EXECUTEES DE 1916 A 1918 A LA STATION GRAVIMETRIQUE DE TOMSK SUR LES DEFORMATIONS DE LA TERRE DUES A L INFLUENCE DE L ATTRACTION LUNI-SOLAIRE IZV.INST.PHYS.MATH.STEKLOV T.2 PP 169-202 — 1927

3.141
ISHIMOTO M. CONSTRUCTION D UN PENDULE HORIZONTAL DE QUARTZ ET OBSERVATIONS SUR LES VARIATIONS DE L INCLINAISON DE LA SURFACE TERRESTRE.
JAP.JOURN. ASTR. AND GEOPH. V.6 N.2 PP 83-118 1928

3.141
EBLE L. NOUVELLES OBSERVATIONS DES DEVIATIONS PERIODIQUES DE LA VERTICALE A PARIS
ANNALES INST.PHYS.DE GLOBE UNIV. PARIS T.VI PP 38-59 1928

3.141
ORLOV A.YA. THE STUDY OF EARTH TIDES ON THE USSR
/TEXTE ANGLAIS EN PP 409-411/
REVUE ASTRON.T.13 PUBL.5 PP 407-409 1936

3.141
SCHAFFERNICHT W. HORIZONTALPENDELBEOBACHTUNGEN UBER LOTSCHWANKUNGEN IN MARBURG/LAHN.
ANNALEN DER PHYS. 5 FOLGE BD. 29 PP 349-357 1937

3.141
ORLOV A.YA. SUR LES DEFORMATIONS DE LA TERRE D APRES LES OBSERVATIONS AU PENDULE HORIZONTAL A TOMSK ET POLTAVA
IZV.ACAD.SC.URSS SER.GEOGR.GEOPH.NO 1 PP 1-29 1939

3.141
AKSENTIEVA Z.N. RESULTATS PRELIMINAIRES DES OBSERVATIONS DE MAREES TERRESTRES A L AIDE DE PENDULES HORIZONTAUX A POLTAVA DE 1930 A 1938
MESSAGER ACAD.SC.UKRAINE NO 2 1940

3.141
BONTCHKOVSKII V.F. INCLINAISON DE LA SURFACE TERRESTRE
TRAV.INST.SEISM.ACAD.SC.URSS NO 99 PP 1-56 1940

3.141
SASSA K.
NISHIMURA E. THE PRELIMINARY REPORT OF THE OBSERVATION OF THE GROUND TILTING.
CHIKYUBUTSURI 5. 1941

3.141
LOMAEV G. SUR LE COEFFICIENT D AMPLIFICATION DU CLINOMETRE AVEC SUSPENSION DE ZOLLNER
TRAV.INST.SEISM.ACAD.SC.URSS NO 106 PP 17-20 1941

3.141
BONTCHKOVSKII V.F. DEFORMATIONS ELASTIQUES DE LA TERRE D APRES LES OBSERVATIONS D INCLINAISONS
INDIC.SUR LA TERRE, NOUV.SER.T.3, 43 PP 3-15 1950

3.141
CALOI P. IL PENDOLO ORIZZONTALE COME CLINOMETRO.
ANNALI DI GEOFISICA 3 PP 451-457 1950

3.141
ZATOPEK K REGISTRACI POHYBU SVISLICE ZOLLNEROVYMI KYVADLY.
CAS.PRO PEST.MAT. A FYS. 79/4 D400-D419 1950

3.141
AKSENTIEVA Z.N. RESULTATS DEFINITIFS DE LA DETERMINATION DE L ONDE M2 DANS LES OSCILLATIONS DE LA VERTICALE A TOMSK DE 1912 A 1920
TRAV.OBS.GRAVIM.POLTAVA ACAD.SC.UKRAINE T.4 PP 3-88 KIEV 1951

3.141
ELLENBERGER H. BESCHREIBUNG DES HORIZONTALPENDEL NACH TOMASCHEK-ELLENBERGER.
OBS.ROY.BELG.COMM.NO 114 S.GEOPH. NO 39 1957

3.141
BONTCHKOVSKII V.F.
KARMALEIEVA R.M. INFLUENCE DE LA TORSION DES FILS SUR LES INDICATIONS D INCLINAISONS DONNEES PAR LE PENDULE
IZV.ACAD.SC.URSS SER.GEOPH.NO 9 PP 1181-1184 1957

3.141
BLUM P.A. SUR UN PENDULE POUR LA MESURE DES DEVIATIONS DE LA VERTICALE EN UN LIEU.
C.R.ACAD.SC.PARIS N.264 P 2389 21 AVRIL 1958

3.141
BLUM P.A.
JOBERT G. OBSERVATION DES DEVIATIONS DE LA VERTICALE A L AIDE DE PENDULES HORIZONTAUX.
OBS.ROY.BELG.COMM.NO 142 S.GEOPH. NO 47 PP 111-113 1958

3.141
FRITZSCH E. DIE EMPFINDLICHKEIT DES ZOLLNERPENDELS UNTER DEM EINFLUSS DES FADENTORSION.
OBS.ROY.BELG.COMM.NO 142 S.GEOPH. NO 47 PP 125-130 1958

3.141
BLUM P.
JOBERT G.
JOBERT N.
PREMIERS RESULTATS OBTENUS A L AIDE D INCLINOMETRES.
C.R.SEANCES ACAD.SC.INST.PHYS.GLOBE T.248 PP 1551-1554 1959

3.141
JOBERT G. THEORIE DU PENDULE DE ZOLLNER ET DU PENDULE DE LETTAU.
GEOFISICA PURA E APPLICATA 44 PP 25-73 1959

3.141
SCHNEIDER M. EINE BEMERKUNG ZUR BESTIMMUNG DER EIGENPERIODE VON HORIZONTALPENDELN.
IVME SYMP.INTERNATIONAL SUR LES MAREES TERRESTRES.
OBS.ROY.BELG.COMM. NO 188 S.GEOPH. NO 58 PP 184-185 1961

3.141
BLUM P.A. CONTRIBUTION A L ETUDE DES VARIATIONS DE LA VERTICALE EN UN LIEU
INST.PHYS.GLOBE PARIS ANN.GEOPH.T.19 N.3 PP 215-243 1963

3.141
PICHA J. OBSERVATIONS OF EARTH TIDES IN CZECHOSLOVAKIA.
STUD.GEOPH.GEOD. 3.11 PP 318-319 1967

3.141
HATZFELD D. MISE EN EVIDENCE DE COMPOSANTES FAIBLES DE LA MAREE TERRESTRE D INCLINAISON.
C.R. AC. SC. PARIS T.270 PP 818-820 1970

3.141
SIMON D. FIRST RESULTS OF EARTH TIDAL OBSERVATIONS IN AIRTIGHT MEASURING BOXES
XV GEN.ASS.UGGI - INT.ASS.GEOD.MOSCOU 1971

3.141
TCHOUPROUNOVA O.V. RESULTATS DES OBSERVATIONS CLINOMETRIQUES A SIMFEROPOL
ROTATION ET DEFORM.MAREES DE LA TERRE,FASC.1
ACAD.SC.UKR.OBS.GRAV.POLTAVA KIEV PP 318-336 1970
* B.I.M. NO 63 PP 3309-3320 1972

3.141
VANICEK P.
LENNON G.W.
THE THEORY OF MOTION OF THE HORIZONTAL PENDULUM WITH A ZOLLNER SUSPENSION AND SOME INDICATIONS FOR THE INSTRUMENTAL DESIGN
STUDIA GEOPH. ET GEOD. N.16 PP 30-50 1972

3.141
AKSENTIEVA Z.N.
TCHOUPROUNOVA O.V.
PREMIERS RESULTATS DES OBSERVATIONS CLINOMETRIQUES AU FLECHISSEMENT FRONTIERE DES CARPATHES
ROTATION ET DEFORM.MAREES DE LA TERRE,FASC.1
ACAD.SC.UKR.OBS.GRAV.POLTAVA KIEV PP 280-282 1970
* B.I.M. NO 65 PP 3422-3425 1973

3.141
OURASSINA I.A. OBSERVATIONS CLINOMETRIQUES A L OBSERVATOIRE ASTRONOMIQUE ENGELHARDT POUR LA PERIODE 1961-1966
ROTATION ET DEFORM.MAREES DE LA TERRE,FASC.1
ACAD.SC.UKR.OBS.GRAV.POLTAVA KIEV PP 206-225 1970
* B.I.M. N.66 PP 3604-3620 1973

3.141
BACHEM H.CH. ZUR AUSZEREN GENAUIGKEIT DER ERDGEZEITENREGISTRIERUNG MIT HORIZONTALPENDELN UNTER BERUCKSICHTIGUNG LOKALER TEKTONISCHER STOREFFEKTE
MITTEILUNGEN NO 52-UNIV.HANNOVER- PP 1-80 1973

3.141
BALENKO V.G.
OVTEKINIKOV V.A.
KOUTNII A.M.
GOLOUBITSKII V.G.
PENDULE HORIZONTAL AVEC SUSPENSION DE ZOLLNER A FILS METALLIQUES
ROT.& DEFORM.DE MAREES DE LA TERRE VOL.VI PP 3-16 KIEV 1974

3.141.1
VERBAANDERT J.
MELCHIOR P.
CONSTUCTION ET ETALONNAGE D UN PENDULE HORIZONTAL EN QUARTZ
OBS.ROY.BELG.COMM.NO 142 S.GEOPH. NO 47 PP 114-115 1958

3.141.1
VERBAANDERT J. MELCHIOR P.J. LES STATIONS GEOPHYSIQUES SOUTERRAINES ET LES PENDULES HORIZONTAUX DE L OBSERVATOIRE ROYAL DE BELGIQUE - RESULTATS DES ANALYSES DES ENREGISTREMENTS PROBLEMES D ETALONNAGES - EFFETS DES CRUES DE LA MEUSE - INTRODUCTION A LA VISITE DE LA STATION DE SCLAIGNEAUX.
IVME SYMP.INTERNATIONAL SUR LES MAREES TERRESTRES.
OBS.ROY.BELG.COMM.NO 188 S.GEOPH. NO 58 PP 192-200 1961

3.141.1
AKHAVAN A. ETUDE COMPARATIVE DES VARIATIONS DU FACTEUR GAMMA D APRES DE LONGS ENREGISTREMENTS SIMULTANES OBTENUS PAR DIFFERENTS PENDULES.
B.I.M. NO 47 PP 2013-2025 1967

3.141.1
VOGEL A. A HORIZONTAL PENDULUM STATION WITH TILT COMPENSATION
REP.1, UNIV.OF UPPSALA, PP 1-8 1970

BEAUMONT C. HYNDMAN R.D. KEEN M.J. A NEW TECHNIQUE FOR THE INSTALLATION OF TILTMETERS
EARTH PLANET.SCI.LETT. VOL 8 PP 337-340 1970

3.141.1
PICHA J. INVESTIGATION OF TIDES OF THE EARTH S CRUST
UPPER MANTLE PROJ.PROGR.CESKOSL.AKAD.VED. PP 8-12 PRAHA 1971

3.141.1
BONATZ M. EIN EXPERIMENT ZUR UNTERSUCHUNG KURZZEITIGER DRIFT-ANDERUNGEN BEI VERBAANDERT-MELCHIOR PENDELN.
B.I.M. N.62 DECEMBRE PP 3196-3201 1971

3.141.1
VOGEL K.A. ANDERSON A.J. AN IMPROVED CONTROLLED TILTMETER SYSTEM AND LATEST MEASUREMENTS IN SWEDEN.
MEASUREM., INTERPRETAT., CHANGES OF STRAIN IN THE EARTH
PHIL.TRANS.ROY.SOC.SER.A, VOL.274, NO 1239 PP 305-310 1973

3.141.1
AKHAVAN SADEGHI A. SUR L ENREGISTREMENT DES VARIATIONS LUNI-SOLAIRE DE LA DIRECTION DE LA VERTICALE PAR UNE METHODE D AMORTISSEMENT ELECTRO-MAGNETIQUE -LA MAREE CLINOMETRIQUE- /LES COMPOSANTES HORIZONTALES/
JOURN.EARTH AND SPACE PHYSICS VOL.3 NO 1 & 2 PP 19-43 1974

3.141.1
ORTIZ R. VIEIRA R. SISTEMA DE DETECCION Y REGISTRO DE INFORMACION APLICADO A LA INSTRUMENTACION DE UNA ESTOCION DE MAREAS TERRESTRES
ASAMBLEA NAC. DE GEOD.Y GEOF. IA 218 VOL 2 PP 181-185 1974

3.141.1
X INSTITUTE OF OCEANOGRAPHIC SCIENCES
ANNUAL REPORT, WORMLEY GREAT BRITAIN 1974

3.141.1
ANDERSON A.J. DYNAMIC RESPONSE OF A QUARTZ HORIZONTAL PENDULUM SYSTEM
PROCEEDINGS 7TH INT.SYMP.EARTH TIDES, SOPRON 1973
AKADEMIAI KIADO BUDAPEST, PP 197-204 1976

3.141.1
VAN RUYMBEKE M. SUR UN PENDULE HORIZONTAL EQUIPE D UN CAPTEUR DE DEPLACEMENT A CAPACITE VARIABLE
BULL.GEOD.VOL 50 NO 3 PP 281-290 1976

3.141.1
VIEIRA R. LAMBAS F. OREJANA M. VIVANCO V. MODIFICATION DE LOS PENDULOS VERBAANDERT-MELCHIOR
2EME ASAMB.NAC.DE GEOD.Y GEOF. P 46 MADRID 197[illegible]

3.141.2
OSTROVSKII A.E. CLINOMETRE SEISMOLOGIQUE AVEC ENREGISTREUR PHOTOELECTRIQUE
BULL.UNION SEISM.ACAD.SC.URSS NO 6 PP 130-134 195[illegible]

3.141.2
OSTROVSKY A.E. HOROMSKY A.V. OBSERVATION OF TIDAL TILTS OF THE EARTH S CRUST BY MEANS OF A PHOTOELECTRIC TILTMETER IN COUNRAD IN OCTOBER DECEMBER 1957.
OBS.ROY.BELG.COMM.NO 142 S.GEOPH. NO 47 PP 96-97 195[illegible]

3.141.2
OSTROVSKII A.E. KHOROMSKII A.V. MIRONOVA L.Y.
RESULTATS DES OBSERVATIONS DES INCLINAISONS DE MAREES DE LA TERRE AVEC INCLINOMETRE PHOTOELECTRIQUE A KONDARA EN OCTOBRE-DECEMBRE 1957
REC.TRAV.MAREES TERR.NO 1 GROUPE 13 GRAV., 1959

3.141.2
OSTROVSKY A.E. KHOROMSKY A. MIRONOVA L.
RESULTS OF OBSERVATIONS OF TIDAL TILTS OF THE EARTH S SURFACE BY MEANS OF A PHOTOELECTRIC TILTMETER IN KOUNRAD IN OCTOBER-DECEMBER 1957
3E SYMP. MAREES TERRESTRES TRIESTE
BOLL.GEOFISICA TEORICA E APPL.VOL.2 NO 5 PP 73-76 1960

3.141.2
OSTROVSKII A.E.
LE CLINOMETRE A ENREGISTREMENT PHOTOELECTRIQUE
GROUPE XIII PROG.AGI GRAV.VOL.2 PP 41-75 MOSCOU 1961
* B.I.M. NO 25 PP 500-536 1961
* B.I.M. NO 26 PP 540-553 1961

3.141.2
OSTROVSKII A.E.
TILTS OF THE EARTH S SURFACE ACCORDING TO OBSERVATIONS IN CENTRAL ASIA DURING THE IGY.
INTERN.ASS.GEOD.VOL.II, XIII GEN.ASS.BERKELEY PP 11-39 1963

3.141.2
OSTROVSKY A.E. FANDUSHINA S.M.
TIDAL TILTS ACCORDING TO TWO-DAY SERIES OF OBSERVATIONS.
5 EME SYMP.INT.SUR LES MAREES TERRESTRES.
OBS.ROY.BELG.COMM.NO 236 S.GEOPH.NO 69 PP 330-332 1964

3.141.2
OSTROVSKY A.E. FANDUSHINA S.M.
TIDAL TILTS OBSERVED WITH TILTMETERS INSTALLED IN THE SAME AZIMUTH.
5 EME SYMP.INT.SUR LES MAREES TERRESTRES.
OBS.ROY.BELG.COMM.NO 236 S.GEOPH.NO 69 PP 333-340 1964

3.141.2
KORBA S.N. MATVEYEV P.S.
QUELQUES RESULTATS D ANALYSE HARMONIQUE DES OBSERVATIONS CLINOMETRIQUES D APRES DES PROGRAMMES GLISSANTS
MAR.TERR.ACAD.SC.UK.SSR.OBS.GRAV.POLTAVA KIEV 1966
* B.I.M. NO 51 PP 2341-2348 1968

3.141.2
OSTROVSKI A.E. MATVEYEV P.S. BALENKO V.G.
SUR LES OBSERVATIONS CLINOMETRIQUES DANS LES SONDAGES.
RAPP.SYMP.INTERN.LENINGRAD 1968 MESURES INCLINAISON COMM.GEOPH.URSS. MOSCOU PP 58-65 1969
* B.I.M. N.57 MAI PP 2729-2733 1970

3.141.2
FANDIOUCHINA S.M.
TIDAL TILTS IN THE TIEN-CHAN.
6 EME SYMPOSIUM MAREES TERRESTRES STRASBOURG
OBS.ROY.BELG.COMM. A9 S.GEOPH. NO 96 PP 73-83 1970

3.141.2
CHIROKOV I.A. OSTROVSKY A.E.
SLOW AND TIDAL DEFORMATIONS OF THE EARTH SURFACE ACCORDING TO TILT OBSERVATIONS AT THE -SAYANY- STATION.
6 EME SYMPOSIUM MAREES TERRESTRES STRASBOURG
OBS.ROY.BELG.COMM. A9 S.GEOPH. NO 96 PP 95-96 1970

3.141.2
CHIROKHOV I.A. OSTROVSKII A.E.
RESULTATS PRELIMINAIRES DES OBSERVATIONS CLINOMETRIQUES A DAGESTAN
ROTATION ET DEFORM.MAREES DE LA TERRE,FASC.1
ACAD.SC.UKR.OBS.GRAV.POLTAVA KIEV PP 165-173 1970

. 3.141.2
BAKHROUCHINE A.B.
CERTAINS RESULTATS D APPLICATION DES CLINOMETRES PHOTOELECTRIQUES POUR L ENREGISTREMENT DES PETITS DEPLACEMENTS ANGULAIRES DANS LES CONDITIONS DES MINES.
ROTATION ET DEFORM.MAREES DE LA TERRE,FASC.1
ACAD.SC.UKR.OBS.GRAV.POLTAVA KIEV PP 234-240 1970

3.141.2
BAKHROUCHINE A.B.
CERTAINS RESULTATS D APPLICATION DES CLINOMETRES PHOTOELECTRIQUES POUR L ENREGISTREMENT DES PETITS DEPLACEMENTS ANGULAIRES DANS LES CONDITIONS DES MINES.
ROTATION ET DEFORM.MAREES DE LA TERRE,FASC.1
ACAD.SC.UKR.OBS.GRAV.POLTAVA KIEV PP 234-240 1970

3.141.2
LATININA L.A.
MESURE DES DEFORMATIONS DE MAREES A LA STATION

STARKOVA E.IA. DJERINO TADJIKISTAN.
KARMALEYEVA R.M. ROTATION ET DEFORM.MAREES DE LA TERRE,FASC.1
RISAEVA S.D. ACAD.SC.UKR.OBS.GRAV.POLTAVA KIEV PP 283-295 1970

3.141.2
MATVEYEV P.S. OBSERVATIONS CLINOMETRIQUES A MOURAKHOVKA
GOLOUBITSKII V.G.
DOUBIK B.S. ROT.ET DEFORM.DE MAREES TERRE VOL.3 PP 64-68 KIEV 1971

3.141.2
EVTOUCHENKO E.Y. RESULTATS DE L ANALYSE HARMONIQUE DES OBSERVATIONS
CLINOMETRIQUES A DONBASS DE 1962 A 1966
ROT.ET DEFORM.DE MAREES TERRE VOL.3 PP 52-60 KIEV 1971

3.141.2
CHIROKOV I.A. RESULTATS PRELIMINAIRES DES OBSERVATIONS
OSTROVSKII A.E. CLINOMETRIQUES A DAGESTAN
ROT.ET DEFORM.MAR.TERR.FASC.1 KIEV 1970 PP 165-173
* B.I.M. NO 61 PP 3077-3083 1971

3.141.2
OSTROVSKII A.E. OBSERVATIONS CLINOMETRIQUES AVEC LA BASE AGRANDIE
CHIROKOV I.A. DU CLINOMETRE.
METH.MES.MAR.TERR.& DEFORM.LENTES SURFACE TERRE
ACAD.SC.URSS.INST.PHYS.T.SCHMIDT MOSCOU PP 160-166 1970
* B.I.M.N.62 PP 3217-3222 1971

3.141.2
IAKOVLIEV V.N. OBSERVATION DES INCLINAISONS DE MAREES PRES DE
SANDLER F.D. L ALIGNEMENT DE LA DIGUE D ANDIGAN EN 1964-1965
LEONTIEVA T.M. * B.I.M. NO 63 PP 3240-3243 1972

3.141.2
BALAVADZE B.K. INVESTIGATION OF RECENT TECTONIC MOVEMENTS OF THE
ABASHIDZE V.G. EARTH S CRUST BY MEANS OF TILTMETER
INST.GEOP.AC.OD SC. GEORGIAN SSR PP 3-9 TBILISI 1973

3.141.2
BAGMET A.L. DETERMINATION DE LA DERIVE INSTRUMENTALE DU POINT ZERO
DU CLINOMETRE AVEC AMPLIFICATION PHOTO-ELECTRIQUE
ROT.ET DEFORM.DE MAREES DE LA TERRE, VOL.5, PP 57-63 1973

3.141.2
BAGMET A.L. AUGMENTATION DE LA STABILITE DE LA SENSIBILITE
MICHATKINE V.N. DES CLINOMETRES AVEC AGRANDISSEMENT PHOTOELECTRIQUE
KOUTNII A.M. ROT.ET DEFORMATIONS DE MAREES DE LA TERRE V.5 PP 56-62 1973

3.141.2
POLIAKOV V.K. RESULTATS DES OBSERVATIONS CLINOMETRIQUES DANS ZEISKAIA
ACAD.NAOUK SSSR, INST.FISIKA ZEMLI PP 119-121 1973
* B.I.M. NO 69 PP 3874-3877 1974

3.141.2
BAGMET A.L. SUR LA PRECISION DE MESURE DES INCLINAISONS DE MAREES
KOLOMIETZ A.S. PAR LES CLINOMETRES AVEC AGRANDISSEMENT PHOTOELECTRIQUE.
OBSERV.GEOPHYSIQUES COMPLEXES A OBNINSK PP 104-108 1974

3.141.2
BAGMET A.L. DETERMINATION DE LA DERIVE INSTRUMENTALE DU CLINOMETRE
AVEC AGRANDISSEMENT PHOTOELECTRIQUE
ROT.ET DEFORM.DE MAREES DE LA TERRE NO 5 PP 63-69 1973
* B.I.M. NO 70 PP 3965-3975 1975

3.141.2
OSTROVSKII A.E. SUR LA POSSIBILITE D UTILISER LE CLINOMETRE
MATVEYEV P.S. PHOTOELECTRIQUE EN METHODE DE ZERO
SCHLIAKHOVOI V.P. ACAD.NAOUK UKRAINSKOI SSR, OBS.GRAV.POLTAVA
ROTAT.DEFOR.MAREES DE LA TERRE VII PP 42-49 1975

3.141.2
CHIROKOV I.A. SUR LES OBSERVATIONS COMPARATIVES DES INCLINAISONS
ANOKHINA K.M. DE MAREES PAR LE CLINOMETRE A FENTE /ASKANIA/
ET LES CLINOMETRES D OSTROVSKII
ACAD.NAOUK UKRAINSKOI SSR, OBS.GRAV.POLTAVA
ROTAT.DEFOR.MAREES DE LA TERRE VII PP 21-24 1975

3.142
V.REBEUR PASCHWITZ RESULTATE AUS BEOBACHTUNGEN AM HORIZONTALPENDEL ZUR
UNTERSUCHUNG DER RELATIVEN VARIATIONEN DER LOTHLINIE.
ASTRONOM.NACHRICHT.BD.126 NO 3001-02 PP 1-18 1891

3.142
V.REBEUR PASCHWITZ UEBER HORIZONTALPENDEL BEOBACHTUNGEN IN WILHELMSHAVEN,
POTSDAM UND PUERTO OROTAVA AUF TENERIFFA.
ASTRONOM.NACHRICHTEN BD.130 NO 3 109-10 PP 194-216 1892

3.142
LEVITSKII G.V. QUELQUES RESULTATS D OBSERVATIONS EFFECTUEES A L OBSERVATOIRE ASTRONOMIQUE DE L UNIVERSITE DE KHARKOV AVEC LES PENDULES REBEUR-PASCHWITZ.
COMM.SOC.MATH.KHARKOV TOME IV NO 5-6 PP 206-208 1894

3.142
KORTASII I.E. OBSERVATIONS A L AIDE DU PENDULE HORIZONTAL REBEUR-PASCHWITZ A L OBSERVATOIRE DE NICOLAIEV.
NELLE.SOC.ASTRON.RUSSE PUBL.IV PP 24-25 1895

3.142
KORTATSII Y. OBSERVATIONS A L AIDE DE PENDULES HORIZONTAUX VON REBEUR-PASCHWITZ A L OBSERVATOIRE DE NIKOLAEV
IZV.OBS.ASTRON.RUSSE PUBL.4 P 24 1895-PUBL.5 P 301 1896

3.142
KORTASII I.E. REBEUR PASCHWITZ.
POSTHUME NELLES.SOC.ASTRON.RUSSE PUBL.V NO 5 PP 200-202 1896

3.142
HECKER O. DAS HORIZONTALPENDEL.
ZEITSCHRIFT FUR INSTRUMENTENKUNDE 16 PP 2-16 1896

3.142
EHLERT R. HORIZONTALPENDELBEOBACHTUNGEN IM MERIDIAN ZU STRASSBURG.I.E.VON APRIL BIS WINTER 1895.
GERL.BEITRAGE ZUR GEOPH.BD.3 PP 131-215 1898

3.142
SCHWEYDAR W. UNTERSUCHUNGEN DER OSCILLATIONEN DER LOTLINIE AUF DEM ASTROMETR.INSTITUT DER GROSSH.STERNWARTE ZU HEIDELBERG.
GERL.BEITRAGE ZUR GEOPH.7 PP 33-120 1905

3.142
HECKER O. BEOBACHTUNGEN AN HORIZONTALPENDELN UBER DIE DEFORMATION DES ERDKORPERS UNTER DEM EINFLUSS VON SONNE UND MOND
VEROFF.K.PREUSS.GEOD.INSTITUTES NF NO 32 1907

3.142
HECKER O. BEOBACHTUNGEN AN HORIZONTALPENDELN UBER DIE DEFORMATION DES ERDKORPERS UNTER DEM EINFLUSS VON SONNE UND MOND
II.VEROFF.D.KONIGL.PREUSS.GEOD.INST.NF 49 1911

3.142
HAID M. GEZEITEN UND STARRHEIDSKOEFFICIENT N DER FESTEN ERDE ABGELEITET AUS DEN REGIESTRIERUNGEN DER HORIZONTALPENDEL IN FREIBURG I.B. UND DURLACH VOM 1 NOVEMBER 1907-1908.
C.R.SCES.17E CONF.GLE.ASSOC.GEOD.INT.HAMBOURG PP 255-258 1912

3.142
SHIDA T.
MATSUYAMA M. NOTE ON HECKER S OBSERVATIONS OF HORIZONTAL PENDULUMS
MEM.OF COLL.SC.&ENG.KYOTO IMP.UNIV.IV PP 187-224 1912

3.142
SHIDA T. HORIZONTAL PENDULUM OBSERVATIONS OF THE CHANGE OF PLUMB LINE AT KAMIGANO,KYOTO.
MEM.OF COLL.SC.& ENG.KYOTO IMP.UNIV.IV PP 23-174 1912

3.142
PICHA J. ERGEBNISSE DER GEZEITENBEOBACHTUNGEN DER FESTEN ERDKRUSTE IN BREZOVE HORY IN DEN JAHREN 1936-1939
B.I.M. NO 5 PP 80-84 1957

3.143
BRAAK C. ON TIDAL FORCES AS DETERMINED BY MEANS OF WIECHERT S ASTATIC SEISMOGRAPH.
PROC.KON.AKAD.WETENSCH.AMSTERDAM SC.XIII,2 PP 1231-1236 1911

3.143
.ENNON G.W. A REPORT ON THE PROGRESS OF WORK ON IGY TILT OBSERVATIONS AT BIDSTON.
3E SYMP. MAREES TERRESTRES TRIESTE
BOLL.GEOFISICA TEORICA E APPL.VOL.2 NO 5 PP 29-37 1960

3.144
ETTAU H. DAS HORIZONTALDOPPELPENDEL.
ZEITSCHRIFT FUR GEOPHYSIK 13 PP 25-33 1937
LEIPZIG-HABILITATIONSSCH. PP 60 1936
VEROFF.GEOPH.INST.LEIPZIG L.WEICKMANN SER.2 PP 142 1937

3.144
OWAK S. WYZNACZANIE CZULOSCI WAHADEL POZIOMYCH PODWOJNYCH /TYPU LETTAU/.
ACTA GEOPH.POLONICA VOL.VIII NO 3 PP 359-369 1960

3.144

3.144
SIMON D. — ANALYSE VON NEIGUNGSGANGEN BEI HORIZONTALDOPPENPENDELN NACH LETTAU AUF METEOROLOGISCHE EFFEKTE HIN. FREIBERG FORSCHUNGSH.REIHE C 194 GEOPH.PP 1-31 — 1966

3.15
SCHULZE R. — DER NEUE ASKANIA-CLINOGRAPH MIT FERNBEDIENUNG UND EICHEINRICHTUNG 3E SYMP. MAREES TERRESTRES TRIESTE BOLL.GEOFISICA TEORICA E APPL.VOL.2 NO 5 PP 99-105 — 1960

3.15
GRAF A. — DAS VERTIKALPENDEL ANSTELLE DES HORIZONTALPENDELS FUR DIE REGISTRIERUNG DER GEZEITEN U.V.KLEINSTEN NEIGUNGEN. B.I.M.NO 34 PP 1069-1075 1963 DEUTSCHE GEOD.KOMM.REIHE B.HEFT NO 104 — 1963

3.15
GRAF A. — ERSTE NEIGUNGSREGISTRIERUNGEN MIT DEM VERTIKALPENDEL IN EINEM 30 M BOHRLOCH. 5 EME SYMP.INT.SUR LES MAREES TERRESTRES. OBS.ROY.BELG.COMM.NO 236 S.GEOPH.NO 69 PP 249-254 — 1964

3.15
JACOBY H.D. — DAS NEUE BOHRLOCH - GEZEITEN - PENDEL NACH GRAF. ASKANIA WARTE 23 HEFT 67 PP 12-17 SONDERDRUCK NO 1468 — 1966

3.15
SCHNEIDER M. — EIN VERTIKAL-EINSTAB-PENDEL ZUR MESSUNG DER ERDGEZEITENBEDINGTEN LOTSCHWANKUNGEN. STUD.GEOPH.GEOD.10 PP 422-436 — 1966

3.15
SCHNEIDER M.S. — DAS VERTIKALPENDEL ALS INDIKATOR FUR LOTSCHWANKUNGEN. GEOF.KOZLEMENYEK XVI, 1-2 PP 91-99 BUDAPEST — 1967

3.15
SIMON I., EMSLIE A.G., MC.CONNELL R.K. — SENSITIVE TILTMETER UTILIZING A DIAMAGNETIC SUSPENSION REV.SCI.INSTRUM. 39 PP 1666-1671 — 1968

3.15
HANSEN S. — A HIGHLY STABLE GEOPHYSICAL TILTMETER /ABSTRACT/ TRANS.AGU 49 664 — 1968

3.15
WOOD M.D., KOVACH R.L. — OBSERVATIONS OF EARTH TIDES AND TILTS IN THE SAN FRANCISCO BAY AREA /ABSTRACT TRANS.AGU 49 PP 114 — 1968

3.15
SIMON I., EMSLIE A.G., STRONG P.F., MC CONNELL R.K. — SENSITIVE TILTMETER UTILIZING A DIAMAGNETIC SUSPENSION REPR.REV.SCIENT.INSTR.VOL.39 NO 11 PP 1666-1671 — 1968

3.15
FLACH D., ROSENBACH O., WILHELM H. — OPERATIONAL TEST OF THE ASKANIA-BOREHOLE TILTMETER / EARTH TIDE PENDULUM/. 6 EME SYMPOSIUM MAREES TERRESTRES STRASBOURG OBS.ROY.BELG.COMM. A9 S.GEOPH. NO 96 PP 167 — 1970

3.15
CABANISS G.H., MC CONNELL R.K., SIMON I. — CRUSTAL TILT IN COASTAL NEW ENGLAND FROM BOREHOLE TILTMETERS /ABSTRACT/ EOS - TRANS. AGU 51, 740 — 1970

3.15
TSUBOKAWA I. — A NEW TYPE ELECTROMAGNETIC TILTMETER. 6 EME SYMPOSIUM MAREES TERRESTRES STRASBOURG OBS.ROY.BELG.COMM. A9 S.GEOPH. NO 96 PP 217-218 — 1970

3.15
TSUBOKAWA I., NAGASAWA K., YANAGISAWA M., MURATA I., TAJIMA H. — NEW ELECTROMAGNETIC TILTOMETER JOURNAL OF GEOD.SOCIETY OF JAPAN, PP 209-231 — 1970

3.15
FLACH D., JENTZSCH G., ROSENBACH O., WILHELM H. — BALL-CALIBRATION OF THE ASKANIA BOREHOLE TILTMETER /EARTH TIDE PENDULUM/ ZEITSCHRIFT GEOPH. V 37 PP 1005-1011 — 197

3.15
KNEISSL M. — EARTH TIDE MEASUREMENTS BY MEANS OF BORE-HOLE

VERTICAL PENDULUMS
XV ASS.GEN. UGGI - ASS.INT.DE GEODESIE MOSCOU AOUT 1971

3.15
SIMON IVAN
EXPERIMENTAL AND THEORETICAL STUDIES OF LONG-PERIOD TILT OF EARTH S CRUST - PART I EXPERIMENTAL
AIR FORCE CAMB.RES.LAB. NO 71-0388/1/ PP 1-66 1971

3.15
ALLEN R.V.
A BOREHOLE TILTMETER FOR MEASUREMENTS AT TIDAL SENSITIVITY
BULL.SEISMOL.SOC.AMER.V.62 N.3 PP 815-821 1972

3.15
BACHEM H.C.
VORSCHLAG ZUM BAU EINES NEIGUNGSGERATES ZUR EICHUNG UND AZIMUTUBERTRAGUNG DES ASKANIA-BOHRLOCH-GEZEITENPENDELS
B.I.M. NO 63 PP 3283-3287 1972

8.15
MAUK F.J.
JOHNSTON M.J.S.
ON THE TRIGGERING OF VOLCANIC ERUPTIONS BY EARTH TIDES
J.G.R. VOL 78 NO 17 PP 3356-3362 1973

3.15
SKALSKY L.
SOUKUP V.
PHOTO-ELECTRIC COMPENSATING TILTMETER
STUDIA GEOPH.ET GEOD. V.18 PP 47-58 1974

3.15
FLACH D.
GROBE W.
AN ELECTRONIC FILTER AND DAMPING SYSTEM FOR THE ASKANIA BOREHOLE TILTMETER
J.GEOPHYS. VOL 41 PP 303-310 1975

3.15
FLACH D.
PRESENT STATE OF THE DEVELOPMENT OF THE ASKANIA BOREHOLE TILTMETER
PROCEEDINGS 7TH INT.SYMP.EARTH TIDES,SOPRON 1973
AKADEMIAI KIADO BUDAPEST, PP 249-258 1976

3.15
FLACH D.
THE ASKANIA BOREHOLE TILTMETER IN FIELD OPERATION
PROCEEDINGS 7TH INT.SYMP.EARTH TIDES, SOPRON 1973
AKADEMIAI KIADO BUDAPEST, PP 243-247 1976

3.15
GROSSE -
BRAUCKMANN W.
HIGH PRECISION BALL CALIBRATION OF THE ASKANIA BOREHOLE TILTMETER / EARTH TIDE PENDULUM/
PROCEEDINGS 7TH INT.SYMP.EARTH TIDES, SOPRON 1973
AKADEMIAI KIADO BUDAPEST, PP 259-265 1976

3.15
SCHMITZ -
HUBSCH H.
SIMULTANEOUS RECORDS WITH TWO BOREHOLE TILTMETERS IN A TEST AREA IN UPPER BAVARIA
PROCEEDINGS 7TH INT.SYMP.EARTH TIDES, SOPRON 1973
AKADEMIAI KIADO BUDAPEST, PP 273-283 1976

3.15
ZSCHAU J.
A CALIBRATION-, COMPENSATION AND DAMPING DEVICE FOR THE ASKANIA BOREHOLE TILTMETER
PROCEEDINGS 7TH INT.SYMP.EARTH TIDES, SOPRON 1973
AKADEMIAI KIADO BUDAPEST, PP 313-326 1976

3.15
SCHLEMMER H.
SOME INSTRUMENTAL EXPERIENCES
PROCEEDINGS 7TH INT.SYMP.EARTH TIDES, SOPRON 1973
AKADEMIAI KIADO BUDAPEST, PP 267-272 1976

3.15
SKALSKY L.
SOUKUP V.
PHOTO-ELECTRIC COMPENSATING TILTMETER
PROCEEDINGS 7TH INT.SYMP.EARTH TIDES, SOPRON 1973
AKADEMIAI KIADO BUDAPEST, PP 291-300 1976

3.16
VERBAANDERT J.
ETALONNAGE DES PENDULES HORIZONTAUX PAR CRAPAUDINE DILATABLE ETUDIEE INTERFEROMETRIQUEMENT.
3E SYMP. MAREES TERRESTRES TRIESTE
BOLL.GEOFISICA TEORICA E APPL.VOL.2 NO 5 PP 81-90 1960

3.16
GRAF A.
VERHALTEN UND EICHUNG DER MUNCHENER HORIZONTALPENDEL.
3E SYMP. MAREES TERRESTRES TRIESTE
BOLL.GEOFISICA TEORICA E APPL.VOL.2 NO 5 PP 91-98 1960

3.16
SCHNEIDER M.
EINE BEMERKUNG ZUR BESTIMMUNG DER EIGENPERIODE VON HORIZONTALPENDELN.
MITT.INST.THEOR.PHYS.GEOPH.BERGAKAD.FREIBERG NO 26
STUDIA GEOPH. ET GEOD. BD.6 NO 1 PP 86-94 1962

3.16
VERBAANDERT J. ETALONNAGE DES PENDULES HORIZONTAUX.
BOLL.GEOFIS.TRIESTE VOL.4 N.16 PP 419-446 1962

3.16
PICHA J.
SKALSKY L. BESTIMMUNG DER SCHWINGUNGSPERIODE UND EMPFINDLICHKEIT EINFACHER HORIZONTALPENDEL FUR GEZEITENMESSUNGEN.
STUD.GEOPH.GEOD.PRAHA 72 PP 156-163 1963

3.16
BLEUMER H. EICHUNG VON HORIZONTALPENDELN MIT EINEM NEIGUNGSGERAT.
WISSENSCH.ARBEIT.INST.GEOD.PHOTOGRAMM.HANNOVER NO 22 1964

3.16
VERBAANDERT J. QUELQUES PROGRES REALISES DANS LA METHODE D ETALONNAGE DES PENDULES HORIZONTAUX.
B.I.M.NO 35 PP 1165-1166 1964

3.16
MELCHIOR P.J. SUR LA QUESTION DE LA DEPENDANCE DE LA PERIODE ET DE L AMPLITUDE POUR LES PENDULES HORIZONTAUX EN QUARTZ.
B.I.M.NO 35 PP 1180-1182 1964

3.16
VERBAANDERT J. NOUVEL INTERFEROMETRE POUR ETALONNAGE DES CRAPAUDINES DILATABLES.
5 EME SYMP.INT.SUR LES MAREES TERRESTRES.
OBS.ROY.BELG.COMM.NO 236 S.GEOPH.NO 69 PP 197-213 1964

3.16
MITTELSTRASS G. DAS HANNOVERSCHE HORIZONTALPENDELEICHGERAET.
5 EME SYMP.INT.SUR LES MAREES TERRESTRES.
OBS.ROY.BELG.COMM.NO 236 S.GEOPH.NO 69 PP 238-244 1964

3.16
MITTELSTRASS G. EINE BEMERKUNG ZUR DIREKTEN EICHUNG VON HORIZONTALPENDELN.
B.I.M.NO 39 PP 1399-1401 NO 1 1965

3.16
MELCHIOR P.J. UN SYSTEME D ETALONNAGE AUTOMATIQUE PROGRAMME POUR LES STATIONS CLINOMETRIQUES SOUTERRAINES.
B.I.M. NO 44 PP 1707-1709 1966

3.16
DUCARME B. ETUDE DU COMPORTEMENT DES CRAPAUDINES DILATABLES.
B.I.M. NO 45 PP 1793-1809 1966

3.16
DUCARME B. COMPARAISON DES CRAPAUDINES AU MOYEN D UN PENDULE ETALON.
B.I.M. N.55 SEPTEMBRE PP 2656-2668 1969

3.16
GOLOUBITSKII V.G. COMPARAISON DES METHODES DYNAMIQUE ET STATIQUE DE L ETALONNAGE DES PENDULES HORIZONTAUX AVEC SUSPENSION DE ZOLLNER
ROTATION ET DEFORMATIONS DE MAREES DE LA TERRE - NO 2
ACAD.NAOUK UKR.SSR - OBS.GRAV.POLTAVA - KIEV PP 108-119 1970

3.16
BALENKO V.G.
KOUTNII A.M. APPLICATION DES NIVEAUX PRECIS POUR LA DETERMINATION DE LA LONGUEUR REDUITE DES PENDULES HORIZONTAUX
ROTATION ET DEFORMATIONS DE MAREES DE LA TERRE - NO 2
ACAD.NAOUK UKR.SSR - OBS.GRAV.POLTAVA - KIEV PP 125-137 1970

3.16
MATVEYEV P.S.
GOLOUBITSKII V.G. SUR LA DETERMINATION DE LA CONSTANTE DU PENDULE HORIZONTAL AVEC SUSPENSION DE ZOLLNER D APRES LE PROCEDE DE GALITZIN - ORLOV
ROTATION ET DEFORMATIONS DE MAREES DE LA TERRE - NO 2
ACAD.NAOUK UKR.SSR - OBS.GRAV.POLTAVA - KIEV PP 92-108 1970

3.16
OKHOTSIMSKAIA M.V.
KHOROMISKII A.V. PLATEFORME POUR L ETALONNAGE DES CLINOMETRES.
INST.PHYS.TERRE SCHMIDT OBS.GEO.OBNINSK-PUB.1 PP 120-127 1970

3.16
LENNON G.W.
VANICEK P. CALIBRATION TESTS AND THE COMPARATIVE PERFORMANCE OF HORIZONTAL PENDULUMS AT A SINGLE STATION.
6 EME SYMPOSIUM MAREES TERRESTRES STRASBOURG
OBS.ROY.BELG.COMM. A9 S.GEOPH. NO 96 PP 183-193 1970

3.16
BONATZ M. IMPROVEMENT OF CLINOMETRIC EARTH-TIDE MEASUREMENTS BY APPLICATION OF A ZERO-COMPENSATION METHOD
XV ASS.GEN. UGGI - ASS.INT.DE GEODESIE MOSCOU AOUT 1971

3.16
VAN RUYMBEKE M. PIEZO-ELECTRICITE ET GEODESIE
CIEL ET TERRE VOL.88, NO 2, 1972
OBS.ROY.BELG.COMM. 69 SERIE GEOPH.NO 106, PP 3-11 1972

3.16
VAN RUYMBEKE M. SUR L ETALONNAGE DE PENDULES HORIZONTAUX A L AIDE
DE CERAMIQUES PIEZO-ELECTRIQUES
BULL.CLASSE SC.ACAD.ROY.BELG.5E SERIE, TOME LVIII, 1972
OBS.ROY.BELG.COMM.B 77-S.GEOPH.NO 112 PP 368-376 1972

3.16
VAN GILS J.M. LA PIEZOELECTRICITE ET L INSTRUMENTATION GEODYNAMIQUE
VAN RUYMBEKE M. CIEL ET TERRE VOL.89, NO 1, 1973
OBS.ROY.BELG.COMM.B 77-S.GEOPH.NO 112 PP 3-12 1973

3.16
VAN GILS J.M. LA PIEZOELECTRICITE ET L INSTRUMENTATION GEODYNAMIQUE
VAN RUYMBEKE M. CIEL ET TERRE VOL.89, NO 1, 1973
OBS.ROY.BELG.COMM.B 77-S.GEOPH.NO 112 PP 3-12 1973

3.16
GOLOUBITSKII V.G. ETUDE DE LA PLATEFORME D ETALONNAGE DE
KOUTNII A.M. L OBSERVATOIRE DE POLTAVA
ROT.ET DEFORM.DE MAREES DE LA TERRE, VOL.5 PP 64-69 1973
* B.I.M. NO 69 PP 3913-3919 1974

3.16
BAGMET A.L. ETALONNAGE DES CLINOMETRES AVEC AMPLIFICATION
KOUTNII A.M. PHOTO-ELECTRIQUE POUR DE PETITS ANGLES D INCLINAISON
ROT.ET DEF.DE MAREES DE LA TERRE,V.5 PP 47-50 1973
* B.I.M. NO 71 PP 4024-4034 1975

3.16
BAGMET A.L. DU PROBLEME DE LA METHODE D ETALONNAGE DES
GOLOUBITSKII V.G. CLINOMETRES A AGRANDISSEMENT PHOTOELECTRIQUE
KOUTNII A.M. ACAD.NAOUK UKRAINSKOI SSR, OBS.GRAV.POLTAVA
ROTAT.DEFOR.MAREES DE LA TERRE VII PP 49-52 1975

3.16
BAGMET A.L. ETUDE DE LA STABILITE DE LA CONSTANTE
GOLOUBITSKII V.G. ELECTRODYNAMIQUE DES CLINOMETRES A AGRANDISSEMENT
KOUTNII A.M. PHOTOELECTRIQUE
NISAMOV R.G. ACAD.NAOUK UKRAINSKOI SSR, OBS.GRAV.POLTAVA
ROTAT.DEFOR.MAREES DE LA TERRE VII PP 53-55 1975

3.16
GOLOUBITSKII V.G. ETALONNAGE DES CLINOMETRES PHOTOELECTRIQUES
KOUTNII A.M. DE L OBSERVATOIRE GRAVIMETRIQUE DE POLTAVA
BALENKO V.G.
BOGDAN I. ACAD.NAOUK UKRAINSKOI SSR, OBS.GRAV.POLTAVA
BAGMET A.L. ROTAT.DEFOR.MAREES DE LA TERRE VII PP 55-58 1975

3.16
GOLOUBITSKII V.G. RESULTATS DE L ETALONNAGE DES PENDULES HORIZONTAUX
KOUTNII A.M. AVEC SUSPENSION DE ZOLLNER
ROT.& DEFORM.DE MAREES DE LA TERRE VOL.VI PP 61-68 1974
* B.I.M. NO 73 PP 4234-4246 1976

3.16
BAGMET A.L. SUR LE PROBLEME DE LA METHODE D ETALONNAGE
GOLOUBITSKII V.G. DES CLINOMETRES A GRANDISSEMENT PHOTOELECTRIQUE
KOUTNII A.M. ROT.ET DEFORM.DE MAREES DE LA TERRE VII PP 49-52 1975
* B.I.M. NO 74 PP 4305-4311 1976

3.16
DUCARME B. SOME CONSIDERATIONS ON THE CRAPAUDINES PERFORMANCES
MELCHIOR P. FOR THE CALIBRATION OF HORIZONTAL PENDULUMS
B.I.M. NO 75 PP 4388-4398 1977

3.17
SKALSKY L. ON SOME PROBLEMS OF TIDAL OBSERVATIONS
PICHA J. WITH TILTMETERS AND THEIR ACCURACY.
STUDIA GEOPH. GEOD. 13 PP 138-172 1969

3.17
BRAGARD L. ETUDE DE LA DERIVE DE PENDULES HORIZONTAUX DE TYPES
DIFFERENTS OBSERVEE A LA STATION DE KANNE /UNIV. DE
LIEGE/ DE 1966 A 1969, EN COMPOSANTE E.-O.
6 EME SYMPOSIUM MAREES TERRESTRES STRASBOURG
OBS.ROY.BELG.COMM. A9 S.GEOPH. NO 96 PP 117-135 1970

3.17
MELCHIOR P. SUR LA PRECISION DES OBSERVATIONS CLINOMETRIQUES DE MAREES TERRESTRES.
6 EME SYMPOSIUM MAREES TERRESTRES STRASBOURG
OBS.ROY.BELG.COMM. A9 S.GEOPH. NO 96 PP 194-196 1970

3.17
ANOKHINA K.M., OSTROVSKY A.E. ON THE EFFECT OF TEMPERATURE ON TIDAL TILTS OF THE EARTH S SURFACE.
6 EME SYMPOSIUM MAREES TERRESTRES STRASBOURG
OBS.ROY.BELG.COMM. A9 S.GEOPH. NO 96 PP 97-100 1970

3.17
VANICEK P. THEORY OF MOTION OF HORIZONTAL PENDULUM WITH ZOLLNER SUSPENSION.
6 EME SYMPOSIUM MAREES TERRESTRES STRASBOURG
OBS.ROY.BELG.COMM. A9 S.GEOPH. NO 96 PP 180-182 1970

3.17
SIMON D. AUFBRAU EINER NEUEN ERDGEZEITENSTATION IM SALZBERGWERK TIEFENORT.
B.I.M. N.59 OCTOBRE PP 2874-2881 1970

3.17
SCHNEIDER M.M. METHODISCHE FRAGEN UND ERFAHRUNGEN BEI ERDGEZEITEN-MESSUNGEN AN DER SOWJETISCHEN UBERWINTERUNGSSTATION WOSTOK IN DER ZENTRALEN ANTARKTIS.
B.I.M. N.59 OCTOBRE PP 2853-2868 1970

3.17
BONATZ M., ROCHOLL W. GESAMTAUSWERTUNG DER MIT HORIZONTALPENDELN IN DER TESTSTATION ERPEL GEWONNENEN MESSUNGSERGEBNISSE 1965 BIS 1971
MITT.INST.FUR THEOR.GEOD.UNIV.BONN NO 21 PP 1-10 1973

3.17
BAGMET A.L., BAGMET M.Y. INFLUENCE DE CERTAINS FACTEURS SUR LA DERIVE INSTRUMENTALE DU CLINOMETRE AVEC AGRANDISSEMENT PHOTOELECTRIQUE.
PROBLEMES DES MESURES DE GRAVITATION - PP 59-61 1974

3.17
BAGMET A.L. ON THE STUDY OF INSTRUMENT DRIFT OF TILTMETER
PROCEEDINGS 7TH INT.SYMP.EARTH TIDES, SOPRON 1973
AKADEMIAI KIADO BUDAPEST, PP 205-222 1976

3.171
BOUNE V.Y. SUR LE PROBLEME DE LA VARIATION DES PERIODES DES CLINOMETRES AU COURS DU TEMPS
IZV.DIVIS.SC.NATUR.DU TADJIKISTAN PUBL.4 PP 3-13 1953

3.171
SCHNEIDER M. ZUR BESTIMMUNG DES AZIMUTS BEI LOTSCHWANKUNGSMESSUNGEN.
B.I.M.NO 32 PP 873-879 1963

3.171
MATVEYEV P.S., GOLOUBITSKII V.G., EVTOUCHENKO E.I. SUR LA DEPENDANCE DE LA PERIODE PROPRE EN FONCTION DE L AMPLITUDE D OSCILLATION POUR LES PENDULES HORIZONTAUX A SUSPENSION ZOLLNER.
TROUDI POLTAVSKOI GRAV.OBS. T.XII PP 100-109 KIEV 1963
* B.I.M. N.35 FEVRIER PP 1134-1141 1964

3.171
SCHNEIDER M. DIE BESTIMMUNG DER TO ZEITEN VON HORIZONTALPENDELN.
MITT.INST.THEOR.PHYS.GEOPH.FREIBERG 43 PP 1710-1713 1966
B.I.M. NO 44 PP 1710-1713 1966

3.171
BONATZ M. ZUR BESTIMMUNG DER SCHWINGUNGSZEIT VON HORIZONTALPENDELN
B.I.M. NO 45 PP 1777-1792 1966

3.171
GOLOUBITSKII V.G. SUR LA QUESTION DE LA DETERMINATION DE LA PERIODE PROPRE DES PENDULES HORIZONTAUX AVEC SUSPENSION DE ZOLLNER.
OBS.GRAV.POLTAVA INF.BULL.NO 11 PP 190-196 KIEV 1967

3.171
BONATZ M. EINE BEMERKUNG ZUR AZIMUTBESTIMMUNG VON HORIZONTALPENDELN
B.I.M. NO 47 PP 153-154 1967

3.171
BONATZ M. DER EINFLUSZ DER SCHWINGUNGSZEIT AUF DEN DAMPFUNGSFAKTOR DES HORIZONTALPENDELS VERBAANDERT-MELCHIOR N.75.
B.I.M. NO 47 PP 2031-2038 1967

3.171
GALOUBITSKII V.G / EVTOUCHENKO E.I. — ETUDE DE LA DEPENDANCE PERIODE-AMPLITUDE, PROPRE AUX PENDULES HORIZONTAUX REPSOLD-LEVITSKII.
MAR.TERR.ACAD.SC.URSS UK.OBS.GRAV.KIEV 1966 PP 138-149
* B.I.M. N.52 DECEMBRE PP 2444-2454 — 1968

3.171
GALOUBITSKII V.G — SUR LA QUESTION DE LA DETERMINATION DE LA PERIODE PROPRE DES PENDULES HORIZONTAUX.
MAR.TERR.ACAD.SC.URSS UK.OBS.GRAV.KIEV 1966 PP 150-154
* B.I.M. N.52 DECEMBRE PP 2455-2459 — 1968

3.171
SKALSKY L. — DETERMINATION OF AZIMUTHS OF SIMPLE HORIZONTAL PENDULA.
STUDIA GEOPH. ET GEOD. 13 4 PP 400-416 — 1969

3.171
BONATZ M. / SCHUSTER O. — AZIMUTMESSUNGEN MIT EINEM THEODOLITKREISEL IN SPITZBERGEN
ALLGEMEINE VERMESSUNGS NACHRICHTEN HEFT 6 PP 215-216 — 1971

3.171
GOLOUBITSKII V.G. — ESTIMATION DE LA PRECISION DE LA DETERMINATION DE LA SENSIBILITE DES PENDULES HORIZONTAUX AVEC SUSPENSION DE ZOLLNER
ROT.ET DEFORM.DE MAREES TERRE VOL.3 PP 128-132 KIEV — 1971

3.171
SASAKI H. / SATO T. — ON THE CALIBRATION OF ORB HORIZONTAL PENDULUMS
PROCEEDINGS OF THE INT.LAT.OBS.MIZUSAWA, PP 21-28 — 1972

3.171
BACHEM H.CHR. — BESTIMMUNG DES AZIMUTS VON HORIZONTALEINFACHPENDELN
B.I.M. NO 64, PP 3366-3380 — 1973

3.171
BACHEM H.C. / WENZEL H.G. — AZIMUTKORREKTION BEI CLINOMETERMESSUNGEN DER ERDGEZEITEN
B.I.M. NO 65 MARS PP 3551-3559 — 1973

3.171
MANDERLIER F. / MOUSSET J. — LE GYROTHEODOLITE ET L ORIENTEMENT DES PENDULES HORIZONTAUX
MIN.DEF.NAT. - INST.GEOGR.MIL. PP 9-54 — 1973

3.171
CHOJNICKI T. — PRZEKSZTALCENIA KIERUNKOWE SKLADOWYCH POZIOMYCH PLYWOW ZIEMSKICH / THE AZIMUTHAL TRANSFORMATIONS OF THE HORIZONTAL TIDAL COMPONENTS/
GEOD.I KARTOGR.,TOM XXIII, 4, PP 251-257 — 1974

3.171
KOUTNII A.M. / BAGMET AD. — SUR L AUGMENTATION DE PRECISION DE LA DETERMINATION DES CARACTERISTIQUES DE PHASE DES INCLINAISONS DE MAREES
ROT.ET DEFORM.DE MAREES DE LA TERRE V. 5 PP 46-50 1973
* B.I.M. NO 70 PP 3957-3964 — 1975

3.171
CHOJNICKI T. — LA PROJECTION DES RESULTATS D ANALYSE DES COMPOSANTES HORIZONTALES DES MAREES SUR LES DIRECTIONS FONDAMENTALES
PUBL INST GEOPH.POL.AC.SCI. VOL 94 PP 3-23 — 1975

3.171
BAKER T.F. / LENNON G.W. — CALIBRATION - CONFIDENCE IN THE PERFORMANCE OF TILTMETERS AND GRAVIMETERS
PROCEEDINGS 7TH INT.SYMP.EARTH TIDES, SOPRON 1973
AKADEMIAI KIADO BUDAPEST, PP 223-230 — 1976

3.175
VERBAANDERT J. — L AMORTISSEMENT ELECTROMAGNETIQUE DES PENDULES HORIZONTAUX.
5 EME SYMP.INT.SUR LES MAREES TERRESTRES.
OBS.ROY.BELG.COMM.NO 236 S.GEOPH.NO 69 PP 214-226 — 1964

3.19
EGEDAL J. — UBER EINE MESSUNG DER BEWEGUNG VON PFEILERN.
ZEITSCHRIFT FUR GEOPH.8 PP 195-196 — 1932

3.19
PREVOT E. — INFLUENCE DES OSCILLATIONS DIURNES DE LA VERTICALE SUR LES RESULTATS DES NIVELLEMENTS DE HAUTE PRECISION.
C.R.ACAD.SCI.PARIS 196 PP 605-607 — 1933

3.19
NORLUND N.E. — HYDROSTATISK NIVELLEMENT OVER STORE BAELT.
DANISH GEODETIC INST.SKRIFTER 3 RAEKKE BD.6 — 1945

3.19
NORLUND N.E. — HYDROSTATISK NIVELLEMENT OVER ORESUND.
DANISH GEODETIC INST.SKRIFTER 3 RAEKKE BD.8 — 1946

3.19
EGEDAL J. ON THE APPLICATION OF THE HYDROSTATIC METHOD TO LEVELLING AND TO DETERMINATION OF VERTICAL MOVEMENTS IN THE EARTH S CRUST.
PUBL.DANSKE METEOROL.INST.MEDDELELSER NO 10 1947

3.19
KUKKAMAKI T.J. TIDAL CORRECTION OF THE LEVELLING.
VEROFF.D.FINN.GEOD.INST.NO 36 1949

3.19
JENSEN H. FORMULES FOR THE ASTRONOMICAL CORRECTION TO THE PRECISE LEVELLING.
DANISH GEODETIC INST.MEDDELELSE NO 23 1949
BULL.GEOD. NO 17 PP 267-277 1950

3.19
VIGNAL J. OSCILLATIONS DIURNES DE LA VERTICALE ET DU SOL.
BULL.GEOD.NO 18 PP 403-426 1950

3.19
RUNE G.A. REPORT ON LEVELLING AND MOON-SUN DIURNAL OSCILLATION OF THE VERTICAL.
BULL.GEOD.NO 18 PP 448-449 1950

3.19
SIMONSEN O. REPORT ON THE ASTRONOMICAL DIURNAL CORRECTION IN THE NEW DANISH PRECISE LEVEL NETWORK.
BULL.GEOD.NO 18 PP 450-451 1950

3.19
SCHULZ G. POSIBLES VARIACIONES DE LA SUPERFICIE DEL GEOIDE Y SUS INFLUENCIAS SOBRE LA NIVELACION DE PRECISION.
BUENOS AIRES 1937 AN.SOC.CT.ARG.UNIV.TUCUMAN PUBL.592 1951

3.19
GERKE K. UBER DIE WIRKUNG VON VERTIKALEN ERDKRUSTENBEWEGUNGEN UND DEFORMATIONEN DER NIVEAUFLACHE AUF NIVELLEMENTS HOHER GENAUIGKEIT.
DEUTSCHE GEOD.KOM.BAYERIS.AKAD.WISSENSCHAFTEN 1953

3.19
GERKE K. UNTERSUCHUNG UBER PERIODISCHE LOTSTORUNGEN IM TIDEGEBIET.
ZEITSCHR.F.VERMESSUNGSWESEN 79JAHRG.S294-299 1954

3.19
JENSEN H. DIURNAL OSCILLATION OF THE VERTICAL OF ASTRONOMIC ORIGIN.
BULL.GEOD.SUPPL.AU NO 34 PP 458-460 1954

3.19
SCHEEL G. SYSTEMATISCHE FEHLER DES HYDROSTATISCHEN NIVELLEMENTS UND VERFAHREN ZU IHRER AUSSCHALTUNG.
MITTEIL.NO 13 INST.ANGEWANDTE GEOD. PP 69-71 FRANKF. 1956

3.19
DECAE A. OPERATIONS METROLOGIQUES RELATIVES A L IMPLANTATION DU SYNCHROTRON A PROTONS DU CERN A MEYRIN.
CERN PS/AED 6 FEVRIER 1957

3.19
DECAE A. IMPLANTATION DE L EUROTRON DU CERN A GENEVE.
BOLL.GEODESIA 1958

3.19
SIMONSEN O. GLOBAL ASPECT OF THE ASTRONOMICAL CORRECTION FOR LEVELLING OF HIGH PRECISION WHEN CONSIDERING THE DEFINITION OF LEVELLING DATUM.
THE DANISH GEODETIC INST.COPENHAGEN. 1965

3.2
LECOLAZET R.
MELCHIOR P.J. POUR UNE COMPARAISON DES OBSERVATIONS DE LA MAREE GRAVIMETRIQUE.
B.I.M.NO 19 PP 322-324 1960

3.21
HARTSOUGH R.C. EFFECT OF LUNAR GRAVITY UPON A QUARTZ THREAD BALANCE.
PHYSICAL REVIEW 19 PP 282 1922

3.21
BROWN E.W.
BROUWER D. ANALYSIS OF RECORDS MADE ON THE LOOMIS CHRONOGRAPH BY THREE SHORTT CLOCKS AND A CRYSTAL OSCILLATOR.
MONTHLY NOT.R.ASTR.SOC.VOL.91 PP 575-591 1930

3.21
SOLLENBERGER P. LUNAR EFFECTS ON CLOCK CORRECTIONS.
CLEMENCE G.M. ASTRON.JOURNAL VOL.XLVIII NO 1107 NO 7 PP 78-80 1939

3.21
STOYKO N. DE L INFLUENCE DE L ATTRACTION LUNI-SOLAIRE SUR LA PESANTEUR.
C.R.ACAD.SCI.PARIS 218 PP 308-310 1944

3.21
STOYKO N. SUR LA DETERMINATION DES TERMES DU JOUR SIDERAL DANS L ATTRACTION LUNI-SOLAIRE PAR L OBSERVATION DES PENDULES A GRAVITE.
C.R.ACAD.SCI.PARIS 220 PP 668-669 1945

3.21
STOYKO N. L ATTRACTION LUNI-SOLAIRE ET LA FORMULE INTERNATIONALE DE LA PESANTEUR.
ANNALES CHRONOMETRIE 16, NO 1 PP 5-29 1946

3.21
HOFFROGGE C. MIT SCHULERPENDELN GEMESSENE ZEITLICHE ANDERUNGENDER SCHWERKRAFT DURCH SONNE UND MOND.
GEOFISICA PURA E APPL.MILANO,VOL.13 PP 118 1948

3.21
STOYKO N. L ATTRACTION LUNI-SOLAIRE ET LES PENDULES.
BULL.ASTRONOM.XIV FASC.I PP 1-36 1949

3.21
STOYKO N. SUR L INFLUENCE DE L ATTRACTION LUNI-SOLAIRE ET DE LA VARIATION DU RAYON TERRESTRE SUR LA ROTATION DE LA TERRE.
C.R.ACAD.SCI.PARIS 230 PP 620-622 1950

3.21
STOYKO N. LA VARIATION DE LA VITESSE DE ROTATION DE LA TERRE.
BULL.ASTR.VOL.15 FASC.3 PP 14 1951

3.21
WOLF H. EINE BESTIMMUNG DES GRAVIMETERFAKTORS AUS PENDELMESSUNGEN.
ZEITSCH.FUR VERMESSUNGSWESEN 85 H2 PP 33-35 1960

3.21
PROVERBIO E. L UTILISATION DES PENDULES ASTRONOMIQUES DANS LA DETERMINATION DES MAREES TERRESTRES.
JOURN.SUISSE D HORLOGERIE NO 9/10 1966
OSS.ASTRONOM.MILANO-MERATE SER.NO 259 PP 1-12 1966

3.22
BREIN R. REPORT ON SOME INVESTIGATIONS CONCERNING GRAVIMETER RECORDINGS IN FRANKFURT A.M.
PROCEEDINGS 7TH INT.SYMP.EARTH TIDES, SOPRON 1973
AKADEMIAI KIADO BUDAPEST, PP 661-666 1976

3.221
BAARS B. GRAVITY EFFECT ON EARTH TIDES.
GEOPH.PROSPECTING I NO 2 PP 82-110 JUIN 1953

3.221
GRAF A. GRAVIMETER, MESSPRINZIPIEN, AUFBAU, MESSTECHNIK.
DEUTSCHE GEOD.KOMM.REIHE B, HEFT NO 30 MUNCHEN 1957

3.221.1
NISHIMURA E. A CONSIDERATION ON EARTH TIDAL CHANGE OF GRAVITY.
ICHINOHE T.
NAKAGAWA I. TELLUS 9 NO 1 PP 118-126 1957

3.221.2
BOLLO R. SUR LA VARIATION PERIODIQUE DE LA GRAVITE EN UN LIEU.
GOUGENHEIM A. C.R.ACAD.SCI.PARIS 229 PP 983-984 1949

3.221.2
BOLLO R. VARIATION PERIODIQUE DE LA GRAVITE EN UN LIEU.
GOUGENHEIM A. ANN.GEOPH.5 NO 2 PP 176-180 1949

3.221.2
BOLLO R. AU SUJET DE LA VARIATION PERIODIQUE DE LA GRAVITE.
GOUGENHEIM A. ANN.DE GEOPH.6 FASC.2 PP 133-135 1950

3.221.2
REFORD M.S. TIDAL VARIATIONS OF GRAVITY.
TRANS.AMER.GEOPH.UNION 32 NO 2 PP 151-156 1951

3.221.2
N. VARIATION PERIODIQUE DE LA GRAVITE /NV.CALEDONIE, GABON, ISLANDE, GROENLAND/.
C.R.COM.NL.FRANCAIS DE GEOD.ET GEOPH.PP 61-63 1951

3.221.2
BREIN R. PHOTOGRAPHISCHE REGISTRIERUNG DER ERDGEZEITEN MIT EINEM GRAVIMETER-BEITRAG ZUR LIBELLENPRUFUNG. DEUTSCHE GEOD.KOMM.BAYERISSCHEN AKAD.WISSENSCHAFTEN 1954

3.221.2
N. VARIATION PERIODIQUE DE LA GRAVITE, ANALYSE /TANANARIVE, STRASBOURG/. C.R.COM.NL.FRANCAIS GEOD.GEOPH.PP 71-72 1954

3.221.2
BREIN R. UNTERSUCHUNG EINIGER FEHLEREINFLUSSE BEI DER MESSUNG VON SCHWEREUNTERSCHIEDEN UND BEI DER REGISTRIERUNG KLEINSTER SCHWEREANDERUNGEN MIT DEM GRAVIMETER. DEUTSCHE GEOD.KOMM.BAYERISCHEN AKAD.WISSENSCHAFTEN 1957

3.221.2
LECOLAZET R. NOTE COMPLEMENTAIRE SUR L EMPLOI DU GRAVIMETRE NORTH-AMERICAN A L ENREGISTREMENT DE LA MAREE GRAVIMETRIQUE. OBS.ROY.BELG.COMM. NO 114 S.GEOPH. NO 39 PP 21-24 1957

3.221.2
BREIN R. ENREGISTREMENT DE LA TEMPERATURE DU NORTH-AMERICAN-GRAVIMETER EN VUE DE DETERMINER LA MARCHE DU GRAVIMETRE. OBS.ROY.BELG.COMM. NO 114 S.GEOPH. NO 39 PP 35-42 1957

3.221.2
GERKE K.
BREIN R. BERICHT UBER DIE VOM INSTITUT FUR ANGEWANDTE GEODASIE /II.ABT.DGFI/ IM RAHMEN DES AGI DURCHGEFUHRTEN GRAVIMETERREGISTRIERUNGEN. OBS.ROY.BELG.COMM. NO 142 S.GEOPH. NO 47 PP 35-57 1958

3.221.2
BREIN R. UBER DIE MESSUNG KLEINSTER SCHWEREANDERUNGEN. ZEITSCH.FUR INSTRUMENTENKUNDE 67 HEFT 11 PP 281-284 1959

3.221.2
BREIN R. LE RESSORT ELECTROMAGNETIQUE UTILISE POUR L ENREGISTREMENT DES MAREES TERRESTRES. IVME SYMP.INTERNATIONAL SUR LES MAREES TERRESTRES. OBS.ROY.BELG.COMM. NO 188 S.GEOPH. NO 58 PP 139-143 1961
INST.ANGEW.GEOD.FRANKFURT A.M.REIHE I HEFT 21 PP 14-17 1962

3.221.2
BREIN R. GEZEITENREGISTRIERUNG MIT HILFE DER ELEKTROMAGNETISCHEN FEDER UN IHRE EICHUNG. MITTEILG.INST.ANGEW.GEOD.BULL.NO 28 PP 648-649 1962

3.221.2
BREIN R. ERGEBNISSE DER SCHWEREREGISTREIRUNGEN MIT VERWENDUNG EINER ELEKTRISCHEN FEDER /1962-1964/. DEUTSCHE GEOD.KOMM.WISS.REIHE B GEOD.HEFT 116 PP 109 1965

3.221.3
WOLF A. TIDAL FORCE OBSERVATIONS /OKLAHOMA/. GEOPHYSICS V NO 4 PP 317-320 OCTOBRE 1940

3.221.3
HOSKINSON A.J. HARMONIC ANALYSIS OF GRAVITY OBSERVATIONS. TRANS.AMER.GEOPH.UNION 32 NO 2 PP 163-165 1951

3.221.3
PETIT J.T.
SLICHTER L.B.
LA COSTE L. EARTH TIDES. TRANS.AM.GEOPH.UNION 34 AVRIL NO 2 PP 174-184 1953

3.221.3
N. DATA ON LA COSTE AND ROMBERG TIDAL GRAVITY METERS. B.I.M. NO 1 PP 2-3 1956

3.221.3
CLARKSON H.N.
LA COSTE L.J.B. AN IMPROVED INSTRUMENT FOR MEASUREMENT OF TIDAL VARIATIONS IN GRAVITY. TRANS.AMER.GEOPH.UNION 37 NO 3 PP 266-272 1956

3.221.3
CLARKSON H.N.
LA COSTE J.B. IMPROVEMENTS IN TIDAL GRAVITY METERS AND THEIR SIMULTANEOUS COMPARISON. TRANS.AMER.GEOPH.UNION 38 NO 1 PP 8-16 1957

3.221.3
NESS N.F. RESULTS AND ANALYSIS OF IGY EARTH TIDE GRAVITY DATA. B.I.M. NO 22 PP 420-426 1960

3.221.3
MELCHIOR P.J. CORRECTIONS A APPORTER AUX PHASES OBTENUES PAR N.F.NESS POUR LES STATIONS DE BIDSTON, TRIESTE, HONOLULU, BERMUDA. B.I.M. NO 24 PP 493-497 1961

3.221.3
BREIN R. — ELEKTRISCHE MESSUNG VON SCHWEREDIFFERENZEN MIT EINEN LA-COSTE-ROMBERG GRAVIMETER. DEUTSCHE GEOD.KOMM.BAYER.AKAD.WISS.PP 16 — 1967

3.221.3
WENZEL H.G. — FILTERPROBLEME BEI DER ERDGEZEITENREGISTRIERUNG MIT LA COSTE-ROMBERG GRAVIMETERN MODELL G. B.I.M. NO 65 MARS PP 3517-3525 — 1973

3.221.3
WENZEL H.G. — ERDGEZEITENREGISTRIERUNG MIT LA COSTE-ROMBERG GRAVIMETERN MODELL G. B.I.M. NO 66 PP 3648-3661 — 1973

3.221.3
SATO T. — ON AN INSTRUMENTAL PHASE-LAG OF THE LACOSTE ROMBERG GRAVIMETER. B.I.M. NO 75 PP 4341-4360 — 1977

3.221.4
ROBINSON ED.S. — TIDAL GRAVITY MEASUREMENTS IN SOUTHEASTERN UNITED STATES. UPPER MANTLE PROJ.UN.STATES PROG.FINAL REP.PP 167-168 — 1971

3.221.6
N. — EIN SERIENMASSIGES ASKANIA-GRAVIMETER GS 9 REGISTRIERT DIE ZEITLICHEN ANDERUNGEN DER ERDBESCHLEUNIGUNG. ASKANIA-WARTE NO 42 PP 29-30 — 1952

3.221.6
HAALCK F. — DIE GENAUIGKEIT EINES MODERNEN GRAVIMETERS. JUBILAUMSBAND D.ZEITSCH.F.GEOPH.BRAUNSCHWEIG PP 21-28 — 1953

3.221.6
SCHULZE R. — GEZEITENREGISTRIERUNG. ASKANIA-WARTE NO 46 SONDERDRUCK GEOPH.935 — 1954

3.221.6
N. — GEZEITENREGISTRIERUNG. ASKANIA-WARTE NO 49 PP 16-17 — 1956

3.221.6
LEFEVRE C. — PERFECTIONNEMENT AUX TECHNIQUES D ENREGISTREMENT DE LA MAREE GRAVIMETRIQUE. CONGRES AFAS — 1956

3.221.6
SCHULZE R. — DAS ASKANIA-GRAVIMETER MIT REGISTRIER EINRICHTUNG. HINWEISE ZUR INBETRIEBNAHME. OBS.ROY.BELG.COMM. NO 114 S.GEOPH. NO 39 PP 15-21 — 1957

3.221.6
LEFEVRE C. — PROGRAMME D ENREGISTREMENT DE LA MAREE GRAVIMETRIQUE AU LABORATOIRE DE GEOPHYSIQUE APPLIQUEE DE LA SORBONNE. B.I.M. NO 4 PP 62-65 — 1957

3.221.6
N. — DIE ASKANIA-GRAVIMETER. ASKANIA-WERKE A.G.BERLIN FRIEDENAU — 1958

3.221.6
HEITZ S. — ZUR INSTRUMENTENGANG - ERMITTLUNG BEI GEZEITENREGISTRIERUNGEN. Z.VERMESS.WES.STUTTGART 83 8 PP 267-271 — 1958

3.221.6
NISHIMURA E. — OBSERVATION OF EARTH TIDAL CHANGE OF GRAVITY DURING I.G.Y. IN JAPAN /PROVISIONAL REPORT/. OBS.ROY.BELG.COMM. NO 142 S.GEOPH. NO 47 PP 64-67 — 1958

3.221.6
NORINELLI A. — CONTRIBUTO ALLO STUDIO DELLA MAREA GRAVITAZIONALE TERRESTRE. RENDIC.CL.SCI.FIS.MAT.NAT.SER.VIII VOL.XXVII PP 212-217 — 1959

3.221.6
NISHIMURA E., ICHINOHE T., NAKAGAWA I., FUNABIKI M. — OBSERVATIONAL RESULTS OF EARTH TIDAL CHANGE OF GRAVITY DURING I.G.Y. IN JAPAN. 3E SYMP. MAREES TERRESTRES TRIESTE. BOLL.GEOFISICA TEORICA E APPL.VOL.2 NO 5 PP 184-185 — 1960

3.221.6
FINSCH K. — UBER DIE WEITERENTWICKLUNG DER GEZEITENREGISTRIERANLAGE. ASKANIA WARTE HEFT 57 APRIL — 1961

3.221.6
OKUDA T. — EARTH S TIDAL CHANGE IN GRAVITY JAPANESE CONTRIBUTION TO THE IGY AND IGC. VOL.III PP 123-140 — 1961

3.221.6
NAKAGAWA I. SOME PROBLEMS ON TIME CHANGE OF GRAVITY PARTS 1 & 2. DIS.PREV.RES.INST.BULL.NO 53 PP 105 PARTS 3,4 & 5 FEBR. DIS.PREV.RES.INST.BULL.NO 57 PP 107 1962

3.221.6
PARIISKI N.N. ETUDE DES VARIATIONS DE MAREES DE LA FORCE DE PESANTEUR EN ASIE CENTRALE/1/ RECH. MAREES TERR. ART. N.3 PUBL.ACAD.SC.URSS MOSCOU 1963
* B.I.M. N.37 SEPTEMBRE PP 1242-1246 1964

3.221.6
BONATZ M. UNTERSUCHUNGEN EINER SYSTEMATISCHEN FEHLERS DER ERDGEZEITENREGISTRIEREINRICHTUNG /GALVANOMETER MIT NACHLAUFSCHREIBER/ ZUM ASKANIA-GRAVIMETER GS 11. B.I.M.NO 38 PP 1342-1346 1964

3.221.6
BONATZ M. EIN VERFAHREN ZUR VERRINGERUNG DES INDUKTIONSEFFEKTES BEI DER ERDGEZEITENREGISTRIERUNG MIT DEM ASKANIA-GRAVIMETER GS 11. B.I.M.NO 38 PP 1347-1352 1964

3.221.6
NAKAGAWA I. ON THE M1 COMPONENT OBTAINED BY GRAVIMETRIC TIDAL OBSERVATION /SCREENING OF GRAVITATIONAL FORCES/. SPEC.CONTR.GEOPH.INST.KYOTO UNIV.4 PP 9-17 1964

3.221.6
BONATZ M. ERFAHRUNGEN MIT DER NEUEN REGISTRIERANLAGE ZU DEN ASKANIA-GRAVIMETERN GS 11 UND GS 12. B.I.M.NO 40 PP 1501-1506 1965

3.221.6
BONATZ M. ZUR BESTIMMUNG DES GUNSTIGSTEN DAMPFUNGWIDERSTANDES BEI DER ERDGEZEITENREGISTRIERUNG MIT DEM ASKANIA-GRAVIMETER GS 11 IN VERBINDUNG MIT LICHTMARKENGALVANOMETER UND NACHLAUFSCHREIBER. B.I.M.NO 40 PP 1507-1511 1965

3.221.6
BONATZ M. DIE GUNSTIGSTE HEIZTEMPERATUR UN HEIZSTUFENEINSTELLUNG BEI DER ERDGEZEITENREGISTRIERUNG MIT DEM ASKANIA-GRAVIMETER GS 11 /12/. B.I.M.NO 41 PP 1555-1563 1965

3.221.6
SCHULZE R. MIKROGRAVIMETRIE. ASKANIA-WARTE. HEFT 26 JAHRGANG 22 PP 7-10 1965

3.221.6
BONATZ I.M. DER EINFLUSZ VON SCHWANKUNGEN DES STROMES FUR DIE BELEUCHTUNG DER PHOTOZELLEN AUF DIE MESZWERTANZEIGE DES ASKANIA-GRAVIMETERS GS 11 NR 116. B.I.M.NO 42 PP 1587-1592 1965

3.221.6
BONATZ M. DER EINFLUSZ DER AUSZENTEMPERATUR AUF DEN GANG DES ASKANIA-GRAVIMETERS GS 11 NO 116. ZEITS.VERMESS.12 90 PP 497-506 1965

3.221.6
HONKASALO T. INVESTIGATION OF AN ASKANIA EARTH TIDE GRAVIMETER. B.I.M. NO 44 PP 1696 1966

3.221.6
CHOJNICKI T. P YWY SKORUPY ZIEMSKICJ. BIUL.MWG NO 3/46 GEOF.POLZKIEZ AKAD.NAUK PP 42-50 1966

3.221.6
BALAKRISHNA S. ESTIMATION OF THE RIGIDITY OF THE EARTH BY EARTH TIDE STUDIES. GEOPH.JOURN.R.ASTR.SOC.14 PP 225-238 1967

3.221.6
DITCHKO I.A. CERTAINES QUESTIONS DE LA THEORIE DU GRAVIMETRE ASKANIA.
* B.I.M. N.52 DECEMBRE PP 2417-2439 1968

3.221.6
GOLDARACENA J.M. RESULTADOS PRELIMINARES DE MEDICION DE MAREAS TERRESTRES EN MEXICO. SIMPOSIO PANAM. MANTO SUPERIOR, MEXICO VOL 1 PP 33-42 1969

3.221.6
NAKAGAWA I. ON CONTINUOUS OBSERVATION OF GRAVITY JOURN.GEOD.SOC.JAPAN VOL.15 PP2-5 1969

3.221.6
VOLKOV V.A. MICHATKINE V.N. — ESSAI DE STABILISATION DE LA TENSION SUR LES THERMOSTATS DES GRAVIMETRES GS-11 POUR LES OBSERVATIONS DE MAREES. INST.PHYS.TERRE SCHMIDT OBS.GEO.OBNINSK-PUB.1 PP 169-173 1970

3.221.6
WORKING GROUP — FOR COMPARING THE GRAVIMETERS IN JAPAN SIMULTANEOUS OBSERVATIONS OF EARTH TIDES WITH FOUR ASKANIA AND TWO LACOSTE & ROMBERG GRAVIMETERS IN MIZUSAWA. JOURN.GEOD.SOC.JAPAN VOL.15 PP 53-67 1969 6 EME SYMPOSIUM MAREES TERRESTRES STRASBOURG OBS.ROY.BELG.COMM. A9 S.GEOPH. NO 96 PP 27-29 1970

3.221.6
NAKAGAWA I. SHIRAKI M. — SOME PROBLEMS ON EARTH TIDES OBSERVED WITH A GRAVIMETER. 6 EME SYMPOSIUM MAREES TERRESTRES STRASBOURG JOURN.GEOD.SOC.JAPAN VOL.15 PP 41-52 1969 OBS.ROY.BELG.COMM. A9 S.GEOPH. NO 96 PP 42-49 1970

3.221.6
PARIISKY N.N. BARSENKOV S.N. VOLKOV V.A. GRIDNEV D.G. KUZNETSOV M.V. KUZNETSOVA L. PERTSEV B.P. SARYTCHEVA J.K. — OBSERVATIONS OF TIDAL GRAVITY VARIATIONS IN ASIAN REGIONS OR THE USSR. 6 EME SYMPOSIUM MAREES TERRESTRES STRASBOURG OBS.ROY.BELG.COMM. A9 S.GEOPH. NO 96 PP 86-89 1970

3.221.6
VARGA P. — HARMONIC ANALYSIS OF EARTH-TIDE OBSERVATIONS IN THE SECOND HALF OF 1967 AS RECORDED IN TIHANY. GEOFIZIKAI KOZLEMENYEK XIX, 1-2 PP 69-75 1970

3.221.6
DITCHKO J.A. — QUELQUES RESULTATS DES OBSERVATIONS DES VARIATIONS DE MAREES DE LA FORCE DE PESANTEUR ROT.ET DEFORM.DE MAREES TERRE VOL.3 PP 68-128 KIEV 1971

3.221.6
VOLKOV V.A. GRIDNIEV D.G. — ETUDE DE LA DEPENDANCE DES INDICATIONS DES GRAVIMETRES GS-11 /124,135/ EN FONCTION DE LA TEMPERATURE. MAR.TERR.ET STRUCT.INTERNE DE LA TERRE ACAD.SC.URSS.INST.PHYS.TERR.SCHMIDT MOSC.PP127-133 1967 * B.I.M. N.62 PP 3164-3173 1971

3.221.6
QUEILLE C. — LA MAREE GRAVIMETRIQUE A GARCHY - ENREGISTREMENT - ANALYSE ET INTERPRETATION DES RESULTATS THESE DE DOCTORAT - UNIV.PARIS - N.CNRS A05926 PP 1-74 1971

3.221.6
SIMON Z. BROZ J. — SBORNIKU VYZKUMNYCH PRACI SVAREK 6 PP 61-83, PRAHA 1972

3.221.6
BONATZ M. CHOJNICKI T. ROCHOLL W. SCHULLER K. — ERDGEZEITENSTATION BONN - GESAMTERGEBNISSE DER SCHWEREREGISTRIERUNGEN MIT DEM ASKANIA - GRAVIMETER GS 11 NR.116 IM ZEITRAUM 1964 BIS 1969 MITTEIL.A.D.INST.THEOR.GEOD.UNIV.BONN NR.11 PP 1-11 1972

3.221.6
BARSENKOV S.N. VOLKOV V.A. KOUZNETSOV M.V. KOUZNETSOVA L.I. PARIISKII N.N. — VARIATIONS DE MAREES DE LA FORCE DE PESANTEUR A TALGAR 2. METH.MES.MAR.TERR.& DEFORM.LENTES SURFACE TERRE ACAD.SC.URSS.INST.PHYS.TERR.SCHMIDT MOSCOU PP 26-73 1970 * B.I.M. NO 66 PP 3576-3590 1973

3.221.6
AKSENTIEVA Z.N. KORBA P.S. LISSENKO G.M. — ORGANISATION ET PREMIERS RESULTATS DES OBSERVATIONS DES VARIATIONS DE MAREES DE LA FORCE DE PESANTEUR A BAKHTCHISSARAE ROTATION ET DEFORMATIONS DE MAREES DE LA TERRE - NO 2 ACAD.NAOUK UKR.SSR - OBS.GR.POLTAVA-KIEV PP 34-41 1970 * B.I.M. NO 64 PP 3335-3339 1973

3.221.6
VOLKOV V.A. GOUSSEVA E.P. DOBROKHOTOV IOU.S. — PREMIERS RESULTATS DES OBSERVATIONS DES VARIATIONS DE MAREES DE LA FORCE DE PESANTEUR A L OBSERVATOIRE GEOPHYSIQUE CENTRAL. INST.PH.T.SCHMIDT OBS.GEO.OBNINSK-PUB.1 PP 148-158 1970 * B.I.M. NO 64 PP 3340-3344 1973

3.221.6
SARITCHEVA YOU.K. PREMIERES OBSERVATIONS DES VARIATIONS DE MAREES DE LA FORCE DE PESANTEUR A NOVOSSIBIRSK. ROTATION ET DEFORM.MAREES DE LA TERRE,FASC.1 ACAD.SC.UKR.OBS.GRAV.POLTAVA KIEV PP 180-192 1970
* B.I.M. NO 64 PP 3388-3397 1973

3.221.6
ROCHOLL W.
WOLF H.
PERIODIZITATEN IN DEN BERECHNETEN AMPLITUDEN-QUOTIENTEN AUS GRAVIMETRISCHEN ERDGEZEITENBEOBACHTUNGEN MITT.INST.THEOR.GEOD.UNIV.BONN NO 30 PP 1-33 1974

3.221.6
AKHAVAN A.
SADEGHI
ETUDE STATISTIQUE DES ERREURS INTERNES ET ETUDES COMPARATIVE DES VARIATIONS DU FACTEUR D AMPLITUDE DANS L OBSERVATION DES MAREES GRAVIMETRIQUE A TEHERAN PROCEEDINGS 7TH INT.SYMP.EARTH TIDES, SOPRON 1973 AKADEMIAI KIADO BUDAPEST, PP 329-343 1976

3.221.6
BONATZ M. A TRANSFORMED ASKANIA-GRAVIMETER GS 12 WITH CAPACITIV TRANSDUCER SYSTEM PROCEEDINGS 7TH INT.SYMP.EARTH TIDES, SOPRON 1973 AKADEMIAI KIADO BUDAPEST, PP 231-233 1976

3.221.6
GRIDNEV D.G.
PARIISKY N.N.
SHIBAYEV YU.
STUDIES OF MAGNETIC FIELD EFFECT ON THE READINGS OF THE ASKANIA GRAVIMETERS PROCEEDINGS 7TH INT.SYMP.EARTH TIDES, SOPRON 1973 AKADEMIAI KIADO BUDAPEST, PP 511-555 1976

3.221.61
BONATZ M. GEZEITENREGISTRIERUNG MIT EINEM AUF KAPAZITIVEN ABGRIFF UMGERUSTETEN ASKANIA-GRAVIMETER GS 12 MITT.INST.FUR THEOR.GEOD.UNIV.BONN NO 22 PP 1-11 1973

3.221.7
MIELBERG J. UBER PERIODISCHE VERANDERUNGEN DER SCHWERKRAFT. PUBL.OBS.TARTU 27 NO 4 1932

3.221.7
LAGRANGE E. LA GRAVITE DANS LE TEMPS, OBSERVATIONS A DORPAT. CIEL ET TERRE 48 NO 6 PP 139-141 1932

3.221.7
WYCKOFF R.D. STUDY OF EARTH TIDES BY GRAVITATIONAL MEASUREMENTS. TRANS.AMER.GEOPH.UNION I PP 46-52 1936

3.221.7
TRUMAN O.H. VARIATIONS OF GRAVITY AT ONE PLACE. ASTROPHYSICAL JOURN.VOL.89 NO 3 PP 445-462 1939

3.221.7
LAGERQVIST P.A. EMPFINDLICHKEITSSTEIGERUNG DES ASTASIERTEN PENDELGRAVIMETERS ZWECKS BEOBACHTUNG DER GEZEITENSCHWANKUNGEN DER SCHWERKRAFT. ARKIV.MATH.ASTRON.PHYS.28 A NO 7 PP 1-20 1942

3.221.7
ECKHARDT E.A. DIURNAL VARIATIONS IN THE ACCELERATION OF GRAVITY. TRANS.AM.GEOPH.UNION VOL.30 NO 2 PP 183-184 194[illegible]

3.221.7
VOIT H. UBER DIE BESTE DIMENSIONIERUNG DES BIFILARGRAVIMETERS. GEOF.PURA ED APPL.XV, PP 90-110 194[illegible]

3.221.7
ZADRO M. NUOVO TIPO DI GRAVIMETRO PER LO STUDIO DELLE MAREE GRAVITAZIONALI. IST.GEOD.TOPOGR.GEOF.UNIV.TRIESTE PUBL.NO 58 PP 30 196[illegible]

3.221.7
BREIN R. DIE ELEKTRISCHE FEDER BEI DER MESSUNG VON SCHWEREDIFFERENZEN MIT HOHER MESZGENAUIGKEIT. INST.ANGEW.GEOD.FRANKFURT A.M. NACHRICHT.KARTEN-VERMESSUNGSW.REIHE I HEFT 27 PP 41-52 196[illegible]

3.221.7
BREIN R. RESULTS OF GRAVIMETER REGISTRATIONS APPLYING AN ELECTRIC-SPRING. 5 EME SYMP.INT.SUR LES MAREES TERRESTRES. OBS.ROY.BELG.COMM.NO 236 S.GEOPH.NO 69 PP 152-157 196[illegible]

3.221.7
MARUSSI A. UN NOUVEAU TYPE DE GRAVIMETRE ASTATISE POUR LA MESURE DES MAREES TERRESTRES.

5 EME SYMP.INT.SUR LES MAREES TERRESTRES.
OBS.ROY.BELG.COMM.NO 236 S.GEOPH.NO 69 PP 163-166 1964

3.221.7
NORINELLI A. UN NUOVO STRUMENTO PER LA MISURA DELLA MAREA GRAVIMETRICA
ACCAD.PATAVINA SSLLAA SC.MAT.NAT.LXXVII PP 21-39 PADOVA 1965

3.221.7
BLOCK B. DILATIONAL MODE OF THE EARTH.
WEISS R. A GRAVIMETER TO MONITOR THE OSO
J.GEOPH.RES. VOL 70 N.22 PP 5615-5627 1965

3.221.7
GARRY C. A GRAVIMETER FOR MARINE, AIRBORNE, AN LUNAR SURFACE
HENDERSON PH.D. MEASUREMENTS.
GENERAL DYNAMICS FORT WORTH TEXAS
APPLIED RESEARCH LABORATORY PP 1-16 1967

3.221.7
PROTHERO JR.W.A. A CRYOGENIC GRAVIMETER
THESIS, UNIV.CALIF.SAN DIEGO 1967

3.221.7
PROTHERO JR.W.A. A SUPERCONDUCTING GRAVIMETER
GOODKIND J.M. PHYS.SUPERCOND.DEV.UNIV.VIRGINIA P. E-1 1967

3.221.7
PROTHERO JR.W.A. PRELIMINARY MEASUREMENTS WITH A SUPERCONDUCTING
GOODKIND J.M. GRAVIMETER
TRANS.AMER.GEOPHYS.UNION VOL.49 NO 4 1968

3.221.7
PROTHERO W.A. A SUPERCONDUCTING GRAVIMETER.
GOODKIND J.M. REV.SCIENT.INST.39 NO 9 PP 1257-1262 1968

3.221.7
BLOCK B. TIDAL TO SEISMIC FREQUENCY INVESTIGATIONS WITH
MOORE R.D. A QUARTZ ACCELEROMETER OF NEW GEOMETRY.
6 EME SYMPOSIUM MAREES TERRESTRES STRASBOURG
OBS.ROY.BELG.COMM. A9 S.GEOPH. NO 96 PP 61-72 1970

3.221.7
BLOCK B. TIDAL TO SEISMIC FREQUENCY INVESTIGATIONS WITH A
MOORE R.D. QUARTZ ACCELEROMETER OF NEW GEOMETRY.
J.GEOPH.RES. VOL 75 N.8 PP 1493-1505 1970

3.221.7
PROTHERO W.A. EARTH TIDE MEASUREMENTS WITH THE SUPERCONDUCTING GRAVI-
GOODKIND J.M. METER
PHYSICS DEP. UNIV.CALIFORNIA SAN DIEGO LA JOLLA 92037 1971

3.221.7
GOODKIND J.M. SUPERCONDUCTING GRAVIMETER
PROTHERO JR.W.A. UPPER MANTLE PROJ.UN.STATES PROG.FINAL REP.PP169-170 1971

3.221.7
GRIDNIEV D.G. ETUDE DE LA DEPENDANCE DE LA DERIVE DU GRAVIMETRE
A QUARTZ EN FONCTION DE LA TEMPERATURE DE
THERMOSTATISATION.
MAR.TERR.ET STRUCT.INTERNE DE LA TERRE
ACAD.SC.URSS.INST.PHYS.T.SCHMIDT MOSCOU PP 123-126 1967
* B.I.M. N.62 PP 3185-3189 1971

3.221.7
PROTHERO W.A. EARTH-TIDE MEASUREMENTS WITH THE SUPERCONDUCTING
GOODKIND J.M. GRAVIMETER
JOURN.OF GEOPH.RESEARCH V.77 N.5 FEBRUARY PP 926-936 1972

3.221.7
GRIDNIEV D.G. OBSERVATIONS DES VARIATIONS DE MAREES DE LA FORCE
PROKHOROVSKII G.S. DE PESANTEUR PAR UN GRAVIMETRE EN QUARTZ ASTATISE
A KRASNAIA PAKHRA EN 1965-1966.
METH.MES.MAR.TERR.& DEFORM.LENTES SURFACE TERRE
ACAD.SC.URSS.INST.PHYS.TERR.SCHMIDT MOSCOU PP100-109 1970
* B.I.M. NO 60 JANVIER PP 3357-3365 1973

3.221.7
WHORF T. NEW GRAVITY TIDE MEASUREMENTS AND AN EXPERIMENT
BERGER J. TO LOOK FOR EVIDENCE OF A - PREFERRED UNIVERSAL
HAUBRICH R. FRAME -
WARBURTON R. EOS VOL.56 NO 12 P 1105 1974

3.221.7
BRODSKII B.I. GRAVIMETRE BALISTIQUE DE L OBSERVATOIRE DE POLTAVA
ACAD.NAOUK UKRAINSKOI SSR, OBS.GRAV.POLTAVA
ROTAT.DEFOR.MAREES DE LA TERRE VII PP 24-26 1975

3.221.9
LASSOVSKY K. OSZLACZKY S. — A NAPES A HOLD GRAVITACIOS HATASA A GRAVIMETERMERESEKRE. GEOF.KOZLEMENYEK BUDAPEST I, 3 PP 13-40 — 1952

3.221.9
LASSOVSKY K. OSZLACZKY S. — GRAVIMETER-REGISTRALASOK GLOBALIS ANALIZISE. GEOF.KOZLEMENYEK BUDAPEST III, 2 PP 27-30 — 1954

3.221.9
LASSOVSKY K. — A FOLD DEFORMACIOS EGYÜTTHATOJANAK MEGHATAROZASO GRAVIMETERESZLELESEKBOL. GEOF.KOZLEMENYEK BUDAPEST V, 1 PP 18-26 — 1956

3.221.9
LASSOVSKY K. — A LUNISZOLARIS HATAS AMPLITUDOVISZONYANAK MEGHATAROZASA A BUDAPESTEN 1954-BEN 37 NAPON AT VEGZETT GRAVIMETERESZLELESEKBOL. GEOF.KOZLEMENYEK BUDAPEST V, 3 PP 9-20 — 1956

3.221.9
OSZLACZKY SZ. TOTH G. — A NAP ES A HOLD GRAVITACIOS HATASANAK MEGFIGYELESE HAZANKBAN. OBSERVATIONS DES EFFETS GRAVITATIONNELS DU SOLEIL ET DE LA LUNE EN HONGRIE. GEOF.KOZLEMENYCK XIII, 1, PP 39-48 — 1964

3.222.1
SCHWEYDAR W. — BEOBACHTUNG DER ANDERUNG DER INTENSITAT DER SCHWERKRAFT DURCH DEN MOND. PREUSS.AK.WISSENSCHAFTEN PHYS.MATH.CL.PP 454-465 — 1914

3.222.1
TOMASCHEK R. SCHAFFERNICHT W. — ZU DEN GRAVIMETRISCHEN BESTIMMUNGSVERSUCHEN DER ABSOLUTEN ERDBEWEGUNG. ASTRONOM.NACHRICHTEN 244 NO 5844 PP 257-266 — 1931

3.222.1
TOMASCHEK R. SCHAFFERNICHT W. — UEBER DIE PERIODISCHEN VERANDERUNGEN DER VERTIKALKOMPONENTE DER SCHWEREBESCHLEUNIGUNG IN MARBURG A.D.LAHN. SITZ.GES.BEFORD.GES.NATURWISS.MARBURG 67,H5 PP 151-174 — 1932

3.222.1
TOMASCHEK R. SCHAFFERNICHT W. — UNTERSUCHUNGEN UBER DIE ZEITLICHEN ANDERUNGEN DER SCHWERKRAFT. MESSUNGEN MIT DEM BIFILARGRAVIMETER. ANN.PHYSIK 5 FOLGE BD.15 PP 787-824 — 1932

3.222.1
TOMASCHEK R. SCHAFFERNICHT W. — TIDAL OSCILLATIONS OF GRAVITY. NATURE 130 PP 165-166 — 1932

3.222.1
TOMASCHEK R. SCHAFFERNICHT W. — UBER DIE MESSUNG DER ZEITLICHEN SCHWANKUNGEN DER SCHWEREBESCHLEUNIGUNG MIT GRAVIMETERN. ZEITSCH.GEOPH.BD.9 PP 125-136 — 1933

3.222.1
TOMASCHEK R. — DIE MESSUNGEN DER ZEITLICHEN ANDERUNGEN DER SCHWERKRAFT. ERGEBNISSE EXAKTEN NATURWISSENSCH.BD 12 PP 36-81 — 1933

3.222.1
TOMASCHEK R. — UBER DIE ZEITLICHEN SCHWANKUNGEN DER SCHWERKRAFT. FORSCHUNGEN-FORTSCHRITTE 9 NO 1 PP 8-9 BERLIN — 1933

3.222.1
TOMASCHEK R. — SCHWERKRAFTMESSUNGEN. NATURWISSENSCH.25 HEFT 12 PP 177-185 — 1937

3.222.1
ELLENBERGER H. — DAS BIFILARE PRINZIP UND SEINE ANWENDUNG ZUM BAU VON HOCHEMPFINDLICHEN UND HANDLICHEN SCHWEREMESSERN. DEUTSCHE GEOD.KOMM.BAYER.AKAD.WISS.REIHE C VER 2 — 1952

3.222.1
ICHINOHE T. — STUDY ON CHANGE OF GRAVITY WITH TIME. PART.1TIDAL VARIATION GRAV.MEM.COLL.SCI.KYOTO PP 289-316 — 1955

3.222.1
ICHINOHE T. — STUDY ON CHANGE OF GRAVITY WITH TIME. MEM.COLL.SC.UNIV.KYOTO SER.A VOL.XXVII NO 3 PP 317-334 — 1955

3.222.1
ICHINOHE T. — STUDY ON CHANGE OF GRAVITY WITH TIME. MEM.COLL.SC.UNIV.KYOTO SER.A VOL.XXVIII NO 1 PP 11-38 — 1956

3.225
KUO J.T. — SPATIAL VARIATIONS OF TIDAL GRAVITY

EWING M. THE EARTH BENEATH THE CONTINENTS
GEOPH.MONOGRAPH 10 AMERICAN GEOPH.UNION 1966

3.225
KUO J.T.
JACHENS R.C.
TRANSCONTINENTAL TIDAL GRAVITY PROFILE ACROSS THE UNITED STATES
SCIENCE VOL.168 PP 968-971 1970

3.225
KUO J.
JACHENS R.C.
WHITE G.
EWING M.
TIDAL GRAVITY MEASUREMENTS ALONG A TRANSCONTINENTAL PROFILE ACROSS THE UNITED STATES.
6 EME SYMPOSIUM MAREES TERRESTRES STRASBOURG
OBS.ROY.BELG.COMM. A9 S.GEOPH. NO 96 PP 50-60 1970

3.225
KUO J.T.
JACHENS R.C.
MELCHIOR P.
EWING M.
A LINK OF THE TRANS-U.S. AND TRANS-EUROPE TIDAL GRAVITY PROFILES
EOS 53 4 P 343 1972

3.225
WOOLLARD G.P.
MARSH H.C.
LONGFIELD R.L.
A COMPARISON OF OBSERVED AND PREDICTED EARTH TIDE OBSERVATIONS BETWEEN ALASKA AND MEXICO AND ACROSS THE ZONE OF CRUSTAL TRANSITION BETWEEN NEBRASKA AND UTAH
HAWAII INST.OF GEOPH. 73-20 PP 1-29 1973

3.225
PARIISKY N.N.
PERTSEV B.P.
SARYCHEVA I.K.
A NOTE CONCERNING REGIONAL VARIATIONS IN THE GRAVITY FACTOR DELTA
DITCHKO I.A. KORBA P.S. IVANOVA M.V.
PROCEEDINGS 7TH INT.SYMP.EARTH TIDES, SOPRON 1973
AKADEMIAI KIADO BUDAPEST, PP 571-576 1976

3.225
BAKER T.F.
LENNON G.W.
THE INVESTIGATION OF MARINE LOADING BY GRAVITY VARIATION PROFILES IN THE U.K.
7TH INTERN.SYMP. EARTH TIDES, SOPRON SEPTEMBER 1973
PROCEEDINGS 7TH INT.SYMP.EARTH TIDES, SOPRON 1973
AKADEMIAI KIADO BUDAPEST, PP 463-476 1976

3.23
SCHULZE R. ZUR EICHUNG VON REGISTRIER-GRAVIMETERN.
OBS.ROY.BELG.NO 114 S.GEOPH. NO 39 1957

3.23
GRAF A. BEMERKUNGEN ZUR INSTRUMENTELLEN AUSRUSTUNG EINER GEZEITENSTATION.
OBS.ROY.BELG.COMM.NO 142 S.GEOPH.NO 47 PP 116-122 1958

3.23
NAKAGAWA I. ON THE CALIBRATIONS OF THE ASKANIA GS 11 GRAVIMETER 111
GEOPH.INST.FAC.SC.KYOTO UNIV.PP 136-150 1960

3.23
MELCHIOR P.J. STATION FONDAMENTALE POUR LA COMPARAISON DES GRAVIMETRES.
B.I.M. NO 18 PP 296-306 1960

3.23
NAKAGAWA I. THE CALIBRATIONS OF THE ASKANIA GS 11 GRAVIMETER NO 111.
IVME SYMP.INTERNATIONAL SUR LES MAREES TERRESTRES.
OBS.ROY.BELG.COMM. NO 188 S.GEOPH. NO 58 PP 307-310 1961

3.23
BREIN R. CALIBRATING RESULTS WITH A REGISTRATION PERFORMED DURING ONE YEAR BY HELP OF THE ELECTROMAGNETIC SPRING.
B.I.M.NO 33 PP 1017-1019 1963

3.23
BONATZ M. FIRST ATTEMPTS TO INCREASE THE ACCURACY OF CALIBRATION OF RECORDING-GRAVIMETERS BY CALIBRATION ALONG A VERTICAL CALIBRATION-BASIS ESTABLISHED WITHIN A LABORATORY.
5 EME SYMP.INT.SUR LES MAREES TERRESTRES.
OBS.ROY.BELG.COMM.NO 236 S.GEOPH.NO 69 PP 158-162 1964

3.23
DOBROCHOTOV Y.S. ERREURS DE DETERMINATION DU COEFFICIENT D ETALONNAGE PENDANT L ENREGISTREMENT DES MAREES, DUES A LA DERIVE DU GRAVIMETRE.
RECH. MAREES TERR. ART.N.3 PUBL.ACAD.SC.URSS MOSCOU 1963
* B.I.M. N.36 MAI PP 1198-1204 1964

3.23
BONATZ M. DIE EICHUNG DER ASKANIA-ERDGEZEITENREGISTRIERANLAGE MIT HILFE EINER VERTIKALEN LABOREICHSTRECKE.
ASKANIA-WARTE.HEDT 66 JAHRGANG 22 PP 11-16 1965

3.23
BONATZ M. — UBER DIE EICHUNG VON REGISTRIERGRAVIMETERN MITTELS EINER VERTIKALEN LABOR-EICHSTRECKE. DEUTSCHE GEOD.KOMM.BAYER.AKAD.WISS.HEFT 84 PP 1-115 — 1965

3.23
SCHULZE R. — EINE NEUE EINRICHTUNG IM GEZEITENGRAVIMETER. ASKANIA-WARTE HEFT 66 22 JAHRGANG PP 17-20 — 1965

3.23
SIMON D.
WALZER U. — EINE BEMERKUNG ZUR EICHUNG VON GRAVIMETRISCHEN GEZEITENREGISTRIERUNGEN. B.I.M. NO 42 PP 1604-1608 — 1965

3.23
SCHULZE R. — EINE NEUE EICHEINRICHTUNG IM GEZEITENGRAVIMETER. ASKANIA-WARTE 22 HEFT 66 PP 17-20 SONDERDRUCK 1442 — 1965

3.23
DUCARME B. — POSSIBILITE D AMELIORATION DES ETALONNAGES DANS LE CAS DES GRAVIMETRES ENREGISTREURS. B.I.M. NO 49 PP 2177-2200 — 1967

3.23
BONATZ M. — ERGEBNISSE DER GRAVIMETEREICHUNG DURCH SPINDELVERSTELLUNG BEI VERWENDUNG EINES ELEKTRONISCHEN REGISTRIERVERSTACKERS. B.I.M. NO 51 PP 2332-2340 — 1968

3.23
BONATZ M. — GRAVIMETEREICHUNG DURCH VERSCHIEBUNG DES VERSTARKERNULLPUNKTES. B.I.M. N.55 SEPTEMBRE PP 2628-2630 — 1969

3.23
KORBA P.S. — NOTE SUR LA FREQUENCE DE LA DETERMINATION DE L ECHELLE DE L ENREGISTREMENT DANS LES OBSERVATIONS DES MAREES TERRESTRES PAR LE GRAVIMETRE ASKANIA. 6 EME SYMPOSIUM MAREES TERRESTRES STRASBOURG OBS.ROY.BELG.COMM. A9 S.GEOPH. NO 96 PP 210-212 — 1970

3.23
MOORE R.D.
FARRELL W.E. — LINEARIZATION AND CALIBRATION OF ELECTROSTATICALLY FEEDBACK GRAVITY METERS. J.G.R. VOL 75 NO 5 PP 928-932 — 1970

3.23
GROTEN E. — CALIBRATION OF A GRAVIMETER BY USING A HEAVY MASS. 6 EME SYMPOSIUM MAREES TERRESTRES STRASBOURG OBS.ROY.BELG.COMM. A9 S.GEOPH. NO 96 PP 197-202 — 1970

3.23
FARRELL W.E. — INSTRUMENTAL ASPECTS OF TIDAL GRAVITY MEASUREMENTS. 6 EME SYMPOSIUM MAREES TERRESTRES STRASBOURG OBS.ROY.BELG.COMM. A9 S.GEOPH. NO 96 PP 216 — 1970

3.23
ZDENEK SIMON
JAROSLAV BROZ — UNTERSUCHUNG ZWEI GRAVIMETRISCHER APPARATUREN FUR DIE ERDGEZEITENREGISTRIERUNG UND DIE BESTIMMUNG IHRER KONSTANTEN VYZKUMNEHO USTAVU GEOD.,TOPO.,KARTOGR.V PRAZE PP57-71 — 1971

3.23
SIMON Z.
SOKOLIK B. — EINE EICHMETHODE FUR DIE ASKANIA ERDGEZEITEN-REGISTRIERANLAGE MIT DEM GALVANOMETER STUDIA GEOPH.ET GEOD.15 PP 92-95 — 1971

3.23
BONATZ M. — ZUR FRAGE DER KONSTANZ DES EICHFAKTORS DES NEUEN ASKANIA-GRAVIMETERS GS 15 B.I.M. N.62 DECEMBRE PP 3214-3216 — 1971

3.23
VALLIANT H.D. — A TECHNIQUE FOR THE PRECISE CALIBRATION OF CONTINUOUSLY RECORDING GRAVIMETERS MEASUREM., INTERPRETAT., CHANGES OF STRAIN IN THE EARTH PHIL.TRANS.ROY.SOC.SER.A, VOL.274, NO 1239 PP 227-230 — 1973

3.23
BONATZ M. — EINE BEMERKUNG ZUM EINFLUSS DER INNENTEMPERATUR DES ASKANIA-GRAVIMETERS GS 15 NR206 AUF DEN EICHFAKTOR B.I.M. NO 64 JANVIER PP 3386-3387 — 1973

3.23
BOGDAN Y.V.
GRIDNIEV D.G.
ETALONNAGE DES GRAVIMETRES -ASKANIA ET DETERMINATION DE L ECHELLE D ENREGISTREMENT DES VARIATIONS DE MAREES DE LA FORCE DE PESANTEUR PAR L INCLINAISON-INCLINAISONS DE TEMPERATURE DES SOCLES
ROT.& DEFORM.DE MAREES DE LA TERRE VOL VI PP 42-54 KIEV 1974
* B.I.M. NO 71 PP 4056-4073 1975

3.23
DITCHKO Y.A.
KORBA P.S.
RESULTATS DE L ETALONNAGE DES GRAVIMETRES ASKANIA PAR LA METHODE DE L INCLINAISON
ROT.ET DEFORM.DE MAREES DE LA TERRE, VOL.5 PP 70-73 1973
* B.I.M. NO 70 PP 3976-3980 1975

3.23
BAKER T.F.
LENNON G.W.
CALIBRATION - CONFIDENCE IN THE PERFORMANCE OF TILTMETERS AND GRAVIMETERS
PROCEEDINGS 7TH INT.SYMP.EARTH TIDES, SOPRON 1973
AKADEMIAI KIADO BUDAPEST, PP 223-230 1976

3.23
BREIN R.
THE INFLUENCE OF THE GRAVIMETRIC SPRINGS ON THE CALIBRATION OF EARTH TIDES RECORDINGS
B.I.M. NO 74 PP 4275-4279 1976

3.24
BREIN R.
UBER HYSTERESIS UND NACHWIRKUNGSERSCHEINUNGEN BEI DER GEZEITENREGISTRIERUNG.
3E SYMP. MAREES TERRESTRES TRIESTE
BOLL.GEOFISICA TEORICA E APPL.VOL.2 NO 5 PP 198-200 1960

3.24
PERTSEV B.
SOME EXPERIENCE IN DETERMINING THE SCALE COEFFICIENT OF THE RECORDINGS IN OBSERVATIONS OF TIDAL VARIATIONS OF GRAVITY.
TRAV.MAR.TERR.NO 1 GROUPE 13 GRAVIM. M.1959
3E SYMP. MAREES TERRESTRES TRIESTE
BOLL.GEOFISICA TEORICA E APPL.VOL.2 NO 5 PP 203-207 1960

3.24
LECOLAZET R.
STEINMETZ L.
SUR UN PROCEDE D ETALONNAGE DES ENREGISTREMENTS GRAVIMETRIQUES AU NORTH AMERICAN
3E SYMP. MAREES TERRESTRES TRIESTE
BOLL.GEOFISICA TEORICA E APPL.VOL.2 NO 5 PP 195-197 1960

3.24
BONATZ M.
EINE EINFACHE MOGLICHKEIT ZUR STEIGERUNG DER BETRIEBSSICHERHEIT DES NACHLAUFSCHREIBERS ZUR ASKANIA-ERDGEZEITENREGISTRIERANLAGE.
B.I.M. NO 43 PP 1674-1676 1966

3.24
BONATZ M.
DER EINFLUSZ DER NETZSPANNUNG AUF DIE SPANNUNGSKONSTANTHALTER ZU DER ASKANIA ERDGEZEITENREGISTRIERANLAGE.
B.I.M. NO 43 PP 1642-1650 1966

3.245
MELCHIOR P.J.
PAQUET P.
L AUTOMATION DANS LES MESURES DE DEFORMATIONS PERIODIQUES DU GLOBE TERRESTRE - REALISATION D UN GRAVIMETRE PERFORATEUR.
ACAD.ROY.BELG.BULL.CL.SC.5E SER.T.XLIX,1963-2
OBS.ROY.BELG.COMM.NO 220 S.GEOPH.NO 65 PP 142-152 1963

3.25
STEINMETZ L.
ETALONNAGE DES ENREGISTREMENTS ET ETUDE EXPERIMENTALE DU TRAINAGE D UN GRAVIMETRE NORTH AMERICAN PAR L EMPLOI D UN DISPOSITIF A ATTRACTION ELECTROSTATIQUE.
INST.PHYS.DU GLOBE A STRASBOURG-BIM NO 21 PP 396-408 1960

3.25
STEINMETZ L.
SUR L ETALONNAGE PERMANENT DES GRAVIMETRES.
IVME SYMP.INTERNATIONAL SUR LES MAREES TERRESTRES.
OBS.ROY.BELG.COMM. NO 188 S.GEOPH. NO 58 PP 302-306 1961

3.25
VENEDIKOV A.P.
SUR L ETALONNAGE DES ENREGISTREURS DES GRAVIMETRES.
B.I.M.NO 33 PP 1020-1028 1963

3.25
VOLKOV V.A.
L INFLUENCE DE LA NON-LINEARITE DE L ECHELLE DU GALVANOMETRE ENREGISTREUR SUR LES RESULTATS DES OBSERVATIONS DE MAREES TERRESTRES.
RECH. MAREES TERR. ART. N.3 PUBL.ACAD.SC.URSS MOSCOU 1963
* B.I.M. N.36 MAI PP 1210-1222 1964

3.25
BONATZ M. DER EINFLUSS DER RAUMTEMPERATUR AUF DAS GALVANOMETER DER ASKANIA ERDGEZEITENREGISTRIEREINRICHTUNG. B.I.M. NO 42 PP 1593-1598 1965

3.25
SIMON D. ELASTISCHEN NACHWIRKUNGEN AN EINEM ASKANIA-GRAVIMETER GS 11. B.I.M. NO 44 PP 1759-1774 1966

3.25
BONATZ M. ZUM PROBLEM DER GRAVIMETERFEHLER BEI DER ERDGEZEITENREGISTRIERUNG. ZEITS.VERMESS.91 4 PP 130-136 1966

3.25
BONATZ M. DER EINFLUSZ DER RAUMTEPERATUR AUF DIE SPANNUNGSKONSTANTHALTER ZU DER ASKANIA-ERDGEZEITENREGISTRIERANLAGE. B.I.M. NO 44 PP 1750 1966

3.25
BONATZ M. ZUR FRAGE DER BETRIEBSSICHERHEIT NETZSPANNUNGSBETRIEBENER REGISTRIERANLAGEN. B.I.M. NO 44 PP 1747 1966

3.25
BONATZ M. DIE ANPASSUNG DES MESSVERSTAKERS AN DEN KOMPENSATIONSSCHREIBER DER NEUEN ASKANIA-ERDGEZEITENREGISTRIERANLAGE. B.I.M. NO 45 PP 1841-1845 1966

3.25
BONATZ M. DER EINFLUSZ VON HELLIGKEITSSCHWANKUNGEN DER GALVANOMETERLICHTMARKE AUF DIE REGISTRIERGENAUIGKEIT DER ASKANIA-ERDGEZEITEN-REGISTRIERANLAGE /GALVANOMETER MIT NACHLAUFSCHREIBER/. B.I.M. NO 45 PP 1846-1850 1966

3.25
SIMON D. DAS VERSCHWINDEN DER EICHSTORUNGEN ALS KRIETERIUM FUR DIE RICHTIGE KORREKTUR DER NICHTLINEARITAT BEI GRAVIMETERREGISTRIERUNGEN. B.I.M. NO 45 PP 1857-1859 1966

3.25
CHOJNICKI METHODS OF CALIBRATING ASKANIA GS 11 GRAVIMETERS. PUBL.GEOD.ASTR.PP 65-144 PAN NO 5 1966

BONATZ M. EINE BEMERKUNG ZU DEM BEITRAG -D.SIMON. DAS VERSCHWINDEN DER EICHSTORUNGEN ALS KRITERIUM FUR DIE RICHTIGE KORREKTUR DER NICHTLINEARITAT BEI GRAVIMETERREGISTRIERUNGEN - IN B.M.TERR. NO 45 S 1857 1966. B.I.M. NO 47 PP 1951-1952 1967

3.25
BONATZ M. DER EINFLUSZ VON RAUMTEMPERATUR UND NESTZSPANNUNGSSCHWANKUNGEN AUF DEN NACHLAUFSCHREIBER ZUR ASKANIA-ERDGEZEITENREGISTRIERANLAGE. B.I.M. NO 47 PP 2026-2030 1967

3.25
BONATZ M. EINFLUSZ VON STORBESCHLEUNIGUNGEN AUF DIE ERGEBNISSE DER ASKANIA-ERDGEZEITEN-REGISTRIERANLAGE. ASKANIA-WARTE HEFT 69 PP 5-12 1967

3.25
BONATZ M. UNTERSUCHUNGEN ELASTISCHER NACHWIRKUNGEN AM ASKANIA-GRAVIMETER GS 11 NO 116. STUDIA GEOPH.GEOD.2, 11 PP 164-182 1967

3.25
BONATZ M. DER EINFLUSZ VON SCHWANKUNGEN DES STROMES FUR DIE PHOTOZELLENBELEUCHTUNG DES ASKANIA-MESZVERSTARKERS AUF DIE MESZWERTANZEIGE. B.I.M. NO 48 PP 2125-2128 1967

3.25
SIMON D. ERWIDERUNG AUF DIE BEMERKUNG VON M.BONATZ /BULL.D INF.NO 47 1951/ ZUM BEITRAG DAS VERSCHWINDEN DER EICHSTORUNGEN ALS KRITERIUM FUR DIE RICHTIGE KORREKTUR DER NICHTLINEARITAT BEI GRAVIMETERREGISTRIERUNG-. /BULL.D INF.45 1857/. B.I.M. NO 48 PP 2129-2132 1967

3.25
BONATZ M. ZUR FRAGE ELASTISCHER NACHWIRKUNGEN AM ASKANIA-GRAVIMETER GS 11 NO 116.
B.I.M. NO 49 PP 2173-2176 1967

3.25
BONATZ M. GENAUIGKEITSTEIGERUNG DER EICHUNG DURCH SPINDELVERSTELLUNG BEI VERWENDUNG EINES ELEKTRONISCHEN REGISTRIERVERSTARKERS ZU DEN ASKANIA-GRAVIMETER GS 11 UND GS 12.
B.I.M. NO 49 PP 2226-2228 1967

3.25
BONATZ M. DER EINFLUSZ DER INNENTEMPERATUR DES ASKANIAGRAVIMETERS GS 11 NO 116 AUF DIE REGISTRIEREMPFINDLICHKEIT.
GRAVIMETERS GS 11 NO 116 AUF DIE REGISTRIEREMPFINDLICHKEIT.
B.I.M. NO 50 PP 2300-2304 1968

3.25
KORBA P.S.
DITCHKO I.A. SUR LES CONSEQUENCES DE L AMORTISSEMENT DANS LES OBSERVATIONS DES VARIATIONS DE LA FORCE DE PESANTEUR AVEC LE GRAVIMETRE ASKANIA.
* B.I.M. N.52 DECEMBRE PP 2440-2443 1968

3.25
BONATZ M. ZUR FRAGE DER STORSIGNALE BEI DER ERDGEZEITENREGISTRIERUNG MIT GRAVIMETERN.
B.I.M. N.52 DECEMBRE PP 2410-2412 1968

3.25
BONATZ M. ERGEBNISSE EINER 100 TAGIGEN GRAVIMETERREGISTRIERUNG BEI VERWENDUNG EINES ELEKTRONISCHEN VERSTARKERS.
B.I.M. N.52 DECEMBRE PP 2413-2416 1968

3.25
WIRTH H. SYSTEMATISCHE AUFZEICHNUNGSFEHLER BEI ERDGEZEITEN-REGISTRIERANLAGEN MIT GRAVIMETERN UND GALVANOMETERN
GERLANDS BEITRAGE ZUR GEOPH.77 HEFT 5 PP 379-384 1968

3.25
BONATZ M. EINE BEMERKUNG ZUR ELIMINATION DES DURCH ANDERUNGEN DES REGISTRIEREMPFINDLICHKEIT VERURSACHTEN DRIFTANTEILS.
B.I.M. NO 51 PP 2358-2359 1968

3.25
PERTSEV B.P. SUR LA QUESTION DU RETARD DE PHASES DANS LES OBSERVATIONS DES MAREES TERRESTRES.
METH.MES.MAR.TERR.& DEFORM.LENTES SURFACE TERRE
ACAD.SC.URSS.INST.PHYS.TERR.SCHMIDT MOSCOU PP 110-112 1970

3.25
VOLKOV V.A.
PARIISKII N.N.
PERTSEV B.P. PREMIERS RESULTATS DE LA DETERMINATION DE LA CARACTERISTIQUE DE PHASE DU SYSTEME ENREGISTREUR DE MAREE GRAVIMETRE-GALVANOMETRE A L AIDE D UNE MASSE MOBILE.
INST.PHYS.TERRE SCHMIDT OBS.GEO.OBNINSK-PUB.1 PP 128-147 1970

3.25
DUCARME B. DETERMINATION EXPERIMENTALE DU DEPHASAGE DES ONDES DE MAREE DANS LE CAS DES GRAVIMETRES ASKANIA.
6 EME SYMPOSIUM MAREES TERRESTRES STRASBOURG
OBS.ROY.BELG.COMM. A9 S.GEOPH. NO 96 PP 203-205 1970

3.25
UEILLE C.
LERC G. THERMOSTAT DE PRECISION AU MILLIEME DE DEGRE CENTIGRADE POUR GRAVIMETRE D ENREGISTREMENT DE LA MAREE TERRESTRE.
B.I.M. N.59 OCTOBRE PP 2882-2920 1970

3.25
OLKOV V.A.
ARIISKY N.N. THE EFFECT OF PHASE CHARACTERISTICS OF ASKANIA GRAVIMETERS UPON MEASUREMENTS OF PHASE LAGS IN TIDAL OBSERVATIONS.
6 EME SYMPOSIUM MAREES TERRESTRES STRASBOURG
OBS.ROY.BELG.COMM. A9 S.GEOPH. NO 96 PP 206-209 1970

3.25
IMON.Z.
OKOLIK B. EINE EICHMETHODE FUR DIE ASKANIA ERDGEZEITEN-REGISTRIERANLAGE MIT DEM GALVANOMETER.

STUDIA GEOPH.ET GEOD.15 PP 92-95 ACAD.PRAHA 1971

3.25
BONATZ M.
ZUR FRAGE DER STORBESCHLEUNIGUNGEN BEI DER EICHUNG VON REGISTRIERGRAVIMETERN AUF EINER VERTIKALEN LABOR-EICHBASIS
B.I.M. NO 61 PP 3066-3069 1971

3.25
STUKENBROKER B.
STEIGERUNG DER RELATIVEN MESSWERTGENAUIGKEIT EINER ERDGEZEITENMESSANLAGE MIT EINEM ASKANIA GS 11 GRAVIMETER UNTER VERWENDUNG EINER VORRICHTUNG ZUR KONTINUIERLICHEN KALIBRIERUNG
B.I.M. NO 61 PP 3051-3065 1971

3.25
BONATZ M.
DER EINFLUSS VON ANDERUNGEN DER VERSTARKEREINGANGSSPANNUNG AUF DIE MESSWERTANZEIGE DES NEUEN ASKANIA-GRAVIMETERS GS 15
B.I.M. NO 61 PP 3070-3072 1971

3.25
VOLKOV V.A.
DETERMINATION DES CARACTERISTIQUES DE PHASES DES SYSTEMES ENREGISTREURS DE MAREES - GRAVIMETRE GALVANOMETRES.
VARIATIONS DE MAREES DE LA FORCE DE PESANTEUR
ACAD.SC.URSS.INST.PHYS.T.SCHMIDT MOSCOU PP 71-82 1964
* B.I.M. N.62 PP 3149-3163 1971

3.25
QUEILLE C.
CLERC G.
THERMOSTAT DE PRECISION AU MILLIEME DE DEGRE CENTIGRADE POUR GRAVIMETRE D ENREGISTREMENT DE LA MAREE TERRESTRE.
B.I.M. N.59 OCTOBRE PP 2882-2920 1970

3.25
VOLKOV V.A.
PARIISKY N.N.
THE EFFECT OF PHASE CHARACTERISTICS OF ASKANIA GRAVIMETERS UPON MEASUREMENTS OF PHASE LAGS IN TIDAL OBSERVATIONS.
6 EME SYMPOSIUM MAREES TERRESTRES STRASBOURG
OBS.ROY.BELG.COMM. A9 S.GEOPH. NO 96 PP 206-209 1970

3.25
SIMON Z.
SOKOLIK B.
EINE EICHMETHODE FUR DIE ASKANIA ERDGEZEITEN-REGISTRIERANLAGE MIT DEM GALVANOMETER.
STUDIA GEOPH.ET GEOD.15 PP 92-95 ACAD.PRAHA 1971

3.25
BONATZ M.
ZUR FRAGE DER STORBESCHLEUNIGUNGEN BEI DER EICHUNG VON REGISTRIERGRAVIMETERN AUF EINER VERTIKALEN LABOR-EICHBASIS
B.I.M. NO 61 PP 3066-3069 1971

3.25
STUKENBROKER B.
STEIGERUNG DER RELATIVEN MESSWERTGENAUIGKEIT EINER ERDGEZEITENMESSANLAGE MIT EINEM ASKANIA GS 11 GRAVIMETER UNTER VERWENDUNG EINER VORRICHTUNG ZUR KONTINUIERLICHEN KALIBRIERUNG
B.I.M. NO 61 PP 3051-3065 1971

3.25
BONATZ M.
DER EINFLUSS VON ANDERUNGEN DER VERSTARKEREINGANGSSPANNUNG AUF DIE MESSWERTANZEIGE DES NEUEN ASKANIA-GRAVIMETERS GS 15
B.I.M. NO 61 PP 3070-3072 197

3.25
VOLKOV V.A.
DETERMINATION DES CARACTERISTIQUES DE PHASES DES SYSTEMES ENREGISTREURS DE MAREES - GRAVIMETRE GALVANOMETRES.
VARIATIONS DE MAREES DE LA FORCE DE PESANTEUR
ACAD.SC.URSS.INST.PHYS.T.SCHMIDT MOSCOU PP 71-82 1964
* B.I.M. N.62 PP 3149-3163 197

3.25
BONATZ M.
UNTERSUCHUNGEN DES EICHFAKTORS DER GRAVIMETRISCHEN GEZEITENME ANLAGE DER UNIVERSITAT BONN
ASKANIA GRAVIMETER GS 15 NR.206
B.I.M. NO 63 PP 3299-3305 19

3.25
GRIDNIEV D.G.
NOUVEAU SYSTEME DE COMPENSATION DE TEMPERATURE DES GRAVIMETRES.
METH.MES.MAR.TERR.& DEFORM.LENTES SURFACE TERRE

ACAD.SC.URSS.INST.PHYS.TERR.SCHMIDT MOSCOU P122-128 1970
* B.I.M. MARS NO 65 PP 3457-3462 1973

3.25
VOLKOV V.A.
ZASSIMOV S.S.
DU PROBLEME DE L AMORTISSEMENT ET DE LA LINEARITE DES SYSTEMES ENREGISTREURS DE MAREES /ASKANIA/ GS15
ACAD.NAOUK UKRAINSKOI SSR, OBS.GRAV.POLTAVA
ROTAT.DEFOR.MAREES DE LA TERRE VII PP 39-42 1975

3.25
BONATZ M.
WILMES H.
ZUR FRAGE INSTRUMENTELLER PHASENFEHLER BEI DER GEZEITENREGISTRIERUNG MIT ASKANIA-GRAVIMETÈRN BN
MITT.INST.THEOR.GEOD.UNIV.BONN N.34 PP 1-9 1975

3.25
VOLKOV V.
ZASIMOV S.S.
A NOTE CONCERNING THE DAMPING AND LINEARITY OF THE TIDE-RECORDING SYSTEM ASKANIA GS 15
PROCEEDINGS 7TH INT.SYMP.EARTH TIDES, SOPRON 1973
AKADEMIAI KIADO BUDAPEST, PP 301-306 1976

3.3
SUGAWA C.
DETERMINATION OF 1 + K - L FROM LATITUDE OBSERVATIONS AT MIZUSAWA.
~~IVME SYMP.INTERNATIONAL SUR LES MAREES TERRESTRES.~~
OBS.ROY.BELG.COMM. NO 188 S.GEOPH. NO 58 PP 76-77 1961

3.3
SUGAWA C.
ON THE LUNAR TIDAL EFFECT IN THE LATITUDE OBSERVATIONS AT MIZUSAWA.
PUBL.INTERN.LAT.OBS.MIZUSAWA VOL.III NO 2 PP 121-135 1961

3.3
SUGAWA C.
EARTH TIDES BY THE MOON DEDUCED FROM LATITUDE OBSERVATIONS.
/EN JAPONAIS/ 1961

3.3
CHUGH R.S.
LATITUDE VARIATION AND EARTH TIDE AT DEHRA DUN.
PROC.IGY SYMP.II NEW DELHI PP 140-153 1961

3.3
SUGAWA CH.
LUNAR EARTH TIDE OBTAINED BY LATITUDE OBSERVATIONS.
/EN JAPONAIS/. 1963

3.31
SHIDA T.
MATSUYAMA M.
CHANGE OF PLUMB LINE REFERRED TO THE AXIS OF THE EARTH AS FOUND FROM THE RESULT OF THE INTERNATIONAL LATITUDE OBSERVATIONS.
MEM.COLL.SC.&ENG.KYOTO IMP.UNIV.IV PP 277-284 1912

3.31
SCHWEYDAR W.
UEBER KURZPERIODISCHE AENDERUNGEN DER GEOGRAPHISCHEN BREITE.
ASTRONOM.NACHRICHTEN BD.193 NO 4627 PP 347-356 1913

3.31
SCHUMANN R.
UBER GEZEITENERSCHEINUNGEN IN DEN SCHWANKUNGEN DER STATIONSPOLHOHEN.
DENKSCH.KAISERLICHEN AKAD.WISSENSCH.WIEN 89 PP 318-400 1913

3.31
CERULLI V.
SU DI UNA PRETESA FORTE VARIAZIONE A CORTO PERIODO DELLA LATITUDINE.
RENDICONTI R.ACCAD.LINCEI VOL.XXVII PP 213-218 1918

3.31
COURVOISIER L.
BEOBACHTUNGEN DES ZENITSTERNS B DRACONIS AM VERTIKALKREISE 1914.6-1918.0.
VEROFF.UNIV.BERLIN BABELSBERG II NO 4 PP 22-24 1919

3.31
CERULLI V.
SOPRA UN ARTICOLO DEL -BULLETIN ASTRONOMIQUE-DI PARIGI.
RENDICONTI R.ACCAD.LINCEI VOL.XXVIII PP 319 1919

3.31
PRZYBYLLOK E.
UBER EINIGE PERIODISCHE ERSCHEINUNGEN IN POLHOHENBEOBACHTUNGEN.
ASTRONOM.NACHRICHTEN BD.213 NO 5101 PP 201-210 1921

3.31
PRZYBYLLOK E.
UBER DIE M2 TIDE DER LOTBEWEGUNG.
ASTRONOM.NACHRICHTEN BD.218 NO 5214 PP 85-88 1923

3.31
BOMFORD G.
VARIATION OF LATITUDE WITH THE MOON S POSITION.
NATURE VOL.123 PP 873 1929

3.31
STETSON H.T. VARIATION OF LATITUDE WITH THE MOON S POSITION.
NATURE VOL.123 PP 127-128 1929

3.31
DROSD A. RESULTATE DES ERWEITERTEN PROGRAMMS DER
BEOBACHTUNGEN FUR DIE BESTIMMUNG DER BREITE DES GROSSEN
PULKOWOER ZENITTELESKOP ANGESTELLT VON A.DROSD UND
FRAU S.ROMANSKAJA IN DEN JAHREN 1917-1920.
PUBL.OBS.CENTRAL A POULKOVO S II VOL.XXXVII 1930

3.31
STETSON H. THE STUDY OF EARTH TIDES FROM THE VARIATION
IN LATITUDE.
TRANS.AMER.GEOPH.UNION PP 148-152 1930

3.31
STETSON H. FURTHER INVESTIGATIONS OF THE MOON S INFLUENCE
ON LATITUDE.
TRANS.AMER.GEOPH.UNION PP 45-46 1931

3.31
STETSON H.T. FURTHER STUDIES ON THE LUNAR CORRELATIONS WITH SMALL
CHANGES IN THE VARIATION OF LATITUDE.
NAT.RES.COUNCIL.TRANS.AMER.GEOPH.UNION PP 85-88 1932

3.31
DOBBIE J.C. A STUDY OF THE EFFECT OF A THEORETICAL TIDAL VARIATION
OF THE ZENITH ON THE ANNUAL VARIATION OF LATITUDE
MONTHLY NOT.R.ASTR.SOC.VOL.93 PP 377-382 1933

3.31
STETSON H.T. VARIATION EFFECT IN LATITUDE CORRELATABLE WITH
THE MOON.
NATURE VOL.131 PP 147 1933

3.31
DOBBIE J.C. EFFECT OF TIDAL VARIATION OF THE VERTICAL ON THE
ANNUAL VARIATION OF LATITUDE.
THE OBSERV.56 NO 707 PP 117-118 1933

3.31
KAWASAKI S. VARIATION IN LATITUDE WITH THE MOON S POSITION.
PROC.IMP.ACAD.TOKYO VOL.11 PP 398-400 1935

3.31
KAWASAKI S. ON MINOR VARIATIONS OF LATITUDE.
MEMOIRS COLL.SC.KYOTO IMP.UNIV.A XX NO 3 PP 87-137 1937

3.31
NISHIMURA E. CHANGE OF PLUMB LINE REFERRED TO THE AXIS OF THE
EARTH AS FOUND FROM THE RESULTS OF THE INTERNATIONAL
LATITUDE OBSERVATIONS.
MEMOIRS COLL.SC.KYOTO IMP.UNIV.A XX NO 5,6 PP 191-206 1937

3.31
SPENCER JONES H. THE TIDAL EFFECT ON THE VARIATION OF LATITUDE
AT GREENWICH.
MONTHLY NOT.R.ASTR.SOC.VOL.99 PP 196-198 1938

3.31
SPENCER JONES H. OBSERVATIONS MADE WITH THE COOKSON FLOATING ZENITH
TELESCOPE IN THE YEARS 1927-1936 AT THE ROYAL
OBSERVATORY, GREENWICH AND THE DETERMINATION OF THE
VARIATION OF LATITUDE AND THE CONSTANT OF NUTATION
FROM THE OBSERVATIONS IN THE YEARS 1911-1936.
CF.PP 95-99 LONDON 1939

3.31
NICOLINI T. VARIAZIONI DELLA LATITUDINE CONNESSE ALL ANGOLO
ORARIO LUNARE.
OSSERV.ASTR.DI CAP.NAPOLI C.A.S.II VOL.III NO 1
REALE ACCAD.D ITALIA FASC.9 S.VII VOL.1 194

3.31
NISHIMURA E. ON EARTH TIDES.
PART.V TIDAL VAR.OF LAT./GEOPH./VOL.8 NO 1 PP 65-72 1947

3.31
CARNERA L. COMMISSION DE LA VARIATION DES LATITUDES. RAPPORT
DU DIRECTEUR DU BUREAU CENTRAL DE 1938 A FIN 1947.
TRANS.INT.ASTR.UNION VOL.VII CF.PP 201-203 CAMBRIDGE 195

3.31
VAN HERK G. AN ATTEMPT TO ANALYSE THE INFLUENCE OF THE MOON ON
THE RESULTS OF THE INTERNATIONAL LATITUDE SERVICE.
BULL.ASTRON.INST.NETHERLANDS XI NO 417 PP 241-243 195

3.31
EVTOUCHENKO E.I. ONDE SEMI-MENSUELLE DANS LES VARIATIONS DE LATITUDE DE LA STATION D UKIAH
CIRC.ASTRON.NO 113-114 P 14 1951

3.31
FEDOROV E.P.
EVTOUCHENKO E.I. VARIATIONS LUNAIRES SEMIMENSUELLES DE LA LATITUDE D APRES LES OBSERVATIONS DANS LES STATIONS DE CARLOFORTE ET UKIAH DE 1899 A 1934
RAPPORTS ACAD.SC.URSS T.85 NO 4 PP 731-732 1952

3.31
SUGAWA CH. DETERMINATION OF 1+K-L FROM LATITUDE OBSERVATIONS
~~MADE AT MIZUSAWA~~
5 EME SYMP.INT.SUR LES MAREES TERRESTRES.
OBS.ROY.BELG.COMM.NO 236 S.GEOPH.NO 69 PP 450-451 1964

3.31
EVTOUCHENKO E.I. L ONDE LUNAIRE SEMI-MENSUELLE DANS LES VARIATIONS DE LATITUDE DE LA STATION DE CARLOFORTE DE 1922 A 1934.
CIRCULAIRE ASTRONOMIQUE N.116 2 JUILLET 1951
* B.I.M. N.39 AVRIL P 1440-1441 1965

3.31
EVTOUCHENKO E.I. L ONDE LUNAIRE SEMI-MENSUELLE DANS LES OBSERVATIONS DE LATITUDE A LA STATION DE MIZUSAWA.
CIRCULAIRE ASTRONOMIQUE N.132 8 DECEMBRE 1952
* B.I.M. N.39 AVRIL P 1443 1965

3.31
FEDOROV E.P. SUR LA NATURE DES VARIATIONS BIMENSUELLES DE LATITUDE.
CURCULAIRE ASTRONOMIQUE N. 110 1951
* B.I.M. N.39 AVRIL PP 1438-1439 1965

3.31
EVTOUCHENKO L ONDE BIMENSUELLE DANS LES VARIATIONS DE LATITUDE DE LA STATION DE UKIAH
CIRC.ASTRON.NO 113-114 MAI 1951
* B.I.M. NO 39 P 1438 1965

3.31
FEDOROV E.P.
EVTOUCHENKO E.I. VARIATIONS LUNAIRES SEMI-MENSUELLES DE LATITUDE D APRES LES OBSERVATIONS DE CARLOFORTE ET UKIAH DE 1899 A 1934
CIRCULAIRE ASTRONOMIQUE N.126 30 AVRIL 1952
* B.I.M. N.39 AVRIL P 1442 1965

3.31
FEDOROV E.P. L ONDE LUNAIRE DIURNE DANS LES VARIATIONS DE LA LATITUDE.
ASTRONOMITCHESKI TSIRKULIAR N.148 P-12 1954
* B.I.M. N.43 MARS PP 1685-1686 1966

3.32
MARKOWITZ W.M.
BESTUL S.M. EFFECT OF THE MOON IN THE DETERMINATION OF LATITUDE AT WASHINGTON.
THE ASTRONOMICAL JOURN.NO 1131 PP 81-86 1940

3.32
WOOLSEY E.G. GRAVITATIONAL EFFECTS ON THE VERTICAL OBSERVED BY THE OTTAWA PZT 1
JOURN.CANADIEN SC.TERRE, VOL.10, NO 3, PP 379-383 1973

3.32
O HORA N.P.J. SEMI-DIURNAL TIDAL EFFECTS IN P.Z.T. OBSERVATIONS
PHYSICS OF THE EARTH V.7 NO 1 APRIL PP 92-96 1973

3.33
GOUGENHEIM A. SUR L EMPLOI DE L ASTROLABE A PRISME POUR L ETUDE DES VARIATIONS DES LATITUDES.
C.R.ACAD.SCI.PARIS 187 PP 281-284 1928

3.33
DEBARBAT S. RECHERCHE DES EFFETS LUNAIRES SUR LA VERTICALE DEDUITE DES OBSERVATIONS FAITES A L ASTROLABE.
BULL.ASTR.S 3 TOME II FASC.4 PP 541-560 1967

3.33
ELSTNER CL.
HOPFNER J. BETRACHTUNGEN ZUM NACHWEIS VON GEZEITENEFFEKTEN IN GEODATISCH-ASTRONOMISCHEN BREITENBESTIMMUNGEN
VEROFF.ZENTRALINST.PHYSIK DER ERDE NO 29 PP 95-107 1974

3.4
OZAWA I. ON SOME COEFFICIENTS OF EARTH TIDAL STRAINS
3E SYMP. MAREES TERRESTRES TRIESTE
BOLL.GEOFISICA TEORICA E APPL.VOL.2 NO 5 PP 119-121 1960

3.4
LATININA L.A. SUR LES DIFFERENCES DANS L ALLURE DES DEPLACEMENTS DES POINTS PROCHES A LA SURFACE DE LA TERRE.
IZVESTIA AC.SC.URSS.SER.GEOPH.NO 4 MOSCOU PP 478-484 1962

3.4
MAJOR M.W. / SUTTON G.H. / OLIVER J. / METSGER R. — ON ELASTIC STRAIN OF THE EARTH IN THE PERIOD RANGE 5 SECONDS TO 100 HOURS. BULL.SEISMOL.SOC.AMERICA 54 NO 1 PP 295-346 — 1964

3.4
OZAWA I. — ON THE EXTENSOMETER WHOSE MAGNIFIER IS A ZOLLNER SUSPENSION TYPE TILTMETER, AND THE OBSERVATION OF THE EARTH S STRAINS BY MEANS OF THE INSTRUMENTS. BOLL.DI GEOF.VOL.XVIII NO 3 ROMA PP 263-278 — 1965

3.4
TORAO TANAKA — ON THE EXTENSOMETER OF A VARIABLE CAPACITOR TYPE. BULL.DISASTER PREV.RES.INST.VOL.15 O NO 100 PP 50-59 — 1966

3.4
BOULATSEN V.G. — QUELQUES QUESTIONS D APPLICATION DES EXTENSOMETRES MAT.TERR.ACAD.SC.URSS.UKRAINE OBS.GRAV.POLTAVA 1966 * B.I.M. NO 51 PP 2366-2385 — 1968

3.4
HADE G. / CONNER M. / KUO J.T. — LASER INTERFEROMETER CALIBRATION SYSTEM FOR EXTENSOMETERS BULL.SEISMOL.SOC.AM. 58 PP 1379 — 1968

3.4
KING G.C.P. / BILHAM R.G. / GERARD V.B. / DAVIES D. / SYDENHAM P.H. — NEW STRAIN METERS FOR GEOPHYSICS. NATURE 223 AUG 23 PP 818-819 — 1969

3.4
BILHAM R.G. — THE MEASUREMENT OF EARTH STRAIN PH.D.THESIS UNIVERSITY CAMBRIDGE — 1970

3.4
STAUDER W. — RESEARCH IN EARTH STRAINS AND FOCAL MECHANISMS UPPER MANTLE PROJ.UN.STATES PROG.FINAL REP.PP 163-164 — 1971

3.4
KING G.G.P. — THE SITING OF STRAINMETERS FOR TELESEISMIC AND TIDAL STUDIES BULL R.SOC.NEW.ZEALAND VOL 9,239, PP 818-819 — 1971

3.4
BOSTROM R.C. / VALI V. — STRAINS AND DETECTION IN A MOSAIC EARTH RECENT CRUSTAL MOVEM. R.S.NEW ZEALAND BULL.9 PP 55-60 — 1972

3.4
BILHAM R. / KING G. — LOCAL ANOMALIES IN STRAIN MEASUREMENT 31TH GEN.AS.EUROPEAN SEISM.COMM.,BRASOV RUMANIA — 1972

3.4
GERARD V.B. — NEW ZEALAND EARTH STRAIN MEASUREMENTS MEASUREM., INTERPRETAT., CHANGES OF STRAIN IN THE EARTH PHIL.TRANS.ROY.SOC.SER.A, VOL.274, NO 1239 PP 311-322 — 1973

3.4
KING G.C.P. / BILHAM R.G. — STRAIN MEASUREMENT INSTRUMENTATION AND TECHNIQUE MEASUREM., INTERPRETAT., CHANGES OF STRAIN IN THE EARTH PHIL.TRANS.ROY.SOC.SER.A, VOL.274, NO 1239 PP 209-218 — 1973

3.4
BERGER J. — THE SPECTRUM OF EARTH STRAIN FROM 10-8 TO 10-2 HZ JOURN.OF GEOPH.RESEARCH V.79 NO 8 PP 1210-1214 MARCH — 1974

3.4
BILHAM R. / KING G. / MCKENZIE D. — INHOMOGENEOUS TIDAL STRAINS IN QUEENSBURY TUNNEL, YORKSHIRE. GEOPH.J.R.ASTR.SOC. VOL 37 PP 217-226 — 1974

3.4
SYDENHAM P.H. — 2000 HR COMPARISON OF 10 M QUARTZ-TUBE AND QUARTZ-CATENARY TIDAL STRAINMETERS GEOPH.J.R.ASTR.SOC. VOL 38 PP 377-387 — 1974

3.4
SYDENHAM P.H. / PRESTON B.C. / MCKNIGHT B.W. / LIMBERT A.R. / GREEN R. / LE BROCQ — SOLID TIDAL MEASUREMENTS IN THE NEW ENGLAND AREA OF AUSTRALIA NATURE, PHYS, SCI., 235 PP 116-118 — 1974

3.4
SYDENHAM P.H. — WHERE IS EXPERIMENTAL ~~RESEARCH ON EARTH STRAIN~~ NATURE, VOL.252,NO 5481, PP 278-280 — 1974

3.4
LEVINE J. STRAINMETER TECHNOLOGY
NATURE VOL.257 P 513 1975

3.4
KING G.C.P.
BEAVAN R.J.
EVANS J.R. STRAINMETER TECHNOLOGY
NATURE VOL 257 P 513 1975

3.4
SYDENHAM P.H. STRAINMETER TECHNOLOGY
NATURE VOL 257 P 514 1975

3.41
BENIOFF H. LIST OF EXTENSOMETERS.
B.I.M. NO 6 PP 96 1957

3.41
BENIOF H. TIDAL STRAIN RECORDER FOR FUSED QUARTZ EXTENSOMETERS.
CR.SEANCES II CONF.TORONTO STRASBOURG 1958 PP 115 1957

3.41
N. MESURES DE L ACCUMULATION DE TENSION EN AMERIQUE DU SUD.
B.I.M. NO 13 PP 204 1958

3.41
BENIOFF H. FUSED-QUARTZ EXTENSOMETER FOR SECULAR, TIDAL AND SEISMIC STRAINS.
BULL.GEOLOGICAL SOC.AMER.VOL.70 PP 1019-1032 1959

3.41
HIERSEMAN J.
MEISSER O. AUFZEICHNUNG LANGPERIODISCHER BODENDEFORMATIONEN MIT EINEM STRAINSEISMOMETER.
IVME SYMP.INTERNATIONAL SUR LES MAREES TERRESTRES.
OBS.ROY.BELG.COMM. NO 188 S.GEOPH. NO 58 PP 293-301 1961

3.41
HIERSEMANN L. FORTLAUFENDE AUFZEICHNUNG VON BODENBEWEGUNGEN DURCH EIN STRAINSEISMOMETER.
FREIBERGER FORSCH.C.135 PP 81 1962

3.41
LATYNINA L.A.
KARMALEYEVA R.M. FIRST RESULTS OF OBSERVATIONS WITH A HORIZONTAL EXTENSOMETER IN THE TIEN SHAN MOUNTAINS.
BULL.AC.SC.URSS.GEOPH.SER.NO 11 PP 982-984 1962

3.41
BALAVADZE B.K.
KARMALEYEVA R.M.
KARTVELISHVILI K.Z.
LATYNINA L.A. HORIZONTAL EXTENSOMETER OBSERVATIONS AT TBILISI /TIFLIS/ ON TIDAL DEFORMATIONS OF THE EARTH.
PHYS.OF SOLID EARTH NR 2 PP 127-130 FEBRUARY 1965

3.41
MAAZ R. ZUR BESTIMMUNG DER SHIDA SCHEN ZAHL AUS EXTENSOMETERAUFZEICHNUNGEN.
INST.GEOD.DEUTSCHEN AKAD.BERLIN B.I.M.NO 39 PP 1402 1965

3.41
KUO J.T.
MAJOR M.
OLIVER J. OBSERVATIONS OF EARTH TIDES WITH STRAIN METERS.
LAMONT GEOLOG.OBS.COLUMBIA UNIV.PALISADES N-Y 1966

3.41
LATININA L.A.
KARMALIEVA R.M. MESURE DES DEPLACEMENTS HORIZONTAUX A LA SURFACE DE LA TERRE PAR UN EXTENSOMETRE EN QUARTZ.
* B.I.M. N.52 DECEMBRE PP 2460-2464 1968

3.41
LATININA L.A.
KARMALIEVA R.M. PREMIERS RESULTATS DES OBSERVATIONS PAR UN EXTENSOMETRE HORIZONTAL A TIAN-SHANE.
* B.I.M. N.55 SEPTEMBRE PP 2637-2643 1969

3.41
LATININA L.A. ANALYSE DU TRAVAIL DU DEFORMOGRAPHE A TIGE
ACAD.SC.URSS, INST.PHYS.TERR.SCHMIDT 1971

3.41
OZAWA I.
MELCHIOR P.
DUCARME B.
FLICK J. OBSERVATION OF THE EARTH TIDAL EXTENSION AT WALFERDANGE
CONTRIBUTIONS GEO.INST.KYOTO UNIV., NO 13, PP 85-92 1973

3.41
KERR GRANT C.
LANGDON J.F.
THOMAS L. A NEW STRAIN METER OBSERVATORY
J.GEOL.SOC.AUSTRALIA VOL.20 NO 1 PP 95-98 1973

3.41
LANGDON J.F. EARTH-TIDE MEASUREMENTS IN SOUTHEASTERN AUSTRALIA

THOMAS L. -BULL.SEISM.SOC.AMERICA V.64 NO 2 PP 457-472 1974

3.41
PETERS J.A.
SYDENHAM P.H.
SOLID TIDES RECORDED WITH 1 M INTERVAL
MECHANICAL STRAINMETER
NATURE VOL.250 NO 5461 PP 43-44 1974

3.41
BOULATSEN V.G.
ETALONNAGE ET SENSIBILITE DES EXTENSOMETRES A TIGE
EN QUARTZ AVEC UN CONVERTISSEUR DU TYPE A TORSION
ROT.&DEFORM.MAREES TERRE VOL.6 PP 68-78 KIEV 1974
* B.I.M. N.72 PP 4123-4142 1975

3.41
BOULATSEN V.G.
ETALONNAGE ET SENSIBILITE DES EXTENSOMETRES A TIGE DE
QUARTZ AVEC UN TRANSFORMATEUR DU TYPE A TORSION
ROTAT.ET DEFORMAT.DE MAREES DE LA TERRE-VI-PP 68-78 1974
* B.I.M. NO 72 PP 4123-4142 1975

3.41
BOULATSEN V.G.
TOKAR V.I.
RESULTATS DES OBSERVATIONS EXTENSOMETRIQUES
DE MAREES EN CRIMEE
ROT ET DEF. DE MAREES DE LA TERRE V.VI PP 16-28 1974
* B.I.M. NO 71 PP PP 4008-4023 1975

3.41
OZAWA I.
MELCHIOR P.
DUCARME B.
FLICK J.
OBSERVATION OF THE EARTH TIDAL EXTENSION
AT WALFERDANGE
CONTRIBUT.OF THE GEOPH.INST.KYOTO UNIV.NO 13 PP 87-92 1975

3.42
OZAWA I.
STUDY ON ELASTIC STRAIN OF THE GROUND IN EARTH TIDES.
DISASTER PREV.RES.INST.BULL.NO 15 1957

3.42
TAKADA M.
ON THE OBSERVING INSTRUMENTS AND THE TELE-METRICAL
DEVICES OF EXTENSOMETERS AND TILTMETERS AT IDE
OBSERVATORY.
3E SYMP. MAREES TERRESTRES TRIESTE
BOLL.GEOFISICA TEORICA E APPL.VOL.2 NO 5 PP 125-126 1960

3.42
OZAWA I.
ON AN EXTENSOMETER OF NEW TYPE.
3E SYMP. MAREES TERRESTRES TRIESTE
BOLL.GEOFISICA TEORICA E APPL.VOL.2 NO 5 PP 117-118 1960

3.42
SYDENHAM P.H.
A TENSIONED WIRE STRAINMETER.
JRN.SC.INST. JRN. PHYS.E SIE.2 VOL.2 PP 1095-1097 1969

3.42
OZAWA I.
NEW TYPES OF HIGHLY SENSITIVE STRAINMETERS
H-70 TYPE EXTENSOMETER AND R-70 TYPE ROTATIONMETER
SPEC.CONTRIB.GEOPH.INST.KYOTO UNIV.NO 10 PP 137-148 1970

3.42
BILHAM R.G.
THE MEASUREMENT OF EARTH STRAIN
DEPT.GEOD.GEOPH.MADINGLEY ROAD CAMBRIDGE 1970

3.42
GERARD V.B.
AN INVAR WIRE EARTH STRAINMETER
J.PHYS.F- SCI.INSTRUM., VOL 4 PP 689-692 1971

3.42
SYDENHAM P.H.
PROGRESS IN THE DESIGN OF TENSIONED-WIRE EARTH
STRAINMETERS
GEOPHYS.J.R.ASTR.SOC.29 PP 319-327 1972

3.42
BILHAM R.
EVANS R.
KING G.
LAWSON A.
MC.KENZIE D.
EARTH STRAIN TIDES OBSERVED IN YORKSHIRE, ENGLAND
WITH A SIMPLE WIRE STRAINMETER
GEOPHYS.J.R. ASTR. SOC. 29 PP 473-485 1972

3.42
SYDENHAM P.H.
STRAIN MEASUREMENT IN AUSTRALIA WITH PARTICULAR
REFERENCE TO THE COONEY OBSERVATORY
PHIL.TRANS.ROY.SOC.SER.A, VOL.274, NO 1239 PP 323-330 1973

3.42
HARWARDT H.
SIMON D.
OBSERVATIONS BY MEANS OF A WIRE EXTENSOMETER AT
THE TIEFENORT STATION
2ND INT.SYMP.GEOD.& PHYS.OF THE EARTH, PP 59-60 1973

3.42
HARWARDT H. SIMON D. — BEOBACHTUNGEN MIT EINEM DRAHTSTRAINMETER IN DER STATION TIEFENORT. 2. INTERN.SYMP.GEODASIE UND PHYSIK DER ERDE, VEROFF.ZENTR.INST.PHYS.ERDE NR 30,TEIL 1,PP 197-206 — 1974

3.43
KNEISSL M. — DER VAISALA-INTERFERENZKOMPARATOR ALS HILFSMITTEL DER ERDGEZEITENMESSUNG. OBS.ROY.BELG.COMM. NO 142 S.GEOPH. NO 47 PP 132-133 — 1958

3.43
BLAYNEY J.L. GILMAN R. — A PORTABLE STRAIN METER WITH CONTINOUS INTERFEROMETRIC CALIBRATION. BULL.SEISM.SOC.AM.VOL.55 PP 955-970 — 1965

3.43
HADE G. CONNER M. KUO J.T. — LASER INTERFEROMETER CALIBRATION SYSTEM FOR EXTENSOMETERS BULL.SEISM.SOC.AM. VOL.58 P 1379 — 1968

3.43
BRODSKII B.Y. — REFLECTEUR MIROIR-LENTILLE POUR LES MESURES INTERFERENTIELLES DES DISTANCES ROTATION ET DEFORMATIONS DE MAREES DE LA TERRE - NO 2 ACAD.NAOUK UKR.SSR - OBS.GRAV.POLTAVA - KIEV PP 124-125 — 1970

3.44
KASAHARA K. — STRAIN MEASUREMENT IN SPACE AND TIME BY USE OF ELECTRONIC DISTANCE MEASURING INSTRUMENTS. PROC.UNITED STATES JAPAN CONF.APP.19 PP 45-46 — 1964

3.44
THIRY C. — CAPTEURS DE DEPLACEMENT ELECTRO-MECANIQUES EMPLOYES POUR MESURER DES MICRODEPLACEMENTS B.I.M. NO 71 PP 4074-4082 — 1975

3.45
VALI V. — MEASUREMENT OF EARTH TIDES AND CONTINENTAL DRIFT WITH LASER INTERFEROMETER.BOEING SCIENTIFIC RESEARCH. LABS.W.VALI-STANFORD UNIV.PROC.INST.ELECT.ELECTRONIC ENG PROC. I.E.E.E. 52 P 7 — 1964

3.45
VALI V. KROGSTAD R.S. MOSS R.W. — LASER INTERFEROMETER FOR EARTH STRAIN MEASUREMENTS. REV.SCI.INSTR.VOL.36 PP 1352-1355 — 1965

3.45
VALI V. KROGSTAD R.S. MOSS R.W. ENGEL R. — MEASUREMENT OF STRAIN RATE AT THE KERN RIVER FAULT USING A LASER INTERFEROMETER. TRANS.AMER.GEOPH.UNION VOL.47 PP 424 — 1966

3.45
VAN VEEN H.J. SAVINO J. ALSOP L.F. — AN OPTICAL MASER STRAINMETER. JOURN.GEOPH.RES.VOL.71 PP 5478-5479 — 1966

3.45
VALI V. KROGSTAD R.S. MOSS R.W. — OBSERVATION OF EARTH TIDES USING A LASER INTERFEROMETER. JOURN.APPL.PHYSICS 37 NO 2 PP 580-582 — 1966

3.45
VALI V. BOSTROM R.C. — ONE THOUSAND METER LASER INTERFEROMETER. REV.SCIENT.INSTR.39 NO 9 PP 1304-1306 — 1968

3.45
KING G.C.P. DAVIES D. — A LASER SEISMOMETER. IN THE APPLICATION OF MODERN PHYSICS TO THE EARTH AND PLANETARY INTERIORS WILEY INTERSCIENCE /ED.RUNCORN/ — 1969

3.45
GERARD V.B. — AN AUTOMATIC FRINGE FOLLOWER FOR A LONG PATH LASER INTERFEROMETER. JRN.SC.INST. JRN. PHYS.E SIE.2 VOL.2 PP 933-935 — 1969

3.45
KING G.C.P. — A LASER INTERFEROMETER. PH.D. THESIS CAMBRIDGE UNIVERSITY — 1969

3.45
KING G.C.P. — A LASER SEISMOMETER THE APPLICATION OF MODERN PHYSICS TO THE EARTH AND PLANETARY INTERIORS WILEY INTERSCIENCE. — 1969

3.45
KING G.C.P. / GERARD V.B.
EARTH TIDES RECORDED BY THE 55-M CAMBRIDGE LASER INTERFEROMETER.
GEOPHYS. J.R. ASTR. SOC. VOL 18 PP 437-438 1969

3.45
BERGER J. / LOVBERG R.H.
A LASER EARTH STRAIN METER.
6 EME SYMPOSIUM MAREES TERRESTRES STRASBOURG
OBS.ROY.BELG.COMM. A9 S.GEOPH. NO 96 PP 154-162 1970

3.45
BERGER J. / LOVBERG R.H.
EARTH STRAIN MEASUREMENTS WITH A LASER INTERFEROMETER
SCIENCE V.170 N.3955 PP 296-303 1970

3.45
VAN VEEN H.J.
A LASER STRAIN SEISMOMETER
VERHAND.ROY.NETH.ACAD.SCI.AMSTERDAM SER.B 36 PP 81 1970

3.45
MANZONI G.
THE LASER STRAINMETER INSTALLED NEAR TRIESTE
BOLL.DI GEODESIA E SCIENZE AFFINI NO 3 PP 231-237 1971

3.45
GERARD V.B.
GEOPHYSICAL STRAINMETER INSTRUMENTATION
RECENT CRUSTAL MOVEM. R.S.NEW ZEALAND BULL.9 PP233-236 1971

3.45
LEVINE J. / HALL J.L.
DESIGN AND OPERATION OF A METHANE ABSORPTION STABILIZED LASER STRAINMETER
J.G.R. VOL 77 PP 2595-2609 1972

3.45
BERGER J. / WYATT F.
SOME OBSERVATIONS OF EARTH STRAIN TIDES IN CALIFORNIA
MEASUREM., INTERPRETAT., CHANGES OF STRAIN IN THE EARTH
PHIL.TRANS.ROY.SOC.SER.A, VOL.274, NO 1239 PP 267-278 1973

3.45
LEVINE J. / STEBBINS R.T.
ULTRA SENSITIVE LASER INTERFEROMETERS AND THEIR APPLICATION TO PROBLEMS OF GEOPHYSICAL INTEREST
MEASUREM., INTERPRETAT., CHANGES OF STRAIN IN THE EARTH
PHIL.TRANS.ROY.SOC.SER.A, VOL.274, NO 1239 PP 279-284 1973

3.45
MANZONI G. / MARCHESINI C.
A 60 M LASER STRAINMETER
MEASUREM., INTERPRETAT., CHANGES OF STRAIN IN THE EARTH
PHIL.TRANS.ROY.SOC.SER.A, VOL.274, NO 1239 PP 285-286 1973

3.45
MANZONI G. / MARCHESINI C.
LA STAZIONE ESTENSOMETRICA DI AURISINA /TRIESTE/
BOLL.DI GEOFISICA VOL XV,N.59 SETTEMBRE PP 255-260 1973

3.45
GOULTY N.R. / KING G.C.P. / WALLARD A.J.
HIGH RESOLUTION LASER STRAINMETER
NATURE VOL.246 PP 470-471 1973

3.45
BERGER J. / LEVINE J.
THE SPECTRUM OF EARTH STRAIN FROM 10-8 HZ TO 10 2 HZ
J.G.R. VOL 79 PP 1210-1214 1974

3.45
GOULTY N.R. / KING G.C.P. / WALLARD A.J.
IODINE STABILIZED LASER STRAINMETER
GEOPH.J.R.ASTR.SOC.V.39 PP 269-282 1974

3.45
MANZONI C. / MARCHESINI C.
FOURIER ANALYSIS OF THE AURISINA /TRIESTE/ LASER STRAINMETER
BOLL.GEOF.TEORICA APPL. VOL XVIII NO 66 PP 123-126 1975

3.46
OZAWA I.
ROTATIONAL STRAIMETER AND THE OBSERVATION OF THE SHEAR STRAIN OF THE EARTH TIDE WITH THE INSTRUMENT.
B.I.M. NO 52 PP 2398-2409 1968

3.47
OZAWA I.
ROTATIONAL STRAINMETER AND THE OBSERVATION OF THE SHEAR STRAIN OF THE EARTH TIDE WITH THIS INSTRUMENT.
* B.I.M. N.52 DECEMBRE PP 2398-2409 1968

3.5
DAVIS S.N. / MOORE G.W.
SEMIDIURNAL MOVEMENT ALONG A BEDROCK JOINT IN WOOL HOLLOW CAVE, CALIFORNIA.
THE NAT.SPELEOLOGICAL SOC.BULL.27 NO 4 PP 133-142 1965

3.5
VARGA P.
INVESTIGATION OF EARTH TIDES BY OBSERVING DILATATIONAL

VARIATIONS OF THE WATER TABLE
B.I.M. NO 74 PP 4319-4332 1976

3.51
PEKERIS C.L. NOTE ON TIDES IN WELLS.
U.S.COAST & GEOD.SURV.SP.PUBL.223, PP 23-24 1940

3.51
MELCHIOR P.J. CONSTRUCTION D UN DILATOMETRE EXPERIMENTAL.
COMPTES RENDUS DU SECOND COLLOQUE INTERNATIONAL
SUR LES MAREES TERRESTRES, MUNICH.
OBS.ROY.BELG.COMM. NO 142 S.GEOPH. NO 47 PP 135-136 1958

3.52
ARAGO F. LES PUITS FORES.
OEUVRES COMPL., TOME 6 CHAP.VII PP 311-314 1856

3.52
KLONNE F.W. DIE PERIODISCHE SCHWANKUNGEN IN DEN INUNDIERTEN
KOHLENSCHACHTEN VON DUX IN DER PERIODE VOM 8 APRIL
BIS 15 SEPTEMBER 1879.
SITZUNGSBER.KAIS.AKAD.WISS.81 PP 101 1880

3.52
GRABLOVITZ G. SUL FENOMENO DI MAREA OSSERVATO NELLE MINIERI
DI DUX IN BOHEMIA.
BOLL.SOC.ADRIATICA SC.NAT.TRIESTE VI PP 34 1880

3.52
LAGRANGE C. LE PHENOMENE DE MAREE SOUTERRAINE DE DUX EN BOHEME.
ANN.DE CHIMIE ET DE PHYS.5E SER.XXV PP 533-546 1882

3.52
ROBERTS I. ON THE ATTRACTIVE INFLUENCE OF THE SUN AND MOON
CAUSING TIDES AND THE VARIATIONS IN ATMOSPHERIC
PRESSURE AND RAINFALL CAUSING OSCILLATIONS IN THE
UNDERGROUND WATER IN POROUS STRATA.
REPORT BRIT.ASSOC. PP 405 1883

3.52
N. LES MAREES SOUTERRAINES.
L ASTRONOMIE V PP 422-425 1886

3.52
VEATCH A.C. FLUCTUATIONS OF THE WATER LEVEL IN WELLS, WITH
SPECIAL REFERENCE TO LONG ISLAND NEW YORK.
UN.STATES GEOL.SURV.WATER-SUPL.&IRR.PAPER NO 155 1906

3.52
YOUNG A. TIDAL PHENOMENA AT INLAND BOREHOLES NEAR CRADOCK.
TRANS.R.SOC.SOUTH AFRICA III, PART 1 PP 61-106 1913

3.52
BILHAM E.G. ON THE RELATION BETWEEN BAROMETRIC PRESSURE AND THE
WATER-LEVEL IN A WELL AT KEW OBSERVATORY, RICHMOND.
PROC.R.SOC.A 94 PP 165-181 1918

3.52
BILHAM E.G. THE LUNAR AND SOLAR DIURNAL VARIATIONS OF WATER-
LEVEL IN A WELL AT KEW OBSERVATORY, RICHMOND.
PROC.R.SOC.A 94 PP 476-478 1918

3.52
FORCHHEIMER ZUR THEORIE DER GRUNDWASSERSTROMUNGEN.
SITZ.AKAD.WISSENSCH.WIEN ABT.IIA, 128 BD.8 PP 1229-1236 1919

3.52
SCHUREMAN P. TIDES IN WELLS.
THE GEOGR.REV.16 PP 479 1926

3.52
KOEHNE W. DAS UNTERIRDISCHE WASSEN WELLEN IM GRUNDWASSER.
HANDB.GEOPH.VII, LIEF 1 PAR.92 PP 214-218 BERLIN 1933

3.52
ROBINSON T.W. EARTH TIDES AS SHOWN BY FLUCTUATIONS OF WATER LEVELS
IN WELLS IN NEW MEXICO AND IOWA.
TRANS.AMER.GEOPH.UNION PART IV PP 656-666 1939

3.52
PEKERIS C.L. NOTE ON TIDES IN WELLS.
TRANS.AMER.GEOPH.UNION PART II PP 212-213 1940

3.52
GEORGE W.O. TIDE PRODUCING FORCES AND ARTESIAN PRESSURES.
ROMBERG F.E. TRANS.AMER.GEOPH.UNION 32 NO 3 PP 369-371 1951

3.52
TOMASCHEK R. SEASONAL VARIATIONS IN THE FLOW OF OIL WELLS.
JOURN.OF THE INST.OF PETROLEUM 38 NO 344 PP 591-605 1952

3.52
SPERLING K. GIBT ES GEZEITENEINFLUSSE IM ERDOLFORDERBETRIEB. ERDOL UND KOHLE 6 NO 8 PP 446-449 1953

3.52
MELCHIOR P.J. COMPOSANTES HARMONIQUES DE LA MAREE DANS LA NAPPE CHAUDE DE KIABUKWA /CONGO BELGE/. ACAD.R.BELG.BULL.CL.SC.XLI PP 204-208 OBS.ROY.BELG.COMM. NO 82 S.GEOPH. NO 31 1955

3.52
MEISSNER R. UNTERSUCHUNGEN UBER DIE ZUSAMMENHANGE ZWISCHEN LUFTDRUCK UND BRUNNENSPIEGEL. ZEITSCHR.FUR GEOPH.21 H2 PP 81-108 1955

3.52
RICHARDSON R.M. TIDAL FLUCTUATIONS OF WATER LEVEL OBSERVED IN WELLS IN EAST TENNESSEE. TRANS.AMER.GEOPH.UNION 37 NO 4 PP 461-462 1956

3.52
CORIN F. A PROPOS DES FLUCTUATIONS DES EAUX DE PUITS EN SYNCHRONISME AVEC LES MAREES - LE SONDAGE DU PALAIS DES TERMES A OSTENDE. BULL.SOC.BELGE GEOLOGIE LXV PP 130-131 1956

3.52
MUGGE R. EXPERIMENTS ON THE MOTION OF WATER IN THE VICINITY OF WELLS. UGGI ASS.INTERN.HYDROL.SC.SYMP.DARCY TOME II PP 255-258 1956

3.52
GULINCK M. CARACTERISTIQUES HYDROGEOLOGIQUES DU SONDAGE DE TURNHOUT OBS.ROY.BELG.COMM. NO 108 S.GEOPH. NO 37 1957

3.52
MELCHIOR P.J. SUR L EFFET DES MAREES TERRESTRES DANS LES VARIATIONS DE NIVEAU OBSERVEES DANS LES PUITS, EN PARTICULIER AU SONDAGE DE TURNHOUT /BELGIQUE/. OBS.ROY.BELG.COMM. NO 108 S.GEOPH. NO 37 1957

3.52
MELCHIOR P.J. ANALYSE HARMONIQUE DES VARIATIONS DE NIVEAU OBSERVEES DANS SIX PUITS /DUCHOV, CARLSBAD, OAK RIDGE, IOWA C., KIABUKWA,TURNHOUT/. 3E SYMP. MAREES TERRESTRES TRIESTE BOLL.GEOFISICA TEORICA E APPL.VOL.2 NO 5 PP 122-124 1960

3.52
MELCHIOR P.J. DIE GEZEITEN IN UNTERIRDISCHEN FLUSSIGKEITEN. ERDOL UN KOHLE 13 PP 312-317 OBS.ROY.BELG.COMM.NO 172 S.GEOPH.NO 55 PP 312-317 1960

3.52
KORB H.G. ~~UEBER DIE ANALYSE DER SCHWANKUNGEN~~ DES GRUNDWASSERSPIE-GELS IN DEM UBERFLUTETEN BERGWERK SONTRA. ZEITSCH.FUR GEOPHYSIK 27 H.2 PP 75-88 1961

3.52
VAN WYK W.L. GROUND-WATER STUDIES IN NORTHERN NATAL, ZULULAND AND SURROUNDING AREAS GEOLOG.SURVEY DEPT.MINES - P.O.BOX 401 PRETORIA R.S.A. MEMOIR 52 PP 1-135 1963

3.52
CREYTZ D.V. DIE GEZEITEN- UND LUFDRUCKWIRKUNG IM UBERFLUTETEN REICHENBERGSCHACHT. ZEITSCH.GEOPH.DEUTSCHEN GEOPH.GESELLS.HEFT 3 PP 140-151 1964

3.52
MELCHIOR P.J.
STERLING A.
WERY A. EFFETS DE DILATATIONS CUBIQUES DUES AUX MAREES TERRESTRES OBSERVES SOUS FORME DE VARIATIONS DE NIVEAU DANS UN PUITS A BASECLES /HAINAUT/. OBS.ROY.BELG.COMM.NO 224 S.GEOPH.NO 66 PP 3-12 1964

3.52
STERLING A. FLUCTUATIONS DE NIVEAU D EAU OBSERVEES DANS LES PUITS. 5 EME SYMP.INT.SUR LES MAREES TERRESTRES. OBS.ROY.BELG.COMM.NO 236 S.GEOPH.NO 69 PP 347-349 1964

3.52
CLARK W.E. COMPUTING THE BAROMETRIC EFFICIENCY OF A WELL. JOURN.HYDRAULICS DIV.VOL.93 NO 4 HY PP 93-98 1967

3.52
BREDEHOEFT J.D. RESPONSE OF WELL-AQUIFER SYSTEMS TO EARTH TIDES. J.GEOPH.RES.VOL.72 NO 12 PP 3075-3087 1967

3.52
WHITE D.E. HYDROLOGY, ACTIVITY, AND HEAT FLOW OF THE STEAMBOAT SPRINGS THERMAL SYSTEM, WASHOE COUNTY, NEVADA GEOL.SURV.PROF.PAP.458-C U.S.GOV.PRINTING OFF. PP 109 1968

3.52
ERNST W. NACHWEIS DER ERDGEZEITEN MIT BODENGASEN. METEOROL. RUNDSCHAU 22, 5, PP 140-142 1969

3.52
BODVARSSON G. CONFINED FLUIDS AS STRAIN METERS. J.GEOPH.RES.VOL.75 NO 14 PP 2711-2718 1970

3.52
SMETS E. EFFETS DYNAMIQUES DU NOYAU LIQUIDE D APRES LES MESURES DANS LE SONDAGE DE HEIBAART. 6 EME SYMPOSIUM MAREES TERRESTRES STRASBOURG OBS.ROY.BELG.COMM. A9 S.GEOPH. NO 96 PP 147 1970

3.52
ROBINSON E.S.
BELL R.TH. TIDES IN CONFINED WELL-AQUIFER SYSTEMS J.GEOPH.RES.VOL.76 NO 8 PP 1857-1869 1971

3.52
STERLING A.
SMETS E. STUDY OF EARTH TIDES BY ANALYSIS OF THE LEVEL FLUCTUATIONS IN A BOREHOLE AT HEIBAART HYDRAULIC LABORATORY-BORGERHOUT ANTW.BELGIUM PP 1-19 1971

3.52
STERLING A.
SMETS E. STUDY OF EARTH TIDES, EARTHQUAKES AND TERRESTRIAL SPECTROSCOPY BY ANALYSIS OF THE LEVEL FLUCTUATIONS IN A BOREHOLE AT HEIBAART /BELGIUM/ GEOPHYS.J.R. ASTR.SOC. NO 23 PP 225-242 1971

3.52
CSABA L. FLUCTATION DE LA NAPPE ET PHENOMENES DE MAREE DANS LE DOMAINE DES EAUX SOUTERRAINES INST.GEOLOGICUM PUBL.HUNGARICUM PP 229-236 1971

3.52
BOWER D.R.
HEATON K.C. RESPONSE OF AN UNCONFINED AQUIFER TO ATMOSPHERIC PRESSURE, EARTH TIDES AND A LARGE EARTHQUAKE PROCEEDINGS 7TH INT.SYMP.EARTH TIDES, SOPRON 1973 AKADEMIAI KIADO BUDAPEST, PP 155-163 1976

3.52
RHOADS G.H. JR. DETERMINATION OF AQUIFER PARAMETERS FROM WELL TIDES THESIS - FAC.VIRGINIA POLYTECHNIC PP 1-77 1976

3.52
BOWER D.R.
HEATON K.C. RESPONSE OF AN AQUIFER NEAR OTTAWA TO THE ALASKAN EARTHQUAKE OF 1964 CONTRIB.FROM THE EARTH PHYS.BRANCH OTTAWA PP 1-14 1977

3.53
JAGGAR T.A. SEISMOMETRIC INVESTIGATION OF THE HAWAIIAN LAVA COLUMN. BULL.SEISM.SOC.AMER.X, NO 4 1920

3.53
IMBO G. SISMICITA DEL PAROSSISMO VESUVIANO DEL MARZO 1944. ANN.OSSERV.VESUVIANO SERIE 6 VOL.1 1954

3.53
IMBO G. AZIONE DELLE MAREE DELLA CROSTA SUI FENOMENI ERUTTIVI. OBS.ROY.BELG.COMM. NO 142 S.GEOPH. NO 47 PP 134 1958

3.53
MACHADO FR. SECULAR VARIATION OF SEISMOVOLCANIC PHENOMENA IN THE AZORES. BULL.VOLCANOLOGIQUE SERIE II TOME XXIII 1960

3.53
SHIMOZURU D.
NAKAGAWA I. TIDAL OSCILLATION OF HALEMAUMAU LAVA LAKE, KILAUEA, HAWAII AND ITS IMPLICATION FOR EXISTENCE OF A MAGMA RESERVOIR. 6 EME SYMPOSIUM MAREES TERRESTRES STRASBOURG OBS.ROY.BELG.COMM. A9 S.GEOPH. NO 96 PP 163-166 1970

3.53
RINEHART J.S. EFFECTS OF 18.6 YEAR EARTH TIDAL COMPONENT ON SEISMIC AND GEYSER ACTIVITY E.O.S. TRANS.AGU 52, P 924 1971

3.53
RINEHART J.S. FLUCTUATIONS IN GEYSER ACTIVITY CAUSED BY

VARIATIONS IN EARTH TIDAL FORCES, BAROMETRIC PRESSURE AND TECTONIC STRESSES
JOURN.GEOPH.RES.VOL.77 NO 2 PP 342-350 1972

3.53
RINEHART J.S. — REPLY-FLUCTUATIONS IN GEYSER ACTIVITY CAUDES BY VARIATIONS IN EARTH TIDAL FORCES, BAROMETRIC PRESSURE AND TECTONIC STRESSES
JOURN.GEOPHYS.RES.VOL.77 NO 29 PP 5830-5831 1972

3.53
WHITE D.F., MARLER G.D. — COMMENTS ON PAPER BY JOHN S.RINEHART, FLUCTUATIONS IN GEYSER ACTIVITY CAUSED BY EARTH TIDAL FORCES, BAROMETRIC PRESSURE AND TECTONIC STRESSES
JOURN.GEOPHYS.RES.VOL.77 NO 29 PP 5825-5829 1972

3.53
RINEHART J.S. — RESPONSE OF GEYSERS TO EARTH TIDAL FORCES
J.COLO.WYO.ACAD.SCI. 7, P 45 1972

3.53
ALEXANDROV Y.M. — COMPARAISON D INTENSITE DU DEGAGEMENT DU GAZ DU PUITS DANS LA MINE NO 1 ARTEMSOL AVEC LES INCLINAISONS DE MAREES
ROTATION ET DEFORMATION DE MAREES DE LA TERRE NO 4
ACAD.NAOUK UKR.SSR.-OBS.GRAV.POLTAVA-KIEV PP 90-93 1972

3.53
RINEHART J.S. — YEAR EARTH TIDE REGULATES GEYSER ACTIVITY
SCIENCE VOL.177, PP 346-347 1972

3.53
JOHNSTON M.J.S., MAUK F.J. — EARTH TIDES AND THE TRIGGERING OF ERUPTIONS FROM MOUNT STROMBOLI ITALY.
NATURE, 239 PP 266-267 1972

3.53
MAUK F.J., JOHNSTON M.J.S. — ON THE TRIGGERING OF VOLCANIC ERUPTIONS BY EARTH TIDES
JOURNAL OF GEOPH.RES. V.78 NO 17 JUNE 10 PP 3356-3362 1973

3.53
HAMILTON W.L. — TIDAL CYCLES OF VOLCANIC ERUPTIONS - FORTNIGHTLY TO 19 YEARLY PERIODS
JOURNAL OF GEOPH.RES. V.78 NO 17 JUNE 10 PP 3363-3375 1973

3.53
MACHADO FR. — THE SEARCH FOR MAGMATIC RESERVOIRS
PHYSICAL VOLCAN./L.CIVETTA & AL.EDITORS PP 255-273 1974

3.53
TENG T.L., MCELRATH R.P. — RESPONSE OF GROUNDWATER RADON CONTENT TO SOLID TIDAL STRAIN
EOS VOL 57 NO 12 P 899 1976

4.1
METZGER J. — MAREE DE L ECORCE - L EFFET DU DEPHASAGE DANS L ANALYSE DES OBSERVATIONS.
GEOFISICA PURA E APPLICATA 28 PP 71-83 1954

4.1
LASSOVSKY K. — A FOLD DEFORMACIOS EGYUTTHATOJANAK MEGHATAROZASA GRAVIMETERESZLELESEKBOL.
GEOFISIKAI KOZLEMENYEK V, 2 PP 18-26 1956

4.1
GOUGENHEIM A. — AU SUJET DE LA DERIVE DES GRAVIMETRES.
B.I.M. NO 2 PP 14-15 1957

4.1
PERTZEV B.P. — ON THE CALCULATION OF THE DRIFT CURVE IN OBSERVATIONS OF BODILY TIDES.
B.I.M. NO 5 PP 71-72 1957

4.1
MELCHIOR P.J. — SUR L INTERPRETATION DES COURBES DE DERIVE DES GRAVIMETRES.
B.I.M. NO 17 PP 279-287 1959

4.1
PERTSEV B.P. — SUR LA DERIVE DANS LES OBSERVATIONS DE MAREES ELASTIQUES
IZV.ACAD.SC.URSS SER.GEOPH.NO 4 PP 547-548 1959

4.1
NAKAGAWA I. — GENERAL CONSIDERATIONS CONCERNING THE ELIMINATION OF THE DRIFT CURVE IN THE EARTH TIDAL OBSERVATIONS.
GEOPH.INST.FAC.SC.KYOTO UNIV.PP 121-135 1960

4.1
ZETLER B.D.
THE EFFECT OF INSTRUMENTAL DRIFT ON THE HARMONIC ANALYSIS OF GRAVITY AT WASHINGTON D.C.
3E SYMP. MAREES TERRESTRES TRIESTE
BOLL.GEOFISICA TEORICA E APPL.VOL.2 NO 5 PP 234-237 1960

4.1
OSZLACZKY S.
NOTE ON THE CORRECTION OF GRAVIMETER READINGS
3E SYMP. MAREES TERRESTRES TRIESTE
BOLL.GEOFISICA TEORICA E APPL.VOL.2 NO 5 PP 201-202 1960

4.1
NAKAGAWA I.
GENERAL CONSIDERATIONS CONCERNING THE ELIMINATION OF THE DRIFT CURVE IN THE EARTH TIDAL OBSERVATIONS.
IVME SYMP.INTERNATIONAL SUR LES MAREES TERRESTRES.
OBS.ROY.BELG.COMM. NO 188 S.GEOPH. NO 58 PP 201-210 1961

4.1
LENNON G.W.
SOME FURTHER COMMENTS UPON THE TREATMENT OF DRIFT AND OTHER NON-TIDAL EFFECTS IN THE ANALYSIS OF EARTH-TIDE OBSERVATIONS.
IVME SYMP.INTERNATIONAL SUR LES MAREES TERRESTRES.
OBS.ROY.BELG.COMM. NO 188 S.GEOPH. NO 58 PP 211-221 1961

4.1
BALENKO V.G.
APPORT DES ONDES A LONGUE PERIODE DANS LES RESULTATS DE L ANALYSE HARMONIQUE D UNE SERIE MENSUELLE D OBSERVATIONS DE MAREES TERRESTRES
TROUDI POLTAVSKOI GRAV.OBS. X PP 44-50 1961

4.1
BALENKO V.G.
CONTRIBUTION DE LA DERIVE DANS R COS ET R SIN DES ONDES DE MAREES TERRESTRES DETERMINEES.
TROUDI POLTAVSKOI GRAV. OBS. XI PP 74-87 1962
* B.I.M. N.32 MAI PP 859-872 1963

4.1
CHOJNICKY T.
BYL J.
ZUR FRAGE DER BESTIMMUNG VON GANGKORRELATIONEN BEI GEZEITENINSTRUMENTEN.
VERMESSUNGSTECHNIK 18 JG.HEFT 10 PP 379-382 1970

4.1
CHOJNICKI T.
DETERMINATION DE LA DERIVE DANS LES MESURES DES MAREES AU MOYEN DE LA METHODE DES POINTS NEUTRES
PUBL.INST.GEOPH.POLISH AC.OF SC. V.71 PP 3-74 1973

4.1
BARTHA G.
EVALUATION OF EARTH-TIDE RECORDS BY SECTIONWISE APPROXIMATION OF THE INSTRUMENTAL DRIFT
ACTA GEODAET.,GEOPHYS.ET MONTANISTICA T.8/3-4/ PP445-450 1973

4.1
CHOJNICKI T.
DETERMINATION DE LA DERIVE DANS LES MESURES DES MAREES AU MOYEN DE LA METHODE DES POINTS NEUTRES
PROCEEDINGS 7TH INT.SYMP.EARTH TIDES, SOPRON 1973
AKADEMIAI KIADO BUDAPEST, PP 351-366 1976

4.15
IVANOVA M.V.
COMPARAISON DES DIFFERENTES METHODES POUR COMBLER LES LACUNES DANS LES OBSERVATIONS DE MAREES TERRESTRES.
RECH. MAREES TERR. ART. N.3 PUBL.ACAD.SC.URSS MOSCOU 1963
* B.I.M. N.36 MAI PP 1205-1209 1964

4.15
DITCHKO Y.A.
TOKAR V.Y.
SUR LA QUESTION DE L INTERPOLATION DES LACUNES DANS L ENREGISTREMENT DES MAREES TERRESTRES.
ACAD. SC. R.S.S. D UKRAINE OBS.GRAV. DE POLTAVA
INST. GEOPH. DE LA RSS D UKRAINE KIEV 1966
* B.I.M. N.49 OCTOBRE PP 2150-2154 1967

4.15
GRIDNIEV D.G.
PROCEDE GRAPHIQUE DE REMPLISSAGE DES LACUNES JUSQU A DEUX JOURS DANS LES OBSERVATIONS DES MAREES TERRESTRES.
MAR.TERR.ET STRUCT.INTERNE DE LA TERRE
ACAD.SC.URSS.INST.PHYS.TERR.SCHMIDT MOSCOU PP 119-122 1967

4.15
MATVEYEV P.S.
BOGDAN Y.D.
INTERPOLATION DES COURTES LACUNES DANS LES OBSERVATIONS DES MAREES TERRESTRES.
ACAD. SC. R.S.S. D UKRAINE OBS.GRAV. DE POLTAVA
INST. GEOPH. DE LA RSS D UKRAINE KIEV 1966
* B.I.M. N.49 OCTOBRE PP 2155-2161 1967

4.15
SCHULLER K. DIE ANWENDUNG DER PRAEDIKTION AUF PERIODISCHE
SCHULZ R.S. PROZESSE, EINE METHODE ZUR UBERBRUCKUNG VON
LUCKEN IN DER ERDGEZEITENREGISTRIERUNG
MITTEIL.INST.THEOR.GEOD.UNIV.BONN, NO 14, PP 1-17 1973

4.2
SIMON D. BEITRAGE ZUR KORREKTUR VON KLINOMETRISCHEN UND
GRAVIMETRISCHEN GEZEITENREGISTRIERUNGEN.
GEOD. UND GEOPH. VEROFF - NAT.KOM. DDR, BERLIN
REIHE III, HEFT 13, PP 110 1969

4.3
NIKITINE M.V. ANALYSE HARMONIQUE DES MAREES
1925

4.3
DOODSON A.T. METHODS OF ANALYSIS OF TIDES.
OBS.ROY.BELG.COMM. NO 114 S.GEOPH. NO 39 1957

4.3
BALENKO V.G. INFLUENCE DES METHODES D ANALYSE HARMONIQUE DES
MAREES TERRESTRES SUR LES ERREURS ACCIDENTELLES
DES ORDONNEES DE DEPART.
TROUDI POLTAVSKOI GRAV. OBS. X PP 38-43 1961
* B.I.M. N.32 MAI PP 902-907 1963

4.3
BALENKO V.G. EVALUATION COMPARATIVE DE LA QUALITE DE L ELIMINATION
DANS LES ONDES DETERMINEES DE L INFLUENCE DES ONDES
PERTURBATRICES POUR LES METHODES D ANALYSE HARMONIQUE
DES MAREES TERRESTRES.
TROUDI POLTAVSKOI GRAV. OBS. XI PP 64-73 1961
* B.I.M. N.32 MAI PP 849-858 1963

4.3
BALENKO V.G. QUELQUES PROBLEMES CONCERNANT LA COMPARAISON DES METHODES
ZACHARTCHENKO S.N. D ANALYSE HARMONIQUE DES MAREES TERRESTRES.
TROUDI POLTAVSKOI GRAV. OBS. X PP20-37 KIEV 1961
* B.I.M. N. 31 FEVRIER PP 801-819 1963

4.3
BALENKO V.G. CONTRIBUTION DES ONDES A LONGUE PERIODE DANS LES
RESULTATS DE L ANALYSE HARMONIQUE D UNE SERIE D UN MOIS
D OBSERVATIONS DE MAREES TERRESTRES.
TROUDI POLTAVSKOI GRAV. OBS. X PP 44-50 1961
* B.I.M. N.32 MAI PP 835-842 1963

4.3
MELCHIOR P.J. A PROPOS DE LA REPARTITION DES ANALYSES HARMONIQUES
MENSUELLES SUR UNE LONGUE SERIE D OBSERVATIONS DE
MAREES TERRESTRES.
B.I.M.NO 35 PP 1150-1156 1964

4.3
MATVEYEV P.S. REMARQUES AU SUJET DU CALCUL DE LA MOYENNE DE RESULTATS
D ANALYSES HARMONIQUES DES MAREES TERRESTRES.
TROUDI POLTAVSKOI GRAV.OBS. T.XII PP 115-124 KIEV 1963
* B.I.M. N.35 FEVRIER PP 1142-1149 1964

4.3
DOS SANTOS FRANCO LA PRECISION RELATIVE DE QUELQUES METHODES D ANALYSE
HARMONIQUE DE LA MAREE.
REV.HYDROGR.INTERN.XLII NO 2 PP 157-166 1965

4.3
KORBA S.N. QUELQUES RESULTATS D ANALYSE HARMONIQUE DES OBSERVATIONS
MATVEYEV P.S. CLINOMETRIQUES D APRES LE PROGRAMME GLISSANT.
ACAD.SC.UKRAINE,OBS.GRAV.POLTAVA-MAR.TERR.PP 94-103 KIEV 1966

4.3
DEUTSCHES TAFELN DER ASTRONOMISCHEN ARGUMENTE UND DER
HYDROGRAPHISCHES KORREKTIONEN ZEIM GEBRAUCH BEI DER HARMONISCHEN
INSTITUT ANALYSE UND VORAUSBERECHNUNG DER GEZEITEN FUR DIE
JAHRE 1900 BIS 1999.
NR 2276 PP 128 HAMBURG 1967

4.3
KRAMER M.V. SUR LA QUESTION DE L INFLUENCE DES ERREURS ACCIDENTELLES
DES OBSERVATIONS SUR LES RESULTATS DE DETERMINATION DES
AMPLITUDES ET DES PHASES DES ONDES DES MAREES ELASTIQUES
ACAD.SC.URSS INST. PHYS.TERR.SCHMIDT MOSCOU 1964
* B.I.M. N.48 JUILLET PP 2041-2056 1967

4.3
PALLAS W. — EINIGE BEMERKUNGEN ZU DEN LABROUSTESCHEN SYMBOLEN YM UND ZM UND DEREN PRODUKTBILDUNG. B.I.M. NO 51 PP 2360-2365 — 1968

4.3
MIKUMO T., NAKAGAWA I. — SOME PROBLEMS ON THE ANALYSIS OF THE EARTH TIDES JOURN.PHYS.EARTH VOL.16 PP 87-95 — 1968

4.3
SCHUSTER O. — UBER DIE RELATIVITAT UND MANGELNDE VERGLEICHBARKEIT DER ERGEBNISSE AUS VERSCHIEDENEN ERDGEZEITEN-ANALYSEN-VERFAHREN GERLANDS BEITRAGE ZUR GEOPHYSIK HEFT 1 — 1972

4.3
MOSETTI F., MANCA B. — SOME METHODS OF TIDAL ANALYSIS INTERN.HYDROGR.REV.VOL.XLIX, NO 2, PP 107-120 — 1972

4.3
GODIN G. — THE ANALYSIS OF TIDES UNIV.OF TORONTO PRESS P 264 — 1972

4.3
MELCHIOR P. — HARMONIC ANALYSIS OF EARTH TIDES METHODS IN COMPUTATIONAL PHYS.V.13 PP 271-343 — 1973

4.3
DE MEYER F. — ESTIMATION AND AUTOMATIC CORRECTION OF THE INDIVIDUAL OBSERVATIONAL ERROR B.I.M. NO 65 MARS PP 3526-3550 — 1973

4.3
VARGA P. — THEORETICAL LIMITATIONS OF THE HARMONIC ANALYSIS OF EARTH TIDES GEOPHYSICAL TRANSACTIONS, VOL.22, PP 51-59, BUDAPEST — 1974

4.3
CHOJNICKI T. — METHODS OF ANALYSIS AND DATA PROCESSING CORRESPONDING TO EARTH TIDES XVI ASS.GEN.DE L UGGI GRENOBLE - AOUT — 1975

4.3
ORZECHOWSKI J. — OBSERVATIONS MODELEES DE MAREE TERRESTRE POUR RECHERCHE ET COMPARAISON DES METHODES D ANALYSE PUBLS.INST.GEOPH.POL.AC.SCI. VOL 94 PP 77-117 — 1975

4.3
SIMON D. — ON SEVERAL EVALUATION PROBLEMS WHICH CAN BE SOLVED BY A MORE ACCURATE TIDE ELIMINATION PROCEEDINGS 7TH INT.SYMP.EARTH TIDES, SOPRON 1973 AKADEMIAI KIADO BUDAPEST, PP 429-452 — 1976

4.3
QUEILLE C. — VARIATION EN FONCTION DU TEMPS DE DELTA PROCEEDINGS 7TH INT.SYMP.EARTH TIDES, SOPRON 1973 AKADEMIAI KIADO BUDAPEST, PP 401-404 — 1976

4.31
DOODSON A.T. — THE ANALYSIS OF TIDAL OBSERVATIONS. PHILOS.TRANS.ROY.SOC.LONDON SER.A VOL.227 PP 223-279 — 1928

4.31
DOODSON A.T., WARBURG H.D. — ADMIRALTY MANUAL OF TIDES. HYDR.DEPART.ADMIR.LONDON — 1941

4.31
DOODSON A.T. — THE ANALYSIS OF TIDAL OBSERVATIONS FOR 29 DAYS. INTERN.HYDROGRAPHIC REVIEW - MAY — 1954

4.31
LENNON G.W. — DIE METHODE DER HARMONISCHEN ANALYSE FUR 29 BEOBACHTUNGS-TAGE DER ERDGEZEITEN DES GEZEITENINSTITUTS LIVERPOOL DT.GEOD.KOMM.WISS.UBERSETZ-DIENST, MUNCHEN 17, 15 S — 1958

4.312
DOODSON A.T., LENNON G.W. — THE ELIMINATION OF DRIFT EFFECTS FROM TIDAL ANALYSES OBS.ROY.BELG.COMM. NO 142 S.GEOPH. NO 47 PP 156-167 — 1958

4.312
LECOLAZET R. — REMARQUES SUR LA NOUVELLE METHODE DE A.T.DOODSON ET G.W.LENNON. 3E SYMP. MAREES TERRESTRES TRIESTE BOLL.GEOFISICA TEORICA E APPL.VOL.2 NO 5 PP 215-217 — 1960

4.312
BALENKO V.G. — SUR LA METHODE D ANALYSE HARMONIQUE DES MAREES

TERRESTRES DE DOODSON-LENNON.
TROUDI POLTAVSKOI GRAV. OBS. X PP 51-56 1961
* B.I.M. N.32 MAI PP 843-848 1963

4.32
LECOLAZET R. APPLICATION A L ANALYSE DES OBSERVATIONS DE LA MAREE GRAVIMETRIQUE, DE LA METHODE DE H. ET Y. LABROUSTE DITE PAR COMBINAISONS LINEAIRES D ORDONNEES
ANNALES DE GEOPH.12, FASC.1 PP 59-71 1956

4.32
LECOLAZET R. NOTE SUR L ANALYSE HARMONIQUE.
OBS.ROY.BELG.COMM. NO 114 S.GEOPH. NO 39 1957

4.32
DOODSON A.T.
LENNON G.W. COMMENTS ON PROFESSOR LECOLAZET S REVISED METHOD OF ANALYSIS.
OBS.ROY.BELG.COMM. NO 142 S.GEOPH. NO 47 PP 161-162 1958

4.32
LECOLAZET R. LA METHODE UTILISEE A STRASBOURG POUR L ANALYSE HARMONIQUE DE LA MAREE GRAVIMETRIQUE.
B.I.M. NO 10 PP 153-178 1958

4.32
LECOLAZET R. LE DEVELOPPEMENT HARMONIQUE DES DEVIATIONS PERIODIQUES THEORIQUES DE LA VERTICALE.
B.I.M. NO 14 PP 211-221 1959

4.32
LECOLAZET R. LA METHODE UTILISEE A STRASBOURG POUR LA SEPARATION DES ONDES K1, P1 ET S1.
B.I.M. NO 19 PP 311-321 1960

4.32
MELCHIOR P.J. VALIDITE DE LA METHODE DE LECOLAZET POUR LES STATIONS CLINOMETRIQUES SOUMISES A D IMPORTANTS EFFETS INDIRECTS BASES D ANALYSE HARMONIQUE DE MAREES TERRESTRES.
IVME SYMP.INTERNATIONAL SUR LES MAREES TERRESTRES.
OBS.ROY.BELG.COMM. NO 188 S.GEOPH. NO 58 PP 222-241 1961

4.32
LECOLAZET R. LE DEVELOPPEMENT HARMONIQUE DE LA MAREE DE DEFORMATION.
B.I.M. NO 31 PP 820-828 1963

4.32
SARITSCHEVA J.K. SUR LA DISPERSION DES VALEURS DU FACTEUR OBTENU PAR LA METHODE DE R. LECOLAZET POUR LA MEME STATION.
5 EME SYMP.INT.SUR LES MAREES TERRESTRES.
OBS.ROY.BELG.COMM.NO 236 S.GEOPH.NO 69 PP 421-425 1964

4.33
PERTZEV B. HARMONIC ANALYSIS OF BODILY TIDES.
OBS.ROY.BELG.COMM. NO 114 S.GEOPH. NO 39 1957

4.33
PERTSEV B.P. ANALYSE HARMONIQUE DES MAREES
IZV.ACAD.SC.URSS SER.GEOPH.NO 8 PP 946-958 1958

4.33
PERTSEV B.P.
PARIISKII N.N.
KRAMER M.V. COMPARAISON DES METHODES DE SEPARATION PAR ANALYSE HARMONIQUE DES DEFORMATIONS DE LA TERRE
IZV.ACAD.SC.URSS SER.GEOPH.NO 2 PP 242-243 1959

4.33
PERTSEV B.P. ANALYSE HARMONIQUE DE SERIES DE 50 JOURS D OBSERVATIONS DE VARIATIONS DES MAREES DE LA FORCE DE PESANTEUR
GROUPE XIII PROGR. AGI GRAV.VOL.2 MOSCOU 1961 PP 20-30
* B.I.M. NO 26 PP 554-571 1961

4.33
BALENKO V.G.
EVTOUCHENKO E.J. SUR LA METHODE DE PERTSEV D ANALYSE HARMONIQUE DE SERIES DE 50 JOURS D OBSERVATIONS DE VARIATIONS DE MAREES DE LA FORCE DE PESANTEUR.
TROUDI POLTAVSKOI GRAV. OBS. T.XII PP 27-47 KIEV 1963
* B.I.M N.34 NOVEMBRE PP 1076-1095 1963

4.33
PERTSEV B.P. DETERMINATION DE ET RELATIFS AUX ONDES SEMI-DIURNES D APRES DES OBSERVATIONS DE MAREES TERRESTRES PORTANT SUR DEUX JOURS.
RECH. MAREES TERR. ART. N.3 PUBL.ACAD.SC.URSS MOSCOU 1963
* B.I.M. N.37 SEPTEMBRE PP 1247-1251 1964

4.33
BALENKO G.V. EVTOUCHENKO E.I. — COMPARAISON DE LA METHODE DE PERTSEV D ANALYSE HARMONIQUE POUR DES SERIES DE 50 ET 29 JOURS DES VARIATIONS DE PESANTEUR
BULL.MATER.ANNEE GEOPH.INTERN.KIEV, 6, PP 86-94 — 1964

4.33
PERTSEV B.P. — SEPARATION DES ONDES DE MAREES DIURNES K1 ET P1
RECH. MAREES TERR. ART. N.3 PUBL.ACAD.SC.URSS MOSCOU 1963
* B.I.M. N.38 NOVEMBRE PP 1337-1341 — 1964

4.331
MATVEEV P.S. — L ANALYSE HARMONIQUE DE LA SERIE DE 29 JOURS DE MAREES TERRESTRES.
B.I.M. NO 31 PP 773-800 — 1962

4.34
IMBERT B. — L ANALYSE DES MAREES PAR LA METHODE DES MOINDRES CARRES.
BULL.INF.COM.CENTRAL OCEANOGR.ET ET.COTES VI — 1954

4.34
KASPAR J. — K STANOVENI KONSTANT PERIODICKEHO POHYBU TIZNICE VYROVNANIM.
TRAV.INST.GEOPH.ACAD.TCHECOSL.SC.NO 43 PRAHA — 1957

4.34
HORN W. — THE HARMONIC ANALYSIS OF TIDAL PHENOMENA THEORY AND PRACTICE.
OBS.ROY.BELG.COMM. NO 142 S.GEOPH. NO 47 PP 152-153 — 1958

4.34
HORN W. — SOME RECENT APPROACHES TO TIDAL PROBLEMS.
CENTRE BGE.OCEANOGR.LIEGE 24-25 FEVRIER PP 111-140 — 1958

4.34
MUNKELT K. — FORMELN ZUR HARMONISCHEN ANALYSE VON GEZEITENERSCHEI-NUNGEN DENEN EIN UNBEKANNTER GANG UBERLAGERT IST
DEUTSCHEN HYDROGR.ZEITSCH.BAND 12 HEFT 5 PP 189-195 — 1959

4.34
HORN W. — THE HARMONIC ANALYSIS, ACCORDING TO THE LEAST SQUARE RULE, OF TIDE OBSERVATIONS UPON WHICH AN UNKNOWN DRIFT IS SUPERPOSED.
3E SYMP. MAREES TERRESTRES TRIESTE
BOLL.GEOFISICA TEORICA E APPL.VOL.2 NO 5 PP 218-222 — 1960

4.34
BREIN R. — ERDGEZEITEN - ANALYSE MIT BERECHNUNG TAGLICHER VERBESSERUNGEN VON CHARAKTERISTISCHEN SCHWEBUNGSKURVEN NACH DER METHODE DER KLEINSTEN QUADRATE.
GEOD.KOMM.BAYER.AK.WISS.HEFT 132 MITTEIL.85 PP 5-23 — 1966

4.34
BREIN R. — UBER DIE ANWENDUNG DER ANALYSEN ZUR BERECHNUNG TAGLICHER VERBESSERUNGEN DER CHARAKTERISTISCHEN SCHWEBUNGSKURVEN NACH DER METHODE DER KLEINSTEN QUADRATE.
B.I.M. NO 43 PP 1624-1632 — 1966

4.34
SCHUSTER O. — STRENGE ZWEIGRUPPEN-ERDGEZEITENANALYSE NACH DER METHODE DER KLEINSTEN QUADRATE.
DEUTSCHE GEOD. KOMM. MUNCHEN C,146 — 1970

4.34
SCHUSTER O. — ANALYSIS OF EARTH-TIDE RECORDINGS ACCORDING TO THE LEAST SQUARE RULE
XV ASS.GEN. UGGI - ASS.INT.DE GEODESIE MOSCOU AOUT — 1971

4.34
CHOJNICKI T. — RESULTATS DES RECHERCHES FAITES SUR LES METHODES D ANALYSE DES MAREES TERRESTRES
PUBLS.INST.GEOPH.POLISH AC.SCI.,F-1/105/ PP 3-40 — 1976

4.34
SCHULZ B.S. — ANALYSIS OF EARTH TIDE RECORDINGS BY MEANS OF ADJUST-MENT COMPUTATION AND CONSIDERING THE MODULATION EFFECTS IN THE FREQUENCY BANDS
PROCEEDINGS 7TH INT.SYMP.EARTH TIDES, SOPRON 1973
AKADEMIAI KIADO BUDAPEST, PP 405-408 — 1976

4.34
BARTHA G. VARGA P. — SOME SUPPLEMENTARY REMARKS TO THE EVALUATION OF EARTH TIDE RECORDS
PROCEEDINGS 7TH INT.SYMP.EARTH TIDES, SOPRON 1973
AKADEMIAI KIADO BUDAPEST, PP 345-365 — 1976

4.34
WENZEL H.G. SOME REMARKS TO THE ANALYSIS METHOD OF CHOJNICKI
B.I.M. NO 73 PP 4187-4191 1976

4.341
VENEDIKOV A.P. APPLICATION A L ANALYSE HARMONIQUE DES OBSERVATIONS DES MAREES TERRESTRES DE LA METHODE DES MOINDRES CARRES.
C.R.ACAD.BULGARE SC.TOME 14 NO 7 PP 671-674 1961

4.341
VENEDIKOV A.P. SUR UNE POSSIBILITE D APPLICATION DE LA METHODE DES MOINDRES CARRES A L ANALYSE DES OBSERVATIONS DES MAREES TERRESTRES.
5 EME SYMP.INT.SUR LES MAREES TERRESTRES.
OBS.ROY.BELG.COMM.NO 236 S.GEOPH.NO 69 PP 412-420 1964

4.341
PICHA J.
VENEDIKOV A.P. COMPARAISON DES METHODES DE PERTSEV ET VENEDIKOV POUR L ANALYSE HARMONIQUE DES OBSERVATIONS DE MAREES.
* B.I.M. N.42 DECEMBRE PP 1581-1586 1965

4.341
VENEDIKOV A.P. UNE METHODE POUR L ANALYSE DES MAREES TERRESTRES A PARTIR D ENREGISTREMENTS DE LONGUEUR ARBITRAIRE.
OBS.ROY.BELG.COMM. NO 250 S.GEOPH. NO 71 PP 437-459 1966

4.341
VENEDIKOV A.P. SUR LA CONSTITUTION DE FILTRES NUMERIQUES POUR LE TRAITEMENT DES ENREGISTREMENTS DES MAREES TERRESTRES.
ACAD.ROY.BELG.BULL.CL.SC.5 LII, 6 PP 827-845 1966

4.341
VENEDIKOV A.P. UNE METHODE POUR L ANALYSE DES MAREES TERRESTRES A PARTIR D ENREGISTREMENTS DE LONGUEUR ARBITRAIRE.
B.I.M. NO 43 PP 1687 1966

4.341
VENEDIKOV A.P.
PAQUET P. SUR L APPLICATION D UNE METHODE POUR L ANALYSE DES MAREES TERRESTRES A PARTIR D ENREGISTREMENTS DE LONGUEURS ARBITRAIRES.
B.I.M. NO 48 PP 2090-2114 1967

4.341
VENEDIKOV A.P. ANALYSE HARMONIQUE DES MAREES TERRESTRES.
RAPP.SYMP.INTERN.LENINGRAD 1968 MESURES INCLINAISON
COMM.GEOPH.URSS.MOSCOU PP 66-91 1969

4.341
SIMON Z. ON SOME PROPERTIES OF VENEDIKOV S METHOD OF TIDAL ANALYSIS
STUDIA GEOPH. ET GEOD. V.18 PP 381-385 1974

4.341
SIMON Z.
WITTINGEROVA E. UEBERPRUEFUNG DER VENEDIKOV-METHODE DER ERDGEZEITEN-ANALYSE AN DEN MODELLEN DER ERDGEZEITENMESSUNGEN
GEOFYSIKALNI SBORNIK XII, NO 417 PP 73-115 1974

4.341
DUCARME B. ON SOME PROPERTIES OF VENEDIKOV S METHOD OF TIDAL ANALYSIS
STUDIA GEOPHYSICA ET GEODAETICA NO 3 PP 301-303 1976

4.341
VENEDIKOV A. NOTE SUR UNE COMPARAISON DE METHODES D ANALYSE DES ENREGISTREMENTS DES MAREES TERRESTRES
B.I.M. NO 75 PP 4371-4380 1977

4.342
MATVEYEV P.S.
ZAKHARTCHENKO S.N. VALEURS DE REDUCTION POUR LE CALCUL DES ONDES DE GROUPE DE LA MAREE TERRESTRE EN 1958-1967
TROUDI POLTAVSKOI GRAV. OBS. T.XII PP59-99 KIEV 1960
* B.I.M. N.34 NOVEMBRE PP 1096-1103 1963

4.342
MATVEYEV P.S. ANALYSE HARMONIQUE DES MAREES TERRESTRES.
TROUDI POLTAVSKOI GRAV. OBS. XI PP 16-63 1962
* B.I.M. N.33 SEPTEMBRE PP 941-986 1963

4.342
MATVEYEV P.S. ANALYSE HARMONIQUE D UNE SERIE DE TROIS JOURS D OBSERVATIONS DE MAREES TERRESTRES.
ACAD.SC.R.S.S. D UKRAINE OBS.GRAV.DE POLTAVA

INST. GEOPH. DE LA RSS D UKRAINE KIEV 1966
* B.I.M. N.48 JUILLET PP 2070-2083 1967

4.342
MATVEYEV P.S. ANALYSE HARMONIQUE D UNE SERIE MENSUELLE D OBSERVATIONS DES MAREES TERRESTRES.
ACAD. SC. R.S.S. D UKRAINE OBS.GRAV. DE POLTAVA
INST. GEOPH. DE LA RSS D UKRAINE KIEV 1966
* B.I.M. N. 50 FEVRIER PP 2231-2261 1968

4.342
MATVEYEV P.S. SEPARATION DES ONDES DIURNES P1, K1 ET S1
ROTATION ET DEFORMATIONS DE MAREES DE LA TERRE - NO 2
ACAD.NAOUK UKR.SSR - OBS.GRAV.POLTAVA - KIEV PP 80-92 1970

4.342
MIRONOVA L.I. COMPARAISON DES RESULTATS DE L ANALYSE HARMONIQUE D APRES LES METHODES DE B.P.PERTSEV ET P.S.MATVEYEV.
ROTATION ET DEFORM.MAREES DE LA TERRE,FASC.1
ACAD.SC.UKR.OBS.GRAV.POLTAVA KIEV PP 225-234 1970

4.343
CHOJNICKI T. DETERMINATION DES PARAMETRES DE MAREES PAR LA COMPENSATION DES OBSERVATIONS.
B.I.M. N.55 SEPTEMBRE PP 2669-2671 1969

4.343
CHOJNICKI T. DETERMINATION DES PARAMETRES DE MAREES PAR LISSAGE DES OBSERVATIONS.
RAPP.SYMP.INTERN.LENINGRAD 1968 MESURES INCLINAISON
COMM.GEOPH.URSS.MOSCOU PP 92-94 1969

4.343
CHOJNICKI T. OBLICZANIE TEORETYCZNYCH PLYWOW SKORUPY ZIEMSKIEJ I ICH DOKLADNOSC.
GEODEZJA I KARTOGRAFIA 19,3 PP 171-202 1970

4.343
CHOJNICKI T. WYZNACZANIE PARAMETROW PLYWOWYCH PRZEZ WYROWNANIE OBSERWACJI METODA NAJMNIEJSZYCH KWADRATOW
GEODEZJA I KARTOGRAFIA 3/70 N.18 PP 152-182 1970

4.343
CHOJNICKI T. WYZNACZANIE PARAMETROW PLYWOWYCH PRZEZ WYROWNANIE OBSERWACJI METODA NAJMNIEJSZYCH KWADRATOW
GEODEZJA I KARTOGRAFIA TOM XX ZESZYT 3 PP 151-182 1971

4.343
CHOJNICKI T. PROBLEMES DES MOYENS D ELABORATION DES OBSERVATIONS DE MAREE EN TENANT COMPTE DES TECHNIQUES CONTEMPORAINES D OBSERVATION ET DE CALCUL.
XV ASS.GEN. UGGI - ASS.INT.DE GEODESIE MOSCOU AOUT 1971

4.343
CHOJNICKI T. DETERMINATION DES PARAMETRES DE MAREE PAR LA COMPENSATION DES OBSERVATIONS AU MOYEN DE LA METHODE DE MOINDRES CARRES
PUBL.INST.GEOPH.POLISH ACAD.SC.MAR.TERR.55 PP 43-80 1972

4.343
CHOJNICKI T. EIN VERFAHREN ZUR ERDGEZEITENANALYSE IN ANLEHNUNG
MITT.INST.FUR THEOR.GEOD.UNIV.BONN.NO.15 PP 1-59 1973

4.344
USANDIVARAS J.C.
DUCARME B. ANALYSE DES ENREGISTREMENTS DE MAREE TERRESTRE PAR LA METHODE DES MOINDRES CARRES
OBS.ROY.BELG.COMM.SER.B NO 45 S.GEOPH.95 PP 560-569 1969

4.344
USANDIVARAS J.C.
DUCARME B. ANALYSE DES ENREGISTREMENTS DE MAREE TERRESTRE PAR LA METHODE DES MOINDRES CARRES.
6 EME SYMPOSIUM MAREES TERRESTRES STRASBOURG
OBS.ROY.BELG.COMM. A9 S.GEOPH. NO 96 PP 174 1970

4.35
DARWIN G.H. THE HARMONIC ANALYSIS OF TIDAL OBSERVATIONS
SCIENT.PAPERS CAMBRIDGE UNIV.PRESS V.I PP 1-69 1907
BRITISH ASSOCIATION REPORT PP 49-118 1883

4.35
DARWIN G.H. ON THE PERIODS CHOSEN FOR HARMONIC ANALYSIS, AND A COMPARISON WITH THE OLDER METHODS BY MEANS OF HOUR-ANGLES AND DECLINATIONS.
SCIENT.PAPERS CAMBRIDGE UNIV.PRESS V.I PP 70-96 1907
BRITISH ASSOCIATION REPORT PP 35-60 1885

4.35
DARWIN G.H. DATUM LEVELS, THE TREATMENT OF A SHORT SERIES OF TIDAL OBSERVATIONS AND ON TIDAL PREDICTION
SCIENT.PAPERS CAMBRIDGE UNIV.PRESS V.I PP 97-118 1907
BRITISH ASSOCIATION REPORT PP 40-58 1886

4.35
DARWIN G.H. A GENERAL ARTICLE ON THE TIDES
SCIENT.PAPERS CAMBRIDGE UNIV.PRESS V.I PP 119-156 1907
ADMIRALTY SCIENTIFIC MANUAL PP 53-91 1886

4.35
DARWIN G.H. ON THE HARMONIC ANALYSIS OF TIDAL OBSERVATIONS OF HIGH AND LOW WATER
SCIENT.PAPERS CAMBRIDGE UNIV.PRESS V.I PP 157-215 1907
PROCEEDINGS OF THE ROY.SOC. XLVIII PP 278-340 1890

4.35
DARWIN G.H. ON TIDAL PREDICTION
SCIENT.PAPERS CAMBRIDGE UNIV.PRESS V.I PP 258-327 1907
PHILOS.TRANSACT.OF THE ROY.SOC. CLXXXII PP 159-229 1891

4.35
DARWIN G.H. ON AN APPARATUS FOR FACILITATING THE REDUCTION OF TIDAL OBSERVATIONS
SCIENT.PAPERS CAMBRIDGE UNIV.PRESS V.I PP 216-257 1907
PROCEEDINGS OF THE ROY.SOC. LII PP 345-389 1892

4.35
SCHWEYDAR W. HARMONISCHE ANALYSE DER LOTSTORUNGEN DURCH SONNE UND MOND.
VEROFF.D.KONIGL.PREUSS.GEOD.INST.N.F.59 1914

4.35
SCHUREMAN P. A MANUAL OF THE HARMONIC ANALYSIS AND PREDICTION OF TIDES.
DEPART.COMM.U.S.COAST & GEOD.SURV.SP.PUBL.NO 98 1924

4.35
VOIT H. UBER EINE EINFACHE METHODE ZUR ERMITTLUNG DER WICHTIGSTEN TIDEN BEI SCHWERE UND LOTSCHWANKUNGEN.
ZEITSCH.GES.NATURWISS.4 PP 443-445 1939

4.35
GOUGENHEIM A. ETUDE PRATIQUE DE LA MAREE GRAVIMETRIQUE.
BULL.GEOD.NO 20 PP 170-187 1951

4.35
SUTHONS C.T. THE SEMI-GRAPHIC METHOD OF HARMONIC ANALYSIS OF TIDAL OBSERVATIONS EXTENDING OVER ABOUT ONE MONTH.
OBS.ROY.BELG.COMM. NO 142 S.GEOPH. NO 47 PP 140-151 1958

4.35
N. THE ADMIRALTY SEMI-GRAPHIC METHOD OF HARMONIC TIDAL ANALYSIS.
ADMIRALTY TIDAL HANDBOOK NO 1 H.D.505 1959

4.35
BROUET J. APPLICATIONS DE LA METHODE SEMIGRAPHIQUE DE L AMIRAUTE BRITANNIQUE AUX ENREGISTREMENTS GRAVIMETRIQUES D UCCLE.
B.I.M. NO 18 PP 293-295 1960

4.35
BROUET J. APPLICATION DE LA METHODE GRAPHIQUE POUR L EXAMEN DES ENREGISTREMENTS DE MAREE TERRESTRE.
B.I.M. NO 22 PP 456-458 1960

4.35
BROUET J. DISCUSSION GRAPHIQUE DES ENREGISTREMENTS DE MAREES TERRESTRES.
OBS.ROY.BELG.COMM. NO 186 S.GEOPH. NO 48 PP 7 1960

4.35
OZAWA I. ON THE MINCING OF THE READING INTERVALS IN THE EARTH TIDE ANALYSES.
PP 15-21 1963

4.38
GROVES G.W. NUMERICAL FILTERS FOR DISCRIMINATION AGAINST TIDAL PERIODICITIES.
TRANS.AMER.GEOPH.UNION 36 NO 6 1073 1955

4.38
CONNES J.
NOZAL V. LE FILTRAGE MATHEMATIQUE DANS LA SPECTROSCOPIE PAR LA TRANSFORMATION DE FOURIER
JOURN.DE PHYS.ET RADIUM VOL.22, PP 359-366 1961

4.38
BLUM P.A. / JOBERT G. — ANALYSE SPECTRALE DE VARIATIONS DE LA PESANTEUR. C.R.AC.SC.PARIS TOME 255 PP 341 — 1962

4.38
JOBERT G. — REMARQUES SUR L ANALYSE SPECTRALE DES VARIATIONS DE LA PESANTEUR. B.I.M. NO 30 PP 740-742 — 1962

JOBERT G. — COMPARAISON DES RESULTATS DE L ANALYSE SPECTRALE DES MAREES TERRESTRES AVEC LES RESULTATS THEORIQUES. B.I.M. NO 33 PP 1013-1016 — 1963

4.38
JEFFREYS H. — NOTE ON FOURIER ANALYSIS. BULL.SEISM.SOC.AMERICA 54 5 A PP 1441-1444 — 1964

4.38
ZETLER B.D. — THE USE OF POWER SPECTRUM ANALYSIS FOR EARTH TIDES. U.S.COAST GEOD.SURVEY-B.I.M.NO 35 PP 1157-1164 — 1964

4.38
JOBERT G. — SUR LES FILTRES NUMERIQUES UTILISES DANS L ANALYSE HARMONIQUE. B.I.M. NO 37 PP 1260-1273 — 1964

4.38
MUNK W.H. / CARTWRIGHT D.E. — TIDAL SPECTROSCOPY AND PREDICTION PHIL.TRANS.ROY.SOC.LONDON A259, 1105, PP 533-581 — 1966

4.38
BARSENKOV S.N. — ANALYSE SPECTRALE DES VARIATIONS DE MAREE DE LA PESANTEUR A TALGAR. AC.SC. URSS PHYS. DE LA TERRE 3 PP 43-51 MOSCOU 1967 * B.I.M. N. 50 FEVRIER PP 2288-2299 — 1968

4.38
VANICEK P. — AN ANALYTICAL TECHNIQUE TO MINIMIZE NOISE IN A SEARCH FOR LINES IN THE LOW FREQUENCY SPECTRUM. 6 EME SYMPOSIUM MAREES TERRESTRES STRASBOURG OBS.ROY.BELG.COMM. A9 S.GEOPH. NO 96 PP 170-173 — 1970

4.38
ZETLER B. / CARTWRIGHT D. — TIDAL CONSTANTS DERIVED FROM RESPONSE ADMITTANCES. 6 EME SYMPOSIUM MAREES TERRESTRES STRASBOURG OBS.ROY.BELG.COMM. A9 S.GEOPH. NO 96 PP 175-178 — 1970

4.38
VARGA P. — FOURIER-TRANSFORMS OF THE TIDAL VARIATIONS IN THE INTENSITY OF GRAVITY GEOFIZIKAI KOZLEMENYEK 19,3-4, PP 13-19 — 1970

4.38
SCHULZ B.-S. — DIE SCHRITTWEISE SPEKTRALAUSGLEICHUNG, EINE METHODE ZUR ERDGEZEITENANALYSE DEUTSCHE GEOD.KOMM. REIHE C HEFT NO 184, PP 5-121 — 1972

4.38
SCHULZ B.S. — ERDGEZEITENANALYSE UNTER BERUCKSICHTIGUNG DER MODULATIONSEFFECTE IN DEN FREQUENZBANDERN MITT.INST.THEOR.GEOD.UNIV.BONN NO 17 PP1-23 — 1973

4.38
NIKOLADZE I.E. — A CONSISTENT ESTIMATE ALGORITHM OF THE TIDAL GRAVITY VARIATION SPECTRUM INST.GEOP.AC.OF SC. GEORGIAN SSR PP 1-26 TBILISI — 1973

4.38
WEBB D.J. — GREEN S FUNCTION AND TIDAL PREDICTION REV.OF GEOPH.& SPACE PHYSICS VOL 12, NO 1 PP 103-115 — 1974

4.38
NIKOLADZE I.YE. — A MINIMUM VARIANCE ESTIMATE ALGORITHM OF THE TIDAL GRAVITY VARIATION SPECTRUM PROCEEDINGS 7TH INT.SYMP.EARTH TIDES, SOPRON 1973 AKADEMIAI KIADO BUDAPEST, PP 367-386 — 1976

4.38
SCHULLER K. — A PROPOSAL OF PREDICTING TIME SERIES BY MEANS OF STATISTICAL METHODS WITH SPECIAL REGARD TO EARTH-TIDE RECORDS. PROCEEDINGS 7TH INT.SYMP.EARTH TIDES, SOPRON 1973 AKADEMIAI KIADO BUDAPEST, PP 411-415 — 1976

4.38
GODIN G. THE USE OF THE ADMITTANCE FUNCTION FOR THE REDUCTION AND INTERPRETATION OF TIDAL RECORDS
MARINE SC.DIR. -MANUSCR.REP.SER.NO 41 PP 145 1976

4.38
SCHULLER K. EIN BEITRAG ZUR AUSWERTUNG VON ERDGEZEITENREGISTRIERUNGEN
D.GEODATISCHE KOM. REIHE C HEFT NO 227 PP 1-86 1976

4.39
MATVEYEV P.S. SUR LA DETERMINATION DE L ONDE O1 ET LE CALCUL DE QUELQUES PETITES ONDES PAR L ANALYSE HARMONIQUE D UNE SERIE D UN MOIS D OBSERVATIONS DE MAREES
TROUDI POLTAVSKOI GRAV.OBS. X PP 57-66 1961
* B.I.M. NO 29 PP 687-698 - NO 30 PP 705-739 1962

4.39
BREIN R. ETUDE DES MAREES TERRESTRES PAR DES VALEURS INSTANTANEES CARACTERISTIQUES.
IVME SYMP.INTERNATIONAL SUR LES MAREES TERRESTRES.
OBS.ROY.BELG.COMM. NO 188 S.GEOPH.NO 58 PP 257-266 1961
INST.ANGEW.GEOD.FRANKFURT A.M.REIHE I H.21 PP 18-25 1962

4.39
MURRAY M.T. A GENERAL METHOD FOR THE ANALYSIS OF HOURLY HEIGHTS OF TIDE.
INT.HYDROGR.REV.VOL.XLI NO 2 PP 91-101 1964

4.39
MUNK W.
HASSELMANN K. SUPER-RESOLUTION OF TIDES.
INST.GEOPH.PLANET.PHYS.UNIV.CALIFORNIA LA JOLLA
STUDIES ON OCEANOGRAPHY TOKYO PP 339-344 1964

4.39
ZETLER B.D.
SCHULDT M.D.
WHIPPLE R.W.
HICKS S.D. TIDAL HARMONIC CONSTANTS OBTAINED FROM OBSERVATIONS IN RANDOM TIME.
5 EME SYMP.INTERNATIONAL SUR LES MAREES TERRESTRES.
OBS.ROY.BELG.COMM.NO 236 S.GEOPH.NO 69 PP 390-402 1964

4.39
MURRAY M.T. OPTIMISATION PROCESSES IN TIDAL ANALYSIS.
INT.HYDROGR.REV.VOL.XLII NO 1 PP 73-81 1965

4.39
BALENKO V.G. SUR LA METHODE D ANALYSE HARMONIQUE DES MAREES FAIBLES
ACAD. SC. R.S.S. D UKRAINE OBS.GRAV. DE POLTAVA
INST. GEOPH. DE LA RSS D UKRAINE KIEV 1966
* B.I.M. N. 49 OCTOBRE PP 2201-2217 1967

4.39
SIMON D. GENAUERE ERFASSUNG DER GEZEITENWELLEN MIT HILFE EINES ITERATIONSVERFAHRENS.
RAPP.SYMP.INTERN.LENINGRAD 1968 MESURES INCLINAISON
COMM.GEOPH.URSS.MOSCOU PP 162-181 1969

4.39
KORBA S.N.
MATVEYEV P.S.
SLAVINSKAYA E.A. SUR LA PRECISION DES RESULTATS OBTENUS D APRES LE SCHEMA D ANALYSE HARMONIQUE DES OBSERVATIONS DES MAREES TERRESTRES
ROT.ET DEFORM.DE MAREES TERRE VOL.3 PP 132-147 KIEV 1971

4.39
NIKOLADZE I.E. ASYMPTOTIC METHOD OF THE ANALYSIS OF THE TIDE GRAVITY VARIATIONS
XV ASS.GEN. UGGI - ASS.INT.DE GEODESIE MOSCOU AOUT 1971

4.39
SLICHTER L.B.
HAGER C.L.
SYPSTAD F.
JACKSON B.V.
ZURN W.E. EARTH TIDES AND EARTH S FREE VIBRATIONS OBSERVATIONS AT THE SOUTH POLE
ANTARTIC JOURN.OF THE UNITED STAT.VOL.VI NO 5 PP 227-228 1971

4.39
GRAFAREND E. NICHTLOKALE GEZEITENANALYSE
INST.FUR THEORET.GEODASIE - UNIV.BONN NO 13 PP 1-14 1973

4.39
LAMBERT A. EARTH TIDE ANALYSIS AND PREDICTION BY THE RESPONSE METHOD
J.G.R. VOL 79 NO 32 PP 4952-4960 1974

4.39
SCHULLER K. EIN VORSCHLAG ZUR BESCHLEUNIGUNG DES ANALYSENVERFAHRENS

CHOJNICKI
B.I.M. NO 70 PP 3976-3981 1975

4.39
PICHA J.
SKALSKY L.
APPLICATION OF THE METHOD OF CHOJNICKI TO THE DETECTION OF THE SYSTEMATIC ERROR OF THE TIDAL OBSERVATIONS.
PROCEEDINGS 7TH INT.SYMP.EARTH TIDES, SOPRON 1973
AKADEMIAI KIADO BUDAPEST, PP 397-400 1976

4.391
KORBA S.N.
PROGRAMME DE REDUCTION D UNE SERIE DE 30 ZONES D OBSERVATIONS DES MAREES TERRESTRES PAR LA METHODE DE MATVEYEV SUR EVM M-220
ROT.ET DEFORM.DE MAREES DE LA TERRE, VOL.5, PP 86-94 1973

4.393
MATEO J.
ANALYSIS OF TIDAL RECORDS
OBS.ASTR. UNIV.NAC.LA PLATA S.GEOF. V.10 PP 1-104 1974

4.5
RINNER K.
VORFUHRUNG EINER PROGRAMMESTEUERTEN HARMONISCHEN ANALYSE AUF DEM ZUSE RECHENAUTOMATEN Z 11.
OBS.ROY.BELG.COMM. NO 142 S.GEOPH. NO 47 PP 169 1958

4.5
MELCHIOR P.J.
PROGRAMMATION DES DIVERSES METHODES D ANALYSE HARMONIQUE SUR ORDINATEUR ELECTRONIQUE /IBM 650/ AU CENTRE INTERNATIONAL DES MAREES TERRESTRES.
B.I.M. NOS 15 PP 249-255/ 16 PP262-266/ 17 PP 288-290 1959

4.5
MELCHIOR P.J.
UTILISATION D UN ORDINATEUR ELECTRONIQUE IBM 650 POUR LES ANALYSES HARMONIQUES, COMPARAISON DES DIVERSES METHODES, POSSIBILITE DE LA REVISION D OBSERVATIONS ANCIENNES. /DEMONSTRATION PRATIQUE
3E SYMP. MAREES TERRESTRES TRIESTE
BOLL.GEOFISICA TEORICA E APPL.VOL.2 NO 5 PP 263-265 1960

4.5
CARROZZO M.T.
A PROGRAMME FOR LECOLAZET METHOD OF HARMONIC ANALYSIS OF THE EARTH TIDES IN THE BELL INTERPRETATIVE SYSTEM FOR AN ELECTRONIC COMPUTER IBM 650
3E SYMP. MAREES TERRESTRES TRIESTE
BOLL.GEOFISICA TEORICA E APPL.VOL.2 NO 5 PP 247-262 1960

4.5
MELCHIOR P.J.
PROGRAMMATION DES DIVERSES METHODES D ANALYSE HARMONIQUE SUR ORDINATEUR ELECTRONIQUE /IBM 650/ AU CENTRE INTERNATIONAL DES MAREES TERRESTRES.
B.I.M. NO 18 PP 307-312 1960

4.5
MELCHIOR P.J.
PROGRAMMATION DES DIVERSES METHODES D ANALYSE HARMONIQUE SUR ORDINATEUR ELECTRONIQUE /IBM 650/ AU CENTRE INTERNATIONAL DES MAREES TERRESTRES.
B.I.M. NO 19 PP 325-326 1960

4.5
LONGMAN I.M.
USE OF DIGITAL COMPUTERS FOR THE REDUCTION AND INTERPRETATION OF EARTH TIDE DATA.
INST GEOPH.UNIV.CALIFORNIA - BIM NO 22 PP 415-419 1960

4.5
SEIFERS H.
BANDPROGRAMME FUR DIE RECHENANLAGE Z 11.
DEUTSCHE GEOD.KOMM.REIHE B NR 80 PP 11-12 1961

4.5
MELCHIOR P.J.
NOUVEAUX DEVELOPPEMENTS APPORTES A L ETUDE DES MAREES TERRESTRES PAR LES PROCEDES DE CALCUL ELECTRONIQUE AU CENTRE INTERNATIONAL.
IVME SYMP.INTERNATIONAL SUR LES MAREES TERRESTRES.
OBS.ROY.BELG.COMM. NO 188 S.GEOPH. NO 58 PP 273--279 1961

4.5
MELCHIOR P.J.
PROGRAMMATION SUR ORDINATEUR ELECTRONIQUE DU TRACE SEMI-AUTOMATIQUE DE COURBES DE NIVEAU A L AIDE D UNE TABLE TRACANTE ELECTRONIQUE.
OBS.ROY.BELG.COMM. NO 198 S.GEOPH. NO 60 1962

4.5
MELCHIOR P.J.
METHODES DE CALCUL ELECTRONIQUE SUR ORDINATEUR IBM 1620 DANS L ANALYSE HARMONIQUE DES MAREES TERRESTRES AU CENTRE INTERNATIONAL.
B.I.M. N.28 PP 645-647 1962

4.5
JOBERT G. — PROGRAMMES D ANALYSE POUR LES MAREES TERRESTRES.
B.I.M.NO 31 PP 829-832 — 1963

4.5
NAKAGAWA I. — A PROGRAM FOR LECOLAZET S METHOD IN HARMONIC ANALYSIS
WRITTEN FOR IBM 7094.
5 EME SYMP.INT.SUR LES MAREES TERRESTRES.
OBS.ROY.BELG.COMM.NO 236 S.GEOPH.NO 69 PP 403 — 1964

4.5
LENNON G.W. — SOME COMPUTER TECHNIQUES FOR THE ANALYTICAL
TREATMENT OF TIDAL DATA.
5 EME SYMP.INT.SUR LES MAREES TERRESTRES.
OBS.ROY.BELG.COMM.NO 236 S.GEOPH.NO 69 PP 404-410 — 1964

4.5
MELCHIOR P.J. — AMELIORATION DES TECHNIQUES DE CALCUL GRACE A
L UTILISATION D UNE ENTREE DISQUES CONNECTES A
L ORDINATEUR IBM 1620.
5 EME SYMP.INT.SUR LES MAREES TERRESTRES.
OBS.ROY.BELG.COMM.NO 236 S.GEOPH.NO 69 PP 411 — 1964

4.5
LENNON G.W. — LE TRAITEMENT DES HAUTEURS HORAIRES DE LA MAREE
AVEC UN IBM 1620.
REV.HYDROGR.INTERN.XLII NO 2 PP 131-155 — 1965

4.5
ORZECHOWSKI J. — ENSEMBLE DES METHODES DE TRANSFORMATION DES DONNEES DE
MAREE, FONDEES SUR LE PRINCIPE DE MOINDRES CARRES
PUBL.INST.GEOPH.POLISH ACAD.SC.MAR.TERR.55 PP 81-166 — 1972

4.5
CHOJNICKI T. — DETERMINATION DE LA COURBE THEORIQUE DE LA MARCHE
DE MAREE DANS LE TEMPS
PUBL.INST.GEOPH.POLISH ACAD.SC.MAR.TERR.55 PP 167-180 — 1972

4.5
STOKES J. — A COMPUTER PROGRAM FOR EARTH TIDE ANALYSIS
REP.7, UNIV.OF UPPSALA, PP 1-41 — 1972

4.5
BACHEM H.C.
WENZEL H.G. — ZUR AUFBEREITUNG DER ERDGEZEITENREGISTRIERUNGEN
FUR DIE HARMONISCHE ANALYSE
B.I.M. NO 67 AOUT, PP 3718-3726 — 1973

4.5
CHOJNICKI T.
ORZECHOWSKI J. — MODIFICATION DU PROGRAMME DE L ANALYSE DES
OBSERVATIONS DES MAREES
PUBL.INST.GEOPH.POLISH AC.OF SC. V.71 PP 117-178 — 1973

4.5
CABANISS G.H.
ECKHARDT D.H. — THE AFCRL EARTH TIDE PROGRAM
AFCRL-TR-73-0084, NO 433, PP 1-27 — 1973

4.5
KORBA S.N. — PROGRAMME DE REDUCTION D UNE SERIE DE 30 JOURS
D OBSERVATIONS DES MAREES TERRESTRES PAR LA METHODE
DE MATVEYEV SUR EVM M 220.
ROTAT.ET DEFORMAT.DE MAREES DE LA TERRE-V-PP 94-98 — 197

4.5
ORZECHOWSKI J. — MOYENS D ACCELERATION ET DE L ACCROISSEMENT DE LA
PRECISION DES CALCULS DANS LES METHODES DES ANALYSES
DES OBSERVATIONS DES MAREES
B.I.M. NO 72 PP 4143-4149 — 197

4.5
CHOJNICKI T. — REMARQUES CONCERNANT L ARTICLE DE K. SCHULLER - EIN
VORSCHLAG ZUR BESCHLEUNIGUNG DES ANALYSENVERFAHRENS
CHOJNICKI -
B.I.M. NO 72 PP 4150-4155 — 197

4.5
DUCARME B. — THE COMPUTATION PROCEDURES AT THE INTERNATIONAL CENTER
FOR EARTH TIDES /I.C.E.T./
B.I.M. NO 72 PP 4156-4181 — 197

4.6
GOUGENHEIM A. — COMMENTAIRES AU SUJET DE L INTERPRETATION DES
OBSERVATIONS.
OBS.ROY.BELG.COMM. NO 100 S.GEOPH. NO 36 — 195

4.6
LECOLAZET R. L INFLUENCE DES ERREURS ACCIDENTELLES DANS L ANALYSE HARMONIQUE. OBS.ROY.BELG.COMM. NO 142 S.GEOPH. NO 47 PP 163-165 1958

4.6
TOMASCHEK R. RINNER K. UEBER DIE GENAUIGKEIT DER ERGEBNISSE DER ERDGEZEITENMESSUNGEN. OBS.ROY.BELG.COMM. NO 142 S.GEOPH. NO 47 PP 166 1958

4.6
NISHIMURA E. METHODICAL COMPARISON OF HARMONIC ANALYSIS OF OBSERVED EARTH TIDAL CHANGE OF GRAVITY. OBS.ROY.BELG.COMM. NO 142 S.GEOPH. NO 47 PP 167-168 1958

4.6
LECOLAZET R. SUR L ESTIMATION DES ERREURS INTERNES AFFECTANT LES RESULTATS D UNE ANALYSE HARMONIQUE MENSUELLE. B.I.M. NO 17 PP 269-278 1959

4.6
SKALSKY L. ZUR FRAGE DER GENAUIGKEIT VON GEZEITENBEOBACHTUNGEN MIT EINFACHEN HORIZONTALPENDELN. 3E SYMP. MAREES TERRESTRES TRIESTE BOLL.GEOFISICA TEORICA E APPL.VOL.2 NO 5 PP 108-116 1960

4.6
PICHA J. VERGLEICH DER METHODEN DER HARMONISCHEN ANALYSE AUF GRUND DES BEOBACHTUNGSMATERIALS DER STATION BREZOVE HORY UND EINIGE BEMERKUNGEN ZU DEN KONTROLLEN DER BERECHNUNGEN 3E SYMP. MAREES TERRESTRES TRIESTE BOLL.GEOFISICA TEORICA E APPL.VOL.2 NO 5 PP 238-246 1960

4.6
DOODSON A.T. LENNON G.W. A MEASURE OF THE ACCURACY OF A SINGLE HARMONIC ANALYSIS 3E SYMP. MAREES TERRESTRES TRIESTE BOLL.GEOFISICA TEORICA E APPL.VOL.2 NO 5 PP 208-214 1960

4.6
PICHA J. VERGLEICH DER METHODEN HARMONISCHER ANALYSE UND EINIGE BEMERKUNGEN ZU DEN KONTROLLEN DER BERECHNUNGEN. STUDIA GEOPH.GEOD.VOL.4 PP 85-93 1960

4.6
BALENKO V.G. QUELQUES PROBLEMES CONCERNANT LA COMPARAISON DES METHODES D ANALYSE HARMONIQUE DES MAREES TERRESTRES. IVME SYMP.INTERNATIONAL SUR LES MAREES TERRESTRES. OBS.ROY.BELG.COMM. NO 188 S.GEOPH. NO 58 PP 242-256 1961

4.6
BALENKO V.G. METHODES D ANALYSE HARMONIQUE DE DOODSON ET LECOLAZET TROUDI POLTAVSKOI GRAV.OBS. X PP 51-56 1961
* B.I.M. NO 30 PP 705-739 1962

4.6
VENEDIKOV A.P. SUR L ESTIMATION DE LA PRECISION DES OBSERVATIONS DES MAREES TERRESTRES. B.I.M.NO 36 PP 1223-1229 1964

4.6
CARROZZO M.T. ANALYSIS AND COMPARAISON OF THE VARIOUS METHODS OF HARMONIC ANALYSIS OF THE EARTH-TIDES. B.I.M.NO 37 PP 1274-1290 1964

4.6
KRAMER M.V. ON THE EFFECT OF CHANCE ERRORS OF OBSERVATIONS UPON THE RESULTS OF DETERMINATION OF AMPLITUDES AND PHASES OF ELASTIC TIDAL WAVES. 5 EME SYMP.INT.SUR LES MAREES TERRESTRES. OBS.ROY.BELG.COMM.NO 236 S.GEOPH.NO 69 PP 426-43P 1964

4.6
VENEDIKOV A.P. ESTIMATION DE LA PRECISION DES OBSERVATIONS DE MAREES TERRESTRES. IZV.GEOPH.INST.BULG.AKAD.NAUK SOFIA 7 PP 173-184 1965

4.6
KHAVAN A. SUR L IMPORTANCE RELATIVE DES ERREURS INTERNES DANS DIVERSES SERIES D OBSERVATIONS DE MAREES TERRESTRES. B.I.M. NO 45 PP 1828-1840 1966

4.6
OUBIK B.S. ESTIMATION DE L ERREUR QUADRATIQUE MOYENNE D UNE

ORDONNEE DE LA COURBE DE MAREE ANALYSEE ET MOYENNE DES RESULTATS DE L ANALYSE HARMONIQUE
ACAD.NAOUK UKRAINSKOI SSR, OBS.GRAV.POLTAVA
ROTAT.DEFOR.MAREES DE LA TERRE VII PP 62-66 1975

4.6
AKHAVAN-SADEGHI A. ETUDE STATISTIQUE ET COMPARATIVE DE LA DISTRIBUTION DES ERREURS INTERNES DANS L OBSERVATION DES MAREES GRAVIMETRIQUES A TEHERAN
JOURN.EARTH AND SPACE PHYSICS VOL 4 NO 1 PP 1-14 1975

4.6
BALENKO V.G.
KORBA P.S. DU PROBLEME DE LA REDUCTION DES OBSERVATIONS DE MAREES TERRESTRES
ACAD.NAOUK UKRAINSKOI SSR, OBS.GRAV.POLTAVA
ROTAT.DEFOR.MAREES DE LA TERRE VII PP 58-62 1975
* B.I.M. NO 73 PP 4225-4233 1976

4.6
ORZECHOVSKI J. OBSERVATIONS MODELEES DE MAREE TERRESTRE POUR RECHERCHE ET COMPARAISON DES METHODES D ANALYSE
PROCEEDINGS 7TH INT.SYMP.EARTH TIDES, SOPRON 1973
AKADEMIAI KIADO BUDAPEST, PP 387-390 1976

4.6
ORZECHOWSKI J. L ANALYSE STATISTIQUE DES RESIDUUMS DES MAREES
PUBLS.INST.GEOPH.POLISH.AC.SCI.,F-1/105/PP 85-98 1976

4.6
NAKAI S. PRE-PROCESSING OF TIDAL DATA
B.I.M. NO 75 PP 4334-4340 1977

4.7
LONGMAN I.M. INTERPOLATION OF EARTH-TIDE.
RECORDS JOURN.GEOPH.RES.65 NO 11 PP 3801-3803 1960

4.7
LECOLAZET R. SUR L INTERPOLATION DE DONNEES MANQUANTES.
IVME SYMP.INTERNATIONAL SUR LES MAREES TERRESTRES.
OBS.ROY.BELG.COMM. NO 188 S.GEOPH. NO 58 PP 267-272 1961

4.7
VENEDIKOV A.P. NOTE SUR LA METHODE DE I.M. LONGMAN POUR L INTERPOLATION DE LA MAREE GRAVIMETRIQUE.
INST.GEOPH.ACAD.BULGARE SCI.-BIM NO 24 PP 487-492 1961

5.1
DARWIN G.H. ON VARIATIONS IN THE VERTICAL DUE TO ELASTICITY OF THE EARTH S SURFACE.
SCIENT.PAPERS I 14 PP 444-459 1882

5.1
HECKER O. DEFORMATIONSBEOBACHTUNGEN IN PRIBRAM IN BOHMEN.
GERL.BEITR.Z.GEOPH.XIII 5-6 1914

5.1
SCHWEYDAR W. DEFORMATION DER ERDE DURCH DIE FLUTKRAFT UNTER BERUCKSICHTIGUNG DER HALBTAGIGEN GEZEITEN DES MEERES.
VEROFF.K.PREUSS.GEOD.INST.66 PP 45-51 1916

5.1
ISHIMOTO M. OBSERVATIONS SUR LES VARIATIONS DE L INCLINAISON DE LA SURFACE TERRESTRE.
BULL.EARTHQU.RES.INST.TOKYO IMP.UNIV.III PP 1-12 1927

5.1
BERROTH A. UEBER DIE MESSUNG DER VARIATION DER SCHWERE DURCH SONNE UND MOND UNTER BERUCKSICHTIGUNG DER DYNAMISCHEN MEERESGEZEITEN.
NACHR.GES.WISS.GOTTINGEN MATH.PHYS.KL. PP 449 1932

TAKAHASI R. TILTING MOTION OF THE EARTH S CRUST OBSERVED AT RYOZYUN /PORT ARTHUR/.
BULL.EARTHQU.RES.INST.TOKYO IMP.UNIV.X, 3 PP 531-559 1932

5.1
TAKAHASI R. A NOTE ON THE TILTING MOTION OF THE EARTH S CRUST OBSERVED AT ZINSEN /CHEMULPO/.
BULL.EARTHQU.RES.INST.TOKYO IMP.UNIV.X, 4 PP 826-843 1932

5.1
NISHIMURA E. OBSERVATION OF GROUND TILTING AT ASO VOLCANOLOGICAL OBSERVATORY.
REPORT JAP.SOC.FOR SCIENCE ADVANCEMT.VOL.14 NO 3 1939

5.1
HAGIWARA T. RIKITAKE T. YAMADA J. — OBSERVATIONS OF THE DEFORMATION OF THE EARTH S SURFACE AT ABURATSUBO, MIURA PENINSULA.
BULL.EARTHQ.RES.INST.TOKYO IMP.UNIV.
PART. I VOL. XXVI PP 23- 27 1948
PART. II VOL.XXVII 1-4 PP 35- 38 1949
PART.III VOL.XXVII 1-4 PP 39- 45 1949
PART. IV VOL. XXIX, 3 PP 455-468 1951
PART. V VOL. XXIX,4 PP 557-561 — 1951

5.1
MELCHIOR P.J. — SUR LES METHODES DE SEPARATION DES EFFETS DIRECTS ET INDIRECTS.
OBS.ROY.BELG.COMM. NO 114 S.GEOPH. NO 39 — 1957

5.1
TOMASCHEK R. — UEBER DER EINFLUSS DER MARITIMEN EFFEKTE IN WINSFORD.
OBS.ROY.BELG.COMM. NO 142 S.GEOPH. NO 47 PP 72 — 1958

5.1
NISHIMURA E. — TIME VARIATION OF CRUSTAL ELASTICITY.
OBS.ROY.BELG.COMM. NO 142 S.GEOPH. NO 47 PP 98-99 — 1958

5.1
TOMASCHEK R. GROTEN E. — THE PROBLEM OF THE RESIDUAL ELLIPSES OF TILT MEASUREMENTS.
IVME SYMP.INTERNATIONAL SUR LES MAREES TERRESTRES.
OBS.ROY.BELG.COMM. NO 188 S.GEOPH. NO 58 PP 78-93 — 1961

5.1
TOMASCHEK R. GROTEN E. — DIE RESIDUALBEWEGUNGEN IN DEN REGISTRIERUNGEN DER HORIZONTALEN GEZEITENKOMPONENTEN.
GEOF.PURA E APPL. MILANO VOL.56 NO 3 PP 1-15 — 1963

5.1
OZAWA I. — ON THE COMBINED OBSERVATIONS OF THE CRUSTAL DEFORMATION AT SOME OBSERVATORIES IN THE SHORT INTERVALS.
GEOPH.PAPERS DEDICATED PROF.K.SASSA PP 427-433 — 1963

5.1
SIMON D. — EMPIRISCHE BESTIMMUNG DER MEERESGEZEITENWIRKUNGEN UND DES EINFLUSSES DER GANZTAGIGEN NUTATION DER ERDE AUF DIE LOTSCHWANKUNGEN IM KALISALZBERGWERK TIEFENORT.
B.I.M. NO 45 PP 1810-1827 — 1966

5.1
PERTSEV B.P. — SUR L INFLUENCE DES MAREES OCEANIQUES SUR LES VARIATIONS DE MAREES DE LA FORCE DE PESANTEUR.
ACAD.SC. URSS INST. DE PHYS. TERRESTRE
* B.I.M. NO 47 AVRIL PP 1955-1961 — 1967

5.1
BUCHHEIM V. — SUR LE PROBLEME DE LA VALEUR DES COEFFICIENTS GAMMA ET DES ECARTS DE PHASES DANS LES OBSERVATIONS CLINOMETRIQUES.
RAPP.SYMP.INTERN.LENINGRAD 1968 MESURES INCLINAISON
COMM.GEOPH.URSS.MOSCOU PP 132-154 — 1969

5.1
NOWROOZI A.A. KUO J. EWING M. — SOLID EARTH AND OCEANIC TIDES RECORDED ON THE OCEAN FLOOR OFF THE COAST OF NORTHERN CALIFORNIA
J.GEOPH.RES.VOL.74 NO 2 PP 605-614 — 1969

5.1
BUCHHEIM W. — DIE KORREKTUR VON ERDGEZEITEN-BEOBACHTUNGEN AUF INDIREKTE EFFEKTE ALS RANDWERTPROBLEM DER MECHANIK.
6 EME SYMPOSIUM MAREES TERRESTRES STRASBOURG
OBS.ROY.BELG.COMM. A9 S.GEOPH. NO 96 PP 137-146 — 1970

5.1
SUGAWA C. HOSOYAMA K. — OBSERVATION OF EARTH TIDES AROUND MIZUSAWA
THE OCEANIC EFFECTS ON EARTH TIDES
XV ASS.GEN. UGGI - ASS.INT.DE GEODESIE MOSCOU AOUT — 1971

5.1
SIMON D. — EINE MOGLICHKEIT DER BESTIMMUNG VON H,K UND I DURCH GEZEITENBEOBACHTUNGEN MIT GRAVIMETERN, EXTENSOMETERN UND KLINOMETERN AN EINEM ORT OHNE VERWENDUNG VON MEERESGEZEITENKARTEN.
B.I.M. N.62 31 DECEMBRE PP 3114-3118 — 1971

5.1
SIMON D. — ZUR FRAGE DER H-, K-, L-BESTIMMUNG DURCH VERWENDUNG EINER KOMBINATION VON GEZEITENGRAVIMETERN, EXTENSOMETERN

UND KLINOMETERN OHNE BENUTZUNG VON MEERESGEZEITENKARTEN
B.I.M. NO 63 PP 3297-3298 1972

5.1
GROTEN E.
BRENNECKE J.
GLOBAL SEA TIDE EFFECTS OF O1 AND K1 CONSTITUENTS
IN EARTH TIDE GRAVITY OBSERVATIONS
RIVISTA ITALIANA DI GEOF.V.XXIII NO 3/4 PP 133-136 1974

5.1
TANAKA T.
ON EFFECTS OF OCEAN TIDES UPON EARTH TIDES
PROCEEDINGS 7TH INT.SYMP.EARTH TIDES, SOPRON 1973
AKADEMIAI KIADO BUDAPEST, PP 607-617 1976

5.11
TERAZAWA K.
ON PERIODIC DISTURBANCE OF LEVEL ARISING FROM THE LOAD
OF NEIGHBORING OCEANIC TIDES
PHIL.TRANS.ROY SOC.LONDON, A 217, P 35 1916

5.11
FEDOROV F.P.
SUR L INFLUENCE DES MAREES OCEANIQUES DANS
L ETUDE DES VARIATIONS LUNI-SOLAIRES DE LA GRAVITE
TRAV.OBS.GRAVIM.DE POLTAVA KIEV T.4 PP 88-102 1951

5.11
JOBERT G.
PERTURBATIONS DES MAREES TERRESTRES.
ANN.GEOPH.16 PP 1-55 1960

5.11
TAKEUCHI H.
SAITO M.
KOBAYASHI N.
STATICAL DEFORMATIONS AND FREE OSCILLATIONS
OF A MODEL EARTH.
J.GEOPH.RES.VOL.67 PP 1141 1962

5.11
GROTEN E.
DER EINFLUSZ DER FREIEN OZEANE AUF DIE BESTIMMUNG DES
VERHALTENS DER ERDKRUSTE AUS HORIZONTALPENDELMESSUNGEN.
GEOF.PURA E APPL.MILANO 54 I, S.6-24 1963

5.11
GROTEN E.
BERECHUNG DER LOTABWEICHUNGEN INFOLGE DER MEERESGEZEITEN-
ATTRAKTION FUR DIE ERDGEZEITENSTATIONEN FREIBERG,
BREZOVE HORY UND SCLAIGNEAUX.
GEOF.PURA E APPL.MILANO BD.55 S.1-15 1963

5.11
GROTEN E.
BERECHNUNG DES EINFLUSSES DER MEERESGEZEITEN AUF DIE
REGISTRIERUNG DER ERDGEZEITEN IN NEUNKIRCHEN SIEGERLAND.
Z.F.GEOPH.WURZBURG 29, 2 PP 57-64 1963

5.11
ZADRO M.
ON THE FREQUENCY DEPENDANCE OF THE LOADING EFFECTS
DUE TO OCEAN TIDES AND SEICHES.
5 EME SYMP.INT.SUR LES MAREES TERRESTRES.
OBS.ROY.BELG.COMM.NO 236 S.GEOPH.NO 69 PP 372-380 1964

5.11
BOZZI ZADRO M.
MAREE TERRESTRI ED EFFETTI DI CARICO.
BOLL.GEOFIS.TEOR.APPL.VIII NO 31 PP 173-195 1966

5.11
PERTSEV B.P.
SUR L INFLUENCE DE L EFFET INDIRECT SUR LES RESULTATS
DES OBSERVATIONS CLINOMETRIQUES.
RAPP.SYMP.INTERN.LENINGRAD 1968 MESURES INCLINAISON
COMM.GEOPH.URSS.MOSCOU PP 182-186 1969

5.11
KUO J.T.
STATIC RESPONSE OF A MULTILAYERED MEDIUM UNDER
INCLINED SURFACE LOADS
J.GEOPH.RES.74, PP 3195-3206 1969

5.11
PERTSEV B.P.
EFFET DES MAREES OCEANIQUES SUR LA FREQUENCE DE L ONDE M2
ROTATION ET DEFORM.MAREES DE LA TERRE.FASC.1
ACAD.SC.UKR.OBS.GRAV.POLTAVA KIEV PP 156-165 1970

5.11
HOSOYAMA K.
THE OCEANIC EFFECT ON THE EARTH TIDES
JOUR.GEOD.SOC.JAPAN 16,3 PP 99-110 1970

5.11
FARRELL W.E.
THE EFFECT OF OCEAN LOADING ON TIDAL GRAVITY.
6 EME SYMPOSIUM MAREES TERRESTRES STRASBOURG
OBS.ROY.BELG.COMM. A9 S.GEOPH. NO 96 PP 116 1970

5.11
LAMBERT A.
THE RESPONSE OF THE EARTH TO LOADING BY THE
OCEAN TIDES AROUND NOVA SCOTIA.
GEOPHYS.J.R.ASTR.SOC. 19, PP 449-477 1970

5.11
LAMBERT A. STUDIES OF OCEAN TIDE LOADING IN NOVA SCOTIA, CANADA. 6 EME SYMPOSIUM MAREES TERRESTRES STRASBOURG OBS.ROY.BELG.COMM. A9 S.GEOPH. NO 96 PP 220 1970

5.11
BOWER D.R. SOME NUMERICAL RESULTS IN THE DETERMINATION OF THE INDIRECT EFFECT. 6 EME SYMPOSIUM MAREES TERRESTRES STRASBOURG OBS.ROY.BELG.COMM. A9 S.GEOPH. NO 96 PP 106-112 1970

5.11
PERTSEV B.P. THE EFFECT OF OCEAN TIDES UPON EARTH-TIDE OBSERVATIONS. 6 EME SYMPOSIUM MAREES TERRESTRES STRASBOURG OBS.ROY.BELG.COMM. A9 S.GEOPH. NO 96 PP 113-115 1970

5.11
BOWER D.R. THE PREDICTION OF OCEAN-TIDE EFFECT IN THE MEASUREMENT OF THE EARTH TIDE XV ASS.GEN. UGGI - ASS.INT.DE GEODESIE MOSCOU AOUT 1971

5.11
FARRELL W.E. GRAVITY TIDES UNIV.CALIF.SAN DIEGO, INST.GEOPH.PLANET.PHYSICS EARTH STRAIN STUDIES 1971

5.11
PERTSEV B.P. SUR L INFLUENCE DE L EFFET INDIRECT SUR LES RESULTATS DES OBSERVATIONS CLINOMETRIQUES. SYMP.INTERN.MESURES INCLIN.PP 182-186 LENINGRAD 1968 * B.I.M. NO 60 PP 2974-2978 1971

5.11
PERTSEV B.P. EFFET INDIRECT DES MAREES OCEANIQUES DANS LA FREQUENCE DE L ONDE M2 ROT.ET DEF.MAREES DE LA TERRE -I PP 156-165 KIEV 1970 * B.I.M. NO 61 PP 3100-3107 1971

5.11
BENHALLOU M.H. REMARQUE SUR LES CORRECTIONS APPORTEES AUX OBSERVATIONS DE MAREES TERRESTRES POUR TENIR COMPTE DES MAREES OCEANIQUES. B.I.M. N.62 DECEMBRE PP 3174-3175 1971

5.11
BEAUMONT CH.
LAMBERT A. CRUSTAL STRUCTURE FROM SURFACE LOAD TILTS, USING A FINITE ELEMENT MODEL GEOPHYS.J.R. ASTR.SOC.29 PP 203-226 1972

5.11
BOZZI ZADRO M. EARTH TIDES AND OCEAN LOAD EFFECTS RECORDED AT TRIESTE BOLL.DI GEOF.TEORICA ED APPL.VOL.XIV, NO 55 PP 192-202 1972

TANAKA T. TIDAL TILTS AND STRAINS AND OCEANIC TIDES /PART 2/ DISASTER PREVENTION RES.INST., KYOTO UNIV., PP 194-201 1972

5.11
TANAKA T. TIDAL TILTS AND STRAINS, AND OCEANIC TIDES /PART 3/ JL.GEODETIC SOC.OF JAPAN, VOL.19, NO 2, PP 85-92 1973

5.11
GROTEN E.
BRENNECKE J. GLOBAL INTERACTION BETWEEN EARTH AND SEA TIDES JOURNAL OF GEOPH.RESEARCH V.78, NO 35 PP 8519-8526 1973

5.11
MERRIAM J.B. COMMENT ON -GLOBAL INTERACTION BETWEEN EARTH AND SEA TIDES- BY E.GROTEN AND J.BRENNECKE J.G.R. VOL.79 NO 29 P 4444 1974

5.11
GROTEN E.
BRENNECKE J. REPLY TO J.B. MERRIAM J.G.R. VOL 79 NO 29 P 4445 1974

5.11
BEAVAN R.J. SOME CALCULATIONS OF OCEAN LOADING STRAIN TIDES IN GREAT BRITAIN GEOPH.J.R.ASTR.SOC.VOL 38 PP 63-82 1974

5.11
PERTSEV B.P. M2 OCEAN-TIDE CORRECTIONS TO TIDAL GRAVITY OBSERVATIONS IN WESTERN EUROPE XVI ASS.GEN.DE L UGGI GRENOBLE - AOUT 1975

5.11
WARBURTON R.J.
BEAUMONT CHR. THE EFFECT OF OCEAN TIDE LOADING ON TIDES OF THE SOLID EARTH OBSERVED WITH THE SUPERCONDUCTING GRAVIMETER

GOODKIND J.M. GEOPHYS.J.R.ASTR.SOC. VOL 43 PP 707-720 1975

5.11
ZSCHAU J. ANALYSE VON AUFLASTGEZEITEN IN LOTSCHWANKUNGSREGISTRIERUNGEN ZUR BESTIMMUNG PHYSIKALISCHER PARAMETER VON ERDKRUSTE UND MANTEL IM BEREICH DER NORDSEE
DEUTSCHEN GEOD.KOM. REIHE B,HEFT NO.211 PP 5-14 1975

5.11
SAITO M. PARTIAL DERIVATIVES OF LOVE NUMBERS AND RELAXATION SPECTRA OF THE EARTH
GEOPHYSICAL FLUID DYNAMICS WHOI II - PP 107-112 1975

5.11
PERTSEV B.P.
IVANOVA M.V. SPATIAL DISTRIBUTION OF OCEAN-TIDE M2 CORRECTIONS TO TIDAL GRAVITY MEASUREMENTS IN THE NORTHERN HEMISPHERE
PROCEEDINGS 7TH INT.SYMP.EARTH TIDES, SOPRON 1973
AKADEMIAI KIADO BUDAPEST, PP 391-400 1976

5.11
BOZZI ZADRO M.
CHIARUTTINI C. LOADING EFFECTS OF THE MEDITERRANEAN TIDES
PROCEEDINGS 7TH INT.SYMP.EARTH TIDES, SOPRON 1973
AKADEMIAI KIADO BUDAPEST, PP 495-502 1976

5.11
PERTSEV B.P. INFLUENCE DES MAREES OCEANIQUES DES ZONES PROCHES SUR LES OBSERVATIONS DE MAREES TERRESTRES
ISVESTIA ACAD.NAOUK SSSR, MOSCOU PP 13-22
*B.I.M. NO 74 PP 4259-4274 1976

5.11
ZSCHAU J. TIDAL SEA LOAD TILT OF THE CRUST, AND ITS APPLICATION TO THE STUDY OF CRUSTAL AND UPPER MANTLE STRUCTURE
GEOPHYS.J.R.ASTR.SOC. VOL 44 PP 577-593 1976

5.11
BRETREGER K.
MATHER R.S. ON THE MODELLING OF THE DEFORMATION OF TIDAL GRAVITY BY OCEAN LOADING
UNISURV G NO 24 PP 71-80 1976

5.11
ANDERSON A.J. SPECTRAL ANALYSIS OF LOW NOISE RECORDING GRAVIMETER STUDIES AND NON-LINEAR TIDAL LOADING IN SCANDINAVIA
UNIV.OF UPPSALA DEPT.SOLID EARTH PHYS. PP 1-36 1976

5.11
MOENS M. SOLID EARTH TIDE AND ARCTIC OCEANIC LOADING TIDE AT LONGYEARBYEN /SPITSBERGEN/
PHYS.OF THE EARTH AND PLAN.INTER.VOL.13 NO 3 PP 197-211 1976

5.11
PERTSEV B.P. INFLUENCE DES MAREES OCEANIQUES DES ZONES PROCHES SUR LES OBSERVATIONS DE MAREES TERRESTRES
PHYSIQUE DE LA TERRE NO 1 PP 13-22 1976
* B.I.M. NO 74 PP 4259-4274 1976

5.11
MOLODENSKI S.M. ABOUT THE CONNECTION OF LOVE NUMBERS WITH LOADING COEFFICIENTS
FISIKA ZEMLI NO 3 PP 3-7 1977

5.11
PERTSEV B.P.
IVANOVA M.V. PRISE EN CONSIDERATION DE L INFLUENCE DES MAREES OCEANIQUES SUR LES OBSERVATIONS DE MAREES TERRESTRES GRAVIMETRIQUES DANS LA PARTIE EST DES ETATS UNIS
PHYS.TERRE NO 11 PP 3-6 MOSCOU 1976
* B.I.M. NO 75 PP 4381-4387 1977

5.111
LENNON G.W. THE ATTRACTION OF OCEAN TIDES AT BIDSTON.
3E SYMP.MAREES TERRESTRES TRIESTE
BOLL. GEOF. TEORICA E APPL. VOL 2 N.5 PP 38-39 1960

5.111
BRAGARD L. DE L INFLUENCE DES ZONES ELOIGNEES SUR LA HAUTEUR DE LA MAREE TERRESTRE.
3E SYMP. MAREES TERRESTRES TRIESTE
BOLL.GEOFISICA TEORICA E APPL.VOL.2 NO 5 PP 190-194 1960

5.111
LENNON G.W. THE DEVIATION OF THE VERTICAL AT BIDSTON IN RESPONSE TO THE ATTRACTION OF OCEAN TIDES.
GEOPH.J.OF THE ROY.ASTRON.SOCIETY VOL.6 NO 1 PP 64-84 1961

5.111
ROCHOLL W. UNTERSUCHUNGEN ZUR ATTRAKTIONSWIRKUNG DER MEERES-

GEZEITEN AUF DIE ERDGEZEITENREGISTRIERUNGEN IN EUROPA
GEODAT.KOMM.REIHE C - HEFT NO 203 PP 3-109 1974

5.112
BOUSSINESQ.J.
EQUILIBRE D ELASTICITE D UN SOL ISOTROPE SANS PESANTEUR SUPPORTANT DIFFERENTS POIDS.
C.R.ACAD.SCI.PARIS 86 PP 1260-1263 1878

5.112
LAMB H.
ON BOUSSINESQ S PROBLEM.
PROC.LONDON MATH.SOC.34 PP 276-284 1902

5.112
TERAZAWA K.
ON PERIODIC DISTURBANCE OF LEVEL ARISING FROM THE LOAD OF NEIGHBOURING OCEANIC TIDES.
PROC.LONDON PHIL.SOC.VOL.217 PP 35-50 1916

5.112
LAMB H.
ON THE DEFLECTION OF THE VERTICAL BY TIDAL LOADING OF THE EARTH S SURFACE.
PROC.ROY.SOC.A 93 PP 293-312 1917

5.112
SEKIGUCHI R.
ON THE TILTING OF THE EARTH AT JINSEN /CHEMULPO/ DUE TO TIDAL LOAD.
MEMOIRS OF IMP.MARINE OBS.KOBE VOL.1 NO 1 PP 1-17 1922

5.112
TAKAHASI R.
TILTING MOTION OF THE EARTH CRUST CAUSED BY TIDAL LOADING
BULL.EARTHQU.RES.INST.TOKYO IMP.UNIV.VOL.VI PP 85-108 1929

5.112
SEZAWA K.
NISHIMURA G.
ELASTIC EQUILIBRIUM OF A SPHERICAL BODY UNDER SURFACE TRACTIONS OF A CERTAIN ZONAL AND AZIMUTHAL DISTRIBUTION.
BULL.EARTHQU.RES.INST.TOKYO IMP.UNIV.VOL.VI PP 47-62 1929

5.112
SEZAWA K.
KANAI K.
ON THE ELASTIC DEFORMATION OF A STRATIFIED BODY SUBJECTED TO VERTICAL SURFACE LOADS.
BULL.EARTHQU.RES.INST.TOKYO IMP.UNIV.VOL.XV PP 359-369 1937

5.112
JOBERT G.
INFLUENCE DU RELIEF SUR LES DEFORMATIONS DE LA TERRE DUES A DES CHARGES SUPERFICIELLES.
C.R.ACAD.SCI.PARIS 239 NO 23 PP 1653-1655 1954

5.112
JOBERT G.
DEFORMATION PLANE D UN SOLIDE ELASTIQUE ISOTROPE ET HETEROGENE.
C.R.ACAD.SCI.PARIS 244 PP 555-558 1957

5.112
TOMASCHEK R.
MEASUREMENTS OF TIDAL GRAVITY AND LOAD DEFORMATIONS ON UNST /SHETLAND/.
OBS.ROY.BELG.COMM. NO 114 S.GEOPH. NO 39 PP 77 1957

5.112
JOBERT G.
INFLUENCE DU RELIEF DANS LES DEFORMATIONS DE LA CROUTE DUES A DES CHARGES SUPERFICIELLES.
OBS.ROY.BELG.COMM. NO 142 S.GEOPH. NO 47 PP 170-171 1958

5.112
SLICHTER L.B.
CAPUTO M.
DEFORMATION OF AN EARTH-MODEL BY SURFACE PRESSURES.
B.I.M.NO 22 PP 427-442 1960

5.112
SLICHTER L.B.
CAPUTO M.
DEFORMATION OF AN EARTH-MODEL BY SURFACE PRESSURES.
IST.GEOD.TOPOGR.E GEOF.UNIV.TRIESTE PUBL.NO 51
JOURN.GEOPH.RES.VOL.65 NO 12 PP 4151-4156 1960

5.112
CAPUTO M.
DEFORMAZIONI DI UN MODELLO DELLA TERRA CAUSATE DA DISTRIBUZIONI SUPERFICIALI DI MASSE GRAVITAZIONALI.
IST.GEOD.TOPOGR.E GEOF.UNIV.TRIESTE PUBL.NO 59
ATTI X ASSOC.GEOF.ITALIANA ROMA PP 157-165 1960

5.112
CAPUTO M.
DEFORMATIONS OF A LAYERED EARTH BY SURFACE FORCES AND BODY FORCES.
4E SYMP.MAR.TERR.COMM.OBS.R.BELG.188 S.GEOPH.58 PP 94-95 1961

5.112
CAPUTO M.
DEFORMAZIONI DI UN MODELLO DELLA TERRA CAUSATE DA MASSE ASSISIMMETRICHE.
IST.GEOD.UNIV.TRIESTE PUBL.NO 62 PP 27-44 1961

5.112
CAPUTO M. ELASTOSTATICA DI UNE SFERA STRATIFICATA E SUE DEFORMAZIONI CAUSATE DA MASSE SUPERFICIALI.
ANNALI GEOFISICA XIV NO 4 PP 363-378 1961

5.112
CAPUTO M. DEFORMATION OF A LAYERED EARTH BY AN AXIALLY SYMMETRIC SURFACE MASS DISTRIBUTION.
JOURN.GEOPH.RES.66 NO 5 PP 1479-1483 1961

5.112
CAPUTO M. TABLES FOR THE DEFORMATION OF AN EARTH MODEL BY SURFACE MASS DISTRIBUTION.
JOURN.GEOPH.RES.67 NO 4 PP 1611-1616 1962

5.113
MATHER R.S.
BRETREGER K. AN EXPERIMENT TO DETERMINE RADIAL DEFORMATION OF EARTH TIDES IN AUSTRALIA BY OCEAN TIDES
UNISURV G NO 23 PP 42-59 1975

5.114
ROSENHEAD L. THE ANNUAL VARIATION OF LATITUDE.
MTLY.NOT.R.ASTR.S.GEOPH.SUPPL.VOL.2 NO 3 PP 140-170 1929

5.12
LONGMAN I.M. A GREEN S FUNCTION FOR DETERMINING THE DEFORMATION OF THE EARTH UNDER SURFACE MASS LOADS. 1, THEORY.
J.GEOPH.RES.67 PP 845-850 1962

5.12
LONGMAN I.M. A GREEN S FUNCTION FOR DETERMINING THE DEFORMATION OF THE EARTH UNDER SURFACE MASS LOADS - 2- COMPUTATIONS AND NUMERICAL RESULTS.
JOURN.GEOPH.RES.68 NO 2 PP 485-496 1963

5.12
BLUM P.A.
CORPEL J.
JOBERT G.
NIERES J. DETERMINATION DES PROPRIETES ELASTIQUES DU MANTEAU A L AIDE DE CHARGES DUES AUX MAREES OCEANIQUES.
GEOPH.JOURN.R.ASTR.SOC.14 PP 219-224 196

5.12
FARREL W.E. GLOBAL CALCULATIONS OF TIDAL LOADING
NATURE NO 238, PP 43-44 197

5.12
FARREL W.E. DEFORMATION OF THE EARTH BY SURFACE LOADS
REV.GEOPHYS.SPACE PHYS.,NO 10, PP 761-797 197

5.12
FARRELL W.E. EARTH TIDES, OCEAN TIDES AND TIDAL LOADING.
MEASUREM., INTERPRETAT., CHANGES OF STRAIN IN THE EARTH
PHIL.TRANS.ROY.SOC.SER.A, VOL.274, NO 1239 PP 253-260 197

5.13
DOODSON A.T.
CORKAN R.H. LOAD TILT AND BODY TILT AT BIDSTON.
MTLY.NOT.R.ASTR.S.GEOPH.SUPPL.VOL.3 NO 6 PP 203-212 193

5.13
EGEDAL J. ON A PSEUDO-ITERATING METHOD FOR REDUCTION OF OBSERVATIONS OF EARTH-TIDE FOR EFFECTS FROM TIDES OF THE SEA.- WITH AN APPLICATION TO OBSERVATIONS MADE AT BERGEN /NORWAY/ 1934.
JOURN.ATMOSPH.TERR.PHYS. SOUS PRESSE 193

5.13
CORKAN R.H. THE ANALYSIS OF TILT RECORDS AT BIDSTON.
MTLY.NOT.R.ASTR.S.GEOPH.SUPPL.VOL.4 NO 7 PP 481-497 193

5.13
CORKAN R.H. A DETERMINATION OF THE EARTH TIDE FROM TILT OBSERVATIONS AT TWO PLACES.
MONTHLY NOT.R.ASTR.SOC.GEOPH.SUPPL.VOL.6 NO 7 PP 431-441 195

5.13
MELCHIOR P.J. APPLICATION DE LA METHODE DE CORKAN A LA DISCUSSION DES OBSERVATIONS DES MAREES TERRESTRES A FREIBERG /SAXE/
ACAD.ROY.BELG.BULL.CL.SC.XL PP 382-388
OBS.ROY.BELG.COMM. NO 74 S.GEOPH. NO 30 19

5.13
MELCHIOR P.J. DISCUSSION DU PROCEDE DE CORKAN POUR LA SEPARATION DES EFFETS DIRECTS ET INDIRECTS.
B.I.M. NO 3 PP 56-60 19

5.13
MELCHIOR P.J. DISCUSSION DU PROCEDE DE CORKAN POUR LA SEPARATION

DES EFFETS DIRECTS ET INDIRECTS DANS LES MAREES TERRESTRES.
OBS.ROY.BELG.COMM. NO 115 S.GEOPH. NO 40 1957

5.13
TCHOUPROUNOVA O. SUR LES EFFETS INDIRECTS DANS LES MAREES TERRESTRES.
* B.I.M. N.55 SEPTEMBRE PP 2611-2625 1969

5.13
BOSSOLASCO M. L INFLUENZA DELLE MAREE OCEANICHE SULLE MAREE TERRESTRI
PANEVA A. NELLE STAZIONI CLINOGRAFICHE DI GENOVA, ROBURENT
CICCONI G. E TOIRANO.
EVA C. GEOFISICA E METEOR. XVIII, 5-6, 117-134. 1969

5.13
BLUM P.A. ETUDE REGIONALE DE L INFLUENCE OCEANIQUE SUR L INCLI-
HATZFELD D. NAISON, PREMIERS RESULTATS A LA STATION DE MOULIS.
6 EME SYMPOSIUM MAREES TERRESTRES STRASBOURG
OBS.ROY.BELG.COMM. A9 S.GEOPH. NO 96 PP 102-105 1970

5.15
VILLAIN C. CARTES DES LIGNES COTIDALS DANS LES OCEANS
ANN.HYDROG. /PARIS/ VOL.3 PP 269-388 1952

5.15
TIMONOV V.V. SUR L ANALYSE CINEMATIQUE DES MAREES - REMARQUES
TRAV.INST.OCEANOGRAPHIQUE DE L ETAT-PUBL.37 PP 185-204 1959

5.15
BOGDANOV K.T. SOLUTION NUMERIQUE DES EQUATIONS HYDRODYNAMIQUES DES
KIM K.V. MAREES SUR ORDINATEUR ELECTRONIQUE BESM.2 POUR
MAGARIK V.A. L EQUATEUR DE L OCEAN PACIFIQUE.
ACAD.SC.URSS.TRAV.INST.OCEANOLOGIE TOME 75 PP 73-87 1964

5.15
MELCHIOR P.J. ANALYSE ~~HARMONIQUE DE VINGT ANNEES D ENREGISTREMENTS~~
PAQUET P. DE MAREES OCEANIQUES A OSTENDE.
VAN CAUWENBERGHE C. BULL.CL.SC.ACAD.ROY.BELG.LIII 1967, 2
OBS.ROY.BELG.SER.B COMM.NO 20 S.GEOPH.NO 82 PP 123-130 1967

5.15
PEKERIS C.L. SOLUTION OF LAPLACE S EQUATION FOR THE M2 TIDE IN THE
ACCAD. Y. WORLD OCEANS
PHIL.TRANS.ROY.SOC.SER.A, VOL.265 PP 413-436 1969

5.15
ZAHEL W. DIE REPRODUKTION GEZEITENBEDINGTER
BEWEGUNGSVORGANGE IN WELTOZEAN
MITTELS DES HYDRODYNAMISCHNUMERISCHEN VERFAHRENS
MITT.INST.MEERESKUNDE UNIV.HAMBURG VOL.17 1970

5.15
FILLOUX J. DEEP-SEA TIDES 1250 KILOMETERS OFF BAJA CALIFORNIA
SCIENCE V.169 N.3948 PP 862-864 1970

5.15
BOGDANOV K.T. SOLUTION NUMERIQUE DU PROBLEME DE LA PROPAGATION
MAGARIK V.A. DES ONDES DE MAREES SEMI-DIURNES /M2 ET S2/ DANS
L OCEAN MONDIAL
DOKLADI AKAD.NAOUK SSSR VOL.172 P 1315 1967
* B.I.M. NO 63 PP 3288-3292 1972

5.15
HENDERSHOTT M.C. THE EFFECTS OF SOLID EARTH DEFORMATION ON GLOBAL
OCEAN TIDES
GEOPHYS.J.R. ASTR.SOC. 29, PP 389-402 1972

5.15
DVORKIN YE.N. CALCULATION OF TIDAL MOTIONS IN ARCTIC SEAS
KAGAN B.A.
KLESHCHEVA G.P. IZV., ATMOSPH.AND OCEANIC PHYSICS V.8 NO 3 PP 298-306 1972

5.15
HENDERSHOTT M.C. OCEAN TIDES
EOS, VOL 54, 2, PP 76-86 1973

5.15
BOGDANOV K.T. NOUVELLES CARTES COTIDALES DES ONDES DE MAREES
NEFEDIEV V.P. DIURNES K1 ET O1 DES MERS AUSTRALO-ASIATIQUES.
RAPP.ACAD.SC.URSS.TOME 144 NO 5 PP 1034-1037 1962
* B.I.M. NO 66 MAI PP 3662-3647 1973

5.15
BOGDANOV K.T. MAREES DE L OCEAN PACIFIQUE.
AC.SC.URSS.TRAV.INST.OCEANOL.TOME 60 PP 142-160 1962
* B.I.M. NO 67 AOUT, PP 3727-3752 1973

5.15
GORDEEV R.G. KAGAN B.A. RIVKIND V.IA. — SOLUTION NUMERIQUE DES EQUATIONS DE LA DYNAMIQUE DES MAREES DANS L OCEAN MONDIAL
DOKLADI AKAD.NAOUK, SSSR, VOL.209, NO 2 PP 340-343 — 1973

5.15
MAXIMOV I.V. — MAREE LUNI-SOLAIRE A LONGUE PERIODE DANS L OCEAN MONDIAL
DOKLADI AKAD.NAOUK SSSR VOL.118 NO 5 PP 888-890 1958
* B.I.M. NO 66 PP 3668-3672 — 1973

5.15
VOROBIEV V.N. — MAREES LUNI-SOLAIRES SEMI-MENSUELLES ET MENSUELLES DANS LES MERS DE L ARCTIQUE SOVIETIQUE
DOKLADI AKAD.NAOUK SSSR VOL.167 NO 5 PP 1039-1041 1966
* B.I.M. NO 67 AOUT, PP 3710-3713 — 1973

5.15
BOULATOV L.V. TITOV V.B. — VARIATIONS SAISONNIERES DU NIVEAU MOYEN ET DES CONSTANTES HARMONIQUES DE LA MAREE A BARENTSBURG
INST.RECH.SC.ARCT.,ANTARCT.LENINGRAD T.269 PP 64-66,1966
* B.I.M. NO 67 AOUT, PP 3714-3717 — 1973

5.15
JACHENS R.C. KUO JOHN T. — THE O1 TIDE IN THE NORTH ATLANTIC OCEAN AS DERIVED FROM LAND-BASED TIDAL GRAVITY MEASUREMENTS
PROCEEDINGS 7TH INT.SYMP.EARTH TIDES, SOPRON 1973
AKADEMIAI KIADO BUDAPEST, PP 165-174 — 1976

5.15
ZAHEL W. — A GLOBAL HYDRODYNAMIC-NUMERICAL 1-MODEL OF THE OCEANTIDE THE OSCILLATION SYSTEM OF THE M2-TIDE AND ITS DISTRIBUTION OF ENERGY DISSIPATION, ANNALES DE GEOPHYSIQUE T.33 — 1977

5.15
BOGDANOV K.T. SEBERKIN B.Y. — GENERATION DES ONDES DE MAREES INTERNES ET INFLUENCE DES MAREES TERRESTRES SUR LES MOUVEMENTS DE MAREES DANS LES OCEANS
PHYS.ATMOSPH.OCEAN XII PP 539-544 MOSCOU 1976
* B.I.M. NO 75 PP 4361-4370 — 1977

5.18
BONTCHKOVSKII V.F. — DEFORMATIONS DE LA SURFACE TERRESTRE SOUS L INFLUENCE D ACTIONS EXTERIEURES
RAPP.ACAD.URSS T.60 NO 6 PP 981-984 — 1948

5.18
BONTCHKOVSKII V.F. — SUR QUELQUES PARTICULARITES DES VARIATIONS PERIODIQUES D INCLINAISON DE LA SURFACE TERRESTRE
BULL.DU CONSEIL DE SEISM.ACAD.SC.URSS NO 6 PP 135-138 — 1957

5.18
SIMON D. — FIRST RESULTS OF EARTH TIDAL OBSERVATIONS IN AIRTIGHT MEASURING BOXES
XV ASS.GEN. UGGI - ASS.INT.DE GEODESIE MOSCOU AOUT — 1971

5.18
SASAKI H. SATO N. — ON THE ANNUAL VARIATION OF THE GROUND TILT AT AKAGANE CRUSTAL MOVEMENT STATION.
PROC.OF THE INTERN.LAT.OBS.MIZUSAWA N.11 PP 81-87 — 1971

5.18
HOLLINGSWORTH A. — THE EFFECT OF OCEAN AND EARTH TIDES ON THE SEMI-DIURNAL AIR TIDE
J. ATMOS. VOL.28 PP 1021-1044 — 1971

5.18
SABURO MIYAHARA — THE EARTH TIDE CORRECTION FOR THE SURFACE BAROMETRIC LUNAR TIDE
JOURN.METEOR.SOC.JAPAN V.50 NO.4 PP 342-345 AUGUST — 1972

5.18
ZSCHAU J. — THE INFLUENCE OF AIR PRESSURE VARIATIONS ON TILT MEASUREMENTS WITH THE ASKANIA BOREHOLE PENDULUM AT THE STATION KIEL-REHMSBERG
PROCEEDINGS 7TH INT.SYMP.EARTH TIDES, SOPRON 1973
AKADEMIAI KIADO BUDAPEST, PP 779-795 — 1976

5.18
SCHNEIDER M.M. SIMON D. — INFLUENCE OF ATMOSPHERIC PRESSURE AND AIR TEMPERATURE FLUCTUATIONS ON TIDAL OBSERVATIONS OF GRAVITY IN CENTRAL ANTARCTICA
PROCEEDINGS 7TH INT.SYMP.EARTH TIDES, SOPRON 1973
AKADEMIAI KIADO BUDAPEST, PP 577-593 — 197

5.2
MELCHIOR P.J. CONCLUSIONS QUE L ON PEUT DEJA TIRER DES OBSERVATIONS DES MAREES TERRESTRES.
OBS.ROY.BELG.COMM.NO 215 S.GEOPH.NO 63 PP 361-371 1963
BOLL.GEOF.TRIESTE VOL.IV NO 16 1962

5.2
GROTEN E. UNTERSUCHUNGEN DER ZEITLICHEN VARIATION DES GRAVIMETERFAKTORS.
ZEITS.GEOPH.WURZBURG HEFT 4 PP 169-172 1963

5.2
MELCHIOR P.J.
PAQUET P. DISCUSSION DES DONNEES CONTENUES DANS LE CATALOGUE GENERAL DES RESULTATS D ANALYSES HARMONIQUES MENSUELLES DE MAREES TERRESTRES.
B.I.M.NO 32 PP 880-894 1963

5.2
BUCHHEIM W. BEMERKUNGEN UBER DIE STREUUNG DER FACTOREN GAMMA UND DELTA UND DIE DAZUGEHORENDEN PHASENWINKEL .
5 EME SYMP.INT.SUR LES MAREES TERRESTRES.
OBS.ROY.BELG.COMM.NO 236 S.GEOPH.NO 69 PP 381-385 1964

5.2
GROTEN E. ON THE CORRELATION OF GRAVITY WITH TIDAL ANOMALIES.
GEOPH.PROSP.VOL.XII NO 4 PP 434-439 OHIO ST.UNIV. 1964

5.2
MELCHIOR P.J. ANALYSE DE LONGS ENREGISTREMENTS DE MAREES TERRESTRES.
B.I.M. NO 46 PP 1862-1930 1966

5.2
MELCHIOR P.J. ANALYSES HARMONIQUES DE LONGUES SERIES D OBSERVATIONS PAR LA METHODE VENEDIKOV.
B.I.M. NO 48 PP 2133-2147 1967

5.2
MELCHIOR P.J. ANALYSES HARMONIQUES DE LONGUES SERIES D OBSERVATIONS PAR LA METHODE VENEDIKOV.
B.I.M. NO 51 PP 2386-2395 1968

5.2
AKSENTIEVA Z.N.
OSTROVSKI A.E.
MATVEYEV P.S. INCLINAISONS DE LA SURFACE DE LA TERRE D APRES LES OBSERVATIONS FAITES EN UNION SOVIETIQUE DE 1957 A 1967.
RAPP.SYMP.INTERN.LENINGRAD 1968 MESURES INCLINAISON
COMM.GEOPH.URSS.MOSCOU PP 36-47 1969
* B.I.M. N.55 SEPTEMBRE PP 2672-2680 1969

5.2
MATVEYEV P.S.
BALENKO V.G.
BOGDAN I.D. SUR LES OBSERVATIONS CLINOMETRIQUES SUR LES PROFILS EN UKRAINE.
RAPP.SYMP.INTERN.LENINGRAD 1968 MESURES INCLINAISON
COMM.GEOPH.URSS.MOSCOU PP 48-57 1969
* B.I.M. N.55 SEPTEMBRE PP 2681-2687 1969

5.2
MELCHIOR P. ANALYSES HARMONIQUES DE LONGUES SERIES D OBSERVATIONS PAR LA METHODE VENEDIKOV.
B.I.M. N.54 MAI P.2578 1969

5.2
HARRISON C. A PRELIMINARY REPORT ON TILT AND GRAVITY-TIDE MESUREMENTS IN THE POORMAN MINE NEAR BOULDER, COLORADO.
ESSA RES. LAB. TECH. MEM. ERLTM-ESL. AUGUST 1969

5.2
SLICHTER L.B.
HAGER C.L.
TAMBURRO M.B. THE LONG-PERIOD EARTH TIDE AT SOUTH POLE.
ANTARCTIC JOURNAL OF THE UNITED STATES 4 N.5 P 214 1969

5.25
OZAWA I. ON OBSERVATIONS OF THE TER-DIURNAL COMPONENT OF THE EARTH S TIDAL STRAIN.
SPEC.CONTR.GEOPH.INST.KYOTO UNIV.NO 4 1964

5.25
MELCHIOR P.J.
VENEDIKOV A. DERIVATION OF THE WAVE M3 /8H.279/ FROM THE PERIODIC TIDAL DEFORMATIONS OF THE EARTH.
PHYS.OF THE EARTH PLANET.INT.1 NO 6 PP 363-372 1968

5.25
BARSENKOV S.N. CALCUL DES MAREES DU TROISIEME ORDRE PAR LES OBSERVATIONS GRAVIMETRIQUES.
AC.SC.URSS PHYS. DE LA TERRE 5 PP 28-32 MOSCOU 1967
* B.I.M. N. 50 FEVRIER PP 2262-2269 1968

5.25
MELCHIOR P.J. DETERMINATION DE L ONDE M3 /RESUME/.
RAPP.SYMP.INTERN.LENINGRAD 1968 MESURES INCLINAISON
COMM.GEOPH.URSS.MOSCOU PP 240 1969

5.25
MELCHIOR P.J. RESULTATS OBTENUS POUR L ONDE M3.
6EME SYMP.MAREES TERRESTRES STRASBOURG.
OBS.ROY.BELG.COMM. A9 S.GEOPH. NO 96 PP 148-149 1970

5.25
KORBA P.S.
KORBA S.N. ONDE TER-DIURNE M3 DANS LES VARIATIONS DE MAREES
DE LA FORCE DE PESANTEUR EN CRIMEE POUR 1964 A 1971
ROT.& DEFORM.DE MAREES DE LA TERRE VOL.VI PP 58-61 KIEV 1974

5.28
LECOLAZET R.
STEINMETZ L. PREMIERS RESULTATS EXPERIMENTAUX CONCERNANT LA VARIATIONS
SEMI-MENSUELLE LUNAIRE DE LA PESANTEUR A STRASBOURG
C.R.ACAD.SCIENCES PARIS B TOME 263 PP 716-719 1966

5.28
SLICHTER L.B.
HAGER C.L. EARTH TIDE GRAVIMETERS AT THE SOUTH POLE.
UPPER MANTLE PROJ.U.S.PROG.REP.PP 131 1967

5.28
SLICHTER L.B.
HAGER C.L. OBSERVATIONS OF EARTH TIDES AND FREE OSCILLATIONS.
SIMP.PANAMERICANO DEL MANTO SUP.MEXICO 1968

5.28
SLICHTER L.B.
HAGER C.L.
TAMBURRO M.
O CONNEL R.V. THE FORTNIGHTLY EARTH TIDE AT SOUTH POLE.
WESTERN NAT.MEETING AMER.GEOPH.UNION SAN FRANCISCO 1968

5.28
VARGA P. EXAMINATION OF WAVE MF WITH REGISTERING GRAVIMETER
PAGEOPH V.97 PP 5-8 1972

5.28
BREIN R. BESTIMMUNG MONATLICHER UND HALBMONATLICHER
SCHWEREVARIATIONEN AUS EINER JAHRESREGISTRIERUNG
B.I.M. NO 63 PP 3275-3282 1972

5.3
MELCHIOR P.J. SUR L EFFET DE L OCEAN ATLANTIQUE DANS LA DIFFERENCE
DE PHASE DES ONDES M2 ET S2 CONSTATEE PAR LES
STATIONS CLINOMETRIQUES D EUROPE OCCIDENTALE.
B.I.M.NO 40 PP 1512-1518 1965

5.4
PAQUET P. PROGRAMME POUR LA COMPARAISON ET LA DISCUSSION D UN
ENSEMBLE DE RESULTATS D ANALYSES HARMONIQUES DE
MAREES TERRESTRES.
B.I.M. NO 29 PP 699-702 1962

5.4
PARIISKII N.N. THE REGIONAL HETEROGENEITY OF THE MANTLE AS
REVEALED BY EARTH TIDES OBSERVATIONS.
INST.TERR.PHYS.MOSCOW-B.I.M.NO 34 PP 1050-1054 1963

5.4
MATVEEV P.S. L ETUDE DES ANOMALIES DES INCLINAISONS DE MAREE DE LA
SURFACE TERRESTRE EN UKRAINE.
OBS.GRAV.ACAD.SC.UKRAINE -B.I.M.NO 34 PP 1055-1059 1963

5.4
MELCHIOR P.J. QUELQUES COMMENTAIRES SUR LES RESULTATS D OBSERVATIONS
EN BELGIQUE ET AU GRAND-DUCHE DE LUXEMBOURG.
5 EME SYMP.INT.SUR LES MAREES TERRESTRES.
OBS.ROY.BELG.COMM.NO 236 S.GEOPH.NO 69 PP 178-186 1964

5.4
PARIISKII N.N. THE REGIONAL HETEROGENEITY OF THE MANTLE AS REVEALED
BY EARTH-TIDE OBSERVATIONS.
TECTONOPHYSICS 1 /5/ PP 439-442 1964

5.4
MATVEYEV P.S. ANOMALIES DES INCLINAISONS DE MAREE DE LA SURFACE
DE LA TERRE A POLTAVA ET TSMAKOVO D APRES LES DONNEES
DES OBSERVATIONS DE 1958-59
OBS.GRAVIMETRIQUE POLTAVA AC. DES SC. DE URSS
GEOFISICA ASTR. INF. BIOULL. NO5 PP 25-32 KIEV 1963
* B.I.M. N.41 SEPTEMBRE PP 1543-1551 1965

5.4
HIERSEMANN L. AUFGABE UND ZIELSTELLUNG DER GEODATISCH-GEOPHYSIKALISCHEN SPEZIALLINIE AM ELBTALGRABEN. GEOL. 15 HEFT 1 AKAD.VERLAG BERLIN PP 19-26 1966

5.4
MELCHIOR P.J. SUR L HETEROGENEITE DE LA CROUTE TERRESTRE EN BELGIQUE MISE EN EVIDENCE PAR LES OBSERVATIONS DES MAREES TERRESTRES A REMOUCHAMPS. OBS.ROY.BELG.COMM. NO 10 S.GEOPH. NO 77 PP 1041-1046 1966

5.4
MELCHIOR P.J. OCEANIC TIDAL LOADS AND REGIONAL HETEROGENEITY IN WESTERN EUROPE. GEOPH.J.R.ASTR.SOC.14 PP 239-244 1967

5.4
CHEH PAN THE GRAVITATIONAL FACTOR FROM MID-CONTINENTAL BODY TIDES AND ITS STATISTICAL ANALYSIS. TECTONOPHYSICS - 9, PP 15-46 1970

5.4
CAPUTO M
PANZA G.F. GEOPHYSICAL CLASSIFICATION OF EARTH TILT. IST. DI FISICA , IST. DI GEOD. BOLOGNA ITALY B.I.M. N.59 OCTOBRE PP 2869-2873 1970

5.4
SUGAWA CH.
HOSOYAMA K.
NAKAI SH.
SATO T. OBSERVATION OF EARTH TIDES AROUND MIZUSAWA PROCEEDINGS OF THE INT.LAT.OBS.OF MIZUSAWA, PP 106-115 1972

5.4
MELCHIOR P. DISTRIBUTION DU FACTEUR GAMMA M2/EW/ EN EUROPE B.I.M. NO 64 P 3420 1973

5.4
WITTLINGER G.
LECOLAZET R. SUR LES OBSERVATIONS DE MAREE CLINOMETRIQUE DANS UN LONG TUNNEL BULL.GEOD.NLLE S. NO 109 SEPTEMBRE PP 293-300 1973

5.4
ROSENMAN M.
JIT SINGH S. STRESS RELAXATION IN A SEMI-INFINITE VISCOELASTIC EARTH MODEL BULL.SEISM.SOC.AMERICA V.63 NO 6 PP 2145-2154 1973

5.4
ROSENMAN M.
JIT SINGH S. QUASI-STATIC STRAINS AND TILTS DUE TO FAULTING IN A VISCOELASTIC HALF-SPACE BULL.SEISM.SOC.AMERICA V.63 NO 5 PP 1737-1752 1973

5.4
BLAIR D. TOPOGRAPHIC EFFECTS ON THE TIDAL STRAIN TENSOR G.J.R. ASTR.SOC. VOL 46 PP 127-140 1976

5.41
ESHELBY J.D. THE DETERMINATION OF THE ELASTIC FIELD OF AN ELLIPSOIDAL INCLUSION AND RELATED PROBLEMS PROC.ROY.SOC., SER.A, 241, P 376 1957

5.41
BREKKE TOR L.
JORSTAD F.A. LARGE PERMANENT UNDERGROUND OPENINGS UNIVERSITETSFORLAGET OSLO-BERGEN-TROMSO PP 1-372 1970

5.41
ADLER LAWRENCE
LIN CHUNG-HSING THE EFFECTS OF BOUNDARY DISCONTINUITIES ON STRESSES ABOUT UNDERGROUND OPENINGS LARGE PERMANENT UNDERGROUND OPENINGS UNIVERSITETSFORLAGET OSLO-BERGEN-TROMSO PP 93-104 1970

5.41
DAHL D.
VOIGHT B. ISOTROPIC AND ANISOTROPIC PLASTIC YIELD ASSOCIATED WITH CYLINDRICAL UNDERGROUND EXCAVATIONS LARGE PERMANENT UNDERGROUND OPENINGS UNIVERSITETSFORLAGET OSLO-BERGEN-TROMSO PP 105-110 1970

5.41
WANG YIH-JIAN
VOIGHT B. A DISCRETE ELEMENT STRESS ANALYSIS MODEL FOR DISCONTINUOUS MATERIALS LARGE PERMANENT UNDERGROUND OPENINGS UNIVERSITETSFORLAGET OSLO-BERGEN-TROMSO PP 111-115 1970

5.41
LECOLAZET R.
WITTLINGER G. SUR L INFLUENCE PERTURBATRICE DE LA DEFORMATION DES CAVITES D OBSERVATION SUR LES MAREES CLINOMETRIQUES C.R.ACAD.SC.PARIS, T.278, SERIE B 663-666 1974

5.41
SIMON D. — EIN BEITRAG ZUR DISKUSSION UBER TEKTONISCHE EINFLUSSE BZW. HOHLRAUMWIRKUNGEN AUF KLINOMETRISCHE GEZEITEN-PARAMETER
ZENTRALINSTITUT FUR PHYSIK DER ERDE MITT.NO 469
ZTSCHR.VERMESSUNGSTECHNIK, BERLIN PP 1-6 — 1974

5.41
ITSUELI U.J., BILHAM R.G., GOULTY N.R., KING G.C.P. — TIDAL STRAIN ENHANCEMENT OBSERVED ACROSS A TUNNEL
GEOPH. J.R. ASTR.SOC. 42 PP 555-564 — 1975

5.41
KING G., ZURN W., EVANS R., EMTER D. — SITE CORRECTION FOR LONG PERIOD SEISMOMETERS, TILTMETERS AND STRAINMETERS
XVI ASS.GEN.DE L UGGI GRENOBLE - AOUT — 1975

5.41
SIMON D. — A CONTRIBUTION TO THE DISCUSSION CORCERNING TECTONIC INFLUENCES OR CAVITY EFFECTS ON CLINOMETRIC TIDAL PARAMETERS
XVI ASS.GEN.DE L UGGI GRENOBLE - AOUT — 1975

5.41
HARRISON J.C. — CAVITY AND TOPOGRAPHIC EFFECTS IN TILT AND STRAIN MEASUREMENT
J.G.R., VOL 81, NO 2, PP 319-328, JANUARY 10 — 1976

5.41
BILHAM R.G., EMTER D., KING G.C.P. — STRAIN MEASUREMENTS IN ROCK CAVITIES
PROCEEDINGS 7TH INT.SYMP.EARTH TIDES, SOPRON 1973
AKADEMIAI KIADO BUDAPEST, P 641 — 1976

5.41
BLAIR DANE, SYDENHAM P. — A TIDAL STRAIN MODEL FOR HILLGROVE GORGE, EASTERN AUSTRALIA
G.J.R. ASTR.SOC. VOL 46 PP 141-153 — 1976

5.41
BLAIR D. — TOPOGRAPHIC, GEOLOGIC AND CAVITY EFFECTS ON THE HARMONIC CONTENT OF TIDAL STRAIN
GEOPHYS.J.R.ASTR.SOC. V.48 — 1977

5.42
TOMASCHEK R. — THE TIDES OF THE SOLID EARTH AND THEIR GEOPHYSICAL AND GEOLOGICAL SIGNIFICANCE.
NATURE 173 NO 4395 PP 143-145 23 JANVIER — 1954

5.42
JOBERT G. — INFLUENCE DE LA STRUCTURE DE LA CROUTE SUR LES DEFORMATIONS CAUSEES PAR LES MAREES OCEANIQUES.
ANN.GEOPH. 12 NO 4 PP 290-295 — 1956

5.42
JOBERT G. — INFLUENCE DE LA STRUCTURE DE LA CROUTE SUR LES DEFORMATIONS CAUSEES PAR LES MAREES OCEANIQUES.
B.I.M.NO 4 PP 66 — 1957

5.42
MATVEEV P.S. — UBER DIE UNSTIMMIGKEITEN IN DEM ERGEBNISSEN DER ERDGEZEITENBESTIMMUNGEN AUS NEIGUNGSBEOBACHTUNGEN AN STATIONEN, DIE AUF DEM EURASISCHEN KONTINENT GELEGEN SIND.
DOPOVIDI AKAD.NAUK.UKR.R S R 5 PP 466-469 — 1957

5.42
JOBERT G. — INFLUENCE DE LA STRUCTURE DE LA CROUTE SUR LES DEFORMATIONS CAUSEES PAR LES MAREES OCEANIQUES.
C.R.ACAD.SCI.PARIS 244 PP 227-230 — 1957

5.42
YOURKEVITCH O.Y. — FACTEUR ENDOGENE DANS LES INCLINAISONS DE LA SURFACE TERRESTRE
TRAV.INST.GEOL.ACAD.SC.UKRAINE SER.GEOPH.PUBL.2 PP 69-78 — 1958

5.42
YOURKEVITCH O.Y. — FACTEUR ENDOGENE DANS LES INCLINAISONS DE LA SURFACE TERRESTRE
THESE DE L AUTEUR DE CANDIDAT EN SCIENCES
PHYS.MATH. M7, 11 ACAD.SC.URSS INST.PHYS.TERRE — 1958

5.42
YOSHIKAWA K.
KIKUCHI S.
ON ANOMALOUS FLUCTUATION OF GROUND TILTING BY OCEANIC TIDAL LOAD RELATED WITH THE VOLCANIC ACTIVITY.
3E SYMP. MAREES TERRESTRES TRIESTE
BOLL.GEOFISICA TEORICA E APPL.VOL.2 NO 5 PP 80 1960

5.42
NISHIMURA E.
TANAKA Y.
TANAKA T.
ON A TILTMETER FOR THE OBSERVATION OF SECULAR GROUND TILTING.
3E SYMP. MAREES TERRESTRES TRIESTE
BOLL.GEOFISICA TEORICA E APPL.VOL.2 NO 5 PP 106-107 1960

5.42
BUCHHEIM W.
UBER DIE AUSWIRKUNG DES GEOTEKTONISCHEN ELBELINEAMENTES AUF DIE ERDGEZEITEN.
B.I.M.NO 22 PP 443-444 1960

5.42
JOBERT G.
LE MOUEL J.
TENSIONS THERMOELASTIQUES DUES A UNE INCLUSION RADIOACTIVE.
ANN.GEOPH.21 NO 3 PP 420-427 1965

5.42
DURNEY B.
STRESSES INDUCED IN A PURELY ELASTIC EARTH MODEL UNDER VARIOUS LOADS.
JOURN.GEOPH.10 NO 2 PP 163-173 1965

5.42
LIOUSTICH E.N.
MAGNITSKII V.A.
LES MOUVEMENTS VERTICAUX DE L ECORCE TERRESTRE ET LES VARIATIONS DU CHAMP GRAVITATIONNEL DUS A DES DEPLACEMENTS DE MASSES SOUS L ECORCE.
RECH.MAREES TERR.ART.NO 1 PUBL.ACAD. SC. URSS MOSCOW 1963
* B.I.M. N.41 SEPTEMBRE PP 1531-1535 1965

5.42
OSTROVSKII A.E.
BAKROUCHINE A.B.
MIRONAVA L.I.
ESSAI DE MESURE DES DEFORMATIONS TECTONIQUES A L AIDE DE CLINOMETRES DANS LA REGION DE DOUCHAMBRE
RECH.MAREES TERR. ART.NO1 PUBL. ACAD.SC.URSS MOSCOU 1963
* B.I.M. N.41 SEPTEMBRE PP 1536-1540 1965

5.42
PICHA J.
USE OF NON-TIDAL EFFECTS FROM RECORDS OF TIDAL OBSERVATIONS IN STUDYING RECENT CRUSTAL MOVEMENTS
STUDIA GEOPH.GEOD.PRAHA 10 PP 101-105 1966

5.42
KUO J.T.
EARTH DEFORMATION IN A TECTONICALLY INACTIVE AREA.
H.KRUMB SCHOOL & LAMONT GEOL.OBS.COLUMBIA UNIV.N-Y 1966

5.42
STARKOV V.I.
STARKOVA E.IA.
RESULTATS DES OBSERVATIONS CLINOMETRIQUES SUR LES FRACTURES /RESUME/.
RAPP.SYMP.INTERN.LENINGRAD 1968 MESURES INCLINAISON COMM.GEOPH.URSS.MOSCOU PP 239 1969

5.42
STARKOV V.I.
STARKOVA E.IA.
INFLUENCE DE LA FRACTURE SUR LA VALEUR Y D APRES LES OBSERVATIONS DE KONDARA
ROTATION ET DEFORM.MAREES DE LA TERRE,FASC.1
ACAD.SC.UKR.OBS.GRAV.POLTAVA KIEV PP 241-249 1970

5.42
WOOD M.D.
ALLEN R.V.
ALLEN S.S.
METHODS FOR PREDICTION AND EVALUATION OF TIDAL TILT DATA FROM BOREHOLE AND OBSERVATORY SITES NEAR ACTIVE FAULTS.
MEASUREM., INTERPRETAT., CHANGES OF STRAIN IN THE EARTH
PHIL.TRANS.ROY.SOC.SER.A, VOL.274, NO 1239 PP 245-252 1973

5.42
KING C.Y.
IOCHER D.
NASON R.D.
KINEMATICS OF FAULT CREEP
MEASUREM., INTERPRETAT., CHANGES OF STRAIN IN THE EARTH
PHIL.TRANS.ROY.SOC.SER.A, VOL.274, NO 1239 PP 355-360
ROYAL SOCIETY - LONDON - P 13 1973

5.42
POLLAK H.L.
TIDAL MODEL OF AN EARTH WITH A LATERAL PETROLOGICAL FACIES CHANGE
B.I.M. NO 66 PP 3621-3667 1973

5.42
BALAVADZE B.K.
BOLOKADZE R.D.
KARTVELISHVILI K.
OBSERVATIONS OF SECULAR DEFORMATIONS AND TILTS OF THE EARTH S SURFACE IN TBILISI
INST.GEOP.AC.OF SC. GEORGIAN SSR PP 3-11 TBILISI 1973

5.42
KOUTNII A.M.   INFLUENCE DES FRACTURES MERIDIENNES SUR LES
                      INCLINAISONS DE MAREES
                      ROT.& DEFORM.DE MAREES DE LA TERRE VOL.VI PP 88-92 KIEV 1974

5.42
KING G.C.P.   DETECTION OF ELASTIC STRAINFIELDS CAUSED BY
BILHAM R.G.   FAULT CREEP EVENTS IN IRAN
CAMPBELL J.W.
MCKENZIE D.P.
NIAZI M.   NATURE VOL.253 NO 5491 PP 420-423   1975
5.42
LATYNINA L.A.   INVESTIGATION OF FAULTS IN THE EARTH S CRUST
                      BY LINEAR TIDE STRAINS
                      PROCEEDINGS 7TH INT.SYMP.EARTH TIDES, SOPRON 1973
                      AKADEMIAI KIADO BUDAPEST, PP 559-568   1976

5.42
BACHEM H.CHR.   EARTH TIDE OBSERVATIONS USING HORIZONTAL PENDULUMS
TORGE W.   INFLUENCED BY LOCAL TECTONIC EFFECTS
                      PROCEEDINGS 7TH INT.SYMP.EARTH TIDES, SOPRON 1973
                      AKADEMIAI KIADO BUDAPEST, PP 453-460   1976

5.42
SIMON D.   ANALYSE DER NICHTPERIODISCHEN BODENDEFORMATIONEN
SCHNEIDER M.M.   AN DER ERDGEZEITENSTATION TIEFENORT 1958-1973
                      PROCEEDINGS 7TH INT.SYMP.EARTH TIDES, SOPRON 1973
                      AKADEMIAI KIADO BUDAPEST, PP 417-428   1976

5.45
TANAKA TORAO   OCEAN TIDES AND TIDAL TILTINGS OF THE GROUND OBSERVED
                      AT THE OURA AND AKIBASAN STATIONS, WAKAYAMA

5.45
MUGGE R.   DAS GRUNDWASSER ALS GEOPHYSIKALISCHER INDIKATOR.
                      ZEITSCHR.FUR GEOPHYSIK 20 H2 PP 65-74   1954

5.45
KIKKAWA K.   THE RELATION BETWEEN THE CYCLIC CHANGE IN GROUNDWATER
                      PRESSURE AND THE TILTING MOTION OF THE GROUND.
                      THE JAPANESE JL.OF LIMNOLOGY 17 NO 3 PP 91-99   1955

5.45
KIKKAWA K.   HYDROLOGICAL RESEARCH IN AMA-GUN, AICHI PREFECTURE II.
                      THE RELATION BETWEEN THE SEA WATER AND THE ARTESIAN
                      GROUNDWATER.
                      THE JAPANESE JL.OF LIMNOLOGY 18 NO 2 PP 67-78   1956

5.45
DIPRIMA R.C.   ON THE DIFFUSION OF TIDES INTO PERMEABLE ROCK OF
                      FINITE DEPTH.
                      QUART.APPLI.MATH. 15 4, 329   1958

5.45
WADATI K.   ON THE SINKING OF THE GROUND.
                      SCIENTIA LII, VOL.XCIII N.DLII PP 93-98 APRIL   1958

5.45
LEONTIEV G.Y.   LES CHARGES ATMOSPHERIQUES ET HYDROLOGIQUES TEMPORAIRES
                      SUR LA SURFACE DE LA TERRE ET LEUR INFLUENCE SUR LE
                      NIVELLEMENT DE HAUTE PRECISION.
                      RECENT CRUSTAL MOV. ART.1 PUBL.ACAD.SC.USSR MOSCOW 1963
                    * B.I.M. N.39 AVRIL PP 1365-1371   1965

5.5
SIMON D.   UBER DEN EINFLUSZ VON LUFTANDERUNGEN UND
                      MEERESGEZEITEN AUF DIE ERGEBNISSE DER HARMONISCHEN
                      ANALYSE VON HORIZONTALPENDELAUFZEICHNUNGEN.
                      B.I.M. NO 44 PP 1720-1746   1966

5.5
SIMON D.   SYSTEMATIC ERRORS OF EARTH-TIDE RESULTS DUE TO BLOCK
BUCHHEIM W.   TILTS INDUCED BY ATMOSPHERIC PRESSURE.
                      MITTEIL.INST.THEOR.PHYS.GEOPH.BERGAK.FREIBERG NO 49   1967

5.5
SIMON D.   ZUM AUFTRETEN LUFTDRUCKBEDINGTER STORUNGEN IN
SCHNEIDER M.   HORIZONTALPENDELAUFZEICHNUNGEN AUF DREI VERSCHIEDENEN
                      ERDGEZEITENSTATIONEN.
                      B.I.M. NO 49 PP 2218-2225   1967

5.5
SIMON D.   METHODEN DER STORPEGELANALYSE UND IHRE BEDEUTUNG FUR DIE

ERDGEZEITENFORSCHUNG.
ERDGEZEITEN SYMP.LENINGRAD PP 95-131 1968

5.6
EGEDAL J. MOTI LENTI DELLA CROSTA TERRESTRE E LORO DETERMINAZIONE.
GEOFISICA PURA E APPLICATA VOL.V FASC.3-4 1943

5.6
DECAE A. ON SOME MOVEMENTS OF THE GROUND IN GENEVA.
GEOPH.JOURN.R.ASTR.SOCIETY VOL.3 NO 1 PP 112-120 MARCH 1960

5.6
MELCHIOR P.J.
BROUET J. SUR LES DIVERSES PERTURBATIONS DANS L ENREGISTREMENT DES MAREES TERRESTRES.
5 EME SYMP.INT.SUR LES MAREES TERRESTRES.
OBS.ROY.BELG.COMM.NO 236 S.GEOPH.NO 69 PP 352-371 1964

5.61
SEZAWA K.
NISHIMURA G. MOVEMENT OF THE GROUND DUE TO ATMOSPHERIC DISTURBANCE IN A SEA REGION.
BUL.EARTHQ.RES.INST.TOKYO I.UNIV.V.IX NO 3 PP 291-309 1931

5.61
LETTAU H. LOTSCHWANKUNG UNTER DEM EINFLUSS VON GEZEITENKRAFTEN UND ATMOSPHARISCHEN KRAFTEN.
GERL.BEITR.ZUR GEOPH.BD. 51 PP 250-269 1937

5.61
LETTAU H. UBER DIE UNMITTELBARE EINWIRKUNG ATMOSPHARISCHER KRAFTE AUF DIE ERDKRUSTE.
METEOR.ZEITSCH. 54 PP 453-457 1937

5.61
LETTAU H. METEOROLOGISCH BEDINGTE ERDKRUSTENDEFORMATIONEN UND IHR NACHWEIS DURCH REGISTRIERUNGEN DES HORIZONTALDOPPELPENDELS.
FORSCH.FORTSCHR.16 PP 223-224 1940

5.61
TOMASCHEK R. NON-ELASTIC TILT OF THE EARTH S CRUST DUE TO METEOROLOGICAL PRESSURE DISTRIBUTIONS.
GEOFISICA PURA E APPLICATA 25 PP 17-25 1953

5.61
JOBERT G. DEFORMATIONS THERMIQUES.
OBS.ROY.BELG.COMM.NO 142 S.GEOPH.NO 47 PP 172-174 1958

5.61
MELCHIOR P.J. PERTURBATION GRAVIMETRIQUE REMARQUABLE OBSERVEE A BRUXELLES LORS DU PASSAGE D UN FRONT FROID ACCOMPAGNE D UNE VARIATION EXCEPTIONNELLE DE LA PRESSION BAROMETRIQUE.
B.I.M.NO 37 PP 1291-1295 1964

5.61
TANAKA T. STUDY ON METEOROLOGICAL AND TIDAL INFLUENCES UPON GROUND DEFORMATIONS.
SPECIAL CONTRIB. GEOPH. INST. KYOTO UNIV. N.9 PP 29-90 1969

5.61
OKHOTSIMSKAIA M.V.
KHOROMSKII A.V. SUR L INFLUENCE DES FACTEURS METEOROLOGIQUES SUR L ALLURE DES VARIATIONS DES INCLINAISONS DE LA SURFACE DE LA TERRE D APRES LES OBSERVATIONS A OBNINSK.
OBSERV.GEOPHYSIQUES COMPLEXES A OBNINSK PP 117-122 1974

5.61
WENZEL H.G. EINFLUSS VON LUFTDRUCK UND LUFTTEMPERATUR AUF GRAVIMETRISCHE ERDGEZEITENBEOBACHTUNGEN
MITTEILUNGEN INST. THEOR.GEODASIE, HANNOVER, PP 17 1974

5.61
BERGER J. A NOTE ON THERMOELASTIC STRAINS AND TILTS
J.G.R., VOL.80, NO 2, PP 274-277, JANUARY 10 1975

5.611
KABOUSENKO S.N. SUR LA THEORIE DES DEFORMATIONS DE TEMPERATURE DE LA SURFACE DE LA TERRE
IZV.ACAD.SC.URSS SER.GEOPH.NO 3 PP 445-449 1959

5.611
SCHNEIDER M. DER EINFLUSS DER LUFTTEMPERATUR AUF DIE BEOBACHTUNG DER LOTSCHWANKUNGEN IN BERGGIESSHUBEL.
B.I.M.NO 22 PP 445-455 1960

5.611
POPOV V.V. THERMAL DEFORMATIONS OF THE EARTH S SURFACE.
BULL./IZVESTIYA/ ACAD.SC.USSR GEOPH.SER.NO 7 PP 611-615 1960

5.611
TANAKA T. STUDY ON THE RELATION BETWEEN LOCAL EARTHQUAKES AND MINUTE GROUND DEFORMATION PART.3 ON EFFECTS OF DIURNAL AND SEMI-DIURNAL FLUCTUATIONS OF THE TEMPERATURE AND ATMOSPHERIC PRESSURE ON GROUND TILTS
BULL.DISAST.PREV.RES.INST.KYOTO UNIV. VOL.16 PP17-36 1967

5.611
OSTROVSKII A.E.
BAGMET A.L. SUR LA VALEUR DE L ONDE DE TEMPERATURE ANNUELLE DES INCLINAISONS DANS LES DIFFERENTES PROFONDEURS.
INST.PHYS.TERRE SCHMIDT OBS.GEO.OBNINSK-PUB.1 PP 148-164 1970

5.611
KORBA P.S. DE LA QUESTION DE L INFLUENCE DES VARIATIONS JOURNALIERES DE LA TEMPERATURE SUR LES RESULTATS DES OBSERVATIONS DES MAREES TERRESTRES
ROTATION ET DEFORM.MAREES DE LA TERRE,FASC.1
ACAD.SC.UKR.OBS.GRAV.POLTAVA KIEV PP 313-317 1970

5.611
BOGDAN Y. ONDE DE TEMPERATURE JOURNALIERE DANS LES INCLINAISONS DE LA SURFACE DE LA TERRE D APRES LES DONNEES DES OBSERVATIONS A DARIEVKA, LIKHOVKA, VELIKIE BOUDICHA ET SAMOTOEVKA
ROT.ET DEFORM.DE MAREES DE LA TERRE, VOL.5, PP 74-76 1973

5.611
BODGAN I.Y. ONDE DIURNE DE TEMPERATURE DANS LES INCLINAISONS DE LA SURFACE DE LA TERRE D APRES LES DONNEES DES OBSERVATIONS A DARIEVKA,LIKHOVKA,VELIKIE,BOUDICHA ET SAMATOEVKA.
ROTAT.ET DEFORMAT.DE MAREES DE LA TERRE-V-PP 76-85 1973

5.611
MATVEYEV P.S.
BOGDAN Y. SUR LA POSSIBILITE D ELIMINER L EFFET DE L ONDE DE TEMPERATURE JOURNALIERE PAR LES RESULTATS DE LA DETERMINATION DES ONDES LES PLUS IMPORTANTES DE LA MAREE TERRESTRE
ROT.ET DEFORM.DE MAREES DE LA TERRE, VOL.5, PP 85-94 1973

5.611
BODGAN I. YOU ONDE DIURNE DE TEMPERATURE DANS LES INCLINAISONS DE LA SURFACE DE LA TERRE D APRES LES DONNEES DES OBSERVATIONS A DARIEVKA,LIKHOVKA,VELIKIE,BOUDICHA ET SAMATOEVKA.
ROTATION ET DEF.DE MAREES DE LA TERRE V.5 PP 76-85 1973

5.611
BOGDAN Y.YOU RESULTATS DE L ETUDE DES CONDIIIONS DE TEMPERATURE DE L OBSERVATION DES INCLINAISONS AU POINT SOUDIEVKA
ROT.& DEFORM.DE MAREES DE LA TERRE VOL.VI PP 78-82 KIEV 1974

5.611
ANOKHINA K.M.
CHIROKOV J.A. QUELQUES RESULTATS DES OBSERVATIONS DES INCLINAISONS PERIODIQUES DE TEMPERATURE DE LA SURFACE DE LA TERRE
ROT.& DEFORM.DE MAREES DE LA TERRE VOL.VI PP 82-88 KIEV 1974

5.611
CHIROKOV I.A.
ANOKHINA K.M. INCLINAISONS LOCALES DE TEMPERATURE DE LA TEMPERATURE DE LA SURFACE DE LA TERRE
ACAD.NAOUK UKRAINSKOI SSR, OBS.GRAV.POLTAVA
ROTAT.DEFOR.MAREES DE LA TERRE VII PP 32-38 1975

5.611
MATVEEV P.S.
BOGDAN I.Y. SUR LA POSSIBILITE D ELIMINER L EFFET RESIDUEL DE L ONDE DIURNE DE TEMPERATURE A PARTIR DES RESULTATS DE LA DETERMINATION DES ONDES LES PLUS IMPORTANTES DE LA MAREE TERRESTRE
ROT.ET DEFORM.DE MAREES DE LA TERRE V. PP 85-94 1973
* B.I.M. NO 72 PP 4086-4100 1975

5.611
SHIROKOV I.A.
ANOKHINA K.M. LOCAL TEMPERATURE TILTS OF THE EARTH S SURFACE
PROCEEDINGS 7TH INT.SYMP.EARTH TIDES, SOPRON 1973
AKADEMIAI KIADO BUDAPEST, PP 592-606 1976

5.612
FIEDLER G. LAS ONDAS PARCIALES DE LA PRESION ATMOSFERICA EN CARACAS Y SUS RELACIONES CONLAS LLUVIAS.
BOL.ACAD.CIENCIAS FIS.MAT.NAT.TOMO XXIV NO 67 PP 42-56 1924

5.612
BONTCHKOVSKII V.F. L INCLINAISON DE LA SURFACE TERRESTRE ET LA DISTRIBUTION DE LA PRESSION ATMOSPHERIQUE
COMM.PRELIM.-METEOR.ET HYDROL.NO 2 PP 9-21 1938

5.612
BROUSSENTSOV G.V. ROLE DE LA PRESSION ATMOSPHERIQUE DANS LES VARIATIONS D INCLINAISON DE LA SURFACE TERRESTRE
INST.TCHERNOVITSK SER.GEOL.GEOGR.T.10 NO 3 PP 103-112 1953

5.612
KHOROCHEVA V.V. INFLUENCE DE LA PRESSION ATMOSPHERIQUE SUR INCLINAISON DE LA SURFACE TERRESTRE
IZV.ACAD.SC.URSS SER.GEOPH.NO 1 PP 131-135 1958

5.612
TOMASCHEK R. SCHWANKUNGEN TEKTONISCHER SCHOLLEN INFOLGE BAROMETRISCHER BELASTUNGANDERUNG.
FREIBERGER FORSCHUNGS.HEFT C.60 PP 35-55 1959

5.612
BONCHKOVSKII V.F. A SPECIFIC CASE OF CONNECTION BETWEEN THE VARIATIONS OF TILTS AND LINEAR DEFORMATIONS AND THE DISTRIBUTION OF ATMOSPHERIC PRESSURE.
BULL./IZVESTIYA/ ACAD.SC.USSR GEOPH.SER.NO 9 PP 832 1963

5.612
SIMON D. ZUM NACHWEIS LUFTDRUCKBEDINGTER KRUSTENBEWEGUNGEN MIT HILFE HORIZONTALPENDELN.
B.I.M.NO 40 PP 1486-1500 1965

5.612
OZAWA I. OBSERVATION OF THE ATMOSPHERIC TIDE EFFECTS ON THE EARTH S DEFORMAT&ONS
SP.CONTR.GEOPH.INST.KYOTO UNIV. VOL.7 PP133-142 1967

5.612
IVANOVA M.V.
PERTSEV B.P. EVALUATION DE L INFLUENCE DES VARIATIONS DE LA PRESSION ATMOSPHERIQUE SUR LES MAREES DE LA FORCE DE PESANTEUR.
METH.MES.MAR.TERR.& DEFORM.LENTES SURFACE TERRE
ACAD.SC.URSS.INST.PHYS.TERR.SCHMIDT MOSCOU PP 113-121 1970

5.612
SIMON D. LUFTDRUCKBEDINGTE SCHOLLENBEWEGUNGEN IN MITTELEUROPA UND IHRE BEDEUTUNG FUR DIE ERFORSCHUNG DER REZENTEN KRUSTENBEWEGUNGEN
ZTSCHR.GEOL.UND GEOPH.LEIPZIG SOUS-PRESSE 1971

5.612
OURASSINA Y.A. LIEN DES VARIATIONS A COURTE PERIODE NON REGULIERES DES INCLINAISONS DE LA SURFACE DE LA TERRE AVEC LA PRESSION ATMOSPHERIQUE D APRES LES OBSERVATIONS CLINO-METRIQUES DANS L OBSERVATOIRE ASTRONOMIQUE ENGELHARDT
IZVESTIA ASTR. ENGELHARDT OBS.NO 40 PP126-130 KAZAN 1973

5.612
IVANOVA M.V.
PERTSEV B.P. EVALUATION DE L INFLUENCE DES VARIATIONS DE LA PRESSION ATMOSPHERIQUE SUR LES MAREES DE LA FORCE DE PESANTEUR
METH.MES.MAR.TERR.& DEFORM.LENTES SURFACE TERRE
ACAD.SC.URSS.INST.PHYS.TERR.SCHMIDT MOSCOU PP113-121 1970
* B.I.M. MARS NO 65 PP 3481-3492 1973

5.612
SIMON D. DETERMINATION OF THE INFLUENCE QUANTITIES OF REGIONAL BAROMETRIC-PRESSURE LOAD BY PARTIAL CURVE FITTING
XVI ASS.GEN.DE L UGGI GRENOBLE - AOUT 1975

5.612
SIMON D. DEFORMATIONS OF UNDERGROUND CAVITIES UNDER THE INFLUENCE OF AIR PRESSURE VARIATIONS INSIDE THEM
XVI ASS.GEN.DE L UGGI GRENOBLE - AOUT 1975

5.614
STEINHAUSER F. UBER DIE ELASTISCHE DEFORMATION DER ERDKRUSTE DURCH LOKALE BELASTUNG MIT BESONDERER BERUCKSICHTIGUNG DER SCHNEEBELASTUNG DER ALPEN.
GERL.BEITR.GEOPH. 41 PP 466-478 1934

5.614
CALOI P. OSCILLAZIONI DEL MARE E PERTURBAZIONI DELLA VERTICALE APPARENTE NEL GOLFO DI TRIESTE DURANTE IL RAPIDO TRANSITARE DI ALCUNI CICLONI ATTRAVERSO L ALTO ADRIATICO.
ATTI R.INST.VENETO 95,PART 2 PP 461-475 1936

5.614
LETTAU H. LOTSCHWANKUNGEN AM GEBIRGSRAND ZUR ZEIT DER SCHNEESCHMELZE.
GERL.BEITR.GEOPH.54 PP 179-193 1939

5.614
ZADRO M. FLESSIONI DELLA CROSTA NELLA ZONA DINARICA INTORNO A TRIESTE PER SBALZI IMPULSIVI DELLA PRESSIONE ATMOSFERICA LOCALE.
ALTI XII CONV.ASSOC.GEOF.ITALIANA ROMA PP 1-18 1962

5.614
HOSOYAMA K.
GOTO T. ON GRAVITATIONAL WAVE OF ATMOSPHERIC PRESSURE AND RELATED GROUND TILTING
PROCEEDINGS OF THE INT.LAT.OBS.MIZUSAWA, PP 162-175 1972

5.7
DENISON F.N. HORIZONTAL PENDULUM MOVEMENTS IN RELATION TO LATITUDE VARIATION.
THE JOURNAL OF THE R.ASTR.SOC.OF CANADA XXXII PP 25-29 1938

5.7
PICHA J. ANNAHERNDE ANALYSE DES NULLPUNKTSGANGES DER HORIZONTALPENDEL AUS DEN JAHREN 1936-1939 FUR DIE GEZEITENSTATION BREZOVE HORY.
OBS.ROY.BELG.COMM.NO 114 S.GEOPH. NO 39 1957

5.7
PICHA J.
SKALSKY L. WEITERE ERGEBNISSE DES NULLPUNKTSGANGESSTUDIUM AN DER GEZEITENSTATION IN BREZOVE HORY.
OBS.ROY.BELG.COMM.142 S.GEOPH.47 PP 92-93 1958

5.7
PICHA J.
SKALSKI L. BEITRAG ZUM STUDIUM DES NULLPUNKTSGANGES DER HORIZONTALPENDEL.
STUDIA GEOPH.ET GEOD. 2 PP 243-260 1958

5.7
WIRT G. ALLURE DE L INCLINAISON ET DE SA CORRELATION POUR DES OBSERVATIONS PARALLELES.
RAPP.SYMP.INTERN.LENINGRAD 1968 MESURES INCLINAISON COMM.GEOPH.URSS.MOSCOU PP 155-161 1969

5.7
BAGMET A.L. MOUVEMENTS LENTS DE L ECORCE TERRESTRE D APRES LES OBSERVATIONS AVEC DES CLINOMETRES PARALLELES
METH.MES.MAR.TERR.& DEFORM.LENTES SURFACE TERRE
ACAD.SC.URSS.INST.PHYS.TERR.SCHMIDT MOSCOU PP 171-180 1970

5.7
LISSENKO G.M.
BOGDAN J.D.
MATVEYEV P.S. SUR LE MOUVEMENT DU ZERO DES CLINOMETRES A LA STATION DE VIELIKIE BOUDICHA
ROTATION ET DEFORMATIONS DE MAREES DE LA TERRE - NO 2
ACAD.NAOUK UKR.SSR - OBS.GRAV.POLTAVA - KIEV PP 120-124 1970

-5.7
BAGMET A.L. MOUVEMENTS LENTS DE L ECORCE TERRESTRE D APRES LES OBSERVATIONS AVEC DES CLINOMETRES PARALLELES
METH.MES.MAR.TERR.& DEFORM.LENTES SURFACE TERRE
ACAD.SC.URSS.INST.PHYS.T. SCHMIDT MOSCOU PP 171-180 1970
* B.I.M. N.62 PP 3223-3231 1971

5.7
BYL J.
CHOJNICKI T. ZUR FRAGE DER BESTIMMUNG VON GANGKORRELATIONEN BEI GEZEITENINSTRUMENTEN
VERMESSUNGSTECHNIK SOUS-PRESSE 1971

5.7
OZAWA I. OBSERVATIONS OF THE SECULAR AND ANNUAL CHANGES OF THE CRUSTAL STRAINS AT OSAKAYAMA
CONTRIBUTIONS, GEOPH.INST.NO 11 PP 205-212 1971

5.7
OZAWA IZUO OBSERVATIONS OF THE SECULAR AND ANNUAL CHANGES OF THE CRUSTAL STRAINS AT OSAKAYAMA
CONTR.GEOPH.INST. KYOTO UNIV. N.11 PP 205-212 1971

5.7
EVTOUCHENKO E.I. INCLINAISONS LENTES DE LA SURFACE DE LA TERRE AU POINT / TORES-1 / D APRES LES DONNEES DES OBSERVATIONS DE 1962 A 1966
ACAD.NAOUK UKRAINSKOI SSR, OBS.GRAV.POLTAVA
ROTAT.DEFOR.MAREES DE LA TERRE VII PP 26-30 1975

6.0
JEFFREYS H. THE EARTH.
5E.EDITION, CAMBRIDGE UN.PRESS, PP 525 1970

6.1
DARWIN G.H. NOTE ON THOMSON S THEORY OF THE TIDES OF

AN ELASTIC SPHERE
SCIENTIFIC PAPERS CAMBRIDGE UNIV.PRESS V.II PP33-35 1908
MESSENGER OF MATHEMATICS,VIII PP 23-26 1879

6.1
DARWIN G.H. A NUMERICAL ESTIMATE OF THE RIGIDITY OF THE EARTH.
NATURE 27 PP 22-23 1882

6.1
CHREE C. THE EQUATIONS OF AN ISOTROPIC ELASTIC SOLID IN POLAR AND
CYLINDRICAL COORDINATES,THEIR SOLUTION AND APPLICATION
TRANS. CAMBRIDGE PHIL.SOC. XIV PP 250-369 1889

6.1
CHREE C. ON THE STRESSES IN ROTATING SPHERICAL SHELLS
TRANS. CAMBRIDGE PHIL.SOC. XIV PP 467-483 1889

6.1
KELVIN LORD DYNAMICAL PROBLEMS REGARDING ELASTIC SPHEROIDAL
SHELLS AND SPHEROIDS OF INCOMPRESSIBLE LIQUID.
OSCILLATIONS OF A LIQUID SPHERE.
PHIL.TRANS.R.SOC.LONDON 153 1863
MATH.& PHYS.PAPERS VOL.III PP 351-386 CAMBRIDGE 1890

6.1
PRESCOTT J. ON THE RIGIDITY OF THE EARTH.
PHIL.MAG.22, PP 481-505 1911

6.1
LIANINE N. SUR LES DEFORMATIONS DE LA TERRE SOUS L INFLUENCE
DE L ATTRACTION LUNI-SOLAIRE EN LIAISON AVEC LES
METHODES DE DETERMINATION DE LA RIGIDITE DE LA TERRE
CAL.CERCLE AMATEURS PHYS.ET ASTRON.P 148 1912

6.1
ORLOV A.YA. DETERMINATION DE LA RIGIDITE MOYENNE DE LA TERRE A
PARTIR DES OBSERVATIONS FAITES A JURJEV, TOMSK
ET POTSDAM
IZV.SOC.GEOGR.RUSSE T.51 PP 479-487 1915

6.1
THOMSON W.
TAIT P.G. A TREATISE ON NATURAL PHILOSOPHY.
PAR.834-838 OXFORD 1867 2D.EDITION CAMBRIDGE 1923

6.1
KELVIN LORD
TAIT P.G. TREATISE ON NATURAL PHILOSOPHY 2 VOL.
CAMBRIDGE UNIV.PRESS 8E ED. 1923

6.2
HOUGH S.S. THE ROTATION OF AN ELASTIC SPHEROID.
PHILOS.TRANS.R.SOC.LONDON A.VOL.187 PP 319-344 1896

6.2
HERGLOTZ G. UBER DIE ELASTIZITAT DER ERDE BEI BERUCKSICHTIGUNG
IHRER VARIABLEN DICHTE.
ZEITSCH.FUR MATH.U.PHYS.52 PP 275-299 1905

6.2
SCHWEYDAR W. EIN BEITRAG ZUR BESTIMMUNG DES STARRHEITSKOEFFIZIENT
DER ERDE.
GERL.BEIT.Z.GEOPHYS.BD.9 PP 41-77 1908

6.2
LOVE A.E.H. THE YIELDING OF THE EARTH TO DISTURBING FORCES.
PROC.R.SOC.LONDON VOL.82 A PP 73-88 1909

6.2
LEIBENSON L.S. DEFORMATION ELASTIQUE D UNE SPHERE EN RELATION AVEC
LA QUESTION DE LA STRUCTURE DE LA TERRE
UNIV.EMP.MOSCOU P 123 1910

6.2
SCHWEYDAR W. THEORIE DER DEFORMATION DER ERDE DURCH FLUTKRAFTE.
VEROFF.PREUSS.GEOD.INST.N.F.66, PP 51 POTSDAM 1916

6.2
HOSKINS L.M. THE STRAIN OF A GRAVITATING SPHERE OF VARIABLE
DENSITY AND ELASTICITY.
TRANS.AMER.MATH.SOC.21 PP 1-43 1920

6.2
HOSKINS L.M. MEMORANDUM CONCERNING EARTH TIDES AND MOVEMENTS
OF THE VERTICAL DUE TO TIDAL FORCES.
BULL.GEOD.NO 1 PP 1-4 1924

6.2
JEFFREYS H. NOTE ON THE BODILY TIDE IN A FLUID EARTH.
MONTHLY NOT.R.ASTR.SOC.VOL.I NO 5 PP 159-160 1924

6.2
ROSENHEAD L. TIDES ON A TWO-LAYERED EARTH.
MONTHLY NOT.R.ASTR.SOC.GEOPH.SUPPL.VOL.2 PP 171-196 1927

6.2
PREY A. UBER DIE ELASTIZITAT DER ERDE I.
GERL.BEIT.Z.GEOPH.23 PP 379-429 1929

6.2
BOAGA G. RICERCHE SULLA RIGIDITA DI UN PIANETA SOLLECITATO
DA FORZE ESTERNE CON SPECIALI IPOTESI SULLA
DISTRIBUZIONE DELLA DENSITA.
ATTI.R.ISTITUTO VENETO DI SCIENCE XC, PP 597-647 1931

6.2
PREY A. UBER DIE ELASTIZITAT DER ERDE II.
GERL.BEIT.Z.GEOPH.44 PP 59-80 1935

6.2
HALES A.L. A WEAK LAYER IN THE MANTLE.
GEOPH.JOURN.ROY.ASTRON.SOC.VOL.4 PP 312-319 1961

6.2
TAKAGI S. EQUATIONS OF MOTION FOR THE EARTH TIDES.
PUBL.INT. LAT.OBS. MIZUSAWA VI N.2 PP 205-223 1968

6.2
MOLODENSKII M.S. THEORIE DES MAREES DANS LA TERRE ELASTIQUE EN TENANT
COMPTE DES TERMES DE L ORDRE DE L APLATISSEMENT
PHYSIQUE DE LA TERRE NO 1 PP 3-8 1974
* B.I.M. NO 69 PP 3904-3912 1974

6.3
TAKEUCHI H. ON THE EARTH TIDE OF THE COMPRESSIBLE EARTH OF
VARIABLE DENSITY AND ELASTICITY.
TRANS.AMER.GEOPH.UNION 31, NO 5 PP 651-689 1950

6.3
TAKEUCHI H. ON THE EARTH TIDE IN THE COMPRESSIBLE EARTH OF
VARYING DENSITY AND ELASTICITY
JL.OF FAC.OF SC.-UNIV.OF TOKYO VOL.VII PART 2 PP 1-153 1951

6.3
MOLODENSKII M.S. ELASTIC TIDES, FREE NUTATION AND SOME PROBLEMS OF
THE EARTH S STRUCTURE
ACAD.SC.URSS TRAV.INST.GEOPH.REC.ART.NO 19 P 146 1953

6.3
ALTERMAN Z.
JAROSCH H.
PEKERIS C.L. OSCILLATIONS OF THE EARTH
PROCEEDINGS ROY.SOC. A 252 PP 80-95 1959

6.3
TACHEUCHI H.
SAITO M.
KOBAYASHI N. STATICAL DEFORMATIONS AND FREE OSCILLATIONS OF A
MODEL EARTH.
JOURN.GEOPH.RES.67 NO 3 PP 1141-1154 1962

6.3
PARIISKII N.N. LES MAREES TERRESTRES ET LA STRUCTURE INTERNE DE LA TERRE
NOUVELLE ACAD. DES SC. URSS /SERIE GEOPH./ N.2 FEVR. 1963
* B.I.M. N.33 SEPTEMBRE PP 913-940 1963

6.3
KAULA W.M. ELASTIC MODELS OF THE MANTLE CORRESPONDING TO
VARIATIONS IN THE EXTERNAL GRAVITY FIELD
J.G.R. VOL 68 NO 17 PP 4967-4978 1963

6.3
ALSOP L.E.
KUO J.T. SEMI-DIURNAL EARTH TIDAL COMPONENTS FOR VARIOUS
EARTH MODELS.
5 EME SYMP.INT.SUR LES MAREES TERRESTRES.
OBS.ROY.BELG.COMM.NO 236 S.GEOPH.NO 69 PP 94-107 1964

6.3
ALSOP L.E.
KUO J.T. THE CHARACTERISTIC NUMBERS OF SEMI-DIURNAL EARTH
TIDAL COMPONENTS FOR VARIOUS EARTH MODELS
ANN.GEOPHYS.NO 20, PP 286-300 1964

6.3
MOLODENSKII M.S.
KRAMER M.V. LES NOMBRES DE LOVE POUR LES MAREES TERRESTRES STATIQUES
DES 2EME ET 3EME ORDRES.
ACAD.SC. URSS MOSCOU INST. PHYS. TERRE O.Y. SCHMIDT 1961
* B.I.M. N. 47 AVRIL PP 1935-1950 1967

6.3
MOLODENSKI M.S. DEPLACEMENTS DUS AUX MAREES DANS UNE TERRE ELASTIQUE
COMPTE TENU DES FORCES DE CORIOLIS

ISVESTIA ACAD.SC.URSS-PHYS.TERR.1970 NO 4 PP 102-107
* B.I.M. NO 61 PP 3002-3008 1971

6.3
TAKEUCHI H. SAITO M. — SEISMIC SURFACE WAVES
METHODS IN COMPUTATIONAL PHYSICS V.11,PP 217-302 1972

6.3
MELCHIOR P. — ON EARTH TIDE MODELS FOR THE REDUCTION OF HIGH PRECISION QUASI-RADIAL MEASUREMENTS
SYMP.EARTH S GRAV.FIELD & SEC.VARIAT.IN POSIT.PP 509-521 1973

6.3
SAITO M. — SOME PROBLEMS OF STATIC DEFORMATION OF THE EARTH
J.PHYS.EARTH VOL 22 PP 123-140 1974

6.3
SAITO M. — PARTIAL DERIVATIVES OF LOVE NUMBERS AND RELAXATION SPECTRA OF THE EARTH
GEOPHYS.FLUID DYNAMICS II PP 107-112 1975

6.3
WILHELM H. — BESTIMMUNG ELASTISCHER PARAMETER DES ERDINNERN DURCH ERDGEZEITEN
DEUTSCHE GEOD.KOM.REIHE C-HEFT NR 222 PP 5-68 1976

6.3
MOLODENSKII S.M. — VARIATION DES NOMBRES DE LOVE POUR UNE VARIATION DU MODELE STRUCTUREL DE LA TERRE
PHYSIQUE DE LA TERRE NO 2 PP 13-21 1976
* B.I.M. NO 74 PP 4293-4304 1976

6.31
KLEINSCHMIDT E. — ZUR FLUTBEWEGUNG DER FESTEN ERDKRUSTE.
ZEITSCHR.FUR GEOPH.9 PP 197-199 1933

6.31
TOMASCHEK R. SCHAFFERNICHT W. — DIE FLUT DER FESTEN ERDE.
ZEITSCHR.FUR GEOPH.IX PP 199-204 1933

6.31
KLEINSCHMIDT E. — ERWIDERUNG AUF DEN AUFSATZ VON R.TOMASCHEK UND W.SCHAFFERNICHT, DIE 6LUT DER FESTEN ERDE.
ZEITSCH.FUR GEOPH.IX PP 308-309 1933

6.31
TOMASCHEK R. SCHAFFERNICHT W. — BEMERKUNG HIERZU.
ZEITSCH.FUR GEOPH.9 PP 309 1933

6.31
MELCHIOR P.J. — SUR L INFLUENCE DE LA LOI DE REPARTITION DES DENSITES A L INTERIEUR DE LA TERRE DANS LES VARIATIONS LUNI-SOLAIRES DE LA GRAVITE EN UN POINT.
GEOFISICA PURA APPL.MILANO XVI FASC.3/4 PP 105-112
OBS.ROY.BELG.COMM. NO 19 S.GEOPH. NO 20 1950

6.31
MELCHIOR P.J. — NOUVELLES RECHERCHES THEORIQUES SUR LES MAREES DE L ECORCE ET LES VARIATIONS DES LATITUDES.
BULL.GEOD.NO 23 PP 59-66
OBS.ROY.BELG.COMM.NO 44 S.GEOPH.23 1952

6.31
JOBERT G. — MAREES TERRESTRES D UN GLOBE FLUIDE HETEROGENE.
ANN.GEOPH.TOME 8 NO 1 PP 106-111 1952

6.31
MOLODENSKII M.S. — DENSITE ET ELASTICITE DE L INTERIEUR DE LA TERRE
TRAV.INST.GEOPH.ACAD.SC.URSS NO 26, 153, PP 121-130 1955

6.31
BRAGARD L. — POTENTIEL DE DEFORMATION SUPERFICIELLE PRODUIT PAR UN BOURRELET DE MAREE. RELATIONS ENTRE LES NOMBRES DE LOVE HN ET KN
XV ASS.GEN. UGGI - ASS.INT.DE GEODESIE MOSCOU AOUT 1971

6.40
HOUGH S.S. — THE OSCILLATIONS OF A ROTATING ELLIPSOIDAL SHELL CONTAINING FLUID
PHIL.TRANS.ROYAL SOC.LONDON VOL.186, 1 PP 469-506 1895

6.40
SLOUDSKY TH. — DE LA ROTATION DE LA TERRE SUPPOSEE FLUIDE A SON INTERIEUR.
BULL.SOC.IMP.NAT.MOSCOU IX PP 285-318 1895

6.40
POINCARE H. — SUR LA PRECESSION DES CORPS DEFORMABLES.
BULL.ASTRON.VOL.27 PP 321-356 1910

6.40
BONDI H.
LYTTLETON R.A.
ON THE DYNAMICAL THEORY OF THE ROTATION OF THE EARTH
1. THE SECULAR RETARDATION OF THE CORE
PROC.CAMBRIDGE PHIL.SOC.VOL.44 PP 345-359 1948

6.40
JEFFREYS H.
THE EARTH CORE AND THE LUNAR NUTATION.
MONTHLY NOT.R.AST.VOL.108 NO 2 PP 206-209 1948

6.40
JEFFREYS H.
DYNAMIC EFFECTS OF A LIQUID CORE.
MONTHLY NOT.R.ASTR.SOC.VOL.109 NO 6 PP 670-687 1949
ET VOL.110 NO 5 PP 460-466 1950

6.40
BONDI H.
LYTTLETON R.A.
ON THE DYNAMICAL THEORY OF THE ROTATION OF THE EARTH
2. THE EFFECT OF PRECESSION ON THE MOTION OF THE LIQUID CORE
PROC.CAMBRIDGE PHIL.SOC.VOL.49 PP 498-515 1953

6.40
JEFFREYS H.
NOTE ON THE THEORY OF THE BODILY TIDE.
OBS.ROY.BELG.COMM.NO 100 S.GEOPH.NO 36 1956

6.40
JEFFREYS H.
VICENTE R.O.
THE THEORY OF NUTATION AND THE VARIATION OF LATITUDE.
MONTHLY NOT.R.ASTR.SOC.VOL.117 NO 2 PP 142-161 1957

6.40
JEFFREYS H.
VICENTE R.O.
THE THEORY OF NUTATION AND THE VARIATION OF LATITUDE
THE-ROCHE MODEL-CORE.
MONTHLY NOT.R.ASTR.SOC.VOL.117 NO 2 PP 162-173 1957

6.40
JEFFREYS H.
THEORETICAL VALUES OF THE BODILY TIDE NUMBERS.
B.I.M. NO 3 PP 29 1957

6.40
PEKERIS C.L.
JAROSCH H.
ALTERMAN Z.
DYNAMICAL THEORY OF THE BODILY TIDE IN THE EARTH.
3E SYMP. MAREES TERRESTRES TRIESTE
BOLL.GEOFISICA TEORICA E APPL.VOL.2 NO 5 PP 17-18 1960

6.40
MOLODENSKII M.S.
THE THEORY OF NUTATIONS AND DIURNAL EARTH TIDES.
IVME SYMP.INTERNATIONAL SUR LES MAREES TERRESTRES.
OBS.ROY.BELG.COMM.NO 188 S.GEOPH.58 PP 25-56 1961

6.40
MOLODENSKII M.
KRAMER M.V.
MAREES TERRESTRES ET NUTATION DE LA TERRE
ACAD.SC.URSS. INST.PHYS.TERRE MOSCOU 1961

6.40
JOBERT G.
THEORIE DE LA NUTATION ET DES MAREES DIURNES D APRES MOLODENSKY. -5 EME SYMP. INT. MAREES TERRESTRES
OBS.ROY.BELG.COMM.NO 236 S.GEOPH.NO 69 PP 64-91 1964

6.40
VICENTE R.O.
THE INFLUENCE ON THE CORE ON EARTH TIDES.
5 EME SYMP.INT.SUR LES MAREES TERRESTRES.
OBS.ROY.BELG.COMM.NO 236 S.GEOPH.NO 69 PP 40-60 1964

6.40
VICENTE R.O.
INVESTIGACOES RECENTES ACERCA DA INFLUENCIA DO NUCLEO NOS MOVIMENTOS DA TERRA.
GAZETA DE MATEMATICA LISBOA 100 PP 91-96 1965

6.40
JEFFREYS H.
VICENTE R.O.
COMPARISON OF FORMS OF THE ELASTIC EQUATIONS FOR THE EARTH
OBS.ROY.BELG.COMM.SER.B NO 3 SER.GEOPH.NO 72 1966
B.I.M. NO 43 PP 1611 1966

6.40
JEFFREYS H.
VICENTE R.O.
THE ENERGY OF ELASTIC STRAIN IN THE EARTH.
ACAD.ROY.BELG.BULL.CL.SC.5E SER.T.LIII PP 926-933 1967

6.40
MELCHIOR P.J.
GEORIS B.
EARTH TIDES, PRECESSION-NUTATION AND THE SECULAR RETARDATION OF EARTH S ROTATION
PHYS.OF THE EARTH PLANET.INT.1 4 PP 267-287 1968

6.40
VICENTE R.A.
A INFLUENCIA DA CONSTITUICAO INTERIOR DA TERRA NO VALOR DAS NUTACOES.
LISBOA, TIPOGRAFIA MATEMATICA LDA 1956
* B.I.M. N.53 PP 2489-2544 1969

6.40
SUESS S.T. SOME EFFECTS OF GRAVITATIONAL TIDES ON A MODEL EARTH CORE
J.GEOPH.RES.75, PP 6650-6661 1970

6.40
KAKUTA CH. THE EFFECT OF A COMPRESSIBLE CORE ON MOLODENSKY S THEORY OF EARTH TIDES
PUBL.ASTR.SOC.JAPAN VOL.22 NO 2 PP 199-222 1970

6.40
MELCHIOR P.J. PRECESSION-NUTATIONS AND TIDAL POTENTIAL
OBS.ROY.BELG. COMM.B NO 62 S.GEOPH 104
CELESTIAL MECHANICS VOL 4, NO 2, PP 190-212 1971

6.40
VICENTE R.O. A DEPENDENCIA DA MARE TERRESTRE DA ESTRUTURA DO NUCLEO DA TERRA.
REV.FAC.SC.LISBOA 2E SER.A VOL.IX FASC.1 PP 45-75 1962
* B.I.M. NO 61 PP 3017-3035 1971

6.40
MOLODENSKII M.S. LES MAREES DANS LA TERRE ELASTIQUE EN ROTATION AVEC UN NOYAU LIQUIDE.
MAREES TERR.STRUCTURE INTERNE TERRE PP 3-9 1967
* B.I.M. NO 60 PP 2979-2987 1971

6.40
VICENTE R.O. INFLUENCE OF THE CORE ON THE NUTATIONS
MANTELLO E NUCLEO NELLA FISICA PLANETARIA, SCUOLA INT. DI FISICA E.FERMI PP 17-26 ACADEMIC PRESS 1971

6.40
PEKERIS C.L.
ACCAD Y. DYNAMICS OF THE LIQUID CORE OF THE EARTH
PHIL.TRANS.ROY.SOC.SER.A, VOL.273, P 1233 1972

6.40
PEKERIS C.L. FINAL REPORT OF RESEARCH ON TIDES IN THE OCEAN - B. EFFECT OF THE LIQUID CORE OF THE EARTH ON THE BODILY TIDE
AMER.COM.WEIZMANN INST.OF SC. 515 PARK AVE. NEW YORK 1973

6.40
SMITH M.L. A PRELIMINARY REPORT OF NUMERICAL RESULTS FOR THE NEARLY DIURNAL LOVE NUMBERS OF THREE EARTH MODELS
COOPERATIVE INST. FOR RESEARCH IN ENVIRONMENTAL SC. UNIV.OF COLORADO/NOAA BOULDER PP 1-6 1974

6.40
WUNSCH C. SIMPLE MODELS OF THE DEFORMATION OF AN EARTH WITH A FLUID CORE-I
GEOPHYS. J.R. ASTR. SOC., 39, PP 413-419 1974

6.4
SHEN
PO-SHEN A THEORY OF TOROIDAL CORE OSCILLATIONS OF THE EARTH
G.J.R.ASTR.SOC.VOL 46. PP 307-318 1976

6.4
MOLODENSKII M.S. MAREES ET OSCILLATIONS PROPRES DE LA TERRE EN TENANT COMPTE DES FORCES DE CORIOLIS
PHYSIQUE DE LA TERRE NO 1 PP 3-12 1976
* B.I.M. NO 74 PP 4280-4292 1976

6.41
BRYAN G.H. THE WAVES ON A ROTATING LIQUID SPHEROID OF FINITE ELLIPTICITY
PHIL.TRANS.ROY.SOC.SER.A, VOL.180, PP 187-219 1888

6.41
BRYAN G.H. THE WAVES ON A ROTATING LIQUID SPHEROID OF FINITE ELLIPTICITY
PHIL.TRANS.ROY.SOC.SER.A, VOL.180, PP 187-219 1888

6.41
ELSASSER W.M. CAUSES OF MOTIONS IN THE EARTH S CORE
TRANS.AMER.GEOPH.UNION VOL.31 NO 3 PP 454-462 1950

6.41
NIGAM S.D. NOTE ON THE BOUNDARY LAYER ON A ROTATING SPHERE
ZEITSCHRIFT ANGEW. MATH. PHYS. V.5 PP 151-155 1954

6.41
FADNIS B.S. BOUNDARY LAYER ON ROTATING SPHEROIDS
ZEITSCHRIFT ANGEW. MATH. PHYS. V.5 PP 156-163 1954

6.41
PROUDMAN I. THE ALMOST-RIGID ROTATION OF VISCOUS FLUID BETWEEN

CONCENTRIC SPHERES
JOURN.FLUID MECHANICS V.1 PP 505-516 1956

6.41
HIDE R. THE HYDRODYNAMICS OF THE EARTH S CORE
PHYSICS AND CHEMISTRY OF THE EARTH, PERGAMON PRESS
VOL 1 PP 94-137 1956

6.41
STEWARTSON K. ROBERTS P.H. ON THE MOTION OF A LIQUID IN A SPHEROIDAL CAVITY OF A PRECESSING RIGID BODY I
JOURN.FLUID MECHANICS V.17 PP 1-20 1963

6.41
GREENSPAN H.P. ON THE TRANSIENT MOTION OF A CONTAINED ROTATING FLUID
J.FLUID MECH.,VOL.20, PART 4, PP 673-696 1964

6.41
GREENSPAN H.P. ON THE GENERAL THEORY OF CONTAINED ROTATING FLUID MOTIONS
J.FLUID MECH. VOL 22, PART 3, PP 449-462 1965

6.41
STEWARTSON K. ROBERTS P.H. ON THE MOTION OF A LIQUID IN A SPHEROIDAL CAVITY OF A PRECESSING RIGID BODY II
PROC. CAMBRIDGE PHIL.SOC. V.61 PP 279-288 1965

6.41
VENEZIAN G. FLOW IN A PRECESSING SPHERICAL ENCLOSURE,2, LOW VISCOSITY
REP.85-34, DIV.ENG.APPL.SCI.CALIF.INST.TECHNOL. 1966

6.41
PEARSON CARL E. A NUMERICAL STUDY OF THE TIME-DEPENDENT VISCOUS FLOW BETWEEN TWO ROTATING SPHERES
J.FLUID MECH. VOL 28, PART 2, PP 323-336 1967

6.41
BARCILON V. PEDLOSKY J. LINEAR THEORY OF ROTATING STRATIFIED FLUID MOTIONS
J.FLUID MECH. VOL 29, PART 1, PP 1-16 1967

6.41
BUSSE F.H. STEADY FLUID FLOW IN A PRECESSING SPHEROIDAL SHELL
J.FLUID MECH. VOL 33, PART 4, PP 739-751 1968

6.41
BUSSE F.H. BEWEGUNGEN IM KERN DER ERDE
ZEITSCH.GEOPH.BAND 37 PHYSICA-VERLAG WURZBURG PP 153-177 1971

6.41
MALKUS W.V.R. MOTIONS IN THE FLUID CORE
MANTELLO E NUCLEO NELLA FISICA PLANETARIA, SCUOLA INT. DI FISICA E.FERMI PP 38-63 ACADEMIC PRESS 1971

6.41
SMYLIE D.E. MANSINHA L. THE ELASTICITY THEORY OF DISLOCATIONS IN REAL EARTH MODELS AND CHANGES IN THE ROTATION OF THE EARTH
GEOPHYS.J.R.ASTR.SOC.NO 23 PP 329-354 1971

6.41
DAHLEN F.A. THE ELASTICITY THEORY OF DISLOCATIONS IN REAL EARTH MODELS AND CHANGES IN THE ROTATION OF THE EARTH
GEOPHYS.J.R.ASTR.SOC.NO 23 PP 355-358 1971

6.41
SMYLIE D.E. MANSINHA L. REPLY TO COMMENTS ON THE ELASTICITY THEORY OF DISLOCATIONS IN REAL EARTH MODELS AND CHANGES IN THE ROTATION OF THE EARTH
GEOPHYS.J.R.ASTR.SOC.NO 23 PP 359-360 1971

6.41
DAHLEN F.A. ON THE STATIC DEFORMATION OF AN EARTH MODEL WITH A FLUID CORE
GEOPH.J.R.ASTR.SOC.VOL 36 PP 461-485 1974

6.41
DENIS C. OSCILLATIONS DE CONFIGURATIONS SPHERIQUES AUTOGRAVITANTES ET APPLICATIONS A LA TERRE
UNIV.DE LIEGE,THESE DE DOCTORAT EN SC. 2 VOL, 359 P 1974

6.41
SMYLIE D.E. DYNAMICS OF THE OUTER CORE
VEROFF.ZENTRALINST.PHYS.ERDE BERLIN VOL 30 PP 91-104 1974

6.42
MELCHIOR P.J. DETERMINATION EXPERIMENTALE DES EFFETS DYNAMIQUES DU NOYAU LIQUIDE DE LA TERRE DANS LES MAREES TERRESTRES DIURNES.
AC.ROY.BELG.COMM.B NO 4 SER.GEOPH.73 PP 93-100 1966
B.I.M. NO 43 PP 1688 1966

6.42
TARADIA V.K. — SUR LE MOUVEMENT ANNUEL DES POLES DE LA TERRE AVEC UN NOYAU LIQUIDE. ASTR.JOURN.ACAD. NAOUK SSSR, T.XLII VOL. 6, PP 1277-1280 * B.I.M. N.43 MARS PP 1618-1623 1966

6.42
MELCHIOR P. — DIURNAL EARTH TIDES AND THE EARTH S LIQUID CORE OBS.ROY.BELG.COMM.B NO 11 S.GEOPH.NO 78 GEOPHYSICAL JOURNAL R.ASTR.SOC.V.12, PP 15-21 1966

6.42
LECOLAZET R. STEINMETZ L. — NOUVEAUX RESULTATS EXPERIMENTAUX CONCERNANT LES ONDES DIURNES DE LA MAREE GRAVIMETRIQUE XV ASS.GEN. UGGI - ASS.INT.DE GEODESIE MOSCOU AOUT 1971

6.42
KORBA P.S. — EFFETS DYNAMIQUES DU NOYAU LIQUIDE DE LA TERRE D APRES LES OBSERVATIONS DES VARIATIONS DE MAREES DE LA FORCE DE PESANTEUR A SIMFEROPOL ET YALTA ROT.ET DEFORM.DE MAREES TERRE VOL.3 PP 30-39 KIEV 1971

6.42
MATVEEV P.S. KORBA P.S. — L EFFET DYNAMIQUE DU NOYAU LIQUIDE DE LA TERRE D APRES LES DONNEES DES OBSERVATIONS DES MAREES TERRESTRES EN CRIMEE ET PRES DE KRIVOI ROG XV ASS.GEN. UGGI - ASS.INT.DE GEODESIE MOSCOU AOUT 1971

6.42
MATVEYEV P.S. BOGDAN Y.D. SLAVINSKAIA F.A. DATSENKO K.N. — LES ONDES DIURNES DANS LES INCLINAISONS DE LA SURFACE DE LA TERRE ENREGISTREES DANS LES POINTS DU PROFIL SOUMI-KHERSON EN 1964 A 1968 ROTATION ET DEFORMATION DE MAREES DE LA TERRE NO 4 ACAD.NAOUK UKR.SSR.-OBS.GRAV.POLTAVA-KIEV PP 75-82 1972

6.42
LECOLAZET R. STEINMETZ L. — NOUVEAUX RESULTATS EXPERIMENTAUX CONCERNANT LES ONDES DIURNES DE LA MAREE GRAVIMETRIQUE BULL.GEOD.NLLE S. NO 109 SEPTEMBRE PP 301-304 1973

6.42
BLUM P.A. HATZFELD D. WITTLINGER G. — RESULTATS EXPERIMENTAUX SUR LA FREQUENCE DE RESONANCE DUE A L EFFET DYNAMIQUE DU NOYAU LIQUIDE C.R. ACAD SC. PARIS 277, PP 241-244 1973

6.42
GRABER M.A. — RESONANCE EFFECTS IN POLAR MOTION MEASURABLE BY RADIO INTERFEROMETRY AND LASER RANGING J.G.P. VOL.79, NO 11, PP 1709-1710 1974

6.42
LECOLAZET R. STEINMETZ L. — SUR LES ONDES DIURNES DE LA MAREE GRAVIMETRIQUE OBSERVEE A STRASBOURG C.R.ACAD.SC.PARIS, T.278, SERIE B 295-297 1974

6.42
LECOLAZET R. STEINMETZ L. — SUR LA STRUCTURE FINE DU SPECTRE DIURNE DE LA MAREE GRAVIMETRIQUE XVI ASS.GEN.DE L UGGI GRENOBLE - AOUT 1975

6.42
LECOLAZET R. MELCHIOR P. — EXPERIMENTAL DETERMINATION OF THE DYNAMICAL EFFECTS OF THE LIQUID CORE OF THE EARTH INTERDISCIPLINARY SYMPOSIUM ON TIDAL INTERACTIONS XVI ASS.GEN.DE L UGGI GRENOBLE - AOUT 1975

6.42
USANDIVARAS J.C. DUCARME B. — ETUDE DE LA STRUCTURE DU SPECTRE DIURNE DES MAREES TERRESTRES PAR LA METHODE DES MOINDRES CARRES BULL.GEOD., VOL.50 NO 2, PP 139-157 1976

6.42
BLUM P. HATZFELD D. WITTLINGER G. — RESULTATS EXPERIMENTAUX SUR LA FREQUENCE DE RESONANCE DUE A L EFFET DYNAMIQUE DU NOYAU LIQUIDE PROCEEDINGS 7TH INT.SYMP.EARTH TIDES, SOPRON 1973 AKADEMIAI KIADO BUDAPEST PP 151-154 1976

6.42
LECOLAZET R. STEINMETZ L. — SUR LES ONDES DIURNES DE LA MAREE GRAVIMETRIQUE OBSERVEE A STRASBOURG PROCEEDINGS 7TH INT.SYMP.EARTH TIDES, SOPRON 1973 AKADEMIAI KIADO BUDAPEST PP 177-180 1976

6.42
DUCARME B. — ABOUT THE FINE STRUCTURE OF THE EARTH TIDES DIURNAL

MELCHIOR P. SPECTRUM
B.I.M. NO 75 PP 4399-4407 1977

6.43
JEFFREYS H. THE RIGIDITY OF THE EARTH CENTRAL CORE.
MONTHLY NOT.R.ASTR.SOC.GEOPH.SUPPL.VOL.1 NO 7 PP 371-383 1926

6.43
VICENTE R.O. THE ELASTICITY OF THE EARTH S OUTER CORE /1/.
REV.FAC.SC.LISBOA 2E SER.A VOL.VIII FASC.2 PP 277-286 1962

6.43
BACKUS G.E. KINEMATICS OF GEOMAGNETIC SECULAR VARIATION IN A
PERFECTLY CONDUCTING CORE
PHIL.TRANSAC.R.SOC.LONDON A.1141 VOL 263 PP 239-266 1968

6.43
JACOBS J.A. ENERGETICS OF THE EARTH S CORE
INTERN.CONFERENCE ON THE CORE-MANTLE INTERFACE,
MELBOURNE, FLORIDA 1972

6.43
JACOBS J.A. THE EARTH S CORE
ACADEMIC PRESS 253 P 1975

6.45
TISSERAND F. TRAITE DE MECANIQUE CELESTE TOME 2
GAUTHIER VILLARS PARIS 1891

6.45
WOOLARD E.W. THEORY OF THE ROTATION OF THE EARTH AROUND ITS CENTER
OF MASS
ASTRON.PAPERS AMER.EPHEM. V.XV PART I PP 11-165 1953

6.45
MOLODENSKII M.S. THE DIRECTIONS OF THE PRINCIPAL AXES IN THE STATE
OF STRESS CAUSED BY EARTH TIDES.
BULL.ACAD.SC.URSS GEOPH.SER.NO 10 PP 896 1963

6.45
MELCHIOR P.J. DEDUCTION DU PHENOMENE DE PRECESSION, NUTATION A PARTIR
DE LA MAREE TERRESTRE.
OBS.ROY.BELG.COMM.NO 240 S.GEOPH.NO 70 1965

6.45
FEDOROV E.P. SUR L ETUDE DU MOUVEMENT DE L AXE INSTANTANE
DE ROTATION DE LA TERRE.
10EME CONF.ASTR. UN. SOV. ACAD.SC.URSS LENINGRAD 1954
* B.I.M. N.42 DECEMBRE PP 1599-1603 1965

6.45
FEDOROV E.P. DETERMINATION DE L AMPLITUDE DU TERME SEMI-MENSUEL DE LA
NUTATION D APRES LES DONNEES DES OBSERVATIONS DE LATITUDE
CIRCULAIRE ASTRONOMIQUE N.116 2 JUILLET 1951
* B.I.M. N.39 AVRIL PP 1440-1441 1965

6.45
PHILIPPOV A.F. ESSAI DE DETERMINATION DE L ONDE LUNAIRE D ABERRATION
DANS LES VARIATIONS DE LATITUDE D APRES LES RESULTATS DES
OBSERVATIONS DES DEUX TELESCOPES ZENITHAUX A POLTAVA
DE 1948.8 A 1954.8.
CIRCULAIRE ASTR.NO 168 PP 14-16 1956
* B.I.M. N.41 SEPTEMBRE PP 1541-1542 1965

6.45
ORLOV A.IA. SUR LES FORMULES DE LA NUTATION EN DECLINAISON.
CIRCULAIRE ASTRONOMIQUE N.116 2 JUILLET 1951
* B.I.M. N.39 AVRIL P 1439 1965

6.45
ORLOV A.IA. CORRECTION DU TERME SEMI-MENSUEL DE LA NUTATION D APRES
LES OBSERVATIONS DE LA LATITUDE A POULKOVO 1915-1928
CIRCULAIRE ASTRONOMIQUE N.126 2 JUILLET 1951
* B.I.M. N.39 AVRIL PP 1442-1443 1965

6.45
EVTOUCHENKO E.I. L ONDE BIMENSUELLE DANS LES VARIATIONS DE LATITUDE
DE LA STATION DE UKIAH
CIRC.ASTRON.NO 113-114 MAI 1951
* B.I.M. NO 39 P 1438 1965

6.45
TARADIA V.K. SUR LA VALEUR DE LA CONSTANTE DE NUTATION POUR LA
TERRE ABSOLUMENT RIGIDE
ASTR.JOURNAL U.D.K. 525.35 PP 227-230 1966

6.45
MATVEYEV P.S. VARIATIONS DE NUTATIONS SEMI-MENSUELLES DE LA LATITUDE D APRES LES OBSERVATIONS DE POLTAVA DE 1949 A 1953. ASTR. TSIRKULIAR N. 143 PP 17-18 1953
* B.I.M. N. 43 MARS PP 1616-1623 1966

6.45
MATVEYEV P.S. LES VARIATIONS SEMI-MENSUELLES DE LA NUTATION EN LATITUDE D APRES LES OBSERVATIONS DE POLTAVA DE 1949 A 1953 CIRCULAIRE ASTR. N.143 PP 17-20 1953
* B.I.M. N.43 MARS PP 1677-1678 1966

6.45
POPOV N.A. SUR LES TERMES A COURTE PERIODE DE LA NUTATION DANS LES OBSERVATIONS DE POLTAVA DES ETOILES ZENITHALES BRILLANTES OBS. GRAV. DE POLTAVA T.IV ACAD.SC. URSS KIEV 1951
* B.I.M. N.43 MARS PP 1651-1673 1966

6.45
POPOV N.A. LE TERME SEMI-ANNUEL DE LA NUTATION TROUDI POLTAVSKOI GRAV. OBS. T.VIII PP 37-42 1959
* B.I.M. N.43 MARS PP 1679-1684 1966

6.45
MELCHIOR P.J. CONTRIBUTION APPORTEE PAR LES MAREES TERRESTRES DANS L ETUDE DE LA ROTATION DE LA TERRE. OBS.ROY.BELG.COMM.B NO 25 S.GEOPH.84 PP 71-76 1968

6.45
NESTEROV V.V. SUR LA DETERMINATION DES CORRECTIONS AUX COEFFICIENTS DU TERME SEMI-ANNUEL DE LA NUTATION EN DECLINAISON.
* B.I.M. N.52 DECEMBRE PP 2482-2485 1968

6.45
FEDOROV E.P. SUR LES FORCES D ACTION RECIPROQUE DU NOYAU ET DE L ENVELOPPE DE LA TERRE APPARAISSANT A CAUSE DE LA NUTATION
* B.I.M. N.55 SEPTEMBRE PP 2631-2636 1969

6.45
TAKAGI S. THEORY OF PRECESSION, NUTATION AND ROTATIONAL VELOCITY OF THE DEFORMABLE EARTH PUBL.INTERN.LATITUDE OBS.MIZUSAWA VOL.VII NO 1 PP 1-95 1969

6.45
WAKO Y. INTERPRETATION OF KIMURA S ANNUAL Z TERM PUBL.ASTR.SOC.JAPAN VOL.22 PP 525-544 1970

6.45
TARADY V.K. DETERMINATION DES TERMES PRINCIPAUX DE LA NUTATION D APRES LES DONNEES DES OBSERVATIONS DE LATITUDE.
* B.I.M. N.57 MAI PP 2778-2780 1970

6.45
WAKO Y. KIMURA S Z-TERM AND THE LIQUID CORE THEORY SYMP.I.A.U..ROTATION OF THE EARTH, MORIOKA PP 1 - 9 1971

6.45
PILNIK G.P. A CORRELATION ANALYSIS OF THE EARTH TIDES AND NUTATION ASTRONOMICHESKII ZHURNAL V.47 N.6 PP 1308-1323 1970
* SOVIET ASTRONOMY - AJ VOL.14,N.6 MAY JUNE PP 1044-1056 1971

6.45
YOKOYAMA K. NOTE ON ERRONEOUS COEFFICIENTS OF SEMI-ANNUAL SOLAR NUTATION TERMS PROC. INTERN.LATITUDE OBS.MIZUSAWA N.11 PP 115-117 1971

6.45
MC CLURE P. DIURNAL POLAR MOTION G.S.F.C. X-592-73-259, PP 1-109 1973

6.45
GOUBANOV V.S. DEFORMATIONS DE MAREES ET NUTATION DE DEUX SEMAINES DE LA TERRE D APRES LES RESULTATS DES OBSERVATIONS DE CINQ SERVICES DE L HEURE. ASTRON.JOURN.ACAD.SC.URSS.TOME 46 FASC.3 PP 671-684 1969
* B.I.M. MARS NO 65 PP 3493-3516 1973

6.45
YOKOYAMA K. STUDY ON SOME NUTATION TERMS BY THE ILS Z-TERM PUBL.INTERN.LAT.OBS.OF MIZUSAWA V.IX NO 1 PP 1-46 1973

6.45
MOCZKO J. ANALYSIS OF THE PERIODIC DEVIATIONS OF ZENITH POINT OF A NUMBER OF OBSERVATORIES - FROM OBSERVATIONAL PUBL.INST.GEOPH.POL.AC.SCI. VOL 94 PP 119-134 1975

6.45
BENDER P.L. FALLER J.E. STOLZ A. SILVERBERG E.C. MULHOLLAND J.D. SHELUS P.J. WILLIAMS J.G. CURRIE D.G. KAULA W.M.
EARTH ROTATION MEASURED BY LUNAR LASER RANGING
SCIENCE VOL.193 PPS 997-999 1976

6.45
FEISSEL M. GUINOT B.
PRELIMINARY STUDY OF THE 18.6 YEAR NUTATION IN POLAR MOTION AND UNIVERSAL TIME
MITTEILUNG DES LOHRMANN-OBSERVATORIUMS NR 33
WISS.Z.TECHN.UNIVERS.DRESDEN 25 H.4 PP 949-950 1976

6.45
HARRIS A.W. WILLIAMS J.G.
EARTH ROTATION STUDY USING LUNAR LASER RANGING DATA
SCIENTIFIC APPLICATIONS LUNAR LASER RANGING,
ASTROPHYSICS AND SPACE SC.LIB.VOL 62 PP 179-190 REIDEL 1976

6.45
WILLIAMS J.G.
PRESENT SCIENTIFIC ACHIEVEMENTS FROM LUNAR LASER RANGING
SCIENTIFIC APPLICATIONS LUNAR LASER RANGING,
ASTROPHYSICS AND SPACE SC.LIB.VOL 62 PP 37-50 REIDEL 1976

6.46
PARIISKII N.N.
SUR LA DECOUVERTE DE LA NUTATION DIURNE DE LA TERRE.
ASTRON.JOURN.ACAD.SC.URSS.TOME 40 FASC.3 PP 556-560 1963

6.46
POPOV N.A.
LA DETERMINATION DE LA NUTATION DIURNE A PARTIR DES OBSERVATIONS DES ETOILES BRILLANTES A POLTAVA.
ASS.BERKELEY DE L UGGI. OBS.GRAV.ACAD.SC.URSS.
B.I.M.NO 34 PP 1062-1065 1963

6.46
PARIISKII N.N.
RESONANCE BETWEEN EARTH-TIDES AND THE NEW DIURNAL NUTATION OF THE EARTH.
INST TERR.PHYS.MOSCOW-B.I.M.NO 34 PP 1047-1049 1963

6.46
POPOV N.
NUTATIONAL MOTION OF THE EARTH S AXIS.
NATURE 198 NO 4886 PP 1153 1963

6.46
VICENTE R.O. JEFFREYS H.
NEARLY DIURNAL NUTATION OF THE EARTH.
NATURE 204 NO 4954 PP 120-121 1964

6.46
POPOV N.
LA DETERMINATION DE LA PERIODE DE NUTATION LIBRE DIURNE DE LA TERRE EN PARTANT DES OBSERVATIONS DE LATITUDE A POLTAVA.
5 EME SYMP.INT.SUR LES MAREES TERRESTRES.
OBS.ROY.BELG.COMM.NO 236 S.GEOPH.NO 69 PP 447-449 1964

6.46
THOMAS D.V.
EVIDENCE FOR A NEARLY DIURNAL TERM IN THE NUTATION OF THE EARTH S AXIS.
NATURE VOL.201 NO 4918 PP 481 1964

6.46
KOULAGINE S.G. KOVBASIOUK L.D.
NUTATION LIBRE DIURNE D APRES LES OBSERVATIONS DE GORKI.
JOURNAL ASTR.AC.SC.URSS T.IXL, 4 MOSCOU PP 758-760 1964
* B.I.M. N.41 SEPTEMBRE PP 1552-1554 1965

6.46
KOVBASSIOUK L.D. KOULAGINE S.G.
TERME DIURNE D APRES LES OBSERVATIONS A GORKI.
ANALYSE DES RESULTATS DES OBSERVATIONS DES LATITUDES
ACAD.SC.UZBEK.SSR.-OBS.ASTR.TACHKENT PP 72-75 1966
* B.I.M. N.52 DECEMBRE PP 2478-2481 1968

6.46
DEBARBAT S.
NUTATION PRESQUE DIURNE ET TERMES PERIODIQUES DES COORDONNEES LOCALES.
ASTR.ET ASTROPHYS.1 NO 3 PP 334-355 1969

6.46
IATSKIV IA S.
NUTATION LIBRE DE LA TERRE D APRES LES OBSERVATIONS DE POULKOVO DE 1915 A 1928
* B.I.M. N.54 MAI PP 2548-2567 1969

6.46
DEBARBAT S.
NUTATION PRESQUE DIURNE ET TERMES PERIODIQUES DES COORDONNEES LOCALES.
6 EME SYMPOSIUM MAREES TERRESTRES STRASBOURG
OBS.ROY.BELG.COMM. A9 S.GEOPH. NO 96 PP 226 1970

6.46
POPOV N.A. SUR LES VARIATIONS DE L AMPLITUDE DE LA
IATSKIV IA.S. NUTATION LIBRE DIURNE DE LA TERRE.
* B.I.M. N.57 MAI PP 2734-2740 1970

6.46
IATSKIV IA. ON THE COMPARISON OF DIURNAL NUTATION DERIVED FROM SEPARATE SERIES OF LATITUDE AND TIME OBSERVATIONS.
SYMP.I.A.U.,ROTATION OF THE EARTH, MORIOKA 1971

6.46
POPOV N.A. AMPLITUDE VARIATIONS IN THE FREE DIURNAL
YATSKIV YA.S. NUTATION OF THE EARTH
ASTRONOMICHESKII ZHURNAL V.47 N.6 PP 1324-1327 1970
* SOVIET ASTRONOMY - AJ VOL14 N.6 MAY-JUNE PP 1057-1059 1971

6.46
CHOLLET F.C. ANALYSE DES OBSERVATIONS DE LATITUDE EFFECTUEES
DEBARBAT S. A L ASTROLABE DANJON DE L OBSERVATOIRE DE PARIS DE 1956.5 A 1970.8
ASTRON. & ASTROPHYS.18, PP 133-142 1972

6.46
OOE M. ON THE NEARLY DIURNAL FREE NUTATION
PUBL.INTERN.LAT.OBS.OF MIZUSAWA V.IX NO 1 PP 133-160 1973

6.46
OOE M. NOTE ON THE RETROGRADE SWAY
PUBL.INTERN.LAT.OBS.OF MIZUSAWA V.IX NO 1 PP 161-166 1973

6.46
TOOMRE A. ON THE -NEARLY DIURNAL WOBBLE- OF THE EARTH.
GEOPH.J.R.ASTR.SOC. VOL.38 PP 335-348 1974

6.46
ROCHESTER M.G. A SEARCH FOR THE EARTH S -NEARLY DIURNAL FREE WOBBLE-
JENSEN O.G.
SMYLIE D.E. GEOPH.J.R.ASTR.SOC. VOL 38 PP 349-363 1974

6.5
PEKERIS C.L. OSCILLATIONS OF THE EARTH.
JAROSCH H.
ALTERMAN Z. DEPT.APPLIED MATH.WEIZMANN INST.REHOVOT ISRAEL JANV. 1959

6.5
CAPUTO M. ELASTODINAMICA ED ELASTOTATICA DI UN MODELLO DELLA TERRA E SUE AUTOOSCILLAZIONI TOROIDALI.
BOLL.GEOF.TEOR.APPL.NO 10 PP 1-20 1961

6.5
CAPUTO M. LE OSCILLAZIONI TORSIONALI LIBERE DELLA TERRA.
ASSOC.GEOF.ITALIANA ROMA PUBL.NO 63 PP 45-50 1961

6.5
CAPUTO M. FREE MODES OF LAYERED OBLATE PLANETS.
JOURN.GEOPH.RES.VOL.68 NO 2 PP 497-500 1963

6.5
ALSOP L.E. FREE SPHEROIDAL VIBRATIONS OF THE EARTH AT VERY LONG PERIODS.
PART I - CALCULATION OF PERIODS FOR SEVERAL EARTH MODELS PP 483-501
PART II - EFFECT OF RIGIDITY OF THE INNER CORE.
BULL.SEISM.SOC.AMER.53,3 PP 503-515 1963

6.5
BLUM P.A. OBSERVATION DES VIBRATIONS PROPRES DE LA TERRE A
GAULON R. L AIDE D INCLINOMETRES.
JOBERT N. 5 EME SYMP.INT.SUR LES MAREES TERRESTRES.
OBS.ROY.BELG.COMM.NO 236 S.GEOPH.NO 69 PP 92 1964

6.5
NAKAGAWA I. FREE OSCILLATIONS OF THE EARTH OBSERVED BY A GRAVIMETER
MELCHIOR P.J. AT BRUSSELS.
TAKEUCHI H. 5 EME SYMP.INT.SUR LES MAREES TERRESTRES.
OBS.ROY.BELG.COMM.NO 236 S.GEOPH.NO 69 PP 108-121 1964

6.5
OZAWA I. ON THE OBSERVATIONS OF THE LONG PERIOD S OSCILLATIONS
ETO OF THE EARTH BY MEANS OF THE EXTENSOMETERS AND THE WATERTUBE TILTMETER.
BULL.DISASTER PREV.RES.INST.VOL.15 P.2 NO 93 PP 43-58 1965

6.5
WON I.J.
KUO J.T.
OSCILLATION OF THE EARTH S INNER CORE AND ITS RELATION TO THE GENERATION OF GEOMAGNETIC FIELD
J.G.R. VOL.78, NO 5, PP 905-911 1973

6.7
DARWIN G.H.
PROBLEMS CONNECTED WITH THE TIDES OF A VISCOUS SPHEROID
SCIENT.PAPERS CAMBRIDGE UNIV.PRESS V.II PP 140-194 1908
PHILOS.TRANSACT.OF THE ROY.SOC.PART II V.170 PP 539-593 1879

6.7
DARWIN G.H.
ON THE BODILY TIDES OF VISCOUS AND SEMI ELASTIC SPHEROIDS AND ON THE OCEAN TIDES UPON A YIELDING NUCLEUS.
SCIENTIFIC PAPERS II, NO 1 PP 1-32 1879

6.7
DARWIN G.H.
ON THE BODILY TIDES OF VISCOUS AND SEMI-ELASTIC SPHEROIDS,AND ON THE OCEAN TIDES UPON A YIELDING NUCLEUS
SCIENTIFIC PAPERS CAMBRIDGE UNIV.PRESS V.II PP 1-32 1908
PHILOS.TRANSACT.OF THE ROY.SOC.PART I V.170 PP 1-35 1879

6.7
DARWIN G.H.
ON THE PRECESSION OF A VISCOUS SPHEROID AND ON THE REMOTE HISTORY OF THE EARTH.
SCIENTIFIC PAPERS II, NO 3 PP 36-139 1879

6.7
JEFFREYS H.
THE VISCOSITY OF THE EARTH I.
MONTHLY NOT.R.ASTR.SOC.VOL.75 NO 8 PP 648-658 1914

6.7
JEFFREYS H.
THE VISCOSITY OF THE EARTH II.
MONTHLY NOT.R.ASTR.SOC.VOL.76 NO 2 PP 84-86 1915

6.7
JEFFREYS H.
THE VISCOSITY OF THE EARTH III.
MONTHLY NOT.R.ASTR.SOC.VOL.77 NO 5 PP 449-456 1917

6.7
JEFFREYS H.
THE VISCOSITY OF THE EARTH IV.
MONTHLY NOT.R.ASTR.SOC.GEOPH.SUPP.I NO 8 PP 412-424 1926

6.7
PREY A.
UEBER FLUTREIBUNG UND KONTINENTALVERSCHIEBUNG.
GERL.BEITR.GEOPH.XV, 4, PP 401-411 1926

6.7
HASKELL N.A.
THE MOTION OF A VISCOUS FLUID UNDER A SURFACE LOAD.
PHYSICS VOL.6 PP 265-269 1935
VOL.7 PP 56- 61 1936

6.7
BUCHHEIM W.
UMRISSE EINER PHANOMENOLOGISCHEN THEORIE DER ELASTISCHEN NACHWIRKUNG UND PLASTIZITAT ISOTROPER GESTEINE.
MITT.INST.THEOR.PHYS.GEOPH.BERGAKAD.FREIBERG NO 14 1958

6.7
MOLODENSKII M.S.
INFLUENCE DE LA VISCOSITE SUR LA PHASE DES MAREES TERRESTRES
BULL.ACAD.SC.URSS SER GEOPH. N.10 PP 1469-1482 1963
* B.I.M. N.39 AVRIL PP 1372-1384 1965

6.7
PEARSON C.F.
A NUMERICAL STUDY OF THE TIME-DEPENDENT VISCOUS FLOW BETWEEN TWO ROTATING SPHERES
J.FLUID MECH. VOL.28 PP 323-336 1967

6.7
SMITH S.W.
JUNGELS P.
PHASE DELAY OF THE SOLID EARTH TIDE.
PHYS. OF THE EARTH & PLANET. INTERIORS VOL2N4 PP233-238 1970

6.7
BODRI B.
EARTH TIDE PHASE LAG AND ANELASTICITY OF THE MANTLE
XVI ASS.GEN.DE L UGGI GRENOBLE - AOUT 1975

6.71
RANALLI G.
SCHEIDEGGER A.E.
RHEOLOGY OF THE TECTONOSPHERE AS INFERRED FROM SEISMIC AFTERSHOCK SEQUENCES.
ANNALI DI GEOFISICA VOL.XXII NO 3 PP 293-305 1969

6.71
SCHEIDEGGER A.E.
THE RHEOLOGY OF THE EARTH IN THE INTERMEDIATE TIME RANGE.
ANNALI DI GEOFISICA VOL.XXIII, NO 1 PP 27-43 1970

6.71
SCHEIDEGGER A.E.
ON THE RHEOLOGY OF ROCK CREEP.
ROCK MECHANICS NO 2 PP 138-145 1970

6.71
PELTIER W.R. THE IMPULSE RESPONSE OF A MAXWELL EARTH
REV.OF GEOPH.AND SPACE PHYSICS VOL 12, NO 4 PP 649-669 1974

6.8
DARWIN G.H. ON THE PRECESSION OF A VISCOUS SPHEROID, AND ON THE REMOTE HISTORY OF THE EARTH
SCIENT.PAPERS CAMBRIDGE UNIV.PRESS V.II PP 36-139 1908
PHILOS.TRANSACT.OF THE ROY.SOC.PART II V.170 PP 447-530 1879

6.8
DARWIN G.H. ON TIDAL FRICTION IN CONNECTION WITH THE HISTORY OF THE SOLAR SYSTEM.
NATURE VOL.23 PP 389-390 1881

6.8
DARWIN G.H. ON THE TIDAL FRICTION OF A PLANET ATTENDED BY SEVERAL SATELLITES,AND ON THE EVOLUTION OF THE SOLAR SYSTEM
PHIL.TRANSACT.ROY.SOCIETY PP 491-535 1881

6.8
DARWIN G. TIDAL FRICTION AND COSMOGONY.
SCIENTIF.PAPERS, VOL.II PP 516 CAMBRIDGE UNIV. 1908

6.8
GERSTERNKORN H. UBER GEZEITENREIBUNG BEIM ZWEIKORPERPROBLEM.
ZEIT.ASTROPH.36 PP 245-274 1955

6.8
GROVES G.
MUNK W. A NOTE ON TIDAL FRICTION.
JOURN.OF MARINE RES.VOL.17 PP 199-211 1958

6.8
ALFVEN H. THE EARLY HISTORY OF THE MOON AND THE EARTH.
ICARUS 1 PP 357-363 1963

6.8
MAC DONALD G.J.F. TIDAL FRICTION.
REVIEW OF GEOPHYSICS VOL.2 NO 3 PP 467-541 1964

6.8
WITKOWSKI J. SUR UN THEOREME DE HENRI POINCARE RELATIF AUX MAREES OCEANIQUES ET SES CONSEQUENCES POUR LES MAREES TERRESTRES
5 EME SYMP.INT.SUR LES MAREES TERRESTRES.
OBS.ROY.BELG.COMM.NO 236 S.GEOPH.NO 69 PP 443-446 1964

6.8
KAULA W.M. TIDAL DISSIPATION BY SOLID FRICTION AND THE RESULTING ORBITAL EVOLUTION
REVIEWS OF GEOPH. NO 2, PP 661-685 1964

6.8
KNOPOFF Q
REVIEWS OF GEOPHYSICS VOL 2 NO 4 PP 625-660 1964

6.8
GOLDREICH P.
SOTER S. Q IN THE SOLAR SYSTEM
ICARUS VOL 5 PP 375-389 1966

6.8
LAGUS P.L.
DON ANDERSON L. TIDAL DISSIPATION IN THE EARTH AND PLANETS.
PHYS.OF EARTH & PLANET.INT.VOL.1 NO 7 PP 505-510 1968

6.8
MUNK W.H. ONCE AGAIN - TIDAL FRICTION.
QUART.J.ROY.ASTR.SOC.VOL.9 PP 352-375 1968

6.8
KAULA W.M. TIDAL FRICTION WITH LATITUDE-DEPENDENT AMPLITUDE AND PHASE ANGLE.
ASTRON.JOURN.T.74 NO 9 NO 1374 PP 1108-1114 1969

6.8
DITCHKO I.A. SUR LE RETARD DES MAREES TERRESTRES
OBS.GRAV.POLTAVA BULL.INF.NO 11 PP 203-208 KIEV 1967
* B.I.M. NO 61 PP 3009-3016 1971

6.8
PERTSEV B.P. SUR LA QUESTION DU RETARD DE PHASE DANS LES OBSERVATIONS DES MAREES TERRESTRES
ACAD.SC.URSS.INST.PHYS.TERR.SCHMIDT PP 110-112 1970
* B.I.M. NO 61 PP 3108-3111 1971

6.8
BOSTROM R.C. WESTWARD DISPLACEMENT OF THE LITHOSPHERE
NATURE VOL.234 PP 536-538 1971

6.8
KNOPOFF L. LITHOSPHERIC MOMENTA AND THE DECELERATION OF THE EARTH
NATURE VOL 237 PP 93-95 1972

6.8
BOSTROM R.C. ARRANGEMENT OF CONVECTION IN THE EARTH BY LUNAR GRAVITY MEASUREM., INTERPRETAT., CHANGES OF STRAIN IN THE EARTH
PHIL.TRANS.ROY.SOC.SER.A, VOL.274, NO 1239 PP 397-408 1973

6.8
WARD W.R. SOLAR TIDAL FRICTION AND SATELLITE LOSS
REID M.J. MNRAS PP.21-32, VOL.164 1973

6.8
MOORE G.W. WESTWARD TIDAL LAG AS THE DRIVING FORCE OF PLATE TECTONICS
GEOLOGY, 1, PP 99-101 1973

6.8
JORDAN TH.H. SOME COMMENTS ON TIDAL DRAG AS A MECHANISM FOR DRIVING PLATE MOTIONS
J.G.R.VOL.79, NO 14, PP 2141-2142 1974

6.8
BOSTROM R.C. EFFECT OF TIDAL FORCES ON THERMAL CONVECTION
UNIVERSITY OF WASHINGTON, SEATTLE, WASH. 1974

6.81
THOMSON W. ON THE OBSERVATIONS AND CALCULATIONS REQUIRED TO FIND THE TIDAL RETARDATION OF THE EARTH S ROTATION.
PHIL.MAG. PP 533-537 1866
MATH.PHYS.PAPERS VOL.III PP 337-341 CAMBRIDGE 1890

6.81
THOMSON W. ON THE THERMODYNAMIC ACCELERATION OF THE EARTH S ROTATION
PROC.R.EDINBURGH VOL.XI PP 396-405 1882
MATH.PHYS.PAPERS VOL.III PP 341-350 CAMBRIDGE 1890

6.81
PARIISKII N.N. IRREGULARITE DE LA ROTATION DE LA TERRE.
TRAV.INST.GEOPH.NO 26/153/ MOSCOU PP 131-152 1955

6.81
PARIISKY N.N. ON THE EFFECT OF EARTH TIDES ON THE SECULAR RETARDATION OF THE EARTH S ROTATION
3E SYMP. MAREES TERRESTRES TRIESTE
BOLL.GEOFISICA TEORICA E APPL.VOL.2 NO 5 PP 19-26 1960

6.81
PARIISKII N.N. THE INFLUENCE OF EARTH TIDES ON THE SECULAR REDARDATION OF THE EARTH S ROTATION.
SOVIET ASTRONOMY AJ VOL 4 N.3 PP 515-522 1960

6.81
SLICHTER L.B. SECULAR EFFECTS OF TIDAL FRICTION UPON THE EARTH S ROTATION.
J.GEOPHYS.RES.VOL.68 PP 4281-4288 1963

6.81
LAMAR D.L. INFLUENCE OF SOLAR TIDAL TORQUE ON LENGTH OF DAY
MERIFIELD P.M. AND SYNODIC MONTH.
J.GEOPH.RES.VOL.72 NO 14 PP 3734-3735 1967

6.81
DICKE R.H. AVERAGE ACCELERATION OF THE EARTH S ROTATION AND THE VISCOSITY OF THE DEEP MANTLE
J.G.R. V.74 NO 25 PP 5895-5901 1969

6.81
PARIISKY N.N. ON THE EFFECT OF OCEANIC TIDES ON THE SECULAR
KUZNETZOV M.V. DECELERATION OF THE EARTH S ROTATION
KUZNETZOVA L.V. SYMP.I.A.U.,ROTATION OF THE EARTH, MORIOKA PP 1 - 16 1971

6.81
NEWTON R.R. ASTRONOMICAL EVIDENCE CORCERNING NON-GRAVITATIONAL FORCES IN THE EARTH-MOON SYSTEM
ASTROPHYSICS AND SPACE SCIENCE V.16 N.2 PP 179-200 1972

6.81
PANNELLA G. PALEONTOLOGICAL EVIDENCE ON THE EARTH S ROTATIONAL HISTORY SINCE EARLY PRECAMBRIAN
ASTROPHYSICS AND SPACE SCIENCE V.16 N.2 PP 212-237 1972

6.81
DITCHKO Y.A. SUR LE RALENTISSEMENT DE LA ROTATION DE LA TERRE
KORBA P.S. DU AUX MAREES
ROT.& DEFORM.DE MAREES DE LA TERRE VOL.VI PP 92-96 KIEV 1973
* B.I.M. NO 71 PP 4002-4007 1975

6.81
VARGA P. POSSIBLE VARIATIONS OF THE MOMENTUM OF INERTIA

AND OF THE ELLIPTICITY OF THE EARTH DURING
THE LAST FIVE HUNDRED MILLION YEARS
GEOFIZIKAI KOZIEMENYEK-HUNGARIAN GEOPH.INST.PP2-101 1975

6.81
BOSTROM R.C. WESTWANDERUNG UND THE LUNAR TIDAL COUPLE - MODULATION
OF CONVECTION BY BULGE STRESS
THE MOON VOL 15 PP 109-117 1976

6.82
GROVES G.W. ON TIDAL TORQUE AND ECCENTRICITY OF A SATELLITE S ORBIT
MON.NAT.ROY.ASTR.SOC.121 PP 497-502 1960

6.82
RUSKOL E.L. ON THE TIDAL CHANGES OF THE ORBITAL INCLINATIONS
OF URANUS SATELLITES RELATIVE TO ITS EQUATORIAL PLANE
ASTRON.VESTNIK PP.150-153, VOL.7 1973

6.82
RUBINCAM D.P. TIDAL FRICTION AND THE EARLY HISTORY OF THE MOON S ORBIT
J.G.R. VOL 80 NO 11 PP 1537-1548 1975

6.82
LAMBECK K. EFFECTS OF TIDAL DISSIPATION IN THE OCEANS ON THE
MOON S ORBIT AND THE EARTH S ROTATION
J.G.R. V.80 NO 20 JULY PP 2917-2925 1975

6.82
BOWER D.R. GEOS-C AND MEASUREMENT OF THE EARTH TIDE
PUBL.EARTH PHYS.BRANCH-OTTAWA VOL.45 NO 3 PP217-223 1976

6.85
DARWIN G.H. ON THE SECULAR CHANGES IN THE ELEMENTS OF THE ORBIT OF
A SATELLITE REVOLVING ABOUT A TIDALLY DISTORTED PLANET
SCIENT.PAPERS CAMBRIDGE UNIV.PRESS V.II PP 208-382 1908
PHILOS.TRANSACT.OF THE ROY.SOC. V.171 PP 713-891 1880

6.85
DARWIN G.H. ON THE ANALYTICAL EXPRESSIONS WHICH GIVE THE HISTORY OF
A FLUID PLANET OF SMALL VISCOSITY, ATTENDED BY A SINGLE
SATELLITE.
SCIENT.PAPERS CAMBRIDGE UNIV.PRESS V.II PP 383-405 1908
PROCEEDINGS OF THE ROY.SOC.V.XXX PP 255-278 1880

6.85
DARWIN G.H. ON THE TIDAL FRICTION OF A PLANET ATTENDED BY SEVERAL
SATELLITES, AND ON THE EVOLUTION OF THE SOLAR SYSTEM.
SCIENT.PAPERS CAMBRIDGE UNIV.PRESS V.II PP 406-458 1908
PHILOS.TRANSACT.OF THE ROY.SOC.V.172 PP 491-535 1881

6.85
DARWIN G.H. TIDAL FRICTION AND THE EVOLUTION OF A SATELLITE.
NATURE VOL.33 PP 367-368 18 FEVRIER
VOL.34 PP 286-288 29 JUILLET
VOL.35 PP 75 25 NOVEMBRE 1886

6.85
JEFFREYS H. TIDAL FRICTION IN SHALLOW SEAS.
PHIL.TRANS.ROY.SOC.LONDON 221 PP 239-264 1920

6.85
HEISKANEN W. UBER DEN EINFLUSS DER GEZEITEN AUF DIE SAKULARE
ACCELERATION DES MONDES.
ANN.ACAD.SCIE.FENNICAE NO 18 PP 1-84 1921

6.85
JEFFREYS H. THE EFFECT OF TIDAL FRICTION ON ECCENTRICITY AND
INCLINATION.
M.N.R.A.S. 122 PP 339-343 1961

6.85
JEFFREYS H. DISSIPATIVE INTERACTION BETWEEN SATELLITES.
M.N.R.A.S. 122 PP 345-347 1961

6.85
MILLER G. THE FLUX OF TIDAL ENERGY OUT OF THE DEEP OCEANS
J.G.R. VOL.71/4 PP 2485-2489 1966

6.85
GOLDREICH P. HISTORY OF THE LUNAR ORBIT.
REV.GEOPHYS.VOL.4 PP 411-439 1966

6.85
BROSCHE P. GEZEITENREIBUNG UND ERDROTATION
SUNDERMANN J. DIE NATURWISSENSCHAFTEN, HEFT 3, S.135, PP 1-2 1969

6.85
NEWTON R.R. SECULAR ACCELERATIONS OF THE EARTH AND MOON.

6 EME SYMPOSIUM MAREES TERRESTRES STRASBOURG
OBS.ROY.BELG.COMM. A9 S.GEOPH. NO 96 PP 235 1970

6.85
BROSCHE P. DIE GEZEITEN DES MEERES UND DIE ROTATION DER ERDE
SUNDERMANN J. PURE & APPLIED GEOPH.VOL.86, III, PP 95-117 1971

6.85
BROSCHE P. DIE BREMSUNG DER ERDROTATION
STERNE UND WELTRAUM, PP 38-40 1971

6.85
BROSCHE P. ON THE TORQUES DUE TO TIDAL FRICTION OF THE
SUNDERMANN J. OCEANS AND ADJACENT SEAS
ROTATION OF THE EARTH PP 235-239 REIDEL 1972

6.85
BROSCHE P. MEERESGEZEITEN BREMSEN DIE ERDRORATION
SUNDERMANN J. UMSCHAU 73 HEFT 7, PP 218-219 1973

6.85
KOUZNETSOV M.V. CALCUL DU RALENTISSEMENT SECULAIRE DE LA ROTATION DE
LA TERRE D APRES LES CARTES COTIDALES ACTUELLES
FIZIKA ZEMLI NO 12 PP 3-10, 1972
* B.I.M. NO 66 PP 3591-3603 1973

6.88
JEFFREYS H. THE RESONANT THEORY OF THE ORIGIN OF THE MOON /2/.
MONTH.NOT.ROY.ASTRON.SOC.91 PP 169-173 1930

6.88
RUBINCAM D.P. THE EARLY HISTORY OF THE LUNAR INCLINATION
G.S.F.C.X-592-73-328, PP 1-167 1973

6.9
LUBIMOVA E.A. THEORY OF THERMAL STATE OF THE EARTH S MANTLE
THE EFFECT OF TIDAL FRICTION
IN GASKELL.THE EARTH S MANTLE, AC.PRESS, P 248 1967

6.9
SHAW H.R. EARTH TIDES, GLOBAL HEAT FLOW, AND TECTONICS
SCIENCE VOL 168 N.3935 PP 1084-1087 1970

6.9
MELCHIOR P. PHYSIQUE ET DYNAMIQUE PLANETAIRES VOL 4
VANDER RUE DEFACQZ 21, BRUXELLES PAGE 28 1973

7.1
ROCHESTER M.G. ON A DYNAMICAL INVARIANT OF EARTH DEFORMATIONS
SMYLIE D.E. AND VARIATIONS IN G.
PROCEEDINGS 7TH INT.SYMP.EARTH TIDES, SOPRON 1973
AKADEMIAI KIADO BUDAPEST PP 187-196 1976

7.1
HEITZ S. MATHEMATISCHE MODELLE DER GEODATISCHEN ASTRONOMIE
DEUTSCHE GEOD.KOM. REIHE A-HEFT NR 85 PP 3-133 1976

7.2
WALKER A.M. THE ANALYSIS OF THE OBSERVATIONS OF THE VARIATION
YOUNG A. OF LATITUDE.
MONTHLY NOT.R.ASTR.SOC.VOL.115 NO 4 PP 443-459 1955

7.2
MELCHIOR P.J. SUR L AMORTISSEMENT DU MOUVEMENT LIBRE DU POLE
INSTANTANE DE ROTATION A LA SURFACE DE LA TERRE.
REND.ACC.NAZ.LINCEI ROMA SER.VIII VOL.XIX PP 137-142
OBS.ROY.BELG.COMM.NO 92 S.GEOPH.NO 34 1955

7.2
ARNOLD K. DER GRAVIMETRISCHE EFFEKT DER POLHOHENSCHWANKUNG.
Z.VERMESS.WES.STUTTGART 83 10 PP 346-348 1958

7.2
HAUBRICH R. THE POLE TIDE.
MUNK W. JOURN.GEOPH.RES.64 NO 12 PP 2373-2388 1959

7.2
MOLODENSKII M.S. LA PERIODE DE CHANDLER ET LA STRUCTURE DU NOYAU DE LA TERRE
11E CONF. ASTR. POULKOVO PP 124-126 1954
* B.I.M. N.43 MARS PP 1612-1615 1966

7.2
GERSTENKORN H. DAMPING OF FREE NUTATION AND RELAXATION TIME OF THE EARTH
ICARUS VOL.6 PP 292-297 1967

7.2
HOLLAND G.L. ON THE POLE TIDE AND RELATED CHANDLER OSCILLATIONS
MURTY T.S. SYMPOSIUM OF COASTAL GEODESY MUNICH 1970

7.2
OKUDA T. LOCAL NON-POLAR VARIATION OF LATITUDE DEDUCED FROM THE ILS DATA FOR THE PERIOD 1933-1965.
6 EME SYMPOSIUM MAREES TERRESTRES STRASBOURG
OBS.ROY.BELG.COMM. A9 S.GEOPH. NO 96 PP 234 1970

7.3
THOMSON W. ON THE RIGIDITY OF THE EARTH, SHIFTINGS OF THE EARTH S INSTANTANEOUS AXIS OF ROTATION, AND IRREGULARITIES OF THE EARTH AS A TIMEKEEPER.
MATH.& PHYS.PAPERS III PP 312-336 CAMBRIDGE U.P. 1890

7.3
JEFFREYS H. POSSIBLE TIDAL EFFECTS ON ACCURATE TIME-KEEPING.
MONTHLY NOT.R.ASTR.SOC.GEOPH.SUPPL.VOL.2 PP 56-58 1928

7.3
ANDERSSON F. BERECHNUNG DER VARIATION DER TAGESLANGE INFOLGE DER DEFORMATION DER ERDE DURCH FLUTERZEUGENDE KRAFTE.
ARKIV MATEMATIK, ASTRON.OCH FYSIK 26A NO 8 PP 1-34 1938

7.3
SEKIGUCHI N. EFFECTS OF THE LONG PERIOD TIDES ON THE ROTATION OF THE EARTH.
PUBL.ASTR.SOC.JAPAN 3 NO 2 PP 94-109 1951

7.3
SEKIGUCHI N. EFFECTS OF THE SHORT PERIOD OCEANIC TIDES ON THE ROTATION OF THE DEFORMABLE EARTH.
PUBL.ASTR.SOC.JAPAN 4 NO 3 PP 139-143 1952

7.3
SEKIGUCHI N. EFFECTS OF THE VISCOSITY OF THE EARTH S CORE ON THE ROTATION OF THE EARTH.
PUBL.ASTR.SOC.JAPAN 4, NO 3 PP 103-114 1952

7.3
MINTZ Y.
MUNK W. THE EFFECT OF WINDS AND BODILY TIDES ON THE ANNUAL VARIATION IN THE LENGTH OF DAY.
MONTHLY NOT.R.ASTR.SOC.GEOPH.SUPPL.VOL.6,9 PP 566-578 1953

7.3
MOLODENSKII M.S.
PARIISKII N.N. ELASTIC TIDES AND IRREGULARITIES OF EARTH S ROTATION IN CONNECTION WITH ITS CONSTITUTION
RAPP.XE ASS.GLE UNION INTERN.GEOD.GEOPH. M. PP 42-43 1954

7.3
MARKOWITZ WM. DETERMINATION OF K FROM THE VARIATIONS OF ROTATION OF THE EARTH DUE TO ZONAL EARTH TIDES
2EME SYMPOSIUM MAREES TERRESTRES - BRUXELLES
OBS.ROY.BELG.COMM.142-S.GEOPH.NO 142 PP 137-139 1958

7.3
WOOLARD E.W. INEQUALITIES IN MEAN SOLAR TIME FROM IDEAL VARIATIONS IN THE ROTATION OF THE EARTH.
THE ASTRON.JOURNAL 64, 4 NO 1269 PP 140-142 1959

7.3
MARKOWITZ WM. VARIATIONS IN ROTATION OF THE EARTH,RESULTS OBTAINED WITH THE DUALRATE MOON CAMERA AND PHOTOGRAPHIC ZENITH TUBES
THE ASTRONOMICAL JOURNAL 64 NO 1268 PP 106-113 1959

7.3
MARKOWITZ W. LUNAR AND SOLAR EARTH TIDES AND THE ROTATION OF THE EARTH
3E SYMP. MAREES TERRESTRES TRIESTE
BOLL.GEOFISICA TEORICA E APPL.VOL.2 NO 5 PP 27-28 1960

7.3
STOYKO A. L INFLUENCE DES MAREES TERRESTRES SUR LA ROTATION DE LA TERRE.
5 EME SYMP.INT.SUR LES MAREES TERRESTRES.
OBS.ROY.BELG.COMM.NO 236 S.GEOPH.NO 69 PP 440-442 1964

7.3
STOYKO A.ET N. AMELIORATION DE L ECHELLE DU TEMPS UNIFORME.
ANN.FRANC.CHRONOM.ET MICROMEC.1 PP 37-41 1966

7.3
PILNIK G.P. MAREES LUNAIRES ET ROTATION DE LA TERRE.
ISV. ACAD. SC. URSS PHYS. DE LA TERRE 8. P 3 1966
* B.I.M. NO 45 PP 1851-1856 1966

7.3
FLIEGEL H.F.
HAWKINS T.P. ANALYSIS OF VARIATIONS IN THE ROTATION OF THE EARTH
THE ASTRONOMICAL JOURNAL VOL.72 NO 4 PP 544-550 1967

7.3
PILNIK G.P. SUR LA DEDUCTION DU NOMBRE DE LOVE K D APRES

L IRREGULARITE DE ROTATION DE LA TERRE
ISVESTIA ACAD.SC.URSS-PHYS.TERR.NO 7 PP 3-13 1967
* B.I.M. NO 51 PP 2318-2331 1968

7.3
STOYKO A.
STOYKO N.
LA DETERMINATION DE L INFLUENCE DES MAREES TERRESTRES A LONGUE PERIODE SUR LA ROTATION DE LA TERRE.
6 EME SYMPOSIUM MAREES TERRESTRES STRASBOURG
OBS.ROY.BELG.COMM. A9 S.GEOPH. NO 96 PP 222-225 1970

7.3
GUINOT B.
SHORT-PERIOD TERMS IN UNIVERSAL TIME
ASTRON. & ASTROPHYS. T 8 PP 26-28 1970

7.3
PARIISKY N.N.
PERTSEV B.
ON THE DETERMINATION OF LOVE NUMBER K FROM VARIATIONS OF THE EARTH S ROTATION
SYMP.I.A.U.,ROTATION OF THE EARTH, MORIOKA PP 1 - 6 1971

7.3
IIJIMA S.
NIIMI Y.
THE EFFECTS OF EARTH TIDES ON TIME AND LATITUDE OBSERVATIONS
JOURN.ASTR.OBS.TOKYO VOL.16 PP 190-202 1971

7.3
DJUROVIC D.
MELCHIOR P.
RECHERCHE DES TERMES DE MAREE DANS LES VARIATIONS DE LA VITESSE DE ROTATION DE LA TERRE
BULL.CLASSE SC.ACAD.ROY.BELG.5E SERIE,TOME LVIII,1972
OBS.ROY.BELG.COMM.B 79-S.GEOPH.NO 115 PP 1248-1257 1972

7.3
PILNIK G.P.
DETERMINATION DE L HEURE ET STRUCTURE INTERNE DE LA TERRE.
ROTATION ET DEFORM.MAREES DE LA TERRE,FASC.1
ACAD.SC.UKR.OBS.GRAV.POLTAVA KIEV PP 56-72 1970
* B.I.M. NO 63 PP 3263-3274 1972

7.3
PILNIK G.P.
OBSERVATIONS ASTRONOMIQUES DES MAREES TERRESTRES.
PHYSIQUE DE LA TERRE 3 PP 3-14 MOSCOU 1970
* B.I.M. NO 63 PP 3249-3262 1972

7.3
PILNIK G.P.
TIDE IRREGULARITY SPECTRA OF THE ROTATION OF THE EARTH
B.I.M. NO 63 PP 3244-3248 1972

7.3
PARIISKII N.N.
PERTSEV B.P.
SUR LA DETERMINATION DU NOMBRE DE LOVE D APRES LES VARIATIONS DUES AUX MAREES DE LA ROTATION D UNE TERRE COMPRESSIBLE
FISIKA ZEMLI 3, PP 11-14 1972
* B.I.M. NO 64 JANVIER PP 3381-3385 1973

7.3
PILNIK G.P.
REMARKS ON THE CALCULATION OF DEFINITIVE UNIVERSAL TIME
REPR.SYMP.NO 48, ROTATION OF THE EARTH, PP 43-45 1973

7.3
PILNIK G.P.
CO-SPECTRA OF EARTH TIDES
BULL.GEODESIQUE, NO 108, PP 211-218 1973

7.3
PILNIK G.P.
SPECTRAL ANALYSIS OF THE LONG PERIODICAL EARTH S TIDES
FISICA ZEMLI NO 4 PP 3-14 1974

7.3
PARIISKII N.N.
PERTSEV B.P.
DETERMINATION DU NOMBRE DE LOVE K D APRES LES VARIATIONS DE MAREES DE LA VITESSE DE ROTATION DE LA TERRE APLATIE
ACAD NAOUK SSSR, INST.PHYS.TERRE SCHMIDT PP 19-33 1973
* B.I.M. NO 69 PP 3840-3849 1974

7.3
GUINOT B.
A DETERMINATION OF THE LOVE NUMBER K FROM THE PERIODIC WAVES OF UT 1
ASTRON.AND ASTROPH. VOL.36 NO 1 PP 1-4 1974

7.3
ROCHESTER M.G.
SMYLIE D.E.
ON CHANGES IN THE TRACE OF THE EARTH S INERTIA TENSOR
J.G.R. VOL.79 NO 32 PP 4948-4951 1974

7.3
DJUROVIC D.
LES TERMES DE MAREE DANS LE TEMPS UNIVERSEL COORDONNEES DU POLE ET TU1-TUC POUR L INTERVALLE 1967-74
BULL.D OBS. O.R.B. VOL.IV, FASC.3 PP 1-64 1975

7.3
PILNIK G.P.
SCIENTIFIC PROBLEMS AND TIME DETERMINATION
2ND CAGLIARI INTERN.MEET.ON TIME DETERMINAT. PP 259-265 — 1975

7.4
DE FAZIO TH.L.
AKI K.
ALBA J.
SOLID EARTH TIDE AND OBSERVED CHANGE IN THE
IN SITU SEISMIC VELOCITY
J.G.R. VOL.78 NO 8, PP 1319-1322 — 1973

7.4
BEAUMONT CH.
BERGER J.
EARTHQUAKE PREDICTION - MODIFICATION OF THE EARTH
TIDE TILTS AND STRAINS BY DILATANCY
GEOPH.J.R.ASTR.SOC. V.39 PP 111-121 — 1974

7.41
DE MONTESSUS
SUR LA REPARTITION HORAIRE DES SEISMES ET LEUR
RELATION SUPPOSEE AVEC LES CULMINATIONS DE LA LUNE.
C.R.ACAD.SCI.PARIS TOME 109 PP 327-330 ERRATUM PP 392 — 1889

7.41
COTTON L.A.
EARTHQUAKE FREQUENCY WITH SPECIAL REFERENCE
TO TIDAL STRESSES IN THE LITHOSPHERE
SEISM.SOC.AM.BULL.NO 12 PP 47 — 1922

7.41
DAVISON C.
THE DIURNAL PERIODICITY OF EARTHQUAKES
J.GEOL. NO 42, PP 449 — 1934

7.41
ALLEN M.W.
THE LUNAR TRIGGERING EFFECT ON EARTHQUAKES IN
SOUTHERN CALIFORNIA.
BULL.SEISMOL.SOC.AMERICA VOL.26 NO 2 PP 147-157 — 1936

7.41
BRAZEE R.J.
EARTH TIDES AND EARTHQUAKES.
EARTHQUAKE NOTES 28, 1, PP 10 — 1957

7.41
TAMRAZYAN G.P.
ON THE SEISMIC ACTIVITY IN THE AREA OF THE NORTHWESTERN
PACIFIC OCEAN MARGIN.
AKAD.NAUK.SSST,GEOPHYS.SER.5. PP 664-668 — 1958

7.41
TAMRAZYAN G.P.
INTERMEDIATE AND DEEP FOCUS EARTHQUAKES IN CONNECTION
WITH THE EARTH S COSMIC SPACE CONDITIONS.
IZV.GEOPHS.SER.4, PP 598-603 — 1959

7.41
HOFFMAN R.B.
AFTERSHOCK-ENERGY RELEASE VERSUS TIDAL EFFECTS.
U.S.GEOL.SURV.PUBL.ART.246 PP C 267 - C 270 — 1961

7.41
GOUGENHEIM A.
CONFIRMATION BY OBSERVATION OF THE NEGLIGIBLE
ROLE OF THE EARTH TIDE IN THE PRODUCTION OF EARTHQUAKES
COMPT.REND.252, PP 3313 — 1961

7.41
MORGAN W.J.
STONER J.O.
DICKE R.H.
PERIODICITY OF EARTHQUAKES AND THE INVARIANCE
OF THE GRAVITATIONAL CONSTANT
J.G.R. VOL.66 PP 3831-3843 — 1961

7.41
TAMRAZYAN G.P.
ON THE PRESENCE OF PULSATIONS OF THE EARTH COUPLED
IN TIME WITH SOLAR ACTIVITY.
DOKL.AKAD.NAUK.SSSR, 147 PP 1361-1364 — 1962

7.41
NESTORENKO P.G.
STOVAS M.V.
VARIATION OF THE GRAVITATIONAL FIELD AS ONE OF THE
CAUSES OF THE EARTH S SEISMICITY
INF.BULL.GEOPH.& ASTR.NO 5, PP 85, UKR.ACAD. — 1963

7.41
KNOPOFF L.
EARTH TIDES AS A TRIGGERING MECHANISM FOR EARTHQUAKES.
PROC.U.S.JAPAN CONF.APPEND.27 PP 61-66 — 1964

7.41
HAGIWARA T.
PROBLEMS IN THE CONTINOUS INSTRUMENTAL OBSERVATIONS
AND EXAMPLES OF THE OBSERVATIONS.
PROC.U.S.-JAPAN CONF. APPEND.18 PP 44-45 — 1964

7.41
KNOPOFF L.
EARTH TIDES AS A TRIGGERING MECHANISM FOR EARTHQUAKES.
UNITED STATES-JAPAN CONFERENCE ON RESEARCH RELATED
TO EARTHQUAKE PREDICTION PROBLEMS PP 61-66 APP.27 1964
BULL.SEISM.SOC.AMER.54 PP 1865-1870 — 1964

7.41
TANAKA T.
STUDY ON THE RELATION BETWEEN LOCAL EARTHQUAKES

AND MINUTE GROUND DEFORMATION.
PART I.- ON SOME STATISTICAL RESULTS FROM LOCAL EARTHQUAKES OCCURED IN THE WAKAYAMA DISTRICT.
BULL.DISASTER PREV.RES.INST.VOL.14 PART.1 PP 55-57 1964

7.41
DIX C.H. TRIGGERING OF SOME EARTHQUAKES.
JAPAN. ACAD. PROC.VOL.40 NO 6, PP 410 1964
GEOPH.INST.TOKYO UNIV.GEOPH.NOTES VOL.17 CONTR.NO 15 1964

7.41
TANAKA T. STUDY ON THE RELATION BETWEEN LOCAL EARTHQUAKES AND MINUTE GROUND DEFORMATION.
PART II.- AN APPLICATION OF THE DIGITAL FILTERING TO THE TILTGRAM FOR THE DETECTION OF THE MINUTE ANOMALOUS TILTING OF THE GROUND.
BULL.DISASTER PREV.RES.INST.KYOTO UNIV.VOL.16PP 57-67 1966

7.41
TAMRAZYAN G.P. ORIGIN TIME OF EARTHQUAKES IN THE KURILE-ADJOINT-TO-KAMCHATKA REGION, AND LOCAL LUNAR AND SOLAR TIME
J.PHYS.EARTH VOL 14 P 41 1966

7.41
TAMRAZYAN G.P. TIDE FORMING FORCES AND EARTHQUAKES
ICARUS 7, N.1 PP 59-65 1967

7.41
SIMPSON J.F. EARTH TIDES AS A TRIGGERING MECHANISM FOR EARTHQUAKES
EARTH & PLANET.SC.LETTERS.2 PP 473-478 1967

7.41
TAMRAZYAN G.P. THE TIDE-GENERATING FORCES AND THE DISTRIBUTION OF THE INTERMEDIATE AND DEEP-FOCI EARTHQUAKES.
GERL.BEITR.GEOPH.77, 3 PP 215-220 1968

7.41
TAMRAZYAN G.P. PRINCIPAL REGULARITIES IN THE DISTRIBUTION OF MAJOR EARTHQUAKES RELATIVE TO SOLAR AND LUNAR TIDES AND OTHER COSMIC FORCES.
ICARUS 9 PP 574-592 1968

7.41
TAMRAZYAN G.P. PRINCIPAL REGULARITIES IN THE DISTRIBUTION OF MAJOR EARTHQUAKES RELATIVE TO SOLAR TIDES, LUNAR TIDES, AND OTHER COSMIC FORCES
INT.J.SOLAR SYSTEM VOL 9 P 574 1968

7.41
RYALL A.
VAN WORMER J.D.
JONES A.F. TRIGGERING OF MICROEARTHQUAKES BY EARTH TIDES AND OTHER FEATURES OF THE TRUCKEE, CALIFORNIA EARTHQUAKE SEQUENCE OF SEPTEMBER 1966
BULL.SEISM.SOC.AMER. VOL.58 NO 1 PP 215-248 1968

7.41
MIDDLEHURST B.M.
CHAPMAN W.B. TIDAL CYCLES AND SEISMIC CAUSES FOR LUNAR EVENTS.
15 TH. ANNUAL MEETING AMERICAN GEOPH. UNION APRIL 1969

7.41
KNOPOFF L. CORRELATION OF EARTHQUAKES WITH LUNAR ORBITAL MOTIONS
THE MOON NO 2 PP 140-143 1970

7.41
SHLIEN S. EARTHQUAKE-TIDE CORRELATION
GEOPH.JOURN.R.ASTR.SOC.,V.28 PP 27-34 1972

7.41
SADEH D.S.
MEIDAV M. SEARCH FOR SIDERAL PERIODICITY IN EARTHQUAKE OCCURENCES
J.G.R.VOL.78, NO 32, PP 7709-7716 1973

7.41
TAMRAZYAN G.P. POSSIBLE COSMIC INFLUENCES ON THE 1966 TASHKENT EARTHQUAKE AND ITS LARGEST AFTERSHOCKS
GEOPH.J.R.ASTR.SOC. VOL 38 PP 423-429 1974

7.41
SHIMSHONI M.
DISHON M. ON THE CAUSE OF SIDERAL PERIODICITIES IN EARTHQUAKE OCCURRENCES
J.G.R. V.80 NO 32 PP 4497-4498 1975

7.41
SADEH DROR
MEIDAV M. A SIDERAL PERIOD IN EARTHQUAKE OCCURENCES RECONFIRMED
J.G.R. V.80 NO 32 P 4499 1975

7.41
HEATON TH.H. TIDAL TRIGGERING OF EARTHQUAKES
GEOPHYS.J.R.ASTR.SOC.NO 43 PP 307-326 1975

7.42
BEAUMONT C. BERGER J. — EARTHQUAKE PREDICTION - MODIFICATION OF THE EARTH TIDE TILTS AND STRAINS BY DILATANCY
G.J.R. ASTR.SOC. VOL 39 PP 111-121 1974

7.42
TANAKA T. — EFFECT OF DILATANCY ON OCEAN LOAD TIDES
PAGEOPH, VOL.114 PP 415-423 1976

7.5
VENING MEINESZ F.A. — THE DETERMINATION OF THE EARTH S PLASTICITY FROM THE POSTGLACIAL UPLIFT OF SCANDINAVIA-ISOSTATIC ADJUSTMENT
PROC.K.AKAD.SC.AMSTERDAM XL, NO 8 PP 654-662 1937

7.5
PREY A. — UEBER POLSCHWANKUNG UND POLWANDERUNG.
GERL.BEITR.GEOPH.56 NO 2 PP 155-202 1940

7.5
NISKANEN E. — ON THE VISCOSITY OF THE EARTH S INTERIOR AND CRUST.
PUBL.ISOSTATIC INST.INTERN.ASSOC.GEODESY NO 20 HELSINKI 1948

7.52
DARWIN G.H. — ON THE STRESSES CAUSED IN THE INTERIOR OF THE EARTH BY THE WEIGHT OF CONTINENTS AND MOUNTAINS.
SCIENT.PAPERS CAMBRIDGE UNIV.PRESS V.II PP 459-514 1908
PHILOS.TRANSACT.OF ROY.SOC.V.173 PP 187-230 1882

7.52
JEFFREYS H. — ON THE STRESSES IN THE EARTH S CRUST REQUIRED TO SUPPORT SURFACE INEQUALITIES
GEOPH.SUPPL.MONTHLY NOT.VOL.III PP 30-41, 60-69 1932

7.52
JEFFREYS H. — THE STRESS DIFFERENCES IN THE EARTH S SHELL
MONTHLY NOT.GEOPH.SUPPL.VOL.5 NO 3 PP 72-89 1943

7.55
MELCHIOR P.J. BROUET J. — CONTRIBUTION DES STATIONS CLINOMETRIQUES DE MAREES TERRESTRES A L ETUDE DES MOUVEMENTS RECENTS DE L ECORCE.
ASSOC.INT.GEOD.COMM.MOUV.REC.CROUTE-2E SYMP.INT.AULANKO 1965

7.55
VICENTE R.O. — THE POSSIBILITY OF DETECTING MOTIONS OF THE CRUST.
GEOPH.JOURN.R.ASTR.SOC.14 NO 5 PP 475-478 1968

7.55
SHAW H.R. KISTLER R.W. EVERNDEN J.F. — SIERRA NEVADA PLUTONIC CYCLE - PART II, TIDAL ENERGY AND A HYPOTHESIS FOR OROGENIC-EPEIROGENIC PERIODICITIES
BULL.VOL.82 PP 869-896 GEOL.SOC.OF AMERICA 1971

7.55
BOSTROM R.C. — ARRANGEMENT OF CONVECTION IN THE EARTH BY LUNAR GRAVITY
PHIL.TRANS.SOC.LOND.A.274, PP 397-407 1973

7.55
GRUNTFEST I.J. — SCALE EFFECTS IN THE STUDY OF EARTH TIDES
TRANSACTION OF THE SOCIETY OF RHEOLOGY 18/2 PP 287-297 1974

7.55
BALAVADZE B.K. ABASHIDZE V.G. — INVESTIGATION OF RECENT TECTONIC MOVEMENTS OF THE EARTH S CRUST BY MEANS OF TILTMETERS
PROCEEDINGS 7TH INT.SYMP.EARTH TIDES, SOPRON 1973
AKADEMIAI KIADO BUDAPEST, PP 621-626 1976

7.6
PEKERIS C.L. — THE MAGNETIC FIELD INDUCED BY THE BODILY TIDE IN THE CORE OF THE EARTH
PROC.NAT.ACAD.SCI.USA VOL.68 NO 6 PP 1111-1113 1971

7.8
NEWTON R.R. — AN OBSERVATION OF THE SATELLITE PERTURBATION PRODUCED BY THE SOLAR TIDE.
GEOPH.JOURN.RES.70 NO 24 PP 5983-5989 1965

7.8
KOZAI Y. — EFFECTS OF THE TIDAL DEFORMATION OF THE EARTH ON THE MOTION OF CLOSE EARTH SATELLITES.
PUBL.ASTR.SOC.JAPAN 17 /4/ PP 395-402 1965

7.8
KOZAI Y. — DETERMINATION OF LOVE S NUMBER FROM SATELLITE OBSERVATIONS.
MEETING ORBITAL ANAL.SPONSORED.ROY.SOC.LONDON PP 97-100 1966

7.8
NEWTON R.R. — TIDAL NUMBERS AND PHASES AS DEDUCED FROM SATELLITE

ORBITS.
J.HOPKINS UNIV.APPL.PHYS.LAB.TECHN.TG 905 PP 80 1967

7.8
NEWTON R.R. A SATELLITE DETERMINATION OF TIDAL PARAMETERS AND EARTH DECELERATION.
GEOPH.JOURN.R.ASTR.SOC.14 NO 5 PP 505-539 1968

7.8
KOZAI Y. LOVE S NUMBER OF THE EARTH DERIVED FROM SATELLITE OBSERVATIONS.
BULL.GEOD.NO 89 PP 355-357 1968

7.8
KOZAI Y. LOVE S NUMBER OF THE EARTH DERIVED FROM SATELLITE OBSERVATIONS.
PUBL.ASTR.SOC.JAPAN 20 NO 1 PP 24-26 1968

7.8
GROTEN E. ON TIDAL EFFECTS IN SATELLITE GRAVITY DATA.
6 EME SYMPOSIUM MAREES TERRESTRES STRASBOURG
OBS.ROY.BELG.COMM. A9 S.GEOPH. NO 96 PP 228-233 1970

7.8
GROTEN E. ON TIDAL EFFECTS IN SATELLITE GRAVITY DATA.
6 EME SYMPOSIUM MAREES TERRESTRES STRASBOURG
OBS.ROY.BELG.COMM. A9 S.GEOPH. NO 96 PP 228-233 1970

7.8
SMITH D.E.
DUNN P.J.
KOLENKIEWICZ R. A LASER POLAR MOTION EXPERIMENT
3 INT.SYMP.-USE ART.SAT.FOR GEOD.-WASHINGTON PP 1-19
MORIOKA-SYMP.ROTATION OF THE EARTH 1971

7.8
LOWREY B.E. TESSERAL RESONANCE ON IMP-4 ORBIT
GODDARD SPACE FLIGHT CENTER X-554-71-198 PP 1-11 1971

7.8
G.R.G.S. PERTURBATIONS DUES AUX DEFORMATIONS DE MAREES TERRESTRES SUR LE MOUVEMENT DES SATELLITES
CENTRE SPATIAL BRETIGNY-AC/GR/1.68/CB/GRGS PP 1-14 1971

7.8
MUSEN PETER
ESTES RONALD ON THE TIDAL EFFECTS IN THE MOTION OF ARTIFICIAL SATELLITES
GSFC X-550-71-341 PP 1-32 1971

7.8
SMITH D.E.
KOLENKIEWICS R.
DUNN P.J. GEODETIC STUDIES BY LASER RANGING TO SATELLITES.
G.S.F.C. X-553-71-360 1971

7.8
MUSEN P.
ESTES R. ON THE TIDAL EFFECTS IN THE MOTION OF ARTIFICIAL SATELLITES
G.S.F.C. GREENBELT,MARYLAND AUGUST PP 1-32 1971

7.8
MUSEN P.
FELSENTREGER TH. ON THE DETERMINATION OF THE LONG PERIOD TIDAL PERTURBATIONS IN THE ELEMENTS OF ARTIFICIAL EARTH SATELLITES
GSFC - X-550-72-192, PP 1-46, MAI 1972

7.8
MUSEN P.
ESTES R. ON THE TIDAL EFFECTS IN THE MOTION OF ARTIFICIAL SATELLITES
CELESTIAL MECHANICS NO 6 PP 4-21 1972

7.8
MUSEN P.
ESTES R. ON THE TIDAL EFFECTS IN THE MOTION OF EARTH SATELLITES AND THE LOVE PARAMETERS OF THE EARTH
GSFC X-550-72-500 PP 1-28 1972

7.8
SMITH DAVID E.
KOLENKIEWICZ R.
DUNN P.J.
PLOTKIN H.H.
JOHNSON TH.S. POLAR MOTION FROM LASER TRACKING OF ARTIFICIAL SATELLITES
G.S.F.C. X-553-72-247 PP 1-5 1972

7.8
DOUGLAS B.C.
KLOSKO S.M.
MARSH J.H.
WILLIAMSON R.G. TIDAL PERTURBATIONS ON THE ORBITS OF GEOS-I AND GEOS-II
G.S.F.C. X-553-72-475 1972

7.8
MUSEN P. / FELSENTREGER TH. — ON THE DETERMINATION OF THE LONG PERIOD TIDAL PERTURBATIONS IN THE ELEMENTS OF ARTIFICIAL EARTH SATELLITES. G.S.F.C., PP 256-279. 1972

7.8
LAMBECK K. / CAZENAVE A. — FLUID TIDAL EFFECTS ON SATELLITE ORBIT AND OTHER TEMPORAL VARIATIONS IN THE GEOPOTENTIAL. G.R.G.S., BULL.NO 7, PP 1-42. 1973

7.8
MUSEN PETER — A SEMI-ANALYTICAL METHOD OF COMPUTATION OF OCEANIC TIDAL PERTURBATIONS IN THE MOTION OF ARTIFICIAL SATELLITES. G.S.F.C. X-590-73-190 JULY PP 1-20. 1973

7.8
DUNN P.J. / SMITH D.E. / KOLENKIEWICZ R. — TECHNIQUES FOR THE ANALYSIS OF GEODYNAMIC EFFECTS USING LASER DATA. GSFC X-592-73-235, PP 1-11. 1973

7.8
KOZAI Y. — A NEW METHOD TO COMPUTE LUNISOLAR PERTURBATIONS IN SATELLITE MOTIONS. RES.IN SPACE SC.-SAO SPEC.REP.NO 349, PP 1-27. 1973

7.8
SMITH D.E. / KOLENKIEWICZ R. / DUNN P.J. — GEODETIC STUDIES BY LASER RANGING TO SATELLITES. GEOPH.MONOGR.SER.VOL.15, PP 187-196. 1973

7.8
SMITH D.E. / KOLENKIEWICZ R. / DUNN P.J. — A DETERMINATION OF THE EARTH TIDAL AMPLITUDE AND PHASE FROM THE ORBITAL PERTURBATIONS OF THE BEACON EXPLORER C SPACECRAFT. G.S.F.C.X-592-73-161, PP 1-5. 1973

7.8
MUSEN P. — ON THE INFLUENCE OF THE SURFACE AND BODY TIDES ON THE MOTION OF A SATELLITE. G.S.F.C.X-920-73-382, PP 1-22. 1973

7.8
BALMINO G. — ANALYTICAL EXPRESSIONS FOR EARTH TIDES PERTURBATIONS ON CLOSE EARTH SATELLITES. PROC.INTERN.SYMP.USE OF ARTIF.SATELL.FOR GEOD.& GEODYN. PUBL.NAT.TECHN.UNIV.OF ATHENS PP 313-348. 1973

7.8
LAMBECK K. / CASENAVE A. / BALMINO G. — SOLID EARTH AND FLUID TIDES FROM SATELLITE ORBIT ANALYSES. PROC.INTERN.SYMP.USE OF ARTIF.SATELL.FOR GEOD.& GEODYN. PUBL.NAT.TECHN.UNIV.OF ATHENS PP 353-394. 1973

7.8
LAMBECK K. — DETERMINATION OF EARTH AND OCEAN TIDES FROM THE ANALYSIS OF SATELLITE ORBITS. PROC.SYMP.ON EARTH S GRAVITATIONAL FIELD AND SECULAR VARIATIONS IN POSITION PP 522-528 SYDNEY. 1973

7.8
SMITH D.E. / KOLENKIEWICZ R. / DUNN P.J. — EARTH TIDAL AMPLITUDE AND PHASE. NATURE, VOL.244, NO 5417, PP 498-499. 1973

7.8
SMITH D.E. / KOLENKIEWICZ R. / AGREEN R.W. / DUNN P.J. — DYNAMIC TECHNIQUES FOR STUDIES OF SECULAR VARIATIONS IN POSITION FROM RANGING TO SATELLITES. PROC.SYMP.ON EARTH S GRAVITATIONAL FIELD & SECULAR VARIATIONS IN POSITIONS PP 291-314. 1973

7.8
MUSEN P. — CONTRIBUTION TO THE THEORY OF TIDAL OSCILLATIONS OF AN ELASTIC EARTH. EXTERNAL TIDAL POTENTIAL. G.S.F.C. X-920-74-202. 1974

7.8
GREENBERG R. — OUTCOMES OF TIDAL EVOLUTION FOR ORBITS WITH ARBITRARY INCLINATION. ICARUS VOL.23 NO 1 PP 51-58. 1974

7.8
BALMINO G. — ANALYTICAL EXPRESSIONS FOR EARTH TIDES PERTURBATIONS ON ARTIFICIAL SATELLITES. GROUPE DE RECH.DE GEDESIE SPATIALE PP 1-85. 1974

7.8
LAMBECK K. — OCEAN TIDE PERTURBATIONS IN THE ORBITS OF GEOS 1 AND GEOS 2
CELESTIAL MECHANICS 10 PP 179-182 — 1974

7.8
LAMBECK K., CAZENAVE A., BALMINO G. — SOLID EARTH AND OCEAN TIDES ESTIMATED FROM SATELLITE ORBIT ANALYSES
REVIEWS OF GEOPH.& SPACE PHYS.VOL.12 NO 3 PP 421-433 — 1974

7.8
CASENAVE A. — ETUDE DES MAREES TERRESTRES ET OCEANIQUES A PARTIR DES PERTURBATIONS D ORBITES DE SATELLITES
ROT.DELLA TERRA E OSSERVAZIONI DI SATELLITI ARTIFICIALI
RENDICONTI FAC.SC.CAGLIARI SUPP.VOL.XLIV, PP 25-29 — 1974

7.8
X. — STARLETTE
CENTRE NAT.D ETUDES SPATIALES
GROUPE DE RECHERCHES DE GEODESIE SPATIALE PP 2-15 — 1975

7.8
MUSEN P. — THE EXTERIOR TIDAL POTENTIAL ACTING ON A SATELLITE
JOURN.ASTRONAUTICAL SC. XXIII, VOL 2 PP 161-178 1975
G.S.F.C. X-920-75-220 PP 1-26 — 1975

7.8
FELSENTREGER T.L., MARSH J.G., AGREEN R.W. — ANALYSES OF THE SOLID EARTH AND OCEAN TIDAL PERTURBATIONS ON THE ORBITS OF THE GEOS-I AND GEOS-II SATELLITES
G.S.F.C. X-921-75-194 PP 1-26 1975
J.G.R. 81, NO 14 PP 2557-2563 — 1976

7.8
LAMBECK K., CAZENAVE A., BALMINO G. — SOLID EARTH AND FLUID TIDES FROM SATELLITE ORBIT ANALYSES
PROCEEDINGS 7TH INT.SYMP.EARTH TIDES, SOPRON 1973
AKADEMIAI KIADO BUDAPEST, PP 517-558 — 1976

7.8
RUBINCAM D. — TIDAL PARAMETERS DERIVED FROM THE PERTURBATIONS IN THE ORBITAL INCLINATIONS OF THE BE-C, GEOS-I, AND GEOS-II SATELLITES
GODDARD SPACE FLIGHT CENTER X-921-76-72 PP 1-98 — 1976

7.8
BOWMAN B.R. — DETERMINATION OF GEOPHYSICAL PARAMETERS FROM LONG TERM ORBIT PERTURBATIONS USING NAVIGATION SATELLITE DOPPLER DERIVED EPHEMERIDES
DEFENSE MAPPING AGENCY TOPOGRAPHIC CENTER PP 1-32 — 1976

7.8
PAUL M.K. — DETERMINATION OF RELATIVE DISPLACEMENT DUE TO EARTH TIDE FROM LASER RANGE DATA FOR A PAIR OF GROUND STATIONS
EOS VOL 57 NO 12 P 896 — 1976

7.9
KAULA W.M. — POTENTIALITIES OF LUNAR LASER RANGING FOR MEASURING TECTONIC MOTIONS.
MEASUREM., INTERPRETAT., CHANGES OF STRAIN IN THE EARTH
PHIL.TRANS.ROY.SOC.SER.A, VOL.274, NO 1239 PP 185-194 — 1973

7.95
PALMER H.P., ANDERSON B. — USEFUL GEODETIC MEASUREMENTS WITH RADIO INTERFEROMETERS.
MEASUREM., INTERPRETAT., CHANGES OF STRAIN IN THE EARTH
PHIL.TRANS.ROY.SOC.SER.A, VOL.274, NO 1239 PP 195-198 — 1973

8.1
RODES L. — THE INFLUENCE OF THE MOON ON THE FREQUENCY OF EARTHQUAKES.
GERL.BEITR.GEOPH.41 H.2 PP 209-212 — 1934

8.1
SEMENOV P.G. — TEMPETE DANS L INCLINAISON DE LA SURFACE TERRESTRE LORS DES TREMBLEMENTS DE TERRE DES 4 MARS - 8 ET 10 JUIN 1949
COMM.TADJIKISTAN ACAD.SC.URSS PUBL.20 PP 29-32 — 1949

8.1
SASSA K., NISHIMURA E. — ON PHENOMENA FORERUNNING EARTHQUAKES.
TRANS.AMER.GEOPH.UNION VOL.32 PP 1-6 — 1951

8.1
HOSOYAMA K. — CHARACTERISTIC TILT OF THE GROUND THAT PRECEDED THE OCCURRENCE OF THE STRONG EARTHQUAKE OF MARCH 7, 1952.
JOURNAL PHYS.EARTH.1 NO 2 PP 75-81 — 1952

8.1
NISHIMURA E.
HOSOYAMA K.
ON TILTING MOTION OF GROUND OBSERVED BEFORE AND AFTER THE OCCURENCE OF AN EARTHQUAKE.
TRANS.AMER.GEOPH.UNION 34 NO 4 PP 597-599 1953

8.1
CALOI P.
OSSERVAZIONI SISMICHE E CLINOGRAFICHE PRESSO GRANDI DIGHE DI SBARRAMENTO.
PUBL.IST.NAZ.DI GEOFISICA NO 276
ANNALI GEOFISICA VI 3 1953

8.1
NISHIMURA E.
ON SOME DESTRUCTIVE EARTHQUAKES OBSERVED WITH THE TILTMETER AT A GREAT DISTANCE.
DISASTER PREV.RES.INST.KYOTO UNIV.BULL.6 PP 15 1953

8.1
SASSA K.
NISHIMURA E.
ON PHENOMENA FORERUNNING EARTHQUAKES.
PUBL.BUR.CENTRAL SEISM.INTERN.A.FASC.19 PP 277-285 ROME 1954

8.1
SAVARENSKII E.F.
REMARQUES AU SUJET DU ROLE DES CONDITIONS SOUTERRAINES DANS LES OBSERVATIONS SEISMIQUES ET CLINOMETRIQUES
TRAV.INST.GEOPH.ACAD.SC.URSS NO 22, 149 PP 102-110 1954

8.1
BONTCHKOVSKII V.F.
INCLINAISONS DE LA SURFACE DE LA TERRE COMME L UN DES SIGNES PRECURSEURS POSSIBLES DES TREMBLEMENTS DE TERRE
TRAV.INST.GEOPH.ACAD.SC.URSS NO 2, 152 PP 134-153 1954

8.1
TOMASCHEK R.
EARTH TILTS IN THE BRITISH ISLES CONNECTED WITH FAR DISTANT EARTHQUAKES.
NATURE 176 PP 24-27 1955

8.1
SOUBBOTINE M.Y.
CLINOMETRE FLUXIOMETRIQUE
TRAV.INST.GEOPH.ACAD.SC.URSS NO 30, 157 PP 198-207 1955

8.1
CALOI P.
SPADEA M.C.
RELAZIONI FRA LENTE VARIAZIONI D INCLINAZIONE E MOTI SISMICI IN ZONA AD ELEVATA SISMICITA.
REND.ACAD.NAZ.LINCEI SER.VIII VOL.XVIII PP 250-256 1955

8.1
CALOI P.
SPADEA M.C.
PRIME INDICAZIONI DI REGISTRAZIONI CLINOGRAFICHE OTTENUTE IN ZONA AD ELEVATA SISMICITA.
PUBL.IST.NAZ.DI GEOFISICA NO 305 ANN.VIII NO 1 1955

8.1
TAMRAZYAN G.P.
TREMBLEMENTS DE TERRE DANS LA REGION DU KASBEK ET MAREES ELASTIQUES
IZV.ACAD.SC.URSS SER.GEOPH.NO 7 PP 840-843 1956

8.1
SEMENOV P.G.
INCLINAISON DE LA SURFACE TERRESTRE EN CONNEXION AVEC LES TREMBLEMENTS DE TERRE
TRAV.ACAD.SC.TADJIKISTAN PUBL.T.54 NO 1 PP 97-101 1956

8.1
NISHIMURA E.
HOSOYAMA K.
ITO Y.
MICRO-TILTING MOTION OF THE GROUND.
TRANS.AMER.GEOPH.UNION 37 NO 5 PP 645-646 1956

8.1
CALOI P.
ABOUT SOME PHENOMENA PRECEDING AND FOLLOWING THE SEISMIC MOVEMENTS IN THE ZONE CHARACTERIZED BY HIGH SEISMICITY.
CONTR.GEOPH.IN HON.OF B.GUTENBERG PERG.PRESS 1958

8.1
CALOI P.
ABOUT SOME PHENOMENA PRECEDING AND FOLLOWING THE SEISMIC MOVEMENTS IN THE ZONE CHARACTERISED BY HIGH SEISMICITY
CONTRIB.GEOPH.B.GUTENBERG PERGAMON PRESS. 1958

8.1
TAKADA M.
ON THE OBSERVING INSTRUMENTS AND TELE-METRICAL DEVICES OF EXTENSOMETERS AND TILTMETERS AT IDE OBSERVATORY AND ON THE CRUSTAL STRAIN ACCOMPANIED BY A GREAT EARTHQUAKE
BULL.DISASTER PREV.RES.INST.NO 27 1959

8.1
TAKADA M.
ON THE CRUSTAL STRAIN ACCOMPANIED BY A GREAT EARTHQUAKE
3E SYMP. MAREES TERRESTRES TRIESTE
BOLL.GEOFISICA TEORICA E APPL.VOL.2 NO 5 PP 127-128 1960

8.1
GOUGENHEIM A. CONFIRMATION PAR L OBSERVATION DU ROLE NEGLIGEABLE DE LA MAREE TERRESTRE DANS LA PRODUCTION DES SEISMES. C.R.ACAD.SC.PARIS3/5/61-BULL.GEOD.NO 20-ANN.GEOF.XIV 1961

8.1
ITO Y. TILTING MOTION OF THE GROUND AS RELATED TO THE VOLCANIC ACTIVITY OF MT.ASO. MICRO-PROCESS OF THE TILTING MOTION OF GROUND AND STRUCTURE. DISASTER PREV.RES.INST.BULL.NO 42 PP 55 1961

8.1
YOSHIKAWA K. ON THE CRUSTAL MOVEMENT ACCOMPANYING WITH THE RECENT ACTIVITY OF THE VOLCANO SAKURAJIMA. DISASTER PREV.RES.INST.BULL.NO 48 1961 ET NO 50 1962

8.1
MATVEYEV P.S.
GOLOUBITSKII V. INFLUENCE DES FORCES GENERATRICES DE MAREES LUNI-SOLAIRES SUR LA FREQUENCE DES TREMBLEMENTS DE TERRE TRANSCAUCASIENS TROUDI POLTAVSKOI GRAV.OBS. X PP 67-74 1961
* B.I.M. NO 28 PP 659-665 1962

8.1
CALOI P. MOUVEMENTS LENTS ET IMPREVUS DANS LA CROUTE TERRESTRE ET LEURS RELATIONS RECIPROQUES. SCIENTIA JANVIER 1962

8.1
BOSSOLASCO M. LE CAUSE METEOROLOGICHE DEI TERREMOTI. GEOF.METEO.BOLL.STA.ITALIANA GENOVA VOL.X.N 5/6 1962

8.1
KARMALEYEVA R.M. ON A CERTAIN COICIDENCE BETWEEN ANOMALOUS TILT AND THE OCCURRENCE OF EARTHQUAKES. BULL.AC.SC.URSS.GEOPH.SER.NO 11 PP 970-978 1962

8.1
PRESS F. DISPLACEMENTS, STRAINS AND TILTS AT TELESEISMIC DISTANCES. J.GEOPH.RES.VOL.70 PP 2395-2412 1965

8.11
SKALSKY L. TILT OBSERVATION BEFORE ROCKBURST. GEOPH.INST.CZECHOSL.ACAD.SC.PRAGUE PP 396-402 1963

8.11
SKALSKY L.
PICHA J. EVALUATION OF ROCKBURSTS OBSERVED IN 1958-1961 AT TIDAL STATIONS OF BREZOVE HORY /PRIBRAM/. 5 EME SYMP.INT.SUR LES MAREES TERRESTRES. OBS.ROY.BELG.COMM.NO 236 S.GEOPH.NO 69 PP 291-317 1964

8.15
MACHADO F. ACTIVITY OF THE ATLANTIC VOLCANOES 1947-1965 CONTROL OF THE ERUPTIONS BY THE EARTH TIDE. BULL.VOLCANOLOGIQUE TOME XXX PP 29-34 1967

8.15
MAUK F.J.
KIENLE J. MICROEARTHQUAKES AT ST.AUGUSTINE VOLCANO, ALASKA, TRIGGERED BY EARTH TIDES SCIENCE VOL.182 PP 386-389 1973

8.15
RINEHART J.S. INFLUENCE OF TIDAL STRAIN ON GEOPHYSICAL PHENOMENA PROCEEDINGS 7TH INT.SYMP.EARTH TIDES, SOPRON 1973 AKADEMIAI KIADO BUDAPEST, PP 181-185 1976

8.2
GROTEN E.
THYSSEN-BORNEMISZ APPROACHES IN DG/G-DETERMINATIONS PURE & APPLIED GEOPH. VOL.99, PP 5-11 1972

8.21
TOMASCHEK R. TIDAL GRAVITY MEASUREMENTS IN THE SHETLANDS. EFFECT OF THE TOTAL ECLIPSE OF JUNE 30, 1954 NATURE 175 PP 937-942 1955

8.21
BREIN R. DIE SCHWERKRAFTREGISTRIERUNGEN BEITRAG ZUR FRAGE EINER ABSORPTION DER SCHWERE. AUS DER VEROFFENTLICHUNG VON R.BREIN - H.S.JELSTRUP - K.NOTTARP - H.U.SANDIG - R.SIGL - BEOBACHTUNGEN ZUR SONNENFINSTERNIS 1954 IN SUDNORWEGEN. DEUTSCHE GEOD.KOMM.BAYER.AKAD.WISSENSCH.REIHE B NO 26 1957

8.21
TOMASCHEK R. CONDITIONS D OBSERVATION DE L ECLIPSE DE SOLEIL DU

15 FEVRIER 1961 POUR LES INSTRUMENTS DE MESURE DES MAREES TERRESTRES.
B.I.M. NO 23 PP 460-465 1961

8.21
CAPUTO M.
OBSERVATIONS FAITES A TRIESTE AVEC LES GRANDS PENDULES HORIZONTAUX LORS DE L ECLIPSE DE SOLEIL DU 15.2.1961.
IVME SYMP.INTERNATIONAL SUR LES MAREES TERRESTRES.
OBS.ROY.BELG.COMM. NO 188 S.GEOPH. NO 58 PP 64-65 1961

8.21
SIGL R.
EBERHARD O.
HORIZONTALPENDELBEOBACHTUNGEN IN BERCHTESGADEN WAHREND DER SONNENFINSTERNIS VOM 15.2.1961.
IVME SYMP.INTERNATIONAL SUR LES MAREES TERRESTRES.
OBS.ROY.BELG.COMM.NO 188 S.GEOPH.NO 58 PP 70-75 1961

8.21
ZADRO M.
-POWER SPECTRUM ANALYSIS - DELLE DEVIAZIONI DELLA VERTICALE REGISTRATE DURANTE L ECLISSI TOTALE DI SOLE DEL 15 FEBBRAIO 1960.
ATTI 11 CONV.ASSOC.GEOF.ITAL.ROMA PUBL.NO 65 PP 63-78 1961

8.21
CAPUTO M.
UN NUOVO LIMITE SUPERIORE PER IL COEFFICIENTE DI ASSORBIMENTO DELLA GRAVITAZIONE.
REND.CL.SC.FIS.MAT.ACCAD.NAZ.LINCEI S.8 V.32 PP 509-51K 1962

8.21
DOBROKHOTOV J.S.
THE SUPPOSED SCREENING EFFECT TO GRAVITY AND THE OBSERVATIONS IN KIEV DURING THE SOLAR ECLIPSE FEBRUARY 15TH 1961.
ASSEMBL.BERKELEY UGGI.-B.I.M.NO 34 PP 1060-1061 1963

8.21
SLICHTER L.B.
CAPUTO M.
HAGER C.L.
AN EXPERIMENT CONCERNING GRAVITATIONAL SHIELDING.
JOURN.GEOPH.RES.VOL.70 NO 6 PP 1541-1551 1965

8.3
TOMASCHEK R.
SCHAFFERNICHT W.
ETHER-DRIFT AND GRAVITY.
NATURE TOME 129 NO 3244 PP 24 1932

8.3
TOMASCHEK R.
SCHAFFERNICHT W.
UBER DIE FRAGE DER NACHWEISBARKEIT EINER LORENTZ-KONTRAKTION DER ERDE.
ASTRONOM.NACHRICHTEN BD.248 NO 5929 PP 1-8 1933

8.3
TOMASCHEK R.
HAT DIE KOSMISCHE BEWEGUNG DER ERDE EINEN EINFLUSS AUF DIE SCHWEREBESCHLEUNIGUNG.
DIE STERNE VOL.13 PP 80-85 1933

8.4
ALTERMAN Z.
JAROSCH H.
PEKERIS C.L.
OSCILLATIONS OF THE EARTH.
PROC.ROY.SOC.A, 252 NO 1268 PP 80-95 1959

8.4
NISHIMURA E.
NAKAGAWA I.
TAKEUCHI H.
FREE OSCILLATIONS OF THE EARTH OBSERVED ON GRAVIMETERS.
IVME SYMP.INTERNATIONAL SUR LES MAREES TERRESTRES.
OBS.ROY.BELG.COMM.NO 188 S.GEOPH.PP 57-63 1961

8.4
BUCHHEIM W.
SMITH S.W.
THE EARTH S FREE OSCILLATIONS OBSERVED ON EARTH TIDE INSTRUMENTS AT TIEFENORT, EAST GERMANY.
J.GEOPH.RES.66 PP 3608-3610 1961

8.4
BOLT B.A.
MARUSSI A.
EIGENVIBRATIONS OF THE EARTH OBSERVED AT TRIESTE.
GEOPH.JOURN.ROY.ASTR.SOC.6 NO 3 PP 299-311 1962

9.0
SUTTON G.H.
NEIDELL N.S.
KOVACH R.L.
THEORETICAL TIDES ON A RIGID, SPHERICAL MOON.
JOURN.GEOPH.RES.VOL.68 NO 14 PP 4261-4267 1963

9.0
HARRISON J.C.
AN ANALYSIS OF THE LUNAR TIDES.
JOURN.GEOPH.RES.VOL.68 NO 14 PP 4269-4280 1963

9.0
KAULA W.M.
TIDAL DISSIPATION IN THE MOON.
JOURN.GEOPH.RES.VOL.68 NO 17 PP 4959-4965 1963

9.0
KAULA W.M.
TIDAL DISSIPATION BY SOLID FRICTION AND ORBITAL

	EVOLUTION. INST.GEOPH.PLANET.PHYS.UNIV.CALIF.LOS ANGELES	1963
9.0 GOLDREICH P.	TIDAL DE-SPIN OF PLANETS AND SATELLITES NATURE VOL.208 PP 375-376	1965
9.0 DERR JOHN S.	TRAVEL TIMES, VARIATIONAL PARAMETERS, AND LOVE NUMBERS FOR MOON MODELS. BULL. SEISM. SOC. AMERICA VOL.60 N.3 PP 697-716	1970
9.0 KNOPOFF L.	CORRELATION OF EARTHQUAKES WITH LUNAR ORBITAL MOTIONS THE MOON 2 PP 140-143	1970
9.0 MEISSNER R. SUTTON G. DUENNEBIER F.	MONDBEBEN UMSCHAU HEFT 4 PP 111-115	1971
9.0 MEISSNER R. SUTTON G. DUENNEBIER F.	POSSIBLE SOURCE MECHANISM OF MOONQUAKES I.U.G.G. MOSKAU - AUGUST PP 2-9	1971
9.0 GOUGENHEIM A.	NOTE SUR L INFLUENCE DES MAREES DE LA LUNE SUR LES OBSERVATIONS TELEMETRIQUES DE CET ASTRE BULL.GEOD. NO 100 PP 225-229	1971
9.0 MEISSNER R. VOSS J. KAESTLE H.J.	MOONQUAKES, PROBLEMS OF DETERMINING THEIR EPICENTERS AND MECHANISMS THE MOON 6, PP 292-303	1973
9.0 PEALE S.J.	SOME EFFECTS OF ELASTICITY ON LUNAR ROTATION THE MOON, VOL.8, NO 4, PP 515-531	1973
9.0 CHAPMAN W.B. MIDDLEHURST B.M. FRISILLO A.L.	MOONQUAKE PREDETERMINATION AND TIDES ICARUS 21, PP 427-436	1974
9.1 ECKHARDT D.H.	COMPUTER SOLUTIONS OF THE FORCED PHYSICAL LIBRATIONS OF THE MOON AFCRL-65-892, PAPERS NO 161, PP 466-470	1965
9.1 ECKHARDT D.H.	COMPUTER SOLUTIONS OF THE FORCED PHYSICAL LIBRATIONS OF THE MOON THE ASTRONOMICAL JL. VOL.70 NO. 7, PP 466-471	1965
9.1 ECKHARDT D.H.	LUNAR PHYSICAL LIBRATION THEORY MEASURE OF THE MOON, PP 40-51, REIDEL PUBL.CO	1967
9.1 ECKHARDT D.H.	LUNAR LIBRATION TABLES THE MOON 1, PP 264-275	1970
9.1 ECKHARDT D.H. DIETER K.	A NONLINEAR ANALYSIS OF THE MOON S PHYSICAL LIBRATION IN LONGITUDE THE MOON 2, PP 309-319	1971
9.1 KAULA W.M. BAXA P.A.	THE PHYSICAL LIBRATIONS OF THE MOON, INCLUDING HIGHER HARMONIC EFFECTS THE MOON 8, PP 287-307	1973
9.1 WILLIAMS J.G. SLADE M.A. ECKHARDT D.H. KAULA W.M.	LUNAR PHYSICAL LIBRATIONS AND LASER RANGING THE MOON V.8 NO 4 OCTOBER PP 469-483	1973
9.91 HAN-SHOU LIU	THERMAL AND TIDAL EFFECT ON THE ROTATION OF MERCURY CELESTIAL MECHANICS INT.JL.SPACE DYN. VOL.2 PP 4-8	1969
9.91 PEALE S.J. BOSS A.P.	A SPIN-ORBIT CONSTRAINT ON THE VISCOSITY OF A MERCURIAN LIQUID CORE J.G.R. VOL 82 NO 5 PP 743-749	1977

9.92
GOLD TH. — ATMOSPHERIC TIDES AND THE RESONANT ROTATION OF VENUS
SOTER ST. — ICARUS 11, PP 356-366 — 1969

9.92
SINGER S.F. — HOW DID VENUS LOSE ITS ANGULAR MOMENTUM.
SCIENCE V.170 N.3963 PP 1196-1198 — 1970

9.93
REDMOND J.C. — THE LUNI-TIDAL INTERVAL IN MARS AND THE SECULAR
FISH F.F. — ACCELERATION OF PHOBOS.
ICARUS VOL.8 NO 2 PP 87-91 — 1964

9.94
HUBBARD W.B. — TIDES IN THE GIANT PLANETS
ICARUS VOL.23 NO 1 PP 42-50 — 1974

10.0
CALOI P. — LA GEOFISICA E LE GRANDI DIGHE.
ENERGIE ELETTRICA VOL.XXXIX FASC. NO 1 — 1962

10.0
CALOI P. — ASPETTI GEODINAMICI DELLA DIGA DELL AMBIESTA.
PUBL.IST.NAZ.GEOF.VOL.XVII NO 3 PP 387-405 — 1964

10.0
DECAE A. — IMPLANTATION DU SYNCHROTON DE 200 GEV DE BERKELEY.
INDUST.ATOMIQUES NO 5.6 /VOL.IX/ PP 51-65 — 1965

10.0
CHIROKOV I.A. — MESURES DES INCLINAISONS DANS LES REGIONS DE
ABACHIDSE V.G. — CONSTRUCTION DE BARRAGES.
BAGMET A.L.
IAKOVLIEV V.N. — RAPP.SYMP.INTERN.LENINGRAD 1968 MESURES INCLINAISON
COMM.GEOPH.URSS.MOSCOU PP 230-238 — 1969

10.0
IAKOVLIEV V.N. — OBSERVATION DES INCLINAISONS DE MAREES PRES DE
SANDLER E.D. — L ALIGNEMENT DE LA DIGUE D ANDIGAN EN 1964-1965
LEONTIEVA T.M. — METHODE DE MESURE DES MAREES TERRESTRES ET DES
DEFORMATIONS LENTES DE LA SURFACE DE LA TERRE 167-170
* B.I.M. 63 PP 3240-3243 — 1972

10.0
WARD W.H. — THE USE OF GROUND STRAIN MEASUREMENTS IN CIVIL
BURLAND J.B. — ENGINEERING
MEASUREM., INTERPRETAT., CHANGES OF STRAIN IN THE EARTH
PHIL.TRANS.ROY.SOC.SER.A, VOL.274, NO 1239 PP 421-428 — 1973

Numerical Results Obtained in the Earth Tide Stations

GEOGRAPHICAL INDEX FOR THE EARTH TIDE STATIONS

001	Arctic	220	Iran
010	United Kingdom	240	India
020	Belgium	245	Nepal
025	Luxemburg	250	Thailand
027	Netherlands	255	Malaysia
030	France	260	China, Hong Kong
040	Spain, Portugal	270	Vietnam
050	Italy	280	Japan
060	Switzerland, Austria	300	North Africa
070	Germany (FGR and DDR)	320	Sahara
080	Iceland	340	Central Africa
082	Denmark	360	Madagascar
084	Normay	400	Philippines
086	Sweden	410	Indonesia
088	Finland	416	Papua New Guinea
090	Poland	420	Australia
092	Czechoslovakia	440	New Zealand
095	Hungary	447	Pacific
100	Rumania	600	USA
101	Bulgaria	680	Canada
110	USSR (Europe)	720	Venezuela
120	USSR (Asia)	990	Antarctic

3.141.1 / 0001
MELCHIOR P. RESULTATS PRELIMINAIRES DE MESURES
BONATZ M. DES MAREES TERRESTRES AU SPITZBERG.
BLANKENBURGH J. 6 EME SYMPOSIUM MAREES TERRESTRES STRASBOURG
OBS.ROY.BELG.COMM. A9 S.GEOPH. NO 96 PP 20 1970

3.141.1 / 0001
BONATZ M. STATION LONGYEARBYEN SPITSBERGEN /ASTRO-GEO PROJECT
MELCHIOR P.J. SPITSBERGEN 1968-1970 -MESURES FAITES DANS LES
DUCARME B. 3 COMPOSANTES AVEC 6 PENDULES HORIZONTAUX VM ET
3 GRAVIMETRES ASKANIA
OBS.ROY.BELG.BULL.OBS.MAR.TERR.VOL.IV,FASC.1 PP 1-110 1971

3.141.1 / 0004
BONATZ M. INTERNATIONAL ASTRO-GEO-PROJECT SPITZBERGEN 1968-70
SCHUSTER O. HORIZONTALPENDELSTATION 4
D.GEOD.KOMM.REIHE B.HEFT N.193 PP 3-69 MUNCHEN 1972

3.141.1 / 0004
BONATZ M. INTERNATIONAL ASTRO-GEO-PROJECT SPITZBERGEN 1968-70
SCHUSTER O. HORIZONTALPENDELSTATION 4
STRENGE ZWEIGRUPPEN-ERDGEZEITENANALYSE NACH DER
METHODE DER KLEINSTEN QUADRATE
ERSTE GRUPPENAUSGLEICHUNG
DEUTSCHE GEOD.KOMM.REIHE B, HEFT NO 193, PP1-69 1972

3.221.6 / 0004
BONATZ M. STATION LONGYEARBYEN SPITSBERGEN /ASTRO-GEO PROJECT
MELCHIOR P.J. SPITSBERGEN 1968-1970 -MESURES FAITES DANS LES
DUCARME B. 3 COMPOSANTES AVEC 6 PENDULES HORIZONTAUX VM ET
3 GRAVIMETRES ASKANIA
OBS.ROY.BELG.BULL.OBS.MAR.TERR.VOL.IV,FASC.1 PP 1-110 1971

3.221.6 / 0004
BONATZ M. BERECHNUNG LANGPERIODISCHER GEZEITENWELLEN FUR DIE
CHOJNICKI T. GRAVIMETERSTATION LONGYEARBYEN
INTERNATIONAL ASTRO-GEO-PROJECT SPITSBERGEN 1969/70
MITT. INST.FUR THEOR.GEOD. UNIV.BONN N.7 PP 1-14 1972

3.221.6 / 0004
BONATZ M. INTERNATIONAL ASTRO-GEO-PROJECT SPITZBERGEN 1968-70
SCHUSTER O. HORIZONTALPENDELSTATION 4
D.GEOD.KOMM.REIHE B.HEFT N.193 PP 3-69 MUNCHEN 1972

3.221.6 / 0004
BONATZ M. INTERNATIONAL ASTRO-GEO-PROJECT SPITZBERGEN 1968-70
SCHULLER KL. STRENGE ZWEIGRUPPEN-ERDGEZEITENANALYSE NACH DER
ERSTE GRUPPENAUSGLEICHUNG-
BAYERISCHEN AK.WISS. REIHE B -ANG.GEOD. H.202 PP 3-54 1973

3.221.9 / 0100
TARRANT L.H. TIDAL GRAVITY EXPERIMENTS AT PEEBLES AND KIRKLINGTON.
MONTHLY NOT.R.ASTR.SOC.GEOPH.SUP.VOL.6 NO 5 PP 278-285 1952

3.221.9 / 0100
TOMASCHEK R. HARMONIC ANALYSIS OF TIDAL GRAVITY EXPERIMENTS AT
PEEBLES AND KIRKLINGTON.
MONTHLY NOT.R.ASTR.SOC.GEOPH.SUP.VOL.6 NO 5 PP 286-302 1952

3.141 / 0100
MELCHIOR P. CLINOMETRIC STATIONS IN EUROPE
PROCEEDINGS 7TH INT.SYMP.EARTH TIDES, SOPRON 1973
AKADEMIAI KIADO BUDAPEST, PP 41-70 1976

3.225 / 0100-0800
MELCHIOR P. EARTH TIDE GRAVITY MAPS FOR WESTERN EUROPE
KUO J.T.
DUCARME B. PHYS.OF THE EARTH AND PLAN.INTER.VOL 13 NO 3 PP 184-196 1976

3.141 / 0101
LENNON G.W. OBSERVATIONS OF BOTH COMPONENTS OF TILT AT BIDSTON.
OBS.ROY.BELG.COMM.NO 114 S.GEOPH. NO 39 1957

3.143 / 0101
LENNON G.W. THE USE OF THE MILNE-SHAW SEISMOGRAPH FOR THE

3.221.9 / 0102
TOMASCHEK R. VARIATIONS OF THE TOTAL VECTOR OF GRAVITY AT
WINSFORD /CHESHIRE/ PART.1 GENERAL RESULTS AND
MARITIME LOAD INFLUENCES.
MONTHLY NOT.R.ASTR.SOC.GEOPH.SUP.VOL.6 NO 9 PP 540-556 1954

3.141 / 0102
TOMASCHEK R. VARIATIONS OF THE TOTAL VECTOR OF GRAVITY AT WINSFORD
/CHESHIRE/

MONTHLY NOT.R.ASTR.SOC.GEOPH.SUPPL.V.6 N.9 PP 540-556 1954

3.141 / 0102
TOMASCHEK R. — ERGEBNISSE DER HORIZONTALPENDELMESSUNGEN IN WINSFORD 1950 BIS 1954.
OBS.ROY.BELG.COMM.NO 142 S.GEOPH. NO 47 PP 70-71 1958

3.141 / 0102
TOMASCHEK R.
GROTEN E. — STATION - WINSFORD /ANGLETERRE/. MESURES FAITES DANS LES COMPOSANTES NORD-SUD ET EST-OUEST AVEC LES PENDULES HORIZONTAUX S T NO 1 ET NO 2 EN 1950, 1951 ET 1952.
OBS.ROY.BELG.BULL.OBS.MAR.TERR.VOL.II FASC.3 PP 85 1964

3.221.3 / 6030
ZURN W.
BEAUMONT CH.
SLICHTER L.B. — GRAVITY TIDES AND OCEAN LOADING IN SOUTHERN ALASKA
J.G.R. VOL 81 NO 26 PP 4923-4932 1976

3.221.9 / 0102
TOMASCHEK R. — TIDAL GRAVITY OBSERVATIONS AT WINSFORD /CHESHIRE/.
MONTHLY NOT.R.ASTR.SOC.GEOPH.SUP.VOL.6 NO 6 PP 372-382 1952

3.221.9 / 0104
TOMASCHEK R. — MEASUREMENTS OF TIDAL GRAVITY AND LOAD DEFORMATIONS ON UNST /SHETLANDS/.
GEOF.PURA E APPLICATA-MILANO VOL.37 PP 55-78 1957

3.22 / 020-090
PICHA J.
SKALSKY L. — PROBLEME DES CORRECTIONS DE MAREE AUX OBSERVATIONS DE HAUTE PRECISION DE LA FORCE DE PESANTEUR
GEOFYSIKALNI SBORNIK XXI NO 399 PP 111-135 1973

3.141.1 / 0200
WERY A. — SITES DES STATIONS GEOPHYSIQUES SOUTERRAINES DE L OBSERVATOIRE ROYAL DE BELGIQUE.
OBS.ROY.BELG.COMM.NO 179 S.GEOPH. NO 56 PP 1-8 1960

3.141.1 / 0200
VERBAANDERT J.
MELCHIOR P.J. — LES STATIONS GEOPHYSIQUES SOUTERRAINES ET LES PENDULES HORIZONTAUX DE L OBSERVATOIRE ROYAL DE BELGIQUE.
OBS.ROY.BELG.MON.7 PP 1-146 1960

3.141.1 / 0200
DUCARME B. — SOME COMMENTS ABOUT THE DISPERSION OF THE CLINOMETRIC RESULTS
PROCEEDINGS 7TH INT.SYMP.EARTH TIDES, SOPRON 1973
AKADEMIAI KIADO BUDAPEST, PP 503-510 1976

3.225 / 0200
MELCHIOR P. — CENTRE INTERNATIONAL DE MAREES TERRESTRES /ICET/
PROCEEDINGS 7TH INT.SYMP.EARTH TIDES, SOPRON 1973
AKADEMIAI KIADO BUDAPEST, PP 71-86 1976

3.225 / 0200
MELCHIOR P. — TRANS-EUROPEAN TIDAL GRAVITY PROFILES. PRELIMINARY EXPERIMENTAL RESULTS
PROCEEDINGS 7TH INT.SYMP.EARTH TIDES, SOPRON 1973
AKADEMIAI KIADO BUDAPEST, PP 87-120 1976

3.221.6 / 0201
MELCHIOR P.J. — RESULTATS DE 8 MOIS D ENREGISTREMENT DE LA MAREE A L OBSERVATOIRE ROYAL DE BELGIQUE /UCCLE/ A L AIDE DU GRAVIMETRE ASKANIA NO 145
3E SYMP. MAREES TERRESTRES TRIESTE
BOLL.GEOFISICA TEORICA E APPL.VOL.2 NO 5 PP 129-139 1960

3.221.6 / 0201
MELCHIOR P.J. — RESULTATS DE SEIZE MOIS D ENREGISTREMENTS DE LA MAREE A L OBSERVATOIRE ROYAL DE BELGIQUE /UCCLE/ A L AIDE DU GRAVIMETRE ASKANIA NO 145 /1958-1959/.
OBS.ROY.BELG.3E SER.T.VIII,FASC.4 S.GEOPH.NO 53 PP 135 1960

3.221.6 / 0201
MELCHIOR P.J. — COMPORTEMENTS DE DEUX GRAVIMETRES ASKANIA PENDANT 48 MOIS D ENREGISTREMENTS CONTINUS A L OBSERVATOIRE ROYAL DE BELGIQUE.
IVME SYMP.INTERNATIONAL SUR LES MAREES TERRESTRES.
OBS.ROY.BELG.COMM. NO 188 S.GEOPH. NO 58 PP 311-334 1961

3.221.6 / 0201
MELCHIOR P.J. — STATION UCCLE-BRUXELLES.MESURES FAITES AVEC LE GRAVIMETRE ASKANIA GS 11 NO 145 EN 1960 ET 1961.
OBS.ROY.BELG.BULL.OBS.MAR.TERR.VOL.1 FASC. 1 PP 87 1962

3.221.6 / 0201
MELCHIOR P.J. — STATION UCCLE-BRUXELLES, MESURES FAITES AVEC LE

GRAVIMETRE ASKANIA GS 11 NO 160 EN 1960, 1961 & 1962.
OBS.ROY.BELG.BULL.OBS.MAR.TERR.VOL.1 FASC.5 PP 117 1962

3.221.6 / 0201
MELCHIOR P. PAQUET P.
CONCLUSION DE QUATRE ANNEES D ENREGISTREMENTS DE MAREES TERRESTRES REALISES A L OBSERVATOIRE ROYAL DE BELGIQUE A BRUXELLES A L AIDE DE DEUX GRAVIMETRES.
ACAD.ROY.BELG.BULL.CL.SC.5E SER.TOME XLVIII, 1962-10
OBS.ROY.BELG.COMM.NO 217 S.GEOPH.NO 64 PP 1115-1127 1963

3.221.6 / 0201
MELCHIOR P.J. PAQUET P.
RESULTATS OBTENUS A BRUXELLES A L AIDE DU GRAVIMETRE PERFORATEUR ASKANIA G.S.11 NO 145 B.
5 EME SYMP.INT.SUR LES MAREES TERRESTRES.
OBS.ROY.BELG.COMM.NO 236 S.GEOPH.NO 69 PP 143-147 1964

3.141.1 / 0202
DOPP S.
RESULTATS DES OBSERVATIONS DE MAREES TERRESTRES AU CENTRE DE PHYSIQUE DU GLOBE DE DOURBES.
INST.ROY.METEO.BELG.PUBL.SER.A NO 47 UCCLE-BXL. 1964

3.141.1 / 0202
DOPP S.
LA STATION DE MAREES TERRESTRES DE DOURBES.
5 EME SYMP.INT.SUR LES MAREES TERRESTRES.
OBS.ROY.BELG.COMM.NO 236 S.GEOPH.NO 69 PP 187-192 1964

3.141.1 / 0202
X.
ANNUAIRE - MAREES TERRESTRES 1964-1965
INST.ROY.METEOROL.BELG. PP 188 1966

3.141.1 / 0204
VERBAANDERT J. MELCHIOR P.J.
LA STATION DE PENDULES HORIZONTAUX DE SCLAIGNEAUX.
ACAD.ROY.BELG.BULL.CL.SC.PP 1084-1086 5 DECEMBRE 1959 ET PP 75-78 6 DECEMBRE 1960
OBS.ROY.BELG.COMM.NO 170 S.GEOPH. NO 54 1960

3.141.1 / 0204
VERBAANDERT J. MELCHIOR P.J.
STATION SCLAIGNEAUX /NAMUR/. MESURES FAITES DANS LA COMPOSANTE NORD-SUD AVEC LE PENDULE HORIZONTAL ORB NO 9 EN 1960, 1961 ET 1962.
OBS.ROY.BELG.BULL.OBS.MAR.TERR.VOL.1 FASC.4 PP 90 1962

3.141.1 / 0204
VERBAANDERT J. MELCHIOR P.J.
STATION SCLAIGNEAUX /NAMUR/. MESURES FAITES DANS LA COMPOSANTE NORD SUD AVEC LE PENDULE HORIZONTAL ORB NO 4 EN 1959 ET 1960.
OBS.ROY.BELG.BULL.OBS.MAR.TERR.VOL.1 FASC.2 PP 37 1962

3.141.1 / 0204
VERBAANDERT J. MELCHIOR P.J.
STATION - SCLAIGNEAUX I /NAMUR/. MESURES FAITES DANS LA COMPOSANTE EST-OUEST AVEC LE PENDULE HORIZONTAL ORB NO 31 EN 1962 ET 1963.
OBS.ROY.BELG.BULL.OBS.MAR.TERR.VOL.22 FASC.4 PP 20 1964

3.141.1 / 0204
VERBAANDERT J. MELCHIOR P.J.
STATION - SCLAIGNEAUX I /NAMUR/. MESURES FAITES DANS LA COMPOSANTE NORD-SUD AVEC LES PENDULES HORIZONTAUX ORB NO 30, NO 42 ET NO 13 EN 1962 ET 1963.
OBS.ROY.BELG.BULL.OBS.MAR.TERR.VOL.II FASC.5 PP 27 1964

3.141.1 / 0204
MELCHIOR P.J.
STATION -SCLAIGNEAUX 1.- MESURES FAITES DANS LA COMPOSANTE E-W AVEC LE PENDULE HORIZONTAL ORB NO 31 EN 1964, 1965 ET 1966.
OBS.ROY.BELG.BULL.OBS.MAR.TERR.VOL.III FASC.III PP 91 1967

3.141.1 / 0205
VERBAANDERT J. MELCHIOR P.J.
STATION SCLAIGNEAUX /NAMUR/. MESURES FAITES DANS LA COMPOSANTE EST-OUEST AVEC LE PENDULE HORIZONTAL ORB NO 1 EN 1960, 1961 ET 1962.
OBS.ROY.BELG.BULL.OBS.MAR.TER.VOL.1 FASC.3 PP 110 1962

3.141.1 / 0205
VERBAANDERT J. MELCHIOR P.
STATION-SCLAIGNEAUX/NAMUR/. MESURES FAITES DANS LA COMPOSANTE NORD-SUD AVEC LE PENDULE HORIZONTAL ORB N.13 EN 1961, 1962 ET 1963.
OBS.ROYAL BELG.BULL.OBS.MAREES TER. V.2 FASC.1,28 1963

3.141.1 / 0205
MELCHIOR P.J.
STATION - SCLAIGNEAUX II, III /NAMUR/. MESURES FAITES DANS LA COMPOSANTE EST-OUEST AVEC LES PENDULES HORIZONTAUX ORB NO 10, NO 23, NO 55 ET NO 56 EN 1961, 1962, 1963 ET 1964.
OBS.ROY.BELG.BULL.OBS.MAR.TERR.VOL.II FASC.6 PP 59 1965

3.141.1 / 0207
VERBAANDERT J.
MELCHIOR P.J.
LA STATION DE PENDULES HORIZONTAUX DE WARMIFONTAINE.
ACAD.ROY.BELG.BULL.CL.SC.PP 427-431, 6 MAI 1961
OBS.ROY.BELG.COMM.NO 190 S.GEOPH. NO 59 PP 427-431 1961

3.141.1 / 0207
VERBAANDERT J.
MELCHIOR P.J.
STATION WARMIFONTAINE /NEUFCHATEAU/. MESURES FAITES
DANS LA COMPOSANTE NORD-SUD AVEC LE PENDULE HORIZONTAL
ORB NO 4 EN 1961 ET 1962.
OBS.ROY.BELG.BULL.OBS.MAR.TERR.VOL.1 FASC.6 1963

3.141.1 / 0207
MELCHIOR P.
DUCARME B.
STATION WARMIFONTAINE - MESURES FAITES DANS LES
COMPOSANTES NORD-SUD ET EST-OUEST AVEC LES
PENDULES HORIZONTAUX VM NO 43, 11, 4 ET 22 DE
1964 A 1969
BULL.OBS.MAR.TERR.VOL.4, FASC.2 PP 1-162 1973

3.141.1 / 0208
VERBAANDERT J.
MELCHIOR P.J.
STATION - WARMIFONTAINE II /NEUFCHATEAU/.
MESURES FAITES DANS LES COMPOSANTES NORD-SUD ET
EST-OUEST AVEC LES PENDULES HORIZONTAUX ORB NO 23
ET NO 9 EN 1962 ET 1963.
OBS.ROY.BELG.BULL.MAR.TERR.VOL.22 FASC.2 PP 35 1963

3.221.6 / 0213
JONES L.
LA STATION GRAVIMETRIQUE SOUTERRAINE DE BATTICE.
5 EME SYMP.INT.SUR LES MAREES TERRESTRES.
OBS.ROY.BELG.COMM.NO 236 S.GEOPH.NO 69 PP 148-151 1964

3.221.6 / 0214
JONES L.
ENREGISTREMENT GRAVIMETRIQUE A LA STATION
SOUTERRAINE DE VEDRIN
3E SYMP. MAREES TERRESTRES TRIESTE
BOLL.GEOFISICA TEORICA E APPL.VOL.2 NO 5 PP 140-144 1960

3.141 / 0215
BRAGARD L.
CONSTRUCTION ET INSTALLATION DE CLINOMETRES A LA STATION
DE KANNE /PROVINCE DE LIMBOURG/.
B.I.M. N.36 PP 1230-1239 1964

3.141 / 0215
BRAGARD L.
NOUVELLES METHODES DE CONSTRUCTION ET D INSTALLATION DE
CLINOMETRES DE HAUTE SENSIBILITE A LA STATION DE KANNE
5 EME SYMP.INT.SUR LES MAREES TERRESTRES.
OBS.ROYAL BELG.COMM.N.236 S.GEOPH. N.69 PP 193-196 1964

3.141 / 0215
BRAGARD L.
RESULTATS D OBSERVATIONS CLINOMETRIQUES A LA
STATION DE KANNE /PROVINCE DE LIMBOURG/.
ACAD.ROY.BELG.BULL.SC.5 LII, PP 1571-1577 1966

3.141 / 0215
BRAGARD L.
RESULTATS COMPARATIFS D ANALYSES HARMONIQUES
D OBSERVATIONS CLINOMETRIQUES EN COMPOSANTE EST-OUEST
A LA STATION DE KANNE.
6 EME SYMPOSIUM MAREES TERRESTRES STRASBOURG
OBS.ROY.BELG.COMM. A9 S.GEOPH. NO 96 PP 136 1970

3.14 / 0215
BRAGARD L.
ETUDE COMPARATIVE DE LA DISTORSION DU FACTEUR D AMPLI-
TUBE GAMMA POUR LES ONDES DIURNES ET COMPOSANTE EST-
OUEST A LA STATION DE KANNE /UNIVERSITE DE LIEGE/
PROCEEDINGS 7TH INT.SYMP.EARTH TIDES, SOPRON 1973
AKADEMIAI KIADO BUDAPEST, PP 653-660 1976

3.221.6 / 0250
FLICK J.
LA STATION GRAVIMETRIQUE DE LUXEMBOURG.
5 EME SYMP.INT.SUR LES MAREES TERRESTRES.
OBS.ROY.BELG.COMM.NO 236 S.GEOPH.NO 69 PP 227-237 1964

3.221.6 / 0250
FLICK J.
MELCHIOR P.J.
MESURES DE LA COMPOSANTE VERTICALE DES MAREES
TERRESTRES AU GRAND DUCHE DE LUXEMBOURG.
ACAD.ROY.BELG.BULL.CL.SC.5-LII, 9 PP 1143-1154 1966

3.221.6 / 0250
FLICK J.
MELCHIOR P.J.
RESULTATS DES OBSERVATIONS DE MAREES TERRESTRES DANS
LES TROIS COMPOSANTES AU GRAND DUCHE DE LUXEMBOURG.
ACAD.ROY.BELG.BULL.CL.SC.TOME 54 PP 1214-1221 1968

3.221.6 / 0250
FLICK J.
MELCHIOR P.
STATION -LUXEMBOURG- MESURES FAITES DANS LA
COMPOSANTE VERTICALE AVEC LE GRAVIMETRE ASKANIA GS11

DUCARME B. N.160 DE 1963 A 1970
BULL.OBS.MAREES TERRESTRES V.III FASC.VI PP 1-94 1971

3.141.1 / 0252
FLICK J.
MELCHIOR P.J.
RESULTATS DES OBSERVATIONS DE MAREES TERRESTRES DANS LES TROIS COMPOSANTES AU GRAND DUCHE DE LUXEMBOURG.
ACAD.ROY.BELG.BULL.CL.SC.TOME 54 PP 1214-1221 1968

3.141.1 / 0252
MELCHIOR P.
FLICK J.
DUCARME B.
STATION -WALFERDANGE- MESURES FAITES DANS LES COMPOSANTES NORD-SUD ET EST-OUEST AVEC LES PENDULES HORIZONTAUX VM N.10,12,23,42,56,65,66 EN 1968-1969
BULL.OBS.MAREES TERRESTRES V.III FASC.V PP 1-78 1970

1.24 / 0252
FLICK J.A.
LE LABORATOIRE SOUTERRAIN DE GEODYNAMIQUE BELGO-LUXEMBOURGEOIS DU GRAND-DUCHE DE LUXEMBOURG ET SA COLLABORATION INTERNATIONALE
ARCHIVES - TOME XXXVI NLLE SERIE PP 374-406 1973

3.141.1 / 0252
TSUBOKAWA I.
YANAGISAWA M.
SUGAWA C.
HOSOYAMA K.
MELCHIOR P.
DUCARME B.
FLICK J.
MOENS M.
RESULTS OF A COMPARISON OF ELECTROMAGNETIC TILTMETER /TEM TYPE/ WITH QUARTZ PENDULUMS /VM TYPE/ AT THE EUROPEAN UNDERGROUND LABORATORY OF GEODYNAMICS OF WALFERDANGE.
JOURNAL GEODETIC SOCIETY OF JAPAN V 20 NO 4 PP 209-220
OBS.ROY.BELG.COMM.NO A31 S.GEOPH. NO 125 1974

3.4 / 0252
MELCHIOR P.
DUCARME B.
THE RATIO L/H - A SIMPLE METHOD FOR THE PROSPECTION OF TIDAL DEFORMATIONS
BULL.GEOD.VOL 50 NO 2 PP 137-138 1976

1.24 / 0252
FLICK J.
DUCARME B.
MELCHIOR P.
VAN GILS J.M.
FIRST RESULTS OF EXTENSOMETRIC REGISTRATIONS AT WALFERDANGE - LUXEMBOURG
PROCEEDINGS 7TH INT.SYMP.EARTH TIDES, SOPRON 1973
AKADEMIAI KIADO BUDAPEST, PP 675-686 1976

3.46 / 0254
MELCHIOR P.
DUCARME B.
VAN GILS J.M.
FLICK J.
DENIS C.
PRELIMINARY RESULTS OBTAINED WITH A VERTICAL STRAINMETER AT THE UNDERGROUND LABORATORY OF GEODYNAMICS AT WALFERDANGE /GRAND-DUCHY OF LUXEMBOURG/
PHYS.OF THE EARTH & PLANETARY INTER.V.9 NO2 PP 97-100 1974

3.142 / 0300
V.REBEUR PASCHWITZ HORIZONTAL-PENDEL BEOBACHTUNGEN AUF DER KAISERLICHEN UNIVERSITAT-STERNWARTE ZU STRASSBURG 1892-1894.
GERL.BEITRAGE ZUR GEOPH.BD.2 PP 211-535 1895

3.221.2 / 0300
LECOLAZET R.
L ENREGISTREMENT DE LA MAREE GRAVIMETRIQUE AVEC UN GRAVIMETRE NORTH-AMERICAN.
B.I.M.NO 1 PP 4-9 1956

3.221.2 / 0300
LECOLAZET R.
ENREGISTREMENT ET ANALYSE HARMONIQUE DE LA MAREE GRAVIMETRIQUE A STRASBOURG.
OBS.ROY.BELG.COMM. NO 114 S.GEOPH. NO 39 PP 76 1957

3.221.2 / 0300
LECOLAZET R.
ENREGISTREMENT ET ANALYSE HARMONIQUE DE LA MAREE GRAVIMETRIQUE A STRASBOURG /8 MOIS D OBSERVATION/.
ANN.GEOPH.TOME 13 PP 186-202 1957

3.221.2/3 / 0300
STEINMETZ L.
ANALYSES COMPAREES DES ENREGISTREMENTS DE LA MAREE GRAVIMETRIQUE OBTENUS, EN NOVEMBRE 1957 A STRASBOURG, AVEC DEUX GRAVIMETRES NORTH AMERICAN ET UN GRAVIMETRE ENREGISTREUR.
OBS.ROY.BELG.COMM. NO 142 S.GEOPH. NO 47 PP 30-32 1958

3.221.2 / 0300
LECOLAZET R.
RESULTATS PROVISOIRES DES ENREGISTREMENTS DE LA MAREE GRAVIMETRIQUE EFFECTUES A STRASBOURG, D AOUT 1957 A FEVRIER 1958.
OBS.ROY.BELG.COMM. NO 142 S.GEOPH. NO 47 PP 33-34 1958

3.221.2 / 0300
LECOLAZET R.
RESULTATS DES ENREGISTREMENTS DE MAREE GRAVIMETRIQUE EFFECTUES A STRASBOURG JUSQU EN 1958

3E SYMP. MAREES TERRESTRES TRIESTE
BOLL.GEOFISICA TEORICA E APPL.VOL.2 NO 5 PP 147-151 1960

3.221.2 / 0300
LECOLAZET R. RAPPORT SUR LES OBSERVATIONS DE MAREE GRAVIMETRIQUE FAITES A STRASBOURG EN 1957, 1958 ET 1959.
B.I.M.NO 21 PP 387-395 1960

3.141 / 0303
BLUM P.A., JOBERT G. RESULTATS OBTENUS A L AIDE D INCLINOMETRES EN 1958 ET 1959
3E SYMP. MAREES TERRESTRES TRIESTE
BOLL.GEOFISICA TEORICA E APPL.VOL.2 NO 5 PP 40-44 1960

3.221.6 / 0303
STANOUDIN B. ENREGISTREMENTS DE LA MAREE GRAVIMETRIQUE.
MEMOIRE PRESENTE FAC.SCIENCES UNIV.PARIS 1953

3.221.3 / 0303
JOBERT G. RESULTATS OBTENUS A L AIDE DU GRAVIMETRE LA COSTE-ROMBERG NO 5-3E SYMP.MAREES TERRESTRES TRIESTE
BOLL.GEOFISICA TEORICA E APPL.VOL.2 NO 5 PP 145-146 1960

3.141 / 0303
BLUM P.A., GAULON R., JOBERT G. TRAVAUX SUR LES MAREES TERRESTRES EFFECTUES EN 1960 A L INSTITUT DE PHYSIQUE DU GLOBE DE PARIS.
IVME SYMP.INTERNATIONAL SUR LES MAREES TERRESTRES.
OBS.ROY.BELG.COMM.NO 188 S.GEOPH. NO 58 PP 161-162 1961

3.141 / 0303
BLUM P.A., GAULON R., JOBERT G. RESULTATS OBTENUS A L AIDE D UN INCLINOMETRE FONCTIONNANT SOUS VIDE.
C.R. SEANCES - ACAD.SC.PARIS 258 PP 283-285 1964

3.141 / 0305
LECOLAZET R., STEINMETZ L., WITTLINGER G. PREMIERS RESULTATS D UNE CAMPAGNE DE MESURE DE LA MAREE CLINOMETRIQUE DANS L EST DE LA FRANCE.
6 EME SYMPOSIUM MAREES TERRESTRES STRASBOURG
OBS.ROY.BELG.COMM. A9 S.GEOPH. NO 96 PP 23-26 1970

3.221.6 / 0315
QUEILLE C. LE GRAVIMETRE ENREGISTREUR DU C.E.G.
5 EME SYMP.INT.SUR LES MAREES TERRESTRES.
OBS.ROY.BELG.COMM.NO 236 S.GEOPH.NO 69 PP 141-142 1964

3.141.1 / 0501
BOSSOLASCO M., CANEVA A., CICCONI G. LA STAZIONE GRAVIMETRICA E GEOMAGNETICA DI ROBURENT /PROV.DI CUNEO/.
RIC.SCI.REND.A SER.2 ROMA 7 2 PP 345-352 1964

3.141.1 / 0501 - 0502 - 0505
BOSSOLASCO M., CANEVA A., CICCONI G., EVA C. LES MAREES TERRESTRES A GENOVA,ROBURENT ET TOIRANO.
B.I.M. N.54 MAI PP 2568-2577 1969

3.140 / 0503
TRUDU R. STAZIONE MONTEPONI / SARDEGNA /.
INST.GEOF.MINERARIA - UNIV. DI CAGLIARI
B.I.M.NO 24 PP 480-484 1961

3.141 / 0503
TRUDU R. LA STAZIONE CLINOGRAFICA DI MONTEPONI NELL ANNO GEOFISICO INTERNAZIONALE.
COMM.NAZ.ITAL.COOP.GEOF.INTERN.PUBL.NO 26 PP 137 1963

3.221.6 / 0504
BOSSOLASCO M. PRIMI RISULTATI DELLE ANALISE DELLE REGISTRAZIONI GRAVIMETRICHE DI ARENZANO
3E SYMP. MAREES TERRESTRES TRIESTE
BOLL.GEOFISICA TEORICA E APPL.VOL.2 NO 5 PP 152-153 1960

3.221.6 / 0508-0509
NORINELLI A., TESI F. RESULTATS DES ENREGISTREMENTS DE LA MAREE GRAVITATIONNELLE TERRESTRE AUX STATIONS DE PADOUE ET DE TRIESTE /GROTTA GIGANTE/.
PUBBL.NUOVA SERIE N 44-63 1958, UNIV.PADOVA IST.GEOD. E GEOF. PP 1-15, 3E SYMP.MAREES TERR. TRIESTE
BOLL.GEOFISICA TEORICA E APPL.VOL.2 NO 5 PP 266-281 1960

3.141 / 0509
MARUSSI A. LES GRANDS PENDULES HORIZONTAUX DE LA STATION DE TRIESTE /GROTTA GIGANTE/ POUR LES MAREES TERRESTRES.
OBS.ROY.BELG.COMM.NO 142 S.GEOPH. NO 47 PP 123-124 1958

3.141 / 0509
MARUSSI A. I PRIMI RISULTATI OTTENUTI NELLA STAZIONE PER LO STUDIO

DELLE MAREE DELLA VERTICALE DELLA GROTTA GIGANTE.
IST.GEOD.GEOF.UNIV.TRIESTE BOLL.NO 4 PP 645-667 1960

3.141 / 0509
MARUSSI A. IN THE UNIVERSITY OF TRIESTE STATION FOR THE STUDY OF THE TIDES OF THE VERTICAL IN THE GROTTA GIGANTE
3E SYMP. MAREES TERRESTRES TRIESTE
BOLL.GEOFISICA TEORICA E APPL.VOL.2 NO 5 PP 45-52 1960

3.141 / 0509
ZADRO M. L ACTIVITE DE LA STATION DE TRIESTE /GROTTA GIGANTE/ PENDANT LA PERIODE DE JUIN 1960 A MARS 1961.
IVME SYMP.INTERNATIONAL SUR LES MAREES TERRESTRES.
OBS.ROY.BELG.COMM.NO 188 S.GEOPH. NO 58 PP 165-168 1961

3.221.6 / 0511
IMBO G., BONASIA V., LO BASCIO A. MAREA GRAVIMETRICA ALL OSSERVATORIO VESUVIANO.
OSSERV.VESUVIANO SER.6 VOL.V PP 161-184 1963

3.221.6 / 0511
IMBO G., BONASIA V., LO BASCIO A. VARIAZIONI DELLA MAREA DELLA CROSTA ALL OSSERVATORIO VESUVIANO.
ANNALI OSS.VESUVIANO 6E SER.VOL.7 PP 181-198 1965

3.141.1 / 0513
JONES L. LA STATION CLINOMETRIQUE SOUTERRAINE DE NICOLOSI AU VOLCAN ETNA.
5 EME SYMP.INT.SUR LES MAREES TERRESTRES.
OBS.ROY.BELG.COMM.NO 236 SER.GEOPH.NO 69 PP 280-288 1964

3.141.1 / 0514
DE FEO A. HORIZONTAL PENDULUMS STATION IN THE CASTELLANA CAVES.
IVME SYMP.INTERNATIONAL SUR LES MAREES TERRESTRES.
OBS.ROY.BELG.COMM.NO 188 S.GEOPH. NO 58 PP 174-183 1961

3.141.1 / 0515
DE FEO A. LA STATION NO 2 POUR L ETUDE DE LA MAREE TERRESTRE DANS LES GROTTES DE CASTELLANA.
5 EME SYMP.INT.SUR LES MAREES TERRESTRES.
OBS.ROY.BELG.COMM.NO 236 S.GEOPH.NO 69 PP 274-279 1964

3.225 / 0610
FLORIN R. REGISTRIERUNG DER ERDGEZEITEN IN CHUR
VERHANDLUNGEN SCHW.NATURFORSCH.GES., PP 155-157 1973

3.141.1 / 0695
RINNER K. ERSTER BERICHT UBER DIE ERDGEZEITENSTATION IM GRASZER SCHLOSSBERG /OSTERREICH/
ASS. BERKELEY U.G.G.I. B.I.M. N.34 PP 1041-1042 1963

3.141.1 / 0695
RINNER K. BERICHT UBER DIT ERDGEZEITENSTATION IM GRAZER SCHLOSSBERG
MITT.GEOD.INST.TECHN.HOCHSCHULE GRAZ FOLGE 9 PP 1-41 1971

3.141.1 / 0695
RINNER K., LICHTENEGGER H. EARTH TIDES REGISTRATIONS IN THE AREA OF GRAZ
XVI ASS.GEN.DE L UGGI GRENOBLE - AOUT 1975
MITTEILUNGEN G.INST.TECH.UNIV.GRAZ V.20 PP 147-172 1975

3.225 / 0700
GERSTENECKER C., GROTEN E., RUMMEL R. REPORT ON TIDAL GRAVITY AND TILT MEASUREMENTS DURING 1969-1973
PROCEEDINGS 7TH INT.SYMP.EARTH TIDES, SOPRON 1973
AKADEMIAI KIADO BUDAPEST, PP 687-706 1976

3.221.6 / 0701
BONATZ M. ERGEBNISSE DER SCHWEREREGISTRIERUNGEN IN DER STATION BONN /1964/1965/.
DEUTSCHE GEOD.KOMM.BAYER.WISS.REIHE B GEOD.NO 133 PP 75 1966

3.221.6 / 0701
BONATZ M. ZUR FRAGE DER STORSIGNALE BEI DER ERDGEZEITENREGISTRIERUNG MIT GRAVIMETERN.
TRAD. B.I.M. NO 52 PP 2410-2412 1968

3.221.6 / 0701
BONATZ M. ERGEBNISSE EINER 100-TAGIGEN GRAVIMETERREGISTRIERUNG BEI VERWENDUNG EINES ELECTRONISCHEN VERSTARKERS.
B.I.M. NO 52 PP 2413-2416 1968

3.221.6 / 0701
BONATZ M. ERGEBNISSE VON PARALLELREGISTRIERUNGEN MIT ZWEI ASKANIA-GRAVIMETERN IN DER ERDGEZEITENSTATION BONN.
GEOD.KOMM.BAYER.AKAD.WISS.REIHE B HEFT.158 PP 1-62 1968

3.221.6 / 0701
BONATZ M. ERGEBNISSE GRAVIMETRISCHER PARALLELREGISTRIERUNGEN IN DER ERDGEZEITENSTATION BONN 1967 /68/ DEUTSCH GEOD. KOM. REIHE B HEFT N.170 1969

3.225 / 0701
BONATZ M. CHOJNICKI T. EUROPAISCHES ERDGEZEITENPROFIL ERGEBNISSE DER GEZEITENREGISTRIERUNGEN MIT DEM ASKANIA-GRAVIMETER GS 15 NR.206 IN DEN STATIONEN BONN, BRUXELLES, WALFERDANGE, STRASBOURG 1970/72 MITT. INST.FUR THEOR.GEOD. UNIV.BONN N.8 PP 1-11 1972

3.141.1 / 0703
BONATZ M. ERGEBNISSE DER HORIZONTALPENDELREGISTRIERUNGEN IN DER ERDGEZEITENSTATION ERPEL BEI BONN 1965 BIS 1967. DEUTSCHE GEOD. KOMM. REIHE B HEFT N.173 1969

3.15 / 0703
BONATZ M. CHOJNICKI T. SCHULLER ERSTE ERGEBNISSE DER MESSUNG KLINOMETRISCHER ERDGEZEITEN MIT DEM ASKANIA-VERTIKALPENDEL GBP 1 NR.12 IN DER TESTSTATION ERPEL 1971/72 MITTEIL.A.D.INST.THEOR.GEOD.UNIV.BONN NR.9 PP 1-9 1972

3.15 / 0703
BONATZ M. CLINOMETRIC MEASUREMENTS WITH ASKANIA VERTICAL PENDULUM BOREHOLE TILTMETER GPB 1 NR 1 IN THE TEST STATION ERPEL PROCEEDINGS 7TH INT.SYMP.EARTH TIDES, SOPRON 1973 AKADEMIAI KIADO BUDAPEST, PP 235-241 1976

3.221.2/6 / 0706
GERKE K. BERICHT UBER DIE GRAVIMETERMESSUNGEN DES INSTITUTS FUR ANGEWANDTE GEODASIE, FRANKFURT AM ZUM GEOPHYSIKALISCHEN JAHR 3E SYMP. MAREES TERRESTRES TRIESTE BOLL.GEOFISICA TEORICA E APPL.VOL.2 NO 5 PP 154-156 1960

3.221.2 / 0706
BREIN R. REPORT ON SOME INVERSTIGATIONS CONCERNING GRAVIMETER RECORDINGS IN FRANKFURT A.M. 6 EME SYMPOSIUM MAREES TERRESTRES STRASBOURG OBS.ROY.BELG.COMM. A9 S.GEOPH. NO 96 PP 30-36 1970

3.221.2 / 0706
BREIN R. REPORT ON SOME INVESTIGATIONS CONCERNING GRAVIMETER RECORDINGS IN FRANKFURT AM MAIN PROCEEDINGS 7TH INT.SYMP.EARTH TIDES, SOPRON 1973 AKADEMIAI KIADO BUDAPEST, PP 661-668 1976

3.221.6 / 0709
BACHEM H.C. WENZEL H.G. ERGEBNISSE DER ERDGEZEITEN REGISTRIERUNG MIT EINEM ASKANIA-GRAVIMETER GS 12 IN DER STATION HANNOVER B.I.M. NO 63 PP 3321-3332 1972

3.225 / 0709
TORGE W. WENZEL H.G. VERGLEICH VON ERDGEZEITENREGISTRIERUNGEN MIT SECHS VERSCHIEDENEN GRAVIMETERN MITTEILUNGEN INST. THEOR.GEODASIE, HANNOVER, PP 19 1974

3.23 / 0709
TORGE W. WENZEL H.G. COMPARISON OF EARTH TIDE OBSERVATIONS WITH SEVEN DIFFERENT GRAVIMETERS AT HANNOVER XVI ASS.GEN.DE L UGGI GRENOBLE - AOUT 1975

3.221.3 / 0709
WENZEL H.G. SIMULTANEOUS OBSERVATIONS WITH LACOSTE-ROMBERG GRAVITYMETERS MODEL G. PROCEEDINGS 7TH INT.SYMP.EARTH TIDES, SOPRON 1973 AKADEMIAI KIADO BUDAPEST, PP 307-311 1976

3.141 / 0711
RINNER K. BERICHT UBER DER HORIZONTALPENDELMESSUNGEN IN BERCHTESGADEN. OBS.ROY.BELG.COMM.NO 142 S.GEOPH. NO 47 PP 82-90 1958

3.141 / 0711
RINNER K. BERICHT UBER DIE HORIZONTALPENDELMESSUNGEN IN BERCHTESGADEN. 3E SYMP. MAREES TERRESTRES TRIESTE BOLL.GEOFISICA TEORICA E APPL.VOL.2 NO 5 PP 53-59 1960

3.141 / 0711
SIGL R. EBERHARD O. HORIZONTALPENDELBEOBACHTUNGEN IM WERK 38 DES SALZBERGWERKES BERCHTESGADEN. IVME SYMP.INTERNATIONAL SUR LES MAREES TERRESTRES.

OBS.ROY.BELG.COMM.NO 188 S.GEOPH. NO 58 PP 171-173 1961

3.141 / 0711
EBERHARD O. ERDGEZEITENBEOBACHTUNGEN. TEIL I. ERGEBNISSE DER ERDGE-ZEITENBEOBACHTUNGEN 1958/1959 DER I. ABT.DES DEUTSCHEN GEOD. FORSCHUNGSINSTITUTS IN BERCHTESGADEN.
DEUTSCHE GEOD.KOMM.REIHE B.HEFT N.70 T.1 MUNCHEN 1962

3.141 / 0711
EBERHARD O. ERDGEZEITENBEOBACHTUNGEN. TEIL II ERGEBNISSE DER HORI-ZONTALPENDELREGISTRIERUNGEN 1960 DER I.ABT.DES DEUTSCHEN GEODATISCHEN FORSCHUNGSINSTITUTS IN BERCHTESGADEN.
DEUTSCHE GEOD.KOMM.REIHE B.HEFT N.70 T.2 MUNCHEN 1963

3.141 / 0711
EBERHARD O. ERDGEZEITENREGISTRIERUNGEN IM SALZBERGWEK BERCHTESGADEN
5 EME SYMP.INT.SUR LES MAREES TERRESTRES.
OBS.ROYAL BELG.COMM.N.236 S.GEOPH.N.69 PP 245-248 1964

3.141 / 0711
EBERHARD O. EARTH TIDE MEASUREMENTS IN THE SALT MINE OF BERCHTESGADEN CARRIED OUT BY DGFI MUNICH WITHIN THE YEARS 1961-1968
XV ASS.GEN. UGGI - ASS.INT.DE GEODESIE MOSCOU AOUT 1971

3.221.6 / 0711
EBERHARD O. ERGEBNISSE DER ERDGEZEITENBEOBACHTUNGEN 1961-1968 IN BERCHTESGADEN DER ABT.I DES DEUTSCHEN GEODATISCHEN FORSCHUNGSINSTITUTS.
DEUTSCHE GEOD.KOMM. - BAYERISCHEN AK. DER WIS.
REIHE B - ANGEW.GEOD. HEFT N.70 TEIL III PP 1-24 1971

3.141 / 0711
EBERHARD O. ERGEBNISSE DER ERDGEZEITENBEOBACHTUNGEN 1961-1968 IN BERCHTESGADEN DER ABT.I DES DEUTSCHEN GEODATISCHEN FORSCHUNGSINSTITUTS.
DEUTSCHE GEOD.KOMM. - BAYERISCHEN AK. DER WIS.
REIHE B - ANGEW.GEOD. HEFT N.70 TEIL III PP 1-24 1971

3.15 / 0712
FLACH D.
ROSENBACH O. DER ASKANIA BOHRLOCH-NEIGUNGSMESSER /GEZEITENPENDEL/ NACH A.GRAF AUF DER TEST-STATION ZELLERFELD-MUHLENHOHE.
B.I.M. NO 60 PP 2934-2943 1971

3.15 / 0712
FLACH D.
ROSENBACH O.
WILHELM H. UNTERSUCHUNGEN DES ASKANIA BOHRLOCH-NEIGUNGSMESSERS /GEZEITENPENDEL/ NACH A.GRAF AUF DER TEST-STATION ZELLERFELD-MUHLENHOHE. /TEIL I/
B.I.M. NO 60 PP 2944-2954 1971

3.15 / 0712
FLACH D.
ROSENBACH O.
WILHELM H. TIDAL ANALYSES OF SIMULTANEOUS RECORDS TAKEN BY TWO ASKANIA BOREHOLE TILTMETERS
ZEITSCHRIFT GEOPH V.37 PP 993-1003 1971

3.221.6 / 0715
STUKENBROKER B. ERGEBNISSE VON ERDGEZEITEN-PARALLELREGISTRIERUNGEN MIT DREI ASKANIA-GRAVIMETERN.
ZEITSCH.FUR GEOPH.BAND 39, PHYSICA-VERLAG PP 1-20 1973

3.15 / 0715
ZSCHAU J. LOTSCHWANKUNGSANOMALIEN IN ERDGEZEITEN-REGISTRIERUNGEN MIT DEM ASKANIA-BOHRLOCH-VERTIKALPENDEL NACH A.GRAF.
DISSERT.DOKTORGRADES CH.A.UNIV.KIEL, PP 1-215 1974

3.225 / 0715
STUKENBROKER B. OBSERVATIONS OF LOADING DEFORMATION BY A GRAVIMETRIC TIDAL PROFILE ACROSS NORTHERN GERMANY
PROCEEDINGS 7TH INT.SYMP.EARTH TIDES, SOPRON 1973
AKADEMIAI KIADO BUDAPEST, PP 137-148 1976

3.141.1 / 0716
KIELSEL H. PRELIMINARY RESULTS OF TIDAL OBSERVATIONS AT SCHILTACH
PROCEEDINGS 7TH INT.SYMP.EARTH TIDES, SOPRON 1973
AKADEMIAI KIADO BUDAPEST, PP 729-734 1976

3.141.1 / 0720
BONATZ M. HORIZONTALPENDELREGISTRIERUNGEN BEI GERINGER GESTEINSUBERDECKUNG.
B.I.M. N.54 MAI PP 2545-2547 1969

3.141.1 / 0720
BONATZ M.
CHOJNICKI T. TROPFSTEINHOHLE WIEHL - ERGEBNISSE VON HORIZONTALPENDEL REGISTRIERUNGEN BEI GERINGER GESTEINSUBERDECKUNG.

ROCHOLL W. MITTEIL.A.D.INST.THEOR.GEOD.UNIV.BONN NR.10 PP 1-11 1972
3.14 / 0762
SIMON D. AUFBAU EINER NEUEN ERDGEZEITENSTATION IM SALZBERGWERK TIEFENORT
VERMESSUNGSTECHNIK 18 HEFT 12 PP 472-474 1970
3.141 / 0762
SIMON D. UBER ERGEBNISSE VON NEIGUNGSBEOBACHTUNGEN IN TIEFENORT
VEROFF.ZENTRALINST.PHYS.ERDE NO 30 TEIL 1 PP 207-233 1974
5.612 / 0762
SIMON D. COHERENCE OF TIDAL PARAMETERS OBSERVED AT TIEFENORT /GDR/ WITH RESPECT TO PSEUDO-TIDAL EFFECTS INDUCED BY ATMOSPHERIC PRESSURE VARIATIONS
XVI ASS.GEN.DE L UGGI GRENOBLE - AOUT 1975
3.221.6 / 0764
BYL J. ERGEBNISSE VON ERDGEZEITENBEOBACHTUNGEN IN POTSDAM /MESSUNG DER ZEITLICHEN VARIATIONEN DER HORIZONTALINTENSITAT/
VERMESSUNGSTECHNIK SOUS-PRESSE 1971
3.221.6 / 0764
DITTFELD H.J. ONE YEAR EARTH TIDE REGISTRATIONS WITH THE GRAVIMETER GS 15 NO 222 IN POTSDAM
XVI ASS.GEN.DE L UGGI GRENOBLE - AOUT 1975
3.221.6 / 0764
DITTFELD H.J. ERSTE ERGEBNISSE MIT DEM GEZEITENGRAVIMETER GS 15 AN DER STATION POTSDAM
B.I.M. NO 69 PP 3850-3853 1974
3.221.6 / 0764
BYL J. ERGEBNISSE HARMONISCHER ANALYSEN VON GRAVIMETER-REGISTRIERUNGEN IN POTSDAM AUS DEM JAHRE 1962.
GERL.BEITR.GEOPH.LEIPZIG 73-3 S.PP 178-183 1964
3.221.6 / 0764
BYL J. ERGEBNISSE DER GRAVIMETERREGISTRIERUNGEN AUS DEN JAHREN 1959, 1960 UND 1961.
ARBEITEN AUS DEM GIP 2 1964
3.141 / 0765
BYL J. RESULTS OF TILT OBSERVATIONS AT POTSDAM
XV ASS.GEN. UGGI - ASS.INT.DE GEODESIE MOSCOU AOUT 1971
3.221.6 / 0764
BYL J. ERGEBNISSE VON ERDGEZEITENBEOBACHTUNGEN IN POTSDAM /MESSUNG DER ZEITLICHEN VARIATIONEN DER VERTIKALINTENSITAT/
VERMESSUNGSTECHNIK 18 1/2 -TEIL I/II 1970
3.221.6 / 0764
DITTFELD H.J. ERSTE ERGEBNISSE DES NEUEN GEZEITENGRAVIMETERS IM GRAVIMETRISCHEN OBSERVATORIUM POTSDAM
AK.DER WISSENSCH.DER DDR,ZENTRALINST.FUR PHYSIK DER ERDE 1974
3.141 / 0767
SCHWEYDAR W. LOTSCHWANKUNG UND DEFORMATION DES ERDE DURCH FLUTKRAFTE /GEMESSEN MIT ZWEI HORIZONTALPENDELN IM BERGWERK IN 189 M TIEFF BEI FREIBERG I.SA./
ZENTRALBUREAU DER INTERN.ERDMESSUNG NF, N.38 1921
3.141 / 0767
MELCHIOR P.J.
PAQUET P. ANALYSE HARMONIQUE DES OBSERVATIONS DE SCHWEYDAR A FREIBERG/SA /1911-1915/ PAR LA METHODE LECOLAZET
B.I.M. NO 32 PP 895-901 1963
3.41 / 0767
HIERSEMANN L. DIE BESTIMMUNG DER SHIDASCHEN ZAHL L AUS FREIBERGER STRAINSEISMOMETERREGISTREIRUNGEN.
VEROFF.INST.ANGEW.GEOPH.BERGAKAD.FREIBERG NR 118 1962
3.141 / 0767
SCHNEIDER M. ERGEBNISSE DER LOTSCHWANKUNGSBEOBACHTUNGEN IN FREIBERG
GERLANDS BEITR.GEOPH.76, 2 PP 106-116 1967
3.14 / 0767
SCHNEIDER M. ERGEBNISSE DER LOTSCHWANKUNGSMESSUNGEN MIT HORIZONTAL- UND VERTIKALPENDELN IN FREIBERG VON 1963 BIS 1966.
GEOD.GEOPH.VEROFF.REIHE III HEFT 10 PP 23 1968
3.15 / 0768
SCHNEIDER M. DIE REGISTRIERUNG VON LOTSCHWANKUNGEN MIT HILFE EINES VERTIKALPENDELS IN FREIBERG /SA/.
BERGAKADEMIE 8 PP 569-570 1962

3.15 / 0768
SCHNEIDER M. LOTSCHWANKUNGSMESSUNGEN MIT VERTIKALPENDELN IN FREIBERG/SA.
5 EME SYMP.INT.SUR LES MAREES TERRESTRES.
OBS.ROY.BELG.COMM.NO 236 S.GEOPH.NO 69 PP 262-273 1964

3.15 / 0768
SCHNEIDER M. MESSUNG DER LOTSCHWANKUNGEN MIT VERTIKALPENDELN IN FREIBERG/SA.
DEUTSCHE GEOD.KOMM.BAYER.AKAD.WISSENSCH.REIHE C HEFT 79 1965

3.221.3 / 0800
TORGE W.
WENZEL H.G. GRAVIMETRIC EARTH TIDE OBSERVATIONS IN ICELAND
B.I.M. NO 74 PP 4312-4318 1976

3.225 / 0840
BONATZ M.
RICHTER B. GRAVIMETRISCHES GEZEITENPROFIL BONN-LONGYEARBYEN /SPITZBERGEN/ HAUPTPARTIALTIDEN FUR DIE STATIONEN OSLO, BERGEN UND TRONDHEIM
MITT.INST.THEOR.GEOD.UNIV.BONN N.33 PP 1-14 1975

3.141.1 / 0862
VOGEL A. THE HORIZONTAL PENDULUMS OF DANNEMORA.
5 EME SYMP.INT.SUR LES MAREES TERRESTRES.
OBS.ROY.BELG.COMM.NO 236 S.GEOPH.NO 69 PP 289-290 1964

3.141.1 / 0862
ANDERSON A.J. SPECTRAL ANALYSIS OF HORIZONTAL TILT FROM FOUR SITES IN SCANDINAVIA
UNIV.UPPSALA DEP.SOLID EARTH PHYSICS NO 21 PP 1-30 1974

1.24 / 0880
HONKASALO T. FINNISH OBSERVATIONS AND RESEARCH ON EARTH TIDES IN 1971 - 1974
XVI ASS.GEN.DE L UGGI GRENOBLE - AOUT 1975

3.225 / 0880
HONKASALO T. FENNOSCANDINAVIAN TIDAL GRAVITY PROFILE, PRELIMINARY EXPERIMENTAL RESULTS
PROCEEDINGS 7TH INT.SYMP.EARTH TIDES, SOPRON 1973
AKADEMIAI KIADO BUDAPEST, PP 25-30 1976

3.141.1 / 0881
HONKASALO T. THE EARTH TIDE STATION LOHJA IN FINLAND.
B.I.M. N.55 SEPTEMBRE PP 2626-2627 1969

3.141.1 / 0881
GROHN P. THE FINNISH EARTH TIDE STATIONS.
6 EME SYMPOSIUM MAREES TERRESTRES STRASBOURG
OBS.ROY.BELG.COMM. A9 S.GEOPH. NO 96 PP 37-39 1970

3.221.6 / 0901
WITKOWSKI J. RESULTATS PROVISOIRES DES ENREGISTREMENTS DES MAREES TERRESTRES A LA STATION DE LATITUDE DE L ACADEMIE POLONAISE DES SCIENCES A BOROWIEC /POZNAN/.
3E SYMP. MAREES TERRESTRES TRIESTE
BOLL.GEOFISICA TEORICA E APPL.VOL.2 NO 5 PP 157-160 1960

3.221.6 / 0901
DOBRZYCKA M. OBSERVATIONS OF TIDAL GRAVITY VARIATIONS AT BOROWIEC.
ACTA GEOPH.POLON.WARSZAWA 8 3 PP 279-280 1960

3.144 / 0901
WITKOWSKI J. RECHERCHES SUR LES PENDULES DOUBLES TYPE LETTAU FAITES A LA STATION DE LATITUDE DE L ACADEMIE POLONAISE DES SCIENCES A BOROWIEC.
IVME SYMP.INTERNATIONAL SUR LES MAREES TERRESTRES.
OBS.ROY.BELG.COMM. NO 188 S.GEOPH. NO 58 PP 186-188 1961

3.144 / 0901
NOWAK S. OBSERVATIONS DE VARIATIONS DE LA VERTICALE A BOROWIEC EN 1961-1962.
ACTA GEOPH.POLON WARSZAWA 14, 1 PP 33-59 1966

3.221.6 / 0901
JAKS W. MAREE GRAVIMETRIQUE A BOROWIEC
B.I.M. N.59 OCTOBRE P 2838 1970

3.221.6 /0905
CHOJNICKI T.
ZARNOWIECKI W. RESULTATS DES MESURES GRAVIMETRIQUES DES MAREES, EXECUTEES AU COURS DES ANNEES 1971/72 A LA STATION VARSOVIE
PUBL.INST.GEOPH.POLISH AC.OF SC. V.71 PP 75-116 1973

3.141. / 0906
CHOJNICKI T. RESULTATS DES MESURES CLINOMETRIQUES DES MAREES EXECU-

WEISS J. — TEES AU COURS DES ANNEES 1973/74 A LA STATION KSIAZ
PUBLS.INST.GEOPH.POL.AC.SCI.VOL 94 PP 25-67 — 1975

3.141 / 0906
CHOJNICKI T.
WEISS J.
RESULTATS DES MESURES CLINOMETRIQUES DES MAREES EXECU-
TEES AU COURS DES ANNEES 1974/75 A LA STATION KSIAZ
PUBLS.INST.GEOPH.POLISH AC.SCI.,F-1/105/PP 42-84 — 1976

3.221.6/ 092
DITTFELD H.J.
SIMON Z.
VARGA P.
VOLKOV V.A.
VENEDIKOV A.P.
BROZ J.
HOLUB S.
EARTH TIDE OBSERVATIONS BY ASKANIA GRAVITYMETERS
AT THE GEODETICAL OBSERVATORY PECNY
/ CZECHOSLOVAKIA /
PUBL.HUNGARIAN GEOPH.INST.-ROLAND EOTVOS

3.141 / 0921
PETR V.
BEOBACHTUNGEN DER GEZEITEN DER ERDKRUSTE IN BREZOVE HORY.
BULL.ASTR.INST.OF CZECHOSLOVAKIA T.VI N.2 PP 27-32 — 1955

3.141 / 0921
PICHA J.
ERGEBNISSE DER GEZEITENBEOBACHTUNGEN DER FESTEN ERDKRUSTE
IN BREZOVE HORY IN DEN JAHREN 1936-1939.
INST.GEOPH.ACAD.TCHECOSL.SC.N.42 GEOPH.SBORNIK 1956
NCSAV PRAHA PP 95-312 — 1957

3.141 / 0921
PICHA J.
GEZEITENBEOBACHTUNGEN IN BREZOVE HORY AUS DEN
JAHREN 1926-1928.
TRAV.INST.GEOPH.AC.TCHEC.SC.NO 64 PP 281-345 — 1957

3.141 / 0921
PICHA J.
ERGEBNISSE DER NEUEN BEARBEITUNG DER ERDGEZEITENBEOBACH-
TUNGENAUS DEN JAHREN 1927-1928 UND 1936-1939 IN BREZOVE
HORY /PRIBRAM/ NACH DER METHODE VON PERTSEV
TRAV.DE L INST GEOPH. DE L AC.TCHECOSL. DES SC. N.198
GEOFYSIKALNI SBORNIK PP 67-95 1964 TRAV. GEOPH. — 1964

3.141 / 0921
PICHA J.
ERGEBNISSE DER GEZEITENBEOBACHTUNGEN AN DER STATION
PRIBRAM-BREZOVE HORY, GRUBE -ANNA- AUS DEN
JAHREN 1958-1961.
GEOPH.SBORNIK PRAHA XIII PP 134 — 1965

3.141 / 0921
PICHA J.
SKALSKII L.
RESULTATS DES OBSERVATIONS COLLECTIVES DES MAREES A
L AIDE DE PENDULES HORIZONTAUX SIMPLES ET DE
CLINOMETRES PHOTOELECTRIQUES A LA STATION DE
PRIBRAM-BREZOVE HORY.
TRAV.INST.GEOPH.TCHEC.SC.GEOF.SBORNIK 243 PP 83-103 1966
B.I.M. NO 52 PP 2465-2477 — 1968

3.14 / 0921
PICHA J.
QUELQUES CONCLUSIONS DES MESURES DE COMPARAISON
INTERNATIONALES FAITES A PRIBRAM
STUDIA GEOPH.ET GEOD. N.16 PP 200-202 — 1972

3.221.6 / 0921
SKALSKY L.
PICHA J.
SKOCH V.
TIDAL MEASUREMENTS WITH THE GS11 GRAVIMETER
IN PRIBRAM
GEOFYSIKALNI SBORNIK XXI, NO 400 PP 137-212 — 1973

3.141.2 / 0924
PICHA J.
SKALSKII L.
RESULTATS DES OBSERVATIONS COLLECTIVES DES MAREES
A L AIDE DE PENDULES HORIZONTAUX SIMPLES ET DE
CLINOMETRES PHOTOELECTRIQUES A LA STATION DE
PRIBRAM - BREZOVE HORY.
* B.I.M. N.52 DECEMBRE PP 2465-2477 — 1968

3.141.2 / 0925
OSTROVSKII A.E.
PICHA J.A.
SKALSKI L.
MIRONOVA L.J.
WITMAN N.G.
INCLINAISONS DE MAREES D APRES LES OBSERVATIONS AVEC LE
CLINOMETRE PHOTOELECTRIQUE A PRIBRAM /PRES DE PRAGUE/
RECH.SUR LES MAR.TERR.ART.XIII SECT.PR.IGY NO 3 1963
* B.I.M. N.35 FEVRIER PP 1167-1179 — 1964

3.141.1 / 0927
MELCHIOR P.
SKALSKY L.
STATION PRIBRAM/BELG. MESURES FAITES DANS LES COMPOSANTES
NORD-SUD ET EST-OUEST AVEC LES PENDULES HORIZONTAUX
VM N.76 77 EN 1966 1967 ET 1968
BULL.OBS. MAREES TERRESTRES III FASC.IV PP 1-12 — 1969

3.221.6 / 0928
BROZ J.
SIMON Z.
DIE REGISTRIERUNG DER GEZEITENVARIATIONEN DER SCHWEREBESCHLEUNIGUNG AUF DER STATION CESKE BUDEJOVICE SBORNIKU VYZKUMNYCH PRACI SVAZEK 9 PP 33-44 1974

3.221.6 / 0928
SIMON Z.
THE OBSERVATION OF TIDAL VARIATIONS OF GRAVITY AT TWO STATIONS IN CZECHOSLOVAKIA
PROCEEDINGS 7TH INT.SYMP.EARTH TIDES, SOPRON 1973
AKADEMIAI KIADO BUDAPEST, PP 767-774 1976

3.221.6 / 0929
SKALSKY L.
PICHA JAN
SKOCH V.
TIDAL MEASUREMENTS WITH THE GS 11 GRAVIMETER IN PRIBRAM
GEOFYSIKALNI SBORNIK XXI PP 137-212 1973

3.221.6 / 0930
DITTFELD H.J.
SIMON Z.
VARGA P.
VOLKOV V.A.
VENEDIKOV A.P.
BROZ J.
HOLUB S.
EARTH TIDE OBSERVATIONS BY ASKANIA GRAVITYMETERS AT THE GEODETICAL OBSERVATORY PECNY /CZECHOSLOVAKIA/
HUNGARIAN GEOPH.INST. -R.EOTVOS-BUDAPEST PP 1-87 1976

3.141.1 / 0951
LICHTENEGGER H.
VORLAUFIGER BERICHT OBER ERDGEZEITENREGISTRIERUNGEN IN SOPRON/UNGARN
PROCEEDINGS 7TH INT.SYMP.EARTH TIDES, SOPRON 1973
AKADEMIAI KIADO BUDAPEST, PP 713-723 1976

3.221.6 / 0954
VARGA P.
HARMONIC ANALYSIS OF EARTH TIDE OBSERVATIONS IN THE SECOND HALF OF 1967 AS RECORDED IN TIHANY.
GEOFIZIKAI KOSLEMENYEK VOL 19 N.1-2 PP 69-75 1969

3.141.1 / 1001
ZUGRAVESCU D.
PENDULUL ORIZONTAL DIN CADRUL OBSERVATORULUI GRAVIMETRIC CALDARUSANI.
STUDII GEOL.GEOF.GEOGRAF.S.GEOF.2 TOMUL 5 PP 231-243 1967

3.221.6 / 1001
ZUGRAVESCU D.
LA STATION POUR L ENREGISTREMENT DES MAREES TERRESTRES DE L OBSERVATOIRE GRAVIMETRIQUE DE CALDARUSANI.
REV.GEOL.GEOPH.GEOGR.SER.11 NO 1 PP 75-85 BUCAREST 1967

3.141.1 / 1002
ZUGRAVESCU D.
STATIILE PENTRU INREGISTRAREA MAREELOR CLINOMETRICE PADES I SI PADES II.
STUD. CERCETARI GEOL.GEOF.GEOG. S.GEOF. V7 N2 PP143-149 1969

3.141.1 / 1002
ZUGRAVESCU D.
STATIA PENTRU INREGISTRAREA MAREELOR CLINOMETRICE, IN CURS DE AMENAJARE LA CLOSANI, VESTUL CARPATILOR MERIDIONALI.
STUD.CERCETARI GEOL.GEOF.GEOG. SERIA GEOF. V7 N1 PP3-8 1969

3.221.6 / 1011
VENEDIKOV A.P.
PREMIERS ENREGISTREMENTS DES MAREES TERRESTRES A SOFIA.
IVME SYMP.INTERNATIONAL SUR LES MAREES TERRESTRES.
OBS.ROY.BELG.COMM. NO 188 S.GEOPH. NO 58 PP 144-148 1961

3.141.1 / 1012
VENEDIKOV A.P.
INSTALLATION D UNE STATION CLINOMETRIQUE PRES DE SOFIA ET NOTE SUR L ORIENTATION DES PENDULES HORIZONTAUX.
5 EME SYMP.INT.SUR LES MAREES TERRESTRES.
OBS.ROY.BELG.COMM.NO 236 S.GEOPH.NO 69 PP 318-322 1964

3.141.2 / 1100
OSTROVSKY A.E.
MATVEEV P.S.
TIDAL TILTS OF THE EARTH BY OBSERVATIONS IN THE USSR.
6 EME SYMPOSIUM MAREES TERRESTRES STRASBOURG
OBS.ROY.BELG.COMM. A9 S.GEOPH. NO 96 PP 90-94 1970

3.141.5 / 1100
OSTROVSKY A.YE.
RESULTS OF OBSERVATIONS OF TIDAL TILTS OF THE EARTH S SURFACE ON THE TERRITORY OF THE USSR FOR THE PERIOD OF PROC 1957-1972
AKADEMIAI KIADO BUDAPEST, PP 121-126 1976

3.221.6 / 1102
DOBROKHOTOV J.S.
BELIKOV B.D.
KRAMER M.V.
BARSENKOV S.
TIDAL VARIATIONS OF GRAVITY IN PULKOVO
3E SYMP. MAREES TERRESTRES TRIESTE
BOLL.GEOFISICA TEORICA E APPL.VOL.2 NO 5 PP 170-176 1960

3.141.2 / 1117
HOROMSKII A.V. PREMIERE SERIE D OBSERVATIONS CLINOMETRIQUES A OBNINSK.
OKHOTSIMSKAIA M.V. METH.MES.MAR.TERR.& DEFORM.LENTES SURFACE TERRE
ACAD.SC.URSS.INST.PHYS.TERR.SCHMIDT MOSCOU PP 155-159 1970
* B.I.M. NO 61 PP 3073-3076 1971

3.221.6 / 1117
VOLKOV V.A. VARIATIONS DE MAREES DE LA PESANTEUR A OBNINSK
GOUSEVA F.P.
DOBROKHOTOV I.S.
IVANOVA M.V. ACAD.SC.URSS-INST.PHYS.TERRE MOSCOU PP 1-9 1970

3.221.6 / 1117
VOLKOV V.A. VARIATIONS DE MAREES DE LA PESANTEUR A OBNINSK
GOUSEVA F.P.
DOBROKHOTOV I.S. ACAD.SC.URSS-INST.PHYS.TERRE MOSCOU PP 1-9 1970
IVANOVA M.V. * B.I.M. MARS NO 65 PP 3468-3480 1973

3.141.2 / 1117
MIRONOVA L.I. RESULTATS DE LA REDUCTION DES INCLINAISONS DE
OKHOTSIMSKAIA M. MAREES A LA STATION D OBNINSK POUR 1967 A 1968
KHOROMSKII ACAD.NAOUK SSSR, INST.FISIKA ZEMLI PP 82-95 1973
* B.I.M. NO 69 PP 3860-3873 1974

3.141.2 / 1117
MIRONOVA L.I. RESULTATS DE LA REDUCTION DES INCLINAISONS DE
KHOROMSKII A.V. OBSERV. GEOPHYSIQUES COMPLEXES A OBNINSK PP 89-103 1974
OKHOTSIMSKAIA M.V. MAREES A TSO OBNINSK.
* B.I.M. NO 72 PP 4101-4115 1975

3.221.6 / 1119
PARIISKY N.N. FIRST OBSERVATIONS AT KRASNAYA PAKHRA NEAR MOSCOW.
PERTZEV B.P.
KRAMER M.V. OBS.ROY.BELG.COMM. NO 142 S.GEOPH. NO 47 PP 58-63 1958

3.221.6 / 1119
PARIISKY N.N. OBSERVATIONS OF TIDAL VARIATIONS OF GRAVITY
DOBROKHOTOV Y.S. IN KRASNAYA PAKHRA /NEAR MOSKOW/
PERTSEV B.P.
KRAMER M.V. TRAV.SUR MAR.TERR.NO 1 GROUPE 13 GRAVIM. M, 1959
BELIKOV B.D. 3E SYMP. MAREES TERRESTRES TRIESTE
BARSENKOV S.N. BOLL.GEOFISICA TEORICA E APPL.VOL.2 NO 5 PP 164-169 1960

3#221.6 / 1119
DOBROCHOTOV Y.S. OBSERVATIONS REITERATIVES DES VARIATIONS DE MAREES DE LA
FORCE DE PESANTEUR A KRASNAYA PAKHRA.
RECH. MAREES TERR. ART. N.3 PUBL.ACAD.SC.URSS MOSCOU1963
* B.I.M. N.38 NOVEMBRE PP 1331-1336 1964

3.221.7 / 1119
GRIDNIEV D.G. ENREGISTREMENT DES VARIATIONS DE MAREES DE
L ACCELERATION DE LA FORCE DE PESANTEUR A
KRASNAIA PAKHRA PAR LE GRAVIMETRE STATIQUE EN
QUARTZ A ENREGISTREMENT PHOTOELECTRIQUE.
INST.PHYS.TERR.MOSCOU 1967 PP 111-118
* B.I.M. NO 60 PP 2988-2998 1971

3.221.6 / 1120
PERTSEV B.P. OBSERVATIONS DES VARIATIONS DE MAREES DE LA FORCE
IVANOVA M.V. DE PESANTEUR A MOSCOU.
ACAD.SC.URSS INST.PHYS. TERR.SCHMIDT MOSCOU 1964
* B.I.M. N.47 AVRIL PP 1996-2000 1967

3.141.2 / 1121
LATININA L.A. RESULTATS DES OBSERVATIONS DE GROUPES DES
KARMALEYEVA R.M. INCLINAISONS DE LA SURFACE DE LA TERRE A LA
STATION OPPOSEE A LA REGION DE MOSCOU.
RAPP.SYMP.INTERN.LENINGRAD 1968 MESURES INCLINAISON
COMM.GEOPH.URSS. MOSCOU PP 217-229 1969
* B.I.M. N.57 MAI PP 2720-2728 1970

3.141.2 / 1121
OSTROVSKII A.E. INCLINAISONS DE MAREES DE LA SURFACE DE LA TERRE D APRES
MIRONOVA L.I. LES OBSERVATIONS DANS LES PUITS VOISINS DE MOSCOU.
METHODE MESURE MAREES TERRESTRES
ACAD.SC.URSS.INST.PHYS.TERR.MOSCOU PP 144-154 1970
* B.I.M. NO 60 PP 2923-2933 1971

3.141.2 / 1121
FANDIOUCHINA S.M. RESULTATS DES OBSERVATIONS CLINOMETRIQUES EFFECTUEES
DANS LA REGION DE MOSCOU EN 1962-1964.
METH.MES.MAR.TERR.& DEFORM.LENTES SURFACE TERRE

ACAD.SC.URSS.INST.PHYS.TERR.SCHMIDT MOSC.PP129-143 1970
* B.I.M. N.62 PP 3202-3213 1971

3.141.2 / 1131
OURASSINA I.A. ANALYSE DES VARIATIONS DE MAREES DE LA VERTICALE A L OBSERVATOIRE ASTRONOMIQUE ENGELHARDT OBSERVEES POUR LA PERIODE DE 1961 A 1966
OBS.ENGELHARDT-IZVESTIA NO 36 -KAZAN- PP 200-203 1968

3.141 / 1133
IVANOVA A.K. STATION POUR L OBSERVATION DES OSCILLATIONS DE LA VERTICALE A L OBSERVATOIRE ASTRONOMIQUE ENGELHARDT
IZV.OBS.ASTRON.ENGELHARDT NO 27 PP 83-87 1951

3.141 / 1133
IVANOVA A.K. RESULTATS PRELIMINAIRES DES OBSERVATIONS SUR LES VARIATIONS LUNI-SOLAIRES DE LA VERTICALE A L OBSERVATOIRE ASTRONOMIQUE ENGELHARDT
CIRCUL.ASTRON.ACAD.SC.URSS NO 157 PP 7 1955

3.141.2 / 1133
OSTROVSKY A.E.
MIRONOVA L.I.
URASINA I.A.
RESULTS OF TIDAL OBSERVATIONS NEAR KAZAN 1960-1962.
5 EME SYMP.INT.SUR LES MAREES TERRESTRES.
OBS.ROY.BELG.COMM.NO 236 S.GEOPH.NO 69 PP 341-343 1964

3.14 / 1133
IVANOVA A.K. RESULTATS DES OBSERVATIONS DES DEVIATIONS DE LA VERTICALE A L OBSERVATOIRE ASTRONOMIQUE ENGELHARDT.
OBS. ASTRON. ENGELHARDT BULL. N.34 PP 29-67 1959
* B.I.M. N.40 JUILLET PP 1455-1485 1965

3 141.2 / 1133
OSTROVSKII A.E.
MIRONOVA L.Y.
OURASINA Y.A.
RESULTATS DES OBSERVATIONS CLINOMETRIQUES DANS LA REGION DE KAZAN POUR 1960-1962
ACAD.SC.URSS INST.PHYS.TERR.SCHMIDT MOSCOU 1964
* B.I.M. N. 47 AVRIL PP 2001-2012 1967

3.141 / 1133
OURASSINA J.A. ANALYSE DES MAREES DE LA VERTICALE A L OBSERVATOIRE ASTRONOMIQUE ENGELHARDT OBSERVEES DE 1961 - 1966.
IZVESTIA NO 36 OBS.ASTR.ENGELHARDT KAZAN 1968

3.141.2 / 1133
OURASSINA I.A. QUELQUES RESULTATS DES OBSERVATIONS DES INCLINAISONS DE LA TERRE A L OBSERVATOIRE D ENGELHARDT POUR 1964
ISVESTIA OBS.ASTRON.ENGELHARDT NO 35 PP 30-32 1966
* B.I.M. NO 51 PP 2337-2340 1968

3.141.2 / 1133
OURASSINA Y.A. LES OBSERVATIONS DES INCLINAISONS A L OBSERVATOIRE ASTRONOMIQUE ENGELHARDT.
RAPP.SYMP.INTERN.LENINGRAD 1968 MESURES INCLINAISON
COMM.GEOPH.URSS.MOSCOU PP 187-196 1969
* B.I.M. N.57 MAI PP 2763-2769 1970

3.141.2 / 1133
OURASSINA Y.A. RESULTATS FONDAMENTAUX DE L ANALYSE DES OBSERVATIONS CLINOMETRIQUES A L OBSERVATOIRE ASTRONOMIQUE ENGELHARDT.
* B.I.M. N.57 MAI PP 2741-2749 1970

3.141 / 1141
TCHOUPROUNOVA O.V. RESULTATS DE L ANALYSE HARMONIQUE DES OBSERVATIONS CLINOMETRIQUES DE 1964 A 1966 DANS LA STATION DE LA REGION DE KALOUCHE IVANO-FRANKOVSKII
ROTATION ET DEFORMATION DE MAREES DE LA TERRE NO 4
AC.NAOUK UKR.SSR-OBS.GRAV.POLTAVA-KIEV-PP 170-178 1972

3.221.6 / 1142
DOBROCHOTOV YOU S.
LISSENKO V.Y.
OBSERVATIONS DES VARIATIONS DE MAREES DE LA FORCE DE PESANTEUR A KIEV
RECH. MAR. TERR. ART. N3 PUBL.ACAD.SC.URSS MOSCOU 1963
* B.I.M. N.39 AVRIL PP 1385-1398 1965

3.252.2 / 1142
BALENKO V.
KOUTNY A.
DETERMINATION DE L ONDE M2 D APRES LES OBSERVATIONS DES INCLINAISONS DE MAREES LE LONG DU PROFIL KIEV - POLTAVA - ARTEMOVSK
PROCEEDINGS 7TH INT.SYMP.EARTH TIDES, SOPRON 1973
AKADEMIAI KIADO BUDAPEST, PP 633-638 1976

3.141.2 / 1143
BALENKO V.G.
KOUTNII A.M.
RESULTATS PRELIMINAIRES DES OBSERVATIONS CLINOMETRIQUES A LA STATION DE LA RESERVE LAVRO-PETCHERSKII DE KIEV
MAR.TERR.ACAD.SC.UK.SSR.OBS.GRAV.POLTAVA KIEV 1966
* B.I.M. NO 51 PP 2349-2357 1968

3.141 / 1152
AKSENTIEVA Z.N. COMPARAISON DES RESULTATS DE LA DETERMINATION DE L ONDE DE MAREE M2 A PARTIR DE DEUX LONGS CYCLES D OBSERVATIONS DES OSCILLATIONS DE LA VERTICALE A POLTAVA /1930-1941 ET 1948-1952/
MESSAGER ACAD.SC.URKAINE NO 9 PP 933-936 1958

3.221.6 / 1152
AKSENTIEVA Z.N. DONNEES PRELIMINAIRES SUR L AMPLITUDE DES VARIATIONS DE MAREE DE L INTENSITE DE LA PESANTEUR A POLTAVA
MESSAGER ACAD.SC.UKRAINE NO 1 PP 29-31 1959

3.141.2 / 1152
OSTROVSKY A.
MATVEIEV P.
FANDUSHINA S.
OBSERVATIONS OF THE TIDAL TILTS OF THE EARTH BY MEANS OF A PHOTOELECTRIC TILTMETER IN POLTAVA, JULY-OCTOBER 1958
3E SYMP. MAREES TERRESTRES TRIESTE
BOLL.GEOFISICA TEORICA E APPL.VOL.2 NO 5 PP 60-63 1960

3.140 / 1152
OSTROVSKII A.E.
MATVEYEV P.S.
LONDAI V.N.
MAREES CLINOMETRIQUES DE LA SURFACE DE LA TERRE A POLTAVA D APRES LES OBSERVATIONS DE 1958-1959
TROUDI POLTAVSKOI GRAV.OBS. X PP 14-19 1961
* B.I.M. NO 28 PP 653-658 1962

3.141 / 1152
AKSENTIEVA Z.N. RESULTATS D UNE SERIE DE ONZE ANNEES D OBSERVATIONS /DE 1930 A 1941/ SUR LES OSCILLATIONS DE LA VERTICALE A POLTAVA.
TROUDI POLTAVSKOI GRAV.OBS. T.II PP 121-138 1948
* B.I.M. N.34 NOVEMBRE PP 1118-1132 1963

3.141 / 1152
AKSENTIEVA Z.N. RESULTATS D UNE SERIE DE ONZE ANNEES D OBSERVATIONS SUR LA MAREE TERRESTRE M2 A L AIDE DES PENDULES HORIZONTAUX A POLTAVA DE 1930 A 1941
TRAVAUX DE LA REUNION SUR LES METHODES D ETUDE DES MOUVEMENTS ET DES DEFORMATIONS DE L ECORCE TERRESTRE
* B.I.M. N.37 SEPTEMBRE PP 1252-1259 1964

3.221.6 / 1152
DYTCHKO I.A.
KORBA S.N.
LES VARIATIONS DE MAREE DE LA GRAVITE A POLTAVA.
5 EME SYMP.INT.SUR LES MAREES TERRESTRES.
OBS.ROY.BELG.COMM.NO 236 S.GEOPH.NO 69 PP 167-169 1964

3.221.6 / 1152
DITCHKO I.A. ORGANISATION ET RESULTATS DES OBSERVATIONS DES VARIATIONS DE LA FORCE DE PESANTEUR AVEC UN GRAVIMETRE GS 11 A POLTAVA.
OBS. GRAV. DE POLTAVA T. XII. ACAD.SC.URSS. KIEV 1963
* B.I.M. N.42 DECEMBRE PP 1567-1580 1965

3.221.6 / 1152
AKSENTIEVA Z.N.
DITCHKO I.A.
RESULTATS PRELIMINAIRES DES OBSERVATIONS SUR LES VARIATIONS DE LA FORCE DE PESANTEUR AVEC UN GRAVIMETRE GS-11 A POLTAVA.
GEOPH. ET ASTR. BULL. INFORM. NO 5
* B.I.M. N.44 JUILLET PP 1754-1758 1966

3.14 / 1152
EVTOUCHENKO E.I. RESULTATS DES OBSERVATIONS CLINOMETRIQUES DANS LE BASSIN DU DONETZ EN 1960.
MAREES TERR.ACAD.SC.R.S.S. D UKRAINE OBS. GRAV. DE POLTAVA INST. GEOPH. DE LA R.S.S. D UKRAINE KIEV 1966
* B.I.M. N.48 JUILLET PP 2066-2068 1967

3.14 / 1152
AKSENTIEVA Z.N.
BOULANIETZ V.G.
TOKAR V.Y.
SUR LA REDUCTION D UNE SERIE DE ONZE ANNEES D OBSERVATIONS /1930 A 1944/ SUR LES OSCILLATIONS DE LA VERTICALE A POLTAVA
ROTATION ET DEFORMATIONS DE MAREES DE LA TERRE - NO 2
ACAD.NAOUK UKR.SSR - OBS.GRAV.POLTAVA - KIEV PP 3-8 1970
* B.I.M. MARS NO 65 PP 3426-3430 1973

3.221.6 / 1150
AKSENTIEVA Z.N.
KORBA P.S.
TCHOUPROUNOVA O.V.
LES OBSERVATIONS SUR LES INCLINAISONS DE MAREES ET LES VARIATIONS DE LA FORCE DE PESANTEUR A SIMFEROPOL.
OBS.GRAV.POLTAVA INF.BULL.NO 11 PP 169-175 KIEV 1967
* B.I.M. NO 60 PP 2955-2961 1971

3.141 / 1150
TCHOUPROUNOVA O.V. QUELQUES PARTICULARITES DANS LE MOUVEMENT DES

PENDULES HORIZONTAUX DANS LA STATION DES MAREES
TERRESTRES -SIMFEROPOL-.
OBS.GRAV.POLTAVA INF.BULL.PP 185 KIEV 1967
* B.I.M. NO 60 PP 2962-2966 1971

3.221.6 / 1150
KORBA P.S. VARIATIONS DE MAREES DE LA FORCE DE PESANTEUR A
SIMFEROPOL EN 1964-1966
ROTATION ET DEFORM.MAREES DE LA TERRE,FASC.1
ACAD.SC.UKR.OBS.GRAV.POLTAVA KIEV PP 199-206 1970
* B.I.M. NO 64 PP 3398-3405 1973

3.41 / 1150
BOULATSEN V.G. RESULTATS PRELIMINAIRES DES OBSERVATIONS DES
DEFORMATIONS DE MAREES LINEAIRES EN CRIMEE
ROTATION ET DEFORMATION DE MAREES DE LA TERRE NO 4
ACAD.NAOUK UKR.SSR.-OBS.GRAV.POLTAVA-KIEV PP 3-20 1972
* B.I.M. NO 66 MAI PP 3560-3575 1973

3.4 / 1150
BOULATSEN V.G.
TOKAR V.Y.
RESULTATS DES OBSERVATIONS DE MAREES
EXTENSOMETRIQUES EN CRIMEE
ROT.& DEFORM.DE MAREES DE LA TERRE VOL.VI PP 16-28 KIEV 1974

3.221.6 / 1150-1151
KORBA S.N.
KORBA P.S.
RESULTATS DE LA REDUCTION DES OBSERVATIONS DES
VARIATIONS DE MAREES DE LA FORCE DE PESANTEUR A
SIMFEROPOL ET YALTA PAR LA METHODE DE VENEDIKOV
ROTATION ET DEFORMATION DE MAREES DE LA TERRE NO 4
ACAD.NAOUK UKR.SSR.-OBS.GRAV.POLTAVA-KIEV PP 54-65 1972

3.221.6 / 1150
KORBA P.S. RESULTATS DES OBSERVATIONS DES VARIATIONS DE
MAREES DE LA PESANTEUR EN CRIMEE
PROCEEDINGS 7TH INT.SYMP.EARTH TIDES, SOPRON 1973
AKADEMIAI KIADO BUDAPEST, PP 707-710 1976

3.41 / 1150
BOULATSEN V.G. PARAMETRES ELASTIQUES L ET H OBTENUS D APRES LES OBSER-
VATIONS EXTENSOMETRIQUES DE MAREES EN CRIMEE
PROCEEDINGS 7TH INT.SYMP.EARTH TIDES, SOPRON 1973
AKADEMIAI KIADO BUDAPEST, PP 647-650 1976

3.41 / 1151
POPOV V.V.
TCHERNIAVKINA M.K.
QUELQUES RESULTATS DES OBSERVATIONS SUR LES
DEFORMATIONS DE LA SURFACE DE LA TERRE A LA
STATION GEOPHYSIQUE DE YALTA.
* B.I.M. N.55 SEPTEMBRE PP 2644-2655 1969

3.221.6 / 1151
KORBA P.S.
KORBA S.N.
VARIATIONS DE MAREES DE LA FORCE DE PESANTEUR A
YALTA DE 1966 A 1968
ROTATION ET DEFORMATIONS DE MAREES DE LA TERRE - NO 2
ACAD.NAOUK UKR.SSR - OBS.GRAV.POLTAVA - KIEV P18-34 1970
* B.I.M. MARS NO 65 PP 3440-3452 1973

3.141 / 1152
AKSENTIEVA Z.N. RESULTATS DE L OBSERVATION DES MAREES TERRESTRES
/ONDE M2/ A L AIDE DE PENDULES HORIZONTAUX A POLTAVA
IZV.ACAD.SC.URSS SER.GEOGR.GEOPH.NO 4 PP 563-567 1940

3.141 / 1152
AKSENTIEVA Z.N. SUR LES OBSERVATIONS CLINOMETRIQUES A POLTAVA DE
1948 A 1952
TRAV.3E CONF.LAT.TOUTES UNIONS POLTAVA MAI 1952
PP 113-118 KIEV 1954

3.141 / 1152
AKSENTIEVA Z.N. SUR LES TRAVAUX DE CLINOMETRIE A POLTAVA DE
1948 A 1952
TRAV.3E CONF.LAT. TOUTES UNIONS POLTAVA MAI 1952
PP 113-118 KIEV 1954

3.141 / 1143
BALENKO V.G.
KOUTNII A.M.
NOVIKOVA A.N.
RESULTATS DES OBSERVATIONS CLINOMETRIQUES A
/LAVRO-PETCHERSKAIA-KIEV/ EN 1964-1966
ROTATION ET DEFORM.MAREES DE LA TERRE,FASC.1
ACAD.SC.UKR.OBS.GRAV.POLTAVA KIEV PP 249-264 1970
* B.I.M. N.62 PP 3129-3138 1971

3.141.2 / 1143
BALENKO V.G.
KOUTNII A.M.
QUELQUES RESULTATS DES OBSERVATIONS CLINOMETRIQUES
SUIVANT LE PROFIL KIEV-POLTAVA-ARTEMOVSK

ROT.ET DEFORM.DE MAREES DE LA TERRE,VOL.5 PP 1-3 1973
* B.I.M. NO 69 PP 3894-3903 1974

3.141 / 1145
BOGDAN Y.D. RESULTATS PRELIMINAIRES DES OBSERVATIONS CLINOMETRIQUES
MATVEYEV P.S. A DARIEVKA.
MAREES TERR. ACAD.SC. R.S.S. D UKRAINE OBS. GRAV. POLTAVA
INST. GEOPH. DE LA R.S.S. D UKRAINE KIEV 1966
* B.I.M. N. 48 JUILLET PP 2057-2061 1967

3.141.2 / 1145
MATVEYEV P.S. RESULTATS DES OBSERVATIONS CLINOMETRIQUES A DARIEVKA
BOGDAN J.D. ROTATION ET DEFORMATION DE MAREES DE LA TERRE NO 4
AC.NAOUK UKR.SSR-OBS.GRAV.POLTAVA-KIEV PP 44-54 1972
* B.I.M. NO 67 AOUT, PP 3702-3709 1973

3.141.2 / 1146-1147
MATVEYEV P.S. RESULTATS DE L ANALYSE HARMONIQUE DES OBSERVATIONS
CLINOMETRIQUES A TSMAKOVO ET INGOULIETS
ROTATION ET DEFORMATION DE MAREES DE LA TERRE NO 4
AC.NAOUK UKR.SSR-OBS.GRAV.POLTAVA-KIEV-PP 105-170 1972

3.141 / 1147
MATVEYEV P.S. RESULTATS PRELIMINAIRES DES OBSERVATIONS CLINOMETRIQUES
DE MAREES DE LA SURFACE DE LA TERRE A TSMAKOVO
TROUDI POLTAVSKOI GRAV.OBS. X PP 3-13 1961
* B.I.M. NO 27 PP 629-640 1962

3.141 / 1148
AKSENTIEVA Z.N. RESULTATS PRELIMINAIRES DES OBSERVATIONS
DITCHKO I.A. CLINOMETRIQUES A INKERMAN
KORBA P.S. ROTATION ET DEFORM.MAREES DE LA TERRE,FASC.1
VAN NIAK TCHAN ACAD.SC.UKR.OBS.GRAV.POLTAVA KIEV PP 300-303 1970
* B.I.M. N.62 PP 3176-3178 1971

3.142 / 1148
DITCHKO I.A. INCLINAISONS DE LA SURFACE DE LA TERRE A INKERMAN
TOKAR V.I. ACAD.NAOUK UKRAINSKOI SSR, OBS.GRAV.POLTAVA
ROTAT.DEFOR.MAREES DE LA TERRE VII PP 30-31 1975
* B.I.M. NO 73 PP 4215-4218 1976

3.42 / 1148
BOULATSEN V.G. MAREES ET DEFORMATIONS LENTES DE L ECORCE TERRESTRE
D APRES LES DONNEES DES OBSERVATIONS
EXTENSOMETRIQUES A INKERMAN / CRIMEE/
ACAD.NAOUK UKRAINSKOI SSR, OBS.GRAV.POLTAVA
ROTAT.DEFOR. MAREES DE LA TERRE VII PP 9-15 1975
* B.I.M. NO 73 PP 4203-4214 1976

3.141 / 1150
AKSENTIEVA Z.N. RESULTATS PRELIMINAIRES DES OBSERVATIONS DES INCLINAISONS
TCHOUPROUNOVA O.V. DUES AUX MAREES DANS LE POLYGONE DE CRIMEE /SIMFEROPOL/
MAREES TERR. ACAD.SC.R.S.S. D UKRAINE OBS. GRAV. DE POLTAVA
INST. GEOPH. DE LA R.S.S. D UKRAINE KIEV 1966
* B.I.M. N.48 JUILLET PP 2062-2065 1967

3.221.6 / 1150
KORBA P.S. VARIATIONS DE LA FORCE DE PESANTEUR POUR DES SERIES
DE TROIS JOURS D OBSERVATIONS A SIMFEROPOL.
ACAD. SC. R.S.S. D UKRAINE OBS.GRAV. DE POLTAVA
INST.GEOPH.DE LA RSS D UKRAINE KIEV 1966
* B.I.M. N.48 JUILLET PP 2084-2089 1967

3.141 / 1150
AKSENTIEVA Z.N. LES OBSERVATIONS SUR LES INCLINAISONS DE MAREES ET
KORBA P.S. LES VARIATIONS DE LA FORCE DE PESANTEUR A SIMFEROPOL.
TCHOUPROUNOVA O.V. OBS.GRAV.POLTAVA INF.BULL.NO 11 PP 169-175 KIEV 1967
* B.I.M. NO 60 PP 2955-2961 1971

3.14 / 1152
MATVEYEV M.S. OBSERVATIONS DES INCLINAISONS DE LA SURFACE DE LA TERRE
BOGDAN Y.D. DANS LES POINTS DU PROFIL SOUMA-KHERSON DE 1964 A 1967
ROTATION ET DEFORMATIONS DE MAREES DE LA TERRE - NO 2
ACAD.NAOUK UKR.SSR - OBS.GRAV.POLTAVA - KIEV PP8-17 1970
* B.I.M. MARS NO 65 PP 3431-3439 1973

3.221.6 / 1152
DITCHKO I.A. VARIATIONS DE MAREES DE LA FORCE DE PESANTEUR A POLTAVA
ROTATION ET DEFORM.MAREES DE LA TERRE,FASC.1
ACAD.SC.UKR.OBS.GRAV.POLTAVA KIEV PP 192-198 1970
* B.I.M. NO 64 PP 3352-3356 1973

3.4 / 1152
BOULANIETS V.G. DEFORMOGRAPHE EXPERIMENTAL DE L OBSERVATOIRE
OVTCHINNIKOV V.A. GRAVIMETRIQUE DE POLTAVA.
ROTATION ET DEFORM.MAREES DE LA TERRE,FASC.1
ACAD.SC.UKR.OBS.GRAV.POLTAVA KIEV PP 295-299 1970
* B.I.M. MARS NO 65 PP 3463-3467 1973

3.142 / 1152
BOULATSEN V.G. ANALYSE D UNE SERIE DE ONZE ANNEES /1930 A 1941/
TOKAR V.Y. D OBSERVATIONS DES INCLINAISONS DE MAREES A POLTAVA
ROT.ET DEFORM.DE MAREES DE LA TERRE, VOL.5 PP 26-33 1973
* B.I.M. NO 70 PP 3937-3956 1975

3.141.2 / 1152
MATVEYEV P.S. DETERMINATION OF THE M2 CONSTITUENT FROM THE TILT
BOGDAN I.Y. OBSERVATIONS ALONG THE PROFILE SUMY - KHERSON.
GOLUBITSKY V.G. PROCEEDINGS 7TH INT.SYMP.EARTH TIDES, SOPRON 1973
AKADEMIAI KIADO BUDAPEST, PP 31-38 1976

3.141 / 1153
BOGDAN I.D. RESULTATS DE L ANALYSE HARMONIQUE DES OBSERVATIONS
LISSENKO G.M. CLINOMETRIQUES A VELIKI BOUDICHA
MATVEYEV P.S. ROTATION ET DEFORM.MAREES DE LA TERRE,FASC.1
ACAD.SC.UKR.OBS.GRAV.POLTAVA KIEV PP 264-279 1970
* B.I.M. N.62 PP 3139-3148 1971

3.141.2 / 1153
MATVEYEV P.S. OBSERVATIONS CLINOMETRIQUES A VELIKIE BOUDICHA.
BOGDAN I.D. OBS.GRAV.POLTAVA INF.BULL.PP 197 KIEV 1967
LISSENKO G.M. * B.I.M. NO 60 PP 2967-2973 1971

3.141.2 / 1153
MATVEYEV P.S. COMPARAISON DES RESULTATS DE L ANALYSE HARMONIQUE
KORBA S.N. DES OBSERVATIONS CLINOMETRIQUES AU POINT -VELIKIE
SLAVINSKAIA E.A. BOUDICHA- OBTENUES PAR DEUX METHODES
ROT.& DEFORM.DE MAREES DE LA TERRE VOL VI PP 54-58 KIEV 1974

3.141.2 / 1153
MATVEYEV P.S. COMPARAISON DES RESULTATS DE L ANALYSE HARMONIQUE
KORBA S.N. DES OBSERVATIONS CLINOMETRIQUES AU POINT DE
BOGDAN I.Y. VELIKIE BOUDICHA, OBTENUES PAR DEUX METHODES
SLAVINSKAYA E.A. ROTAT.ET DEFORMAT.DE MAREES DE LA TERRE-VI-PP 54-58 1974
* B.I.M. NO 72 PP 4116-4122 1975

3.141 / 1149 - 1154
MATVEYEV P.S. RESULTATS PRELIMINAIRES DES OBSERVATIONS CLINOMETRIQUES
BOGDAN I.D. A LIKHOVKA ET SAMATOEVKA.
ROTATION ET DEFORM.MAREES DE LA TERRE,FASC.1
ACAD.SC.UKR.OBS.POLTAVA KIEV PP 303-313 1970
* B.I.M. N.62 PP 3179-3184 1971

3.141.2 / 1154
MATVEYEV P.S. RESULTATS DE L ANALYSE HARMONIQUE DES OBSERVATIONS
BOGDAN Y.D. CLINOMETRIQUES A SAMOTOEVKA ET LIKHOVKA
DOUBIK B.S.
SLAVINSKAYA E.A. ROT.ET DEFORM.DE MAREES TERRE VOL.3 PP 39-52 KIEV 1971

3.14 / 1155
BALENKO V.G. OBSERVATIONS CLINOMETRIQUES A LA STATION
KOUTNII A.M. CHEVTCHENKOVO DE LA REGION DE POLTAVA
NOVIKOVA A.N. ROTATION ET DEFORMATIONS DE MAREES DE LA TERRE - NO 2
ACAD.NAOUK UKR.SSR - OBS.GR.POLTAVA-KIEV PP41-57 1970
* B.I.M. NO 64 PP 3405-3419 1973

3.141.2 / 1156
BALENKO V.G. RESULTATS PRELIMINAIRES DES OBSERVATIONS DES
KOUTNII A.M. INCLINAISONS DE MAREES A LA STATION CATHERINOVKA
NOVIKOVA A.N. DE LA REGION DE KHARKOV
ROT.ET DEFORM.DE MAREES TERRE VOL.3 PP 60-63 KIEV 1971

Subject Index

391 - 405